Non-Linear Elastic Deformations

by

R. W. Ogden
George Sinclair Professor of Mathematics
University of Glasgow

DOVER PUBLICATIONS, INC.
Mineola, New York

To
My parents

Copyright

Published in Canada by General Publishing Company, Ltd., 30 Lesmill Road, Don Mills, Toronto, Ontario.

Published in the United Kingdom by Constable and Company, Ltd., 3 The Lanchesters, 162–164 Fulham Palace Road, London W6 9ER.

Bibliographical Note

This Dover edition, first published in 1997, is an unabridged and corrected republication of the work first published by Ellis Harwood Ltd., Chichester, England and Halsted Press/John Wiley & Sons, New York, in 1984.

Library of Congress Cataloging-in-Publication Data

Ogden, R. W., 1943–
Non-linear elastic deformations / by R. W. Ogden.
p. cm.
Originally published: Chichester : E. Horwood ; New York : Halsted Press, 1984.
Includes bibliographical references (p.).
ISBN 0-486-69648-0 (pbk.)
1. Deformations (Mechanics) 2. Elasticity. I. Title.
TA17.6.034 1997
620.1'1232—dc21 97–16162
CIP

Manufactured in the United States of America
Dover Publications, Inc., 31 East 2nd Street, Mineola, N.Y. 11501

Table of Contents

Acknowledgements

The content and style of the book owes much to the influence of many friends and colleagues over several years, but I should like to record my particular gratitude to Professor Rodney Hill, FRS, University of Cambridge, who introduced me to the subject of non-linear elasticity when I was a research student (1967–70), and Professor Peter Chadwick, FRS, University of East Anglia, with whom I spent two years as a research fellow (1970–72).

Thanks are due to Drs. Gareth Parry and Keith Walton, University of Bath, who undertook the task of reading and criticizing the whole of the manuscript, to Michael Warby, Brunel University, on whose calculations Figs. 6.7 and 6.9 were based, and Molly Demmar for her excellent typing. Finally, I should like to thank my publisher Ellis Horwood for his patience in waiting for the delivery of a long-overdue manuscript.

R.W. Ogden
Brunel University
May 1983

Preface

This book is concerned with the mathematical theory of non-linear elasticity, the application of this theory to the solution of boundary-value problems (including discussion of bifurcation and stability) and the analysis of the mechanical properties of solid materials capable of large elastic deformations. The setting is purely isothermal and no reference is made to thermodynamics. For the most part attention is restricted to the quasi-static theory, but some brief relevant discussion of time-dependent problems is included.

Apart from much basic material the book includes many previously unpublished results and also provides new approaches to some problems whose solutions are known. In part the book can be regarded as a research monograph but, at the same time, parts of it should also be suitable as a postgraduate text. Problems designed to develop further the text material are given throughout and some of these contain statements of new results.

Because so much of the theory depends on the use of tensors, Chapter 1 concentrates on the development of much of the tensor algebra and analysis which is used in subsequent chapters. However, there are parts of the book (in particular, Sections 4.4 and 7.2) which do not rely on a knowledge of tensors and can be read accordingly. Chapter 2 provides a detailed development of the basic kinematics of deformation and motion. Chapter 3 deals with the balance laws for a general continuum and the concept of stress. Prominence is given to the nominal stress tensor and the notion of conjugate stress and strain tensors is examined in detail.

In Chapter 4 the properties of the constitutive laws of both Cauchy- and Green-elastic materials are studied and, in particular, the implications of objectivity and material symmetry are assessed. Considerable attention is devoted to isotropic constitutive laws for both (internally) constrained and unconstrained materials. The basic boundary-value problems of non-linear elasticity are formulated in Chapter 5 and the governing equations are solved for a selection of problems in respect of unconstrained and incompressible isotropic materials. A section dealing with variational aspects of boundary-value problems is included along with a short discussion of conservation laws.

Chapter 6, the longest chapter, is concerned with incremental deformations superposed on an underlying finite deformation. The resulting (linearized) boundary-value problem is formulated and its structure discussed in relation to the analysis of uniqueness, stability and bifurcation. The role of the strong ellipticity inequality is examined. Constitutive inequalities are discussed and the implications of their failure in relation to bifurcation (or branching) is assessed from the local (i.e. incremental) viewpoint. Global aspects of non-uniqueness are also considered. The incremental theory is then applied to some representative problems whose bifurcation behaviour is studied in detail.

In the final chapter, Chapter 7, the theory of elasticity is applied to certain deformations and geometries associated with simple experimental tests, in particular the pure homogeneous biaxial deformation of a rectangular sheet. The relevant theory is provided in a concise form as a background for comparison with experimental results, isotropic materials being considered for simplicity of illustration. This is then used to assess the elastic response of certain rubberlike materials. The incremental theory governing the change in deformation due to a small change in material properties is developed and applied to the case of a slightly compressible material and this in turn is illustrated by means of rubberlike materials.

The book concentrates on 'exact' theories in the sense that no discussion of 'special' theories, such as shell, rod or membrane theories, or of numerical methods is included. (Excellent separate accounts of these topics are available elsewhere.) Within this framework a broad spectrum of topics has been covered and a balanced overview attempted (although this is, not surprisingly, influenced by the areas of the subject on which the writer has been actively engaged). Attention is confined to twice-continuously differentiable deformations on the whole, with discontinuities being touched on only briefly, in Chapter 6, in relation to failure of ellipticity.

References to standard works for background reading are given throughout the text but historical attributions and detailed lists of references to papers are not provided. Only where further development of the textual material might be required are references to the more recent papers cited, but the list of references is not intended to be exhaustive. References are indicated by the author's name followed by the year of publication in the text and gathered together at the end of each chapter.

CHAPTER 1

Tensor Theory

The use of vector and tensor analysis is of fundamental importance in the development of the theory which describes the deformation and motion of continuous media. In non-linear elasticity theory, in particular, little progress can be made or insight gained without the use of tensor formulations. This first chapter is therefore devoted to an account of the vector and tensor algebra and analysis which underlies the requirements of subsequent chapters. Some theorems of tensor algebra, however, are not dealt with here but postponed until the later chapters in which they are needed.

It is assumed that the reader is familiar with elementary vector and matrix algebra, including determinants, with the concept of a vector space, including linear independence and the notion of a basis, and with linear mappings. Also some familiarity with the index (or suffix) notation and the summation convention is assumed. Nevertheless, certain basic ideas are summarized in the early part of this chapter, primarily to establish notations but also for convenience of reference.

1.1 EUCLIDEAN VECTOR SPACE

The set of real numbers is denoted by $\mathbb{R}$. A *scalar* is a member of $\mathbb{R}$.

A (real) vector space V is a set of elements (called *vectors*[†]) such that (a) $\mathbf{u}+\mathbf{v}\in V$, $\mathbf{u}+\mathbf{v}=\mathbf{v}+\mathbf{u}$, $\mathbf{u}+(\mathbf{v}+\mathbf{w})=(\mathbf{u}+\mathbf{v})+\mathbf{w}$ for all $\mathbf{u},\mathbf{v},\mathbf{w}\in V$, (b) V contains the *zero vector* $\mathbf{0}$ such that $\mathbf{u}+\mathbf{0}=\mathbf{u}$ for all $\mathbf{u}\in V$ and for every $\mathbf{u}\in V$ there is an inverse element, denoted $-\mathbf{u}$, such that $\mathbf{u}+(-\mathbf{u})=\mathbf{0}$, (c) $\alpha\mathbf{u}\in V$, $1\mathbf{u}=\mathbf{u}$, $\alpha(\beta\mathbf{u})=(\alpha\beta)\mathbf{u}$, $(\alpha+\beta)\mathbf{u}=\alpha\mathbf{u}+\beta\mathbf{u}$, $\alpha(\mathbf{u}+\mathbf{v})=\alpha\mathbf{u}+\alpha\mathbf{v}$ for all $\alpha,\beta\in\mathbb{R}$, $\mathbf{u},\mathbf{v}\in V$, where 1 denotes unity.

A Euclidean vector space $\mathbb{E}$ is a real vector space such that, for any pair of vectors $\mathbf{u},\mathbf{v}\in\mathbb{E}$, there is defined a scalar, denoted $\mathbf{u}\cdot\mathbf{v}$, with the properties

$$\mathbf{u}\cdot\mathbf{v}=\mathbf{v}\cdot\mathbf{u} \tag{1.1.1}$$

$$\mathbf{u}\cdot\mathbf{u}\geq 0 \tag{1.1.2}$$

[†]In this chapter vectors are denoted by bold-face, lower-case letters, e.g. $\mathbf{t},\mathbf{u},\mathbf{v},\ldots$.

for all $\mathbf{u}, \mathbf{v} \in \mathbb{E}$, equality in (1.1.2) holding if and only if $\mathbf{u} = \mathbf{0}$. The *scalar product* (or '*dot*' *product*) $\mathbf{u} \cdot \mathbf{v}$ of $\mathbf{u}$ and $\mathbf{v}$ is bilinear (that is linear in each element of the product.[†] Thus

$$(\alpha\mathbf{u} + \beta\mathbf{v}) \cdot \mathbf{w} = \alpha(\mathbf{u} \cdot \mathbf{w}) + \beta(\mathbf{v} \cdot \mathbf{w}) \tag{1.1.3}$$

for all $\alpha, \beta \in \mathbb{R}$ and all $\mathbf{u}, \mathbf{v}, \mathbf{w} \in \mathbb{E}$, and dually by (1.1.1).

The *magnitude* (or *modulus*) of $\mathbf{u}$ is denoted by $|\mathbf{u}|$ and defined as the *positive* square root of

$$|\mathbf{u}|^2 = \mathbf{u} \cdot \mathbf{u}. \tag{1.1.4}$$

If $|\mathbf{u}| = 1$ then $\mathbf{u}$ is said to be a *unit vector*.

If $\mathbf{u} \cdot \mathbf{v} = 0$ then $\mathbf{u}$ and $\mathbf{v}$ are said to be *orthogonal*.

The above discussion applies to a vector space of arbitrary finite dimension. With a view to the application in subsequent chapters to continuous bodies occupying three-dimensional physical space we confine the remaining development in this chapter to an underlying three-dimensional Euclidean space. (Generalization to n dimensions, however, is for the most part a straightforward matter.)

In three dimensions we denote the *vector product* of $\mathbf{u}$ and $\mathbf{v}$ by $\mathbf{u} \wedge \mathbf{v}$. It is a vector with the properties

$$\mathbf{u} \wedge \mathbf{v} = -\mathbf{v} \wedge \mathbf{u}, \tag{1.1.5}$$

$$(\mathbf{u} \wedge \mathbf{v}) \cdot (\mathbf{u} \wedge \mathbf{v}) = (\mathbf{u} \cdot \mathbf{u})(\mathbf{v} \cdot \mathbf{v}) - (\mathbf{u} \cdot \mathbf{v})^2, \tag{1.1.6}$$

$$\mathbf{u} \cdot (\mathbf{u} \wedge \mathbf{v}) = 0, \tag{1.1.7}$$

$$(\alpha\mathbf{u} + \beta\mathbf{v}) \wedge \mathbf{w} = \alpha(\mathbf{u} \wedge \mathbf{w}) + \beta(\mathbf{v} \wedge \mathbf{w}) \tag{1.1.8}$$

for all $\alpha, \beta \in \mathbb{R}$, $\mathbf{u}, \mathbf{v}, \mathbf{w} \in \mathbb{E}$.

It follows immediately from (1.1.5) that

$$\mathbf{u} \wedge \mathbf{u} = \mathbf{0} \tag{1.1.9}$$

for each $\mathbf{u} \in \mathbb{E}$.

If $\mathbf{u}$ and $\mathbf{v}$ are unit vectors, (1.1.6) can be written

$$|\mathbf{u} \wedge \mathbf{v}|^2 + (\mathbf{u} \cdot \mathbf{v})^2 = 1$$

and this, together with (1.1.9), leads naturally to the following geometrical

[†] More precisely, the scalar product is a bilinear mapping from $\mathbb{E} \times \mathbb{E}$ to $\mathbb{R}$, where the *Cartesian product* $\mathbb{E} \times \mathbb{E}$ is defined to be the set of ordered pairs $(\mathbf{u}, \mathbf{v})$ for all $\mathbf{u}, \mathbf{v} \in \mathbb{E}$.

interpretations of the scalar and vector products. We write

$$\mathbf{u}\cdot\mathbf{v} = |\mathbf{u}|\,|\mathbf{v}|\cos\theta, \tag{1.1.10}$$

$$\mathbf{u}\wedge\mathbf{v} = |\mathbf{u}|\,|\mathbf{v}|\sin\theta\,\mathbf{k}, \tag{1.1.11}$$

where (1.1.10) defines the angle θ between the directions **u** and **v**, and **k** is a unit vector in the direction $\mathbf{u}\wedge\mathbf{v}$ $(0 \le \theta \le \pi)$.

So far the discussion has been in *invariant* (or *absolute*) notation, that is no reference to a 'basis', 'axes' or 'components' is made, implied or required. It turns out that much of the theory to be developed is more concise and transparent in such notation than in terms of corresponding component notations. However, there are circumstances in which use of the component forms of vectors (and tensors) is helpful. In particular, practical ideas can often be fixed more readily by reference to components (and associated basis vectors) and algebraic manipulations can be made more convincing for the beginner. The component representation of vectors is therefore examined in the following two sub-sections.

1.1.1 Orthonormal bases and components

A *basis* for $\mathbb{E}$ is a set of three linearly independent vectors. An *orthonormal basis* is a set of three vectors, here denoted $\mathbf{e}_1$, $\mathbf{e}_2$, $\mathbf{e}_3$ and collectively by $\{\mathbf{e}_i\}$, such that

$$\mathbf{e}_i\cdot\mathbf{e}_j = \delta_{ij} = \begin{cases} 1 & i=j \\ 0 & i\neq j \end{cases} \tag{1.1.12}$$

for any pair of indices i,j. (Italic letters $i,j,\ldots,p,q,\ldots$ are used for indices running over values $1,2,3$.) The *Krönecker delta* symbol, δ_{ij}, is defined by the right-hand equality in (1.1.12).

With reference to the basis $\{\mathbf{e}_i\}$ a vector **u** is decomposed as

$$\mathbf{u} = u_1\mathbf{e}_1 + u_2\mathbf{e}_2 + u_3\mathbf{e}_3, \tag{1.1.13}$$

where u_1, u_2, u_3 are called the *components* of **u** relative to the given basis.

This summation convention allows (1.1.13) to be written in the compact form

$$\mathbf{u} = u_j\mathbf{e}_j \tag{1.1.14}$$

in which summation over an index (in this case j) from 1 to 3 is implied by its repetition. This convention is followed throughout the book without further comment except where an explicit statement is made to the contrary.

If the dot product of equation (1.1.14) with $\mathbf{e}_i$ is taken, use of (1.1.12) leads to

$$u_i = \mathbf{u} \cdot \mathbf{e}_i.$$

Thus, for an arbitrary choice of orthonormal basis $\{\mathbf{e}_i\}$ the component u_i of a vector $\mathbf{u}$ is defined as the scalar product of $\mathbf{u}$ with the basis vector $\mathbf{e}_i$.

It is left as an exercise for the reader to show that

$$\mathbf{u} \cdot \mathbf{v} = u_i v_i \equiv u_1 v_1 + u_2 v_2 + u_3 v_3,$$
$$|\mathbf{u}|^2 = u_1^2 + u_2^2 + u_3^2.$$

It is assumed here that the orthonormal basis $\{\mathbf{e}_i\}$ forms a *right-handed triad* of unit vectors, that is

$$\mathbf{e}_2 \wedge \mathbf{e}_3 = \mathbf{e}_1, \qquad \mathbf{e}_3 \wedge \mathbf{e}_1 = \mathbf{e}_2, \qquad \mathbf{e}_1 \wedge \mathbf{e}_2 = \mathbf{e}_3.$$

In summation notation these are put jointly as

$$\mathbf{e}_i \wedge \mathbf{e}_j = \varepsilon_{ijk} \mathbf{e}_k, \tag{1.1.15}$$

where ε_{ijk}, which is called the *alternating symbol*, is defined by

$$\varepsilon_{ijk} = \begin{cases} 1 & \text{if } (ijk) \text{ is a cyclic permutation of } (123) \\ -1 & \text{if } (ijk) \text{ is an anticyclic permutation of } (123) \\ 0 & \text{otherwise.} \end{cases} \tag{1.1.16}$$

We note, in particular, the cyclic properties

$$\varepsilon_{ijk} = \varepsilon_{kij} = \varepsilon_{jki} \tag{1.1.17}$$

and the antisymmetry

$$\varepsilon_{ijk} = -\varepsilon_{ikj} \tag{1.1.18}$$

(on any pair of indices) which follow immediately from the definition (1.1.16).

In this notation the vector product $\mathbf{u} \wedge \mathbf{v}$ becomes

$$\mathbf{u} \wedge \mathbf{v} = u_i v_j \mathbf{e}_i \wedge \mathbf{e}_j = \varepsilon_{ijk} u_i v_j \mathbf{e}_k \tag{1.1.19}$$

and the three components of this are

$$u_2 v_3 - u_3 v_2, \qquad u_3 v_1 - u_1 v_3, \qquad u_1 v_2 - u_2 v_1.$$

The *triple scalar product* $(\mathbf{u} \wedge \mathbf{v})\cdot\mathbf{w}$ is written

$$(\varepsilon_{ijp}u_i v_j \mathbf{e}_p)\cdot(w_k \mathbf{e}_k) = \varepsilon_{ijk}u_i v_j w_k \tag{1.1.20}$$

on use of (1.1.12) and (1.1.19). This leads to the convenient determinantal representation

$$(\mathbf{u} \wedge \mathbf{v})\cdot\mathbf{w} \equiv \begin{vmatrix} u_1 & v_1 & w_1 \\ u_2 & v_2 & w_2 \\ u_3 & v_3 & w_3 \end{vmatrix} \tag{1.1.21}$$

from which the properties

$$(\mathbf{u} \wedge \mathbf{v})\cdot\mathbf{w} = (\mathbf{w} \wedge \mathbf{u})\cdot\mathbf{v} = (\mathbf{v} \wedge \mathbf{w})\cdot\mathbf{u}$$

follow immediately together with their anticyclic counterparts. These properties can, of course, be established without reference to the basis $\{\mathbf{e}_i\}$; see, for example, Chadwick (1976, p. 13).† A further result which can be seen immediately from (1.1.21) is that $\mathbf{u}, \mathbf{v}, \mathbf{w}$ are linearly dependent if and only if $(\mathbf{u} \wedge \mathbf{v})\cdot\mathbf{w} = 0$.

Problem 1.1.1 Let $\det \mathsf{A}$ denote the determinant of the 3×3 matrix A which has elements A_{ij}. Use (1.1.20) and (1.1.21) to show that

$$\det \mathsf{A} = \varepsilon_{ijk}A_{i1}A_{j2}A_{k3} = \varepsilon_{ijk}A_{1i}A_{2j}A_{3k}. \tag{1.1.22}$$

Deduce that

$$\varepsilon_{ijk}A_{ip}A_{jq}A_{kr} = (\det \mathsf{A})\varepsilon_{pqr} \tag{1.1.23}$$

and hence show that

$$\det \mathsf{A} = \tfrac{1}{6}\varepsilon_{ijk}\varepsilon_{pqr}A_{ip}A_{jq}A_{kr}. \tag{1.1.24}$$

Problem 1.1.2 If A and B denote two 3×3 matrices, use (1.1.22) and (1.1.23) to show that

$$\det \mathsf{AB} = (\det \mathsf{A})(\det \mathsf{B}). \tag{1.1.25}$$

(The reader is reminded that $(\mathsf{AB})_{ij} = A_{ip}B_{pj}$.)

† References are listed at the end of the chapter.

From the definitions of δ_{ij} and ε_{ijk} it is easily seen that

$$\varepsilon_{ijk} = \begin{vmatrix} \delta_{1i} & \delta_{1j} & \delta_{1k} \\ \delta_{2i} & \delta_{2j} & \delta_{2k} \\ \delta_{3i} & \delta_{3j} & \delta_{3k} \end{vmatrix} = \begin{vmatrix} \delta_{1i} & \delta_{2i} & \delta_{3i} \\ \delta_{1j} & \delta_{2j} & \delta_{3j} \\ \delta_{1k} & \delta_{2k} & \delta_{3k} \end{vmatrix}.$$

Use of this with (1.1.25) leads to the representation

$$\varepsilon_{ijk}\varepsilon_{pqr} = \begin{vmatrix} \delta_{ip} & \delta_{iq} & \delta_{ir} \\ \delta_{jp} & \delta_{jq} & \delta_{jr} \\ \delta_{kp} & \delta_{kq} & \delta_{kr} \end{vmatrix} \tag{1.1.26}$$

and on setting $r = k$ and summing over k from 1 to 3 in (1.1.26) we obtain the useful identity

$$\varepsilon_{kij}\varepsilon_{kpq} = \varepsilon_{ijk}\varepsilon_{pqk} = \delta_{ip}\delta_{jq} - \delta_{iq}\delta_{jp}. \tag{1.1.27}$$

With reference to the basis $\{\mathbf{e}_i\}$ the *triple vector product* $\mathbf{u} \wedge (\mathbf{v} \wedge \mathbf{w})$ is expanded as

$$u_s\mathbf{e}_s \wedge (\varepsilon_{kpq}v_p w_q \mathbf{e}_k) = \varepsilon_{kpq}\varepsilon_{krs}v_p w_q u_s \mathbf{e}_r$$

by use of (1.1.15), (1.1.17) and (1.1.19). Application of (1.1.27) reduces this to

$$u_q w_q v_p \mathbf{e}_p - u_p v_p w_q \mathbf{e}_q$$

and the identity

$$\mathbf{u} \wedge (\mathbf{v} \wedge \mathbf{w}) = (\mathbf{u}\cdot\mathbf{w})\mathbf{v} - (\mathbf{u}\cdot\mathbf{v})\mathbf{w}$$

follows.

1.1.2 Change of basis

We now consider a second (right-handed) orthonormal basis $\{\mathbf{e}'_i\}$ oriented with respect to $\{\mathbf{e}_i\}$ as depicted in Fig. 1.1.

Since $\{\mathbf{e}_i\}$ is a basis, each of $\mathbf{e}'_1, \mathbf{e}'_2, \mathbf{e}'_3$ is expressible as a linear combination of $\mathbf{e}_1, \mathbf{e}_2, \mathbf{e}_3$. We therefore write

$$\mathbf{e}'_i = Q_{ip}\mathbf{e}_p \qquad (i = 1, 2, 3), \tag{1.1.28}$$

and, on taking the dot product of (1.1.28) with $\mathbf{e}_j$, it is seen that the coefficients Q_{ij} are given by

$$Q_{ij} = \mathbf{e}'_i \cdot \mathbf{e}_j. \tag{1.1.29}$$

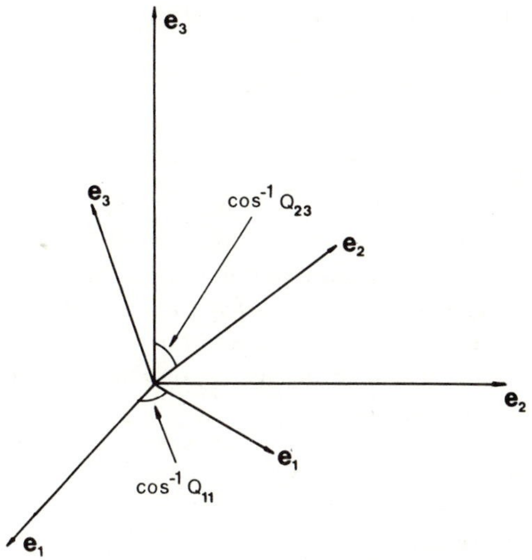

Fig. 1.1 Orientation of the basis vectors $\mathbf{e}_i'$ relative to $\mathbf{e}_i$

The definition (1.1.10) with (1.1.29) shows that the Q_{ij}'s are the direction cosines of the vectors $\mathbf{e}_i'$ relative to the $\mathbf{e}_j$, as indicated in Fig. 1.1.

By orthonormality and (1.1.28), we have

$$\delta_{ij} = \mathbf{e}_i' \cdot \mathbf{e}_j' = Q_{ik}\mathbf{e}_k \cdot \mathbf{e}_j' = Q_{ik}Q_{jk}. \tag{1.1.30}$$

It is convenient to represent the collection of coefficients Q_{ij} as a matrix Q, with transpose Q^T. Then (1.1.30) shows that Q^T is the inverse matrix of Q and so

$$\mathsf{Q}\mathsf{Q}^T = \mathsf{I} = \mathsf{Q}^T\mathsf{Q}, \tag{1.1.31}$$

where I is the identity matrix, or, in component notation,

$$Q_{ik}Q_{jk} = \delta_{ij} = Q_{ki}Q_{kj}. \tag{1.1.32}$$

A matrix Q satisfying (1.1.31) is said to be an *orthogonal matrix.*

Premultiplication of (1.1.28) by Q_{ij} and use of (1.1.32) leads to the dual connections

$$\mathbf{e}_i' = Q_{ij}\mathbf{e}_j, \qquad \mathbf{e}_j = Q_{ij}\mathbf{e}_i' \tag{1.1.33}$$

between the basis vectors.

From (1.1.22), (1.1.25) and (1.1.31) we obtain

$$(\det \mathsf{Q})^2 = (\det \mathsf{Q}^{\mathrm{T}})(\det \mathsf{Q}) = \det \mathsf{Q}^{\mathrm{T}}\mathsf{Q} = 1$$

and hence

$$\det \mathsf{Q} = \pm 1. \tag{1.1.34}$$

Here we have restricted attention to the situation in which $\det \mathsf{Q} = +1$, this corresponding to maintenance of right-handedness of the basis vectors. In this case Q is said to be *proper orthogonal* (it may be interpreted as a rotation which takes $\{\mathbf{e}_i\}$ into $\{\mathbf{e}_i'\}$; see Section 1.3.5). For a change of basis in which right-handedness is not preserved, on the other hand, $\det \mathsf{Q} = -1$ and Q is *improper orthogonal.*

Let v_i, v_i' be the components of a vector $\mathbf{v}$ with respect to bases $\{\mathbf{e}_i\}$, $\{\mathbf{e}_i'\}$ respectively. Then, by use of (1.1.33),

$$v_k'\mathbf{e}_k' = v_j\mathbf{e}_j = v_jQ_{kj}\mathbf{e}_k'.$$

Taking the dot product of this with $\mathbf{e}_i'$ and applying (1.1.32), we obtain

$$v_i' = Q_{ij}v_j, \qquad v_j = Q_{ij}v_i' \tag{1.1.35}$$

which show that the components of $\mathbf{v}$ transform under change of (orthonormal) basis according to the same rule (1.1.33) applicable to the basis vectors themselves.

As a specific example we consider a change of basis for which $\mathbf{e}_3' = \mathbf{e}_3$ and

$$\left.\begin{aligned} \mathbf{e}_1' &= \cos\theta\,\mathbf{e}_1 + \sin\theta\,\mathbf{e}_2, \\ \mathbf{e}_2' &= -\sin\theta\,\mathbf{e}_1 + \cos\theta\,\mathbf{e}_2, \end{aligned}\right\} \tag{1.1.36}$$

corresponding to a positive (i.e. anticlockwise) rotation through an angle θ about $\mathbf{e}_3$.

Then

$$\mathsf{Q} = \begin{bmatrix} \cos\theta & \sin\theta & 0 \\ -\sin\theta & \cos\theta & 0 \\ 0 & 0 & 1 \end{bmatrix} \tag{1.1.37}$$

and it is easily confirmed that (1.1.32) are satisfied and $\det \mathsf{Q} = 1$.

Problem 1.1.3 Write down (a) the Q corresponding to a rotation θ about $\mathbf{e}_2$, (b) the Q corresponding to a rotation ϕ about $\mathbf{e}_1$, (c) the Q corresponding to (b) followed by (a).

Problem 1.1.4 Show that

$$\mathrm{Q} = \begin{bmatrix} \cos 2\theta & \sin 2\theta & 0 \\ \sin 2\theta & -\cos 2\theta & 0 \\ 0 & 0 & 1 \end{bmatrix}$$

is an improper orthogonal matrix that represents a change of basis equivalent to a reflection in the plane through $\mathbf{e}_3$ inclined at a positive angle θ to $\mathbf{e}_1$.

1.1.3 Euclidean point space: Cartesian coordinates

The mechanical behaviour of continuous media is most conveniently described in terms of scalars, vectors and tensors which in general vary from point to point in the material, and may therefore be regarded as functions of position in the physical space occupied by the material. In order to express this formally in mathematical terms the notion of a Euclidean point space is required.

Let $\mathscr{E}$ be a set of elements which we refer to as *points*. If, for each pair (x, y) of points x, y of $\mathscr{E}$ there exists a vector,† denoted $\mathbf{v}(x, y)$, in $\mathbb{E}$ such that

$$\text{(a)}\quad \mathbf{v}(x, y) = \mathbf{v}(x, z) + \mathbf{v}(z, y) \tag{1.1.38}$$

for all x, y, z in $\mathscr{E}$, and

$$\text{(b)}\quad \mathbf{v}(x, y) = \mathbf{v}(x, z) \quad \text{if and only if } y = z \tag{1.1.39}$$

for each x in $\mathscr{E}$, then $\mathscr{E}$ is said to be a *Euclidean point space* (it is *not* a vector space).

From (a) it is easily shown that

$$\mathbf{v}(x, x) = \mathbf{0} \quad \text{for all } x \text{ in } \mathscr{E} \tag{1.1.40}$$

and hence that

$$\mathbf{v}(y, x) = -\mathbf{v}(x, y) \quad \text{for all } x,\ y \text{ in } \mathscr{E}. \tag{1.1.41}$$

For what follows it is convenient to adopt the notation $\mathbf{x}(y)$ for $\mathbf{v}(x, y)$. Then, if a fixed (but arbitrary) point o of $\mathscr{E}$ is chosen for reference $\mathbf{x}(o)$ is called the *position vector* of the point x relative to o, and o is referred to as the *origin*.

† Which may be thought of as a bilinear mapping from $\mathscr{E} \times \mathscr{E}$ to $\mathbb{E}$.

By (1.1.38) and (1.1.41), we obtain

$$\mathbf{x}(y) = \mathbf{x}(o) - \mathbf{y}(o),$$

and this is *independent* of the choice of o. It is therefore convenient to write this in the conventional form $\mathbf{x} - \mathbf{y}$ and to use the abbreviated notation $\mathbf{x}$ in place of $\mathbf{x}(o)$.

The distance $d(x, y)$ between two points x, y of $\mathscr{E}$ is defined by means of the dot product on $\mathbb{E}$ according to

$$d(x, y) = |\mathbf{x} - \mathbf{y}| = \{(\mathbf{x} - \mathbf{y})\cdot(\mathbf{x} - \mathbf{y})\}^{1/2}. \tag{1.1.42}$$

It is straightforward to establish that the bilinear mapping d from $\mathscr{E} \times \mathscr{E}$ to $\mathbb{R}$ is a *metric*, that is

(a) $d(x, y) = d(y, x)$

(b) $d(x, y) \leq d(x, z) + d(z, y)$

(c) $d(x, y) \geq 0$ with equality if and only if $x = y$

for all x, y, z in $\mathscr{E}$. (a) follows from the definition (1.1.42), (b) by use of the inequality $(\mathbf{x} - \mathbf{z})\cdot(\mathbf{z} - \mathbf{y}) \leq d(x, z)d(z, y)$, which can be obtained from (1.1.10) with (1.1.42), and (c) follows from (a) and (b) by putting $x = y$ and using (1.1.42).

Since $\mathscr{E}$ is endowed with a metric it is a *metric space.*

The angle θ between the lines joining o to x and o to y in $\mathscr{E}$ is also defined through the scalar product on $\mathbb{E}$. Thus, by (1.1.10),

$$\cos\theta = \mathbf{x}\cdot\mathbf{y}/|\mathbf{x}|\cdot|\mathbf{y}|$$

for an arbitrary choice of origin o.

With an origin o fixed in $\mathscr{E}$, an arbitrary point x of $\mathscr{E}$ corresponds to a unique position vector $\mathbf{x}$ in $\mathbb{E}$. Let $\{\mathbf{e}_i\}$ be an orthonormal basis for $\mathbb{E}$. Then the components x_i of $\mathbf{x}$ are given by $x_i = \mathbf{x}\cdot\mathbf{e}_i$ (Section 1.1.1). They may alternatively be defined by three mappings† from $\mathscr{E}$ to $\mathbb{R}$, e_i say, such that $e_i(x) = \mathbf{x}\cdot\mathbf{e}_i$ ($i = 1$, 2, 3) for every x in $\mathscr{E}$. The origin o, together with the collection of mappings e_i is denoted $\{o, e_i\}$ and this is said to form a *(rectangular) Cartesian coordinate system* on $\mathscr{E}$. The components x_i are called *(rectangular) Cartesian coordinates* of the point x in the coordinate system $\{o, e_i\}$. The distinction between e_i and $\mathbf{e}_i$ can be ignored for most of the applications envisaged in this book, and, moreover, when o is fixed the point

† In the terminology of Section 1.5, $e_i : \mathscr{E} \to \mathbb{R}$ ($i = 1, 2, 3$) is a *scalar field* over $\mathscr{E}$.

x may be identified with its position vector $\mathbf{x}$ relative to o. We shall emphasize this identification in later sections.

With respect to rectangular Cartesian coordinate systems $\{o, e_i\}$ and $\{o', e_i'\}$ the point x has coordinates x_i and x_i' respectively, where the basis vectors are related by (1.1.33). Let $\mathbf{o}'(o)$ be denoted by $\mathbf{c}$. Then

$$\mathbf{x}' = \mathbf{x}'(o') = \mathbf{x}(o) + \mathbf{o}(o') = \mathbf{x} - \mathbf{c} \tag{1.1.43}$$

and it follows that

$$x_i' = Q_{ij}(x_j - c_j), \quad x_i = Q_{ji}(x_j' + c_j'), \tag{1.1.44}$$

where c_i, c_i' are the components of $\mathbf{c}$ relative to the bases $\{\mathbf{e}_i\}$, $\{\mathbf{e}_i'\}$ respectively. When $\mathbf{c} = \mathbf{0}$ the transformation law (1.1.44) for the components of $\mathbf{x}$ is equivalent to that given in (1.1.35).

From (1.1.44) we obtain

$$Q_{ij} = \frac{\partial x_i'}{\partial x_j} = \frac{\partial x_j}{\partial x_i'} \tag{1.1.45}$$

and the chain rule for partial derivatives may be used to show that

$$Q_{ik}Q_{jk} = \frac{\partial x_i'}{\partial x_k}\frac{\partial x_k}{\partial x_j'} = \delta_{ij},$$

thus confirming (1.1.30).

Equation (1.1.45) describes the elements of the Jacobian matrix $(\partial x_i'/\partial x_j)$ corresponding to the (linear) coordinate transformation (1.1.44) and, in that such a matrix plays an important role for more general (non-linear) coordinate transformations, the above provides a lead in to our discussion of curvilinear coordinates in Section 1.5.4.

Until Section 1.5, however, we shall not consider basis vectors defined over $\mathscr{E}$ and in the meantime therefore we take the terms 'orthonormal' and 'rectangular Cartesian' to be equivalent.

1.2 CARTESIAN TENSORS

1.2.1 Motivation: stress in a continuum

Consider an infinitesimal element of surface area dS in a continuous medium. Let $\mathbf{n}$ be the unit normal to dS (by convention we take $\mathbf{n}$, rather than $-\mathbf{n}$, to be the *positive* unit normal; see Section 1.5.5 for a definition of positive unit normal). In general, the material on one side of dS exerts a force on the

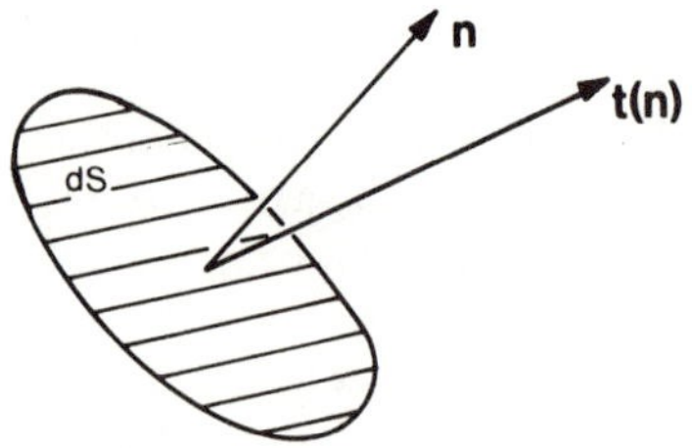

Fig. 1.2 The force **t**(**n**) per unit area acting on an infinitesimal surface area dS with unit normal **n**.

material on the other side (that into which **n** points in our convention). As indicated in Fig. 1.2, we denote the force by $\mathbf{t}(\mathbf{n})\,dS$, where $\mathbf{t}(\mathbf{n})$ is called the *stress vector* and is such that $\mathbf{t}(-\mathbf{n}) = -\mathbf{t}(\mathbf{n})$. It has dimensions of force per unit area and it depends on the orientation of dS, that is on **n**. In fact, as is shown in Chapter 3, $\mathbf{t}(\mathbf{n})$ depends linearly on **n**. We express this dependence by writing

$$\mathbf{t}(\mathbf{n}) = \mathbf{T}\mathbf{n}, \tag{1.2.1}$$

where **T**, which is independent of **n**, is a *linear mapping* from $\mathbb{E}$ into $\mathbb{E}$.

Equation (1.2.1) is in *invariant form* since it does not depend on any preferred choice of basis for $\mathbb{E}$. When referred to the orthonormal basis $\{\mathbf{e}_i\}$, however, (1.2.1) is decomposed as

$$t_i(\mathbf{n}) = T_{ij}n_j, \tag{1.2.2}$$

where T_{ij} are called the *components* of **T** relative to the basis $\{\mathbf{e}_i\}$.

If, for example, we take $\mathbf{n} = \mathbf{e}_1$ then the components of **n** are $n_i = \delta_{i1}$ and (1.2.2) becomes $t_i(\mathbf{e}_1) = T_{i1}$, and similarly $t_i(\mathbf{e}_2) = T_{i2}, t_i(\mathbf{e}_3) = T_{i3}$. The nine components T_{ij} can therefore be thought of as representing the components $t_i(\mathbf{e}_j)$ of the vectors $\mathbf{t}(\mathbf{e}_j), j = 1, 2, 3$, corresponding to the forces per unit area on three mutually perpendicular planes (at a point in the material). In other words, the *state of stress* in the material can be represented by the components T_{ij}, for any chosen basis $\{\mathbf{e}_i\}$, or, in invariant form, by the linear mapping **T**. This provides a physical interpretation of **T** which will be elaborated in Chapter 3.

The linear mapping **T** is also called a *tensor* on $\mathbb{E}$ (or more specifically a second-order tensor). In the present context it is a *stress tensor*, but, as will be seen in Chapter 3, many different tensor measures of stress can be constructed so we do not refer to **T** as *the* stress tensor. Unless indicated otherwise tensors are denoted by bold-face, upper-case letters such as $\mathbf{T}, \mathbf{U}, \mathbf{V}, \mathbf{W}, \ldots$ in this chapter.

As a specific example, we take $\mathbf{T} = -p\mathbf{I}$, where p is a scalar and $\mathbf{I}$ is the identity mapping on $\mathbb{E}$ (in components $T_{ij} = -p\delta_{ij}$), so that $\mathbf{t}(\mathbf{n}) = -p\mathbf{n}$. Physically, this corresponds to the situation in an inviscid fluid for which p acts in the direction normal to an arbitrary surface and no shearing (viscous) forces act parallel to the surface.

We now determine how the components of $\mathbf{T}$ transform under change of orthonormal basis. Let t_i, T_{ij}, n_j and t'_i, T'_{ij}, n'_j be the components of $\mathbf{t}, \mathbf{T}, \mathbf{n}$ with respect to bases $\{\mathbf{e}_i\}$ and $\{\mathbf{e}'_i\}$ respectively, the basis vectors being related by (1.1.33). Then

$$t_i = T_{ij}n_j, \quad t'_i = T'_{ip}n'_p. \tag{1.2.3}$$

But, since $\mathbf{t}$ and $\mathbf{n}$ are vectors, their components transform according to (1.1.35); thus

$$t'_i = Q_{ip}t_p, n'_p = Q_{pq}n_q. \tag{1.2.4}$$

Combining (1.2.3) and (1.2.4) we obtain

$$Q_{ip}T_{pq}n_q = T'_{ip}Q_{pq}n_q.$$

But this holds for arbitrary $\mathbf{n}$ so that

$$Q_{ip}T_{pq} = T'_{ip}Q_{pq},$$

and use of (1.1.32) after post-multiplication of this equation by Q_{jq} leads to

$$T'_{ij} = Q_{ip}Q_{jq}T_{pq}. \tag{1.2.5}$$

This is the transformation rule for the (rectangular Cartesian) components of the second-order tensor $\mathbf{T}$ under change of basis (1.1.33). Indeed, (1.2.5) provides the basis for the characterization of second- and higher-order Cartesian tensors as defined in the following section.

Let T, T' respectively denote the matrices (T_{ij}) and (T'_{ij}). Then (1.2.5) may be represented

$$\mathsf{T}' = \mathsf{Q}\mathsf{T}\mathsf{Q}^{\mathsf{T}}, \quad \mathsf{T} = \mathsf{Q}^{\mathsf{T}}\mathsf{T}'\mathsf{Q}.$$

We emphasize that it is important to distinguish between the tensor $\mathbf{T}$ and its component representation T. Although this matrix representation is useful for manipulative purposes in respect of second-order tensors it does not generalize conveniently to tensors of higher order.

1.2.2 Definition of a Cartesian tensor

An entity† **T** which has components $T_{ijk}\cdots$ (n indices) relative to a rectangular Cartesian basis $\{\mathbf{e}_i\}$ and transforms like

$$T'_{ijk}\cdots = Q_{ip}Q_{jq}Q_{kr}\cdots T_{pqr}\cdots \tag{1.2.6}$$

under a change of basis $\mathbf{e}_i \to \mathbf{e}'_i = Q_{ij}\mathbf{e}_j$, where $\mathsf{Q} \equiv (Q_{ij})$ is a (proper) orthogonal matrix, is called a *Cartesian tensor of order n*. We abbreviate this phrase as CT(n) for convenience. All indices run over values 1, 2, 3 so that a CT(n) has 3^n components.

Examples

(a) The tensor described in Section 1.2.1 is a CT(2), and its components transform according to (1.2.5).

(b) A vector is a CT(1) whose components transform in accordance with (1.1.35).

(c) A scalar is a CT(0) and is unchanged by a transformation of basis.

(d) The Krönecker delta is a CT(2) since

$$\delta'_{ij} = \mathbf{e}'_i\cdot\mathbf{e}'_j = (Q_{ip}\mathbf{e}_p)\cdot(Q_{jq}\mathbf{e}_q) = Q_{ip}Q_{jq}\delta_{pq},$$

use having been made of (1.1.12) and (1.1.33). Furthermore, it follows from the definition of δ_{pq} and (1.1.32) that $\delta'_{ij} = \delta_{ij}$, that is the components of the identity tensor **I** are unaffected by a change of basis. In fact, **I** is a member of an important special class of Cartesian tensors, referred to as *isotropic tensors*, which are discussed more fully in Section 1.2.5.

Problem 1.2.1 If **S** is a CT(3) and **T** is a CT(2) with components S_{ijk}, T_{lm} respectively with respect to the basis $\{\mathbf{e}_i\}$, show that $S_{ijk}T_{lm}$ are the components of a CT(5). Generalize this result to the situation where **S** is a CT(m) and **T** is a CT(n). (An invariant notation to represent such a product is introduced in Section 1.2.3.)

Deduce that $U_{ijk} \equiv S_{ijp}T_{kp}$, $v_i = S_{ijk}T_{jk}$ are respectively the components of a CT(3) and a CT(1).

Problem 1.2.2 If **T** is a CT(2) and $\mathbf{Tn} = \mathbf{0}$ for arbitrary vectors $\mathbf{n}$ show that

†This is an unsatisfactory word to describe what is more formally called a multilinear mapping. The abstract definition is postponed until Section 1.3.

$\mathbf{T} = \mathbf{0}$ (the zero second-order tensor), that is $T_{ij} = 0$ with respect to an arbitrary basis. (This result was used in the derivation of equation (1.2.5).)

Problem 1.2.3 If $\mathbf{S}$ and $\mathbf{T}$ are each CT(n)'s and α, β are scalars, prove that $\alpha\mathbf{S} + \beta\mathbf{T}$ is also a CT(n). This shows that Cartesian tensors of the *same order* may be added, but no meaning can be attached to the sum of tensors of different orders.

Problem 1.2.4 Show that $A^{\pm}_{ijkl} = \frac{1}{2}(\delta_{ik}\delta_{jl} \pm \delta_{il}\delta_{jk})$ are the components of a CT(4) and prove that $A^{+}_{ijkk} = \delta_{ij}$, $A^{-}_{ijkk} = 0$.

Problem 1.2.5 If E_{ij} are the components of a CT(2) and λ, μ are scalars, deduce that

$$T_{ij} = 2\mu E_{ij} + \lambda E_{kk}\delta_{ij}$$

are also the components of a CT(2), and show that

$$\begin{aligned} T_{kk} &= (3\lambda + 2\mu)E_{kk}, \\ T_{ij}E_{ij} &= 2\mu E_{ij}E_{ij} + \lambda(E_{kk})^2 \\ T_{ij}T_{ij} &= 4\mu^2 E_{ij}E_{ij} + \lambda(3\lambda + 4\mu)(E_{kk})^2. \end{aligned}$$

If $\bar{E}_{ij}$ and $\bar{T}_{ij}$ are defined by

$$\bar{E}_{ij} = E_{ij} - \tfrac{1}{3}E_{kk}\delta_{ij}, \quad \bar{T}_{ij} = T_{ij} - \tfrac{1}{3}T_{kk}\delta_{ij},$$

deduce that $\bar{E}_{kk} = \bar{T}_{kk} = 0$, $\bar{T}_{ij} = 2\mu\bar{E}_{ij}$, and hence show that

$$\begin{aligned} T_{ij}E_{ij} &= \bar{T}_{ij}\bar{E}_{ij} + \tfrac{1}{3}T_{kk}E_{ll}, \\ \bar{E}_{ij}\bar{E}_{ij} &= \bar{E}_{ij}E_{ij} = E_{ij}E_{ij} - \tfrac{1}{3}(E_{kk})^2. \end{aligned}$$

1.2.3 The tensor product

Consider two vectors $\mathbf{u}$ and $\mathbf{v}$. The product $u_i v_j$ of their components transforms according to

$$u_i' v_j' = Q_{ip}Q_{jq}u_p v_q,$$

so the tensor with components $u_i v_j$ with respect to the basis $\{\mathbf{e}_i\}$ is a CT(2). It is denoted by $\mathbf{u} \otimes \mathbf{v}$, and is called the *tensor product* (or dyadic product) of $\mathbf{u}$ and $\mathbf{v}$, so that $(\mathbf{u} \otimes \mathbf{v})_{ij} = u_i v_j$. It follows that $\mathbf{v} \otimes \mathbf{u} \neq \mathbf{u} \otimes \mathbf{v}$ in general. In fact, $\mathbf{v} \otimes \mathbf{u}$ is the transpose of $\mathbf{u} \otimes \mathbf{v}$, denoted $(\mathbf{u} \otimes \mathbf{v})^{\mathrm{T}}$ by analogy with the notation used for matrix transposes in Section 1.1.2.

Since $(\mathbf{u}\otimes\mathbf{v})_{ij}w_j = v_j w_j u_i$ we have, in invariant notation,

$$(\mathbf{u}\otimes\mathbf{v})\mathbf{w} = (\mathbf{v}\cdot\mathbf{w})\mathbf{u} \tag{1.2.7}$$

for all $\mathbf{u}$, $\mathbf{v}$, $\mathbf{w}$ in $\mathbb{E}$.

In respect of the basis vectors $\mathbf{e}_i$ equation (1.2.7) gives

$$(\mathbf{e}_i\otimes\mathbf{e}_j)\mathbf{n} = n_j\mathbf{e}_i \tag{1.2.8}$$

for all vectors $\mathbf{n}$ in $\mathbb{E}$. If $\mathbf{T}$ is an arbitrary CT(2) with components T_{ij} with respect to the basis $\{\mathbf{e}_i\}$ then multiplication of (1.2.8) by T_{ij} with summation over i and j leads to

$$(T_{ij}\mathbf{e}_i\otimes\mathbf{e}_j)\mathbf{n} = T_{ij}n_j\mathbf{e}_i = \mathbf{Tn}$$

since, by definition, $\mathbf{Tn}$ has components $T_{ij}n_j$ (see Section 1.2.1).

Since $\mathbf{n}$ is arbitrary, it follows from the result of Problem 1.2.2 that

$$\mathbf{T} = T_{ij}\mathbf{e}_i\otimes\mathbf{e}_j. \tag{1.2.9}$$

Thus (1.2.9) provides a representation for an arbitrary CT(2) $\mathbf{T}$ with respect to an arbitrarily chosen basis $\{\mathbf{e}_i\}$. (The tensors $\mathbf{e}_i\otimes\mathbf{e}_j$ in fact form a basis for the nine-dimensional vector space of linear mappings from $\mathbb{E}$ to $\mathbb{E}$. This will be discussed in more detail in Section 1.3.)

In passing we remark that in general a pair of vectors $\mathbf{u}$, $\mathbf{v}$ cannot be found such that an arbitrary CT(2) can be represented in the form $\mathbf{u}\otimes\mathbf{v}$.

For the identity tensor $\mathbf{I}$, with components δ_{ij}, we have

$$\mathbf{I} = \delta_{ij}\mathbf{e}_i\otimes\mathbf{e}_j = \mathbf{e}_i\otimes\mathbf{e}_i \tag{1.2.10}$$

for an arbitrary (orthonormal) basis $\{\mathbf{e}_i\}$.

The tensor product may be applied repeatedly to any number of vectors or tensors. For example, $\mathbf{u}\otimes\mathbf{v}\otimes\mathbf{w}$ is a third-order tensor with components $u_i v_j w_k$ with respect to the basis $\{\mathbf{e}_i\}$. There is no distinction here between this and either of $\mathbf{u}\otimes(\mathbf{v}\otimes\mathbf{w})$ or $(\mathbf{u}\otimes\mathbf{v})\otimes\mathbf{w}$ each of which may be decomposed as $u_i v_j w_k \mathbf{e}_i\otimes\mathbf{e}_j\otimes\mathbf{e}_k$. The most general CT(3) may be represented in the form $T_{ijk}\mathbf{e}_i\otimes\mathbf{e}_j\otimes\mathbf{e}_k$.

More generally, a CT(n) may be represented

$$\mathbf{T} = T_{i_1 i_2 \ldots i_n}\mathbf{e}_{i_1}\otimes\mathbf{e}_{i_2}\otimes\cdots\otimes\mathbf{e}_{i_n}$$

and the tensor product $\mathbf{S}\otimes\mathbf{T}$ of a CT(m) $\mathbf{S}$ and a CT(n) $\mathbf{T}$ is a CT($m+n$)

such that

$$\mathbf{S}\otimes\mathbf{T}=S_{i_1i_2\ldots i_m}T_{j_1j_2\ldots j_n}\mathbf{e}_{i_1}\otimes\mathbf{e}_{i_2}\otimes\cdots\otimes\mathbf{e}_{i_m}\otimes\mathbf{e}_{j_1}\otimes\mathbf{e}_{j_2}\otimes\cdots\otimes\mathbf{e}_{j_n}.$$

1.2.4 Contraction

Consider the components $T_{i_1i_2\ldots i_p\ldots i_q\ldots i_n}$ of a CT(n). Set any two indices equal, $i_q=i_p$ say, and sum over i_p from 1 to 3. These indices are then said to be *contracted* and the order of the tensor is reduced by two. This result follows from the transformation rule (1.2.6) on use of (1.1.32). The general proof is left as an exercise for the reader, but we examine here some special cases.

(a) The tensor product $\mathbf{u}\otimes\mathbf{v}$ of two vectors becomes their dot product $\mathbf{u}\cdot\mathbf{v}$ on contraction since u_iv_j becomes u_iv_i.

(b) If $\mathbf{T}$ is a CT(2) with components T_{ij} then this contracts to the scalar T_{ii}. Since

$$T'_{ij}=Q_{ip}Q_{jq}T_{pq}$$

it follows by (1.1.32) that

$$T'_{ii}=\delta_{pq}T_{pq}=T_{pp}.$$

This scalar is called the *trace* of $\mathbf{T}$, and is denoted by tr $\mathbf{T}$. It is a particular example of a *scalar invariant* of $\mathbf{T}$ (full discussion of scalar invariants is given in Section 1.3 and Chapter 4).

(c) If $\mathbf{S}$ and $\mathbf{T}$ are CT(2)'s then $\mathbf{S}\otimes\mathbf{T}$ is a CT(4).

Let $S_{ij}T_{kl}$ be the components of $\mathbf{S}\otimes\mathbf{T}$ with respect to the basis $\{\mathbf{e}_i\}$. Then the indices can be contracted in a number of ways.

For example $S_{ij}T_{jl}$ are the components of the CT(2) which is written $\mathbf{ST}$ and this in turn contracts to the scalar $\operatorname{tr}(\mathbf{ST})=S_{ij}T_{ji}=\operatorname{tr}(\mathbf{TS})$. Similarly $S_{ij}T_{kj}$ are the components of $\mathbf{ST}^{\mathrm{T}}$ when $\mathbf{T}^{\mathrm{T}}$ denotes the transpose of $\mathbf{T}$ (that is $\mathbf{T}^{\mathrm{T}}=T_{ij}\mathbf{e}_j\otimes\mathbf{e}_i$). The results $\operatorname{tr}(\mathbf{ST})=\operatorname{tr}(\mathbf{S}^{\mathrm{T}}\mathbf{T}^{\mathrm{T}})$, $\operatorname{tr}(\mathbf{ST}^{\mathrm{T}})=\operatorname{tr}(\mathbf{S}^{\mathrm{T}}\mathbf{T})$ are then easily established.

If $\mathbf{S}=\mathbf{T}$ then $\mathbf{TT}$ is denoted by $\mathbf{T}^2$, $\mathbf{TT}^2$ by $\mathbf{T}^3$ and so on, and $\operatorname{tr}(\mathbf{T}^2)$, $\operatorname{tr}(\mathbf{T}^3)$ are further examples of scalar invariants of $\mathbf{T}$.

For the products of higher-order tensors there are many possible contractions available but we do not need to go into details here.

Problem 1.2.6 If $\mathbf{T}$ is an arbitrary CT(2) and $\mathbf{I}$ is the identity tensor on $\mathbb{E}$, deduce that $\mathbf{T}=\mathbf{TI}$. Hence use the representation (1.2.10) for $\mathbf{I}$ to show that

T has the representation (1.2.9). The decomposition $\mathbf{Tn} = T_{ik}n_k\mathbf{e}_i$ may be used.

1.2.5 Isotropic tensors

If the components of a Cartesian tensor **T** are unchanged under an arbitrary (subject to right-handedness being maintained) transformation of rectangular Cartesian basis then **T** is said to be an *isotropic tensor.*

Examples

(a) CT(0): all scalars are isotropic.

(b) CT(1): there are no non-trivial isotropic vectors.

If **v** is an isotropic vector then its components must satisfy $Q_{ij}v_j = v_i$ for arbitrary proper orthogonal matrices Q. The choice

$$\mathsf{Q} = \begin{bmatrix} 0 & 1 & 0 \\ -1 & 0 & 0 \\ 0 & 0 & 1 \end{bmatrix},$$

corresponding to $\theta = \pi/2$ in (1.1.37), leads to $v_1 = v_2 = 0$ and, similarly, another choice of Q give $v_3 = 0$. Hence $\mathbf{v} = \mathbf{0}$ is the only isotropic vector.

(c) CT(2): scalar multiples of the identity **I** are the only isotropic CT(2). For this particular example we work through the proof in detail. Let **T** be an isotropic CT(2). Then its components T_{ij} must satisfy

$$Q_{ip}Q_{jq}T_{pq} = T_{ij}$$

or, in matrix form,

$$\mathsf{QTQ^T} = \mathsf{T} \tag{1.2.11}$$

for all proper orthogonal Q.

The choice of Q used in (b) above leads to

$$\begin{bmatrix} T_{22} & -T_{21} & T_{23} \\ -T_{12} & T_{11} & -T_{13} \\ T_{32} & -T_{31} & T_{33} \end{bmatrix} = \begin{bmatrix} T_{11} & T_{12} & T_{13} \\ T_{21} & T_{22} & T_{23} \\ T_{31} & T_{32} & T_{33} \end{bmatrix}$$

so that $T_{22} = T_{11}$, $T_{12} = -T_{21}$, $T_{23} = T_{13} = T_{31} = T_{32} = 0$.

The choice

$$\mathsf{Q} = \begin{bmatrix} 1 & 0 & 0 \\ 0 & 0 & 1 \\ 0 & -1 & 0 \end{bmatrix}$$

then yields $T_{12}=0$, $T_{33}=T_{11}$, so that $T_{ij}=T_{11}\delta_{ij}$. Since (1.2.11) is unaffected by multiplication of T by an arbitrary scalar, the required result follows.

(d) CT(3): scalar multiples of the tensor which has components ε_{ijk} (defined by (1.1.16)) are the only isotropic CT(3)'s.

Firstly, we note from (1.1.23) that

$$\varepsilon'_{ijk} \equiv Q_{ip}Q_{jq}Q_{kr}\varepsilon_{pqr} = (\det \mathsf{Q})\varepsilon_{ijk}.$$

If Q is proper orthogonal then $\det \mathsf{Q}=1$ and it follows that $\varepsilon'_{ijk}=\varepsilon_{ijk}$ in accordance with the definition of isotropy. If $\det \mathsf{Q} = -1$, on the other hand, then $\varepsilon'_{ijk} = -\varepsilon_{ijk}$. This problem does not arise in respect of even-order tensors, but explains why proper orthogonal changes of basis are used in the definition of isotropy. It follows that scalar multiples of ε_{ijk} are isotropic, but it is left to the reader to prove that these are the only ones (follow the method used for CT(2)'s above).

(e) CT(4): if the tensor **T** has components T_{ijkl} with respect to an arbitrary basis $\{\mathbf{e}_i\}$ then the only independent forms of T_{ijkl} are scalar multiples of

$$\delta_{ij}\delta_{kl}, \qquad \delta_{ik}\delta_{jl}, \qquad \delta_{il}\delta_{jk}.$$

The most general isotropic CT(4) is therefore expressible in the component form

$$T_{ijkl} = \alpha\delta_{ij}\delta_{kl} + \beta\delta_{ik}\delta_{jl} + \gamma\delta_{il}\delta_{jk}, \tag{1.2.12}$$

where α, β, γ are scalars. The proof of this is omitted, but details can be found in Jeffreys (1952), for example.

As an example of the application of the above we note from (d) that, since $\varepsilon'_{ijk}=\varepsilon_{ijk}$ then $\varepsilon'_{ijk}\varepsilon'_{pqk}=\varepsilon_{ijk}\varepsilon_{pqk}$ are the components of a fourth-order isotropic tensor. Hence it is expressible in the form

$$\varepsilon_{ijk}\varepsilon_{pqk} = \alpha\delta_{ij}\delta_{pq} + \beta\delta_{ip}\delta_{jq} + \gamma\delta_{iq}\delta_{jp},$$

where α, β, γ are to be determined.

But, because of the antisymmetry property (1.1.18) of ε_{ijk} it follows immediately that $\alpha=0$ and $\gamma=-\beta$. The choice $ij=pq=12$ gives $\beta=1$ and the result (1.1.27) is obtained.

(f) Isotropic tensors of all higher orders have components expressible as linear combinations of products of Krönecker deltas and alternating symbols†.

†A proof is given by Jeffreys (1973).

For CT(5), for example, with *ijklm* as indices, there are ten sets of components (not all independent), namely

$$\delta_{ij}\varepsilon_{klm}, \delta_{ik}\varepsilon_{jlm}, \delta_{il}\varepsilon_{jkm}, \delta_{im}\varepsilon_{jkl}, \delta_{jk}\varepsilon_{ilm},$$
$$\delta_{jl}\varepsilon_{ikm}, \delta_{jm}\varepsilon_{ikl}, \delta_{kl}\varepsilon_{ijm}, \delta_{km}\varepsilon_{ijl}, \delta_{lm}\varepsilon_{ijk}.$$

Every isotropic CT(6) has components expressible as linear combinations of products of Krönecker deltas alone. Amongst the indices *ijklmn* there are fifteen independent products of the form $\delta_{ij}\delta_{kl}\delta_{mn}$. A consequence of this is that $\varepsilon_{ijk}\varepsilon_{lmn}$, being the components of an isotropic CT(6), is expressible as

$$\begin{aligned}\varepsilon_{ijk}\varepsilon_{lmn} = \delta_{il}\delta_{jm}\delta_{kn} + \delta_{im}\delta_{jn}\delta_{kl} + \delta_{in}\delta_{jl}\delta_{km}\\ - \delta_{il}\delta_{jn}\delta_{km} - \delta_{im}\delta_{jl}\delta_{km} - \delta_{in}\delta_{jm}\delta_{kl},\end{aligned} \tag{1.2.13}$$

which is equivalent to (1.1.26) The reader is invited to consider the consequences of these results for isotropic CT(n)'s of all higher even and odd orders.

Finally in this section we remark (i) for isotropic tensors of even order eight and higher not all the products of Krönecker deltas are independent and (ii) the above results apply if $\mathbb{E}$ is three-dimensional; if $\mathbb{E}$ has dimension two or greater than three parallel results can be obtained.

1.3 TENSOR ALGEBRA

Following the discussion of Cartesian tensors in terms of component transformations under change of basis in Section 1.2 we now develop the theory in its equivalent invariant form. Since second-order tensors are most important in applications attention is confined to these in Sections 1.3.1 to 1.3.5, while Section 1.3.6 is devoted to a short discussion of higher-order tensors.

1.3.1 Second-order tensors

As we indicated in Section 1.2 a second-order Cartesian tensor $\mathbf{T}$ can be regarded as a linear mapping of the vector space $\mathbb{E}$ into itself. We write $\mathbf{T}:\mathbb{E}\to\mathbb{E}$ to indicate this, and the element (vector) in $\mathbb{E}$ to which the vector $\mathbf{u}$ maps is denoted $\mathbf{Tu}$. This viewpoint embodies a fairly general definition of a tensor which does not involve a choice of basis for $\mathbb{E}$ (either orthonormal or otherwise) and this is the starting point for the development in this section.

The tensor $\mathbf{T}$ is said to be *linear* if

$$\mathbf{T}(\alpha\mathbf{u} + \beta\mathbf{v}) = \alpha\mathbf{Tu} + \beta\mathbf{Tv} \tag{1.3.1}$$

for all $\mathbf{u}, \mathbf{v} \in \mathbb{E}$ and all $\alpha, \beta \in \mathbb{R}$. If an orthonormal basis $\{\mathbf{e}_i\}$ is chosen for $\mathbb{E}$ then

(1.3.1) has the component form

$$T_{ij}(\alpha u_j + \beta v_j) = \alpha T_{ij} u_j + \beta T_{ij} v_j.$$

For the most part, however, we avoid using such component representations in this section.

Let $\mathscr{L}(\mathbb{E}, \mathbb{E})$ denote the set of all linear mappings from $\mathbb{E}$ to $\mathbb{E}$. Then $\mathscr{L}(\mathbb{E}, \mathbb{E})$ is itself a vector space with the element $\alpha\mathbf{S} + \beta\mathbf{T}$ defined according to

$$(\alpha\mathbf{S} + \beta\mathbf{T})\mathbf{u} = \alpha(\mathbf{Su}) + \beta(\mathbf{Tu}) \tag{1.3.2}$$

for all $\mathbf{S}, \mathbf{T} \in \mathscr{L}(\mathbb{E}, \mathbb{E})$, all $\alpha, \beta \in \mathbb{R}$ and all $\mathbf{u} \in \mathbb{E}$.

The *inner product* $\mathbf{ST}$ is defined by

$$(\mathbf{ST})\mathbf{u} = \mathbf{S}(\mathbf{Tu}) \tag{1.3.3}$$

for all $\mathbf{S}, \mathbf{T} \in \mathscr{L}(\mathbb{E}, \mathbb{E})$ and all $\mathbf{u} \in \mathbb{E}$. Note that $\mathbf{ST}$ has the Cartesian component form $S_{ik}T_{kj}$ which represents a contraction of $\mathbf{S} \otimes \mathbf{T}$, as discussed in Section 1.2.4(c).

The *zero tensor* $\mathbf{0}$ maps every vector in $\mathbb{E}$ to the zero vector $\mathbf{o}$ and the *identity tensor* $\mathbf{I}$ maps every vector in $\mathbb{E}$ to itself. Thus

$$\mathbf{0u} = \mathbf{o}, \qquad \mathbf{Iu} = \mathbf{u} \tag{1.3.4}$$

for all $\mathbf{u} \in \mathbb{E}$.

The set of all bilinear functions over $\mathbb{E} \times \mathbb{E}$ forms a vector space over $\mathbb{R}$, denoted by $\mathscr{L}(\mathbb{E} \times \mathbb{E}, \mathbb{R})$, which may be identified† with $\mathscr{L}(\mathbb{E}, \mathbb{E})$.

If an orthonormal basis $\{\mathbf{e}_i\}$ is chosen for $\mathbb{E}$ then the bilinear function $(\mathbf{e}_i \otimes \mathbf{e}_j)$ is defined so that

$$(\mathbf{e}_i \otimes \mathbf{e}_j)(\mathbf{u}, \mathbf{v}) = u_i v_j \tag{1.3.5}$$

for all $\mathbf{u}, \mathbf{v} \in \mathbb{E}$.

For an arbitrary member $\mathbf{T}$ of $\mathscr{L}(\mathbb{E} \times \mathbb{E}, \mathbb{R})$ equation (1.3.5) leads to

$$\mathbf{T}(\mathbf{u}, \mathbf{v}) = u_i v_j \mathbf{T}(\mathbf{e}_i, \mathbf{e}_j) \tag{1.3.6}$$

and hence

$$\mathbf{T}(\mathbf{u}, \mathbf{v}) = \mathbf{T}(\mathbf{e}_i, \mathbf{e}_j)(\mathbf{e}_i \otimes \mathbf{e}_j)(\mathbf{u}, \mathbf{v}).$$

Since this holds for all $\mathbf{u}, \mathbf{v} \in \mathbb{E}$, we obtain the representation

$$\mathbf{T} = \mathbf{T}(\mathbf{e}_i, \mathbf{e}_j)\mathbf{e}_i \otimes \mathbf{e}_j$$

† Strictly $\mathscr{L}(\mathbb{E} \times \mathbb{E}, \mathbb{R})$ is isomorphic to $\mathscr{L}(\mathbb{E}, \mathbb{E})$, i.e. there is a one-to-one correspondence between the elements of the two spaces. We use the same symbol, $\mathbf{T}$ for example, to denote either an element of $\mathscr{L}(\mathbb{E} \times \mathbb{E}, \mathbb{R})$ or the corresponding element of $\mathscr{L}(\mathbb{E}, \mathbb{E})$.

for $\mathbf{T}$ with respect to the basis $\{\mathbf{e}_i\}$. This is equivalent to (1.2.9) and we identify $\mathbf{T}(\mathbf{e}_i, \mathbf{e}_j)$ as the component T_{ij} of $\mathbf{T}$ relative to the basis $\{\mathbf{e}_i\}$. Equation (1.3.6) then becomes

$$\mathbf{T}(\mathbf{u}, \mathbf{v}) = u_i T_{ij} v_j \tag{1.3.7}$$

and this can also be written as the scalar product $\mathbf{u}$ with $\mathbf{Tv}$. Thus

$$\mathbf{T}(\mathbf{u}, \mathbf{v}) = \mathbf{u}\cdot(\mathbf{Tv}). \tag{1.3.8}$$

In general, however, this is not equal to $\mathbf{v}\cdot(\mathbf{Tu})$. This leads us to define the transpose $\mathbf{T}^{\mathrm{T}}$, also a member of $\mathscr{L}(\mathbb{E} \times \mathbb{E}, \mathbb{R})$, of $\mathbf{T}$ by

$$\mathbf{v}\cdot(\mathbf{T}^{\mathrm{T}}\mathbf{u}) = \mathbf{u}\cdot(\mathbf{Tv}) \tag{1.3.9}$$

for all $\mathbf{u}, \mathbf{v} \in \mathbb{E}$, or, equivalently,

$$\mathbf{T}^{\mathrm{T}}(\mathbf{v}, \mathbf{u}) = \mathbf{T}(\mathbf{u}, \mathbf{v}).$$

(This generalizes the definition given in Section 1.2.4 in respect of Cartesian tensors.) With respect to an arbitrary orthonormal basis the components of $\mathbf{T}^{\mathrm{T}}$ are given by

$$(\mathbf{T}^{\mathrm{T}})_{ij} = T_{ji}.$$

The properties

$$\left.\begin{aligned} (\mathbf{T}^{\mathrm{T}})^{\mathrm{T}} &= \mathbf{T}, \\ (\alpha\mathbf{S} + \beta\mathbf{T})^{\mathrm{T}} &= \alpha\mathbf{S}^{\mathrm{T}} + \beta\mathbf{T}^{\mathrm{T}}, \\ (\mathbf{ST})^{\mathrm{T}} &= \mathbf{T}^{\mathrm{T}}\mathbf{S}^{\mathrm{T}}, \end{aligned}\right\} \tag{1.3.10}$$

for all $\mathbf{S}$, $\mathbf{T} \in \mathscr{L}(\mathbb{E} \times \mathbb{E}, \mathbb{R})$ and all $\alpha, \beta \in \mathbb{R}$ follow immediately from the definitions.

A second-order tensor $\mathbf{T}$ is said to be *symmetric* if $\mathbf{T}^{\mathrm{T}} = \mathbf{T}$. In components, this implies that $T_{ij} = T_{ji}$ with respect to an arbitrary basis $\{\mathbf{e}_i\}$. Note that the identity tensor $\mathbf{I}$ is symmetric and

$$\mathbf{I}(\mathbf{u}, \mathbf{v}) = \mathbf{u}\cdot\mathbf{v}.$$

A second-order tensor $\mathbf{T}$ is said to be *skew-symmetric* (or *antisymmetric*) if $\mathbf{T}^{\mathrm{T}} = -\mathbf{T}$, or, in components $T_{ij} = -T_{ji}$.

The reader should confirm that symmetry or antisymmetry of the components T_{ij} is preserved under change of orthonormal basis. Note that a symmetric tensor has six independent components while an antisymmetric one has three, and that a general second-order tensor (with nine independent components) can be written as the sum of a symmetric part and an antisymmetric part. Thus

$$\mathbf{T} = \tfrac{1}{2}(\mathbf{T} + \mathbf{T}^{\mathrm{T}}) + \tfrac{1}{2}(\mathbf{T} - \mathbf{T}^{\mathrm{T}}).$$

The *trace* of $\mathbf{T}$ may be defined with respect to an orthonormal basis $\{\mathbf{e}_i\}$ as in Section 1.2.4(b). Thus

$$\operatorname{tr}\mathbf{T} = T_{ii} = \mathbf{e}_i\cdot(\mathbf{T}\mathbf{e}_i) = \mathbf{T}(\mathbf{e}_i, \mathbf{e}_i).$$

Likewise, the determinant of $\mathbf{T}$ is defined as the determinant of the matrix T of components of $\mathbf{T}$ with respect to an orthonormal basis. Thus, by (1.1.22),

$$\det\mathbf{T} = \varepsilon_{ijk}T_{i1}T_{j2}T_{k3}.$$

Using the fact that ε_{ijk} are the components of an isotropic tensor it is easy to establish, by (1.1.23) applied to Q_{ij} and the transformation rule (1.2.5), that $\det\mathbf{T}$ is independent of the choice of basis. In other words, like $\operatorname{tr}\mathbf{T}$, $\det\mathbf{T}$ is a scalar invariant of $\mathbf{T}$.

If $\det\mathbf{T} \neq 0$ then there exists a unique *inverse tensor*, denoted $\mathbf{T}^{-1}$, such that

$$\left.\begin{aligned} \mathbf{T}\mathbf{T}^{-1} &= \mathbf{I} = \mathbf{T}^{-1}\mathbf{T}, \\ \det(\mathbf{T}^{-1}) &= (\det\mathbf{T})^{-1}, \\ (\mathbf{S}\mathbf{T})^{-1} &= \mathbf{T}^{-1}\mathbf{S}^{-1} \end{aligned}\right\} \tag{1.3.11}$$

for all $\mathbf{S}, \mathbf{T} \in \mathscr{L}(\mathbb{E}\times\mathbb{E}, \mathbb{R})$ such that $\det\mathbf{S} \neq 0$.

The *adjugate tensor* of $\mathbf{T}$, denoted $\operatorname{adj}\mathbf{T}$, is then defined by

$$\operatorname{adj}(\mathbf{T}^{\mathrm{T}}) = (\det\mathbf{T})\mathbf{T}^{-1}, \tag{1.3.12}$$

although it may also be defined when $\mathbf{T}^{-1}$ does not exist. It is easily shown that

$$\det(\operatorname{adj}\mathbf{T}) = (\det\mathbf{T})^2.$$

Problem 1.3.1 Suppose $A^{\pm}_{ijkl}$ are as defined in Problem 1.2.4. Show that (a) if T_{ij} are the components of a symmetric CT(2) then

$$A^{+}_{ijkl}T_{kl} = T_{ij}, \qquad A^{-}_{ijkl}T_{kl} = 0$$

and (b) if T_{ij} are the components of a skew-symmetric CT(2) then

$$A^+_{ijkl}T_{kl}=0, \qquad A^-_{ijkl}T_{kl}=T_{ij}.$$

Problem 1.3.2 If T_{ij} are the components of an arbitrary CT(2), show that $\varepsilon_{ijk}T_{jk}$ are the components of a CT(1). Deduce that T_{ij} is symmetric if and only if $\varepsilon_{ijk}T_{jk}=0$.

Problem 1.3.3 If W_{ij} are the components of an antisymmetric CT(2) $\mathbf{W}$ then the vector $\mathbf{w}$ with components $w_i=\frac{1}{2}\varepsilon_{ijk}W_{kj}$ is called the *axial vector* of $\mathbf{W}$. Show that $\varepsilon_{ipq}w_i=W_{qp}$ and that, for an arbitrary vector $\mathbf{a}$, $\mathbf{w}\wedge\mathbf{a}=\mathbf{Wa}$.

Deduce that $\mathbf{u}\wedge\mathbf{v}$ is the axial vector of $\mathbf{v}\otimes\mathbf{u}-\mathbf{u}\otimes\mathbf{v}$.

Problem 1.3.4 If

$$\mathsf{T}\equiv\begin{bmatrix}0 & T_{12} & 0\\ -T_{12} & 0 & 0\\ 0 & 0 & 0\end{bmatrix}$$

is the matrix representing the components of an antisymmetric CT(2) $\mathbf{T}$ with respect to basis $\{\mathbf{e}_i\}$, show that, for any change of basis $\mathbf{e}_i\to\mathbf{e}'_i=Q_{ij}\mathbf{e}_j$ such that $\mathbf{e}'_3=\mathbf{e}_3$, the matrix representing the components of $\mathbf{T}$ is unchanged.

1.3.2 Eigenvalues and eigenvectors of a second-order tensor

Let $\mathbf{T}$ be a second-order tensor. A vector $\mathbf{v}$ in $\mathbb{E}$ is called an *eigenvector*† of $\mathbf{T}$ if there is a scalar, λ say, such that

$$\mathbf{Tv}=\lambda\mathbf{v}. \tag{1.3.13}$$

Then λ is called the *eigenvalue*† of $\mathbf{T}$ corresponding to the eigenvector $\mathbf{v}$.

The set of homogeneous equations (1.3.13) has non-trivial solutions for $\mathbf{v}$ if and only if

$$\det(\mathbf{T}-\lambda\mathbf{I})=0. \tag{1.3.14}$$

This is called the *characteristic equation for* $\mathbf{T}$, which may also be written

$$\det(T_{ij}-\lambda\delta_{ij})=0 \tag{1.3.15}$$

in terms of Cartesian components.

†The terms *proper vector* (*proper number*) or *characteristic vector* (*characteristic value*) are also in common use.

Expansion of the determinant (1.3.15) leads to the equation

$$\lambda^3 - I_1(\mathbf{T})\lambda^2 + I_2(\mathbf{T})\lambda - I_3(\mathbf{T}) = 0, \tag{1.3.16}$$

where

$$\left.\begin{aligned} I_1(\mathbf{T}) &= \operatorname{tr}\mathbf{T}, \\ I_2(\mathbf{T}) &= \tfrac{1}{2}\{(\operatorname{tr}\mathbf{T})^2 - \operatorname{tr}\mathbf{T}^2\}, \\ I_3(\mathbf{T}) &= \det\mathbf{T} = \tfrac{1}{6}\{(\operatorname{tr}\mathbf{T})^3 - 3(\operatorname{tr}\mathbf{T})(\operatorname{tr}\mathbf{T}^2) + 2\operatorname{tr}\mathbf{T}^3\}. \end{aligned}\right\} \tag{1.3.17}$$

These are called the *principal invariants of* $\mathbf{T}$. We have already noted in Sections 1.2.4 and 1.3.1 that $\operatorname{tr}\mathbf{T}$, $\operatorname{tr}\mathbf{T}^2$ and $\det\mathbf{T}$ are scalar invariants of $\mathbf{T}$.

To each real solution λ of the cubic equation (1.3.16) corresponds a real eigenvector. Later we shall see an example in which (1.3.16) yields three real eigenvalues and examples in which there is only one.

By repeated application of $\mathbf{T}$ to equation (1.3.13), we obtain

$$\mathbf{T}^r\mathbf{v} = \lambda^r\mathbf{v} \tag{1.3.18}$$

for any positive integer r. Multiplication of $\mathbf{v}$ by equation (1.3.16) and use of (1.3.18) then leads to

$$\mathbf{T}^3 - I_1(\mathbf{T})\mathbf{T}^2 + I_2(\mathbf{T})\mathbf{T} - I_3(\mathbf{T})\mathbf{I} = \mathbf{0}. \tag{1.3.19}$$

This is the mathematical expression of the *Cayley–Hamilton theorem* which states that a (second-order) tensor satisfies its own characteristic equation. (The above proof is valid only if (1.3.18) is applied to three linearly independent eigenvectors of $\mathbf{T}$.)

Problem 1.3.5 If $\det\mathbf{T} \neq 0$ deduce from (1.3.14) that

$$\det(\mathbf{T}^{-1} - \lambda^{-1}\mathbf{I}) = 0$$

and hence

$$\lambda^{-3} - I_1(\mathbf{T}^{-1})\lambda^{-2} + I_2(\mathbf{T}^{-1})\lambda^{-1} - I_3(\mathbf{T}^{-1}) = 0.$$

Show that

$$\left.\begin{aligned} I_1(\mathbf{T}^{-1}) &= I_2(\mathbf{T})/I_3(\mathbf{T}), \\ I_2(\mathbf{T}^{-1}) &= I_1(\mathbf{T})/I_3(\mathbf{T}), \\ I_3(\mathbf{T}^{-1}) &= 1/I_3(\mathbf{T}). \end{aligned}\right\} \tag{1.3.20}$$

Problem 1.3.6 Show that

$$\mathbf{T}^{-1} = (\mathbf{T}^2 - I_1(\mathbf{T})\mathbf{T} + I_2(\mathbf{T})\mathbf{I})/I_3(\mathbf{T}) \tag{1.3.21}$$

and deduce that $\mathbf{T}^r$ is expressible in terms of $\mathbf{I}$, $\mathbf{T}$ and $\mathbf{T}^2$, the coefficients being invariants of $\mathbf{T}$, for any positive or negative integer r.

1.3.3 Symmetric second-order tensors

For a symmetric second-order tensor $\mathbf{T}$ the results of Section 1.3.2 can be made more specific. In particular, the eigenvectors of $\mathbf{T}$ are mutually orthogonal and the eigenvalues are real. We now prove these results.

Let λ_i be the eigenvalue of $\mathbf{T}$ associated with the eigenvector $\mathbf{v}^{(i)}$ $(i = 1, 2, 3)$. Then

$$\mathbf{T}\mathbf{v}^{(i)} = \lambda_i \mathbf{v}^{(i)} \qquad (i = 1, 2, 3;\ \text{no summation}).$$

It follows that

$$\mathbf{v}^{(j)} \cdot (\mathbf{T}\mathbf{v}^{(i)}) = \lambda_i \mathbf{v}^{(j)} \cdot \mathbf{v}^{(i)}$$

and similarly

$$\mathbf{v}^{(i)} \cdot (\mathbf{T}\mathbf{v}^{(j)}) = \lambda_j \mathbf{v}^{(i)} \cdot \mathbf{v}^{(j)}.$$

On use of the symmetry (1.1.1) with $\mathbf{T}^{\mathrm{T}} = \mathbf{T}$ and the result (1.3.9), subtraction of these two equations gives

$$(\lambda_i - \lambda_j)\mathbf{v}^{(i)} \cdot \mathbf{v}^{(j)} = 0. \tag{1.3.22}$$

Firstly, if $\lambda_i \neq \lambda_j$ then the required result follows, that is

$$\mathbf{v}^{(i)} \cdot \mathbf{v}^{(j)} = 0 \qquad (i \neq j). \tag{1.3.23}$$

Secondly, if $\lambda_i = \lambda_j \neq \lambda_k$ $(i \neq j \neq k \neq i)$ then it follows from (1.3.22) that

$$\mathbf{v}^{(i)} \cdot \mathbf{v}^{(k)} = \mathbf{v}^{(j)} \cdot \mathbf{v}^{(k)} = 0.$$

There is no restriction of the relation between $\mathbf{v}^{(i)}$ and $\mathbf{v}^{(j)}$ in the plane normal to $\mathbf{v}^{(k)}$ so they can be chosen arbitrarily in that plane so that (1.3.23) is satisfied.

Thirdly, if $\lambda_i = \lambda_j = \lambda_k$ $(i \neq j \neq k \neq i)$ then no restriction is imposed on the relative orientations of the eigenvectors and they can be chosen arbitrarily in accordance with (1.3.23).

In order to prove that λ_i is real we assume that $\bar{\lambda}_i \neq \lambda_i$, where the overbar denotes complex conjugate. Let $\bar{\mathbf{v}}^{(i)}$ be the complex conjugate of the corresponding eigenvector $\mathbf{v}^{(i)}$. Then, since $\mathbf{T}$ is real we obtain

$$\mathbf{T}\bar{\mathbf{v}}^{(i)} = \bar{\lambda}_i \bar{\mathbf{v}}^{(i)}$$

on taking the complex conjugate of the equation for $\mathbf{v}^{(i)}$. Thus $\bar{\mathbf{v}}^{(i)}$ is an eigenvector of $\mathbf{T}$ corresponding to the eigenvalue $\bar{\lambda}_i$ which, by hypothesis, is distinct from λ_i. Substituting $\bar{\lambda}_i$ and $\bar{\mathbf{v}}^{(i)}$ for λ_j and $\mathbf{v}^{(j)}$ in (1.3.22), we obtain

$$(\lambda_i - \bar{\lambda}_i)\mathbf{v}^{(i)} \cdot \bar{\mathbf{v}}^{(i)} = 0.$$

But $\mathbf{v}^{(i)} \cdot \bar{\mathbf{v}}^{(i)} = |\mathbf{v}^{(i)}|^2$, this defining the modulus of a complex vector, which is positive provided $\mathbf{v}^{(i)} \neq \mathbf{0}$. Hence $\bar{\lambda}_i = \lambda_i$, contrary to hypothesis, and therefore λ_i is real $(i = 1, 2, 3)$. It also follows that $\mathbf{v}^{(i)}$ is real. The proof of this is left to the reader.

Let $\mathbf{v}^{(i)}$ be the normalized eigenvectors of $\mathbf{T}$ so that the identity tensor $\mathbf{I}$ is given by

$$\mathbf{I} = \mathbf{v}^{(i)} \otimes \mathbf{v}^{(i)} \qquad \text{(summed over } i\text{)}.$$

Hence

$$\mathbf{T} = \mathbf{T}\mathbf{I} = (\mathbf{T}\mathbf{v}^{(i)}) \otimes \mathbf{v}^{(i)} = \sum_{i=1}^{3} \lambda_i \mathbf{v}^{(i)} \otimes \mathbf{v}^{(i)}. \tag{1.3.24}$$

Thus $\mathbf{T}$ has the representation (1.3.24) with respect to the basis $\{\mathbf{v}^{(i)}\}$, and the matrix of its components T' with respect to this basis is diagonal with components $\lambda_1, \lambda_2, \lambda_3$. The expansion (1.3.24) is called the *spectral representation* of $\mathbf{T}$.

With respect to a general basis $\{\mathbf{e}_i\}$ the component matrix of $\mathbf{T}$ is T, say. Then

$$\mathsf{T} = \mathsf{Q}^{\mathrm{T}}\mathsf{T}'\mathsf{Q},$$

where the matrix Q has components $Q_{ij} = \mathbf{v}^{(i)} \cdot \mathbf{e}_j$.

If $\lambda_2 = \lambda_1$ it is easily deduced from (1.3.24) that

$$\mathbf{T} = \lambda_1 \mathbf{I} + (\lambda_3 - \lambda_1)\mathbf{v}^{(3)} \otimes \mathbf{v}^{(3)},$$

while

$$\mathbf{T} = \lambda_1 \mathbf{I}$$

if $\lambda_3 = \lambda_2 = \lambda_1$.

The eigenvectors of $\mathbf{T}$ are also called *principal axes* (or *principal directions*) and the eigenvalues the *principal values* of $\mathbf{T}$, a terminology used frequently in subsequent chapters.

Problem 1.3.7 Show that if $\lambda_1, \lambda_2, \lambda_3$ are the eigenvalues of a symmetric second-order tensor $\mathbf{T}$ then the principal invariants of $\mathbf{T}$ are expressible as

$$I_1(\mathbf{T}) = \lambda_1 + \lambda_2 + \lambda_3,$$
$$I_2(\mathbf{T}) = \lambda_2\lambda_3 + \lambda_3\lambda_1 + \lambda_1\lambda_2,$$
$$I_3(\mathbf{T}) = \lambda_1\lambda_2\lambda_3.$$

Problem 1.3.8 Two symmetric second-order tensors are said to be *coaxial* if their principal axes coincide (in some order). Prove that $\mathbf{S}$ and $\mathbf{T}$ are coaxial if and only if $\mathbf{ST} = \mathbf{TS}$.

Problem 1.3.9 Suppose $\{\mathbf{e}_i\}$ is an orthonormal basis such that $\mathbf{e}_3$ is the eigenvector of a symmetric second-order tensor $\mathbf{T}$ corresponding to eigenvalue λ_3. If λ_1, λ_2 are the other two eigenvalues, show that there exists an angle θ such that $\mathbf{T}$ has the representation

$$\mathbf{T} = (\lambda_1 \cos^2\theta + \lambda_2 \sin^2\theta)\mathbf{e}_1 \otimes \mathbf{e}_1 + (\lambda_1 \sin^2\theta + \lambda_2 \cos^2\theta)\mathbf{e}_2 \otimes \mathbf{e}_2$$
$$+ (\lambda_1 - \lambda_2)\sin\theta\cos\theta(\mathbf{e}_1 \otimes \mathbf{e}_2 + \mathbf{e}_2 \otimes \mathbf{e}_1) + \lambda_3 \mathbf{e}_3 \otimes \mathbf{e}_3. \quad (1.3.25)$$

A second-order tensor $\mathbf{T}$ is said to be *positive definite* if $\mathbf{v}\cdot(\mathbf{Tv}) > 0$ for all non-zero $\mathbf{v}$ in $\mathbb{E}$, and *positive semi-definite if* $\mathbf{v}\cdot(\mathbf{Tv}) \geq 0$ for all $\mathbf{v}$ in $\mathbb{E}$ with equality for at least one non-zero $\mathbf{v}$.

If $\mathbf{T}$ is symmetric and positive definite then it follows from the spectral decomposition (1.3.24) that $\lambda_i > 0$ $(i = 1, 2, 3)$, while $\lambda_i \geq 0$ with equality for at least one index i for the case of positive semi-definiteness. From (1.3.24) we then define the symmetric tensor $\mathbf{T}^{1/2}$ as the *positive square root* of $\mathbf{T}$ according to

$$\mathbf{T}^{1/2} = \sum_{i=1}^{3} \lambda_i^{1/2} \mathbf{v}^{(i)} \otimes \mathbf{v}^{(i)},$$

where $\lambda_i^{1/2}$ is the positive square root of λ_i. This tensor is uniquely determined by $\mathbf{T}$.

When $\lambda_i > 0$ $(i = 1, 2, 3)$, the inverse of $\mathbf{T}$ exists and is uniquely defined by

$$\mathbf{T}^{-1} = \sum_{i=1}^{3} \lambda_i^{-1} \mathbf{v}^{(i)} \otimes \mathbf{v}^{(i)}.$$

We note that $\mathbf{T}^{1/2}$ and $\mathbf{T}^{-1}$ are necessarily coaxial with $\mathbf{T}$.

1.3.4 Antisymmetric second-order tensors

Let $\mathbf{W}$ be an antisymmetric second-order tensor, so that

$$\mathbf{W}^{\mathrm{T}} = -\mathbf{W}. \tag{1.3.26}$$

From the definitions (1.3.17) of the principal invariants it follows on use of (1.3.26) that

$$I_1(\mathbf{W}) = 0, \qquad I_2(\mathbf{W}) = W_{23}^2 + W_{31}^2 + W_{12}^2, \qquad I_3(\mathbf{W}) = 0,$$

where W_{ij} are the components of $\mathbf{W}$ relative to an arbitrary choice of orthonormal basis $\{\mathbf{e}_i\}$. Hence, the characteristic equation (1.3.16) reduces to

$$\lambda^3 + I_2(\mathbf{W})\lambda = 0.$$

Since $I_2(\mathbf{W}) > 0$ for $\mathbf{W} \neq \mathbf{0}$, $\mathbf{W}$ possesses only one real eigenvalue, namely $\lambda = 0$. Let the corresponding eigenvector be $\mathbf{w}$ so that

$$\mathbf{W}\mathbf{w} = \mathbf{0}. \tag{1.3.27}$$

With respect to the basis $\{\mathbf{e}_i\}$ we may write

$$\mathbf{W} = W_{ji}\mathbf{e}_j \otimes \mathbf{e}_i = \tfrac{1}{2} W_{ji}(\mathbf{e}_j \otimes \mathbf{e}_i - \mathbf{e}_i \otimes \mathbf{e}_j)$$

on making use of (1.3.26).

For an arbitrary vector $\mathbf{a}$ we have

$$(\mathbf{e}_j \otimes \mathbf{e}_i - \mathbf{e}_i \otimes \mathbf{e}_j)\mathbf{a} = (\mathbf{a}\cdot\mathbf{e}_i)\mathbf{e}_j - (\mathbf{a}\cdot\mathbf{e}_j)\mathbf{e}_i = (\mathbf{e}_i \wedge \mathbf{e}_j) \wedge \mathbf{a}$$

in terms of the cross product (see Section 1.1.1).

Hence

$$\mathbf{W}\mathbf{a} = \tfrac{1}{2}(W_{ji}\mathbf{e}_i \wedge \mathbf{e}_j) \wedge \mathbf{a}.$$

But, from equation (1.1.15) and Problem 1.3.3 we have

$$\tfrac{1}{2} W_{ji}\mathbf{e}_i \wedge \mathbf{e}_j = \tfrac{1}{2}\varepsilon_{kij} W_{ji}\mathbf{e}_k = \mathbf{w},$$

where $\mathbf{w}$ is the axial vector of $\mathbf{W}$.

Since

$$\mathbf{W}\mathbf{a} = \mathbf{w} \wedge \mathbf{a} \tag{1.3.28}$$

reduces to (1.3.27) on choosing $\mathbf{a} = \mathbf{w}$ it follows that the axial vector is the only (real) eigenvector of $\mathbf{W}$.

Problem 1.3.10 Let $\mathbf{u}, \mathbf{v}$ and $\hat{\mathbf{w}} = \mathbf{w}/|\mathbf{w}|$ form an orthonormal basis, where $\mathbf{w}$ is the axial vector of $\mathbf{W}$. Show that $\mathbf{W}$ has the representation

$$\mathbf{W} = \mathbf{W}(\mathbf{v}, \mathbf{u})(\mathbf{v} \otimes \mathbf{u} - \mathbf{u} \otimes \mathbf{v}),$$

where $\mathbf{W}(\mathbf{v}, \mathbf{u})$ is the bilinear expression defined in accordance with equation (1.3.8). Deduce that

$$(\mathbf{W}\mathbf{u}) \otimes \mathbf{u} - \mathbf{u} \otimes (\mathbf{W}\mathbf{u}) = \mathbf{W}.$$

1.3.5 Orthogonal second-order tensors

In general, the scalar product $\mathbf{v} \cdot \mathbf{u}$ is not preserved under a linear mapping $\mathbf{T}: \mathbb{E} \rightarrow \mathbb{E}$, for, by the definition (1.3.9) of the transpose $\mathbf{T}^{\mathrm{T}}$, we have

$$(\mathbf{T}\mathbf{u}) \cdot (\mathbf{T}\mathbf{v}) = \mathbf{v} \cdot (\mathbf{T}^{\mathrm{T}}\mathbf{T}\mathbf{u})$$

and this is not equal to $\mathbf{v} \cdot \mathbf{u}$ for all $\mathbf{u}, \mathbf{v} \in \mathbb{E}$ in general

A tensor $\mathbf{Q}$ which does preserve the scalar product must satisfy the equations

$$\mathbf{Q}^{\mathrm{T}}\mathbf{Q} = \mathbf{I} = \mathbf{Q}\mathbf{Q}^{\mathrm{T}}. \tag{1.3.29}$$

Such a $\mathbf{Q}$ is said to be an *orthogonal tensor* by analogy with the definition of an orthogonal matrix given in Section 1.1.2. As in the matrix situation it follows that

$$\det \mathbf{Q} = \pm 1, \tag{1.3.30}$$

and if $\det \mathbf{Q} = +1(-1)$, $\mathbf{Q}$ is said to be a *proper* (*improper*) orthogonal tensor.

At this point we emphasize the important distinction between the orthogonal tensor $\mathbf{Q}$, which is a linear mapping from $\mathbb{E}$ to $\mathbb{E}$, and the matrix Q which represents the direction cosines of the change of basis given by (1.1.28).

However, $\mathbf{Q}$ and Q can be related in the following manner. Let $\mathbf{P}: \mathbb{E} \rightarrow \mathbb{E}$ be an orthogonal tensor which rotates an arbitrary vector $\mathbf{u}$ into $\mathbf{v} = \mathbf{P}\mathbf{u}$. With respect to the basis $\{\mathbf{e}_i\}$, we have $v_i = P_{ij}u_j$, and if the basis is changed to $\{\mathbf{e}'_i\}$ according to (1.1.33) then

$$u'_i = Q_{ij}u_j$$

It follows that $u'_i = v_i$ (respectively $v'_i = u_i$) for all vectors $\mathbf{u}$ in $\mathbb{E}$ if and only if $\mathsf{P} = \mathsf{Q}$ (respectively $\mathsf{P} = \mathsf{Q}^{\mathrm{T}}$), where $\mathsf{Q} = (Q_{ij})$ and P is the matrix (P_{ij}) of the components of $\mathbf{P}$ relative to $\{\mathbf{e}_i\}$. Thus, we may interpret a change of basis as equivalent to the orthogonal transformation that rotates all vectors with a rotation *equal and opposite* (respectively *equal*) to that which takes $\{\mathbf{e}_i\}$ to $\{\mathbf{e}'_i\}$.

We rewrite equation (1.3.29) as

$$\mathbf{Q}^{\mathrm{T}}(\mathbf{Q}-\mathbf{I}) = -(\mathbf{Q}^{\mathrm{T}}-\mathbf{I}),$$

and on taking the determinant of this equation we obtain

$$\det(\mathbf{Q}-\mathbf{I}) = 0.$$

This implies that $\lambda = 1$ is an eigenvalue of $\mathbf{Q}$ and therefore there exists a vector, $\mathbf{u}$ say, such that

$$\mathbf{Qu} = \mathbf{u}.$$

Let $\mathbf{u}, \mathbf{v}, \mathbf{w}$ form an orthonormal basis. Then

$$0 = \mathbf{v}\cdot\mathbf{u} = \mathbf{v}\cdot(\mathbf{Q}^{\mathrm{T}}\mathbf{Qu}) = (\mathbf{Qv})\cdot(\mathbf{Qu}) = (\mathbf{Qv})\cdot\mathbf{u}$$

and hence $\mathbf{Qv}$, and similarly $\mathbf{Qw}$, is normal to $\mathbf{u}$. Also $\mathbf{Qv}$ and $\mathbf{Qw}$ are orthogonal. By use of (1.1.36) we see that there exists an angle θ such that

$$\mathbf{Qv} = \mathbf{v}\cos\theta + \mathbf{w}\sin\theta,$$

$$\mathbf{Qw} = -\mathbf{v}\sin\theta + \mathbf{w}\cos\theta$$

and hence $\mathbf{Q}$ has the representation

$$\mathbf{Q} = \mathbf{u}\otimes\mathbf{u} + (\mathbf{v}\otimes\mathbf{v} + \mathbf{w}\otimes\mathbf{w})\cos\theta + (\mathbf{w}\otimes\mathbf{v} - \mathbf{v}\otimes\mathbf{w})\sin\theta. \quad (1.3.31)$$

Problem 1.3.11 Use (1.3.31) to show that for an arbitrary vector $\mathbf{a}$

$$\mathbf{Qa} = \mathbf{a}\cos\theta + (\mathbf{a}\cdot\mathbf{u})\mathbf{u}(1-\cos\theta) + \mathbf{u}\wedge\mathbf{a}\sin\theta.$$

Problem 1.3.12 Show that the principal invariants of $\mathbf{Q}$ are given by

$$I_1(\mathbf{Q}) = I_2(\mathbf{Q}) = 1 + 2\cos\theta, \quad I_3(\mathbf{Q}) = 1.$$

Hence obtain the characteristic equation for $\mathbf{Q}$ and deduce that $\mathbf{Q}$ has only one real eigenvalue.

1.3.6 Higher-order tensors

Let $\mathbb{E}^n \equiv \underbrace{\mathbb{E}\times\mathbb{E}\times\cdots\times\mathbb{E}}_{n\ \text{times}}$ denote the set of n-tuples $(\mathbf{u}^{(1)},\ldots,\mathbf{u}^{(n)})$ of vectors $\mathbf{u}^{(i)}\in\mathbb{E}$ $(i = 1,\ldots,n)$. Then a tensor $\mathbf{T}$ of order n is a *real-valued multilinear function* over

$\mathbb{E}^n$. That is $\mathbf{T}: \mathbb{E}^n \to \mathbb{R}$ such that

$$\mathbf{T}(\mathbf{u}^{(1)}, \ldots, \alpha\mathbf{u}^{(i)} + \beta\mathbf{v}^{(i)}, \ldots, \mathbf{u}^{(n)}) = \alpha\mathbf{T}(\mathbf{u}^{(1)}, \ldots, \mathbf{u}^{(i)}, \ldots, \mathbf{u}^{(n)}) + \beta\mathbf{T}(\mathbf{u}^{(1)}, \ldots, \mathbf{v}^{(i)}, \ldots, \mathbf{u}^{(n)}) \quad (1.3.32)$$

for all $\mathbf{u}^{(i)}, \mathbf{v}^{(i)} \in \mathbb{E}$, $\alpha, \beta \in \mathbb{R}$ and for $i = 1, \ldots, n$. This generalizes the definition of a second-order tensor given in Section 1.3.1.

The set of all real-valued multilinear functions over $\mathbb{E}^n$ forms a real vector space of dimension 3^n, denoted $\mathscr{L}(\mathbb{E}^n, \mathbb{R})$, with respect to addition and scalar multiplication. These two operations are independent of choice of basis for $\mathbb{E}$, that is they are invariant. A third such operation is that of the *tensor product*: if $\mathbf{S} \in \mathscr{L}(\mathbb{E}^m, \mathbb{R})$ and $\mathbf{T} \in \mathscr{L}(\mathbb{E}^n, \mathbb{R})$ then the tensor product of $\mathbf{S}$ and $\mathbf{T}$, denoted $\mathbf{S} \otimes \mathbf{T}$, is an element of $\mathscr{L}(\mathbb{E}^{m+n}, \mathbb{R})$ defined according to

$$(\mathbf{S} \otimes \mathbf{T})(\mathbf{u}^{(1)}, \ldots, \mathbf{u}^{(m)}, \mathbf{v}^{(1)}, \ldots, \mathbf{v}^{(n)}) = \mathbf{S}(\mathbf{u}^{(1)}, \ldots, \mathbf{u}^{(m)})\mathbf{T}(\mathbf{v}^{(1)}, \ldots, \mathbf{v}^{(n)}) \quad (1.3.33)$$

for all $\mathbf{u}^{(i)}$, $\mathbf{v}^{(\alpha)} \in \mathbb{E}$ $(i = 1, \ldots, m; \alpha = 1, \ldots, n)$. Note that $\mathscr{L}(\mathbb{E}^2, \mathbb{R}) = \mathscr{L}(\mathbb{E} \times \mathbb{E}, \mathbb{R})$.

In respect of an orthonormal basis $\{\mathbf{e}_i\}$, we define

$$(\mathbf{e}_{i_1} \otimes \mathbf{e}_{i_2} \otimes \cdots \otimes \mathbf{e}_{i_n})(\mathbf{u}^{(1)}, \mathbf{u}^{(2)}, \ldots, \mathbf{u}^{(n)}) = u_{i_1}^{(1)} u_{i_2}^{(2)} \ldots u_{i_n}^{(n)},$$

where $u_{i_k}^{(k)}$ are the components of the vector $\mathbf{u}^{(k)}$.

Then, for a general $\mathbf{T} \in \mathscr{L}(\mathbb{E}^n, \mathbb{R})$, we obtain

$$\begin{aligned}\mathbf{T}(\mathbf{u}^{(1)}, \ldots, \mathbf{u}^{(n)}) &= u_{i_1}^{(1)} u_{i_2}^{(2)} \ldots u_{i_n}^{(n)} \mathbf{T}(\mathbf{e}_{i_1}, \ldots, \mathbf{e}_{i_n}) \\ &= \mathbf{T}(\mathbf{e}_{i_1}, \ldots, \mathbf{e}_{i_n})(\mathbf{e}_{i_1} \otimes \mathbf{e}_{i_2} \otimes \ldots \otimes \mathbf{e}_{i_n})(\mathbf{u}^{(1)}, \mathbf{u}^{(2)}, \ldots, \mathbf{u}^{(n)})\end{aligned}$$

for all $\mathbf{u}^{(i)} \in \mathbb{E}$ $(i = 1, \ldots, n)$. This leads to the representation

$$\mathbf{T} = T_{i_1 i_2 \ldots i_n} \mathbf{e}_{i_1} \otimes \mathbf{e}_{i_2} \otimes \ldots \otimes \mathbf{e}_{i_n} \quad (1.3.34)$$

for $\mathbf{T}$, where we have written $T_{i_1 i_2 \ldots i_n}$ for $\mathbf{T}(\mathbf{e}_{i_1}, \ldots, \mathbf{e}_{i_n})$, the (Cartesian) components of $\mathbf{T}$ relative to the basis $\{\mathbf{e}_i\}$, thus generalizing a result derived in Section 1.3.1 for second-order tensors.

The operation of contraction on any pair of indices of $\mathbf{T}$ may be defined as in Section 1.2.4 with respect to an orthonormal basis. However, contraction is in fact independent of choice of basis. In other words, it is an invariant operation. This will be discussed further in Section 1.4.

Symmetry or antisymmetry of $\mathbf{T}$ with respect to any pair of spaces $\mathbb{E}$ in the Cartesian product $\mathbb{E}^n$ may be defined following the corresponding definitions for a second-order tensor. Thus, $\mathbf{T}$ is symmetric $(+)$, antisymmetric $(-)$. in the

pth and qth places if

$$\mathbf{T}(\mathbf{u}^{(1)},\ldots,\mathbf{u}^{(p)},\ldots,\mathbf{u}^{(q)},\ldots,\mathbf{u}^{(n)}) = \pm\mathbf{T}(\mathbf{u}^{(1)},\ldots,\mathbf{u}^{(q)},\ldots,\mathbf{u}^{(p)},\ldots,\mathbf{u}^{(n)}) \tag{1.3.35}$$

for all $\mathbf{u}^{(i)} \in \mathbb{E}$ $(i = 1,\ldots,n)$.

Clearly, these definitions may be extended to cover several pairs of places. Thus, a tensor may be partially symmetric and partially antisymmetric at one and the same time. If $\mathbf{T}(\mathbf{u}^{(1)},\ldots\mathbf{u}^{(n)})$ is indifferent to any pairwise intercharge of the $\mathbf{u}^{(i)}$ then $\mathbf{T}$ is said to be *completely symmetric*. On the other hand, $\mathbf{T}$ is *completely antisymmetric* if $\mathbf{T}(\mathbf{u}^{(\sigma_1)},\ldots,\mathbf{u}^{(\sigma_n)}) = \pm\mathbf{T}(\mathbf{u}^{(1)},\ldots,\mathbf{u}^{(n)})$ with the plus (minus) sign if $\sigma_1,\ldots,\sigma_n$ is a cyclic (anticyclic) permutation of $1,\ldots,n$.

1.4 CONTRAVARIANT AND COVARIANT TENSORS

1.4.1 Reciprocal basis. Contravariant and covariant components

So far we have considered only orthonormal bases for $\mathbb{E}$. We now let $\{\mathbf{e}_i\}$ be an arbitrary basis for $\mathbb{E}$ so that in general $\mathbf{e}_1, \mathbf{e}_2, \mathbf{e}_3$ are neither unit vectors nor mutually orthogonal. A vector $\mathbf{u}$ in $\mathbb{E}$ is then written

$$\mathbf{u} = u^k \mathbf{e}_k, \tag{1.4.1}$$

where u^k are the *contravariant* components of $\mathbf{u}$ with respect of the basis $\{\mathbf{e}_i\}$, distinguished by use of a *super*script.

The *reciprocal basis* of $\{\mathbf{e}_i\}$, denoted by $\{\mathbf{e}^i\}$, is also a basis for $\mathbb{E}$ and the reciprocal basis vectors $\mathbf{e}^i$ $(i = 1, 2, 3)$ are defined uniquely by the equations

$$\mathbf{e}^i \cdot \mathbf{e}_j = \delta^i_j = \begin{cases} 1 & i = j \\ 0 & i \neq j, \end{cases} \tag{1.4.2}$$

where δ^i_j is the *Krönecker delta*, defined by the right-hand equality in (1.4.2). We distinguish between δ^i_j and the Krönecker delta, δ_{ij}, used in Section 1.1.1 which is appropriate only in respect of orthonormal bases $\{\mathbf{e}_i\}$, which are self-reciprocal, when the distinction between subscripts and superscripts is unnecessary.

With respect to the reciprocal basis $\{\mathbf{e}^i\}$ a vector $\mathbf{u}$ in $\mathbb{E}$ may be decomposed as

$$\mathbf{u} = u_k \mathbf{e}^k, \tag{1.4.3}$$

where u_k are called the *covariant* components of $\mathbf{u}$, distinguished from the contravariant components by use of a *sub*script.

On taking the dot product of (1.4.1) and (1.4.3) respectively with $\mathbf{e}^i$ and

$\mathbf{e}_i$ we obtain

$$u^i = \mathbf{u}\cdot\mathbf{e}^i, \qquad u_i = \mathbf{u}\cdot\mathbf{e}_i \tag{1.4.4}$$

on use of (1.4.2). On the other hand, the dot product of (1.4.1) and (1.4.3) with $\mathbf{e}_i$ and $\mathbf{e}^i$ respectively leads to

$$u_i = I_{ik}u^k, \qquad u^i = I^{ik}u_k, \tag{1.4.5}$$

where we have introduced the symbols I_{ik}, I^{ik} defined by

$$I_{ik} = \mathbf{e}_i\cdot\mathbf{e}_k = I_{ki},$$
$$I^{ik} = \mathbf{e}^i\cdot\mathbf{e}^k = I^{ki}.$$

We note that I_{ik} and I^{ik} are reciprocal in the sense that

$$I_{ij}I^{jk} = \delta_i^k. \tag{1.4.6}$$

It follows from (1.4.1) and (1.4.3)–(1.4.5) that

$$\mathbf{e}^i = I^{ik}\mathbf{e}_k, \qquad \mathbf{e}_i = I_{ik}\mathbf{e}^k. \tag{1.4.7}$$

Since $(\mathbf{e}_i \otimes \mathbf{e}_j)\mathbf{u} = u_j\mathbf{e}_i$, it follows that

$$I^{ij}(\mathbf{e}_i \otimes \mathbf{e}_j)\mathbf{u} = I^{ij}u_j\mathbf{e}_i = u^i\mathbf{e}_i = \mathbf{u}$$

for all $\mathbf{u}\in\mathbb{E}$. Thus $I^{ij}\mathbf{e}_i\otimes\mathbf{e}_j$ defines the identity mapping on $\mathbb{E}$, denoted by $\mathbf{I}$, and we have

$$\mathbf{I} = I^{ij}\mathbf{e}_i\otimes\mathbf{e}_j = I_{ij}\mathbf{e}^i\otimes\mathbf{e}^j = \mathbf{e}^i\otimes\mathbf{e}_i = \mathbf{e}_i\otimes\mathbf{e}^i \tag{1.4.8}$$

as alternative representations for $\mathbf{I}$.

We say that the second-order tensor $\mathbf{I}$ has contravariant components I^{ij}, covariant components I_{ij} and mixed components δ^i_j, all of which reduce to the Krönecker delta when the basis is orthonormal. This generalizes the terminology used for the components of vectors.

The dot product $\mathbf{u}\cdot\mathbf{v}$ is easily seen to have the component representations

$$\mathbf{u}\cdot\mathbf{v} = u^i v_i = u_i v^i = I_{ij}u^i v^j = I^{ij}u_i v_j. \tag{1.4.9}$$

A second-order tensor $\mathbf{T}$ has the representations

$$\mathbf{T} = T^{ij}\mathbf{e}_i\otimes\mathbf{e}_j = T^i{}_j\mathbf{e}_i\otimes\mathbf{e}^j = T_i{}^j\mathbf{e}^i\otimes\mathbf{e}_j = T_{ij}\mathbf{e}^i\otimes\mathbf{e}^j, \tag{1.4.10}$$

where $T^{ij}, T^i{}_j, T_i{}^j$ and T_{ij} are respectively the *contravariant*, *right-covariant mixed*, *left-covariant mixed* and *covariant* components of $\mathbf{T}$ with respect to the basis $\{\mathbf{e}_i\}$ and its reciprocal. It is easily shown by use of (1.4.7) that the different types of components are related by

$$T^{ij} = I^{ik}T_k{}^j = I^{jk}T^i{}_k = I^{ik}I^{jl}T_{kl},$$

for example.

With the interpretation of $\mathbf{T}$ as a bilinear function or a linear mapping as in Section 1.3.1, we see that (1.3.7) is replaced by

$$\mathbf{T}(\mathbf{u}, \mathbf{v}) = \mathbf{u}\cdot(\mathbf{T}\mathbf{v}) = T_{ij}u^i v^j, \tag{1.4.11}$$

and this has the alternative component forms

$$T_i{}^j u^i v_j = T^i{}_j u_i v^j = T^{ij}u_i v_j.$$

From (1.3.9) it follows that

$$(\mathbf{T}^{\mathrm{T}})_{ij} = T_{ji}, \qquad (\mathbf{T}^{\mathrm{T}})^{ij} = T^{ji}, \qquad (\mathbf{T}^{\mathrm{T}})^i{}_j = T_j{}^i, \qquad (\mathbf{T}^{\mathrm{T}})_i{}^j = T^j{}_i \tag{1.4.12}$$

so that if $\mathbf{T}$ is symmetric

$$T_{ij} = T_{ji}, \qquad T^{ij} = T^{ij}, \qquad T_j{}^i = T^i{}_j \tag{1.4.13}$$

and the ordering of the indices is irrelevant. Note that $\mathbf{I}$ is an example of a symmetric tensor and we have set $I_i{}^j = I^j{}_i = \delta^i_j$.

Problem 1.4.1 If $\mathbf{T}$ is a second-order tensor show that the equation $\mathbf{v} = \mathbf{T}\mathbf{u}$ has component representations

$$v^i = T^{ij}u_j = T^i{}_j u^j,$$
$$v_i = T_{ij}u^j = T_i{}^j u_j.$$

Problem 1.4.2 If $\mathbf{S}$ and $\mathbf{T}$ are second-order tensors show that the composition $\mathbf{ST}$ has covariant components

$$(\mathbf{ST})_{ij} = S_{ik}T^k{}_j = S_i{}^k T_{kj}$$

and derive corresponding expressions for its contravariant and mixed components.

Problem 1.4.3 If $\mathbf{T}$ is a second-order tensor, show that its components satisfy

$$T^k{}_k = T_k{}^k = I_{kl}T^{kl} = I^{kl}T_{kl}.$$

The representations (1.4.10) for a second-order tensor are generalized in an obvious way to higher-order tensors. Let $\mathbf{T}$ be a tensor of order n. Then $\mathbf{T}$ has the completely contravariant representation

$$\mathbf{T} = T^{i_1 i_2 \ldots i_n} \mathbf{e}_{i_1} \otimes \mathbf{e}_{i_2} \otimes \cdots \otimes \mathbf{e}_{i_n},$$

the completely covariant representation

$$\mathbf{T} = T_{i_1 i_2 \ldots i_n} \mathbf{e}^{i_1} \otimes \mathbf{e}^{i_2} \otimes \cdots \otimes \mathbf{e}^{i_n}$$

or any representation intermediate between these involving mixed components; for example

$$\mathbf{T} = T^{i_1 i_2 \ldots i_p}{}_{i_{p+1} \ldots i_{q-1}}{}^{i_q \ldots i_n} \mathbf{e}_{i_1} \otimes \ldots \otimes \mathbf{e}_{i_p}$$
$$\otimes \mathbf{e}^{i_{p+1}} \otimes \ldots \otimes \mathbf{e}^{i_{q-1}} \otimes \mathbf{e}_{i_q} \otimes \ldots \otimes \mathbf{e}_{i_n}.$$

These all reduce to (1.3.34) in respect of an orthonormal basis.

1.4.2 Change of basis

Suppose that a second basis, $\{\mathbf{e}'_i\}$ say, is chosen for $\mathbb{E}$ so that

$$\mathbf{e}'_i = P^j_i \mathbf{e}_j, \tag{1.4.14}$$

where the coefficients P^j_i are the elements of a non-singular matrix. We write the inverse of this relationship in the form

$$\mathbf{e}_i = Q^j_i \mathbf{e}'_j, \tag{1.4.15}$$

where the coefficients Q^j_i satisfy

$$P^j_i Q^k_j = \delta^k_i = P^k_j Q^j_i. \tag{1.4.16}$$

It is easily shown from the definition (1.4.2) together with (1.4.14) to (1.4.16) that the corresponding reciprocal basis vectors are related according to

$$\mathbf{e}'^i = Q^i_j \mathbf{e}^j, \qquad \mathbf{e}^i = P^i_j \mathbf{e}'^j. \tag{1.4.17}$$

By use of these relations in conjunction with (1.4.1) and (1.4.3), it follows that the contravariant and covariant components of a vector $\mathbf{u}$ transform according to

$$u'^i = Q^i_k u^k, \qquad u'_i = P^k_i u_k \tag{1.4.18}$$

respectively under change of basis.

Problem 1.4.4 Show that a change of basis (1.1.14) such that $\mathbf{e}'_i = \mathbf{e}^i (i = 1, 2, 3)$ to change of basis.

Problem 1.4.5 Show that a change of basis (1.4.14) such that $\mathbf{e}'_i = \mathbf{e}^i (i = 1,2,3)$ reverse the roles of contravariance and covariance.

From (1.4.10), (1.4.15) and (1.4.17), it is easily established that the components of a second-order tensor $\mathbf{T}$ transform according to

$$\left.\begin{aligned} T'^{ij} &= Q^i_m Q^j_n T^{mn}, \quad T'^i{}_j = Q^i_m P^n_j T^m{}_n, \\ T'_i{}^j &= P^m_i Q^j_n T_m{}^n, \quad T'_{ij} = P^m_i P^n_j T_{mn}. \end{aligned}\right\} \tag{1.4.19}$$

With the help of (1.4.16) we obtain from (1.4.19) $T'^k{}_k = T^k{}_k$ so that the quantity $T^k{}_k$, which has the alternative expressions given in Problem 1.4.3, is a scalar invariant of $\mathbf{T}$. It is called the trace of $\mathbf{T}$, tr $\mathbf{T}$, as in Section 1.2.4, but is written T_{kk} only when the basis is orthonormal. It follows from (1.4.19) that neither T_{kk} nor T^{kk} is invariant in general.

For tensors of higher order each contravariant superscript (respectively covariant subscript) transforms by means of Q (respectively P). Thus

$$\left.\begin{aligned} T'^{i_1 i_2 \ldots i_n} &= Q^{i_1}_{j_1} Q^{i_2}_{j_2} \ldots Q^{i_n}_{j_n} T^{j_1 j_2 \ldots j_n}, \\ T'_{i_1 i_2 \ldots i_n} &= P^{j_1}_{i_1} P^{j_2}_{i_2} \ldots P^{j_n}_{i_n} T_{j_1 j_2 \ldots j_n}, \\ T'^{i_1 \ldots i_m}{}_{i_{m+1} \ldots i_n} &= Q^{i_1}_{j_1} \ldots Q^{i_m}_{j_m} P^{j_{m+1}}_{i_{m+1}} \ldots P^{j_n}_{i_n} T^{j_1 \ldots j_m}{}_{j_{m+1} \ldots j_n}, \end{aligned}\right\} \tag{1.4.20}$$

for example.

We note that the contravariant and covariant components of the identity tensor $\mathbf{I}$ transform as

$$I'^{ij} = Q^i_m Q^j_n I^{mn}, \qquad I'_{ij} = P^m_i P^n_j I_{mn}$$

while the mixed components are unchanged since $Q^i_m P^n_j \delta^m_n = \delta^i_j$.

It follows from (1.4.20) and (1.4.16) that

$$I'_{i_p i_q} T'^{i_1 \ldots i_p \ldots i_q \ldots i_n}$$

$$= Q^{i_1}_{j_1} \ldots Q^{i_{p-1}}_{j_{p-1}} Q^{i_{p+1}}_{j_{p+1}} \ldots Q^{i_{q-1}}_{j_{q-1}} Q^{i_{q+1}}_{j_{q+1}} \ldots Q^{i_n}_{j_n} I_{j_p j_q} T^{j_1 \ldots j_p \ldots j_q \ldots j_n}, \tag{1.4.21}$$

that is

$$I_{i_p i_q} T^{i_1 \dots i_p \dots i_q \dots i_n} \equiv T^{i_1 \dots i_p \dots i_{q-1}}{}_{i_p}{}^{i_{q+1} \dots i_n}$$
$$\equiv T^{i_1 \dots i_{p-1}}{}_{i_q}{}^{i_{p+1} \dots i_q \dots i_n}$$
$$\equiv I^{i_p i_q} T^{i_1 \dots i_{p-1}}{}_{i_p}{}^{i_{p+1} \dots i_{q-1}}{}_{i_q}{}^{i_{q+1} \dots i_n} \qquad (1.4.22)$$

are the contravariant components of a tensor of order $n-2$.

The above manipulation is referred to as contraction of the p and qth indices of $\mathbf{T}$. It is the counterpart in respect of a general basis of the operation described in Section 1.2.4 for Cartesian components and is equivalent to it. We emphasize that in component form contraction is achieved by setting equal a covariant and a contravariant index. It is an *invariant* operation in that the end result represents a *tensor* (of order $n-2$), as distinct from the operation of setting two covariant or two contravariant indices equal.

Contraction may be expressed invariantly in the following way. If $\mathbf{T}$ is a tensor of order n then we denote by $\mathbf{C}_{pq}(\mathbf{T})$ the tensor of order $n-2$ resulting from contracting on the p and q th indices. For example, by writing

$$\mathbf{T} = T^{i_1 \dots i_n} \mathbf{e}_{i_1} \otimes \cdots \otimes \mathbf{e}_{i_n}$$

we obtain

$$\mathbf{C}_{pq}(\mathbf{T}) = T^{i_1 \dots i_p \dots i_q \dots i_n} (\mathbf{e}_{i_p} \cdot \mathbf{e}_{i_q}) \mathbf{e}_{i_1} \otimes \cdots \otimes \mathbf{e}_{i_{p-1}}$$
$$\otimes \mathbf{e}_{i_{p+1}} \otimes \cdots \otimes \mathbf{e}_{i_{q-1}} \otimes \mathbf{e}_{i_{q+1}} \otimes \cdots \otimes \mathbf{e}_{i_n}$$
$$= T^{i_1 \dots i_p \dots i_q \dots i_n} I_{i_p i_q} \mathbf{e}_{i_1} \otimes \cdots \otimes \mathbf{e}_{i_{p-1}} \otimes \mathbf{e}_{i_{p+1}} \otimes \cdots \otimes \mathbf{e}_{i_{q-1}}$$
$$\otimes \mathbf{e}_{i_{q+1}} \otimes \cdots \otimes \mathbf{e}_{i_n} \qquad (1.4.23)$$

or equivalently in any other component representation.

Particular examples of contraction are

$$\mathbf{C}_{12}(\mathbf{u} \otimes \mathbf{v}) = \mathbf{u} \cdot \mathbf{v},$$
$$\mathbf{C}_{12}(\mathbf{T}) = \operatorname{tr} \mathbf{T},$$
$$\mathbf{C}_{23}(\mathbf{S} \otimes \mathbf{T}) = \mathbf{ST},$$

where $\mathbf{u}$ and $\mathbf{v}$ are vectors and $\mathbf{S}$, $\mathbf{T}$ second-order tensors.

1.4.3 Dual space. General tensors

The notion of a tensor is more general than we have described in Sections 1.3, 1.4.1 and 1.4.2, so, in order to set the theory within a broader framework, this section deals with the definition of tensors based on the vector space and its dual. The results of this section, however, are not used in later chapters so omission of this section will not detract from the subsequent development.

The *dual space* of $\mathbb{E}$, denoted $\mathbb{E}^*$, is the vector space consisting of the set of linear mappings from $\mathbb{E}$ to $\mathbb{R}$, that is $\mathscr{L}(\mathbb{E}, \mathbb{R})$. It has dimension equal to that of $\mathbb{E}$ (three in the present context) but in general cannot be identified with $\mathbb{E}$.

Vectors in $\mathbb{E}$ are called *contravariant vectors* to distinguish them from the elements of $\mathbb{E}^*$ which are called *covariant vectors*. The connection between this terminology and that used in Section 1.4.1 to describe *components* of a vector in $\mathbb{E}$ will be discussed shortly.

If $\mathbf{u}\in\mathbb{E}$ and $\mathbf{v}^*\in\mathbb{E}^*$ then we denote by $\mathbf{v}^*(\mathbf{u})$ the scalar to which $\mathbf{u}$ maps under the *linear function* $\mathbf{v}^*: \mathbb{E}\to\mathbb{R}$. Since $\mathbb{E}^*$ is a vector space we have

$$(\alpha\mathbf{v}^* + \beta\mathbf{w}^*)(\mathbf{u}) = \alpha\mathbf{v}^*(\mathbf{u}) + \beta\mathbf{w}^*(\mathbf{u})$$

for all $\mathbf{v}^*, \mathbf{w}^*\in\mathbb{E}^*, \alpha, \beta\in\mathbb{R}$ and $\mathbf{u}\in\mathbb{E}$. Also $\mathbf{v}^*(\mathbf{u})$ can be regarded as the result of a bilinear function operating on the pair $(\mathbf{v}^*, \mathbf{u})$ in the Cartesian product $\mathbb{E}^* \times \mathbb{E}$. We denote this bilinear function by $\langle\ ,\ \rangle: \mathbb{E}^* \times \mathbb{E}\to\mathbb{R}$, so that

$$\langle\mathbf{v}^*, \mathbf{u}\rangle = \mathbf{v}^*(\mathbf{u}) \tag{1.4.24}$$

for all $\mathbf{u}\in\mathbb{E}, \mathbf{v}^*\in\mathbb{E}^*$.

Since the result is a scalar $\langle\ ,\ \rangle$ is called a *scalar product* over $\mathbb{E}^* \times \mathbb{E}$. It is important at this point to distinguish between the scalar product over $\mathbb{E}^* \times \mathbb{E}$ and the inner product (or dot product) which is a scalar-valued bilinear function over $\mathbb{E} \times \mathbb{E}$. As we see later, however, the two products can be identified by means of an isomorphism between $\mathbb{E}$ and $\mathbb{E}^*$.

From the vector space axioms it follows that $\langle\mathbf{v}^*, \mathbf{u}\rangle = 0$ for all $\mathbf{u}\in\mathbb{E}$ if and only if $\mathbf{v}^* = \mathbf{0}$ and, dually, $\langle\mathbf{v}^*, \mathbf{u}\rangle = 0$ for all $\mathbf{v}^*\in\mathbb{E}^*$ if and only if $\mathbf{u} = \mathbf{0}$. If $\langle\mathbf{v}^*, \mathbf{u}\rangle = 0$ then $\mathbf{u}\in\mathbb{E}$ and $\mathbf{v}^*\in\mathbb{E}^*$ are said to be *orthogonal*, following the terminology used in respect of the dot product.

The dual space of $\mathbb{E}^*$, denoted $\mathbb{E}^{**}$, may also be defined but can be identified with $\mathbb{E}$ in the following way. If $\mathbf{u}\in\mathbb{E}, \mathbf{v}^*\in\mathbb{E}^*$ then we define a unique element $\mathbf{u}^{**}\in\mathbb{E}^{**}$ by means of

$$\mathbf{u}^{**}(\mathbf{v}^*) = \mathbf{v}^*(\mathbf{u})$$

for all $\mathbf{v}^*\in\mathbb{E}^*$ and each $\mathbf{u}\in\mathbb{E}$. In terms of the scalar product, over $\mathbb{E}^{**} \times \mathbb{E}^*$ and $\mathbb{E}^* \times \mathbb{E}$ respectively, we rewrite this as

$$\langle\mathbf{u}^{**}, \mathbf{v}^*\rangle = \langle\mathbf{v}^*, \mathbf{u}\rangle.$$

The identification $\mathbb{E}^{**} = \mathbb{E}$, or $\mathbf{u}^{**} = \mathbf{u}$ for each $\mathbf{u} \in \mathbb{E}$, then leads to

$$\langle \mathbf{u}, \mathbf{v}^* \rangle = \langle \mathbf{v}^*, \mathbf{u} \rangle, \tag{1.4.25}$$

showing that the scalar product is then symmetric.

Since $\mathbf{v}^* \in \mathscr{L}(\mathbb{E}, \mathbb{R})$ we then have the dual interpretation that $\mathbf{v} \in \mathscr{L}(\mathbb{E}^*, \mathbb{R})$ for any $\mathbf{v} \in \mathbb{E}$, that is contravariant vectors can be regarded as linear functions over $\mathbb{E}^*$. We make no use of $\mathbb{E}^{**}$ here but the interested reader can find further discussion in Bowen and Wang (1976), for example.

Let $\{\mathbf{e}_i\}$ be an arbitrary basis for $\mathbb{E}$, so that an arbitrary $\mathbf{u} \in \mathbb{E}$ may be written in the component form

$$\mathbf{u} = u^i \mathbf{e}_i \tag{1.4.26}$$

as in (1.4.1). Thus, in the present terminology, the *contravariant* vector $\mathbf{u}$ has *contravariant* components u^i.

Let $\{\mathbf{e}^i_*\}$ denote the basis in $\mathbb{E}^*$ that is uniquely determined by the equations

$$\langle \mathbf{e}^i_*, \mathbf{e}_j \rangle = \delta^i_j. \tag{1.4.27}$$

It is called the *dual basis* of $\{\mathbf{e}_i\}$. Recall equation (1.4.2) but note the distinction between *reciprocal* basis vectors in $\mathbb{E}$ and *dual* basis vectors in $\mathbb{E}^*$.

A covariant vector $\mathbf{v}^*$ in $\mathbb{E}^*$ may be decomposed on the dual basis according to

$$\mathbf{v}^* = v^*_i \mathbf{e}^i_*, \tag{1.4.28}$$

where v^*_i are the *covariant components* of $\mathbf{v}^*$.

In this sense a contravariant (respectively covariant) vector has only contravariant (respectively covariant) components. It is only when $\mathbb{E}^*$ is identified with $\mathbb{E}$ and hence $\mathbf{v}$ with $\mathbf{v}^*$ and $\mathbf{e}^i_*$ with $\mathbf{e}^i$ that a vector $\mathbf{v}$ can be said to have both contravariant components (v^i) and covariant components ($v^*_i = v_i$). This identification is made shortly.

From (1.4.26)–(1.4.28) we obtain

$$\langle \mathbf{v}^*, \mathbf{e}_i \rangle = v^*_i, \quad \langle \mathbf{e}^i_*, \mathbf{v} \rangle = v^i \tag{1.4.29}$$

and hence

$$\langle \mathbf{v}^*, \mathbf{v} \rangle = v^*_i v^i \tag{1.4.30}$$

for arbitrary $\mathbf{v} \in \mathbb{E}$ and $\mathbf{v}^* \in \mathbb{E}^*$.

By drawing analogy with (1.2.7) consideration of (1.4.24) leads us to define the tensor product $\mathbf{v}\otimes\mathbf{v}^*$ by

$$(\mathbf{v}\otimes\mathbf{v}^*)(\mathbf{u}) = \langle\mathbf{v}^*,\mathbf{u}\rangle\mathbf{v} \tag{1.4.31}$$

for all $\mathbf{u},\mathbf{v}\in\mathbb{E}$ and all $\mathbf{v}^*\in\mathbb{E}^*$.

We can therefore regard $\mathbf{v}\otimes\mathbf{v}^*$ as a linear mapping from $\mathbb{E}$ to $\mathbb{E}$. Alternatively, it can be regarded as a bilinear function over $\mathbb{E}^*\times\mathbb{E}$ according to

$$(\mathbf{v}\otimes\mathbf{v}^*)(\mathbf{u}^*,\mathbf{u}) = \langle\mathbf{u}^*,(\mathbf{v}\otimes\mathbf{v}^*)(\mathbf{u})\rangle = \langle\mathbf{v}^*,\mathbf{u}\rangle\langle\mathbf{u}^*,\mathbf{v}\rangle$$

for all $\mathbf{u},\mathbf{v}\in\mathbb{E}$ and all $\mathbf{u}^*,\mathbf{v}^*\in\mathbb{E}^*$, use having been made of (1.4.31) and linearity.

From (1.4.26), (1.4.29) and (1.4.31) we find that

$$(\mathbf{e}_i\otimes\mathbf{e}_*^i)(\mathbf{u}) = \mathbf{u}$$

for all $\mathbf{u}\in\mathbb{E}$ and hence we may write

$$\mathbf{e}_i\otimes\mathbf{e}_*^i = \mathbf{I}, \tag{1.4.32}$$

where $\mathbf{I}:\mathbb{E}\to\mathbb{E}$ is the identity mapping. With the bilinear function interpretation, we have

$$\mathbf{I}(\mathbf{u}^*,\mathbf{u}) = \langle\mathbf{u}^*,\mathbf{u}\rangle$$

for all $\mathbf{u}\in\mathbb{E},\mathbf{u}^*\in\mathbb{E}^*$.

More generally, if $\mathbf{T}:\mathbb{E}\to\mathbb{E}$ is a linear mapping it may also be regarded as a bilinear function over $\mathbb{E}^*\times\mathbb{E}$ according to

$$\mathbf{T}(\mathbf{v}^*,\mathbf{u}) = \langle\mathbf{v}^*,\mathbf{T}\mathbf{u}\rangle, \tag{1.4.33}$$

and $\mathscr{L}(\mathbb{E},\mathbb{E})$ is identified with $\mathscr{L}(\mathbb{E}^*\times\mathbb{E},\mathbb{R})$. (The reader is reminded of the definition given in Section 1.3.1 of $\mathbf{T}$ as a bilinear function over $\mathbb{E}\times\mathbb{E}$. The present definition is more general since it does not require the existence of an inner (or dot) product.)

From (1.4.29), (1.4.31) and (1.4.33) we find

$$(\mathbf{e}_i\otimes\mathbf{e}_*^j)(\mathbf{v}^*,\mathbf{u}) = u^j v_i^*$$

so that, by use of linearity, we may write

$$\begin{aligned}\mathbf{T}(\mathbf{v}^*,\mathbf{u}) &= \mathbf{T}(\mathbf{e}_*^i,\mathbf{e}^j)u^j v_i^* \\ &= \mathbf{T}(\mathbf{e}_*^i,\mathbf{e}_j)(\mathbf{e}_i\otimes\mathbf{e}_*^j)(\mathbf{v}^*,\mathbf{u})\end{aligned}$$

for all $\mathbf{u} \in \mathbb{E}$ and $\mathbf{v}^* \in \mathbb{E}^*$. Hence $\mathbf{T}$ has the representation

$$\mathbf{T} = T^i{}_j \mathbf{e}_i \otimes \mathbf{e}_*^j \tag{1.4.34}$$

which generalizes the second form given in (1.4.10), where $T^i{}_j$ are the 'mixed' components of $\mathbf{T}$ defined as $\mathbf{T}(\mathbf{e}_*^i, \mathbf{e}_j)$, and $\{\mathbf{e}_i \otimes \mathbf{e}_*^j\}$ is a basis for the vector space $\mathscr{L}(\mathbb{E}^* \times \mathbb{E}, \mathbb{R})$.

Given $\mathbf{T} \in \mathscr{L}(\mathbb{E}, \mathbb{E})$ the *dual mapping* $\mathbf{T}^* \in \mathscr{L}(\mathbb{E}^*, \mathbb{E}^*)$ is defined uniquely by

$$\langle \mathbf{v}^*, \mathbf{T}\mathbf{u} \rangle = \langle \mathbf{T}^*\mathbf{v}^*, \mathbf{u} \rangle \tag{1.4.35}$$

for all $\mathbf{u} \in \mathbb{E}, \mathbf{v}^* \in \mathbb{E}^*$. Analogously to (1.4.34) $\mathbf{T}^*$ has the representation

$$\mathbf{T}^* = T_i^{*j} \mathbf{e}_*^i \otimes \mathbf{e}_j \tag{1.4.36}$$

and $\{\mathbf{e}_*^i \otimes \mathbf{e}_j\}$ forms a basis for $\mathscr{L}(\mathbb{E} \times \mathbb{E}^*, \mathbb{R})$. Note that $\mathbf{I}^*$, the identity mapping on $\mathbb{E}^*$, has the representation

$$\mathbf{I}^* = \mathbf{e}_*^i \otimes \mathbf{e}_i. \tag{1.4.37}$$

Equally, $\mathbf{T}^*$ may be regarded as a bilinear mapping over $\mathbb{E} \times \mathbb{E}^*$ such that

$$\mathbf{T}^*(\mathbf{u}, \mathbf{v}^*) = \langle \mathbf{T}^*\mathbf{v}^*, \mathbf{u} \rangle \tag{1.4.38}$$

and hence by (1.4.33) and (1.4.35)

$$\mathbf{T}^*(\mathbf{u}, \mathbf{v}^*) = \mathbf{T}(\mathbf{v}^*, \mathbf{u}). \tag{1.4.39}$$

In addition to the vector spaces $\mathscr{L}(\mathbb{E}^* \times \mathbb{E}, \mathbb{R})$ and $\mathscr{L}(\mathbb{E} \times \mathbb{E}^*, \mathbb{R})$ just considered there are two other such spaces of bilinear functions, namely $\mathscr{L}(\mathbb{E} \times \mathbb{E}, \mathbb{R})$ and $\mathscr{L}(\mathbb{E}^* \times \mathbb{E}^*, \mathbb{R})$. The first of these may be identified with $\mathscr{L}(\mathbb{E}, \mathbb{E}^*)$, for if $\mathbf{T} \in \mathscr{L}(\mathbb{E}, \mathbb{E}^*)$ then a unique bilinear function over $\mathbb{E} \times \mathbb{E}$ is defined by

$$\mathbf{T}(\mathbf{u}, \mathbf{v}) = \langle \mathbf{T}\mathbf{u}, \mathbf{v} \rangle \tag{1.4.40}$$

for all $\mathbf{u}, \mathbf{v} \in \mathbb{E}$, and such a $\mathbf{T}$ has the representation

$$\mathbf{T} = T_{ij} \mathbf{e}_*^i \otimes \mathbf{e}_*^j \tag{1.4.41}$$

with respect to the basis $\{\mathbf{e}_*^i \otimes \mathbf{e}_*^j\}$ of $\mathscr{L}(\mathbb{E} \times \mathbb{E}, \mathbb{R})$. In general, this does not have a dual in a sense similar to that described by (1.4.39), but its transpose

$\mathbf{T}^{\mathrm{T}}$, also in $\mathscr{L}(\mathbb{E} \times \mathbb{E}, \mathbb{R})$, is defined by

$$\langle \mathbf{T}\mathbf{u}, \mathbf{v} \rangle = \langle \mathbf{T}^{\mathrm{T}}\mathbf{v}, \mathbf{u} \rangle \tag{1.4.42}$$

for all $\mathbf{u}, \mathbf{v} \in \mathbb{E}$, analogously to (1.3.9).

Similarly, if $\mathbf{T} \in \mathscr{L}(\mathbb{E}^*, \mathbb{E})$ then the identity

$$\mathbf{T}(\mathbf{u}^*, \mathbf{v}^*) = \langle \mathbf{u}^*, \mathbf{T}\mathbf{v}^* \rangle \tag{1.4.43}$$

for all $\mathbf{u}^*, \mathbf{v}^* \in \mathbb{E}^*$ defines a unique bilinear function over $\mathbb{E}^* \times \mathbb{E}^*$, which also has a transpose defined analogously to (1.4.42). It has the representation

$$\mathbf{T} = T^{ij}\mathbf{e}_i \otimes \mathbf{e}_j \tag{1.4.44}$$

relative to the basis $\{\mathbf{e}_i \otimes \mathbf{e}_j\}$ of $\mathscr{L}(\mathbb{E}^* \times \mathbb{E}^*, \mathbb{R})$.

Recalling that a covariant vector is a linear function over $\mathbb{E}$ and that we may interpret a contravariant vector as a linear function over $\mathbb{E}^*$, we now extend this terminology to cover bilinear functions.

A bilinear function over $\mathbb{E}^* \times \mathbb{E}^*$ (respectively $\mathbb{E} \times \mathbb{E}$) is called a *second-order contravariant* (respectively *covariant*) *tensor*. A bilinear function over $\mathbb{E} \times \mathbb{E}^*$ (respectively $\mathbb{E}^* \times \mathbb{E}$) is called a *second-order left-covariant* (respectively *right-covariant*) *mixed tensor*. In their component representations a superscript (subscript) corresponds to a contravariant (covariant) index.

Alternatively, elements of $\mathscr{L}(\mathbb{E}^* \times \mathbb{E}^*, \mathbb{R})$, $\mathscr{L}(\mathbb{E} \times \mathbb{E}, \mathbb{R})$, $\mathscr{L}(\mathbb{E} \times \mathbb{E}^*, \mathbb{R})$ and $\mathscr{L}(\mathbb{E}^* \times \mathbb{E}, \mathbb{R})$ are called tensors of type $(2,0)$, $(0,2)$, $(1,1)$ and $(1,1)$ respectively, a notation which is generalized to multilinear functions shortly. This notation does not distinguish $\mathscr{L}(\mathbb{E} \times \mathbb{E}^*, \mathbb{R})$ from $\mathscr{L}(\mathbb{E}^* \times \mathbb{E}, \mathbb{R})$. In other words, the order of $\mathbb{E}$ and $\mathbb{E}^*$ in the Cartesian product is not important, this being justified by means of the duality between $\mathscr{L}(\mathbb{E} \times \mathbb{E}^*, \mathbb{R})$ and $\mathscr{L}(\mathbb{E}^* \times \mathbb{E}, \mathbb{R})$ expressed in (1.4.39).

In general, a *tensor* on $\mathbb{E}$ is a *multilinear function of order* (r, s), otherwise called an $(r+s)$-linear function. We write

$$\mathbf{T}: \underbrace{\mathbb{E}^* \times \cdots \times \mathbb{E}^*}_{r\text{ times}} \times \underbrace{\mathbb{E} \times \cdots \times \mathbb{E}}_{s\text{ times}} \to \mathbb{R},$$

where r is the *contravariant order* and s is the *covariant order* of $\mathbf{T}$. A contravariant vector is of order $(1,0)$, a covariant vector of order $(0,1)$ and scalars are classified as $(0,0)$.

The set $\mathscr{L}(\underbrace{\mathbb{E}^* \times \cdots \times \mathbb{E}^*}_{r\text{ times}} \times \underbrace{\mathbb{E} \times \cdots \times \mathbb{E}}_{s\text{ times}}, \mathbb{R})$ is a vector space with product basis $\{\mathbf{e}_{i_1} \otimes \cdots \otimes \mathbf{e}_{i_r} \otimes \mathbf{e}_*^{j_1} \otimes \cdots \otimes \mathbf{e}_*^{j_s}\}$, and a general tensor of type (r, s) has the component representation

$$\mathbf{T} = T^{i_1\ldots i_r}{}_{j_1\ldots j_s}\mathbf{e}_{i_1}\otimes\cdots\otimes\mathbf{e}_{i_r}\otimes\mathbf{e}_*^{j_1}\otimes\cdots\otimes\mathbf{e}_*^{j_s}. \tag{1.4.45}$$

Under change of basis $\{\mathbf{e}_i\}\to\{\mathbf{e}'_i\}$ and dual basis $\{\mathbf{e}_*^i\}\to\{\mathbf{e}_*'^i\}$ we have

$$\mathbf{e}'_i = P_i^j\mathbf{e}_j, \qquad \mathbf{e}_*'^i = Q_j^i\mathbf{e}_*^j, \tag{1.4.46}$$

where the matrices P and Q are as defined in Section 1.4.2, and the components of $\mathbf{T}$ transform according to

$$T'^{i_1\ldots i_r}{}_{j_1\ldots j_s} = Q^{i_1}_{m_1}\cdots Q^{i_r}_{m_r}P^{n_1}_{j_1}\cdots P^{n_s}_{j_s}T^{m_1\ldots m_r}{}_{n_1\ldots n_s}. \tag{1.4.47}$$

The operation of contraction is defined analogously to that described in Section 1.4.2. It reduces a tensor of type (r,s) to one of type $(r-1,s-1)$ according to

$$\begin{aligned}\mathbf{C}_q^p(\mathbf{T}) = T^{i_1\ldots i_p\ldots i_r}{}_{j_1\ldots j_q\ldots j_s}\langle\mathbf{e}_*^{j_q},\mathbf{e}_{i_p}\rangle\mathbf{e}_{i_1}\otimes\cdots\otimes\mathbf{e}_{i_{p-1}}\\ \otimes\mathbf{e}_{i_{p+1}}\otimes\cdots\otimes\mathbf{e}_{i_r}\otimes\mathbf{e}_*^{j_1}\otimes\cdots\otimes\mathbf{e}_*^{j_{q-1}}\\ \otimes\mathbf{e}_*^{j_{q+1}}\otimes\cdots\otimes\mathbf{e}_*^{j_s}.\end{aligned} \tag{1.4.48}$$

Symmetry or antisymmetry with respect to a pair of contravariant (or covariant) indices may be defined as in Section 1.3.6, but not between contravariant and covariant indices. A completely symmetric or completely antisymmetric tensor must therefore be either purely contravariant $(r,0)$ or purely covariant $(0,s)$.

For any given inner product on $\mathbb{E}$ (in particular the dot product) there is a unique tensor $\mathbf{g}:\mathbb{E}\to\mathbb{E}^*$ of type $(0,2)$ induced by the inner product (in this case the dot product) through the identity

$$\langle\mathbf{gu},\mathbf{v}\rangle = \mathbf{u}\cdot\mathbf{v} \tag{1.4.49}$$

for all $\mathbf{u},\mathbf{v}\in\mathbb{E}$ such that $\mathbf{g}$ is an isomorphism. Equation (1.1.1) ensures that $\langle\mathbf{gv},\mathbf{u}\rangle = \langle\mathbf{gu},\mathbf{v}\rangle$ and hence symmetry of $\mathbf{g}$ follows from (1.4.42). Since $\mathbf{g}$ is an isomorphism it possesses an inverse (or reciprocal) $\mathbf{g}^{-1}$ which is a tensor of type $(2,0)$. In general $\mathbf{g}$ is called the metric tensor associated with the inner product, but in respect of the dot product it is referred to as the Euclidean metric tensor. It is a straightforward matter to associate with $\mathbf{g}$ a bilinear function $d:\mathbb{E}\times\mathbb{E}\to\mathbb{R}$ having the properties (a), (b), (c) ascribed to a *metric* in Section 1.1.3.

By reference to (1.4.2) and (1.4.27) it is easy to see that (1.4.49) yields the connection

$$\mathbf{e}_*^i = \mathbf{g}\mathbf{e}^i, \tag{1.4.50}$$

where $\{\mathbf{e}^i\}$ is the reciprocal basis of $\{\mathbf{e}_i\}$ in $\mathbb{E}$.

If $\mathbf{u} \in \mathbb{E}$ then we may write

$$\mathbf{u} = u^i \mathbf{e}_i = u_i \mathbf{e}^i,$$

where u^i are the contravariant components of $\mathbf{u}$ and u_i are the components of $\mathbf{u}$ on the reciprocal basis (we shall call them the reciprocal components of $\mathbf{u}$ here). Hence, by use of (1.4.50), the image $\mathbf{u}^*$ of $\mathbf{u}$ under $\mathbf{g}$ is given by

$$u_i^* \mathbf{e}_*^i = \mathbf{u}^* = \mathbf{g}\mathbf{u} = u_i \mathbf{e}_*^i. \tag{1.4.51}$$

It follows that $u_i^* = u_i$, that is the covariant components of $\mathbf{u}^*$ in $\mathbb{E}^*$ are also the reciprocal components of $\mathbf{u}$ in $\mathbb{E}$, but since $\mathbf{u}$ is a contravariant vector they are not in general called covariant components of $\mathbf{u}$. The terminology used in Section 1.4.1 is, however, now justified by the identification of $\mathbf{u}$ and $\mathbf{u}^*$ through $\mathbf{g}$.

With $\mathbf{g}$ regarded as a bilinear function over $\mathbb{E} \times \mathbb{E}$, its (covariant) components are given by

$$g_{ij} \equiv \mathbf{g}(\mathbf{e}_i, \mathbf{e}_j) = \langle \mathbf{g}\mathbf{e}_i, \mathbf{e}_j \rangle = \mathbf{e}_i \cdot \mathbf{e}_j = I_{ij} \tag{1.4.52}$$

on use of (1.4.49) and the definition of I_{ij} given in Section 1.4.1. Similarly, with $\mathbf{g}^{-1}$ as a bilinear function over $\mathbb{E}^* \times \mathbb{E}^*$, we have

$$(\mathbf{g}^{-1})^{ij} = \mathbf{g}^{-1}(\mathbf{e}_*^i, \mathbf{e}_*^j) = \langle \mathbf{e}_*^i, \mathbf{g}^{-1}\mathbf{e}_*^j \rangle = \langle \mathbf{g}\mathbf{e}^i, \mathbf{e}^j \rangle = I^{ij}. \tag{1.4.53}$$

Since, by (1.4.41), we may write $\mathbf{g} = g_{ij}\mathbf{e}_*^i \otimes \mathbf{e}_*^j$, it follows from (1.4.51) on use of (1.4.27) that

$$u_i = g_{ij} u^j$$

and dually

$$u^i = (\mathbf{g}^{-1})^{ij} u_j.$$

In view of (1.4.52) and (1.4.53) these are equivalent to (1.4.5).

Thus, through the components g_{ij} and $(\mathbf{g}^{-1})^{ij}$ the tensors $\mathbf{g}$ and $\mathbf{g}^{-1}$ respectively correspond to lowering and raising of an index.

More generally, $\mathbf{g}$ may be applied to a tensor of order (r, s) to convert it into a tensor of order $(r-1, s+1)$ while $\mathbf{g}^{-1}$ converts a tensor of order (r, s) into one of order $(r+1, s-1)$. For example, if $\mathbf{g}$ is applied to the pth element of the tensor

$$\mathbf{e}_{i_1} \otimes \cdots \otimes \mathbf{e}_{i_p} \otimes \cdots \otimes \mathbf{e}_{i_r}$$

of order $(r, 0)$ it becomes a tensor of order $(r-1, 1)$, namely

$$g_{i_p i_q} \mathbf{e}_{i_1} \otimes \cdots \otimes \mathbf{e}_{i_{p-1}} \otimes \mathbf{e}_*^{i_q} \otimes \mathbf{e}_{i_{p+1}} \otimes \cdots \otimes \mathbf{e}_{i_r}.$$

The same operation on the tensor

$$\mathbf{T} = T^{i_1 \ldots i_r} \mathbf{e}_{i_1} \otimes \cdots \otimes \mathbf{e}_{i_r}$$

converts it to

$$T^{i_1 \ldots i_{p+1}}{}_{i_p}{}^{i_{p+1} \ldots i_r} \mathbf{e}_{i_1} \otimes \cdots \otimes \mathbf{e}_{i_{p-1}} \otimes \mathbf{e}_*^{i_p} \otimes \mathbf{e}_{i_{p+1}} \otimes \cdots \mathbf{e}_{i_r},$$

the pth index being lowered according to

$$T^{i_1 \ldots i_p \ldots i_r} g_{i_p i_q} = T^{i_1 \ldots i_{p-1}}{}_{i_q}{}^{i_{p+1} \ldots i_r}.$$

The operation of lowering or raising an index may be carried out on as many indices as required; for example, a tensor of type $(r, 0)$ may be converted to one of type $(0, r)$ by means of the component transformation

$$g_{i_1 j_1} g_{i_2 j_2} \cdots g_{i_r j_r} T^{i_1 i_2 \ldots i_r} = T_{j_1 j_2 \ldots j_r}$$

and, conversely,

$$(\mathbf{g}^{-1})^{i_1 j_1} (\mathbf{g}^{-1})^{i_2 j_2} \ldots (\mathbf{g}^{-1})^{i_r j_r} T_{i_1 i_2 \ldots i_r} = T^{j_1 j_2 \ldots j_r}.$$

Problem 1.4.6 If $\mathbf{T} \in \mathscr{L}(\mathbb{E}, \mathbb{E})$ and $\mathbf{T}^* \in \mathscr{L}(\mathbb{E}^*, \mathbb{E}^*)$ are as defined in equations (1.4.34)–(1.4.36) show that $T^*{}_j{}^i = T^i{}_j$.

Given that $\mathbf{S}$ is a tensor of type $(2, 0)$, show that $\mathbf{gS}$ and $\mathbf{Sg}$ are each of type $(1, 1)$, and that $(\mathbf{Sg})^* = \mathbf{gS}$ if and only if $\mathbf{S}^{\mathsf{T}} = \mathbf{S}$.

Since, by (1.4.52) and (1.4.53), we have $g_{ij} = I_{ij}$ and $(\mathbf{g}^{-1})^{ij} = I^{ij}$ and also $u_i^* = u_i$ we are led to make the identification of $\mathbb{E}^*$ with $\mathbb{E}$ by setting $\mathbf{gu} = \mathbf{u}$ for each $\mathbf{u}$ in $\mathbb{E}$. Thus $\mathbf{g} = \mathbf{I}$, the identity on $\mathbb{E}$ and the components of $\mathbf{g}$ have all the properties ascribed to those of $\mathbf{I}$ in Section 1.4.1. In particular we note that

$$(\mathbf{g}^{-1})^{ij} = g^{ij}, \quad (\mathbf{g}^{-1})_{ij} = g_{ij}$$

and

$$g^{ik}g_{kj} = \delta^i_j.$$

The dual basis is then identified with the reciprocal basis and the scalar product with the inner product. Further discussion of these matters is given in Bowen and Wang (1976).

For the purposes of the use of tensors in elasticity theory, sufficient generality is available when $\mathbb{E}^*$ is identified with $\mathbb{E}$ and the theory described in Sections 1.4.1 and 1.4.2 is as general as we require. Henceforth, we assume that this identification is made.

1.5 TENSOR FIELDS

1.5.1 The gradient of a tensor field

In general, tensors associated with physical quantities have values which depend on position in space. In mathematical terms, we therefore regard tensors as defined over a Euclidean point space $\mathscr{E}$ or some (open) subset $\mathscr{D}$ of $\mathscr{E}$, and such tensors are called *fields*. More specifically, a *scalar field* $\phi:\mathscr{D} \to \mathbb{R}$ is a real-valued function over $\mathscr{D}$, a *vector field* $\mathbf{v}$ is a mapping $\mathbf{v}:\mathscr{D} \to \mathbb{E}$ and a *tensor field* $\mathbf{T}$ of order n is a mapping $\mathbf{T}:\mathscr{D} \to \mathscr{L}(\underbrace{\mathbb{E} \times \cdots \times \mathbb{E}}_{n\,\text{times}}, \mathbb{R}) = \mathscr{L}(\mathbb{E}^n, \mathbb{R})$.

At each point x in $\mathscr{D}$ the scalar $\phi(x)$, vector $\mathbf{v}(x)$ and tensor $\mathbf{T}(x)$ have the algebraic properties discussed in previous sections in which the notations α, $\mathbf{v}$, $\mathbf{T}$ were used respectively for scalars, vectors and tensors without the need for the argument x. Our purpose now is to discuss the changes in $\phi(x)$, $\mathbf{v}(x)$ and $\mathbf{T}(x)$ as x varies over $\mathscr{D}$, but in order to simplify this task we first of all make a small notational change.

In Section 1.1.3 an arbitrary pair of points x, y in $\mathscr{E}$ was associated with a vector in $\mathbb{E}$ denoted by $\mathbf{x} - \mathbf{y}$, where $\mathbf{x}$ and $\mathbf{y}$ respectively are the position vectors in $\mathbb{E}$ of the points x and y relative to an arbitrarily chosen origin (point) o in $\mathscr{E}$. The difference $\mathbf{x} - \mathbf{y}$ is independent of the choice of o. When o is fixed an arbitrary point x in $\mathscr{E}$ is associated uniquely with a vector $\mathbf{x}$ in $\mathbb{E}$, the position vector of x relative to o. This one-to-one correspondence may be described in terms of the vector field $\mathbf{r}:\mathscr{E} \to \mathbb{E}$, where $\mathbf{r}(x) = \mathbf{x}$ for each x in $\mathscr{E}$. A choice of origin is implicit in the definition of $\mathbf{r}$ which is called the *position vector field relative to o*.

Because of this correspondence, we may identify x with $\mathbf{x}$ for each x in $\mathscr{E}$ so that (given o) $\mathbf{r}$ becomes the identity mapping according to $\mathbf{r}(\mathbf{x}) = \mathbf{x}$. However, this does not enable $\mathscr{E}$ to be identified with $\mathbb{E}$ since $\mathbf{x} \in \mathscr{E}$ corresponds to a different vector in $\mathbb{E}$ when a different origin is chosen in $\mathscr{E}$. Thus, if $\mathbf{r}'$ is the position vector field relative to the origin o', we have $\mathbf{r}'(\mathbf{x}) = \mathbf{x} - \mathbf{c}$, where $\mathbf{c}$ is the position vector of o' relative to o.

But, we emphasize that when o is fixed $\mathbf{x}$ has a dual role—as a point in $\mathscr{E}$

and as a vector in $\mathbb{E}$. *Henceforth, we denote points in $\mathscr{E}$ by bold-face, lower-case letters and a given* $\mathbf{x}$ *in* $\mathscr{E}$ *has the dual interpretation described above.*

The reader is assumed to be familiar with basic notions of continuity and differentiability, so here we merely summarize some definitions relating to a (scalar, vector or tensor) field over $\mathscr{D}$. We represent a general field by the notation f so that $f:\mathscr{D}\rightarrow\mathscr{I}$, where $\mathscr{I}$ represents one of the (inner product) spaces $\mathbb{R}$, $\mathbb{E}$ or $\mathscr{L}(\mathbb{E}^n, \mathbb{R})$.

The field f is said to be *continuous* at $\mathbf{x}\in\mathscr{D}$ if, for any given positive number ε, there exists a scalar $\delta(\varepsilon, \mathbf{x}) > 0$ such that $d_{\mathscr{I}}(f(\mathbf{y}), f(\mathbf{x})) < \varepsilon$ whenever $d(\mathbf{y}, \mathbf{x}) < \delta(\varepsilon, \mathbf{x})$, where d is the metric defined in Section 1.1.3 and $d_{\mathscr{I}}$ is the metric associated with the inner product on $\mathscr{I}$. We note, in passing, that on $\mathscr{L}(\mathbb{E} \times \mathbb{E}, \mathbb{R})$, for example, a suitable inner product is given by $\mathrm{tr}(\mathbf{S}\mathbf{T}^{\mathrm{T}})$, where $\mathbf{S}$ and $\mathbf{T}$ are second-order tensors. In rectangular Cartesian components this is $S_{ij}T_{ij}$, an expression which is easily generalized to tensors of higher order.

If f is continuous at each point of $\mathscr{D}$ it is said to be continuous in $\mathscr{D}$.

For any two points $\mathbf{x}$ and $\mathbf{y}$ in $\mathscr{E}$ the *point difference* $\mathbf{y} - \mathbf{x}$ is a vector in $\mathbb{E}$. Let $\mathbf{a}$ be an arbitrary vector in $\mathbb{E}$. Then, given $\mathbf{x}\in\mathscr{D}$, we define $\mathbf{y}$ by $\mathbf{y} - \mathbf{x} = t\mathbf{a}$, where $t\in\mathbb{R}$. Provided t is restricted to a suitable (open) subset of $\mathbb{R}$ the equation $\mathbf{y} = \mathbf{x} + t\mathbf{a}$ defines a set of points in $\mathscr{D}$. Geometrically, it describes a segment in $\mathscr{D}$ of the straight line through $\mathbf{x}$ parallel to $\mathbf{a}$. We are now in a position to define differentiability at a point of $\mathscr{D}$.

The field $f:\mathscr{D}\rightarrow\mathscr{I}$ is said to be *differentiable at* $\mathbf{x}\in\mathscr{D}$ if there exists a (unique) linear mapping $\mathbf{G}_{\mathbf{x}}\in\mathscr{L}(\mathbb{E}, \mathscr{I})$ such that

$$\mathbf{G}_{\mathbf{x}}\mathbf{a} = \lim_{t\rightarrow 0}\left\{\frac{f(\mathbf{x} + t\mathbf{a}) - f(\mathbf{x})}{t}\right\} = \frac{\mathrm{d}}{\mathrm{d}t}f(\mathbf{x} + t\mathbf{a})\bigg|_{t=0}$$

for every $\mathbf{a}\in\mathbb{E}$.

If f is differentiable at each point of $\mathscr{D}$ it is said to be differentiable in $\mathscr{D}$ and there is a mapping $\mathbf{G}:\mathscr{D}\rightarrow\mathscr{L}(\mathbb{E}, \mathscr{I})$ such that $\mathbf{G}(\mathbf{x}) = \mathbf{G}_{\mathbf{x}}$, $\mathbf{x}\in\mathscr{D}$. If f is a tensor field of order n then $\mathbf{G}$ is a tensor field of order $n + 1$. It is called the *gradient* of f and we write $\mathbf{G} = \mathrm{grad}\, f$. Thus, $\mathrm{grad}\, f$ is defined by

$$(\mathrm{grad}\, f(\mathbf{x}))\mathbf{a} = \frac{\mathrm{d}}{\mathrm{d}t}f(\mathbf{x} + t\mathbf{a})\bigg|_{t=0} \tag{1.5.1}$$

for arbitrary $\mathbf{a}$ in $\mathbb{E}$, and this equation is the basis for our subsequent discussion.

We remark that, for any given *unit* vector $\mathbf{a}$, $(\mathrm{grad}\, f(\mathbf{x}))\mathbf{a}$ is called the *directional derivative* of f in the direction $\mathbf{a}$ at $\mathbf{x}$.

1.5.2 Symbolic notation for differential operators

In vector calculus it is common practice to use the symbol ∇ for the differential operator grad, and we adopt this notation here in accordance with the following conventions.

For a scalar field ϕ, grad ϕ is a vector field, written symbolically as $\nabla\phi$, and (1.5.1) becomes

$$(\nabla\phi(\mathbf{x}))\cdot\mathbf{a} = \frac{\mathrm{d}}{\mathrm{d}t}\phi(\mathbf{x}+t\mathbf{a})\bigg|_{t=0}. \tag{1.5.2}$$

In view of the symmetry of the dot product, the left-hand side of (1.5.2) may also be written as $(\mathbf{a}\cdot\nabla)\phi(\mathbf{x})$, thereby defining the *scalar* differential operator $\mathbf{a}\cdot\nabla$ for any $\mathbf{a}$ in $\mathbb{E}$.

For a tensor field T of order n, grad T is written $\nabla\otimes\mathsf{T}$ and this is a tensor field of order $n+1$. In this case (1.5.1) becomes

$$(\nabla\otimes\mathsf{T}(\mathbf{x}))\mathbf{a} = \frac{\mathrm{d}}{\mathrm{d}t}\mathsf{T}(\mathbf{x}+t\mathbf{a})\bigg|_{t=0} \tag{1.5.3}$$

and the left-hand side of this may also be put as $(\mathbf{a}\cdot\nabla)\mathsf{T}(\mathbf{x})$ in terms of the scalar differential operator defined above.

In particular, we note the important special case in which we have a vector field $\mathbf{v}(n=1)$. Then $\nabla\otimes\mathbf{v}$ is a second-order tensor field which enables us to define two differential operators closely related to grad.

Firstly, if $\mathbf{v}$ is a vector field, the scalar field div $\mathbf{v}$ (the *divergence* of $\mathbf{v}$) is defined by

$$\operatorname{div}\mathbf{v} \equiv \nabla\cdot\mathbf{v} = \operatorname{tr}(\nabla\otimes\mathbf{v}). \tag{1.5.4}$$

The divergence operation can be regarded as the contraction of the second-order tensor (field) $\nabla\otimes\mathbf{v}$ (see Sections 1.2.4, 1.3.1 and 1.4.2 for discussion of contraction). Equally the operation of contraction may be applied to the tensor field $\nabla\otimes\mathsf{T}$ of order $n+1$ to reduce it to a tensor field of order $n-1$. There are n possible contractions between ∇ and T in addition to $\frac{1}{2}n(n-1)$ contractions of T itself.

If, for example, T is a second-order tensor field then $\nabla\otimes\mathsf{T}$ has three possible contractions. Two of these are between ∇ and T, and we write the results as div T and div T^{T}, while the third is grad(tr T). In symbolic notation these are written as $\nabla\cdot\mathsf{T}$, $\nabla\cdot\mathsf{T}^{\mathrm{T}}$ and $\nabla(\operatorname{tr}\mathsf{T})$ respectively, each being a vector field. It is a matter of convention which of the two contractions between ∇ and T is denoted by $\nabla\cdot\mathsf{T}$. For example, we may define $\nabla\cdot\mathsf{T}$ by

$$(\nabla\cdot\mathsf{T}(\mathbf{x}))\cdot\mathbf{a} = \nabla\cdot(\mathsf{T}(\mathbf{x})\mathbf{a}) \tag{1.5.5}$$

for all $\mathbf{a}$ in $\mathbb{E}$, but alternatively $\mathbf{T}$ could be replaced by $\mathbf{T}^T$ on one side of this identity. In respect of higher-order tensors analogous conventions may be adopted but we shall not require them here.

Secondly, the *vector* field curl $\mathbf{v}$, denoted symbolically by $\nabla \wedge \mathbf{v}$, is defined by

$$(\nabla \wedge \mathbf{v}(\mathbf{x}))\cdot\mathbf{a} = \nabla\cdot(\mathbf{v}(\mathbf{x}) \wedge \mathbf{a}) \tag{1.5.6}$$

for all $\mathbf{a}$ in $\mathbb{E}$. In fact, $\nabla \wedge \mathbf{v}$ is the axial vector (field) of the skew-symmetric second-order tensor (field) $\nabla \otimes \mathbf{v} - (\nabla \otimes \mathbf{v})^T$ and may alternatively be defined by the identity

$$\{\nabla \otimes \mathbf{v}(\mathbf{x}) - (\nabla \otimes \mathbf{v}(\mathbf{x}))^T\}\mathbf{a} = \{\nabla \wedge \mathbf{v}(\mathbf{x})\} \wedge \mathbf{a} \tag{1.5.7}$$

for all $\mathbf{a}$ in $\mathbb{E}$ in accordance with equation (1.3.28) of Section 1.3.4. The equivalence of (1.5.6) and (1.5.7) is established by, for example, replacing $\mathbf{a}$ by $\mathbf{a} \wedge \mathbf{k}$ in (1.5.6), where $\mathbf{k}$ is an arbitrary vector in $\mathbb{E}$. Use of the properties of triple scalar and vector products (described in Section 1.1.1) in the resulting equation leads to (1.5.7) on removal of the dot product with arbitrary $\mathbf{k}$.

Second- and higher-order derivatives of fields are defined in an obvious way. For example, if grad f is differentiable in $\mathscr{D}$ then grad grad f is defined by (1.5.1) with f replaced by grad f. If ϕ is a scalar field, grad grad ϕ is denoted symbolically by $\nabla \otimes \nabla\phi$, while grad grad $\mathbf{T}$ is written $\nabla \otimes \nabla \otimes \mathbf{T}$ in respect of a tensor field $\mathbf{T}$, and so on for third- and higher-order derivatives. Note that each operation by grad raises the order of the tensor field by one.

The definitions of the differential operators described above do not depend on any choice of basis for $\mathbb{E}$ and are therefore tensorially (or vectorially) *invariant*. Accordingly, we have thus far refrained from making such a choice of basis. In Sections 1.5.3 and 1.5.4, however, we discuss differentiation relative to Cartesian and general (or curvilinear) coordinate systems and their associated basis vectors (which are themselves vector fields in general).

Problem 1.5.1 †Show that

$$\nabla \otimes (\phi\mathbf{T}) = \phi\nabla \otimes \mathbf{T} + \mathbf{T} \otimes \nabla\phi.$$

If $\mathbf{T}$ is a second-order tensor field, deduce that

$$\nabla\cdot(\phi\mathbf{T}) = \phi\nabla\cdot\mathbf{T} + \mathbf{T}^T\nabla\phi.$$

† If ϕ is a scalar field and $\mathbf{S}, \mathbf{T}$ are tensor fields the product fields $\phi\mathbf{T}$ and $\mathbf{S} \otimes \mathbf{T}$ are defined by $(\phi\mathbf{T})(\mathbf{x}) = \phi(\mathbf{x})\mathbf{T}(\mathbf{x})$ and $(\mathbf{S} \otimes \mathbf{T})(\mathbf{x}) = \mathbf{S}(\mathbf{x}) \otimes \mathbf{T}(\mathbf{x})$.

Show further that

$$\nabla\cdot(\mathbf{T}\mathbf{v}) = (\nabla\cdot\mathbf{T})\cdot\mathbf{v} + \operatorname{tr}\{\mathbf{T}(\nabla\otimes\mathbf{v})\},$$

where $\mathbf{v}$ is a vector field.

Problem 1.5.2 Use equation (1.5.3) to show that $\nabla\otimes\mathbf{r}(\mathbf{x}) = \mathbf{I}$, where $\mathbf{r}$ is the position vector field defined in Section 1.5.1 and $\mathbf{I}$ is the identity. Deduce that $\nabla\cdot\mathbf{r}(\mathbf{x}) = 3$.

Problem 1.5.3 If $r = |\mathbf{r}(\mathbf{x})|$ show that $\nabla r = \mathbf{r}(\mathbf{x})/r$. Hence simplify $\nabla\cdot(r^\alpha\mathbf{r}(\mathbf{x}))$, $\nabla\cdot(r^\alpha\mathbf{a})$ and $\nabla\cdot\{r^\alpha(\mathbf{a}\wedge\mathbf{r}(\mathbf{x}))\}$, where $\alpha\in\mathbb{R}$ and $\mathbf{a}\in\mathbb{E}$.

Problem 1.5.4 Show that $\nabla\wedge\mathbf{r}(\mathbf{x}) = \mathbf{0}$ and

$$\nabla\wedge(\phi\mathbf{v}) = \phi\nabla\wedge\mathbf{v} + \nabla\phi\wedge\mathbf{v}.$$

Problem 1.5.5 Show that $\nabla\otimes\nabla\otimes\mathbf{v}$ has contractions $\nabla(\nabla\cdot\mathbf{v})$ and $(\nabla\cdot\nabla)\mathbf{v}$. (The so-called Laplacian operator $\nabla\cdot\nabla$ is also denoted ∇^2.)

Use equation (1.5.6), with $\mathbf{v}$ replaced by $\nabla\wedge\mathbf{v}$, together with (1.5.7) to establish the identity

$$\nabla\wedge(\nabla\wedge\mathbf{v}) = \nabla(\nabla\cdot\mathbf{v}) - \nabla^2\mathbf{v}.$$

Problem 1.5.6 Prove that

$$\nabla\cdot(\mathbf{u}\wedge\mathbf{v}) = (\nabla\wedge\mathbf{u})\cdot\mathbf{v} - (\nabla\wedge\mathbf{v})\cdot\mathbf{u},$$

$$\nabla\wedge(\mathbf{u}\wedge\mathbf{v}) = (\nabla\cdot\mathbf{v})\mathbf{u} + (\mathbf{v}\cdot\nabla)\mathbf{u} - (\nabla\cdot\mathbf{u})\mathbf{v} - (\mathbf{u}\cdot\nabla)\mathbf{v},$$

$$\nabla(\mathbf{u}\cdot\mathbf{v}) = (\mathbf{u}\cdot\nabla)\mathbf{v} + (\mathbf{v}\cdot\nabla)\mathbf{u} - (\nabla\wedge\mathbf{v})\wedge\mathbf{u} - (\nabla\wedge\mathbf{u})\wedge\mathbf{v}$$

for any pair of vector fields $\mathbf{u}$ and $\mathbf{v}$.

Problem 1.5.7 If ϕ is a scalar field, show that $\nabla\otimes\nabla\phi$ is a second-order tensor field and hence show that $\nabla\wedge(\nabla\phi) = \mathbf{0}$ for arbitrary ϕ.

Problem 1.5.8 Show that, for any vector field $\mathbf{v}$, $\nabla\cdot(\nabla\wedge\mathbf{v}) = 0$.

1.5.3 Diffeıentiation in Cartesian coordinates

For simplicity we consider first the representation of the differential operators in a rectangular Cartesian coordinate system. This serves to illustrate the theory in terms of the familiar partial derivatives, and such representations are sufficiently general for use in most of the applications to elasticity theory in subsequent chapters.

We recall from Section 1.1.3 that a Cartesian coordinate system on $\mathscr{E}$ consists of an arbitrary choice of origin in $\mathscr{E}$ together with three scalar fields $e_i:\mathscr{E} \to \mathbb{R}$ which define the Cartesian coordinates x_i by $e_i(\mathbf{x}) = x_i (i = 1, 2, 3)$. In the notation adopted in Section 1.5.1 the position vector $\mathbf{x}$ of the point $\mathbf{x}$ relative to origin $\mathbf{o}$ has the representation $\mathbf{x} = x_i\mathbf{e}_i$ with respect to an (orthonormal) basis $\{\mathbf{e}_i\}$ of $\mathbb{E}$.

We now use the notation $\mathbf{e}$ to represent the triad (e_1, e_2, e_3) of scalar fields so that $\mathbf{e}:\mathscr{E} \to \mathbb{R}^3$ (where $\mathbb{R}^3 \equiv \mathbb{R} \times \mathbb{R} \times \mathbb{R}$) is defined by

$$\mathbf{e}(\mathbf{x}) = (e_1(\mathbf{x}), e_2(\mathbf{x}), e_3(\mathbf{x})) = (x_1, x_2, x_3) \tag{1.5.8}$$

for all $\mathbf{x}$ in $\mathscr{E}$. Then the mapping $\mathbf{e}$ is referred to as a *(rectangular) Cartesian coordinate system* on $\mathscr{E}$, an origin $\mathbf{o}$ being implicit in this definition.

Given $\mathbf{o}$, $\mathbf{e}$ is a one-to-one mapping on $\mathscr{E}$ and therefore invertible with inverse, denoted by $\mathbf{e}^{-1}$, defined over the whole of $\mathbb{R}^3$ so that

$$\mathbf{x} = \mathbf{e}^{-1}(x_1, x_2, x_3). \tag{1.5.9}$$

By definition $\mathbf{e}$ and $\mathbf{e}^{-1}$ are linear mappings and therefore continuous and differentiable.

For any field $f:\mathscr{D} \to \mathscr{J}$ we use (1.5.9) to write

$$f(\mathbf{x}) = f\{\mathbf{e}^{-1}(x_1, x_2, x_3)\} = f_{\mathbf{e}}(x_1, x_2, x_3),$$

wherein $f_{\mathbf{e}}:\mathbf{e}(\mathscr{D}) \to \mathscr{J}$ is defined. Thus, for any given Cartesian coordinate system $\mathbf{e}$ a field over $\mathscr{D}$ can be interpreted as a mapping over a subset $\mathbf{e}(\mathscr{D})$ of $\mathbb{R}^3$, and

$$f(\mathbf{x} + t\mathbf{a}) = f_{\mathbf{e}}(x_1 + ta_1, x_2 + ta_2, x_3 + ta_3)$$

for $\mathbf{x} \in \mathscr{D}$, arbitrary $\mathbf{a} \in \mathbb{E}$, and any $t \in \mathbb{R}$ such that $\mathbf{x} + t\mathbf{a} \in \mathscr{D}$. From equation (1.5.1) we therefore obtain

$$\begin{aligned}(\operatorname{grad} f(\mathbf{x}))\mathbf{a} &= \frac{\mathrm{d}}{\mathrm{d}t} f_{\mathbf{e}}(x_1 + ta_1, x_2 + ta_2, x_3 + ta_3)\bigg|_{t=0} \\ &= \frac{\partial f_{\mathbf{e}}}{\partial x_i}(x_1, x_2, x_3)a_i \\ &= \frac{\partial f}{\partial x_i}(\mathbf{x})a_i \end{aligned} \tag{1.5.10}$$

by use of the chain rule for partial derivatives.

Since $\mathbf{a}$ is arbitrary, we conclude that

$$\operatorname{grad} f(\mathbf{x}) = \frac{\partial f}{\partial x_i}(\mathbf{x})\mathbf{e}_i \tag{1.5.11}$$

and the field grad f has the representation $(\partial f/\partial x_i)\mathbf{e}_i$ with respect to the (orthonormal) basis $\{\mathbf{e}_i\}$.

For a scalar field ϕ, specialization of (1.5.11) gives

$$\nabla\phi(\mathbf{x}) = \frac{\partial \phi}{\partial x_i}(\mathbf{x})\mathbf{e}_i \tag{1.5.12}$$

and, for each i, $\partial\phi/\partial x_i : \mathscr{D} \to \mathbb{R}$ is a scalar field whose value at $\mathbf{x} \in \mathscr{D}$ is given by

$$\frac{\partial \phi}{\partial x_i}(\mathbf{x}) = \frac{\partial \phi_{\mathbf{e}}}{\partial x_i}(x_1, x_2, x_3). \tag{1.5.13}$$

As a particular example we choose $\phi = e_i$ so that $\phi_{\mathbf{e}}(x_1, x_2, x_3) = x_i$ and hence

$$\operatorname{grad} e_i(\mathbf{x}) \equiv \nabla e_i(\mathbf{x}) = \mathbf{e}_i \tag{1.5.14}$$

for all $\mathbf{x}$ in $\mathscr{E}$, recalling that $e_i(\mathbf{x}) = x_i$.

Next we consider the vector field $\mathbf{v}$, which, by (1.5.3), gives

$$\begin{aligned} \operatorname{grad} \mathbf{v}(\mathbf{x}) \equiv \nabla \otimes \mathbf{v}(\mathbf{x}) &= \frac{\partial}{\partial x_j}\mathbf{v}(\mathbf{x}) \otimes \mathbf{e}_j \\ &= \frac{\partial v_i}{\partial x_j}(\mathbf{x})\mathbf{e}_i \otimes \mathbf{e}_j, \end{aligned} \tag{1.5.15}$$

where the components v_i of $\mathbf{v}$ are scalar fields with values at $\mathbf{x}$ given by $\mathbf{e}_i \cdot \mathbf{v}(\mathbf{x}) = \mathbf{e}_i \cdot \mathbf{v}_{\mathbf{e}}(x_1, x_2, x_3)$.

In respect of the position vector field $\mathbf{r}$, for example, we have $\mathbf{v}_{\mathbf{e}}(x_1, x_2, x_3) = \mathbf{x}$ and hence equation (1.5.15) yields

$$\nabla \otimes \mathbf{r}(\mathbf{x}) = \mathbf{e}_i \otimes \mathbf{e}_i = \mathbf{I},$$

recalling the result (1.2.10), to confirm the (invariant) formula stated in Problem 1.5.2.

Contraction of (1.5.15) gives the component expression for div $\mathbf{v}$. Thus

$$\nabla \cdot \mathbf{v}(\mathbf{x}) = \frac{\partial v_i}{\partial x_i}(\mathbf{x}). \tag{1.5.16}$$

Use of equation (1.1.15) leads to the corresponding component form of curl **v** as follows:

$$\nabla \wedge \mathbf{v}(\mathbf{x}) = \mathbf{e}_i \wedge \mathbf{e}_j \frac{\partial v_j}{\partial x_i}(\mathbf{x}) = \varepsilon_{kij} \frac{\partial v_j}{\partial x_i}(\mathbf{x})\mathbf{e}_k. \tag{1.5.17}$$

For a tensor field **T** of order n equation (1.5.15) generalizes to

$$\begin{aligned}\nabla \otimes \mathbf{T} &= \frac{\partial \mathbf{T}}{\partial x_i} \otimes \mathbf{e}_i \\ &= \frac{\partial T_{i_1 i_2 \ldots i_n}}{\partial x_i} \mathbf{e}_{i_1} \otimes \mathbf{e}_{i_2} \otimes \cdots \otimes \mathbf{e}_{i_n} \otimes \mathbf{e}_i,\end{aligned} \tag{1.5.18}$$

where the components $T_{i_1 i_2 \ldots i_n}$ are scalar fields.

As discussed in Section 1.5.2, $\nabla \otimes \mathbf{T}$ has n possible contractions which can be regarded as the operation of div on **T**. Equally, the operation of curl may be applied to any one of the n basis vectors in the product basis on which **T** is decomposed. We do not require the results of such operations here, but details may be worked out using a simple generalization of (1.5.17).

Suppose the coordinate system **e** is changed to **e**′ in accordance with equations (1.1.33) and (1.1.43) so that the *point* **x** has coordinates x_i' in **e**′ (but has position vector $\mathbf{x} - \mathbf{c}$). On use of (1.1.45) it follows that

$$\frac{\partial \phi}{\partial x_i'}(\mathbf{x}) = \frac{\partial x_j}{\partial x_i'} \frac{\partial \phi}{\partial x_j}(\mathbf{x}) = Q_{ij} \frac{\partial \phi}{\partial x_j}(\mathbf{x})$$

so that the components of $\nabla \phi(\mathbf{x})$ conform with the rule for the transformation of components of a CT(1) (see Section 1.2.2).

Similarly, for a tensor field **T** of order n whose (rectangular Cartesian components transform according to equation (1.2.6) it is easy to show that the components of $\nabla \otimes \mathbf{T}$ are those of a CT $(n + 1)$.

For practical purposes, it is common practice to ignore the distinction between a field f, its value $f(\mathbf{x})$ at a point **x** and its representation $f_{\mathbf{e}}(x_1, x_2, x_3)$ in a Cartesian coordinate system. For most applications of the theory in this book it is convenient to disregard this distinction and for the most part we omit the argument **x** or (x_1, x_2, x_3) and use the same notation for a field and its value in any coordinate system.

Problem 1.5.9 Use Cartesian component representations to derive the results stated in Problems 1.5.2, 1.5.5–1.5.8.

Problem 1.5.10 If ϕ is a scalar field, show that the components of $\nabla \otimes \nabla \phi$ transform like those of a CT(2). Deduce that $\nabla^2 \phi$ is a CT(0).

Show that $\nabla \wedge \mathbf{v}$ is a CT(1).

Problem 1.5.11 If ϕ, $\mathbf{v}$ and $\mathbf{S}$ are scalar, vector and tensor fields, such that $\nabla\cdot\mathbf{v}=0$ and

$$\mathbf{S}=-\phi\mathbf{I}+\alpha\{\nabla\otimes\mathbf{v}+(\nabla\otimes\mathbf{v})^{\mathrm{T}}\},$$

where $\alpha\in\mathbb{R}$, show that

$$\nabla\cdot\mathbf{S}=-\nabla\phi+\alpha\nabla^2\mathbf{v}$$

and

$$\nabla(\operatorname{tr}\mathbf{S})=-3\nabla\phi.$$

Given that A_{ijk} and B_{ijkl} respectively are the components in the Cartesian coordinate system $\mathbf{e}$ of an isotropic CT(3) and CT(4), write down their most general forms. Deduce that if T_{ij} are the components of a symmetric CT(2) then there are scalars β, γ such that

$$A_{ijk}T_{jk}=0, \quad B_{ijkl}T_{jl}=\beta T_{ik}+\gamma T_{jj}\delta_{ik}.$$

Hence show that if

$$v_i=A_{ijk}S_{jk}+B_{ijkl}\frac{S_{jl}}{\partial x_k}$$

then

$$\mathbf{v}=\alpha\beta\nabla^2\mathbf{v}-(\beta+3\gamma)\nabla\phi$$

for some scalars α, β, γ.

1.5.4 Differentiation in curvilinear coordinates

We now generalize the analysis given in Section 1.5.3 in respect of a Cartesian coordinate system to a curvilinear coordinate system as defined below.

Let $\boldsymbol{\psi}:\mathscr{D}\to\mathbb{R}^3$ be a *continuous, one-to-one* mapping whose inverse $\boldsymbol{\psi}^{-1}$, defined on $\boldsymbol{\psi}(\mathscr{D})$, is also continuous.† If $\mathbf{x}\in\mathscr{D}$ we write

$$\boldsymbol{\psi}(\mathbf{x})=(x^1,x^2,x^3), \quad \mathbf{x}=\boldsymbol{\psi}^{-1}(x^1,x^2,x^3). \tag{1.5.19}$$

Further, suppose that $\boldsymbol{\psi}$ and $\boldsymbol{\psi}^{-1}$ have continuous derivatives up to any desired order (which is often taken to be infinity).‡

† Such a $\boldsymbol{\psi}$ is called a *homeomorphism*.

‡ If $\boldsymbol{\psi}$ and $\boldsymbol{\psi}^{-1}$ are n-times continuously differentiable $\boldsymbol{\psi}$ is called a *C^n-diffeomorphism* on $\mathscr{D}$. A *C^n-chart* at $\mathbf{x}\in\mathscr{E}$ is a pair $(\mathscr{D},\boldsymbol{\psi})$, where $\mathscr{D}$ is an open subset of $\mathscr{E}$ containing $\mathbf{x}$ and $\boldsymbol{\psi}:\mathscr{D}\to\mathbb{R}^3$ is a C^n-diffeomorphism. A *C^n-atlas* on $\mathscr{E}$ is a family of C^n-charts $(\mathscr{D}_\mu,\boldsymbol{\psi}_\mu)$ such that $\mathscr{E}=\bigcup_\mu\mathscr{D}_\mu$, where μ runs over some index set. A *C^n-Euclidean manifold* is a Euclidean point space equipped with a C^n-atlas. A Euclidean manifold has the important property that it can be equipped with an atlas consisting of a single chart $(\mathscr{E},\mathbf{e})$, $\mathbf{e}$ being a Cartesian coordinate system on $\mathscr{E}$.

Given $\boldsymbol{\psi}$ and $\mathscr{D}$ there are three scalar fields $\psi^i:\mathscr{D}\rightarrow\mathbb{R}$ such that

$$\boldsymbol{\psi}(\mathbf{x})=(\psi^1(\mathbf{x}),\psi^2(\mathbf{x}),\psi^3(\mathbf{x}))\quad \mathbf{x}\in\mathscr{D}.$$

The fields ψ^i are called the *coordinate functions* of $\boldsymbol{\psi}$ on $\mathscr{D}$, $\boldsymbol{\psi}$ is a *coordinate system* on $\mathscr{D}$ and $\mathscr{D}$ is a *coordinate neighbourhood.* The coordinates x^i of the point $\mathbf{x}$ in the coordinate system $\boldsymbol{\psi}$ are given by

$$x^i=\psi^i(\mathbf{x}). \tag{1.5.20}$$

They are also called *curvilinear* (or general) coordinates covering $\mathscr{D}$, and it is important to distinguish the curvilinear coordinates x^i (superscripts) from the Cartesian coordinates x_i (subscripts).

The equation

$$x^i\equiv\psi^i(\mathbf{x})=\text{constant} \tag{1.5.21}$$

defines a subset of $\mathscr{D}$ called an x^i-*coordinate surface* of ψ in $\mathscr{D}$. The intersection of coordinate surfaces corresponding to two different values of i defines a *coordinate curve* in $\mathscr{D}$. For example, suppose $\boldsymbol{\psi}^{-1}(x^1,x^2,x^3)=\mathbf{x}$ is a given point in $\mathscr{D}$. Then, with x^2 and x^3 fixed, the mapping $\mathbf{x}_1:\mathbb{D}_1\rightarrow\mathscr{D}$ defined by $\mathbf{x}_1(t)=\boldsymbol{\psi}^{-1}(t,x^2,x^3)$, where $\mathbb{D}_1$ is an open subset of $\mathbb{R}$ containing x^1, is called an x^1-coordinate curve of $\boldsymbol{\psi}$ in $\mathscr{D}$ passing through $\mathbf{x}$. The coordinate curves $\mathbf{x}_2$ and $\mathbf{x}_3$ are defined in a similar way.

The *tangent vector* to the curve $\mathbf{x}_i$ at the point $\mathbf{x}_i(t)$ is defined as $\mathrm{d}\mathbf{x}_i(t)/\mathrm{d}t$. Equivalently, the tangent vector to the x^i-coordinate curve at $\mathbf{x}=\boldsymbol{\psi}^{-1}(x^1,x^2,x^3)$ may be written as $\partial\mathbf{x}/\partial x^i$ and this is the notation we use henceforth.

For a field $f:\mathscr{D}\rightarrow\mathscr{J}$, we define the mapping $f_{\psi}:\boldsymbol{\psi}(\mathscr{D})\rightarrow\mathscr{J}$ by

$$f(\mathbf{x})=f\{\boldsymbol{\psi}^{-1}(x^1,x^2,x^3)\}=f_{\psi}(x^1,x^2,x^3)\quad \mathbf{x}\in\mathscr{D}.$$

From (1.5.1) it then follows that

$$\begin{aligned}(\operatorname{grad} f(\mathbf{x}))\mathbf{a}&=\frac{\mathrm{d}}{\mathrm{d}t}f_{\psi}(x^1+ta^1,x^2+ta^2,x^3+ta^3)\\&=\frac{\partial f_{\psi}}{\partial x^i}(x^1,x^2,x^3)a^i=\frac{\partial f}{\partial x^i}(\mathbf{x})a^i\end{aligned} \tag{1.5.22}$$

where the 'components' a^i are defined by $x^i+ta^i=\psi^i(\mathbf{x}+t\mathbf{a})$, recalling (1.5.20). Application of (1.5.22) to the scalar field ψ^i shows that

$$a^i=(\operatorname{grad}\psi^i(\mathbf{x}))\cdot\mathbf{a}, \tag{1.5.23}$$

while in respect of the position vector field $\mathbf{r}$ relative to some chosen origin we obtain

$$(\operatorname{grad}\mathbf{r}(\mathbf{x}))\mathbf{a} = \frac{\partial \mathbf{x}}{\partial x^i} a^i \tag{1.5.24}$$

since $\mathbf{r}_\psi(x^1, x^2, x^3) = \mathbf{r}(\mathbf{x}) = \mathbf{x}$.

But, by the result of Problem 1.5.2, we have $\operatorname{grad}\mathbf{r}(\mathbf{x}) = \mathbf{I}$, where $\mathbf{I}$ is the identity on $\mathbb{E}$. Hence (1.5.24) gives

$$\mathbf{a} = \frac{\partial \mathbf{x}}{\partial x^i} a^i, \tag{1.5.25}$$

and, since $\mathbf{a}$ is an arbitrary vector in $\mathbb{E}$, the three vectors $\partial\mathbf{x}/\partial x^i$ form a basis $\{\partial\mathbf{x}/\partial x^i\}$ at each point $\mathbf{x}$ of $\mathscr{D}$. It is called the *natural basis* of ψ at $\mathbf{x}$ and we write

$$\mathbf{g}_i(\mathbf{x}) = \frac{\partial \mathbf{x}}{\partial x^i} \qquad (i = 1, 2, 3), \tag{1.5.26}$$

where $\mathbf{g}_i$ is a vector field on $\mathscr{D}$. Note that $\mathbf{g}_i(\mathbf{x})$ is tangent to the x^i-coordinate curve of ψ.

The 'components' a^i are therefore the contravariant components of $\mathbf{a}$ with respect to the natural basis of ψ.

We follow Section 1.4.1 and define the reciprocal basis $\{\mathbf{g}^i(\mathbf{x})\}$ of $\{\mathbf{g}_i(\mathbf{x})\}$ at each point $\mathbf{x}$ of $\mathscr{D}$ such that

$$\mathbf{g}^i(\mathbf{x}) \cdot \mathbf{g}_j(\mathbf{x}) = \delta^i_j. \tag{1.5.27}$$

Since (1.5.23) and (1.5.25) hold for arbitrary $\mathbf{a} \in \mathbb{E}$ we have

$$\mathbf{g}^i(\mathbf{x}) = \operatorname{grad}\psi^i(\mathbf{x}).$$

Note that $\mathbf{g}^i(\mathbf{x})$ is normal to the surface $\psi^i(\mathbf{x}) = \text{constant}$ at an arbitrary point $\mathbf{x}$ on that surface.

From the results of Section 1.4.1, we have

$$\mathbf{I} = \mathbf{g}^i(\mathbf{x}) \otimes \mathbf{g}_i(\mathbf{x}) = \mathbf{g}_i(\mathbf{x}) \otimes \mathbf{g}^i(\mathbf{x}) \tag{1.5.28}$$

for each $\mathbf{x}$ in $\mathscr{D}$. The contravariant and covariant components of $\mathbf{I}$ are now denoted by $g^{ij}(\mathbf{x})$ and $g_{ij}(\mathbf{x})$ respectively, so that

$$\begin{aligned} g^{ij}(\mathbf{x}) &= \mathbf{g}^i(\mathbf{x}) \cdot \mathbf{g}^j(\mathbf{x}), \qquad & g_{ij}(\mathbf{x}) &= \mathbf{g}_i(\mathbf{x}) \cdot \mathbf{g}_j(\mathbf{x}), \\ \mathbf{g}^i(\mathbf{x}) &= g^{ij}(\mathbf{x})\mathbf{g}_j(\mathbf{x}), \qquad & \mathbf{g}_i(\mathbf{x}) &= g_{ij}(\mathbf{x})\mathbf{g}^j(\mathbf{x}), \end{aligned} \tag{1.5.29}$$

and the mixed components are δ^i_j.

Since **a** is arbitrary it follows from (1.5.22) that

$$\operatorname{grad} f(\mathbf{x}) = \frac{\partial f}{\partial x^i}(\mathbf{x})\mathbf{g}^i(\mathbf{x}) \tag{1.5.30}$$

and

$$(\operatorname{grad} f(\mathbf{x}))\mathbf{g}_i(\mathbf{x}) = \frac{\partial f}{\partial x^i}(\mathbf{x}) = \frac{\partial f_\psi}{\partial x^i}(x^1, x^2, x^3), \qquad i = 1, 2, 3, \tag{1.5.31}$$

the fields $\partial f/\partial x^i : \mathscr{D} \to \mathscr{I}$ thereby being defined.

At a point $\mathbf{x} = \psi^{-1}(x^1, x^2, x^3)$ let $\mathrm{d}x^i$ denote an increment or differential in x^i. These are the contravariant components with respect to the basis $\{\mathbf{g}_i(\mathbf{x})\}$ of the vector $\mathrm{d}\mathbf{x} = (\partial\mathbf{x}/\partial x^i)\,\mathrm{d}x^i = \mathbf{g}_i(\mathbf{x})\,\mathrm{d}x^i$. However, the coordinates x^i cannot be interpreted as the contravariant components of $\mathbf{x}$ itself since $\mathrm{d}\mathbf{x} = \mathbf{g}_i(\mathbf{x})\,\mathrm{d}x^i \neq \mathrm{d}(\mathbf{g}_i(\mathbf{x})x^i)$ in general. Moreover, the covariant components $(\mathrm{d}\mathbf{x})_i$ of $\mathrm{d}\mathbf{x}$, defined by $(\mathrm{d}\mathbf{x})_i = g_{ij}(\mathbf{x})\,\mathrm{d}x^j$, are not in general differentials. Clearly, $\mathrm{d}\mathbf{x}$ is *naturally contravariant* in character. On the other hand, $\operatorname{grad} \equiv \mathbf{g}^i(\mathbf{x})(\partial/\partial x^i)$ is *naturally covariant.*

From (1.5.26) we obtain

$$\frac{\partial \mathbf{g}_i}{\partial x^j}(\mathbf{x}) = \frac{\partial^2 \mathbf{x}}{\partial x^i \partial x^j} = \frac{\partial \mathbf{g}_j}{\partial x^i}(\mathbf{x}).$$

This leads us to introduce the coefficients $\Gamma^k_{ij}(\mathbf{x}) = \Gamma^k_{ji}(\mathbf{x})$ so that

$$\frac{\partial \mathbf{g}_i}{\partial x^j}(\mathbf{x}) = \Gamma^k_{ij}(\mathbf{x})\mathbf{g}_k(\mathbf{x}) \tag{1.5.32}$$

and therefore, by (1.5.27),

$$\Gamma^k_{ij}(\mathbf{x}) = \frac{\partial \mathbf{g}_i}{\partial x^j}(\mathbf{x}) \cdot \mathbf{g}^k(\mathbf{x}) = -\,\mathbf{g}_i(\mathbf{x}) \cdot \frac{\partial \mathbf{g}^k}{\partial x^j}(\mathbf{x}). \tag{1.5.33}$$

It also follows from (1.5.27) and (1.5.29) that

$$\frac{\partial \mathbf{g}^i}{\partial x^j}(\mathbf{x}) = -\,\Gamma^i_{jk}(\mathbf{x})\mathbf{g}^k(\mathbf{x}). \tag{1.5.34}$$

In symbolic notation, we use (1.5.30) to rewrite (1.5.32) and (1.5.34) in the forms

$$\begin{aligned} \nabla \otimes \mathbf{g}_i(\mathbf{x}) &= \Gamma^k_{ij}(\mathbf{x})\mathbf{g}_k(\mathbf{x}) \otimes \mathbf{g}^j(\mathbf{x}), \\ \nabla \otimes \mathbf{g}^i(\mathbf{x}) &= -\,\Gamma^i_{jk}(\mathbf{x})\mathbf{g}^k(\mathbf{x}) \otimes \mathbf{g}^j(\mathbf{x}). \end{aligned} \tag{1.5.35}$$

The coefficients Γ^i_{jk} are called *Christoffel symbols* associated with $\mathscr{D}$ and ψ. They are not the components of a tensor, as is seen in the result of Problem 1.5.17 below.

For a tensor field $\mathbf{T}$ with, for example, covariant component fields $T_{i_1 \dots i_n}$ with respect to the basis $\{\mathbf{g}^i(\mathbf{x})\}$, application of (1.5.30) and (1.5.35) to $T_{i_1 \dots i_n}\mathbf{g}^{i_1} \otimes \cdots \otimes \mathbf{g}^{i_n}$ yields

$$\nabla \otimes \mathbf{T} = \left\{ \frac{\partial T_{i_1 \dots i_n}}{\partial x^i} - \sum_{p=1}^{n} \Gamma^j_{i i_p} T_{i_1 \dots j \dots \dots i_n} \right\} \mathbf{g}^{i_1} \otimes \cdots \otimes \mathbf{g}^{i_n} \otimes \mathbf{g}^i, \tag{1.5.36}$$

the argument $\mathbf{x}$ having been omitted. The index j occurs as the pth subscript of $\mathbf{T}$ in the summation.

More compactly, we write

$$(\nabla \otimes \mathbf{T})_{i_1 \dots i_n i} = T_{i_1 \dots i_n, i},$$

where $,i$ denotes the *covariant derivative* defined by

$$T_{i_1 \dots i_n, i} = \frac{\partial T_{i_1 \dots i_n}}{\partial x^i} - \sum_{p=1}^{n} \Gamma^j_{i i_p} T_{i_1 \dots j \dots i_n}, \tag{1.5.37}$$

in respect of the covariant components of $\mathbf{T}$.

Similarly, $T^{i_1 \dots i_n}{}_{,i}$ denotes the covariant derivative of the contravariant components of $\mathbf{T}$ and is defined by

$$T^{i_1 \dots i_n}{}_{,i} = \frac{\partial T^{i_1 \dots i_n}}{\partial x^i} + \sum_{p=1}^{n} \Gamma^{i_p}_{ij} T^{i_1 \dots j \dots i_n}. \tag{1.5.38}$$

The covariant derivative of mixed components of $\mathbf{T}$ is defined in an obvious way intermediate between (1.5.37) and (1.5.38). For example, if $\mathbf{T}$ is a second-order tensor field $\nabla \otimes \mathbf{T}$ has four possible component representations, given by

$$\left.\begin{aligned}
T^{pq}{}_{,i} &= \frac{\partial T^{pq}}{\partial x^i} + \Gamma^p_{ir} T^{rq} + \Gamma^q_{ir} T^{pr}, \\
T_{pq,i} &= \frac{\partial T_{pq}}{\partial x^i} - \Gamma^r_{ip} T_{rq} - \Gamma^r_{iq} T_{pr}, \\
T^p{}_{q,i} &= \frac{\partial T^p{}_q}{\partial x^i} + \Gamma^p_{ir} T^r{}_q - \Gamma^r_{iq} T^p{}_r, \\
T_p{}^q{}_{,i} &= \frac{\partial T_p{}^q}{\partial x^i} - \Gamma^r_{pi} T_r{}^q + \Gamma^q_{ir} T_p{}^r.
\end{aligned}\right\} \tag{1.5.39}$$

Problem 1.5.12 If $\mathbf{I}$ is the second-order identity tensor show that $\nabla \otimes \mathbf{I} = \mathbf{0}$ and deduce that $g^{pq}{}_{,i} = 0, g_{pq,i} = 0$.

Problem 1.5.13 If $\mathbf{T}$ is a second-order tensor field, obtain alternative expressions for all the contractions of $\nabla \otimes \mathbf{T}$. Deduce that $\nabla \cdot \mathbf{T} = T^{ip}{}_{,i}\mathbf{g}_p = T^i{}_{p,i}\mathbf{g}^p$.

Problem 1.5.14 If $\mathbf{v}$ is a vector field, show that

$$\nabla \otimes \mathbf{v} = v^k{}_{,j}\mathbf{g}_k \otimes \mathbf{g}^j = v_{k,j}\mathbf{g}^k \otimes \mathbf{g}^j,$$

where

$$v^k{}_{,j} = \frac{\partial v^k}{\partial x^j} + \Gamma^k_{ij}v^i$$

$$v_{k,j} = \frac{\partial v_k}{\partial x^j} - \Gamma^i_{jk}v_i.$$

Deduce an expression for $\nabla \cdot \mathbf{v}$.

Problem 1.5.15 If g denotes the determinant of the components g_{ij}, use the results of Problem 1.1.1 to show that

$$\frac{1}{g}\frac{\partial g}{\partial g_{pq}} = g^{pq}$$

and hence prove that

$$\frac{1}{\sqrt{g}}\frac{\partial}{\partial x^i}(\sqrt{g}) = \Gamma^k_{ik}.$$

Use this result to show that

$$\nabla \cdot \mathbf{v} = \frac{1}{\sqrt{g}}\frac{\partial}{\partial x^i}(\sqrt{g}v^i).$$

Problem 1.5.16 Show that $(\mathbf{g}_i \wedge \mathbf{g}_j) \cdot \mathbf{g}_k = \sqrt{g}\varepsilon_{ijk}$ and deduce that

$$\mathbf{g}_i \wedge \mathbf{g}_j = \sqrt{g}\varepsilon_{ijk}\mathbf{g}^k,$$

where ε_{ijk} is the alternating symbol defined in (1.1.16).

Show similarly that

$$\mathbf{g}^i \wedge \mathbf{g}^j = \frac{1}{\sqrt{g}} \varepsilon^{ijk} \mathbf{g}_k,$$

where $\varepsilon^{ijk} = \varepsilon_{ijk}$.

Use these results to derive component expressions for $\nabla \wedge \mathbf{v}$, where $\mathbf{v}$ is a vector field.

Suppose $\boldsymbol{\psi}' : \mathscr{D}' \to \mathbb{R}^3$ is a second coordinate system so that a point $\mathbf{x} \in \mathscr{D}'$ has coordinates x'^i. Then, if $\mathbf{x}$ is also in $\mathscr{D}$ (the intersection $\mathscr{D} \cap \mathscr{D}'$ of $\mathscr{D}$ and $\mathscr{D}'$ is assumed to be non-empty), we have $x^i = \psi^i(\mathbf{x})$ and $x'^i = \psi'^i(\mathbf{x})$.

The equation

$$(x'^1, x'^2, x'^3) = \boldsymbol{\psi}'(\mathbf{x}) = \boldsymbol{\psi}'\{\boldsymbol{\psi}^{-1}(x^1, x^2, x^3)\} = \boldsymbol{\psi}' \circ \boldsymbol{\psi}^{-1}(x^1, x^2, x^3)$$

then defines a coordinate transformation from (x^1, x^2, x^3) to (x'^1, x'^2, x'^3), the composite mapping $\boldsymbol{\psi}' \circ \boldsymbol{\psi}^{-1} : \boldsymbol{\psi}(\mathscr{D} \cap \mathscr{D}') \to \boldsymbol{\psi}'(\mathscr{D} \cap \mathscr{D}')$ thereby being defined. For convenience we now write the coordinate transformation and its inverse compactly as

$$x'^i = x'^i(x^1, x^2, x^3), \qquad x^i = x^i(x'^1, x'^2, x'^3) \tag{1.5.40}$$

ignoring the distinction between functions and function values.

The matrix of components $\partial x'^i(x^1, x^2, x^3)/\partial x^j$ is called the Jacobian matrix of the coordinate transformation (1.5.40). The properties of $\boldsymbol{\psi}$ and $\boldsymbol{\psi}'$ ensure that this matrix is non-singular, and its inverse has components $\partial x^i(x'^1, x'^2, x'^3)/\partial x'^j$ such that

$$\frac{\partial x'^i}{\partial x^k}(x^1, x^2, x^3) \frac{\partial x^k}{\partial x'^j}(x'^1, x'^2, x'^3) = \delta^i_j,$$

$$\frac{\partial x^i}{\partial x'^k}(x'^1, x'^2, x'^3) \frac{\partial x'^k}{\partial x^j}(x^1, x^2, x^3) = \delta^i_j. \tag{1.5.41}$$

Under the coordinate transformation (1.5.40) the basis vectors transform according to

$$\mathbf{g}'_i(\mathbf{x}) = \frac{\partial x^j}{\partial x'^i}(x'^1, x'^2, x'^3) \mathbf{g}_j(\mathbf{x}), \qquad \mathbf{g}'^i(\mathbf{x}) = \frac{\partial x'^i}{\partial x^j}(x^1, x^2, x^3) \mathbf{g}^j(\mathbf{x}),$$

while the contravariant and covariant components of a vector field $\mathbf{v}$ satisfy

$$v'^i = \frac{\partial x'^i}{\partial x^j} v^j, \qquad v'_i = \frac{\partial x^j}{\partial x'^i} v_j.$$

Clearly $\partial x'^i/\partial x^j$ and $\partial x^j/\partial x'^i$ have adopted the roles played respectively by Q^i_j and P^j_i in Section 1.4.2, and the components of tensor fields transform according to (1.4.20).

Problem 1.5.17 Show that the Christoffel symbols Γ^k_{ij} and Γ'^k_{ij} associated with $(\mathscr{D}, \psi)$ and $(\mathscr{D}', \psi')$ respectively are related by

$$\Gamma'^k_{ij} = \frac{\partial x^p}{\partial x'^i}\frac{\partial x^q}{\partial x'^j}\frac{\partial x'^k}{\partial x^r}\Gamma^r_{pq} + \frac{\partial^2 x^p}{\partial x'^i \partial x'^j}\frac{\partial x'^k}{\partial x^p}.$$

For a (rectangular) Cartesian coordinate system $\mathbf{e}$ the whole of $\mathscr{E}$ is a coordinate neighbourhood, the coordinate curves are straight lines parallel to the basis vectors $\mathbf{e}_i$ and the coordinate surfaces are planes defined by pairs of these lines. If we take $\psi' = \mathbf{e}, \mathscr{D}' = \mathscr{E}$ then the coordinate transformation (1.5.40) becomes

$$x_i = x_i(x^1, x^2, x^3), \qquad x^i = x^i(x_1, x_2, x_3) \tag{1.5.42}$$

for all $\mathbf{x}$ in $\mathscr{D}$, where (x_1, x_2, x_3) are the (rectangular) Cartesian coordinates of $\mathbf{x}$ as discussed in Section 1.5.3.

Such a Cartesian coordinate system provides a convenient starting point for generating a general curvilinear coordinate system as described in the following.

Relative to the choice of origin implicit in $\mathbf{e}$, let $\mathbf{x}$ be the position vector of the point $\mathbf{x}$. With respect to an orthonormal basis $\{\mathbf{e}_i\}$ it has the representation $\mathbf{x} = x_j\mathbf{e}_j$. Thus, from (1.5.26), (1.5.28) and (1.5.42), we obtain

$$\left.\begin{aligned} \mathbf{g}_i(\mathbf{x}) &= \frac{\partial x_j}{\partial x^i}(x^1, x^2, x^3)\mathbf{e}_j, \\ \mathbf{g}^i(\mathbf{x}) &= \frac{\partial x^i}{\partial x_j}(x_1, x_2, x_3)\mathbf{e}_j \end{aligned}\right\} \tag{1.5.43}$$

and hence

$$g_{ij} = \frac{\partial x_k}{\partial x^i}\frac{\partial x_k}{\partial x^j}, \qquad g^{ij} = \frac{\partial x^i}{\partial x_k}\frac{\partial x^j}{\partial x_k},$$

while from (1.5.33) expressions for Γ^k_{ij} may be obtained.

Problem 1.5.18 Show that

$$\sqrt{g} = \det\left(\frac{\partial x_i}{\partial x^j}\right),$$

where g is defined in Problem 1.5.15.

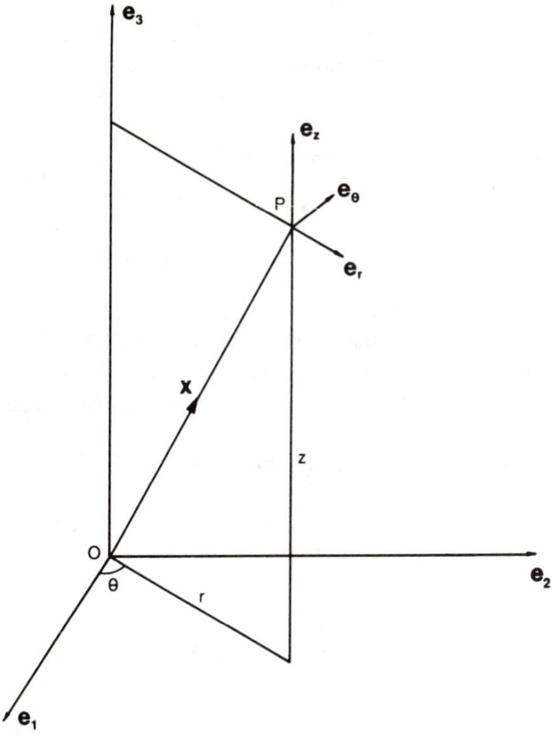

Fig. 1.3 Position vector **x** of a point P relative to origin O, showing cylindrical polar coordinates of P with respect to the orthonormal basis $\{\mathbf{e}_i\}$ through O. The coordinate surfaces are circular cylinders coaxial with $\mathbf{e}_3$ ($r = \text{constant}$), planes containing $\mathbf{e}_3$ ($\theta = \text{constant}$) and planes normal to $\mathbf{e}_3$ ($z = \text{constant}$). The unit basis vectors $\mathbf{e}_r$, $\mathbf{e}_\theta$, $\mathbf{e}_z$ at P are also shown.

We now apply the above theory to a cylindrical polar coordinate system with coordinates $(x^1, x^2, x^3) = (r, \theta, z)$ defined in Fig. 1.3.

The coordinate transformation and its inverse are written

$$(x_1, x_2, x_3) = (r\cos\theta, r\sin\theta, z)$$

$$(r, \theta, z) = \left((x_1^2 + x_2^2)^{1/2}, \tan^{-1}\left(\frac{x_2}{x_1}\right), x_3\right)$$

there being a one-to-one correspondence between coordinate neighbourhoods provided (r, θ, z) is in the open subset of $\mathbb{R}^3$ defined by

$$0 < r < \infty, \qquad 0 < \theta < 2\pi, \qquad -\infty < z < \infty.$$

Under the transformation this set of points is mapped onto the whole of $\mathbb{R}^3$ excluding the plane defined by $x_2 = 0, x_1 \geq 0$. Then both transformations are continuously differentiable infinitely many times.

Since $\mathbf{x} = r\cos\theta\mathbf{e}_1 + r\sin\theta\mathbf{e}_2 + z\mathbf{e}_3$, equation (1.5.43) leads to

$$\left.\begin{aligned} \mathbf{g}_1 &= \mathbf{e}_1\cos\theta + \mathbf{e}_2\sin\theta = \mathbf{e}_r = \mathbf{g}^1, \\ \mathbf{g}_2 &= r(-\mathbf{e}_1\sin\theta + \mathbf{e}_2\cos\theta) = r\mathbf{e}_\theta = r^2\mathbf{g}^2, \\ \mathbf{g}_3 &= \mathbf{e}_3 = \mathbf{e}_z = \mathbf{g}^3, \end{aligned}\right\} \tag{1.5.44}$$

the *unit* basis vectors $\mathbf{e}_r, \mathbf{e}_\theta, \mathbf{e}_z$ thereby being defined. The values of the components g_{ij} and g^{ij} are then found by using (1.5.29). Note that $g_{ij} = 0$ $(i \neq j)$, a condition which is necessary and sufficient for the coordinate curves to be orthogonal, and therefore defining an *orthogonal curvilinear coordinate system.*

The only non-zero Christoffel symbols are given by

$$\Gamma^2_{21} = \Gamma^2_{12} = \frac{1}{r}, \qquad \Gamma^1_{22} = -r \tag{1.5.45}$$

(use equation (1.5.33) with (1.5.44)).

For a vector field $\mathbf{v}$ with covariant components v_i we define the *physical components* $\hat{v}_i$ by

$$\hat{v}_i = v_i|\mathbf{g}^i| = v_i\sqrt{g^{ii}} \qquad \text{(no sum)} \tag{1.5.46}$$

so that

$$\mathbf{v} = v_i\mathbf{g}^i = \hat{v}_i\hat{\mathbf{g}}^i, \tag{1.5.47}$$

where $\hat{\mathbf{g}}^i = \mathbf{g}^i/\sqrt{g^{ii}}$ is a unit vector. Thus, physical components correspond to unit basis vectors and all such components of a given vector (or tensor) have the same physical dimensions.

With respect to the cylindrical polar coordinate system

$$(\hat{\mathbf{g}}^1, \hat{\mathbf{g}}^2, \hat{\mathbf{g}}^3) = (\mathbf{e}_r, \mathbf{e}_\theta, \mathbf{e}_z) \tag{1.5.48}$$

and we write

$$(\hat{v}_1, \hat{v}_2, \hat{v}_3) = (v_r, v_\theta, v_z). \tag{1.5.49}$$

It follows from (1.5.44) that $v_1 = v_r, v_2 = rv_\theta, v_3 = v_z$. This enables us to calculate the components of $\nabla \otimes \mathbf{v}$ with respect to the orthonormal basis

$(\mathbf{e}_r, \mathbf{e}_\theta, \mathbf{e}_z)$ as follows. From Problem 1.5.14, we have

$$\nabla \otimes \mathbf{v} = \left(\frac{\partial v_k}{\partial x^j} - \Gamma^i_{jk} v_i\right)\mathbf{g}^k \otimes \mathbf{g}^j$$

which, with the help of (1.5.44)–(1.5.49), yields

$$\begin{aligned}\nabla \otimes \mathbf{v} = {} & \frac{\partial v_r}{\partial r}\mathbf{e}_r \otimes \mathbf{e}_r + \frac{1}{r}\left(\frac{\partial v_r}{\partial \theta} - v_\theta\right)\mathbf{e}_r \otimes \mathbf{e}_\theta + \frac{\partial v_r}{\partial z}\mathbf{e}_r \otimes \mathbf{e}_z \\ & + \frac{\partial v_\theta}{\partial r}\mathbf{e}_\theta \otimes \mathbf{e}_r + \frac{1}{r}\left(\frac{\partial v_\theta}{\partial \theta} + v_r\right)\mathbf{e}_\theta \otimes \mathbf{e}_\theta + \frac{\partial v_\theta}{\partial z}\mathbf{e}_\theta \otimes \mathbf{e}_z \\ & + \frac{\partial v_z}{\partial r}\mathbf{e}_z \otimes \mathbf{e}_r + \frac{1}{r}\frac{\partial v_z}{\partial \theta}\mathbf{e}_z \otimes \mathbf{e}_\theta + \frac{\partial v_z}{\partial z}\mathbf{e}_z \otimes \mathbf{e}_z. \end{aligned} \tag{1.5.50}$$

Contraction immediately gives

$$\nabla \cdot \mathbf{v} = \frac{\partial v_r}{\partial r} + \frac{v_r}{r} + \frac{1}{r}\frac{\partial v_\theta}{\partial \theta} + \frac{\partial v_z}{\partial z}. \tag{1.5.51}$$

In the following problems we use the *physical components* of a tensor $\mathbf{T}$, defined on each subscript by analogy with (1.5.46) or on a superscript by making use of the unit vectors $\hat{\mathbf{g}}_i = \mathbf{g}_i/\sqrt{g_{ii}}$ (no sum). Thus, for a second-order tensor, for example,

$$\begin{aligned}\mathbf{T} &= \hat{T}_{ij}\hat{\mathbf{g}}^i \otimes \hat{\mathbf{g}}^j = \hat{T}^{ij}\hat{\mathbf{g}}_i \otimes \hat{\mathbf{g}}_j \\ &= T_{ij}\mathbf{g}^i \otimes \mathbf{g}^j = T^{ij}\mathbf{g}_i \otimes \mathbf{g}_j.\end{aligned}$$

For an orthogonal coordinate system $\hat{T}_{ij} = \hat{T}^{ij}$ and, in particular, for a cylindrical polar coordinate system these are written as $T_{rr}, T_{r\theta}, \ldots, T_{zz}$.

Problem 1.5.19 If $\mathbf{T}$ is a second-order tensor field, obtain the components of $\nabla \otimes \mathbf{T}$ with respect to the basis $\mathbf{e}_r, \mathbf{e}_\theta, \mathbf{e}_z$ in terms of the physical components of $\mathbf{T}$.

Hence show that the r, θ and z components of $\nabla \cdot \mathbf{T}$ are respectively

$$\left.\begin{aligned} &\frac{\partial T_{rr}}{\partial r} + \frac{1}{r}\frac{\partial T_{\theta r}}{\partial \theta} + \frac{\partial T_{zr}}{\partial z} + \frac{1}{r}(T_{rr} - T_{\theta\theta}), \\ &\frac{\partial T_{r\theta}}{\partial r} + \frac{1}{r}\frac{\partial T_{\theta\theta}}{\partial \theta} + \frac{\partial T_{z\theta}}{\partial z} + \frac{1}{r}(T_{r\theta} + T_{\theta r}), \\ &\frac{\partial T_{rz}}{\partial r} + \frac{1}{r}\frac{\partial T_{\theta z}}{\partial \theta} + \frac{\partial T_{zz}}{\partial z} + \frac{1}{r}T_{rz}. \end{aligned}\right\} \tag{1.5.52}$$

Problem 1.5.20 Spherical polar coordinates (r, θ, ϕ) are defined by the coordinate transformation

$$x_1 = r \sin \theta \cos \phi, \qquad x_2 = r \sin \theta \sin \phi, \qquad x_3 = r \cos \theta.$$

Find the subset of $\mathbb{R}^3$ on which the transformation is one-to-one, and obtain the inverse transformation.

Show that

$$\mathbf{g}_1 = \mathbf{g}^1, \qquad \mathbf{g}_2 = r^2 \mathbf{g}^2, \qquad \mathbf{g}_3 = r^2 \sin^2 \theta \mathbf{g}^3$$

and that the only non-zero Christoffel symbols are given by

$$\Gamma^2_{12} = \Gamma^3_{13} = \frac{1}{r}, \qquad \Gamma^1_{22} = -r, \qquad \Gamma^3_{23} = \cot \theta,$$

$$\Gamma^1_{33} = -r \sin^2 \theta, \qquad \Gamma^2_{33} = -\sin \theta \cos \theta.$$

If $\mathbf{v}$ is a vector field with physical components v_r, v_θ, v_ϕ find the physical components of $\nabla \otimes \mathbf{v}$ with respect to the orthonormal basis $(\mathbf{e}_r, \mathbf{e}_\theta, \mathbf{e}_\phi)$ and deduce that

$$\nabla \cdot \mathbf{v} = \frac{\partial v_r}{\partial r} + \frac{2}{r} v_r + \frac{1}{r} \frac{\partial v_\theta}{\partial \theta} + \frac{\cot \theta v_\theta}{r} + \frac{1}{r \sin \theta} \frac{\partial v_\phi}{\partial \phi}. \tag{1.5.53}$$

Alternatively, calculate g_{ij} and g and use the expression in Problem 1.5.15 to find $\nabla \cdot \mathbf{v}$.

Problem 1.5.21 Repeat Problem 1.5.19 for spherical polar coordinates, showing that the r, θ, ϕ components of $\nabla \cdot \mathbf{T}$ are respectively

$$\left.\begin{array}{l}
\dfrac{\partial T_{rr}}{\partial r} + \dfrac{2}{r} T_{rr} + \dfrac{1}{r} \dfrac{\partial T_{\theta r}}{\partial \theta} + \dfrac{1}{r \sin \theta} \dfrac{\partial T_{\phi r}}{\partial \phi} + \dfrac{\cot \theta}{r} T_{\theta r} - \dfrac{1}{r}(T_{\theta\theta} + T_{\phi\phi}), \\[2ex]
\dfrac{\partial T_{r\theta}}{\partial r} + \dfrac{2}{r} T_{r\theta} + \dfrac{1}{r} T_{\theta r} + \dfrac{1}{r} \dfrac{\partial T_{\theta\theta}}{\partial \theta} + \dfrac{1}{r \sin \theta} \dfrac{\partial T_{\phi\theta}}{\partial \phi} + \dfrac{\cot \theta}{r}(T_{\theta\theta} - T_{\phi\phi}), \\[2ex]
\dfrac{\partial T_{r\phi}}{\partial r} + \dfrac{2}{r} T_{r\phi} + \dfrac{1}{r} T_{\phi r} + \dfrac{1}{r} \dfrac{\partial T_{\theta\phi}}{\partial \theta} + \dfrac{1}{r \sin \theta} \dfrac{\partial T_{\phi\phi}}{\partial \phi} + \dfrac{\cot \theta}{r}(T_{\theta\phi} + T_{\phi\theta}).
\end{array}\right\} \tag{1.5.54}$$

1.5.5 Curves and surfaces

To enable evaluation of the covariant derivative of a field on a curve or surface in $\mathscr{E}$ appropriate smoothness must be imposed and with this in mind we now define a regular curve and a regular surface. These definitions[†] are also relevant for the discussion of integrals of fields over curves and surfaces in Section 1.5.6. *Coordinate* curves and surfaces have already been defined in Section 1.5.4.

A *curve* is a mapping $\boldsymbol{\lambda}: I \rightarrow \mathscr{E}$, where I is an open interval in $\mathbb{R}$. Suppose that for any point $\mathbf{x} \in \boldsymbol{\lambda}(I)$ there exists a (coordinate) neighbourhood $\mathscr{D}$ in $\mathscr{E}$ and an open interval $I' \subset I$ such that $\boldsymbol{\lambda}(I') = \mathscr{D} \cap \boldsymbol{\lambda}(I)$. Then, if $\boldsymbol{\psi}$ is a coordinate system on $\mathscr{D}$, $\boldsymbol{\lambda}(t)$ has the coordinate representation

$$\boldsymbol{\psi}\{\boldsymbol{\lambda}(t)\} = (\lambda^1(t), \lambda^2(t), \lambda^3(t)) \quad t \in I'.$$

The curve is said to be *smooth* if I can be covered with subintervals $\mathscr{D} \cap \boldsymbol{\lambda}(I)$ for each of which the coordinate functions λ^i $(i = 1, 2, 3)$ are continuously differentiable infinitely many times.

The *tangent vector* to $\boldsymbol{\lambda}$ at $t \in I$ is defined as $\dot{\boldsymbol{\lambda}}(t) \equiv \mathrm{d}\boldsymbol{\lambda}(t)/\mathrm{d}t$, and this is smooth if $\boldsymbol{\lambda}$ is smooth. A smooth curve $\boldsymbol{\lambda}$ is said to be *regular* if $\dot{\boldsymbol{\lambda}}(t) \neq \mathbf{0}$ for all $t \in I$. A coordinate curve is clearly a regular curve.

A subset $\mathscr{S}$ of $\mathscr{E}$ is called a *regular surface* if, for each $\mathbf{x} \in \mathscr{S}$, there exists a (coordinate) neighbourhood $\mathscr{D}$ in $\mathscr{E}$ and a mapping $\mathbf{a}: D \rightarrow \mathscr{D} \cap \mathscr{S}$ from an open set D in $\mathbb{R}^2$ onto $\mathscr{D} \cap \mathscr{S}$ such that (a) $\mathbf{a}$ is smooth in D, (b) $\mathbf{a}$ is one-to-one and its inverse is continuous and (c) if $(x^1, x^2) \in D$ are coordinates covering $\mathscr{D} \cap \mathscr{S}$ then $\mathbf{a}_1 \wedge \mathbf{a}_2 \neq \mathbf{0}$ for all $(x^1, x^2) \in D$, where $\mathbf{a}_\alpha = \partial \mathbf{a}(x^1, x^2)/\partial x^\alpha$ $(\alpha = 1, 2)$. Note that in general $\mathscr{D} \cap \mathscr{S} \neq \mathscr{S}$, that is $\mathscr{S}$ cannot be covered by a single coordinate neighbourhood.

The (positive) unit normal $\mathbf{n}$ to $\mathscr{D} \cap \mathscr{S}$ is given by

$$\mathbf{n} = (\mathbf{a}_1 \wedge \mathbf{a}_2)/|\mathbf{a}_1 \wedge \mathbf{a}_2|. \tag{1.5.55}$$

A coordinate surface is clearly a regular surface since, for example, $\mathbf{x} = \boldsymbol{\psi}^{-1}(x^1, x^2, 0) \equiv \mathbf{a}(x^1, x^2)$ satisfies (a), (b), (c) by definition of a coordinate system. Conversely, a given regular surface $\mathscr{S}$ can be identified as a coordinate surface locally by choosing a coordinate neighbourhood $\mathscr{D}$ and a coordinate system $\boldsymbol{\psi}$ so that $\mathscr{D} \cap \mathscr{S}$ is characterized by $\psi^3(\mathbf{x}) = 0$. The positive normal to $\mathscr{D} \cap \mathscr{S}$ is then defined to have the sense of grad $\psi^3(\mathbf{x})$.

Suppose $\mathscr{D}'$ is a second (coordinate) neighbourhood containing the point $\mathbf{x}$ of $\mathscr{S}$ so that $\mathscr{D}' \cap \mathscr{D} \cap \mathscr{S}$ contains $\mathbf{x}$. Let $\mathbf{x}$ have coordinates $(x^1, x^2) \in D$ and $(x'^1, x'^2) \in D'$ and let the mapping $\mathbf{a}': D' \rightarrow \mathscr{D}' \cap \mathscr{S}$ satisfy the requirements (a), (b), (c) in respect of D' and $\mathscr{D}'$. Then, by means of (1.5.55) it is easily shown that the unit normal $\mathbf{n}'$ to $\mathscr{D}' \cap \mathscr{S}$ is related to $\mathbf{n}$ on $\mathscr{D}' \cap \mathscr{D} \cap \mathscr{S}$ by

[†] For full discussion of the definitions and their implications see, for example, do Carmo (1976).

$\mathbf{n}' = \pm\mathbf{n}$, the sign being + or − according as the Jacobian

$$\frac{\partial x'^1}{\partial x^1}\frac{\partial x'^2}{\partial x^2} - \frac{\partial x'^1}{\partial x^2}\frac{\partial x'^2}{\partial x^1}$$

of the coordinate transformation is positive or negative.

If a regular surface $\mathscr{S}$ can be covered by a family of coordinate neighbourhoods $\mathscr{D} \cap \mathscr{S}$ such that the Jacobian of the coordinate transformation is positive wherever two such neighbourhoods overlap on $\mathscr{S}$ then $\mathscr{S}$ is said to be *orientable.* A choice of such a family of coordinate neighbourhoods is called an *orientation* of $\mathscr{S}$ and when such a choice has been made $\mathscr{S}$ is said to be *oriented.* For an orientable surface a smooth unit normal (field) covering the whole surface may be defined and either $\mathbf{n}$ or $-\mathbf{n}$ may be described as the *positive* unit normal (field).

Suppose that $\boldsymbol{\lambda}(t), t \in I$, defines a regular curve in the domain $\mathscr{D} \subset \mathscr{E}$ of a differentiable field f. Then the (covariant) derivative of f along $\boldsymbol{\lambda}$ is given by

$$\frac{\mathrm{d}}{\mathrm{d}t} f\{\boldsymbol{\lambda}(t)\} = (\operatorname{grad} f(\mathbf{x}))\frac{\mathrm{d}\boldsymbol{\lambda}(t)}{\mathrm{d}t}, \tag{1.5.56}$$

where grad $f(\mathbf{x})$, given by (1.5.30), is evaluated for $\mathbf{x} = \boldsymbol{\lambda}(t)$. Essentially, (1.5.56) is locally equivalent to the definition of grad given by (1.5.1).

Suppose next that $\mathbf{a}(u^1, u^2)$, $(u^1, u^2) \in D$, defines a regular surface $\mathscr{S}$ in $\mathscr{D}$. Then

$$\frac{\partial f}{\partial u^\alpha}\{\mathbf{a}(u^1, u^2)\} = (\operatorname{grad} f(\mathbf{x}))\frac{\partial \mathbf{a}}{\partial u^\alpha}(u^1, u^2) \quad \alpha = 1, 2,$$

evaluated for $\mathbf{x} = \mathbf{a}(u^1, u^2)$. The surface covariant derivative of f on $\mathscr{S}$ is defined as

$$\frac{\partial f}{\partial u^\alpha}\{\mathbf{a}(u^1, u^2)\}\mathbf{a}^\alpha,$$

where $\mathbf{a}^\alpha(\alpha = 1, 2)$ are the reciprocals of $\mathbf{a}_\alpha = \partial\mathbf{a}(u^1, u^2)/\partial u^\alpha$ on $\mathscr{S}$. Thus, if $\mathbf{T}$ is a tensor field, for example, the surface covariant derivative of $\mathbf{T}$ is written symbolically as

$$[(\mathbf{a}_\alpha \cdot \nabla)\mathbf{T}\{\mathbf{a}(u^1, u^2)\}] \otimes \mathbf{a}^\alpha,$$

while the covariant derivative of $\mathbf{T}$ along $\boldsymbol{\lambda}$ is just

$$(\dot{\boldsymbol{\lambda}} \cdot \nabla)\mathbf{T}\{\boldsymbol{\lambda}(t)\}$$

and in each case $\nabla \otimes \mathbf{T}$ is determined in accordance with (1.5.36) or its equivalent in other components.

1.5.6 Integration of tensor fields

In order to define integrals of fields over curves, surfaces and volumes in $\mathscr{E}$ we require expressions for the differential elements of arclength, surface area and volume. These are now provided.

For a regular curve $\boldsymbol{\lambda}: I \to \mathscr{E}$ the element of arclength $\mathrm{d}s$ at $\boldsymbol{\lambda}(t)$ is defined by

$$\mathrm{d}s = |\dot{\boldsymbol{\lambda}}(t)|\,\mathrm{d}t \qquad t \in I \tag{1.5.57}$$

and this is independent of the choice of parameter t.

For an oriented surface $\mathscr{S}$ defined by $\mathbf{x} = \mathbf{a}(u^1, u^2)$, $(u^1, u^2) \in D$, the *vector* element of surface area $\mathbf{d}\boldsymbol{\sigma}$ at $\mathbf{a}(u^1, u^2)$ is defined as

$$\mathbf{n}\,\mathrm{d}\sigma \equiv \mathbf{d}\boldsymbol{\sigma} = (\mathbf{a}_1 \wedge \mathbf{a}_2)\,\mathrm{d}u^1\,\mathrm{d}u^2, \tag{1.5.58}$$

where the unit normal $\mathbf{n}$ is given by (1.5.55) and

$$\mathrm{d}\sigma = |\mathbf{a}_1 \wedge \mathbf{a}_2|\,\mathrm{d}u^1\,\mathrm{d}u^2. \tag{1.5.59}$$

Each of the expressions (1.5.58) and (1.5.59) is independent of the choice of coordinates for $\mathscr{S}$.

Note that

$$|\mathbf{a}_1 \wedge \mathbf{a}_2| = \sqrt{a} = \sqrt{\det(a_{\alpha\beta})}, \tag{1.5.60}$$

where $a_{\alpha\beta} = \mathbf{a}_\alpha \cdot \mathbf{a}_\beta$ ($\alpha, \beta = 1$ or 2), this defining a.

At a point $\mathbf{x}$ in a coordinate neighbourhood $\mathscr{D} \subset \mathscr{E}$ corresponding to coordinates (x^1, x^2, x^3) the volume element $\mathrm{d}\tau$ is defined as

$$\mathrm{d}\tau = \sqrt{g}\,\mathrm{d}x^1\,\mathrm{d}x^2\,\mathrm{d}x^3, \tag{1.5.61}$$

where

$$\sqrt{g} = \mathbf{g}_1 \cdot (\mathbf{g}_2 \wedge \mathbf{g}_3) = \sqrt{\det(g_{ij})}. \tag{1.5.62}$$

It is an easy matter to show that $\mathrm{d}\tau$ is independent of the choice of coordinates for $\mathscr{D}$.

If f is a continuous field on $\mathscr{D}$ then the integral of f over $\mathscr{D}$ is defined by

$$\int_{\mathscr{D}} f\,\mathrm{d}\tau = \int_{\psi(\mathscr{D})} f_\psi(x^1, x^2, x^3)\sqrt{g}\,\mathrm{d}x^1\,\mathrm{d}x^2\,\mathrm{d}x^3, \tag{1.5.63}$$

this being independent of the choice of coordinate system ψ. It is assumed here that $\mathscr{D}$ is covered by ψ, but the definition is extended in a straightforward manner to domains not covered by a single coordinate neighbourhood.[†]

If, for example, $\mathbf{T}$ is a second-order tensor field, we may write

$$\int_{\mathscr{D}} \mathbf{T}\,\mathrm{d}\tau = \left(\int_{\mathscr{D}} T_{ij}\,\mathrm{d}\tau\right)\mathbf{e}_i \otimes \mathbf{e}_j$$

independently of the choice of (rectangular) Cartesian coordinate system $\mathbf{e}$. In order to evaluate integrals of tensor fields it is sufficient therefore to know how to evaluate integrals of scalar fields—the (rectangular) Cartesian components of vectors and tensors. Note that the total volume of $\mathscr{D}$ is obtained by setting $f \equiv 1$ in (1.5.63).

Problem 1.5.22 Calculate $\sqrt{g}$ for cylindrical and spherical polar coordinate systems, and show that $g = 1$ for a rectangular Cartesian coordinate system.

If $\mathscr{S}$ is an oriented surface and $\mathscr{D}$ is a coordinate neighbourhood of a point $\mathbf{x}$ on $\mathscr{S}$ such that $\mathscr{D} \cap \mathscr{S}$ has a smooth bounding closed curve then the (*surface*) integral of f over $\mathscr{D} \cap \mathscr{S}$ is defined by

$$\int_{\mathscr{D}\cap\mathscr{S}} f\,\mathrm{d}\sigma = \int_{D} f\{\mathbf{a}(u^1, u^2)\}\sqrt{a}\,\mathrm{d}u^1\,\mathrm{d}u^2 \qquad (1.5.64)$$

independently of the choice of coordinate system (or parametrization) of $\mathscr{D} \cap \mathscr{S}$, where $\mathbf{a}(D) = \mathscr{D} \cap \mathscr{S}$. The total surface area of $\mathscr{D} \cap \mathscr{S}$ is obtained by putting $f = 1$ in (1.5.64).

As in the case of volume integrals the definition (1.5.64) can be extended to surfaces not covered by a single coordinate neighbourhood.

For a regular curve $\boldsymbol{\lambda}: I \to \mathscr{E}$ the (*line*) integral of f along $\boldsymbol{\lambda}$ is defined by

$$\int_{\boldsymbol{\lambda}} f\,ds = \int_{I} f\{\boldsymbol{\lambda}(t)\}\,|\dot{\boldsymbol{\lambda}}(t)|\,\mathrm{d}t, \qquad (1.5.65)$$

the total length of $\boldsymbol{\lambda}$ being obtained from (1.5.65) on setting $f = 1$.

If $\mathscr{D}$ is a domain in $\mathscr{E}$ with oriented closed boundary $\mathscr{S}$ then for any field f which is continuously differentiable in $\mathscr{D}$ and continuous on $\mathscr{S}$

$$\int_{\mathscr{D}} \operatorname{grad} f\,\mathrm{d}\tau = \int_{\mathscr{S}} f\mathbf{n}\,\mathrm{d}\sigma, \qquad (1.5.66)$$

[†]See, for example, Bowen and Wang (1976) or Flanders (1963).

where the positive unit normal $\mathbf{n}$ is the *outward* normal to $\mathscr{S}$. Equation (1.5.66) is a generalized statement of *Gauss's theorem*, which is of fundamental importance in the derivation of the field equations governing the motion of continuous media.

For a tensor field $\mathbf{T}$, for example, (1.5.66) is written symbolically as

$$\int_{\mathscr{D}} \nabla \otimes \mathbf{T}\, \mathrm{d}\tau = \int_{\mathscr{S}} \mathbf{T} \otimes \mathbf{n}\, \mathrm{d}\sigma \tag{1.5.67}$$

or, in rectangular Cartesian components,

$$\int_{\mathscr{D}} \frac{\partial T_{ijk}}{\partial x_p} \ldots \mathrm{d}\tau = \int_{\mathscr{S}} T_{ijk} \ldots n_p\, \mathrm{d}\sigma.$$

The 'usual' form of Gauss's theorem (or the divergence theorem) is recovered from (1.5.67) by taking $\mathbf{T}$ to be a vector field and contracting.

In the context of elasticity theory Stokes's theorem is far less useful than Gauss's theorem, but we include a statement of it here for completeness. If $\mathscr{S}$ is an oriented surface whose bounding edge $\mathscr{C}$ is a regular closed curve $\boldsymbol{\lambda}$ then, for any field f continuously differentiable on $\mathscr{S}$ and continuous on $\mathscr{C}$, we have

$$\int_{\mathscr{S}} (\mathbf{d}\boldsymbol{\sigma} \wedge \mathrm{grad}) f = \int_{\mathscr{C}} f\, \mathbf{d}\boldsymbol{\lambda}. \tag{1.5.68}$$

For full discussion of these theorems reference may be made to Apostol (1965), Kellogg (1954), Flanders (1963), Spivak (1965) and Bowen and Wang (1976).

Problem 1.5.23 If $\mathbf{T}$ is a second-order tensor field on the domain $\mathscr{D}$ which has an oriented closed bounding surface $\mathscr{S}$, show that

$$\int_{\mathscr{D}} \nabla \cdot \mathbf{T}\, \mathrm{d}\tau = \int_{\mathscr{S}} \mathbf{T}^{\mathrm{T}} \mathbf{n}\, \mathrm{d}\sigma,$$

$$\int_{\mathscr{D}} \{(\nabla \cdot \mathbf{T}) \otimes \mathbf{x} + \mathbf{T}^{\mathrm{T}}\}\, \mathrm{d}\tau = \int_{\mathscr{S}} (\mathbf{T}^{\mathrm{T}} \mathbf{n}) \otimes \mathbf{x}\, \mathrm{d}\sigma,$$

$$\int_{\mathscr{D}} \{(\mathbf{x} \cdot \mathbf{x})(\nabla \cdot \mathbf{T}) + 2\mathbf{T}^{\mathrm{T}} \mathbf{x}\}\, \mathrm{d}\tau = \int_{\mathscr{S}} (\mathbf{x} \cdot \mathbf{x}) \mathbf{T}^{\mathrm{T}} \mathbf{n}\, \mathrm{d}\sigma,$$

where $\mathbf{x}$ is the position vector relative to an arbitrary origin.

Problem 1.5.24 If $\mathbf{W}$ is a skew-symmetric second-order tensor field with axial vector field $\mathbf{w}$, show that

$$\int_{\mathscr{D}} \{\mathbf{x} \wedge (\nabla \cdot \mathbf{W}) - 2\mathbf{w}\}\, d\tau = \int_{\mathscr{S}} (\mathbf{W}\mathbf{n}) \wedge \mathbf{x}\, d\sigma,$$

where $\mathscr{D}, \mathscr{S}$ and $\mathbf{x}$ are as defined in Problem 1.5.23.

Problem 1.5.25 Show that for any closed oriented surface $\mathscr{S}$ with unit outward normal $\mathbf{n}$ and enclosed volume τ

$$\int_{\mathscr{S}} \mathbf{x} \otimes \mathbf{n}\, d\sigma = \tau \mathbf{I},$$

where $\mathbf{x}$ is the position vector relative to an origin within $\mathscr{S}$ and $\mathbf{I}$ is the identity tensor. Deduce that if $\mathscr{S}$ is the unit sphere, denoted Ω, then

$$\int_{\Omega} \mathbf{n} \otimes \mathbf{n}\, d\sigma = \tfrac{4}{3}\pi \mathbf{I}.$$

If n_i are rectangular Cartesian components of $\mathbf{n}$ find

$$\int_{\Omega} n_i n_j n_k\, d\sigma$$

and show that

$$\int_{\Omega} n_i n_j n_k n_l\, d\sigma = \frac{4\pi}{15}(\delta_{ij}\delta_{kl} + \delta_{ik}\delta_{jl} + \delta_{il}\delta_{jk}).$$

Express the latter result in invariant form.

REFERENCES

T.M. Apostol, *Mathematical Analysis*, Addison Wesley (1965).
R.M. Bowen and C.-C. Wang, *Introduction to Vectors and Tensors* (2 volumes), Plenum (1976).
P. Chadwick, *Continuum Mechanics*, George Allen & Unwin (1976).
M.P. do Carmo, *Differential Geometry of Curves and Surfaces*, Prentice-Hall (1976).
H. Flanders, *Differential Forms with Applications to the Physical Sciences*, Academic Press (1963).
H. Jeffreys, *Cartesian Tensors*, Cambridge University Press (1952).
H. Jeffreys, On Isotropic Tensors, *Proceedings of the Cambridge Philosophical Society*, Vol. 73, pp. 173–176 (1973).
O.D. Kellogg, *Foundations of Potential Theory*, Dover (1954).
M. Spivak, *Calculus on Manifolds*, Benjamin (1965).

CHAPTER 2

Analysis of Deformation and Motion

2.1 KINEMATICS

2.1.1 Observers and frames of reference

In order to describe qualitatively and quantitatively phenomena which occur in the physical world we need some framework which can be used as a basis for this description. Such a framework is encapsulated in the notion of an *observer*. In essence an observer embodies a means of measuring physical quantities, and, in particular, is equipped to monitor the relative positions of points in space and the progress of time.

We suppose that the space in which physical phenomena are observed is a (three-dimensional) Euclidean point space $\mathscr{E}$ and that the times at which they are observed to occur are measured on the real line $\mathbb{R}$. An *event* in the physical world, as perceived by a given observer, manifests itself at a point of $\mathscr{E}$ at a time in $\mathbb{R}$. For an observer O a certain event is recorded as a pair $(\mathbf{x}, t)$ in the Cartesian product $\mathscr{E} \times \mathbb{R}$, where $\mathbf{x} \in \mathscr{E}$ and $t \in \mathbb{R}$. According to O the event occurs at the *place* $\mathbf{x}$ at *time* t. Formally, an observer is a one-to-one mapping which assigns a pair $(\mathbf{x}, t) \in \mathscr{E} \times \mathbb{R}$ to an event in the physical world. For full discussion of this abstract concept the reader is referred to, for example, Wang and Truesdell (1973) and Truesdell (1977). It is not our intention to give prominence to the underlying abstract ideas here.

If $\mathbf{x}$ and $\mathbf{x}_0$ are distinct points of $\mathscr{E}$ and t and t_0 are distinct times in $\mathbb{R}$ then the two events which are observed by O as $(\mathbf{x}, t)$ and $(\mathbf{x}_0, t_0)$ in $\mathscr{E} \times \mathbb{R}$ are separated by a distance $|\mathbf{x} - \mathbf{x}_0|$ in $\mathscr{E}$ and a time interval $t - t_0$ in $\mathbb{R}$. (We recall the interpretation, given in Section 1.1.3, of the point difference $\mathbf{x} - \mathbf{x}_0$ as an element of the vector space $\mathbb{E}$.)

Suppose there is a mapping of $\mathscr{E} \times \mathbb{R}$ into itself such that the pairs $(\mathbf{x}_0, t_0)$ and $(\mathbf{x}, t)$ are mapped to $(\mathbf{x}_0^*, t_0^*)$ and $(\mathbf{x}^*, t^*)$ respectively. We ask what is the nature of the mapping under which $|\mathbf{x} - \mathbf{x}_0|$ and $t - t_0$ are preserved for arbitrary $(\mathbf{x}, t)$, given any fixed, but arbitrary, pair $(\mathbf{x}_0, t_0)$. By reference to equation (1.1.42) and the results of Section 1.3.5, we deduce that the most general such mapping is characterized by the equations

$$\mathbf{x}^* - \mathbf{x}_0^* = \mathbf{Q}(t)(\mathbf{x} - \mathbf{x}_0), \quad t^* = t - \mathrm{a}, \tag{2.1.1}$$

where $a \in \mathbb{R}$ is a constant and $\mathbf{Q}(t)$ is a (second-order) orthogonal tensor which depends on t. The mapping is clearly one-to-one. Notice that its spatial specification involves point differences and can therefore be interpreted as a linear transformation of vectors in $\mathbb{E}$, no preferred choice of origin for $\mathscr{E}$ being implied.

The description of distances and time intervals is patently not unique, being invariant under (infinitely many possible) transformations of the type (2.1.1). *We stipulate* that different observers should agree about distance and time intervals between events† and it is therefore appropriate to refer to (2.1.1) as an *observer transformation.* It represents the most general one-to-one mapping of $\mathscr{E} \times \mathbb{R}$ to itself which preserves (a) the distance between an arbitrary pair of points of $\mathscr{E}$, (b) time intervals between events and (c) the order in which events occur.

We interpret the mapping which takes $(\mathbf{x}, t)$ to $(\mathbf{x}^*, t^*)$ as a *change of observer*, from O to O^* say, so that the event which is recorded at the place $\mathbf{x}$ at time t by O is the *same* event as that observed to occur at $\mathbf{x}^*$ at time t^* by O^*. A change of observer merely changes the description in $\mathscr{E} \times \mathbb{R}$ of an event.

If $\mathbf{Q}(t)$ is a *proper* orthogonal tensor then orientation is preserved and (2.1.1) can be interpreted as a transformation which *rotates* vectors in $\mathbb{E}$. It is sufficient for our purposes to confine attention to orientation preserving observer transformations so that $\mathbf{Q}(t)$ is henceforth to be regarded as proper orthogonal.

With the notational change $\mathbf{c}(t) = \mathbf{x}_0^* - \mathbf{Q}(t)\mathbf{x}_0$, equations (2.1.1) take on the standard forms

$$\mathbf{x}^* = \mathbf{c}(t) + \mathbf{Q}(t)\mathbf{x}, \quad t^* = t - a \tag{2.1.2}$$

which appear in many texts (for example, Truesdell and Noll, 1965), $\mathbf{c}(t)$ being an arbitrary vector.

If observers O and O^* respectively choose origins $\mathbf{o}$ and $\mathbf{o}^*$ arbitrarily in $\mathscr{E}$ then $\mathbf{x}$ and $\mathbf{x}^*$ may be interpreted as the position vectors of the points $\mathbf{x}$ and $\mathbf{x}^*$ respectively relative to $\mathbf{o}$ and $\mathbf{o}^*$, but, of course, a choice of origin is not implicit in the definition of an observer.

As indicated above a change of observer affects the points of $\mathscr{E}$ (and also vectors in $\mathbb{E}$). A coordinate transformation, on the other hand, affects the coordinates of points but not the points themselves. This distinction is of the same nature as that made in Section 1.3.5 between an orthogonal tensor

† We are not concerned with relativistic effects here.

(transformation) and an orthogonal matrix representing a change of coordinates. We need to make the distinction clear at this stage prior to selecting coordinate representations for points of $\mathscr{E}$.

The definitions of an observer and an observer transformation are independent of any preferred choice of basis for the vector space $\mathbb{E}$. Nevertheless, any observer is free to select such a basis. Let O choose a rectangular Cartesian coordinate system such that the point $\mathbf{x}$ has coordinates (x_1, x_2, x_3) relative to an origin $\mathbf{o}$ and basis† $\{\mathbf{e}_i\}$ which are *at rest* relative to O (i.e. do not depend on t). Then $\mathbf{x} = x_i\mathbf{e}_i$. Similarly, suppose O^* chooses an origin $\mathbf{o}^*$ and basis $\{\mathbf{e}_i^*\}$ at rest relative to himself so that the point $\mathbf{x}^*$ has rectangular Cartesian coordinates (x_1^*, x_2^*, x_3^*), and hence $\mathbf{x}^* = x_i^*\mathbf{e}_i^*$.

Because of the (length-preserving) transformation of vectors in $\mathbb{E}$ defined by (2.1.1) it is convenient to identify $\mathbf{Q}(t)\mathbf{e}_i$ with $\mathbf{e}_i^*$ (although $\mathbf{Q}(t)\mathbf{e}_i$ could be allowed to differ from $\mathbf{e}_i^*$ by a time-independent orthogonal transformation). Then the observer transformation (2.1.2) has the component form

$$x_i^* = c_i^*(t) + x_i, \tag{2.1.3}$$

where $c_i^*(t) = \mathbf{c}(t)\cdot\mathbf{e}_i^*$. Thus, relative to their respective choices of basis the two observers assign, apart from a shift in origin, the same coordinates to the places at which they observe a given event.

In the literature the phrases 'frame of reference' and 'change of frame (of reference)' are frequently used as the equivalents of 'observer' and 'change of observer' respectively as we have defined them above. But, the reader should be aware that the terms 'observer' and 'frame of reference' are also used by some authors to mean simply 'the choice of a rectangular Cartesian coordinate system'.‡ In this book, however, we attach a separate meaning to 'frame of reference' to that which we have attributed to 'observer'. Indeed, once an observer has been (arbitrarily) chosen, we refer to an assigned rectangular Cartesian coordinate system for that observer as a *frame of reference of the observer*, the time coordinate being implied by the selection of observer.

We emphasize that so far we have considered different (but equivalent) descriptions of the same event as perceived by two observers in relative motion, viewed from their respective frames of reference. The basis vectors $\mathbf{e}_i$ and $\mathbf{e}_i^* = \mathbf{Q}(t)\mathbf{e}_i$ are fixed relative to O and O^* respectively. But $\mathbf{e}_i^*$ can also be viewed by O as a rotating vector and this consideration leads to a legitimate alternative assesment of events which entails a single (but arbitrary) observer. Specifically, we examine how the description of an event by a given

† We restrict attention to positively oriented (i.e. right-handed) bases here.

‡ Then $\mathscr{E} \times \mathbb{R}$ itself essentially has the role of the event world and the transformation (2.1.2) is not considered. For an abstract generalization see Wang and Truesdell (1973).

observer is subject to a change in his frame of reference, rotating frames being admitted. As we see below, a change of frame of reference merely changes the coordinate representations of points of $\mathscr{E}$, not the points themselves, as distinct from what happens in an observer transformation.

Let O choose a second frame of reference involving an origin $\mathbf{o}'$ and (rectangular Cartesian) basis $\{\mathbf{e}'_i(t)\}$, so that the place $\mathbf{x}$ of an event observed by O has position vectors $\mathbf{x}$ and $\mathbf{x}'$ relative to $\mathbf{o}$ and $\mathbf{o}'$ respectively. We may write

$$x'_i \mathbf{e}'_i = \mathbf{x}' = \mathbf{a}(t) + \mathbf{x}, \tag{2.1.4}$$

where $\mathbf{a}(t) = \mathbf{o} - \mathbf{o}'$ represents the translation of origin and x'_i are (rectangular Cartesian) coordinates.

The quantities $Q'_{ij}(t) = \mathbf{e}'_i(t)\cdot\mathbf{e}_j$ define the direction cosines of the basis $\{\mathbf{e}'_i(t)\}$ relative to $\{\mathbf{e}_i\}$ in accordance with the convention used in Section 1.1.2. From (2.1.4) the component transformation

$$x'_i = a'_i(t) + Q'_{ij}(t)x_j = Q'_{ij}(t)\{x_j + a_j(t)\} \tag{2.1.5}$$

is then obtained, where $a_i(t)$ and $a'_i(t)$ are the components of $\mathbf{a}(t)$ relative to the bases $\{\mathbf{e}_i\}$ and $\{\mathbf{e}'_i(t)\}$ respectively. Also $Q'_{ij}(t)$ can be regarded as the components of a proper orthogonal tensor $\mathbf{Q}'(t)$ with respect to the basis $\{\mathbf{e}_i\}$, so that

$$\mathbf{e}'_i(t) = \mathbf{Q}'(t)^{\mathrm{T}}\mathbf{e}_i.$$

The identification $\mathbf{Q}'(t) = \mathbf{Q}(t)^{\mathrm{T}}$ then ensures that $\{\mathbf{e}'_i(t)\}$ coincides with the basis $\{\mathbf{e}^*_i\}$ associated with the observer O^*. If also $\mathbf{Q}(t)\mathbf{a}(t)$ is equated to $\mathbf{c}(t)$ then $\mathbf{o}'$ becomes $\mathbf{o}^*$ and the frames of reference of the two observers coincide. Thus, the transformation (of basis vectors) corresponding to a change of frame of reference for a given observer is equivalent to the spatial part of the transformation associated with a change of observer. A change of frame of reference thereby defines the spatial part of an observer transformation. (Note that, as far as O is concerned, $\mathbf{x}^*$ corresponds to a different event from $\mathbf{x}$ so that it serves no useful purpose to decompose $\mathbf{x}^*$ on $\{\mathbf{e}_i\}$ or $\{\mathbf{e}'_i(t)\}$.) That coordinates transform according to (2.1.5) under a change of frame of reference is equivalent to the preservation of distances under (the spatial part of) the observer transformation thereby defined.

In general, measurements of physical quantities depend on the choice of observer. For example, two observers in relative motion will naturally ascribe different values to the speed of a moving point. However, physical phenomena, as distinct from their kinematical descriptions, do not depend on the choice of observer, and the mathematical formulation of physical laws must therefore reflect this independence. For example, the elastic modulus of a spring is a material constant and all observers should attribute to it the same value. This aspect of the theory is discussed in Section 2.4 and Chapter 4.

2.1.2 Configurations and motions

We are concerned with the deformation and motion of 'bodies' consisting of continuously distributed material. Formally, a body B is a set of points,† referred to as *particles* (or *body points* or *material points*), which can be put into one-to-one correspondence with some region, $\mathscr{B}$ say, of the Euclidean point space $\mathscr{E}$. The body is said to *occupy* $\mathscr{B}$. As the body moves the region it occupies in $\mathscr{E}$ changes continuously.

A *configuration* of B is a one-to-one mapping $\chi: B \to \mathscr{E}$ which takes particles of B to the places they occupy in $\mathscr{E}$. In this book we assume that χ and its inverse χ^{-1} are twice continuously differentiable.‡ This degree of regularity is not too restrictive for our purposes since it is not our intention to discuss singularities such as shock waves and acceleration waves.

We identify a generic particle of B by the label X. Then, for $X \in B$, we write

$$\mathbf{x} = \chi(X), \quad X = \chi^{-1}(\mathbf{x}), \tag{2.1.6}$$

where $\mathbf{x}$ is the place occupied by the particle X in the configuration χ. The configuration χ clearly depends on the choice of observer since $\mathbf{x} \in \mathscr{E}$ is the spatial recognition of an event. The effect of a change of observer on a configuration will be examined shortly.

We write $\chi(B) = \{\chi(X), X \in B\}$ for the set of places occupied by the particles of B in the configuration χ, i.e. $\mathscr{B}$. (Often, $\chi(B)$ is itself referred to as a configuration of B.)

A *motion* of B (as seen by an arbitrary observer O)§ is a one-parameter family of configurations $\chi_t: B \to \mathscr{E}$, where the subscript identifies the time t as parameter. We write

$$\mathbf{x} = \chi_t(X) = \chi(X, t) \tag{2.1.7}$$

for the place of the particle X at time t. For a given particle (fixed X) equation (2.1.7) describes a curve in $\mathscr{E}$ called the *path* of X in the motion. The parameter t may be restricted to some subset of $\mathbb{R}$.

The observer O of the motion records the place $\mathbf{x}$ of the particle X at the current time t. According to a second observer O^* of the same motion,

† In abstract language, a body is a smooth three-dimensional manifold homeomorphic to the closure of a connected open subset of $\mathscr{E}$.

‡ That is, χ is a C^2-diffeomorphism.

§ The abstract definition of a motion does not involve an observer (Wang and Truesdell, 1973).

X is seen to be at $\mathbf{x}^*$ at current t^* related to $(\mathbf{x}, t)$ by (2.1.2). We write

$$\mathbf{x}^* = \boldsymbol{\chi}^*(X, t^*), \tag{2.1.8}$$

where $\boldsymbol{\chi}^*(X, t^*)$ is defined through the observer transformation by

$$\boldsymbol{\chi}^*(X, t^*) = \mathbf{c}(t) + \mathbf{Q}(t)\boldsymbol{\chi}(X, t), \quad t^* = t - a, \tag{2.1.9}$$

use having been made of (2.1.7). Equation (2.1.8) describes the same motion as does (2.1.7), and two observers O and O^* therefore have equivalent mathematical characterizations of that motion provided $\boldsymbol{\chi}$ and $\boldsymbol{\chi}^*$ are related by (2.1.9).

The velocity and acceleration of the particle X as observed by O are defined by

$$\frac{\partial \boldsymbol{\chi}(X, t)}{\partial t}, \quad \frac{\partial^2 \boldsymbol{\chi}(X, t)}{\partial t^2} \tag{2.1.10}$$

respectively, where the partial derivative indicates differentiation with respect to t for a given material particle X.

From (2.1.9), we calculate the velocity and acceleration observed by O^* as

$$\frac{\partial \boldsymbol{\chi}(X, t^*)}{\partial t^*} = \mathbf{Q}(t)\frac{\partial \boldsymbol{\chi}(X, t)}{\partial t} + \dot{\mathbf{c}}(t) + \dot{\mathbf{Q}}(t)\boldsymbol{\chi}(X, t) \tag{2.1.11}$$

and

$$\frac{\partial^2 \boldsymbol{\chi}^*(X, t^*)}{\partial t^{*2}} = \mathbf{Q}(t)\frac{\partial^2 \boldsymbol{\chi}(X, t)}{\partial t^2} + \ddot{\mathbf{c}}(t)$$

$$+ 2\dot{\mathbf{Q}}(t)\frac{\partial \boldsymbol{\chi}(X, t)}{\partial t} + \ddot{\mathbf{Q}}(t)\boldsymbol{\chi}(X, t) \tag{2.1.12}$$

respectively, where the superposed dot indicates differentiation with respect to t.

Clearly, the two observers' perceptions of the velocity and acceleration of a given motion are different even though the prescription (rate of change at fixed X) is the same in each case. This is, of course, to be expected since O and O^* are in relative motion by virtue of the time-dependence of $\mathbf{c}(t)$ and $\mathbf{Q}(t)$.

Suppose the observer O selects an origin $\mathbf{o}$ and basis $\{\mathbf{e}_i\}$. Then the motion (2.1.7) has the component representation

$$x_i = \chi_i(X, t) \quad i = 1, 2, 3. \tag{2.1.13}$$

Under a change of frame of reference for O defined by the coordinate transformation (2.1.5), we obtain

$$x'_i = \chi'_i(X,t) = a'_i(t) + Q'_{ij}(t)\chi_j(X,t), \tag{2.1.14}$$

and successive differentiations of this with respect to t lead to

$$\left.\begin{aligned} \frac{\partial \chi'_i(X,t)}{\partial t} &= Q'_{ij}(t)\frac{\partial \chi_j(X,t)}{\partial t} + \dot{Q}'_{ij}(t)\chi_j(X,t) + \dot{a}'_i(t), \\ \frac{\partial^2 \chi'_i(X,t)}{\partial t^2} &= Q'_{ij}(t)\frac{\partial^2 \chi_j(X,t)}{\partial t^2} + 2\dot{Q}'_{ij}(t)\frac{\partial \chi_j(X,t)}{\partial t} \\ &\quad + \ddot{Q}'_{ij}(t)\chi_j(X,t) + \ddot{a}'_i(t). \end{aligned}\right\} \tag{2.1.15}$$

The left-hand sides of (2.1.15) are the components respectively of velocity and acceleration *relative* to the rotating basis $\{\mathbf{e}'_i(t)\}$. The components of the actual velocity and acceleration (as observed by O) on the two bases are, on the other hand, related according to the usual transformation rule for the components of a vector. Thus

$$\mathbf{e}'_i \cdot \frac{\partial \boldsymbol{\chi}(X,t)}{\partial t} = Q'_{ij}(t)\mathbf{e}_j \cdot \frac{\partial \boldsymbol{\chi}(X,t)}{\partial t},$$

the left-hand side of which is

$$\frac{\partial \chi'_i(X,t)}{\partial t} + \dot{Q}'_{jk}(t)Q'_{ik}(t)\chi'_j(X,t),$$

and similarly for the acceleration.

The distinctions between the transformation laws for the components of vectors and tensors recorded by a given observer and those of their *relative* counterparts (as illustrated above) have important consequences for the formulation of physical laws.

2.1.3 Reference configurations and deformations

Physical observations of the body B are made in specific configurations. For reference purposes, it is therefore convenient to identify a certain fixed (but arbitrarily chosen) configuration of B so that the particles of B may then be labelled during the motion by their places in that configuration. Let $\boldsymbol{\chi}_0$ denote some fixed (i.e. time-independent) configuration and write

$$\mathbf{X} = \boldsymbol{\chi}_0(X), \quad X = \boldsymbol{\chi}_0^{-1}(\mathbf{X}), \tag{2.1.16}$$

where $\mathbf{X}$ is the place of the particle X in configuration $\boldsymbol{\chi}_0$. The subscript zero may, but need not, correspond to $t = 0$ (in 2.1.7). (Indeed, χ_0 need not be a configuration actually occupied by the body in a given motion.) If, however, the body occupies $\chi_0(B)$ at some time $t_0 \le t$ during the motion (2.1.7) then we have $\mathbf{X} = \chi(X, t_0) = \chi_0(X)$ for all $X \in B$.

A fixed configuration χ_0 is called a *reference configuration*, and we denote by $\mathscr{B}_0$ the region of $\mathscr{E}$ occupied by B in this configuration. The configuration χ_t, on the other hand, is called the *current configuration*, and we denote $\chi_t(B)$ by $\mathscr{B}_t$. Of course, it may sometimes be convenient to choose a reference configuration which coincides with the current configuration, as we shall see in Chapter 6 in particular.

Elimination of X between (2.1.7) and (2.1.16) leads to

$$\mathbf{x} = \chi\{\chi_0^{-1}(\mathbf{X}), t\}. \tag{2.1.17}$$

Through (2.1.16) we have identified X with its place $\mathbf{X}$ in the configuration χ_0, and for all practical purposes the distinction between X and $\mathbf{X}$ need not be made. Henceforth, therefore, it will often be convenient to ignore the distinction and, in order to simplify the notation, we rewrite (2.1.17) as

$$\mathbf{x} = \chi(\mathbf{X}, t). \tag{2.1.18}$$

Its inverse is written

$$\mathbf{X} = \chi^{-1}(\mathbf{x}, t), \tag{2.1.19}$$

and, for a fixed place $\mathbf{x}$, this equation (or, equivalently, the inverse of (2.1.7)) identifies the particles which pass through $\mathbf{x}$ during the motion.

This new definition of χ, now to be regarded as a mapping from $\mathscr{B}_0$ to $\mathscr{B}_t$, depends implicitly on the choice of reference configuration (the notations in (2.1.7) and (2.1.18) should not be confused). For any fixed t, χ is called a *deformation* from the reference configuration, and equation (2.1.18) specifies a one-parameter family of such deformations. We emphasize that, as distinct from 'configuration', the word 'deformation' implies that a reference configuration is involved.

With X replaced by $\mathbf{X}$ in (2.1.8) and (2.1.9), χ and χ^* are now interpreted as deformations from the respective reference configurations of the observers O and O^*. Note that the definitions (2.1.10) of velocity and acceleration do not depend on any preferred choice of reference configuration, although such a choice may be entered into (2.1.10) from (2.1.16).

As we have indicated in Section 2.1.2 the manifestation of a motion through the family of configurations χ_t depends on the choice of observer, and the observer transformation (2.1.9) must be satisfied for all t for which the motion is defined. Equally, different observers are at liberty to choose different reference configurations. However, it is sometimes convenient to select a reference configuration which is independent of observer so that all observers allocate the particle X to the same reference place $\mathbf{X}$ in $\mathscr{E}$. Under a change of observer the place $\mathbf{x}$ transforms according to (2.1.2), but $\mathbf{X}$ is then unaffected.

If the reference configuration χ_0 is occupied by the body at some time t_0 during the considered motion (as observed by O) then (2.1.18) specifies the deformation at time t *relative* to that at time t_0 (the dependence on t_0 being implicit in (2.1.18)). Such a deformation is sometimes referred to as a *relative deformation.*

The reference places $\mathbf{X}$ and $\mathbf{X}^*$ that X is observed to occupy by O and O^* respectively are then related through the observer transformation (2.1.2) according to

$$\mathbf{X}^* = \mathbf{c}(t_0) + \mathbf{Q}(t_0)\mathbf{X}. \qquad (2.1.20)$$

The reference configuration is independent of observer if and only if

$$\mathbf{c}(t_0) = \mathbf{0}, \quad \mathbf{Q}(t_0) = \mathbf{I}, \qquad (2.1.21)$$

initial conditions we are free to impose. Henceforth, any choice of reference configuration is understood to be independent of observer.

The places $\mathbf{X}$ and $\mathbf{x}$ occupied by the particle X in the reference and current configurations respectively may also be regarded as position vectors in $\mathbb{E}$ relative to fixed origins $\mathbf{O}$ and $\mathbf{o}$ respectively in $\mathscr{E}$. In general $\mathbf{O}$ and $\mathbf{o}$ need not coincide. This is depicted in Fig. 2.1.

A given observer O may choose different frames of reference for different

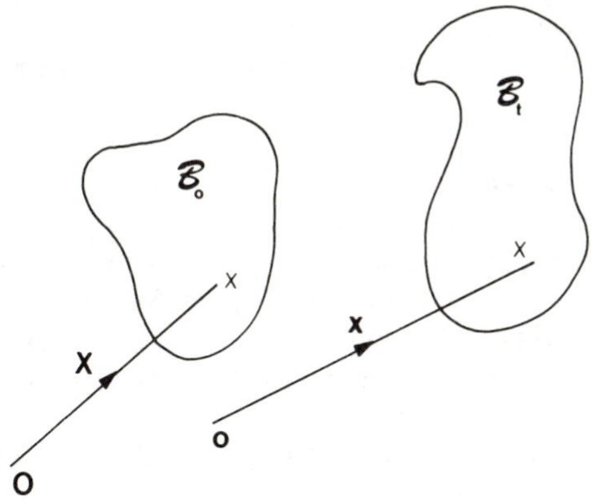

Fig. 2.1 The body B depicted in configurations $\mathscr{B}_0$ and $\mathscr{B}_t$. Particle X has position vectors $\mathbf{X}$ in $\mathscr{B}_0$ and $\mathbf{x}$ in $\mathscr{B}_t$ relative to origins $\mathbf{O}$ and $\mathbf{o}$ respectively.

configurations of the body B. In particular, his choice of origin $\mathbf{o}$ and basis $\{\mathbf{e}_i\}$ at time t can be made independently of that in the reference configuration. The component form of (2.1.18) is written

$$x_i = \chi_i(\mathbf{X}, t) \quad i = 1, 2, 3. \tag{2.1.22}$$

which transforms to

$$x_i' = \chi_i'(\mathbf{X}, t) \quad i = 1, 2, 3, \tag{2.1.23}$$

under a change of frame defined by the coordinate transformation (2.1.5). (These are essentially (2.1.13) and (2.1.14) with X replaced by $\mathbf{X}$.)

If O now chooses an origin $\mathbf{O}$ and basis $\{\mathbf{E}_\alpha\}$ in the reference configuration so that

$$\mathbf{X} = X_\alpha \mathbf{E}_\alpha, \tag{2.1.24}$$

where X_α ($\alpha = 1, 2, 3$) are rectangular Cartesian components of $\mathbf{X}$, then (2.1.22) and (2.1.23) respectively are replaced by

$$x_i = \chi_i(X_\alpha, t), \quad x_i' = \chi_i'(X_\alpha, t) \quad i, \alpha = 1, 2, 3. \tag{2.1.25}$$

We emphasize that the reference coordinates X_α are unaffected by a change of frame at current time t, but the coordinates x_i transform according to (2.1.5).

We use Greek character indices $\alpha, \beta, \ldots$ for the components of vectors and tensors associated with a basis $\{\mathbf{E}_\alpha\}$ in the reference configuration. In particular, X_α are called *referential* or *Lagrangean* coordinates of the particle X. On the other hand, italic letter indices $i, j, \ldots$ are associated with the basis $\{\mathbf{e}_i\}$ in the current configuration and x_i are called *current* or *Eulerian* coordinates of X.

2.1.4 Rigid-body motions

As we saw in Section 2.1.1 the spatial part of the observer transformation (2.1.1) or (2.1.2) is the most general mapping of points of $\mathscr{E}$ which preserves the distance between an arbitrary pair of points. An equivalent mathematical statement defines a rigid-body motion seen by a single observer O: a *rigid-body motion* of a body B is one during which the distance between an arbitrary pair of particles of B is preserved.

Consider the motion observed by O as $\mathbf{x} = \boldsymbol{\chi}(X, t)$. This motion is rigid if and only if

$$\mathbf{x} = \mathbf{c}(t) + \mathbf{Q}(t)\mathbf{X} \tag{2.1.26}$$

for all $\mathbf{X} \in \mathscr{B}_0$ with an arbitrary choice of reference configuration (2.1.16) such that $\mathscr{B}_0 = \chi_0(B)$.

In (2.1.26), $\mathbf{c}(t)$ is a vector corresponding to a pure translation of the body, while $\mathbf{Q}(t)$ is a proper orthogonal tensor defining a pure rotation of the body. This terminology has been anticipated in Section 2.1.1.

More generally, if two distinct motions of B are given by $\mathbf{x} = \chi(X, t)$ and $\bar{\mathbf{x}} = \bar{\chi}(X, t)$, $X \in B$, then they differ by a rigid-body motion if and only if

$$\bar{\mathbf{x}} = \mathbf{c}(t) + \mathbf{Q}(t)\mathbf{x}. \tag{2.1.27}$$

This can be interpreted as a rigid-body motion superimposed on the motion defined by $\mathbf{x} = \chi(X, t)$.

Clearly (2.1.27) has the same form as the spatial part of (2.1.2). But, it must be distinguished from an observer transformation since $\mathbf{x}$ and $\bar{\mathbf{x}}$ are the places of two distinct events observed by O, whereas $\mathbf{x}$ and $\mathbf{x}^*$ are the places of one event recorded by two different observers.

2.2 ANALYSIS OF DEFORMATION AND STRAIN

2.2.1 The deformation gradient

Our immediate purpose is to analyse the deformation from some reference configuration $\mathscr{B}_0$ of the body B to a generic configuration $\mathscr{B}_t$ described by (2.1.18). In particular, we wish to compare the two configurations without requiring knowledge of the intermediate stages in the motion so that the time dependence in (2.1.18) is not then needed. We therefore suppress the dependence on t and replace (2.1.18) by

$$\mathbf{x} = \chi(\mathbf{X}) \tag{2.2.1}$$

and $\mathscr{B}_t$ by $\mathscr{B}$.

For the most part we restrict attention to the view of a single observer O, although the effect of a change of observer on the analysis will be discussed briefly in Section 2.2.8. Let O choose origins $\mathbf{O}$ and $\mathbf{o}$ and bases $\{\mathbf{E}_\alpha\}$ and $\{\mathbf{e}_i\}$ in the reference and current configurations respectively so that the (rectangular Cartesian) component form of (2.2.1) may be written

$$x_i = \chi_i(X_\alpha). \tag{2.2.2}$$

Either of (2.2.1) or (2.2.2) describes the deformation from $\mathscr{B}_0$ to the current configuration $\mathscr{B}$. How the body achieves its current configuration and/or is maintained there is a question which is not our concern in this section but will be addressed in Chapter 3.

In order to analyse the deformation locally (i.e. in a neighbourhood of a material particle X) we require that the derivatives $\partial x_i/\partial X_\alpha$ are continuous (this is ensured by the assumptions made about the regularity of $\boldsymbol{\chi}$ at the beginning of Section 2.1.2). Then, on taking the differential of (2.2.2), we obtain

$$\mathrm{d}x_i = \frac{\partial x_i}{\partial X_\alpha}\mathrm{d}X_\alpha. \tag{2.2.3}$$

The invariant form of (2.2.3), obtainable directly from (2.2.1), is written

$$\mathbf{dx} = \mathbf{A}\,\mathbf{dX}, \tag{2.2.4}$$

where $\mathbf{A}$ is a second-order tensor defined by

$$\mathbf{A} = \mathrm{Grad}\,\boldsymbol{\chi}(\mathbf{X}) \equiv \boldsymbol{\nabla} \otimes \boldsymbol{\chi}(\mathbf{X}). \tag{2.2.5}$$

We distinguish Grad (with upper-case G) from grad (with lower-case g); respectively they refer to the gradient operation taken with respect to $\mathbf{X}$ and $\mathbf{x}$. An alternative notation for $\mathbf{A}$ is $\partial\mathbf{x}/\partial\mathbf{X} \equiv \partial\boldsymbol{\chi}(\mathbf{X})/\partial\mathbf{X}$.

Let $\mathbf{A}$ have components $A_{i\alpha}$ with respect to bases $\{\mathbf{E}_\alpha\}$ and $\{\mathbf{e}_i\}$. Then $A_{i\alpha} = \partial x_i/\partial X_\alpha$ and

$$\mathbf{A} = A_{i\alpha}\mathbf{e}_i \otimes \mathbf{E}_\alpha,$$

a representation which is easily generalized to the case of curvilinear coordinates if required.

A vector $\mathbf{dX}$ at the point $\mathbf{X}$ in the reference configuration $\mathscr{B}_0$ is called a *material line element* (or *fibre*)†. Equation (2.2.4) describes how an arbitrary line element $\mathbf{dX}$ at $\mathbf{X}$ transforms under the deformation to the line element $\mathbf{dx}$ (consisting of the same material as $\mathbf{dX}$) at the point $\mathbf{x}$ in the configuration $\mathscr{B}$. The transformation is linear locally since $\mathbf{A}$ depends on $\mathbf{X}$ (in general) but *not* on $\mathbf{dX}$. The local nature of the deformation is therefore embodied in the properties of the tensor $\mathbf{A}$, which is called the *deformation gradient* relative to the reference configuration $\mathscr{B}_0$. (The effect on $\mathbf{A}$ of a change in reference configuration will be considered in Section 2.2.8.)

Consider the equation $\mathbf{AdX} = \mathbf{0}$. If $\mathbf{dX} \neq \mathbf{0}$ this implies that there exists at

†Strictly, it is the set of material particles lying along $\mathbf{dX}$ at the point $\mathbf{X}$ to which this name should be attached. Since $\mathbf{dX}$ (respectively $\mathbf{dx}$) is associated with the reference (respectively current) configuration, an appropriate terminology is *Lagrangean* (respectively *Eulerian*) *line element*.

The set of line elements $\mathbf{dX}$ at $\mathbf{X}$ forms a vector space $T_\mathbf{X}(\mathscr{B}_0)$ called the *tangent space* of $\mathscr{B}_0$ at $\mathbf{X}$; $T_\mathbf{x}(\mathscr{B})$ is defined similarly. The second-order tensor $\mathbf{A}$ is an example of a *two-point tensor* since it can be regarded as a bilinear mapping from $T_\mathbf{X}(\mathscr{B}_0) \times T_\mathbf{x}(\mathscr{B})$ to $\mathbb{R}$, involving points in two distinct configurations.

least one line element of material in the reference configuration whose length is reduced to zero by the deformation. In other words, it is annihilated. This is regarded as physically unrealistic and is therefore ruled out from consideration. Thus $\mathbf{A}\mathbf{dX} \neq \mathbf{0}$ for all $\mathbf{dX} \neq \mathbf{0}$, i.e. $\mathbf{A}$ is a *non-singular* tensor. This imposes the restriction

$$\det \mathbf{A} \neq 0$$

and, for future reference, we introduce the standard notation

$$J = \det \mathbf{A}. \tag{2.2.6}$$

Since $\mathbf{A}$ is non-singular it has an inverse $\mathbf{A}^{-1}$, and we write

$$\mathbf{B} = (\mathbf{A}^{-1})^{\mathrm{T}} \tag{2.2.7}$$

so that (2.2.4) has inverse

$$\mathbf{dX} = \mathbf{B}^{\mathrm{T}}\mathbf{dx} \tag{2.2.8}$$

and

$$\mathbf{A}\mathbf{B}^{\mathrm{T}} = \mathbf{I} = \mathbf{B}^{\mathrm{T}}\mathbf{A}. \tag{2.2.9}$$

In (rectangular Cartesian) components the inverse deformation gradient $\mathbf{B}$ has the representation

$$\mathbf{B} = B_{i\alpha}\mathbf{e}_i \otimes \mathbf{E}_\alpha, \qquad B_{i\alpha} = \frac{\partial X_\alpha}{\partial x_i}, \tag{2.2.10}$$

where X_α are the components of

$$\mathbf{X} = \chi^{-1}(\mathbf{x}), \tag{2.2.11}$$

the inverse of (2.2.1).

Since $(\mathbf{A}^{\mathrm{T}}\mathbf{A})^{\mathrm{T}} = \mathbf{A}^{\mathrm{T}}(\mathbf{A}^{\mathrm{T}})^{\mathrm{T}} = \mathbf{A}^{\mathrm{T}}\mathbf{A}$ (on use of the results (1.3.10)) it follows that $\mathbf{A}^{\mathrm{T}}\mathbf{A}$ is *symmetric*, as also are $\mathbf{A}\mathbf{A}^{\mathrm{T}}$ and the respective inverses $\mathbf{B}^{\mathrm{T}}\mathbf{B}$ and $\mathbf{B}\mathbf{B}^{\mathrm{T}}$ of $\mathbf{A}^{\mathrm{T}}\mathbf{A}$ and $\mathbf{A}\mathbf{A}^{\mathrm{T}}$. Moreover, because $\mathbf{A}$ is non-singular, we have

$$\mathbf{dX}\cdot(\mathbf{A}^{\mathrm{T}}\mathbf{A})\mathbf{dX} = (\mathbf{A}\mathbf{dX})\cdot(\mathbf{A}\mathbf{dX}) = |\mathbf{A}\mathbf{dX}|^2 > 0$$

for $\mathbf{dX} \neq \mathbf{0}$. Hence $\mathbf{A}^{\mathrm{T}}\mathbf{A}$ is *positive definite* in accordance with the definition given in Section 1.3.3. So also are $\mathbf{A}\mathbf{A}^{\mathrm{T}}$, $\mathbf{B}\mathbf{B}^{\mathrm{T}}$ and $\mathbf{B}^{\mathrm{T}}\mathbf{B}$.

The tensors $\mathbf{A}^{\mathrm{T}}\mathbf{A}$ and $\mathbf{A}\mathbf{A}^{\mathrm{T}}$ respectively are called the *right* and *left Cauchy–Green deformation tensors*. Their properties of symmetry and positive definiteness have important consequences for the geometrical interpretation of the deformation, as we see in Sections 2.2.3 to 2.2.5.

Problem 2.2.1 If ϕ is a scalar field defined on $\mathscr{B}$ and

$$\Phi(\mathbf{X}) = \phi(\mathbf{x}) \equiv \phi\{\boldsymbol{\chi}(\mathbf{X})\}$$

through equation (2.2.1), show that

$$\text{Grad}\,\Phi = \mathbf{A}^{\mathrm{T}}\,\text{grad}\,\phi.$$

For a vector field

$$\mathbf{V}(\mathbf{X}) = \mathbf{v}(\mathbf{x}) = \mathbf{v}\{\boldsymbol{\chi}(\mathbf{X})\}$$

derive the parallel result

$$\text{Grad}\,\mathbf{V} = (\text{grad}\,\mathbf{v})\mathbf{A}.$$

Problem 2.2.2 If $\mathsf{A} \equiv (A_{i\alpha})$ denotes the matrix of components of the deformation gradient tensor $\mathbf{A}$ relative to the reference and current bases $\{\mathbf{E}_\alpha\}$ and $\{\mathbf{e}_i\}$, show that AP^{T} denotes the matrix of components of $\mathbf{A}$ relative to bases $\{\mathbf{E}'_\alpha\}$ and $\{\mathbf{e}_i\}$, where $\mathbf{E}'_\alpha = P_{\alpha\beta}\mathbf{E}_\beta$ and $\mathsf{P} \equiv (P_{\alpha\beta})$ is a proper orthogonal matrix.

On the other hand, if $\{\mathbf{E}_\alpha\}$ is fixed while $\{\mathbf{e}_i\}$ is changed according to $\mathbf{e}'^i = Q_{ij}\mathbf{e}_j$, where $\mathsf{Q} \equiv (Q_{ij})$ is a proper orthogonal matrix, show that the component matrix of $\mathbf{A}$ transforms to QA.

Deduce that the components of $\mathbf{A}$ transform like those of a CT(2) if the two bases are changed simultaneously in such a way that $\mathsf{P} = \mathsf{Q}$. Note, however, that the components of $\mathbf{A}$ associated with $\{\mathbf{E}_\alpha\}$ and $\{\mathbf{e}_i\}$ respectively should in general be treated independently, and individually they transform like the components of *vectors*.

In Problem 2.2.1 we saw that a scalar or vector field could be defined over either $\mathscr{B}$ or $\mathscr{B}_0$. In general, if f is a (scalar, vector or tensor) field defined on $\mathscr{B}$ the corresponding field over $\mathscr{B}_0$, denoted by F, is defined by

$$F(\mathbf{X}) = f(\mathbf{x}) \equiv f\{\boldsymbol{\chi}(\mathbf{X})\}$$

through (2.2.1) for all $\mathbf{x}$ in $\mathscr{B}$.

Conversely, if F is defined over $\mathscr{B}_0$, we define f over $\mathscr{B}$ by

$$f(\mathbf{x}) = F(\mathbf{X}) = F\{\boldsymbol{\chi}^{-1}(\mathbf{x})\}$$

for all $\mathbf{X}$ in $\mathscr{B}_0$, the inverse deformation (2.2.11) having been used.

The *Lagrangean* (or *referential*) *description* of a physical phenomenon associated with the deformation (2.2.1) of a body B involves fields defined over $\mathscr{B}_0$, while the *Eulerian* (or *spatial*) *description* requires fields defined over $\mathscr{B}$.

2.2.2 Deformation of volume and surface

Consider three non-coplanar line elements $\mathbf{dX}^{(1)}, \mathbf{dX}^{(2)}, \mathbf{dX}^{(3)}$ at the point $\mathbf{X}$ in $\mathscr{B}_0$ so that

$$\mathbf{dx}^{(i)} = \mathbf{A}\mathbf{dX}^{(i)} \qquad i = 1, 2, 3 \tag{2.2.12}$$

in accordance with (2.2.4). We assume that the triad $\mathbf{dX}^{(1)}$, $\mathbf{dX}^{(2)}$, $\mathbf{dX}^{(3)}$ is positively oriented, i.e. $\mathbf{dX}^{(1)}\cdot(\mathbf{dX}^{(2)} \wedge \mathbf{dX}^{(3)}) > 0$, and we write

$$\mathrm{d}V = \mathbf{dX}^{(1)}\cdot(\mathbf{dX}^{(2)} \wedge \mathbf{dX}^{(3)})$$

for the volume of the infinitesimal parallepiped whose edges are $\mathbf{dX}^{(1)}$, $\mathbf{dX}^{(2)}$, $\mathbf{dX}^{(3)}$. Alternatively, by (1.1.21), we have

$$\mathrm{d}V = \det(\mathbf{dX}^{(1)}, \mathbf{dX}^{(2)}, \mathbf{dX}^{(3)}) \tag{2.2.13}$$

in which $\mathbf{dX}^{(i)}$ denotes a column vector ($i = 1, 2, 3$).

The corresponding volume $\mathrm{d}v$ in the deformed configuration is

$$\mathrm{d}v = \det(\mathbf{dx}^{(1)}, \mathbf{dx}^{(2)}, \mathbf{dx}^{(3)})$$

which, on use of (2.2.12), (1.1.25), (2.2.13) and (2.2.6), becomes

$$\mathrm{d}v = (\det \mathbf{A})\,\mathrm{d}V \equiv J\,\mathrm{d}V. \tag{2.2.14}$$

By convention we define volume elements to be positive so that relative orientation of the line elements is preserved under deformation. It follows from (2.2.14) that

$$J \equiv \det \mathbf{A} > 0, \tag{2.2.15}$$

and we adopt this constraint henceforth. Equation (2.2.14) then provides J with a physical interpretation: it is the local ratio of current to reference volume of a material volume element. In components J represents the Jacobian determinant $\det(\partial x_i/\partial X_\alpha)$ of the transformation (2.2.2). Note that $J = 1$ if the current and reference configurations coincide or if the deformation is just a rigid rotation.

If the volume does not change locally during the deformation then

$$J \equiv \det \mathbf{A} = 1 \tag{2.2.16}$$

at $\mathbf{X}$, and the deformation is then said to be *isochoric* (volume preserving) at $\mathbf{X}$. If (2.2.16) is satisfied for all $\mathbf{X}$ in $\mathscr{B}_0$ then the deformation of the body is isochoric.

The mathematical constraint (2.2.16) is of considerable assistance in the solution of certain boundary-value problems and, although it is an idealization as far as real materials are concerned, there are, as we see in later chapters, circumstances in which (2.2.16) provides a good first approximation to material behaviour.

The local connection (2.2.14) may also be put in the global form

$$\int_{\mathscr{B}} \mathrm{d}v = \int_{\mathscr{B}_0} J\,\mathrm{d}V \tag{2.2.17}$$

in which $\mathscr{B}$ and $\mathscr{B}_0$ may be replaced respectively by an arbitrary subset of $\mathscr{B}$ and the corresponding subset of $\mathscr{B}_0$. Note that a deformation of a body B may be globally volume preserving without (2.2.16) being satisfied for all $\mathbf{X}$ in $\mathscr{B}_0$.

Consider next an infinitesimal vector element of material surface $\mathbf{dS}$ in a neighbourhood of the point $\mathbf{X}$ in $\mathscr{B}_0$ such that $\mathbf{dS} = \mathbf{N}\,\mathrm{d}S$, where $\mathbf{N}$ is the (positive) unit normal to the surface. Let $\mathbf{dX}$ be an arbitrary material line element cutting the edge of $\mathbf{dS}$ such that $\mathbf{dX}\cdot\mathbf{dS} > 0$. Then the cylinder with base $\mathrm{d}S$ and generators $\mathbf{dX}$ has volume $\mathrm{d}V = \mathbf{dX}\cdot\mathbf{dS}$. Suppose that $\mathbf{dX}$ and $\mathbf{dS}$ respectively become $\mathbf{dx}$ and $\mathbf{ds}$ under the deformation (2.1.1), where $\mathbf{ds} = \mathbf{n}\,\mathrm{d}s$ and $\mathbf{n}$ is the (positive) unit normal to the surface $\mathrm{d}s$. The material of the volume $\mathrm{d}V$ forms a cylinder of volume $\mathrm{d}v = \mathbf{dx}\cdot\mathbf{ds}$ in the current configuration and so, by (2.2.14). we have

$$\mathbf{dx}\cdot\mathbf{ds} = J\,\mathbf{dX}\cdot\mathbf{dS}.$$

On use of (2.2.4) we obtain

$$\mathbf{A}^{\mathrm{T}}\,\mathbf{ds} = J\,\mathbf{dS}$$

on removal of the arbitrary $\mathbf{dX}$. With the help of (2.2.9) this becomes

$$\mathbf{ds} = J\mathbf{B}\mathbf{dS}, \qquad \mathbf{n}\,\mathrm{d}s = J\mathbf{B}\mathbf{N}\,\mathrm{d}S. \tag{2.2.18}$$

Either of the connections (2.2.18) between the reference and current area elements is known as *Nanson's formula.* Note that the surface normal is *not* embedded in the material, i.e. does not transform according to the rule (2.2.4) appropriate to line elements. Thus, in general, the material lying along $\mathbf{n}$ in the neighbourhood of the point $\mathbf{x}$ of $\mathscr{B}$ is not the same as that along $\mathbf{N}$ in a neighbourhood of $\mathbf{X}$ in $\mathscr{B}_0$.

Let $\partial\mathscr{B}$ and $\partial\mathscr{B}_0$ respectively denote the bounding surfaces of $\mathscr{B}$ and $\mathscr{B}_0$.

On use of the divergence theorem and (2.2.18) we have

$$\mathbf{0} = \int_{\partial\mathscr{B}} \mathbf{ds} = \int_{\partial\mathscr{B}_0} J\mathbf{BN}\,\mathrm{d}S = \int_{\mathscr{B}_0} \mathrm{Div}(J\mathbf{B}^{\mathrm{T}})\mathrm{d}V, \tag{2.2.19}$$

where Div (with upper-case D) denotes the divergence operation with respect to $\mathbf{X}$, as distinct from div (with lower-case d) which refers to $\mathbf{x}$.

The result (2.2.19) is equally valid for an arbitrary subregion of $\mathscr{B}_0$ and therefore the local result

$$\mathrm{Div}(J\mathbf{B}^{\mathrm{T}}) = \mathbf{0} \tag{2.2.20}$$

follows because of the assumed regularity of (2.2.1) (the integrand in (2.2.19) is continuous). In components (2.2.20) takes the form

$$\frac{\partial}{\partial X_\alpha}(JB_{i\alpha}) = 0 \qquad i = 1, 2, 3. \tag{2.2.21}$$

Problem 2.2.3 Use (2.2.20) to show that $\mathrm{div}(J^{-1}\mathbf{A}) = \mathbf{0}$.

Problem 2.2.4 Prove (2.2.21) directly by use of the component form of (2.2.9) and the formula $\partial J/\partial A_{j\beta} = JB_{j\beta}$ obtainable from the results of Problem 1.1.1.

Problem 2.2.5 Let S be a material surface in the reference configuration $\mathscr{B}_0$ of a body B. Suppose that $\mathbf{N}$ is the (positive) unit normal at an arbitrary point of S and let $\mathbf{L}, \mathbf{M}$ be mutually orthogonal unit tangent vectors to S at $\mathbf{X}$ such that $\mathbf{N} = \mathbf{L} \wedge \mathbf{M}$. If $\mathbf{A}$ is the gradient of the deformation (2.2.1), S deforms to the surface s in $\mathscr{B}$ with (positive) unit normal $\mathbf{n}$ at $\mathbf{x}$, $\mathbf{l} = \mathbf{AL}$, $\mathbf{m} = \mathbf{AM}$ and $\mathbf{n} = (\mathbf{l} \wedge \mathbf{m})/|\mathbf{l} \wedge \mathbf{m}|$, show that

$$\mathbf{A}^{\mathrm{T}}(\mathbf{l} \wedge \mathbf{m}) = J\mathbf{N}$$

and

$$\mathbf{n}\,\mathrm{d}s = (\mathbf{l} \wedge \mathbf{m})\,\mathrm{d}S.$$

Reconcile these results with Nanson's formula. [The results of Problem 1.1.1 may be used.]

2.2.3 Strain, stretch, extension and shear

The material is said to be *unstrained* at $\mathbf{X}$ if the length of an arbitrary line element $\mathbf{dX}$ based on $\mathbf{X}$ is unchanged after deformation. From (2.2.4) we obtain

$$|\mathbf{dx}|^2 - |\mathbf{dX}|^2 = \mathbf{dX}\cdot(\mathbf{A}^{\mathrm{T}}\mathbf{A} - \mathbf{I})\mathbf{dX}, \tag{2.2.22}$$

which represents the difference between the squared lengths of a line element in the current and reference configurations. The material is unstrained at **X** if the right-hand side of (2.2.22) vanishes for arbitrary **dX**. This leads to the tensorial restriction

$$\mathbf{A}^{\mathrm{T}}\mathbf{A} = \mathbf{I}. \tag{2.2.23}$$

The condition (2.2.23), which is necessary and sufficient for the material to be unstrained at **X**, allows the possibility that **A** is simply a rigid rotation since a proper orthogonal tensor satisfies (2.2.23) and (2.2.15) jointly. Thus, we regard a rigid rotation as a local deformation which corresponds to zero strain. The most general deformation which yields zero strain and is consistent with (2.2.15) comprises a rigid translation combined with a rigid rotation, as described by equation (2.1.26) *locally*.

If (2.2.23) is satisfied for all **X** in $\mathscr{B}_0$ then the body is said to be unstrained. Then the deformation gradient is a proper orthogonal tensor **Q** at every **X** in $\mathscr{B}_0$. It is left as an exercise for the reader to prove that **Q** is necessarily independent of **X**; in other words the deformation of the body is described by (2.1.26) *globally* (i.e. for all **X** in $\mathscr{B}_0$).

If (2.2.23) is not satisfied at **X** then the material is said to be *strained* at **X**, and the tensor $\mathbf{A}^{\mathrm{T}}\mathbf{A} - \mathbf{I}$ can be regarded as a *strain tensor* in that it provides, through (2.2.22), a measure of the change in length of an arbitrary line element of material. We define the *Green* (or *Lagrangean*†) strain tensor **E** by

$$\mathbf{E} = \tfrac{1}{2}(\mathbf{A}^{\mathrm{T}}\mathbf{A} - \mathbf{I}), \tag{2.2.24}$$

the factor $\frac{1}{2}$ being a normalization factor (whose significance is explained in a more general context in Section 2.2.7). Clearly, the material is unstrained at **X** if and only if $\mathbf{E} = \mathbf{0}$ at **X**.

In components, (2.2.24) has the form

$$E_{\alpha\beta} = \tfrac{1}{2}(A_{i\alpha}A_{i\beta} - \delta_{\alpha\beta}) = \frac{1}{2}\left(\frac{\partial x_i}{\partial X_\alpha}\frac{\partial x_i}{\partial X_\beta} - \delta_{\alpha\beta}\right). \tag{2.2.25}$$

From (2.2.8) we obtain an alternative for (2.2.22), namely

$$|\mathbf{dx}|^2 - |\mathbf{dX}|^2 = \mathbf{dx}\cdot(\mathbf{I} - \mathbf{B}\mathbf{B}^{\mathrm{T}})\mathbf{dx},$$

and the tensor

$$\tfrac{1}{2}(\mathbf{I} - \mathbf{B}\mathbf{B}^{\mathrm{T}}) \tag{2.2.26}$$

†The word *Lagrangean* is used since **E** is associated with the reference configuration by virtue of being a bilinear mapping from $T_{\mathbf{X}}(\mathscr{B}_0) \times T_{\mathbf{X}}(\mathscr{B}_0)$ to $\mathbb{R}$, where $T_{\mathbf{X}}(\mathscr{B}_0)$ is the tangent space defined in the footnote on *p.* 84.

is called the *Eulerian* strain tensor. A general discussion of strain measures is postponed until Section 2.2.7.

The *displacement* of a particle X from the reference to the current configuration is defined by the point difference $\mathbf{u} = \mathbf{x} - \mathbf{X}$, and we write

$$\mathbf{u}(\mathbf{X}) = \boldsymbol{\chi}(\mathbf{X}) - \mathbf{X}. \tag{2.2.27}$$

The *displacement gradient*, denoted by $\mathbf{D}$, is a (two-point) tensor given by

$$\mathbf{D} = \mathrm{Grad}\, \mathbf{u}(\mathbf{X}) \equiv \mathbf{A} - \mathbf{I}. \tag{2.2.28}$$

It follows from (2.2.24) that

$$\mathbf{E} = \tfrac{1}{2}(\mathbf{D} + \mathbf{D}^{\mathrm{T}} + \mathbf{D}^{\mathrm{T}}\mathbf{D}), \tag{2.2.29}$$

or, in component form,

$$E_{\alpha\beta} = \frac{1}{2}\left(\frac{\partial u_\alpha}{\partial X_\beta} + \frac{\partial u_\beta}{\partial X_\alpha} + \frac{\partial u_\gamma}{\partial X_\alpha}\frac{\partial u_\gamma}{\partial X_\beta}\right), \tag{2.2.30}$$

where u_α are the components of $\mathbf{u}$ with respect to the reference basis $\{\mathbf{E}_\alpha\}$.

Let $\mathbf{M}$ and $\mathbf{m}$ be unit vectors along $\mathbf{dX}$ and $\mathbf{dx}$ respectively. Then, from (2.2.4), we obtain

$$\mathbf{m}|\mathbf{dx}| = \mathbf{AM}|\mathbf{dX}|$$

and hence

$$|\mathbf{dx}|^2 = \mathbf{M}\cdot(\mathbf{A}^{\mathrm{T}}\mathbf{AM})|\mathbf{dX}|^2.$$

We write

$$\frac{|\mathbf{dx}|}{|\mathbf{dX}|} \equiv |\mathbf{AM}| \equiv \{\mathbf{M}\cdot(\mathbf{A}^{\mathrm{T}}\mathbf{AM})\}^{1/2} = \lambda(\mathbf{M}), \tag{2.2.31}$$

which defines $\lambda(\mathbf{M})$, the *stretch* in the direction $\mathbf{M}$ at $\mathbf{X}$, as the ratio of current to reference lengths of a line element which was in the direction $\mathbf{M}$ in the reference configuration. Because of the restriction (2.2.15), we have

$$0 < \lambda(\mathbf{M}) < \infty \tag{2.2.32}$$

for all non-zero vectors $\mathbf{M}$.

The quantity $\lambda(\mathbf{M}) - 1$ is called the *extension ratio* in the direction $\mathbf{M}$, while $|\mathbf{dx}| - |\mathbf{dX}|$ is the *extension*.

Note that $\lambda(\mathbf{M}) = 1$ for all unit vectors $\mathbf{M}$ if and only if $\mathbf{E} = \mathbf{0}$.

If $\mathbf{dX}$ and $\mathbf{dX}'$ are two line elements based on the point $\mathbf{X}$ in the reference configuration, the corresponding line elements in the current configuration are given by

$$\mathbf{dx} = \mathbf{A}\mathbf{dX}, \quad \mathbf{dx}' = \mathbf{A}\mathbf{dX}'.$$

Let $\mathbf{M}$ and $\mathbf{M}'$ respectively be unit vectors along $\mathbf{dX}$ and $\mathbf{dX}'$, with $\mathbf{m}$ and $\mathbf{m}'$ similarly defined in the current configuration.

If Θ denotes the angle between the directions $\mathbf{M}$ and $\mathbf{M}'$ and θ that between $\mathbf{m}$ and $\mathbf{m}'$, then

$$\cos\Theta = \mathbf{M}\cdot\mathbf{M}', \quad \cos\theta = \mathbf{m}\cdot\mathbf{m}' = \mathbf{M}\cdot(\mathbf{A}^{\mathrm{T}}\mathbf{A}\mathbf{M}')/\lambda(\mathbf{M})\lambda(\mathbf{M}'), \tag{2.2.33}$$

the latter following on use of (2.2.31).

The *decrease* $\Theta - \theta$ in the angle between the pairs of line elements is called the *angle of shear* of the directions $\mathbf{M}$ and $\mathbf{M}'$ in the *plane of shear* defined by $\mathbf{M}$ and $\mathbf{M}'$.

Each of the quantities *stretch* and *shear* depends on the deformation only through the right Cauchy–Green deformation tensor $\mathbf{A}^{\mathrm{T}}\mathbf{A}$ and is therefore independent of any rigid-body constituent of $\mathbf{A}$.

2.2.4 Polar decomposition of the deformation gradient

The polar decomposition theorem is of considerable assistance in the geometrical interpretation of the deformation and in order to prove it we require the following preliminary lemma.

Lemma 2.2.1 To any symmetric, positive definite second-order tensor $\mathbf{T}$ there corresponds a unique symmetric, positive definite second-order tensor $\mathbf{U}$ (the positive square root of $\mathbf{T}$) such that $\mathbf{U}^2 = \mathbf{T}$.

Proof The result follows from a simple application of the spectral representation for $\mathbf{T}$ discussed in Section 1.3.3, and we do not give details here.

Theorem 2.2.1 The polar decomposition theorem: for any non-singular second-order tensor $\mathbf{A}$ there exist unique positive definite symmetric second-order tensors $\mathbf{U}$ and $\mathbf{V}$, and an orthogonal second-order tensor $\mathbf{R}$ such that

$$\mathbf{A} = \mathbf{R}\mathbf{U} = \mathbf{V}\mathbf{R}. \tag{2.2.34}$$

Proof It follows from Lemma 2.2.1 that there exist symmetric, positive definite second-order tensors $\mathbf{U}$ and $\mathbf{V}$ such that

$$\mathbf{A}^{\mathrm{T}}\mathbf{A} = \mathbf{U}^2, \quad \mathbf{A}\mathbf{A}^{\mathrm{T}} = \mathbf{V}^2 \tag{2.2.35}$$

since $\mathbf{A}^T\mathbf{A}$ and $\mathbf{A}\mathbf{A}^T$ are symmetric and positive definite.

Define

$$\mathbf{R} = \mathbf{A}\mathbf{U}^{-1}, \quad \mathbf{S} = \mathbf{V}^{-1}\mathbf{A}.$$

Then, on use of (2.2.35) and the symmetry of $\mathbf{U}$, we obtain

$$\mathbf{R}^T\mathbf{R} = \mathbf{U}^{-1}\mathbf{A}^T\mathbf{A}\mathbf{U}^{-1} = \mathbf{I}.$$

Hence $\mathbf{R}$ is orthogonal, $\mathbf{A} = \mathbf{R}\mathbf{U}$ and similarly $\mathbf{A} = \mathbf{V}\mathbf{S}$, where $\mathbf{S}$ is orthogonal.

To prove uniqueness, we suppose that there exist second-order tensors $\mathbf{R}'$ and $\mathbf{U}'$, orthogonal and symmetric, positive definite respectively, such that $\mathbf{A} = \mathbf{R}\mathbf{U} = \mathbf{R}'\mathbf{U}'$. It follows that $\mathbf{R} = \mathbf{R}'\mathbf{U}'\mathbf{U}^{-1}$ and hence $\mathbf{R}^{-1} = \mathbf{U}\mathbf{U}^{-1}\mathbf{R}'^T$. But since $\mathbf{R}^{-1} = \mathbf{R}^T = \mathbf{U}^{-1}\mathbf{U}'\mathbf{R}'^T$ we obtain $\mathbf{U}'^2 = \mathbf{U}^2$ and the result $\mathbf{U}' = \mathbf{U}$ follows from Lemma 2.2.1, and therefore $\mathbf{R}' = \mathbf{R}$. Similarly for $\mathbf{V}\mathbf{S}$.

It remains to show that $\mathbf{S} = \mathbf{R}$. Since $\mathbf{R}\mathbf{U} = \mathbf{A} = \mathbf{S}(\mathbf{S}^T\mathbf{V}\mathbf{S})$, the uniqueness result proved above implies that $\mathbf{S} = \mathbf{R}$ and hence

$$\mathbf{U} = \mathbf{R}^T\mathbf{V}\mathbf{R}. \tag{2.2.36}$$

In respect of the deformation gradient, we have $\det \mathbf{A} > 0$, as imposed in (2.2.15). It then follows that $\mathbf{R}$ is *proper* orthogonal and, by (2.2.34),

$$\det \mathbf{A} = \det \mathbf{U} = \det \mathbf{V}. \tag{2.2.37}$$

The deformation gradient $\mathbf{A}$ represents a rigid rotation if and only if $\mathbf{U} = \mathbf{V} = \mathbf{I}$. If $\mathbf{R} = \mathbf{I}$, on the other hand, then $\mathbf{A} = \mathbf{U} = \mathbf{V}$ and the deformation is said to be a *pure strain*. This is consistent with the terminology 'strain' introduced in Section 2.2.3 since, by (2.2.24), (2.2.26) and (2.2.35), the Lagrangean and Eulerian strain tensors become

$$\tfrac{1}{2}(\mathbf{U}^2 - \mathbf{I}), \quad \tfrac{1}{2}(\mathbf{I} - \mathbf{V}^{-2})$$

respectively.

The tensors $\mathbf{U}$ and $\mathbf{V}$ respectively are called the *right* and *left stretch tensors*. Their connection with the stretch defined in (2.2.31) is as follows. Firstly, from (2.2.35) and (2.2.31), we obtain

$$\lambda(\mathbf{M}) = \{\mathbf{M}\cdot(\mathbf{U}^2\mathbf{M})\}^{1/2} = |\mathbf{U}\mathbf{M}|. \tag{2.2.38}$$

Let λ_i be the principal values of $\mathbf{U}$ corresponding to the principal directions

$\mathbf{u}^{(i)}$ so that

$$\mathbf{U}\mathbf{u}^{(i)} = \lambda_i \mathbf{u}^{(i)} \quad i = 1, 2, 3.$$

Substitution of $\mathbf{u}^{(i)}$ for $\mathbf{M}$ in (2.2.38) shows that $\lambda_i = \lambda(\mathbf{u}^{(i)})$ and hence provides justification for referring to $\lambda_i (i = 1, 2, 3)$ as the *principal stretches.*

Secondly, use of (2.2.34) leads to

$$\mathbf{V}\mathbf{R}\mathbf{u}^{(i)} = \mathbf{R}\mathbf{U}\mathbf{u}^{(i)} = \lambda_i \mathbf{R}\mathbf{u}^{(i)}.$$

This shows that λ_i *are also the principal values of* $\mathbf{V}$, corresponding to principal directions $\mathbf{R}\mathbf{u}^{(i)}$ $(i = 1, 2, 3)$. Thus, the deformation rotates the principal directions of $\mathbf{U}$ into those of $\mathbf{V}$.

Note that in terms of the principal stretches, equation (2.2.37) gives

$$J \equiv \det \mathbf{A} = \lambda_1 \lambda_2 \lambda_3. \tag{2.2.39}$$

Problem 2.2.6 Show that $\mathbf{B} = \mathbf{R}\mathbf{U}^{-1} = \mathbf{V}^{-1}\mathbf{R}$, $\mathbf{B}^{\mathrm{T}}\mathbf{B} = \mathbf{U}^{-2}$, $\mathbf{B}\mathbf{B}^{\mathrm{T}} = \mathbf{V}^{-2}$ and prove that $\mathbf{B}^{\mathrm{T}}\mathbf{B}, \mathbf{A}^{\mathrm{T}}\mathbf{A}$ and $\mathbf{U}$ have the same principal directions. Show also that $\mathbf{B}\mathbf{B}^{\mathrm{T}}$ is coaxial with $\mathbf{A}\mathbf{A}^{\mathrm{T}}$ and $\mathbf{V}$. Deduce that $\mathbf{B}\mathbf{B}^{\mathrm{T}}$ and $\mathbf{B}^{\mathrm{T}}\mathbf{B}$ have principal values $\lambda_i^{-2} (i = 1, 2, 3)$.

Problem 2.2.7 Use the spectral decomposition

$$\mathbf{A}^{\mathrm{T}}\mathbf{A} = \sum_{i=1}^{3} \lambda_i^2 \mathbf{u}^{(i)} \otimes \mathbf{u}^{(i)}$$

to show that the stretch $\lambda(\mathbf{M})$ defined in (2.2.31) is given by

$$\lambda(\mathbf{M})^2 = \sum_{i=1}^{3} \lambda_i^2 M_i^2,$$

where $M_i = \mathbf{M} \cdot \mathbf{u}^{(i)} (i = 1, 2, 3)$.

Show that if $\mathbf{M} = \cos\phi \mathbf{u}^{(1)} + \sin\phi \mathbf{u}^{(2)}$ and $\mathbf{M}' = -\sin\phi \mathbf{u}^{(1)} + \cos\phi \mathbf{u}^{(2)}$ then, from (2.2.33), it follows that $\Theta = \pi/2$ and

$$\cos\theta = (\lambda_2^2 - \lambda_1^2) \sin 2\phi \{4\lambda_1^2\lambda_2^2 + (\lambda_1^2 - \lambda_2^2)^2 \sin^2 2\phi\}^{-1/2}.$$

Find the value of ϕ for which the angle of shear of the directions $\mathbf{M}$ and $\mathbf{M}'$ has maximum magnitude.

2.2.5 Geometrical interpretations of the deformation

The set of line elements $\mathbf{dX}$ based on the point $\mathbf{X}$ in $\mathscr{B}_0$ and satisfying the equation

$$\mathbf{dX} \cdot \mathbf{dX} \equiv |\mathbf{dX}|^2 = \text{constant} \tag{2.2.40}$$

defines a sphere of radius $|\mathbf{dX}|$ and centre $\mathbf{X}$. It follows from (2.2.8) that

$$\mathbf{dx}\cdot(\mathbf{BB}^{\mathrm{T}}\mathbf{dx}) = |\mathbf{dX}|^2 = \text{constant}. \tag{2.2.41}$$

Since $\mathbf{BB}^{\mathrm{T}}$ is positive definite equation (2.2.41) defines an ellipsoid centred on the point $\mathbf{x}$ in $\mathscr{B}$, and the material contained within the sphere (2.2.40) in the reference configuration is deformed into the ellipsoid (2.2.41) in the current configuration.

Since it is based on the current configuration (2.2.41) is referred to as the *Eulerian strain ellipsoid.* Its principal axes coincide with those of $\mathbf{BB}^{\mathrm{T}}$ and hence with those of $\mathbf{AA}^{\mathrm{T}}$ and $\mathbf{V}$, and, since $\mathbf{BB}^{\mathrm{T}}$ has principal values $\lambda_i^{-2}(i=1,2,3)$, the semi-axes of the ellipsoid have lengths proportional to $\lambda_1, \lambda_2, \lambda_3$. The *reciprocal* Eulerian strain ellipsoid, on the other hand, is based on $\mathbf{AA}^{\mathrm{T}}$ and has semi-axes proportional to $\lambda_1^{-1}, \lambda_2^{-1}, \lambda_3^{-1}$.

A sphere

$$\mathbf{dx}\cdot\mathbf{dx} = |\mathbf{dx}|^2 = \text{constant}$$

formed from line elements $\mathbf{dx}$ based at the point $\mathbf{x}$ in $\mathscr{B}$ comes from the ellipsoid

$$\mathbf{dX}\cdot(\mathbf{A}^{\mathrm{T}}\mathbf{A}\mathbf{dX}) = |\mathbf{dx}|^2 = \text{constant}$$

centred on $\mathbf{X}$ in the reference configuration $\mathscr{B}_0$. This is called the *reciprocal Lagrangean strain ellipsoid* to distinguish it from the Lagrangean strain ellipsoid which is based on $\mathbf{B}^{\mathrm{T}}\mathbf{B}$ and coaxial with $\mathbf{B}^{\mathrm{T}}\mathbf{B}$ and $\mathbf{U}$. When $\mathbf{A}$ is known both sets of principal axes can be calculated on the basis of (2.2.35).

The unit eigenvectors $\mathbf{u}^{(i)}\,(i=1,2,3)$ of $\mathbf{U}$ define the *principal axes of the Lagrangean strain ellipsoid.* For brevity we refer to these simply as the *Lagrangean principal axes* or, when the context is clear, just the *Lagrangean axes.* Similarly, the *principal axes* $\mathbf{Ru}^{(i)}$ *of the Eulerian strain ellipsoid* is abbreviated to *Eulerian (principal) axes.*

Perpendicular diameters of the sphere (2.2.40) along the line elements $\mathbf{dX}$ and $\mathbf{dX}'$ are such that $\mathbf{dX}'\cdot\mathbf{dX}=0$. It follows from (2.2.8) that the corresponding diameters along $\mathbf{dx}$ and $\mathbf{dx}'$ of the ellipsoid (2.2.41) satisfy

$$\mathbf{dx}'\cdot(\mathbf{BB}^{\mathrm{T}}\mathbf{dx}) = 0. \tag{2.2.42}$$

Note that since $\mathbf{BB}^{\mathrm{T}}\mathbf{dx}$ is normal† to the surface of the ellipsoid (2.2.41) equation (2.2.42) is a statement that $\mathbf{dx}$ and $\mathbf{dx}'$ are conjugate diameters of (2.2.41). Thus, perpendicular diameters of a sphere at $\mathbf{X}$ are deformed into

†Recall that if $\mathbf{T}$ is a symmetric second-order tensor then $\mathbf{Tx}$ is normal to the surface $\mathbf{x}.(\mathbf{Tx}) \equiv \phi = \text{constant}$ since $2\mathbf{Tx} = \operatorname{grad}\phi$.

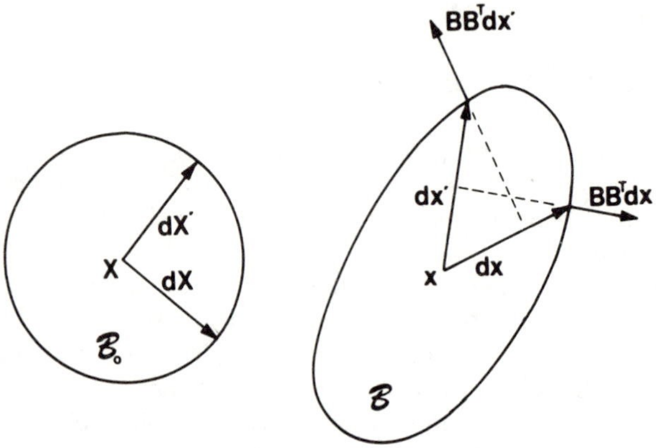

Fig. 2.2 Perpendicular line elements **dX**, **dX**′ along radii of a sphere based on **X** in $\mathscr{B}_0$ are deformed into conjugate diameters **dx**, **dx**′ of an ellipsoid centred on **x** in $\mathscr{B}$.

conjugate diameters of the Eulerian strain ellipsoid at **x**. This is illustrated in Fig. 2.2.

If, an addition to (2.2.42), we have $\mathbf{dx}'\cdot\mathbf{dx}=0$ then **dx** and **dx**′ are necessarily along the principal axes of the Eulerian strain ellipsoid. It follows that at any point **X** in $\mathscr{B}_0$ there exist three† mutually orthogonal line elements which are mutually orthogonal in $\mathscr{B}$, namely those lying along the Eulerian axes. But in general the angle between a pair of line element **dX** and **dX**′ changes during deformation in accordance with (2.2.33).

In the light of the polar decomposition (2.2.34) the deformation may be viewed simplistically in the following two ways:

(a) First apply the right stretch tensor so that the line elements lying along the Lagrangean axes are stretched to their final lengths according to

$$\mathbf{U}\mathbf{u}^{(i)}=\lambda_i\mathbf{u}^{(i)}\quad i=1,2,3.$$

Next rotate the Lagrangean axes to their final positions $\mathbf{R}\mathbf{u}^{(i)}$, so that

$$\mathbf{A}\mathbf{u}^{(i)}=\mathbf{R}\mathbf{U}\mathbf{u}^{(i)}=\lambda_i\mathbf{R}\mathbf{u}^{(i)}\quad i=1,2,3.$$

(b) First rotate the line elements lying along the Lagrangean axes $\mathbf{u}^{(i)}$ to their final positions $\mathbf{R}\mathbf{u}^{(i)}$ and then apply the left stretch tensor to take them to

† If two or more principal stretches are equal there are infinitely many such line elements.

their final lengths. Thus

$$\mathbf{Au}^{(i)} = \mathbf{VRu}^{(i)} = \lambda_i \mathbf{Ru}^{(i)}.$$

Any rigid translation is not taken into account in (a) and (b).

As we have seen, the mutually orthogonal triads of line elements lying along the Lagrangean axes before deformation are mutually orthogonal after deformation and lie along the Eulerian axes. But, they are not in general mutually orthogonal at intermediate stages of the deformation and, as the deformation proceeds, both the Lagrangean and Eulerian axes change their orientation, with different triads of line elements continuously being aligned along these axes. The following examples illustrate the effect of specific deformations on line elements of various orientations.

Example 2.2.1 If a rectangular strip of material (rubber, for example) is extended uniformly (with or without accompanying transverse contraction) then the Lagrangean and Eulerian axes coincide and remain fixed in direction parallel to the edges of the strip during deformation. Line elements parallel to the edges of the strip remain parallel but other line elements rotate during deformation (and also stretch or contract in general). This is illustrated in Fig. 2.3 which shows a circle described on the surface of the strip in the reference configuration $\mathscr{B}_0$ and the ellipse which it becomes in the current configuration $\mathscr{B}$.

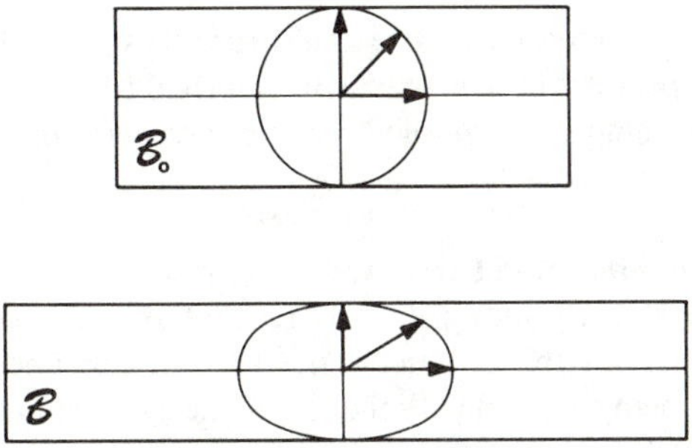

Fig. 2.3 The effect on various line elements of the uniform extension (with lateral contraction) of a rectangular strip from $\mathscr{B}_0$ to $\mathscr{B}$.

Example 2.2.2 In the (simple) shear deformation of a rectangular block the orientation of the principal axes changes with the magnitude of the deformation (full details are given in Section 2.2.6). But, as illustrated in Fig. 2.4 certain line elements remain fixed in direction (the horizontal ones in the figure) although they do not lie along a principal axis. This contrasts with what happens in the uniform extension of a strip.

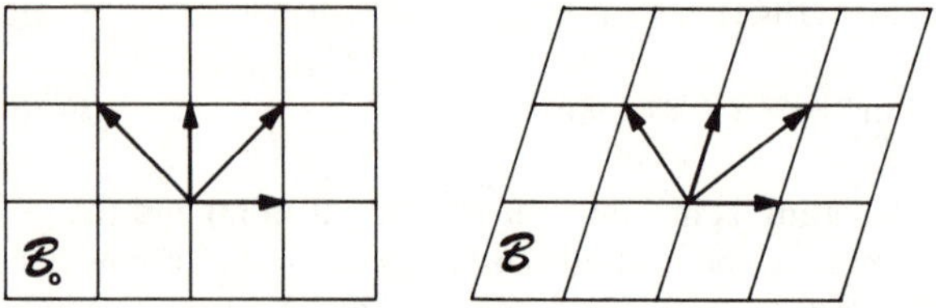

Fig. 2.4 The effect on various line elements of the simple shear of a rectangular block from $\mathscr{B}_0$ to $\mathscr{B}$.

The mathematical description of these two deformations amongst others is examined in Section 2.2.6.

2.2.6 Examples of deformations

In general the deformation gradient $\mathbf{A}$ depends on $\mathbf{X}$, i.e. varies from point to point in the material of the body B. Such a deformation is said to be *inhomogeneous*. If, on the other hand, $\mathbf{A}$ is independent of $\mathbf{X}$ for the body in question then the deformation of the body is said to be *homogeneous*. The most general homogeneous deformation of the body B from its reference configuration $\mathscr{B}_0$ is given by

$$\mathbf{x} = \mathbf{A}\mathbf{X} + \mathbf{c} \tag{2.2.43}$$

for $\mathbf{X} \in \mathscr{B}_0$, where the vector $\mathbf{c}$ and second-order tensor $\mathbf{A}$ (the deformation gradient) are independent of $\mathbf{X}$, $\mathbf{c}$ representing a rigid translation of the whole body. We now examine some simple homogeneous deformations for specific bodies.

Uniform extension with lateral contraction Consider a circular cylindrical body extended uniformly along its axis so that the circular symmetry is maintained. Because of the symmetry the Lagrangean and Eulerian axes coincide, one axis being aligned with the axis of the cylinder and the other pair orientated arbitrarily normal to the axis so as to form a mutually orthogonal triad.

Let L and l be the reference and current lengths of the cylinder. Then the principal stretch λ_1 along the axis is $\lambda_1 = l/L$, and $\lambda_2 = \lambda_3$ for those normal to the axis.

If reference and current basis vectors are taken to coincide with the principal axes then (2.2.43) has the component form

$$x_1 = \lambda_1 X_1, \quad x_2 = \lambda_2 X_2, \quad x_3 = \lambda_3 X_3, \tag{2.2.44}$$

c having been set equal to zero for convenience, and **A** has component matrix

$$\mathsf{A}=\begin{bmatrix}\lambda_1 & 0 & 0\\ 0 & \lambda_2 & 0\\ 0 & 0 & \lambda_3\end{bmatrix} \tag{2.2.45}$$

with $\lambda_3=\lambda_2$.

Expressed in terms of reference and current cylindrical polar coordinates (R,Θ,Z) and (r,θ,z) respectively (2.2.43) becomes

$$r=\lambda_2 R,\quad \theta=\Theta,\quad z=\lambda_1 Z, \tag{2.2.46}$$

where $X_1=R\cos\Theta$, $X_2=R\sin\Theta$, $X_3=Z$ and similarly for the current coordinates.

If the deformation is isochoric then from (2.2.16) and (2.2.39) we obtain $\lambda_1\lambda_2\lambda_3=1$, and (2.2.45) reduces to

$$\mathsf{A}=\begin{bmatrix}\lambda_1 & 0 & 0\\ 0 & \lambda_1^{-1/2} & 0\\ 0 & 0 & \lambda_1^{-1/2}\end{bmatrix}. \tag{2.2.47}$$

The deformation associated with the strip extension considered at the end of Section 2.2.5 also has the form (2.2.45) when referred to basis vectors along and normal to the axis of the strip despite the lack of geometrical symmetry. But in general the condition $\lambda_3=\lambda_2$ cannot then be invoked.

Pure dilatation Suppose a cube of material has edges of lengths L and l in the reference and current configurations. Then, by symmetry, the principal stretches are given by $\lambda_1=\lambda_2=\lambda_3=l/L$, and **A** has matrix representation

$$\mathsf{A}=\begin{bmatrix}\lambda_1 & 0 & 0\\ 0 & \lambda_1 & 0\\ 0 & 0 & \lambda_1\end{bmatrix} \tag{2.2.48}$$

with respect to reference and current basis vectors of arbitrary orientation (i.e. the Lagrangean and Eulerian axes). This deformation is isochoric if and only if $\lambda_1=1$.

Plane deformations Assume that, relative to the reference configuration, no deformation takes place in the direction defined by the (coincident) reference and current basis vectors $\mathbf{E}_3$ and $\mathbf{e}_3$. The deformation confined to the (1, 2)-plane defined by the basis vectors $\mathbf{e}_1=\mathbf{E}_1$ and $\mathbf{e}_2=\mathbf{E}_2$ is said to be a *plane deformation*† (the body is sometimes said to be in a state of *plane strain*).

The deformation gradient associated with a plane deformation has matrix representation

$$\mathsf{A}=\begin{bmatrix} A_{11} & A_{12} & 0 \\ A_{21} & A_{22} & 0 \\ 0 & 0 & 1 \end{bmatrix} \tag{2.2.49}$$

relative to the chosen basis, and $\mathbf{A}$ itself may be written

$$\mathbf{A}=A_{11}\mathbf{e}_1\otimes\mathbf{e}_1+A_{12}\mathbf{e}_1\otimes\mathbf{e}_2+A_{21}\mathbf{e}_2\otimes\mathbf{e}_1 \\ +A_{22}\mathbf{e}_2\otimes\mathbf{e}_2+\mathbf{e}_3\otimes\mathbf{e}_3. \tag{2.2.50}$$

From the polar decomposition theorem (2.2.34) we have $\mathbf{A}=\mathbf{RU}$, where $\mathbf{R}$ and $\mathbf{U}$ respectively have matrix representations

$$\mathsf{R}=\begin{bmatrix} \cos\theta & \sin\theta & 0 \\ -\sin\theta & \cos\theta & 0 \\ 0 & 0 & 1 \end{bmatrix},\quad \mathsf{U}=\begin{bmatrix} \lambda_1 & 0 & 0 \\ 0 & \lambda_2 & 0 \\ 0 & 0 & 1 \end{bmatrix} \tag{2.2.51}$$

with respect to the Lagrangean axes (which in general do not coincide with the chosen basis vectors). The angle θ and the principal stretches λ_1 and λ_2 may be calculated from A_{11}, A_{12}, A_{21} and A_{22} as we illustrate shortly in the case of a simple shear deformation.

If the plane deformation is isochoric then the components of $\mathbf{A}$ satisfy the constraint

$$A_{11}A_{22}-A_{12}A_{21}=1 \tag{2.2.52}$$

or, equivalently,

$$\lambda_1\lambda_2=1. \tag{2.2.53}$$

Pure shear In general the orientation of the (Lagrangean and Eulerian) principal axes changes with the magnitude of the deformation, in particular with the magnitude of λ_1 in the isochoric plane deformation defined above. If, however, the principal axes are fixed in direction and $\mathbf{R}=\mathbf{I}$ the resulting isochoric plane deformation is said to be a *pure shear*.

Such a deformation can be achieved in practice by clamping a rectangular

†This term is also used for a plane deformation superposed on a uniform extension normal to the plane.

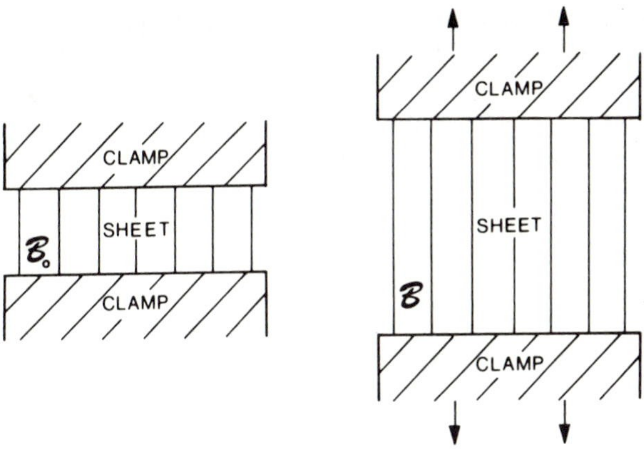

Fig. 2.5 Pure shear of a plane rectangular sheet normal to a pair of clamped edges. The arrows show the direction of extension. Reference configuration: $\mathscr{B}_0$. Current configuration: $\mathscr{B}$.

sheet of material along a parallel pair of edges (in the $\mathbf{E}_3$ direction) and extending the sheet in its plane normal to the clamped edges. If λ_1 is the principal stretch in the direction of extension then $\lambda_2 = \lambda_1^{-1}$ is that normal to the sheet and $\lambda_3 = 1$ parallel to the clamps. The reference and current configurations of the sheet are depicted in Fig. 2.5. Strictly, the deformation has the form of a pure shear only in the central part of the sheet because the free edges of the sheet tend to curve on deformation and this edge effect propagates inwards to a certain extent (this is not illustrated in Fig. 2.5). Detailed experimental results were first obtained by Rivlin and Saunders (1951), and a discussion of more recent work is given by Treloar (1975).

Simple shear This deformation was referred to briefly at the end of Section 2.2.5. It is defined as an isochoric plane deformation for which there exists a set of line elements whose orientation is such that they are unchanged in length and direction by the deformation. Let $\mathbf{e}_1$ define this orientation. Then we must have $\mathbf{A}\mathbf{e}_1 = \mathbf{e}_1$ so that application of (2.2.50) leads to $A_{11} = 1$ and $A_{21} = 0$, and (2.2.52) yields $A_{22} = 1$. We write $A_{12} = \gamma$ so that (2.2.49) becomes

$$\mathsf{A} = \begin{bmatrix} 1 & \gamma & 0 \\ 0 & 1 & 0 \\ 0 & 0 & 1 \end{bmatrix}, \tag{2.2.54}$$

γ being called the *amount of shear*. It follows that $\mathbf{A}\mathbf{e}_2 = \mathbf{e}_2 + \gamma\mathbf{e}_1$.

The deformation is illustrated in Fig. 2.4 in respect of a rectangular block, $\mathbf{e}_1$ and $\mathbf{e}_2$ corresponding to the horizontal and vertical axes respectively.

Horizontal line elements remain horizontal and the perpendicular distance between any pair of horizontal line elements is unchanged by the deformation. The angle of shear of a pair of line elements parallel to an orthogonal (in $\mathscr{B}_0$) pair of edges of the block is $\tan^{-1}\gamma$.

On use of equation (1.2.10) the representation (2.2.50) becomes

$$\mathbf{A} = \mathbf{I} + \gamma \mathbf{e}_1 \otimes \mathbf{e}_2. \tag{2.2.55}$$

With $\mathbf{c} = \mathbf{0}$ equation (2.2.43) then reduces to

$$\mathbf{x} = \mathbf{X} + \gamma(\mathbf{X}\cdot\mathbf{e}_2)\mathbf{e}_1 \tag{2.2.56}$$

or, in components,

$$x_1 = X_1 + \gamma X_2, \quad x_2 = X_2, \quad x_3 = X_3.$$

From (2.2.55) we calculate the right Cauchy–Green deformation tensor as

$$\mathbf{U}^2 = \mathbf{A}^{\mathrm{T}}\mathbf{A} = \mathbf{I} + \gamma(\mathbf{e}_1 \otimes \mathbf{e}_2 + \mathbf{e}_2 \otimes \mathbf{e}_1) + \gamma^2 \mathbf{e}_2 \otimes \mathbf{e}_2, \tag{2.2.57}$$

and, on removal of the factor corresponding to $\lambda = \lambda_3 = 1$, the characteristic equation

$$\det(\mathbf{A}^{\mathrm{T}}\mathbf{A} - \lambda^2 \mathbf{I}) = 0$$

reduces to

$$\lambda^2 + \lambda^{-2} = 2 + \gamma^2$$

or, equivalently,

$$\lambda - \lambda^{-1} = \gamma \tag{2.2.58}$$

since we may take $\lambda_2^{-1} \equiv \lambda_1 = \lambda \geq 1$ without loss of generality.

Let θ denote the orientation (in the anticlockwise sense relative to $\mathbf{e}_1$ and $\mathbf{e}_2$) of the in-plane Lagrangean principal axes, and let

$$\mathsf{A}^{\mathrm{T}}\mathsf{A} \equiv \mathsf{U}^2 = \begin{bmatrix} \lambda_1^2 & 0 & 0 \\ 0 & \lambda_2^2 & 0 \\ 0 & 0 & 1 \end{bmatrix} \equiv \begin{bmatrix} \lambda^2 & 0 & 0 \\ 0 & \lambda^{-2} & 0 \\ 0 & 0 & 1 \end{bmatrix}$$

be the matrix representation of the tensor $\mathbf{A}^{\mathrm{T}}\mathbf{A}$ relative to the Lagrangean axes.

Then

$$\begin{bmatrix} 1 & \gamma & 0 \\ \gamma & 1+\gamma^2 & 0 \\ 0 & 0 & 1 \end{bmatrix} = \begin{bmatrix} \cos\theta & -\sin\theta & 0 \\ \sin\theta & \cos\theta & 0 \\ 0 & 0 & 1 \end{bmatrix} \begin{bmatrix} \lambda^2 & 0 & 0 \\ 0 & \lambda^{-2} & 0 \\ 0 & 0 & 1 \end{bmatrix} \times$$

$$\times \begin{bmatrix} \cos\theta & \sin\theta & 0 \\ -\sin\theta & \cos\theta & 0 \\ 0 & 0 & 1 \end{bmatrix}$$

which leads to

$$\lambda^2 \cos^2\theta + \lambda^{-2}\sin^2\theta = 1$$

$$\lambda^2 \sin^2\theta + \lambda^{-2}\cos^2\theta = 1 + \gamma^2$$

$$(\lambda^2 - \lambda^{-2})\sin\theta\cos\theta = \gamma$$

and hence (2.2.58) together with

$$\tan 2\theta = -2/\gamma.$$

The orientation of the Eulerian axes is found by replacing $\mathbf{A}^{\mathrm{T}}\mathbf{A}$ by $\mathbf{A}\mathbf{A}^{\mathrm{T}}$ in the above. In order to distinguish the orientations of the two sets of principal axes we use the notations θ_{L} and θ_{E}. Then

$$\tan 2\theta_{\mathrm{L}} = -2/\gamma = -\tan 2\theta_{\mathrm{E}} \tag{2.2.59}$$

and the angles are restricted according to

$$\frac{\pi}{4} \leq \theta_{\mathrm{L}} < \frac{\pi}{2}, \qquad 0 < \theta_{\mathrm{E}} \leq \frac{\pi}{4}. \tag{2.2.60}$$

In the reference configuration $\gamma = 0$ and $\theta_{\mathrm{L}} = \theta_{\mathrm{E}} = \pi/4$, the other limits in the inequalities being approached as $\gamma \to \infty$.

The changing orientation of the Eulerian axes as γ increases from zero is illustrated in Fig. 2.6 in respect of a square block whose diagonals coincide in direction with the Eulerian axes for $\gamma = 0$.

If, as indicated in Fig. 2.6, θ denotes the orientation of the lengthening diagonal of the block then simple geometrical considerations show that

$$\tan\theta = (1+\gamma)^{-1}.$$

It is straightforward to deduce from this and (2.2.59) that $\theta < \theta_{\mathrm{E}}$ with θ_{E}

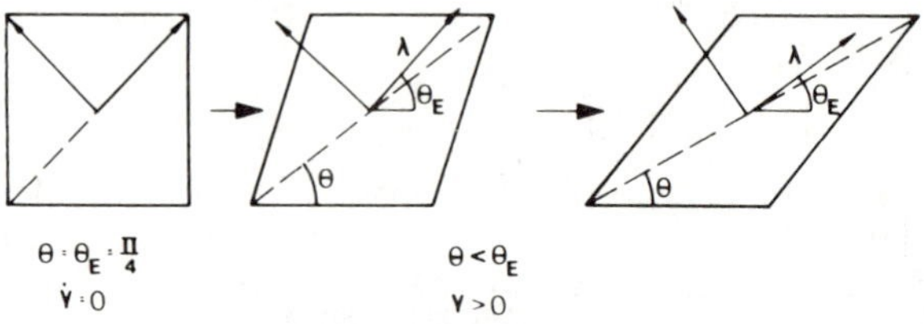

Fig. 2.6 Simple shear of a square block, showing orientation of the Eulerian axes compared with the orientation of the lengthening diagonal.

approaching θ as $\gamma \to \infty$. For large γ, we have

$$\tan 2\theta \simeq \frac{2}{\gamma}\left(1 - \frac{1}{\gamma}\right)$$

and the stretching effectively takes place along the diagonal with principal stretch λ. The shortening diagonal is associated with the principal stretch λ^{-1}.

Inextensibility Simple shear is an example of a deformation in which line elements aligned in a certain direction do not change their lengths (although they are capable of extension). However, if such line elements are constrained so that their lengths do not change on deformation then the material is said to be *inextensible* in the direction concerned.

Thus, if the unit vector **L** in the reference configuration defines a direction of inextensibility we must have

$$|\mathbf{AL}| = 1$$

or, equivalently,

$$\mathbf{L}\cdot(\mathbf{A}^{\mathrm{T}}\mathbf{AL}) = 1. \tag{2.2.61}$$

If **L** and **M** are two directions of inextensibility (not necessarily orthogonal) then

$$\mathbf{L}\cdot(\mathbf{A}^{\mathrm{T}}\mathbf{AL}) = \mathbf{M}\cdot(\mathbf{A}^{\mathrm{T}}\mathbf{AM}) = 1. \tag{2.2.62}$$

For a plane deformation taking place in the plane defined by **L** and **M**, for example, reference to the Lagrangean axes reduces (2.2.62) to

$$(\lambda_1^2 - \lambda_2^2)\cos^2\theta = 1 - \lambda_2^2 = (\lambda_1^2 - \lambda_2^2)\cos^2\phi,$$

where θ and ϕ are the orientations of the (fixed) directions **L** and **M** respectively relative to the Lagrangean axes in the plane of deformation. It follows that *either* $\phi = \theta$ or $\theta + \pi$, i.e. $\mathbf{M} = \pm\mathbf{L}$, *or* $\lambda_1 = \lambda_2 = 1$, i.e. there is no deformation, *or* $\theta + \phi = \pi$ or 2π.

If the single constraint (2.2.61) is combined with an isochoric plane deformation with $\mathbf{L}$ in the plane of deformation then $\lambda_1\lambda_2 = 1$ and the orientation θ of the Lagrangean axes relative to $\mathbf{L}$ is given by

$$\cos^2\theta = (\lambda_1^2 + 1)^{-1}.$$

It is easy to show that this deformation is necessarily a simple shear.

Orientation constraints In simple shear certain line elements retain their orientation throughout the deformation. This *orientation constraint* is generalized as follows. If a line element in the direction $\mathbf{L}$ in the reference configuration is constrained to maintain its direction during deformation of the material then

$$\mathbf{AL} = \lambda\mathbf{L}, \tag{2.2.63}$$

for some positive scalar λ. This allows the line elements to change their lengths. If there are two such directions $\mathbf{L}$ and $\mathbf{M}$ then

$$\mathbf{AL} = \lambda\mathbf{L}, \qquad \mathbf{AM} = \mu\mathbf{M} \tag{2.2.64}$$

jointly, where $\mu > 0$.

For a plane deformation in the plane defined by $\mathbf{L}$ and $\mathbf{M}$ it is easy to show that $\mathbf{A}$ has the representation

$$\begin{aligned}\mathbf{A} = \{1 - (\mathbf{L}\cdot\mathbf{M})^2\}^{-1}[&\lambda\mathbf{L}\otimes\mathbf{L} - \lambda(\mathbf{L}\cdot\mathbf{M})\mathbf{L}\otimes\mathbf{M} \\ &- \mu(\mathbf{L}\cdot\mathbf{M})\mathbf{M}\otimes\mathbf{L} + \mu\mathbf{M}\otimes\mathbf{M}] + \mathbf{N}\otimes\mathbf{N},\end{aligned}$$

where $\mathbf{N}$ is a unit vector normal to the plane of $\mathbf{L}$ and $\mathbf{M}$. If, additionally, the deformation is isochoric then it follows that $\lambda\mu = 1$ and further, if $\mathbf{L}\cdot\mathbf{M} = 0$, $\mathbf{A}$ reduces to

$$\mathbf{A} = \lambda\mathbf{L}\otimes\mathbf{L} + \mu\mathbf{M}\otimes\mathbf{M} + \mathbf{N}\otimes\mathbf{N},$$

i.e. the deformation is necessarily a pure shear.

Problem 2.2.8 A pure strain is defined by the deformation gradient

$$\mathbf{A} = \lambda_1\mathbf{e}_1\otimes\mathbf{e}_1 + \lambda_2\mathbf{e}_2\otimes\mathbf{e}_2 + \mathbf{e}_3\otimes\mathbf{e}_3,$$

where $\mathbf{e}_1, \mathbf{e}_2, \mathbf{e}_3$ are fixed mutually orthogonal unit vectors. If two line elements are aligned with $\mathbf{M} = \cos\phi\mathbf{e}_1 + \sin\phi\mathbf{e}_2$ and $\mathbf{M}' = -\sin\phi\mathbf{e}_1 + \cos\phi\mathbf{e}_2$ in the reference configuration calculate the change in the angle between them due to the deformation. Deduce that the maximum change among all such pairs of line elements is

$$\sin^{-1}\left\{\frac{\lambda_1^2 - \lambda_2^2}{\lambda_1^2 + \lambda_2^2}\right\}.$$

Problem 2.2.9 If $\mathbf{M}$ and $\mathbf{M}'$ are as defined in Problem 2.2.8 show that for the simple shear deformation

$$\mathbf{A} = \mathbf{I} + \gamma\mathbf{e}_1 \otimes \mathbf{e}_2$$

the largest change in angle between line elements aligned with $\mathbf{M}$ and $\mathbf{M}'$ in the reference configuration is

$$\tan^{-1}\{\gamma(1 + \tfrac{1}{4}\gamma^2)\}.$$

Show that this value corresponds to $\tan 2\phi = \frac{1}{2}\gamma$ and deduce that $\phi = \frac{1}{4}\pi - \theta_{\mathrm{E}}$, where θ_{E} is given by (2.2.59).

Problem 2.2.10 Let the simple shear deformation (2.2.55) have matrix representation

$$\mathsf{A} = \begin{bmatrix} 1 & \gamma & 0 \\ 0 & 1 & 0 \\ 0 & 0 & 1 \end{bmatrix}.$$

Obtain the corresponding matrix polar decomposition $\mathsf{A} = \mathsf{RU}$ by writing

$$\mathsf{U} = \begin{bmatrix} \cos\theta & -\sin\theta & 0 \\ \sin\theta & \cos\theta & 0 \\ 0 & 0 & 1 \end{bmatrix} \begin{bmatrix} \lambda & 0 & 0 \\ 0 & \lambda^{-1} & 0 \\ 0 & 0 & 1 \end{bmatrix} \begin{bmatrix} \cos\theta & \sin\theta & 0 \\ -\sin\theta & \cos\theta & 0 \\ 0 & 0 & 1 \end{bmatrix},$$

using the connection $\mathsf{A}^{\mathrm{T}}\mathsf{A} = \mathsf{U}^2$ to show that $\tan 2\theta = -2/\gamma$, and deducing that

$$\mathsf{R} = (4 + \gamma^2)^{-1/2} \begin{bmatrix} 2 & \gamma & 0 \\ -\gamma & 2 & 0 \\ 0 & 0 & 1 \end{bmatrix}.$$

Problem 2.2.11 If the deformation gradient $\mathbf{A} = \mathbf{R}$ represents a rotation through an angle θ about the direction of the unit vector $\mathbf{n}$, show that the displacement gradient $\mathbf{D} = \mathbf{A} - \mathbf{I}$ may be expressed

$$\mathbf{D} = -\mathbf{W}\sin\theta + \mathbf{W}^2(1 - \cos\theta),$$

where $\mathbf{W}$ is the antisymmetric tensor whose axial vector is $\mathbf{n}$ (see Section 1.3.4).

Deduce that

$$1 - \cos\theta = -\tfrac{1}{2}\operatorname{tr}(\mathbf{D}), \qquad \mathbf{W}\sin\theta = -\tfrac{1}{2}(\mathbf{D} - \mathbf{D}^{\mathrm{T}}).$$

Problem 2.2.12 From Nanson's formula (2.2.18) deduce that

$$(\mathrm{d}s/\mathrm{d}S)^2 = J^2\mathbf{N}\cdot(\mathbf{B}^{\mathrm{T}}\mathbf{B}\mathbf{N}).$$

For the simple shear deformation

$$\mathbf{A} = \mathbf{I} + \gamma\mathbf{e}_1 \otimes \mathbf{e}_2$$

defined in (2.2.55) show that the inverse deformation is given by

$$\mathbf{B} = \mathbf{I} - \gamma\mathbf{e}_2 \otimes \mathbf{e}_1$$

(see (2.2.7)). Hence find $\mathbf{B}^{\mathrm{T}}\mathbf{B}$, and show that the maximum value of $(\mathrm{d}s/\mathrm{d}S)^2$ is

$$1 + \tfrac{1}{2}\gamma^2 + \tfrac{1}{2}\gamma(4 + \gamma^2)^{1/2}$$

and corresponds to $N_3 = 0$ and $N_1^2 - N_2^2 = -2\gamma N_1 N_2$. What is the minimum value of $(\mathrm{d}s/\mathrm{d}S)^2$?

Problem 2.2.13 A body contains two families of inextensible line elements (or reinforcing fibres) aligned along the directions

$$\mathbf{M} = \cos\phi\,\mathbf{e}_1 + \sin\phi\,\mathbf{e}_2, \qquad \mathbf{M}' = \cos\phi\,\mathbf{e}_1 - \sin\phi\,\mathbf{e}_2$$

$(0 \leq \phi \leq \pi/2)$ in the reference configuration, where the unit vectors $\mathbf{e}_1, \mathbf{e}_2, \mathbf{e}_3$ form a fixed orthonormal basis.

If the body is subjected to the pure strain

$$\mathbf{A} = \lambda^{-1/2}\mu\mathbf{e}_1 \otimes \mathbf{e}_1 + \lambda^{-1/2}\mu^{-1}\mathbf{e}_2 \otimes \mathbf{e}_2 + \lambda\mathbf{e}_3 \otimes \mathbf{e}_3$$

show that the inextensibility constraints lead to

$$\mu^2 = \{\lambda \pm (\lambda^2 - \sin^2 2\phi)^{1/2}\}/2\cos^2\phi.$$

Prove that (a) if $\phi = \frac{1}{4}\pi$ then no contraction along $\mathbf{e}_3$ is possible, (b) if $\lambda > \sin 2\phi$ then two deformed configurations are possible provided $\lambda \neq 1$, (c) the maximum contraction in the $\mathbf{e}_3$ direction is achieved when $\mu^2 = \tan\phi$ which corresponds to the two families of fibres being mutually orthogonal in the *current* configuration.

Problem 2.2.14 A body contains two families of fibres whose orientations $\mathbf{L}$ and $\mathbf{M}$ are preserved during deformation, i.e. there are positive scalars λ and μ such that

$$\mathbf{AL} = \lambda\mathbf{L}, \qquad \mathbf{AM} = \mu\mathbf{M}.$$

In the reference configuration a line element is parallel to $\mathbf{L} + \alpha\mathbf{M}$. Show that its length increases in the ratio

$$[\lambda^2 + 2\lambda\mu\alpha(\mathbf{L}\cdot\mathbf{M}) + \mu^2\alpha^2]^{1/2}[1 + 2\alpha(\mathbf{L}\cdot\mathbf{M}) + \alpha^2]^{-1/2}$$

and that this value is stationary provided α satisfies

$$\lambda(\mathbf{L}\cdot\mathbf{M}) + (\lambda + \mu)\alpha + \mu\alpha^2(\mathbf{L}\cdot\mathbf{M}) = 0$$

$(\lambda \neq \mu)$.

If the two solutions of this equation are denoted by α_1 and α_2 show that $\mathbf{L} + \alpha_1\mathbf{M}$ and $\mathbf{L} + \alpha_2\mathbf{M}$ respectively have orientations $\alpha_2\mathbf{L} + \mathbf{M}$ and $\alpha_1\mathbf{L} + \mathbf{M}$ in the current configuration. Deduce that the angles of rotation of line elements along $\mathbf{L} + \alpha_1\mathbf{M}$ and $\mathbf{L} + \alpha_2\mathbf{M}$ in the reference configuration are equal.

Problem 2.2.15 If

$$\mathbf{U} = \lambda_1\mathbf{u}^{(1)} \otimes \mathbf{u}^{(1)} + \lambda_1^{-1}\mathbf{u}^{(2)} \otimes \mathbf{u}^{(2)} + \mathbf{u}^{(3)} \otimes \mathbf{u}^{(3)},$$

where the unit vectors $\mathbf{u}^{(i)}$ $(i = 1, 2, 3)$ define the Lagrangean principal axes, show that $\mathbf{U}$ is expressible in the form $\mathbf{R}^{\mathrm{T}}\mathbf{A}$, where $\mathbf{R}$ is a rotation and $\mathbf{A}$ is a simple shear in the plane of $\mathbf{u}^{(1)}$ and $\mathbf{u}^{(2)}$.

Combinations and decompositions of deformations Suppose that a body B is deformed from a reference configuration $\mathscr{B}_0$ to the configuration $\mathscr{B}$ and then to a new configuration $\mathscr{B}'$. Let a typical material point have places $\mathbf{X}, \mathbf{x}$ and $\mathbf{x}'$ in $\mathscr{B}_0$, $\mathscr{B}$ and $\mathscr{B}'$ respectively so that

$$\mathbf{x} = \boldsymbol{\chi}(\mathbf{X}), \qquad \mathbf{x}' = \boldsymbol{\chi}'(\mathbf{X}),$$

where $\boldsymbol{\chi}$ and $\boldsymbol{\chi}'$ represent the deformations from $\mathscr{B}_0$ to $\mathscr{B}$ and $\mathscr{B}'$ respectively.

We may also regard $\mathscr{B}$ as a reference configuration and define the deformation $\boldsymbol{\kappa}: \mathscr{B} \rightarrow \mathscr{B}'$ by

$$\mathbf{x}' = \boldsymbol{\kappa}(\mathbf{x})$$

so that

$$\boldsymbol{\chi}'(\mathbf{X}) = \boldsymbol{\kappa}\{\boldsymbol{\chi}(\mathbf{X})\}$$

defines the composite deformation from $\mathscr{B}_0$ to $\mathscr{B}'$.

The deformation gradients

$$\mathbf{A} = \operatorname{Grad} \boldsymbol{\chi}(\mathbf{X}), \qquad \mathbf{A}' = \operatorname{Grad} \boldsymbol{\chi}'(\mathbf{X})$$

of $\mathscr{B}$ and $\mathscr{B}'$ relative to $\mathscr{B}_0$ are related by

$$\mathbf{A}' = (\operatorname{grad} \boldsymbol{\kappa}(\mathbf{x}))\mathbf{A} = \bar{\mathbf{A}}\mathbf{A}, \tag{2.2.65}$$

where $\bar{\mathbf{A}} = \operatorname{grad} \boldsymbol{\kappa}(\mathbf{x})$ defines the deformation gradient from $\mathscr{B}$ to $\mathscr{B}'$. Thus, the deformation gradients of successive deformations are *multiplied* together in accordance with (2.2.65).

For homogeneous deformations, we have $\mathbf{x} = \mathbf{A}\mathbf{X} + \mathbf{c}$ and

$$\mathbf{x}' = \bar{\mathbf{A}}\mathbf{x} + \bar{\mathbf{c}} = \bar{\mathbf{A}}\mathbf{A}\mathbf{X} + \bar{\mathbf{c}} + \bar{\mathbf{A}}\mathbf{c} \equiv \mathbf{A}'\mathbf{X} + \mathbf{c}'.$$

It should be emphasized that the order in which successive deformations are applied is crucial since deformations do not in general commute, i.e. $\bar{\mathbf{A}}\mathbf{A} \neq \mathbf{A}\bar{\mathbf{A}}$ in general. This is illustrated in the case of two pure strains $\mathbf{A} = \mathbf{U}$ and $\bar{\mathbf{A}} = \bar{\mathbf{U}}$, when $\mathbf{U}$ and $\bar{\mathbf{U}}$ are right stretch tensors. From the result of Problem 1.3.8 it follows that $\mathbf{U}\bar{\mathbf{U}} = \bar{\mathbf{U}}\mathbf{U}$ if and only if $\mathbf{U}$ and $\bar{\mathbf{U}}$ are coaxial, i.e. the Lagrangean principal axes of the two deformations coincide.

A more instructive example is afforded by the combination of two simple shears. Let

$$\mathbf{A} = \mathbf{I} + \gamma \mathbf{m} \otimes \mathbf{n}, \qquad \mathbf{m} \cdot \mathbf{n} = 0 \tag{2.2.66}$$

be a simple shear, where $\mathbf{m}$ and $\mathbf{n}$ are unit vectors (replacing $\mathbf{e}_1$ and $\mathbf{e}_2$ in (2.2.55)), $\mathbf{m}$ is the *direction of shear* and $\mathbf{m}$ and $\mathbf{n}$ define the *plane of shear*. The planes normal to $\mathbf{n}$ are called *glide planes*.

If

$$\bar{\mathbf{A}} = \mathbf{I} + \bar{\gamma} \bar{\mathbf{m}} \otimes \bar{\mathbf{n}}, \qquad \bar{\mathbf{m}} \cdot \bar{\mathbf{n}} = 0$$

is a second simple shear then, provided γ and $\bar{\gamma}$ are non-zero, it is easily shown that $\mathbf{A}\bar{\mathbf{A}} = \bar{\mathbf{A}}\mathbf{A}$ if and only if

$$(\mathbf{n}\cdot\bar{\mathbf{m}})\mathbf{m} \otimes \bar{\mathbf{n}} = (\bar{\mathbf{n}}\cdot\mathbf{m})\bar{\mathbf{m}} \otimes \mathbf{n}.$$

Pre- and post-multiplication of this by $\mathbf{m}, \bar{\mathbf{m}}, \mathbf{n}, \bar{\mathbf{n}}$ in turn leads to the conclusions

$$\mathbf{n}\cdot\bar{\mathbf{m}} = 0 = \bar{\mathbf{n}}\cdot\mathbf{m}$$

and hence *either* (a) $\bar{\mathbf{m}} = \pm\mathbf{m}$ *or* (b) $\bar{\mathbf{n}} = \pm\mathbf{n}$.

In case (a) the composite deformation represents a simple shear of amount $(\gamma^2 + \bar{\gamma}^2 \pm 2\gamma\bar{\gamma}\mathbf{n}\cdot\bar{\mathbf{n}})^{1/2}$ in the direction $\mathbf{m}$ with glide planes normal to $\gamma\mathbf{n} \pm \bar{\gamma}\bar{\mathbf{n}}$ while in the case (b) a simple shear of amount $(\gamma^2 + \bar{\gamma}^2 \pm 2\bar{\gamma}\mathbf{m}\cdot\bar{\mathbf{m}})^{1/2}$ in the direction $\gamma\mathbf{m} \pm \bar{\gamma}\bar{\mathbf{m}}$ with glide planes normal to $\mathbf{n}$ results from the combination of $\mathbf{A}$ and $\bar{\mathbf{A}}$. In essence we have established the following:

Theorem 2.2.2 Two simple shears combine to give a third simple shear if and only if each direction of shear is parallel to both families of glide planes.

From the polar decomposition $\mathbf{A} = \mathbf{R}\mathbf{U}$ we see that a general deformation comprises a pure strain $\mathbf{U}$ followed by a rigid rotation $\mathbf{R}$, but we may make further decompositions as follows. Since $\det \mathbf{A} = J$, we define

$$\mathbf{A}^* = J^{-1/3}\mathbf{A} \tag{2.2.67}$$

so that

$$\det \mathbf{A}^* = 1. \tag{2.2.68}$$

Thus $\mathbf{A} = J^{1/3}\mathbf{A}^*$ is composed of a *pure dilatation* $J^{1/3}\mathbf{I}$ and an *isochoric deformation* $\mathbf{A}^*$. We now restrict attention to isochoric deformations and write

$$\mathbf{A}^* = \mathbf{R}^*\mathbf{U}^*,$$

where $\mathbf{R}^*$ is proper orthogonal and $\mathbf{U}^*$ is positive definite and symmetric and $\det \mathbf{U}^* = 1$. The principal values of $\mathbf{U}^*$ are given by

$$\lambda_i^* = J^{-1/3}\lambda_i, \tag{2.2.69}$$

where $\lambda_i\,(i = 1, 2, 3)$ are the actual principal stretches, and hence

$$\lambda_1^*\lambda_2^*\lambda_3^* = 1. \tag{2.2.70}$$

If the unit vectors $\mathbf{u}^{(i)}\,(i = 1, 2, 3)$ correspond to the Lagrangean principal axes then we may write

$$\mathbf{U}^* = \lambda_1^* \mathbf{u}^{(1)} \otimes \mathbf{u}^{(1)} + \lambda_2^* \mathbf{u}^{(2)} \otimes \mathbf{u}^{(2)} + \lambda_3^* \mathbf{u}^{(3)} \otimes \mathbf{u}^{(3)},$$

and this may be decomposed as

$$\mathbf{U}^* = [\bar{\lambda}_1^* \mathbf{u}^{(1)} \otimes \mathbf{u}^{(1)} + \bar{\lambda}_2^* \mathbf{u}^{(2)} \otimes \mathbf{u}^{(2)} + \mathbf{u}^{(3)} \otimes \mathbf{u}^{(3)}] \times$$
$$\times [\lambda_3^{*-1/2}(\mathbf{u}^{(1)} \otimes \mathbf{u}^{(1)} + \mathbf{u}^{(2)} \otimes \mathbf{u}^{(2)}) + \lambda_3^* \mathbf{u}^{(3)} \otimes \mathbf{u}^{(3)}],$$

where $\bar{\lambda}_1^* = \lambda_1^* \lambda_3^{*1/2}$, $\bar{\lambda}_2^* = \lambda_2^* \lambda_3^{*1/2}$ and hence $\bar{\lambda}_1^* \bar{\lambda}_2^* = 1$.

But from the result of Problem 2.2.15 the first element in the above product may be written as $\mathbf{Q}^{\mathrm{T}}\mathbf{S}$, where $\mathbf{Q}$ is a rotation and $\mathbf{S}$ is a simple shear in the plane of $\mathbf{u}^{(1)}$ and $\mathbf{u}^{(2)}$, while the second element of the product represents a simple extension in the $\mathbf{u}^{(3)}$ direction with associated lateral contraction. Thus

$$\mathbf{A}^* = \mathbf{R}^*\mathbf{Q}^{\mathrm{T}}\mathbf{S}\mathbf{E} = \mathbf{R}^*\mathbf{Q}^{\mathrm{T}}\mathbf{E}\mathbf{S},$$

where

$$\mathbf{E} = \lambda_3^{*-1/2}(\mathbf{u}^{(1)} \otimes \mathbf{u}^{(1)} + \mathbf{u}^{(2)} \otimes \mathbf{u}^{(2)}) + \lambda_3^* \mathbf{u}^{(3)} \otimes \mathbf{u}^{(3)}$$

is a simple extension with accompanying lateral contraction and $\mathbf{R}^*\mathbf{Q}^{\mathrm{T}}$ is a rotation. These simple constituent deformations have been discussed individually earlier in this section. We have proved the following:

Theorem 2.2.3 An arbitrary (homogeneous) isochoric deformation is composed of a *simple extension* (with lateral contraction), a *simple shear* in the plane *normal* to the direction of extension, and a *rotation*.

So far we have concentrated on homogeneous deformations but for the remainder of this section we turn attention to a discussion of certain inhomogeneous isochoric deformations.

Combined extension and inflation of a circular tube Suppose that in some reference configuration a circular cylindrical tube has internal and external radii A and B respectively and length L. Then, in terms of cylindrical polar coordinates (R, Θ, Z), the tube is defined by

$$A \le R \le B, \quad 0 \le \Theta \le 2\pi, \quad 0 \le Z \le L.$$

Let the tube be extended uniformly to length $l = \lambda_z L$ and its internal radius changed to a so that the circular symmetry is maintained. The current configuration of the body satisfies

$$a \leq r \leq b, \quad 0 \leq \theta \leq 2\pi, \quad 0 \leq z \leq l,$$

where (r, θ, z) are cylindrical polar coordinates and the resulting (isochoric) deformation is given by

$$r^2 - a^2 = \lambda_z^{-1}(R^2 - A^2), \quad \theta = \Theta, \quad z = \lambda_z Z \tag{2.2.71}$$

(the material volume $\pi(R^2 - A^2)L$ contained between cylinders of radii A and R is unchanged by the deformation and therefore equal to $\pi(r^2 - a^2)l$).

Because of the symmetry the Lagrangean (and Eulerian) principal axes coincide locally with the cylindrical polar basis vectors $\mathbf{e}_r$, $\mathbf{e}_\theta$, $\mathbf{e}_z$. The corresponding principal stretches are respectively

$$\lambda_1 = (\lambda\lambda_z)^{-1}, \quad \lambda_2 = \frac{r}{R} \equiv \lambda, \quad \lambda_3 = \lambda_z, \tag{2.2.72}$$

where λ is defined as the azimuthal principal stretch. On setting $\lambda_a = a/A$ and $\lambda_b = b/B$, we obtain from (2.2.71) the connections

$$\lambda_a^2\lambda_z - 1 = \frac{R^2}{A^2}(\lambda^2\lambda_z - 1) = \frac{B^2}{A^2}(\lambda_b^2\lambda_z - 1) \tag{2.2.73}$$

and it follows that

$$\lambda_a \geq \lambda \geq \lambda_b, \tag{2.2.74}$$

with equality holding if and only if $\lambda_a = \lambda_b = 1$ $(a \neq b)$, provided $\lambda_a^2\lambda_z \geq 1$.

Combined extension and torsion of a solid cylinder We consider a circular cylindrical body which has radius A and length L in some reference configuration. In terms of cylindrical polar coordinates (R, Θ, Z) the body is defined by

$$0 \leq R \leq A, \quad 0 \leq \Theta \leq 2\pi, \quad 0 \leq Z \leq L.$$

Let the cylinder be extended uniformly to length $l = \lambda L$ so that its radius becomes $a = \lambda^{-1/2}A$ and in terms of cylindrical polar coordinates (r, θ, z) the current configuration of the body is such that

$$0 \leq r \leq a, \quad 0 \leq \theta \leq 2\pi, \quad 0 \leq z \leq l.$$

The deformation is defined by

$$r=\lambda^{-1/2}R,\quad \theta=\Theta,\quad z=\lambda Z.$$

The plane face $z=l$ is then rotated, while that at $z=0$ is fixed, in such a way that plane sections of the cylinder normal to its axis remain plane and the radius r turns through an angle τz, where $\tau(>0)$ represents the twist (or torque) per unit length of the deformed cylinder. The combined deformation is given by

$$r=\lambda^{-1/2}R,\quad \theta=\Theta+\tau\lambda Z,\quad z=\lambda Z. \tag{2.2.75}$$

By writing

$$\mathbf{X}=R\mathbf{E}_R+Z\mathbf{E}_Z,\quad \mathbf{x}\equiv\boldsymbol{\chi}(\mathbf{X})=\lambda^{-1/2}R\mathbf{e}_r+\lambda Z\mathbf{e}_z$$

for the position vectors of a material point in the reference and current configurations, and

$$\mathrm{Grad}=\mathbf{E}_R\frac{\partial}{\partial R}+\mathbf{E}_\Theta\frac{1}{R}\frac{\partial}{\partial\Theta}+\mathbf{E}_Z\frac{\partial}{\partial Z}$$

(see Section 1.5.4), where $(\mathbf{E}_R,\mathbf{E}_\Theta,\mathbf{E}_Z)$ and $(\mathbf{e}_r,\mathbf{e}_\theta,\mathbf{e}_z)$ are the reference and current cylindrical polar unit basis vectors, we obtain

$$\begin{aligned}\mathbf{A}\equiv\mathrm{Grad}\,\mathbf{x}&=(\mathbf{I}+\tau r\mathbf{e}_\theta\otimes\mathbf{e}_z)\times\\&\times(\lambda^{-1/2}\mathbf{e}_r\otimes\mathbf{e}_r+\lambda^{-1/2}\mathbf{e}_\theta\otimes\mathbf{e}_\theta+\lambda\mathbf{e}_z\otimes\mathbf{e}_z)\times\\&\times(\mathbf{e}_r\otimes\mathbf{E}_R+\mathbf{e}_\theta\otimes\mathbf{E}_\Theta+\mathbf{e}_z\otimes\mathbf{E}_z).\end{aligned} \tag{2.2.76}$$

Note that $\mathbf{e}_z=\mathbf{E}_Z$. Thus the deformation is composed locally of a rotation $\mathbf{e}_r\otimes\mathbf{E}_R+\mathbf{e}_\theta\otimes\mathbf{E}_\Theta+\mathbf{e}_z\otimes\mathbf{E}_Z$, which takes the reference basis vectors $(\mathbf{E}_R,\mathbf{E}_\Theta,\mathbf{E}_Z)$ to $(\mathbf{e}_r,\mathbf{e}_\theta,\mathbf{e}_z)$, a simple extenstion $\lambda^{-1/2}\mathbf{e}_r\otimes\mathbf{e}_r+\lambda^{-1/2}\mathbf{e}_\theta\otimes\mathbf{e}_\theta+\lambda\mathbf{e}_z\otimes\mathbf{e}_z$ and a simple shear $\mathbf{I}+\tau r\mathbf{e}_\theta\otimes\mathbf{e}_z$ of amount τr in planes normal to $\mathbf{e}_r$.

The left Cauchy–Green deformation tensor is

$$\begin{aligned}\mathbf{A}\mathbf{A}^{\mathrm{T}}=&\lambda^{-1}\mathbf{e}_r\otimes\mathbf{e}_r+(\lambda^{-1}+\lambda^2\tau^2r^2)\mathbf{e}_\theta\otimes\mathbf{e}_\theta\\&+\lambda^2\mathbf{e}_z\otimes\mathbf{e}_z+\lambda^2\tau r(\mathbf{e}_z\otimes\mathbf{e}_\theta+\mathbf{e}_\theta\otimes\mathbf{e}_z).\end{aligned}$$

One of the Eulerian principal axes is along $\mathbf{e}_r$ and corresponds to the principal stretch $\lambda_1=\lambda^{-1/2}$. A slight modification of the calculation which gives the orientation (2.2.59) of the principal axes in a simple shear to take account of the pre-stretch λ shows that the Eulerian axes are given by

$$\cos\psi\mathbf{e}_\theta + \sin\psi\mathbf{e}_z, \quad -\sin\psi\mathbf{e}_\theta + \cos\psi\mathbf{e}_z,$$

where

$$\tan 2\psi = 2\lambda^3\tau r/(1 - \lambda^3 + \lambda^3\tau^2 r^2) \tag{2.2.77}$$

$(0 < \psi \le \pi/4)$. The corresponding principal stretches λ_2 and λ_3 satisfy

$$\lambda_2^2\lambda_3^2 = \lambda, \quad \lambda_2^2 + \lambda_3^2 = \lambda^{-1} + \lambda^2 + \lambda^2\tau^2 r^2. \tag{2.2.78}$$

For the case of *simple torsion*, which corresponds to $\lambda = 1$, we obtain $\tan 2\psi = 2/\tau r$ and the deformation is just a simple shear locally.

Bending of a rectangular block Consider a rectangular block defined by

$$-A \le X_1 \le A, \qquad -B \le X_2 \le B, \qquad -C \le X_3 \le C$$

in some reference configuration, where (X_1, X_2, X_3) are rectangular Cartesian coordinates. We suppose that the block is deformed into a sector of a circular cylindrical tube so that the planes $X_1 = \text{constant}$ become sectors of the cylindrical surface $r = \text{constant}$, the planes $X_2 = \text{constant}$ become planes $\theta = \text{constant}$ and $X_3 = \text{constant}$ becomes $z = \text{constant}$, where (r, θ, z) are cylindrical polar coordinates as indicated in Fig. 2.7.

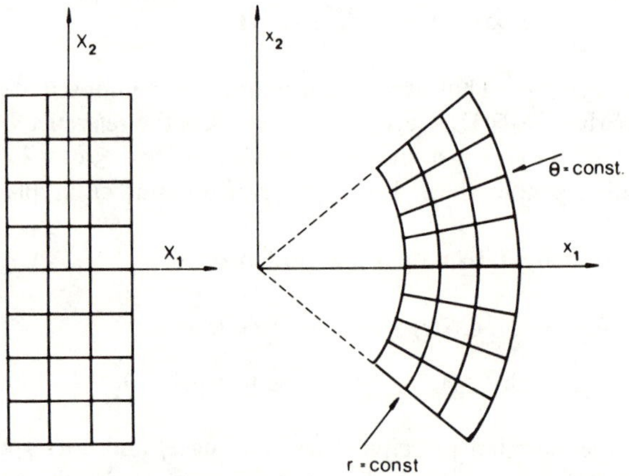

Fig. 2.7 Bending of a rectangular block into a sector of a circular cylindrical tube.

The deformation is described by

$$r = f(X_1), \qquad \theta = g(X_2), \qquad z = \lambda X_3 \tag{2.2.79}$$

in general, where f and g are functions to be determined. With respect to a reference rectangular Cartesian basis $\{\mathbf{E}_\alpha\}$ and a current cylindrical polar bases $\{\mathbf{e}_r, \mathbf{e}_\theta, \mathbf{e}_z\}$ the deformation gradient has the form

$$\begin{aligned}&[f'(X_1)\mathbf{e}_r \otimes \mathbf{e}_r + f(X_1)g'(X_2)\mathbf{e}_\theta \otimes \mathbf{e}_\theta + \lambda \mathbf{e}_z \otimes \mathbf{e}_z] \times \\ &\quad \times [\mathbf{e}_r \otimes \mathbf{E}_1 + \mathbf{e}_\theta \otimes \mathbf{E}_2 + \mathbf{e}_z \otimes \mathbf{E}_3],\end{aligned}$$

where $\mathbf{e}_z = \mathbf{E}_3$ and the prime denotes differentiation.

If the deformation is isochoric it follows that $\lambda f f' g' = 1$ and separation of variables then yields

$$r^2 = \frac{2X_1}{\alpha} + \beta, \qquad \theta = \frac{\alpha X_2}{\lambda}, \tag{2.2.80}$$

where α and β are constants which may be determined in terms of A, B and C when $f(A)$ and $g(B)$ are specified, symmetry about $\theta = 0$ having been assumed.

The explicit forms of the above inhomogeneous deformations arise because it is assumed that the deformations are isochoric. Other such examples are listed in Truesdell and Noll (1965) and Spencer (1970), for example, and we do not provide details here. If a material body is such that all deformations are *constrained* to be isochoric at each point of the body then the material is said to be *incompressible* and the constraint is referred to as *incompressibility*. The assumption of incompressibility is not always sufficient on its own to give the deformation an explicit form, and in general one or more arbitrary functions are needed for the specification of the deformation. Such functions are (in principle) obtainable ultimately from the solution of differential equations (generally non-linear) which derive from the governing equations of motion or equilibrium when the material properties have been specified (see Chapters 3 and 4). The following example, which involves two arbitrary functions, illustrates the point.

Combined axial and azimuthal shear of a circular tube Axial (or telescopic) shear of a circular cylindrical tube is defined by

$$r = R, \qquad \theta = \Theta, \qquad z = Z + w(R)$$

and azimuthal (or torsional) shear by

$$r = R, \qquad \theta = \Theta + \phi(R), \qquad z = Z$$

in terms of reference and current cylindrical polar coordinates. The combined (isochoric) deformation

$$r = R, \qquad \theta = \Theta + \phi(R), \qquad z = Z + w(R) \tag{2.2.81}$$

has deformation gradient

$$\text{Grad}\left[R\mathbf{e}_r + \{Z + w(R)\}\mathbf{e}_z\right]$$

which may be decomposed as

$$[\mathbf{I} + w'(R)\mathbf{e}_z \otimes \mathbf{e}_r][\mathbf{I} + R\phi'(R)\mathbf{e}_\theta \otimes \mathbf{e}_r] \times$$
$$\times [\mathbf{e}_r \otimes \mathbf{E}_R + \mathbf{e}_\theta \otimes \mathbf{E}_\Theta + \mathbf{e}_z \otimes \mathbf{E}_Z].$$

The axial and azimuthal shears correspond to local simple shears of amounts $w'(R)$ and $R\phi'(R)$ in planes normal to $\mathbf{e}_\theta$ and $\mathbf{e}_z$ respectively. The resultant deformation consists of a rotation, which takes $(\mathbf{E}_R, \mathbf{E}_\Theta, \mathbf{E}_Z)$ to $(\mathbf{e}_r, \mathbf{e}_\theta, \mathbf{e}_z)$, and, by Theorem 2.2.2, a local simple shear of amount $v = [w'(R)^2 + R^2\phi'(R)^2]^{1/2}$ in the direction $\sin\chi\mathbf{e}_\theta + \cos\chi\mathbf{e}_z$ with $\mathbf{e}_r$ normal to the glide planes, where $\tan\chi = R\phi'(R)/w'(R)$, $0 \le \chi \le \frac{1}{2}\pi$. The Eulerian principal axes are

$$\cos\psi\mathbf{e}_r - \sin\psi(\sin\chi\mathbf{e}_\theta + \cos\chi\mathbf{e}_z),$$
$$\cos\chi\mathbf{e}_\theta - \sin\chi\mathbf{e}_z,$$
$$\sin\psi\mathbf{e}_r + \cos\psi(\sin\chi\mathbf{e}_\theta + \cos\chi\mathbf{e}_z),$$

where $\tan 2\psi = 2/v$, and the corresponding principal stretches are respectively λ^{-1}, 1, $\lambda(\ge 1)$, where $\lambda - \lambda^{-1} = v$.

For inhomogeneous deformations which are not isochoric it is not possible to find explicit forms from purely kinematical considerations unless other constraints are imposed. For example, the spherically symmetric deformation of a thick-walled spherical shell is described by $\mathbf{x} = f(R)\mathbf{X}$, where $R = |\mathbf{X}|$ and $\mathbf{X}$ and $\mathbf{x}$ are the position vectors of a material point in the reference and current configurations relative to the centre of the sphere. If the material of the shell is inextensible in the radial direction it follows that $f(R) = 1 + (a - A)/R$, where A and a are the reference and current internal radii. The change in total volume of the shell is $4\pi(B - A)(ab - AB)$.

Problem 2.2.16 Show that the gradient of the spherically symmetric deformation $\mathbf{x} = f(R)\mathbf{X}$, where $R = |\mathbf{X}|$, is

$$f(R)\mathbf{I} + \frac{1}{R}f'(R)\mathbf{X} \otimes \mathbf{X}$$

and find $f(R)$ if the deformation is isochoric.

A spherical shell is defined by

$$A \le R \le B, \qquad 0 \le \Theta \le \pi, \qquad 0 \le \Phi \le 2\pi$$

in some reference configuration, where (R, Θ, Φ) are spherical polar coordinates. If the material of the shell is incompressible and the shell deformed so that spherical symmetry is maintained, show that

$$r^3 = R^3 + a^3 - A^3, \qquad \theta = \Theta, \qquad \phi = \Phi,$$

where the current configuration of the shell is defined in terms of spherical polar coordinates (r, θ, ϕ) such that $a \le r \le b$.

Deduce that the principal stretches are λ^{-2}, λ, λ, where $\lambda = r/R$, and show that if $\lambda_a = a/A$, $\lambda_b = b/B$ then

$$\left(\frac{B}{A}\right)^3 (\lambda_b^3 - 1) = \lambda_a^3 - 1$$

and hence either $\lambda_a \ge \lambda_b \ge 1$ or $\lambda_a \le \lambda_b \le 1$.

Problem 2.2.17 Obtain the form of the isochoric deformation which takes a sector of a circular cylindrical tube into a rectangular block. Compare the result with the inverse of (2.2.80).

Problem 2.2.18 The combination of the axial shear and torsion of a circular cylindrical tube is defined by

$$r = R, \qquad \theta = \Theta + \tau[Z + w(R)], \qquad z = Z + w(R).$$

Show that the deformation gradient may be written

$$[\mathbf{I} + w'(R)\mathbf{e}_z \otimes \mathbf{e}_r + \tau r \mathbf{e}_\theta \otimes \mathbf{e}_z + \tau r w'(R)\mathbf{e}_\theta \otimes \mathbf{e}_r] \times$$
$$\times [\mathbf{e}_r \otimes \mathbf{E}_R + \mathbf{e}_\theta \otimes \mathbf{E}_\Theta + \mathbf{e}_z \otimes \mathbf{E}_Z]$$

and deduce that it does *not* represent a simple shear locally.

Problem 2.2.19 By rearranging the deformation (2.2.76) in accordance with Theorem 2.2.3 show that it corresponds to an extension $\lambda^{-1/2}\mathbf{e}_r \otimes \mathbf{e}_r + \lambda^{1/4}(\mathbf{e}_\theta \otimes \mathbf{e}_\theta + \mathbf{e}_z \otimes \mathbf{e}_z)$, a simple shear of amount $[\lambda^{3/2}\tau^2 r^2 + (\lambda^{3/4} - \lambda^{-3/4})^2]^{1/2}$ in planes normal to $\mathbf{e}_r$ locally, a rotation about $\mathbf{e}_r$ through an angle ϕ, where $(\lambda^{3/4} + \lambda^{-3/4})\cos\phi - \lambda^{3/4}\tau r \sin\phi = 2$, and a rotation $\mathbf{e}_r \otimes \mathbf{E}_R + \mathbf{e}_\theta \otimes \mathbf{E}_\Theta + \mathbf{e}_z \otimes \mathbf{E}_Z$.

Problem 2.2.20 Find the deformation gradient, the principal stretches and the orientations of the Eulerian principal axes for the deformation

$$r^2 = \lambda_z^{-1}(R^2 - A^2) + a^2, \qquad \theta = \Theta + \psi(R), \qquad z = \lambda_z Z + w(R)$$

defined in terms of cylindrical polar coordinates.

2.2.7 Strain tensors In Section 2.2.3 we defined the Lagrangean and Eulerian strain tensors

$$\tfrac{1}{2}(\mathbf{A}^{\mathrm{T}}\mathbf{A} - \mathbf{I}) \equiv \tfrac{1}{2}(\mathbf{U}^2 - \mathbf{I}), \qquad \tfrac{1}{2}(\mathbf{I} - \mathbf{B}\mathbf{B}^{\mathrm{T}}) \equiv \tfrac{1}{2}(\mathbf{I} - \mathbf{V}^{-2}).$$

The fact that the strain vanishes if and only if $\mathbf{U} = \mathbf{V} = \mathbf{I}$ enables many other tensor measures of strain based on $\mathbf{U}$ or $\mathbf{V}$ to be defined. For example, the second-order tensors, respectively Lagrangean and Eulerian, defined as

$$\left.\begin{array}{lll} \dfrac{1}{m}(\mathbf{U}^m - \mathbf{I}), & \dfrac{1}{m}(\mathbf{V}^m - \mathbf{I}) & m \neq 0, \\ \ln \mathbf{U}, & \ln \mathbf{V} & m = 0, \end{array}\right\} \tag{2.2.82}$$

where m is an integer (positive or negative), are suitable measures of strain.

They are coaxial with $\mathbf{U}$, $\mathbf{V}$ respectively and have principal values

$$\left.\begin{array}{ll} \dfrac{1}{m}(\lambda_i^m - 1) & m \neq 0, \\ \ln \lambda_i & m = 0, \end{array}\right\} \tag{2.2.83}$$

($i = 1, 2, 3$), where λ_i are the principal stretches.

In fact, the validity of (2.2.82) may be extended to non-integral real values of m, as suggested by Doyle and Ericksen (1956), if $\dfrac{1}{m}(\mathbf{U}^m - \mathbf{I})$ and $\dfrac{1}{m}(\mathbf{V}^m - \mathbf{I})$ respectively are defined to be the tensors coaxial with $\mathbf{U}$ and $\mathbf{V}$ having principal values $\dfrac{1}{m}(\lambda_i^m - 1)$. The following generalization is due to Hill (1968).

Let $f: (0, \infty) \to \mathbb{R}$ be a scalar function such that

$$f(1) = 0, \qquad f'(1) = 1, \tag{2.2.84}$$

where f is a monotonic increasing function, i.e. $f'(\lambda) > 0$ for all $\lambda > 0$. We denote by $\mathbf{F}(\mathbf{U})$ the second-order tensor function of $\mathbf{U}$ which is coaxial with $\mathbf{U}$ and has principal values $f(\lambda_i)$ which are functions of the principal values λ_i of $\mathbf{U}$. Thus, if $\mathbf{u}^{(i)}$ ($i = 1, 2, 3$) are unit vectors along the Lagrangean principal axes then in terms of the spectral decomposition, we have

$$\left.\begin{aligned}\mathbf{U} &= \sum_{i=1}^{3} \lambda_i \mathbf{u}^{(i)} \otimes \mathbf{u}^{(i)},\\ \mathbf{F}(\mathbf{U}) &= \sum_{i=1}^{3} f(\lambda_i) \mathbf{u}^{(i)} \otimes \mathbf{u}^{(i)}.\end{aligned}\right\} \tag{2.2.85}$$

Equally, we may define the corresponding Eulerian strain measures relative to the spectral decomposition of $\mathbf{V}$, so that

$$\left.\begin{aligned}\mathbf{V} &= \sum_{i=1}^{3} \lambda_i (\mathbf{R}\mathbf{u}^{(i)}) \otimes (\mathbf{R}\mathbf{u}^{(i)}) \equiv \mathbf{R}\mathbf{U}\mathbf{R}^{\mathrm{T}},\\ \mathbf{F}(\mathbf{V}) &= \sum_{i=1}^{3} f(\lambda_i)(\mathbf{R}\mathbf{u}^{(i)}) \otimes (\mathbf{R}\mathbf{u}^{(i)}) \equiv \mathbf{R}\mathbf{F}(\mathbf{U})\mathbf{R}^{\mathrm{T}},\end{aligned}\right\} \tag{2.2.86}$$

where $\mathbf{R}$ is the proper orthogonal tensor arising in the polar decomposition theorem

The monotonicity ascribed to f ensures that all measures of strain defined above are such that an increase in the length of a line element of material (and therefore the stretch in that direction) is always accompanied by an increase in the corresponding strain in that direction. The normality condition $f'(1) = 1$ is imposed so that for 'small' strains all strain measures are equivalent to the first order in some suitable measure of 'smallness'. This will be discussed further in Chapter 6.

More general measures of strain $\mathbf{F}(\mathbf{U})$ may be defined, which are not coaxial with $\mathbf{U}$ for example, but we do not consider them here.

Some specific members of the class (2.2.82) deserve special mention. We confine the discussion to the Lagrangean tensors, however. The special cases $m = 0$ and $m = 2$ have already been mentioned, but we remark that $\ln \mathbf{U}$ is associated with the name of *Hencky*. For $m = 1$, we have

$$\mathbf{F}(\mathbf{U}) = \mathbf{U} - \mathbf{I} \tag{2.2.87}$$

which is called the *Biot* strain tensor, while for $m = -2$

$$\mathbf{F}(\mathbf{U}) = \tfrac{1}{2}(\mathbf{I} - \mathbf{U}^{-2}) \equiv \tfrac{1}{2}(\mathbf{I} - \mathbf{B}^{\mathrm{T}}\mathbf{B}) \tag{2.2.88}$$

and this is referred to as the *Almansi* strain tensor, as also is $\frac{1}{2}(\mathbf{I} - \mathbf{V}^{-2})$.

It turns out, as we see in subsequent chapters, that the Green ($m = 2$), Biot ($m = 1$) and Hencky ($m = 0$) strain tensors are of central importance in applications to elasticity.

2.2.8 Change of reference configuration or observer

With respect to a fixed choice of reference configuration $\mathscr{B}_0$ it was shown in Section 2.2.6 that the deformation gradients $\mathbf{A}$ and $\mathbf{A}'$ associated with the deformations $\chi:\mathscr{B}_0\to\mathscr{B}$ and $\chi':\mathscr{B}_0\to\mathscr{B}'$ respectively are related by

$$\mathbf{A}' = \mathbf{P}\mathbf{A}, \tag{2.2.89}$$

where $\mathbf{P} = \text{grad}\,\boldsymbol{\kappa}(\mathbf{x})$ and $\boldsymbol{\kappa}:\mathscr{B}\to\mathscr{B}'$ is the deformation defined by $\boldsymbol{\kappa}(\mathbf{x}) = \chi'\{\chi^{-1}(\mathbf{x})\}$ for all $\mathbf{x}$ in $\mathscr{B}$.

Change of reference configuration We now consider the effect on the deformation gradient of a change of reference configuration. Let χ be as defined above with respect to the reference configuration $\chi_0:B\to\mathscr{B}_0$ so that

$$\mathbf{X} = \chi_0(X)$$

for all X in B. Suppose that $\chi_0':B\to\mathscr{B}_0'$ is another reference configuration so that

$$\mathbf{X}' = \chi_0'(X), \qquad X\in B.$$

The deformation $\chi':\mathscr{B}_0'\to\mathscr{B}$ is defined by

$$\mathbf{x} = \chi'(\mathbf{X}')$$

for all $\mathbf{X}'$ in $\mathscr{B}_0'$ and the change of reference configuration satisfies

$$\mathbf{X}' = \chi_0'\{\chi_0^{-1}(\mathbf{X})\} \equiv \boldsymbol{\kappa}_0(\mathbf{X})$$

wherein $\boldsymbol{\kappa}_0:\mathscr{B}_0\to\mathscr{B}_0'$ is specified.

It follows that the deformation gradients $\mathbf{A} = \text{Grad}\,\chi(\mathbf{X})$ and $\mathbf{A}' = \text{Grad}'\,\chi'(\mathbf{X}')$ are related by

$$\mathbf{A}' = \mathbf{A}\mathbf{P}_0^{-1}, \tag{2.2.90}$$

where $\mathbf{P}_0 = \text{Grad}\,\boldsymbol{\kappa}_0(\mathbf{X})$. The formulae (2.2.89) and (2.2.90) should be compared bearing in mind that the notation $\mathbf{A}'$ describes different quantitities in the separate equations.

Change of observer In order to discuss the effect on the deformation gradient of a change of observer we need to replace the time dependence which has been omitted from explicit consideration so far in this section. Under a change of observer the motion described as (2.1.7) by O is described by O^*

as (2.1.8) subject to the transformation rule (2.1.9). For a fixed reference configuration the corresponding deformation is given by (2.1.18) according to O.

If the reference configuration is independent of observer then the observer transformation may be written

$$\chi^*(\mathbf{X}, t^*) = \mathbf{c}(t) + \mathbf{Q}(t)\chi(\mathbf{X}, t), \qquad t^* = t - a \tag{2.2.91}$$

and it follows that the deformation gradients $\mathbf{A} = \mathrm{Grad}\,\chi(\mathbf{X}, t)$ and $\mathbf{A}^* = \mathrm{Grad}\,\chi^*(\mathbf{X}, t^*)$ are related by[†]

$$\mathbf{A}^* = \mathbf{Q}(t)\mathbf{A}. \tag{2.2.92}$$

Thus, comparison of (2.2.92) and (2.2.89) shows that the effect of a change of observer on the deformation gradient is equivalent to that of a superposed rigid-body rotation $\boldsymbol{\kappa}(\mathbf{x}) = \mathbf{c}(t) + \mathbf{Q}(t)\mathbf{x}$, for which $\mathbf{P} = \mathbf{Q}(t)$.

In the case of a relative deformation $\mathbf{X}$ is replaced by $\mathbf{X}^*$ on the left-hand side of (2.2.91), where, in view of (2.1.20), we have

$$\mathbf{X}^* = \mathbf{c}(t_0) + \mathbf{Q}(t_0)\mathbf{X}$$

and (2.2.92) is replaced by

$$\mathbf{A}^* = \mathbf{Q}(t)\mathbf{A}\mathbf{Q}(t_0)^{\mathrm{T}}, \tag{2.2.93}$$

where $\mathbf{A}^* = \mathrm{Grad}^*\,\chi^*(\mathbf{X}^*, t^*)$.

The tensor transformations (2.2.89), (2.2.90), (2.2.92) and (2.2.93) should be compared with the transformation rules for the component matrices of $\mathbf{A}$ under changes of (reference and current) frames of reference for a single observer O, as discussed in Problem 2.2.2.

2.3 ANALYSIS OF MOTION

2.3.1 Deformation and strain rates

We now return to the deformation expressed in the form (2.1.18), i.e.

$$\mathbf{x} = \chi(\mathbf{X}, t) \tag{2.3.1}$$

for an arbitrary choice of reference configuration, with the time dependence included.

[†] It follows from (2.2.92) that the restriction to proper orthogonal $\mathbf{Q}(t)$ ensures that the sign of det $\mathbf{A}$ is preserved under observer transformations.

The velocity of the material particle which is allotted the place **X** in the reference configuration is denoted by $\mathbf{V}(\mathbf{X}, t)$ in the Lagrangean specification and defined by

$$\mathbf{V}(\mathbf{X}, t) = \frac{\partial \chi}{\partial t}(\mathbf{X}, t) \tag{2.3.2}$$

following (2.1.10). In the Eulerian specification we write $\mathbf{v}(\mathbf{x}, t)$ for the velocity of the material particle which occupies the place **x** at time t, where

$$\mathbf{V}(\mathbf{X}, t) = \mathbf{v}\{\chi(\mathbf{X}, t), t\}. \tag{2.3.3}$$

It will sometimes make for conciseness to represent $\partial/\partial t$ at fixed **X** by a superposed dot so that

$$\mathbf{V}(\mathbf{X}, t) = \dot{\chi}(\mathbf{X}, t)$$

and the acceleration $\ddot{\chi}(\mathbf{X}, t)$.

The gradient of the deformation (2.3.1) is written

$$\mathbf{A} = \text{Grad}\, \chi(\mathbf{X}, t).$$

Since **X** and t are independent variables the partial derivatives with respect to **X** and t commute and hence

$$\dot{\mathbf{A}} = \text{Grad}\, \dot{\chi}(\mathbf{X}, t) = (\text{grad}\, \mathbf{v}(\mathbf{x}, t))\mathbf{A}$$

on use of the result quoted in Problem 2.2.1.

We write

$$\boldsymbol{\Gamma} = \text{grad}\, \mathbf{v}(\mathbf{x}, t) \tag{2.3.4}$$

for the *velocity gradient* tensor, noting that this is the Eulerian velocity gradient as distinct from the Lagrangean velocity gradient Grad $\mathbf{V}(\mathbf{X}, t)$. Thus

$$\dot{\mathbf{A}} = \boldsymbol{\Gamma}\mathbf{A}, \qquad \dot{\mathbf{A}}^{\mathrm{T}} = \mathbf{A}^{\mathrm{T}}\boldsymbol{\Gamma}^{\mathrm{T}} \tag{2.3.5}$$

and it follows by differentiation of (2.2.9) with respect to t at fixed **X** that

$$\dot{\mathbf{B}} = -\boldsymbol{\Gamma}^{\mathrm{T}}\mathbf{B}, \qquad \dot{\mathbf{B}}^{\mathrm{T}} = -\mathbf{B}^{\mathrm{T}}\boldsymbol{\Gamma}. \tag{2.3.6}$$

It is appropriate to refer to $\partial/\partial t$ at fixed **X** (respectively **x**) as the *Lagrangean* (respectively *Eulerian*) *time derivative*. The connection

$$\left.\frac{\partial}{\partial t}\right|_{\mathbf{X}} = \left.\frac{\partial}{\partial t}\right|_{\mathbf{x}} + \mathbf{v}\cdot\text{grad}$$

follows from an application of the chain rule for partial derivatives and use of (2.3.1)–(2.3.3). The Lagrangean time derivative is also referred to as the *intrinsic derivative* or the *material time derivative* or, in the fluid-mechanical context, as the *rate of change following the flow*, but we do not use these alternative terminologies in this book.

On taking the Lagrangean time derivative of the Green strain tensor (2.2.24) and making use of (2.3.5) we obtain

$$\dot{\mathbf{E}} = \mathbf{A}^{\mathrm{T}}\mathbf{\Sigma}\mathbf{A}, \tag{2.3.7}$$

where

$$\mathbf{\Sigma} = \tfrac{1}{2}(\mathbf{\Gamma} + \mathbf{\Gamma}^{\mathrm{T}}), \tag{2.3.8}$$

the symmetric part of $\mathbf{\Gamma}$, is called the *Eulerian strain rate* tensor. Note that it is not in general expressible as either the Lagrangian or Eulerian time derivative of a strain tensor so it is not a rate of strain.

We now consider the physical interpretations of $\mathbf{\Sigma}$ and $\mathbf{\Gamma}$. From (2.2.22) and (2.2.4) we obtain

$$\frac{1}{2}\frac{\partial}{\partial t}\{|\mathbf{dx}|^2 - |\mathbf{dX}|^2\} = \mathbf{dX}\cdot(\dot{\mathbf{E}}\mathbf{dX}) = \mathbf{dx}\cdot(\mathbf{\Sigma}\mathbf{dx}) \tag{2.3.9}$$

or, equivalently,

$$\frac{1}{|\mathbf{dx}|}\frac{\partial}{\partial t}|\mathbf{dx}| = \mathbf{m}\cdot(\mathbf{\Sigma}\mathbf{m}), \tag{2.3.10}$$

where $\mathbf{m} = \mathbf{dx}/|\mathbf{dx}|$.

Thus $\mathbf{\Sigma}$ is a measure of the rate at which line elements of material are changing their lengths. The motion is rigid if and only if $\mathbf{\Sigma} = \mathbf{0}$.

Relative to a rectangular Cartesian basis $\{\mathbf{e}_i\}$ $\mathbf{\Gamma}$ and $\mathbf{\Sigma}$ have components

$$\Gamma_{ij} = \frac{\partial v_i}{\partial x_j}, \qquad \Sigma_{ij} = \frac{1}{2}\left(\frac{\partial v_i}{\partial x_j} + \frac{\partial v_j}{\partial x_i}\right)$$

and if, for example, $\mathbf{m} = \mathbf{e}_i$ it follows from (2.3.10) that

$$\Sigma_{ii} = \frac{1}{|\mathbf{dx}|}\frac{\partial}{\partial t}|\mathbf{dx}| \qquad (i = 1, 2, 3;\ \text{no summation}).$$

This provides an interpretation for the *normal components* Σ_{ii} of $\mathbf{\Sigma}$ but not for the *shear components* Σ_{ij} $(i \neq j)$.

From (2.2.4) and (2.3.5) we obtain

$$\frac{\partial}{\partial t}(\mathbf{dx}) = \dot{\mathbf{A}}\,\mathbf{dX} = \mathbf{\Gamma}\,\mathbf{dx}$$

so that $\mathbf{\Gamma}$ itself measures not only the changing lengths of line elements but also their changing orientations. Use of (2.3.10) in this equation shows that

$$\dot{\mathbf{m}} = \mathbf{\Gamma m} - [\mathbf{m}\cdot(\mathbf{\Sigma m})]\mathbf{m}, \tag{2.3.11}$$

use having been made of the result $\mathbf{m}\cdot(\mathbf{\Gamma m}) = \mathbf{m}\cdot(\mathbf{\Sigma m})$ which follows from symmetry.

If $\mathbf{dx}' = \mathbf{m}'|\mathbf{dx}'|$ denotes a second Eulerian line element based at $\mathbf{x}$ such that $\mathbf{m}\cdot\mathbf{m}' = \cos\theta$ differentiation of this leads to

$$\dot{\theta} = [\{\mathbf{m}\cdot(\mathbf{\Sigma m}) + \mathbf{m}'\cdot(\mathbf{\Sigma m}')\}(\mathbf{m}\cdot\mathbf{m}') - 2\mathbf{m}\cdot(\mathbf{\Sigma m}')]/|\mathbf{m} \wedge \mathbf{m}'|. \tag{2.3.12}$$

Thus $\mathbf{\Sigma}$ is also a measure of the rate of change of the angle of shear between the directions $\mathbf{m}$ and $\mathbf{m}'$. With $\mathbf{m} = \mathbf{e}_i$ and $\mathbf{m}' = \mathbf{e}_j$, for example, this equation reduces to

$$\Sigma_{ij} = -\tfrac{1}{2}\dot{\theta},$$

and the shear components $\Sigma_{ij}(i \neq j)$ are interpreted as the rates at which instantaneously orthogonal line elements are changing their relative orientations.

We have dealt with the symmetric part of $\mathbf{\Gamma}$. It remains to discuss the antisymmetric part, denoted by

$$\mathbf{\Omega} = \tfrac{1}{2}(\mathbf{\Gamma} - \mathbf{\Gamma}^{\mathrm{T}}). \tag{2.3.13}$$

Since $\mathbf{\Gamma} = \mathbf{\Sigma} + \mathbf{\Omega}$ we obtain from (2.3.11)

$$\dot{\mathbf{m}} = \mathbf{\Omega m} + \mathbf{\Sigma m} - [\mathbf{m}\cdot(\mathbf{\Sigma m})]\mathbf{m}. \tag{2.3.14}$$

If $\mathbf{\Sigma} = \mathbf{0}$ then the motion is rigid and (2.3.14) reduces to $\dot{\mathbf{m}} = \mathbf{\Omega m} = \boldsymbol{\omega} \wedge \mathbf{m}$, where $\boldsymbol{\omega}$ is the axial vector of $\mathbf{\Omega}$ defined in accordance with (1.3.28). Thus the motion represents a rigid-body rotation (or spin) with angular velocity $\boldsymbol{\omega} = \tfrac{1}{2}\operatorname{curl}\mathbf{v}$. If $\mathbf{\Sigma} \neq \mathbf{0}$ then the term $\mathbf{\Omega m}$ contributes a rigid-body spin to the motion, while the term $\mathbf{\Sigma m} - [\mathbf{m}\cdot(\mathbf{\Sigma m})]\mathbf{m}$ in (2.3.14) describes the rotation of line elements superposed on the rigid-body spin.

If $\mathbf{m}$ lies along a principal axis of $\mathbf{\Sigma}$ at some instant then $\mathbf{\Sigma m} - [\mathbf{m}\cdot(\mathbf{\Sigma m})]\mathbf{m} = \mathbf{0}$ *at that instant* and $\mathbf{\Omega}$ then has precisely the interpretation that it is the instantaneous rate of rotation (or spin) of the triad of line elements directed along the principal axes of $\mathbf{\Sigma}$ at the instant in question. The term *body spin* is used to describe $\mathbf{\Omega}$. At different instants different triads of line elements are directed along the principal axes of $\mathbf{\Sigma}$ in general and, in particular, such triads do not rotate rigidly with these axes. Moreover, although the line elements are mutually orthogonal at a considered instant they do not in general remain mutually orthogonal as the motion proceeds.

Suppose the line element $\mathbf{dx} = \mathbf{m}|\mathbf{dx}|$ lies along a principal axis of $\mathbf{\Sigma}$ at a given instant. Then, by (2.3.10), $\mathbf{m}\cdot(\mathbf{\Sigma}\,\mathbf{dx})$ is its rate of increase in length and we may therefore summarize the motion in the neighbourhood of a point $\mathbf{x}$ as comprising a triaxial stretching (along the principal axes of $\mathbf{\Sigma}$) and a (body) spin $\mathbf{\Omega}$.

To illustrate the above theory we consider the simple shear deformation defined in Section 2.2.6 with the amount of shear γ depending on t. We restrict attention to the Cartesian coordinate representation of the deformation so that $\mathbf{A}$ and $\mathbf{\Gamma}$ have component matrices A and Γ, where

$$\mathsf{A} = \begin{bmatrix} 1 & \gamma & 0 \\ 0 & 1 & 0 \\ 0 & 0 & 1 \end{bmatrix}$$

as in (2.2.54) and hence

$$\Sigma = \tfrac{1}{2}(\Gamma + \Gamma^{\mathrm{T}}) = \begin{bmatrix} 0 & \tfrac{1}{2}\dot{\gamma} & 0 \\ \tfrac{1}{2}\dot{\gamma} & 0 & 0 \\ 0 & 0 & 0 \end{bmatrix}$$

$$\Omega = \tfrac{1}{2}(\Gamma - \Gamma^{\mathrm{T}}) = \begin{bmatrix} 0 & \tfrac{1}{2}\dot{\gamma} & 0 \\ -\tfrac{1}{2}\dot{\gamma} & 0 & 0 \\ 0 & 0 & 0 \end{bmatrix}.$$

It is easily shown that the in-plane principal axes of $\mathbf{\Sigma}$ bisect the Cartesian coordinate directions $\mathbf{e}_1$ and $\mathbf{e}_2$ for all values of $\dot{\gamma}$, i.e. they do not rotate. With $\mathbf{m} = \cos\theta\,\mathbf{e}_1 + \sin\theta\,\mathbf{e}_2$ we obtain from (2.3.14) the formula

$$\dot{\theta} = -\dot{\gamma}\sin^2\theta.$$

In particular, for $\theta = \tfrac{1}{4}\pi$, this gives the body spin $-\tfrac{1}{2}\dot{\gamma}$. If $\dot{\gamma} > 0$ with the direction of shear as indicated in Fig. 2.6 then the spin is in the clockwise sense, as is to be expected.

From the polar decomposition (2.2.34) we obtain

$$\dot{\mathbf{A}} = \dot{\mathbf{R}}\mathbf{U} + \mathbf{R}\dot{\mathbf{U}}$$

and the combination of this with (2.3.5) yields

$$\boldsymbol{\Gamma} = \dot{\mathbf{R}}\mathbf{R}^{\mathrm{T}} + \mathbf{R}\dot{\mathbf{U}}\mathbf{U}^{-1}\mathbf{R}^{\mathrm{T}}.$$

Hence

$$\boldsymbol{\Sigma} = \tfrac{1}{2}\mathbf{R}(\dot{\mathbf{U}}\mathbf{U}^{-1} + \mathbf{U}^{-1}\dot{\mathbf{U}})\mathbf{R}^{\mathrm{T}} \tag{2.3.15}$$

and

$$\boldsymbol{\Omega} = \dot{\mathbf{R}}\mathbf{R}^{\mathrm{T}} + \tfrac{1}{2}\mathbf{R}(\dot{\mathbf{U}}\mathbf{U}^{-1} - \mathbf{U}^{-1}\dot{\mathbf{U}})\mathbf{R}^{\mathrm{T}}. \tag{2.3.16}$$

We note, in particular, that the interpretation of the decomposition $\boldsymbol{\Gamma} = \boldsymbol{\Omega} + \boldsymbol{\Sigma}$ into its spin and strain-rate constituents is more complicated than the corresponding polar decomposition $\mathbf{A} = \mathbf{R}\mathbf{U}$ comprising rotation and pure strain since $\boldsymbol{\Omega}$ has contributions not only from the spin $\dot{\mathbf{R}}\mathbf{R}^{\mathrm{T}}$ associated with the rate of rotation $\dot{\mathbf{R}}$ but also from $\dot{\mathbf{U}}$.

Problem 2.3.1 Show that the Lagrangean time derivative of the strain tensor $\tfrac{1}{2}(\mathbf{I} - \mathbf{B}^{\mathrm{T}}\mathbf{B})$ is $\mathbf{B}^{\mathrm{T}}\boldsymbol{\Sigma}\mathbf{B}$.

Problem 2.3.2 For the rigid-body motion

$$\mathbf{x} = \boldsymbol{\chi}(\mathbf{X}, t) \equiv \mathbf{c}(t) + \mathbf{Q}(t)\mathbf{X}$$

show that the velocity and acceleration may be written

$$\dot{\mathbf{x}} = \dot{\mathbf{c}} + \boldsymbol{\omega} \wedge (\mathbf{x} - \mathbf{c})$$

and

$$\ddot{\mathbf{x}} = \ddot{\mathbf{c}} + \dot{\boldsymbol{\omega}} \wedge (\mathbf{x} - \mathbf{c}) + \boldsymbol{\omega} \wedge \{\boldsymbol{\omega} \wedge (\mathbf{x} - \mathbf{c})\}$$

respectively, where $\boldsymbol{\omega}(t)$ is the axial vector associated with the antisymmetric tensor $\dot{\mathbf{Q}}\mathbf{Q}^{\mathrm{T}}$.

Problem 2.3.3 If $J = \det \mathbf{A}$, and $\partial J/\partial \mathbf{A}$ denotes the second-order tensor whose rectangular Cartesian components are $\partial J/\partial A_{i\alpha}$ deduce from the result of Problem 2.2.4 that

$$\frac{\partial J}{\partial \mathbf{A}} = J\mathbf{B}^{\mathrm{T}}.$$

Hence show that

$$\dot{\mathbf{A}}\frac{\partial J}{\partial \mathbf{A}} = J\boldsymbol{\Gamma}$$

and

$$\frac{\dot{J}}{J} = \operatorname{tr}(\boldsymbol{\Gamma}). \tag{2.3.17}$$

Problem 2.3.4 Use Nanson's formula (2.2.18) and the result (2.3.17) to show that

$$\frac{\partial}{\partial t}(\mathbf{ds}) = \operatorname{tr}(\boldsymbol{\Gamma})\,\mathbf{ds} - \boldsymbol{\Gamma}^{\mathrm{T}}\mathbf{ds}.$$

If $\mathbf{n}$ is the unit normal to the (Eulerian) surface area element $\mathrm{d}s = |\mathbf{ds}|$ derive the formulae

$$\frac{1}{\mathrm{d}s}\frac{\partial}{\partial t}(\mathrm{d}s) = \operatorname{tr}(\boldsymbol{\Gamma}) - \mathbf{n}\cdot(\boldsymbol{\Gamma}\mathbf{n})$$

and

$$\dot{\mathbf{n}} = [\mathbf{n}\cdot(\boldsymbol{\Gamma}\mathbf{n})]\mathbf{n} - \boldsymbol{\Gamma}^{\mathrm{T}}\mathbf{n}.$$

Use the latter result to show that

$$\dot{\mathbf{m}} = \boldsymbol{\Gamma}\mathbf{m} - [\mathbf{m}\cdot(\boldsymbol{\Gamma}\mathbf{m})]\mathbf{m} + \alpha\mathbf{m}\wedge\mathbf{n}$$

for any unit vector $\mathbf{m}$ normal to $\mathbf{n}$, where α is a scalar (possibly dependent on t). Reconcile this with the result (2.3.11) and explain why α should be zero.

Problem 2.3.5 For the deformation

$$\mathbf{A} = \lambda\mathbf{e}_1\otimes\mathbf{e}_1 + \lambda^{-1/2}(\mathbf{e}_2\otimes\mathbf{e}_2 + \mathbf{e}_3\otimes\mathbf{e}_3)$$

show that $\boldsymbol{\Omega} = \mathbf{0}$ and

$$\boldsymbol{\Sigma} = \frac{\dot{\lambda}}{\lambda}[\mathbf{e}_1\otimes\mathbf{e}_1 - \tfrac{1}{2}(\mathbf{e}_2\otimes\mathbf{e}_2 + \mathbf{e}_3\otimes\mathbf{e}_3)].$$

If $\mathbf{m} = \cos\theta\mathbf{e}_1 + \sin\theta\mathbf{e}_2$ use the formula (2.3.11) to show that

$$\dot{\theta} = -\frac{3}{2}\frac{\dot{\lambda}}{\lambda}\sin\theta\cos\theta.$$

2.3.2 Spins of the Lagrangean and Eulerian axes

Let $\mathbf{u}^{(i)}$ $(i = 1, 2, 3)$ be unit vectors defining the Lagrangean principal axes at a point $\mathbf{X}$ in some reference configuration. In general $\mathbf{u}^{(i)}$ depends on t so we define a fixed rectangular Cartesian basis $\{\mathbf{E}_i\}$ so that

$$\mathbf{u}^{(i)} = \mathbf{P}\mathbf{E}_i \qquad (i = 1, 2, 3),$$

where $\mathbf{P}$ is a proper orthogonal tensor which describes the orientation of the Lagrangean axes relative to the fixed basis.

On taking the Lagrangean time derivative of the above equation and then eliminating $\mathbf{E}_i$ we obtain

$$\dot{\mathbf{u}}^{(i)} = \mathbf{\Omega}^{(\mathrm{L})}\mathbf{u}^{(i)} \qquad (i = 1, 2, 3), \tag{2.3.18}$$

where the antisymmetric tensor $\mathbf{\Omega}^{(\mathrm{L})}$ is defined as $\dot{\mathbf{P}}\mathbf{P}^{\mathrm{T}}$. It is called the *spin of the Lagrangean principal axes.* The representation

$$\mathbf{\Omega}^{(\mathrm{L})} = \sum_{i=1}^{3} \dot{\mathbf{u}}^{(i)} \otimes \mathbf{u}^{(i)} \tag{2.3.19}$$

follows from (2.3.18) and the components $\Omega_{ij}^{(\mathrm{L})}$ of $\mathbf{\Omega}^{(\mathrm{L})}$ on the basis $\{\mathbf{u}^{(i)}\}$ are given by

$$\Omega_{ij}^{(\mathrm{L})} = \mathbf{u}^{(i)} \cdot \dot{\mathbf{u}}^{(j)} = -\Omega_{ji}^{(\mathrm{L})}. \tag{2.3.20}$$

From the spectral decomposition

$$\mathbf{U} = \sum_{i=1}^{3} \lambda_i \mathbf{u}^{(i)} \otimes \mathbf{u}^{(i)} \tag{2.3.21}$$

we then obtain

$$\dot{\mathbf{U}} = \sum_{i=1}^{3} \dot{\lambda}_i \mathbf{u}^{(i)} \otimes \mathbf{u}^{(i)} + \mathbf{\Omega}^{(\mathrm{L})}\mathbf{U} - \mathbf{U}\mathbf{\Omega}^{(\mathrm{L})} \tag{2.3.22}$$

on use of (2.3.18). It follows that with respect to the basis $\{\mathbf{u}^{(i)}\}$ the normal and shear components of $\dot{\mathbf{U}}$ are given by

$$\left.\begin{aligned} (\dot{\mathbf{U}})_{ii} &= \dot{\lambda}_i \qquad i = 1, 2, 3, \\ (\dot{\mathbf{U}})_{ij} &= \Omega_{ij}^{(\mathrm{L})}(\lambda_j - \lambda_i) \qquad i \neq j \end{aligned}\right\} \tag{2.3.23}$$

respectively.

More generally, for the strain tensor $\mathbf{F}(\mathbf{U})$ given by (2.2.85)

$$\dot{\mathbf{F}} = \sum_{i=1}^{3} f'(\lambda_i)\dot{\lambda}_i \mathbf{u}^{(i)} \otimes \mathbf{u}^{(i)} + \mathbf{\Omega}^{(L)}\mathbf{F} - \mathbf{F}\mathbf{\Omega}^{(L)} \tag{2.3.24}$$

and this has components

$$\left.\begin{aligned} (\dot{\mathbf{F}})_{ii} &= f'(\lambda_i)\dot{\lambda}_i \qquad i=1,2,3, \\ (\dot{\mathbf{F}})_{ij} &= \Omega_{ij}^{(L)}\{f(\lambda_j) - f(\lambda_i)\} \qquad i \neq j. \end{aligned}\right\} \tag{2.3.25}$$

From (2.3.15) and (2.3.21)–(2.3.23) it is easily shown that

$$\mathbf{R}^{\mathrm{T}}\mathbf{\Sigma}\mathbf{R} = \sum_{i=1}^{3} \frac{\dot{\lambda}_i}{\lambda_i} \mathbf{u}^{(i)} \otimes \mathbf{u}^{(i)} + \sum_{i \neq j} \Omega_{ji}^{(L)} \frac{(\lambda_i^2 - \lambda_j^2)}{2\lambda_i\lambda_j} \mathbf{u}^{(i)} \otimes \mathbf{u}^{(j)}. \tag{2.3.26}$$

If we denote the components of $\mathbf{\Sigma}$ on the Eulerian principal axes, i.e. the components of $\mathbf{R}^{\mathrm{T}}\mathbf{\Sigma}\mathbf{R}$ on the Lagrangean principal axes, by Σ_{ij} then it follows from (2.3.26) that

$$\left.\begin{aligned} \Sigma_{ii} &= \dot{\lambda}_i/\lambda_i \qquad i=1,2,3, \\ \Sigma_{ij} &= \Omega_{ij}^{(L)}(\lambda_j^2 - \lambda_i^2)/2\lambda_i\lambda_j \qquad i \neq j. \end{aligned}\right\} \tag{2.3.27}$$

The second equation in (2.3.27) defines the components $\Omega_{ij}^{(L)}$ when $\mathbf{\Sigma}$ is known and provided $\lambda_i \neq \lambda_j (i \neq j)$. If $\lambda_i = \lambda_j$ $(i \neq j)$ then $\Omega_{ij}^{(L)}$ is indeterminate from the kinematics.

In respect of the Eulerian principal axes $\mathbf{v}^{(i)}$ $(i=1,2,3)$, the spin $\mathbf{\Omega}^{(E)}$ of the Eulerian principal axes, the left stretch tensor $\mathbf{V}$ and $\mathbf{F}(\mathbf{V})$ defined by (2.2.86), equations analogous to (2.3.18)–(2.3.25) may be written down but there is no easily interpretable analogue of (2.3.26). Since $\mathbf{v}^{(i)} = \mathbf{R}\mathbf{u}^{(i)}$ it follows with the help of (2.3.19) that

$$\mathbf{\Omega}^{(E)} = \dot{\mathbf{R}}\mathbf{R}^{\mathrm{T}} + \mathbf{R}\mathbf{\Omega}^{(L)}\mathbf{R}^{\mathrm{T}} \tag{2.3.28}$$

and hence by (2.3.16) that

$$\mathbf{\Omega}^{(E)} = \mathbf{\Omega} + \mathbf{R}\mathbf{\Omega}^{(L)}\mathbf{R}^{\mathrm{T}} - \tfrac{1}{2}\mathbf{R}(\dot{\mathbf{U}}\mathbf{U}^{-1} - \mathbf{U}^{-1}\dot{\mathbf{U}})\mathbf{R}^{\mathrm{T}}. \tag{2.3.29}$$

Further use of (2.3.21)–(2.3.23) reduces this to

$$\mathbf{\Omega}^{(E)} = \mathbf{\Omega} - \sum_{i \neq j} \Omega_{ji}^{(L)} \frac{(\lambda_i^2 + \lambda_j^2)}{2\lambda_i\lambda_j} \mathbf{v}^{(i)} \otimes \mathbf{v}^{(j)}.$$

If $\Omega_{ij}^{(E)}$ and Ω_{ij} denote the components of $\mathbf{\Omega}^{(E)}$ and $\mathbf{\Omega}$ respectively on the basis $\{\mathbf{v}^{(i)}\}$ then

$$\Omega_{ij}^{(E)} = \Omega_{ij} + (\lambda_i^2 + \lambda_j^2)\Omega_{ij}^{(L)}/2\lambda_i\lambda_j \tag{2.3.30}$$

or, from (2.3.27),

$$\Omega_{ij}^{(E)} = \Omega_{ij} + (\lambda_i^2 + \lambda_j^2)\Sigma_{ij}/(\lambda_j^2 - \lambda_i^2) \tag{2.3.31}$$

provided $\lambda_i \neq \lambda_j \quad (i \neq j)$.

Problem 2.3.6 Use the results (2.3.17) and (2.3.25) with $f(\lambda_i) = \frac{1}{2}(\lambda_i^2 - 1)$ to derive (2.3.26).

Problem 2.3.7 Show that $\dot{\mathbf{R}}\mathbf{R}^{\mathrm{T}}$ has components

$$\Omega_{ij} + (\lambda_j - \lambda_i)\Sigma_{ij}/(\lambda_i + \lambda_j)$$

on the Eulerian principal axes.

2.4 OBJECTIVITY OF TENSOR FIELDS

2.4.1 Eulerian and Lagrangean objectivity

The description of a physical quantity associated with the motion of a body depends in general on the choice of observer. Our purpose here is to distinguish those scalar, vector and tensor fields which depend intrinsically on the observer from those which are essentially independent of the observer. We begin by examining how an observer transformation affects certain fields.

From (2.1.2) we recall that an observer transformation $(\mathbf{x}, t) \rightarrow (\mathbf{x}^*, t^*)$ is specified by

$$\mathbf{x}^* = \mathbf{Q}(t)\mathbf{x} + \mathbf{c}(t), \qquad t^* = t - a \tag{2.4.1}$$

so that for the motion (2.1.7) we have

$$\chi^*(X, t^*) = \mathbf{Q}(t)\chi(X, t) + \mathbf{c}(t), \qquad t^* = t - a. \tag{2.4.2}$$

It follows that the velocity $\dot{\chi}(X, t)$ and acceleration $\ddot{\chi}(X, t)$ transform according to

$$\dot{\chi}^*(X, t^*) = \mathbf{Q}(t)\dot{\chi}(X, t) + \dot{\mathbf{c}}(t) + \dot{\mathbf{Q}}(t)\chi(X, t). \tag{2.4.3}$$

$$\ddot{\chi}^*(X, t) = \mathbf{Q}(t)\ddot{\chi}(X, t) + \ddot{\mathbf{c}}(t) + \ddot{\mathbf{Q}}(t)\chi(X, t) + 2\dot{\mathbf{Q}}(t)\dot{\chi}(X, t). \tag{2.4.4}$$

Notice in particular that the definitions of velocity and acceleration are relative and therefore inextricably linked to the kinematics through the relative motion of the observers implied by the quantities $\dot{\mathbf{c}}(t)$, $\ddot{\mathbf{c}}(t)$, $\dot{\mathbf{Q}}(t)$ and $\ddot{\mathbf{Q}}(t)$ occurring in (2.4.3) and (2.4.4).

According to (2.2.92) the transformation rule for the deformation gradient is

$$\mathbf{A}^*(\mathbf{X}, t^*) = \mathbf{Q}(t)\mathbf{A}(\mathbf{X}, t) \tag{2.4.5}$$

when the reference configuration is independent of observer. For a relative deformation, on the other hand, this is replaced by (2.2.93) and in either case

$$J^* \equiv \det \mathbf{A}^* = \det \mathbf{A} \equiv J. \tag{2.4.6}$$

Thus the scalar quantity J is unaffected by an observer transformation and this is consistent with the intuitive notion that local volume ratio is independent of its kinematical description.

In what follows we restrict attention to reference configurations independent of observer so that (2.4.5) holds, but definitions and results are easily modified to account for relative deformations as in (2.2.93) if required.

It follows from (2.4.5) that

$$(\mathbf{A}\mathbf{A}^{\mathrm{T}})^* \equiv \mathbf{A}^*\mathbf{A}^{*\mathrm{T}} = \mathbf{Q}\mathbf{A}\mathbf{A}^{\mathrm{T}}\mathbf{Q}^{\mathrm{T}} \tag{2.4.7}$$

and similarly

$$(\mathbf{B}\mathbf{B}^{\mathrm{T}})^* \equiv \mathbf{B}^*\mathbf{B}^{*\mathrm{T}} = \mathbf{Q}\mathbf{B}\mathbf{B}^{\mathrm{T}}\mathbf{Q}^{\mathrm{T}}, \tag{2.4.8}$$

where, for convenience, we now omit the arguments from the tensors. Also we obtain

$$(\mathbf{A}^{\mathrm{T}}\mathbf{A})^* = \mathbf{A}^{\mathrm{T}}\mathbf{A}, \qquad (\mathbf{B}^{\mathrm{T}}\mathbf{B})^* = \mathbf{B}^{\mathrm{T}}\mathbf{B} \tag{2.4.9}$$

and

$$\mathbf{E}^* = \mathbf{E}, \tag{2.4.10}$$

where $\mathbf{E}$ is the Green strain tensor defined in (2.2.24). Moreover, application of the polar decomposition theorem (2.2.34) to $\mathbf{A}$ and $\mathbf{A}^*$ in (2.4.5) yields

$$\mathbf{V}^* = \mathbf{Q}\mathbf{V}\mathbf{Q}^{\mathrm{T}}, \qquad \mathbf{R}^* = \mathbf{Q}\mathbf{R}, \qquad \mathbf{U}^* = \mathbf{U}. \tag{2.4.11}$$

From (2.4.1) it follows that Eulerian line elements transform according to $\mathbf{dx}^* = \mathbf{Q}\,\mathbf{dx}$ while Lagrangean line elements $\mathbf{dX}$ are unaffected by the observer transformation. Hence

$$\mathbf{dX}'\cdot(\mathbf{E}\,\mathbf{dX}) = \tfrac{1}{2}\mathbf{dx}'\cdot(\mathbf{I} - \mathbf{B}\mathbf{B}^{\mathrm{T}})\,\mathbf{dx} = \tfrac{1}{2}\mathbf{dx}^{*\prime}\cdot(\mathbf{I} - \mathbf{B}^*\mathbf{B}^{*\mathrm{T}})\,\mathbf{dx}^*, \tag{2.4.12}$$

where $\mathbf{dx}' = \mathbf{A}\,\mathbf{dX}'$.

On taking the Lagrangean time derivative of (2.4.12) we obtain

$$\mathbf{dX}'\cdot(\dot{\mathbf{E}}\,\mathbf{dX}) = \mathbf{dx}'\cdot(\mathbf{\Sigma}\,\mathbf{dx}) = \mathbf{dx}^{*\prime}\cdot(\mathbf{\Sigma}^*\,\mathbf{dx}^*) \tag{2.4.13}$$

by use of (2.3.7) and hence the transformation rule for the Eulerian strain-rate $\mathbf{\Sigma}$ is

$$\mathbf{\Sigma}^* = \mathbf{Q}\mathbf{\Sigma}\mathbf{Q}^{\mathrm{T}}. \tag{2.4.14}$$

We note in passing that $\dot{\mathbf{E}}, \ddot{\mathbf{E}}$ and higher derivatives of $\mathbf{E}$ are not affected by an observer transformation.

From (2.4.5) we have

$$\dot{\mathbf{A}}^* = \mathbf{Q}\dot{\mathbf{A}} + \dot{\mathbf{Q}}\mathbf{A} \tag{2.4.15}$$

and we note the distinction between the transformation rule (2.4.5) for $\mathbf{A}$ and that for $\dot{\mathbf{A}}$. Use of (2.3.5) in (2.4.15) yields

$$\mathbf{\Gamma}^* = \mathbf{Q}\mathbf{\Gamma}\mathbf{Q}^{\mathrm{T}} + \dot{\mathbf{Q}}\mathbf{Q}^{\mathrm{T}}. \tag{2.4.16}$$

The symmetric part of this gives (2.4.14) while its antisymmetric part has the transformation rule

$$\mathbf{\Omega}^* = \mathbf{Q}\mathbf{\Omega}\mathbf{Q}^{\mathrm{T}} + \dot{\mathbf{Q}}\mathbf{Q}^{\mathrm{T}}. \tag{2.4.17}$$

Clearly, the body spin $\mathbf{\Omega}$, as a measure of the rigid rotation of a triad of line elements, is influenced by the rotation of observers through their relative spin $\dot{\mathbf{Q}}\mathbf{Q}^{\mathrm{T}}$, as intuition demands.

The tensors of strain and strain-rate involved in (2.4.12) and (2.4.13), on the other hand, do not depend on $\dot{\mathbf{Q}}$ (or indeed $\dot{\mathbf{c}}$) and can therefore be regarded as intrinsic to the material in that they incorporate measures of the changing lengths of line elements of material and the changing angles between pairs of line elements. Every observer attaches the same value to each such quantity and this 'observer indifference' is reflected in the tensor transformation rules (2.4.8), (2.4.10) and (2.4.14). It is important to notice the distinction between the transformation rules for the *Eulerian tensors* $\mathbf{B}\mathbf{B}^{\mathrm{T}}$

and Σ, for example, the *Lagrangean tensors* such as $\mathbf{E}$ and $\mathbf{B}^T\mathbf{B}$ and the *two-point* (or *mixed Eulerian–Lagrangean*) *tensor* $\mathbf{A}$.† We use the word *objective* to describe a (second-order) tensor which is independent of the kinematics in the sense that it satisfies transformation rules of the type (2.4.5), (2.4.7)–(2.4.11) and (2.4.14), but this word must be qualified by *Lagrangean*, *Eulerian* or *two-point* as we indicate in the general definition given below.

We distinguish between a Lagrangean (or Eulerian) tensor field and the Lagrangean (or Eulerian) description of a tensor field since any Lagrangean, Eulerian or two-point tensor field may be given either a Lagrangean or an Eulerian description through the motion (2.1.18) or its inverse, as shown at the end of Section 2.2.1.

Firstly we consider scalar fields. An Eulerian scalar field ϕ, defined over the current configuration $\mathscr{B}$ of the body B, is said to be *objective* if it transforms according to

$$\phi^*(\mathbf{x}^*, t^*) = \phi(\mathbf{x}, t) \tag{2.4.18}$$

under the observer transformation (2.4.1), where ϕ^* is the corresponding scalar field observed by O^*. Equally, a Lagrangean scalar field Φ, defined over the reference configuration $\mathscr{B}_0$, is said to be objective if

$$\Phi^*(\mathbf{X}, t) = \Phi(\mathbf{X}, t). \tag{2.4.19}$$

Since $\mathbf{x} = \boldsymbol{\chi}(\mathbf{X}, t)$ each Eulerian scalar field is associated with a Lagrangean scalar field through

$$\Phi(\mathbf{X}, t) = \phi\{\boldsymbol{\chi}(\mathbf{X}, t), t\}. \tag{2.4.20}$$

Conversely

$$\phi(\mathbf{x}, t) = \Phi\{\boldsymbol{\chi}^{-1}(\mathbf{x}, t), t\} \tag{2.4.21}$$

converts a Lagrangean to an Eulerian scalar field. Clearly, a Lagrangean (respectively Eulerian) scalar field is simply the Lagrangean (respectively Eulerian) description of the associated Eulerian (respectively Lagrangean) scalar field. It follows from (2.4.20) or (2.4.21) that the definitions (2.4.18) and (2.4.19) are equivalent. Equation (2.4.6) demonstrates the objectivity of the

† A second-order Lagrangean (respectively Eulerian) tensor is a member of the space $\mathscr{L}(T_{\mathbf{X}}(\mathscr{B}_0) \times T_{\mathbf{X}}(\mathscr{B}_0), \mathbb{R})$ (respectively $\mathscr{L}(T_{\mathbf{x}}(\mathscr{B}) \times T_{\mathbf{x}}(\mathscr{B}), \mathbb{R})$) of bilinear mappings, where $T_{\mathbf{X}}(\mathscr{B}_0)$ and $T_{\mathbf{x}}(\mathscr{B})$ are the tangent spaces defined in the footnote on p. 84. Higher-order tensors are defined similarly as multilinear mappings over $T_{\mathbf{X}}(\mathscr{B}_0)$ and $T_{\mathbf{x}}(\mathscr{B})$ while Lagrangean (respectively Eulerian) vectors constitute the space $\mathscr{L}(T_{\mathbf{X}}(\mathscr{B}_0), \mathbb{R})$ (respectively $\mathscr{L}(T_{\mathbf{x}}(\mathscr{B}), \mathbb{R})$). Note that $T_{\mathbf{X}}(\mathscr{B}_0)$ is not affected by an observer transformation but $T_{\mathbf{x}}(\mathscr{B})$ changes to $T_{\mathbf{x}}(\mathscr{B}^*)$. We continue to omit the subscript t from $\mathscr{B}$ and also t^* from $\mathscr{B}^*$.

scalar field J while from (2.4.3) it is seen that the speed $|\dot{\chi}|$ is not objective.

Secondly, let $\mathbf{T}$ denote an *Eulerian* tensor field of order $n (n = 1, 2, \ldots)$.† Then $\mathbf{T}$ is said to be an *objective Eulerian tensor field* of order n if the associated multilinear form transforms in accordance with

$$\mathbf{T}^*(\mathbf{x}^*, t^*)(\mathbf{dx}^{(1)*}, \ldots, \mathbf{dx}^{(n)*}) = \mathbf{T}(\mathbf{x}, t)(\mathbf{dx}^{(1)}, \ldots, \mathbf{dx}^{(n)}) \tag{2.4.22}$$

under the observer transformation (2.4.1), where $\mathbf{dx}^{(k)*} = \mathbf{Q}(t)\,\mathbf{dx}^{(k)}$, for arbitrary $\mathbf{dx}^{(k)}\,(k = 1, .., n)$.

For a vector field $\mathbf{v}\,(n = 1)$, equation (2.4.22) is reducible to

$$\mathbf{v}^*(\mathbf{x}^*, t^*) = \mathbf{Q}(t)\mathbf{v}(\mathbf{x}, t) \tag{2.4.23}$$

and for a second-order tensor field $(n = 2)$ to

$$\mathbf{T}^*(\mathbf{x}^*, t^*) = \mathbf{Q}(t)\mathbf{T}(\mathbf{x}, t)\mathbf{Q}(t)^{\mathrm{T}} \tag{2.4.24}$$

but there is no corresponding simple representation for the tensor transformation rule when $n \geq 3$. However, it is instructive to examine the (rectangular Cartesian) component representations

$$\mathbf{T}(\mathbf{x}, t) = T_{ijk\ldots}(\mathbf{x}, t)\mathbf{e}_i \otimes \mathbf{e}_j \otimes \mathbf{e}_k \otimes \ldots,$$
$$\mathbf{T}^*(\mathbf{x}^*, t^*) = T^*_{ijk\ldots}(\mathbf{x}^*, t^*)\mathbf{e}_i^* \otimes \mathbf{e}_j^* \otimes \mathbf{e}_k^* \otimes \ldots$$

for, with $\mathbf{e}_i^* = \mathbf{Q}(t)\mathbf{e}_i$, the objectivity statement (2.4.22) becomes

$$T^*_{ijk\ldots}(\mathbf{x}^*, t^*) = T_{ijk\ldots}(\mathbf{x}, t). \tag{2.4.25}$$

By means of the motion described by $\mathbf{x} = \chi(\mathbf{X}, t)$ and $\mathbf{x}^* = \chi^*(\mathbf{X}, t^*)$ by the respective observers the definition (2.4.22) is expressible in a Lagrangean description. For example, (2.4.23) becomes

$$\mathbf{v}^*\{\chi^*(\mathbf{X}, t^*), t^*\} = \mathbf{Q}(t)\mathbf{v}\{\chi(\mathbf{X}, t), t\},$$

but we emphasize that $\mathbf{v}\{\chi(\mathbf{X}, t), t\}$ is *not* a Lagrangean vector. Similarly for tensors.

Thirdly, if $\mathbf{T}$ is a *Lagrangean* tensor field (of order n)‡ then it is said to be an *objective Lagrangean tensor field* (of order n) if it is unaffected by an

†Thus $\mathbf{T}$ is a member of the space $\mathscr{L}(T_{\mathbf{x}}(\mathscr{B})^n, \mathbb{R})$ of real n-multilinear mappings over $T_{\mathbf{x}}(\mathscr{B})$, and the Eulerian line elements $\mathbf{dx}^{(k)}$ are members of $T_{\mathbf{x}}(\mathscr{B})$. We emphasize that we do not distinguish $T_{\mathbf{x}}(\mathscr{B})$ from its dual space here.

‡That is, a real n-multilinear mapping over $T_{\mathbf{X}}(\mathscr{B}_0)$.

observer transformation. Symbolically this is expressed

$$\mathbf{T}^*(\mathbf{X}, t^*) = \mathbf{T}(\mathbf{X}, t) \tag{2.4.26}$$

(in terms of the Lagrangean description) or, equivalently.

$$\mathbf{T}^*(\mathbf{X}, t^*)(\mathbf{dX}^{(1)}, \dots, \mathbf{dX}^{(n)}) = \mathbf{T}(\mathbf{X}, t)(\mathbf{dX}^{(1)}, \dots, \mathbf{dX}^{(n)}) \tag{2.4.27}$$

for arbitrary Lagrangean line elements $\mathbf{dX}^{(\alpha)}$ $(\alpha = 1, \dots, n)$.

The definition of an *objective two-point tensor field* is intermediate between (2.4.22) and (2.4.27). Thus, in the case of the deformation gradient $\mathbf{A}$, for example, equation (2.4.5) is equivalent to the bilinear form equality

$$\mathbf{A}^*(\mathbf{X}, t^*)(\mathbf{dx}^*, \mathbf{dX}) = \mathbf{A}(\mathbf{X}, t)(\mathbf{dx}, \mathbf{dX}) \tag{2.4.28}$$

for arbitrary Lagrangean and Eulerian line elements $\mathbf{dX}$ and $\mathbf{dx}$ respectively, where $\mathbf{dx}^* = \mathbf{Q}(t)\,\mathbf{dx}$. It follows that $\mathbf{A}$ is an objective two-point tensor, and more generally equation (2.4.28) defines an objective two-point tensor of the second order, a definition which is easily extended to two-point tensors of higher order if required.

In standard texts such as Truesdell and Noll (1965) no distinction is made between Eulerian and Lagrangian fields in the definition of objectivity. In such work a field is said to be objective if it satisfies the appropriate one of the Eulerian rules (2.4.18), (2.4.22)–(2.4.24), but vector and tensor fields for which (2.4.27) holds are not regarded as objective (e.g. $\mathbf{A}^{\mathrm{T}}\mathbf{A}$). On the other hand, only fields for which (2.4.26) holds does Hill (1978) refer to as objective. The distinction we have made between Eulerian and Lagrangean fields reconciles these two views. We remark that the definition of an objective Eulerian field is independent of reference configuration although the value of the field itself, for example $\mathbf{A}\mathbf{A}^{\mathrm{T}}(\mathbf{x}, t)$, will in general depend on the choice of reference configuration. Equally, Lagrangean objectivity is invariant under a change of reference configuration.

The terminology 'observer frame indifferent', or simply 'frame indifferent' is often used as the equivalent of 'objective', but some authors attach a (subtly) different meaning to it. In this book we adopt the definition of 'objectivity' described above.

At this point we remark that the definitions of Eulerian and Lagrangean objectivity may equally well be considered in relation to a single observer O if the observer transformation is interpreted as a superposed rigid-body motion given by

$$\bar{\mathbf{x}} = \mathbf{Q}(t)\mathbf{x} + \mathbf{c}(t), \tag{2.4.29}$$

as described in Section 2.1.4, recalling that attention is restricted to proper orthogonal $\mathbf{Q}(t)$. The definitions (2.4.18) and (2.4.24) are equivalent to

$$\left.\begin{aligned} \phi(\bar{\mathbf{x}}, t) &= \phi(\mathbf{x}, t), \\ \mathbf{v}(\bar{\mathbf{x}}, t) &= \mathbf{Q}(t)\mathbf{v}(\mathbf{x}, t), \\ \mathbf{T}(\bar{\mathbf{x}}, t) &= \mathbf{Q}(t)\mathbf{T}(\mathbf{x}, t)\mathbf{Q}(t)^{\mathrm{T}} \end{aligned}\right\} \tag{2.4.30}$$

respectively, where $\mathbf{x}$ and $\bar{\mathbf{x}}$ are related by (2.4.29). Equally, a Lagrangean field observed by O is objective if and only if it is unaffected by a superposed rigid-body motion.

A direct correspondence between the two views is provided by setting $\mathbf{x}^* = \bar{\mathbf{x}}$ and introducing the time shift $t^* = t - a$ so that, for example,

$$\mathbf{T}^*(\mathbf{x}^*, t^*) = \mathbf{T}(\mathbf{x}^*, t).$$

We have seen in (2.4.20) and (2.4.21) that, through the motion $\mathbf{x} = \chi(\mathbf{X}, t)$ and its inverse, each Eulerian (respectively Lagrangean) scalar field is associated with a unique Lagrangean (respectively Eulerian) scalar field. Vector and tensor Lagrangean and Eulerian fields, on the other hand, may be associated through the deformation gradient $\mathbf{A}$ and its inverse, but the association is not unique. If, for example, $\mathbf{v}$ is an Eulerian vector field then $\mathbf{A}^{\mathrm{T}}\mathbf{v}$ and $\mathbf{B}^{\mathrm{T}}\mathbf{v}$ are each Lagrangean vector fields. Respectively, they are covariant and contravariant in character in the sense that

$$(\mathbf{A}^{\mathrm{T}}\mathbf{v})\cdot\mathbf{dX} = \mathbf{v}\cdot(\mathbf{A}\,\mathbf{dX}) = \mathbf{v}\cdot\mathbf{dx} \equiv v_i\,\mathrm{d}x^i,$$

$$(\mathbf{B}^{\mathrm{T}}\mathbf{v})\cdot\mathrm{Grad} = \mathbf{v}\cdot(\mathbf{B}\,\mathrm{Grad}) = \mathbf{v}\cdot\mathrm{grad} \equiv v^i\frac{\partial}{\partial x^i},$$

where v_i and v^i respectively are covariant and contravariant components of $\mathbf{v}$ with respect to a general curvilinear basis.

We call $\mathbf{A}^{\mathrm{T}}\mathbf{v}$ and $\mathbf{B}^{\mathrm{T}}\mathbf{v}$ the (covariant and contravariant) *induced Lagrangean fields* of $\mathbf{v}$. Similarly, if $\mathbf{V}$ is a Lagrangean vector field then $\mathbf{AV}$ and $\mathbf{BV}$ are the *induced Eulerian fields* of $\mathbf{V}$. Note that $\mathbf{A}^{\mathrm{T}}\mathbf{AV}$ and $\mathbf{B}^{\mathrm{T}}\mathbf{BV}$ are Lagrangean while $\mathbf{AA}^{\mathrm{T}}\mathbf{v}$ and $\mathbf{BB}^{\mathrm{T}}\mathbf{v}$ are Eulerian vector fields.

If $\mathbf{T}$ is a second-order Eulerian tensor field then there are four induced second-order Lagrangean tensor fields, namely $\mathbf{A}^{\mathrm{T}}\mathbf{TA}$, $\mathbf{A}^{\mathrm{T}}\mathbf{TB}$, $\mathbf{B}^{\mathrm{T}}\mathbf{TA}$, $\mathbf{B}^{\mathrm{T}}\mathbf{TB}$, and also four induced second-order two-point tensor fields: $\mathbf{A}^{\mathrm{T}}\mathbf{T}$, $\mathbf{B}^{\mathrm{T}}\mathbf{T}$, $\mathbf{TA}$, $\mathbf{TB}$. The generalization to higher-order tensors is straightforward. If $\mathbf{T}^{(\mathrm{E})}$ (respectively $\mathbf{T}^{(\mathrm{L})}$) is an Eulerian (respectively Lagrangean) tensor field of order n then 2^n induced Lagrangean (respectively Eulerian) tensor fields $\mathbf{T}^{(\mathrm{L})}$ (respectively $\mathbf{T}^{(\mathrm{E})}$) are defined through the equation

$$\mathbf{T}^{(\mathrm{E})}(\mathbf{x}, t)(\mathbf{v}^{(1)}, \ldots, \mathbf{v}^{(n)}) = \mathbf{T}^{(\mathrm{L})}(\mathbf{X}, t)(\mathbf{V}^{(1)}, \ldots, \mathbf{V}^{(n)}),$$

where $\mathbf{x} = \boldsymbol{\chi}(\mathbf{X}, t)$ and the vectors $\mathbf{v}^{(k)}$ and $\mathbf{V}^{(k)}$ are related by either $\mathbf{v}^{(k)} = \mathbf{A}\mathbf{V}^{(k)}$ or $\mathbf{v}^{(k)} = \mathbf{B}\mathbf{V}^{(k)}$ for each $k\,(k = 1, \ldots, n)$. It is left as an exercise for the reader to enumerate the number of possible induced two-point fields of either $\mathbf{T}^{(\mathrm{E})}$ or $\mathbf{T}^{(\mathrm{L})}$.

From the definitions it follows immediately that *a field is objective if and only if each of its induced fields is objective*. Thus,

$$\mathbf{v}^* = \mathbf{Q}\mathbf{v} \Leftrightarrow \mathbf{A}^{*\mathrm{T}}\mathbf{v}^* = \mathbf{A}^{\mathrm{T}}\mathbf{v} \Leftrightarrow \mathbf{B}^{*\mathrm{T}}\mathbf{v}^* = \mathbf{B}^{\mathrm{T}}\mathbf{v}, \tag{2.4.31}$$

for example, while for the non-objective velocity field $\dot{\boldsymbol{\chi}}$ it follows from (2.4.3) and (2.4.5) that $\mathbf{A}^{*\mathrm{T}}\dot{\boldsymbol{\chi}}^* \neq \mathbf{A}^{\mathrm{T}}\dot{\boldsymbol{\chi}}$ since $\dot{\boldsymbol{\chi}}^* \neq \mathbf{Q}\dot{\boldsymbol{\chi}}$.

Conclusions concerning the objectivity or otherwise of the Lagrangean time derivative of an objective field may now be drawn. For example, if $\mathbf{v}$ is an objective Eulerian vector field then differentiation of the left-hand equality in (2.4.31) shows that

$$\dot{\mathbf{v}}^* = \mathbf{Q}\dot{\mathbf{v}} + \dot{\mathbf{Q}}\mathbf{v} \tag{2.4.32}$$

so that $\dot{\mathbf{v}}$ is not objective. This is a particular case of the general statement

$$\frac{\partial}{\partial t}\,(\text{objective Eulerian field}) \neq \text{objective Eulerian field.} \tag{2.4.33}$$

On the other hand, since $\mathbf{A}^{\mathrm{T}}\mathbf{v}$ is an objective Lagrangean vector field it follows that

$$\frac{\partial}{\partial t}(\mathbf{A}^{\mathrm{T}}\mathbf{v}) = \mathbf{A}^{\mathrm{T}}(\dot{\mathbf{v}} + \boldsymbol{\Gamma}^{\mathrm{T}}\mathbf{v}) \tag{2.4.34}$$

is objective and, more generally,

$$\frac{\partial}{\partial t}(\text{objective Lagrangean field}) = \text{objective Lagrangean field.} \tag{2.4.35}$$

Since $\dot{\mathbf{v}} + \boldsymbol{\Gamma}^{\mathrm{T}}\mathbf{v}$ is an induced Eulerian field of $\partial(\mathbf{A}^{\mathrm{T}}\mathbf{v})/\partial t$ it is objective. This result may be confirmed directly by using (2.4.16) to show that

$$\dot{\mathbf{v}}^* + \boldsymbol{\Gamma}^{*\mathrm{T}}\mathbf{v}^* = \mathbf{Q}(\dot{\mathbf{v}} + \boldsymbol{\Gamma}^{\mathrm{T}}\mathbf{v}).$$

Problem 2.4.1 If $\mathbf{E}$ is the Green strain field show that

$$\ddot{\mathbf{E}} = \mathbf{A}^{\mathrm{T}}(\dot{\boldsymbol{\Sigma}} + \boldsymbol{\Gamma}^{\mathrm{T}}\boldsymbol{\Sigma} + \boldsymbol{\Sigma}\boldsymbol{\Gamma})\mathbf{A}$$

and deduce that $\dot{\boldsymbol{\Sigma}} + \boldsymbol{\Gamma}^{\mathrm{T}}\boldsymbol{\Sigma} + \boldsymbol{\Sigma}\boldsymbol{\Gamma}$ is an objective Eulerian tensor field.

Problem 2.4.2 Given that $\mathbf{T}$ is an objective second-order Eulerian tensor field find the induced objective Eulerian tensor field of the Lagrangean time derivative of each of $\mathbf{A}^T\mathbf{T}\mathbf{A}$, $\mathbf{B}^T\mathbf{T}\mathbf{B}$, $\mathbf{A}^T\mathbf{T}\mathbf{B}$, $\mathbf{B}^T\mathbf{T}\mathbf{A}$. Deduce that

$$\dot{\mathbf{T}} + \boldsymbol{\Gamma}^T\mathbf{T} + \mathbf{T}\boldsymbol{\Gamma} + \alpha\boldsymbol{\Sigma}\mathbf{T} + \beta\mathbf{T}\boldsymbol{\Sigma}$$

is an objective Eulerian tensor field provided α and β are objective scalar fields.

Problem 2.4.3 Suppose $\mathbf{R}$ is the proper orthogonal tensor arising from the polar decomposition of $\mathbf{A}$ and that $\mathbf{v}$ (respectively $\mathbf{T}$) is an Eulerian vector (respectively second-order tensor) field. Show that $\mathbf{R}^T\mathbf{v}$ (respectively $\mathbf{R}^T\mathbf{T}\mathbf{R}$) is an objective Lagrangean vector (respectively tensor) if and only if $\mathbf{v}$ (respectively $\mathbf{T}$) is objective.

Show that

$$\dot{\mathbf{v}} - \dot{\mathbf{R}}\mathbf{R}^T\mathbf{v}, \qquad \dot{\mathbf{T}} - \dot{\mathbf{R}}\mathbf{R}^T\mathbf{T} + \mathbf{T}\dot{\mathbf{R}}\mathbf{R}^T$$

are objective Eulerian fields. If $\boldsymbol{\Omega}$ is the body spin deduce that

$$\dot{\mathbf{v}} - \boldsymbol{\Omega}\mathbf{v}, \qquad \dot{\mathbf{T}} - \boldsymbol{\Omega}\mathbf{T} + \mathbf{T}\boldsymbol{\Omega}$$

are objective.

Hence find the Lagrangean tensor field associated with $\mathbf{T}$ whose Lagrangean time derivative is expressible as

$$\mathbf{P}^T(\dot{\mathbf{T}} - \boldsymbol{\Omega}\mathbf{T} + \mathbf{T}\boldsymbol{\Omega})\mathbf{P},$$

where $\mathbf{P}$ is a proper orthogonal tensor whose physical interpretation should be investigated.

2.4.2 Embedded components of tensors

At this point it is instructive to examine the component representations for various tensors with respect to general curvilinear bases, noting first that

$$\mathbf{A} = A^i{}_\alpha \mathbf{e}_i \otimes \mathbf{E}^\alpha, \qquad \mathbf{B} = B_i{}^\alpha \mathbf{e}^i \otimes \mathbf{E}_\alpha,$$

where

$$A^i{}_\alpha = \frac{\partial x^i}{\partial X^\alpha}, \qquad B_i{}^\alpha = \frac{\partial X^\alpha}{\partial x^i}.$$

Material line elements along the basis vectors $\mathbf{E}_\alpha$ transform to $\mathbf{A}\mathbf{E}_\alpha$ under the deformation. These form a basis, $\{\boldsymbol{\varepsilon}_\alpha\}$ say, in the current configuration and we refer to this as an *embedded basis*. Its reciprocal basis $\{\varepsilon^\alpha\}$ is such that $\varepsilon^\alpha = \mathbf{B}\mathbf{E}^\alpha$ (note that these vectors are not embedded).

If $\mathbf{v}$ is an Eulerian vector field then

$$(\mathbf{A}^{\mathrm{T}}\mathbf{v})\cdot\mathbf{E}_\alpha = \mathbf{v}\cdot\boldsymbol{\varepsilon}_\alpha, \qquad (\mathbf{B}^{\mathrm{T}}\mathbf{v})\cdot\mathbf{E}^\alpha = \mathbf{v}\cdot\boldsymbol{\varepsilon}^\alpha \tag{2.4.36}$$

are respectively the covariant and contravariant components of $\mathbf{v}$ on the embedded basis. Similarly, for a second-order Eulerian tensor $\mathbf{T}$,

$$\left.\begin{aligned}
\boldsymbol{\varepsilon}_\alpha\cdot(\mathbf{T}\boldsymbol{\varepsilon}_\beta) &= \mathbf{E}_\alpha\cdot(\mathbf{A}^{\mathrm{T}}\mathbf{T}\mathbf{A})\mathbf{E}_\beta = A^i{}_\alpha T_{ij} A^j{}_\beta \\
\boldsymbol{\varepsilon}^\alpha\cdot(\mathbf{T}\boldsymbol{\varepsilon}^\beta) &= \mathbf{E}^\alpha\cdot(\mathbf{B}^{\mathrm{T}}\mathbf{T}\mathbf{B})\mathbf{E}^\beta = B_i{}^\alpha T^{ij} B_j{}^\beta \\
\boldsymbol{\varepsilon}_\alpha\cdot(\mathbf{T}\boldsymbol{\varepsilon}^\beta) &= \mathbf{E}_\alpha\cdot(\mathbf{A}^{\mathrm{T}}\mathbf{T}\mathbf{B})\mathbf{E}^\beta = A^i{}_\alpha T_i{}^j B_j{}^\beta \\
\boldsymbol{\varepsilon}^\alpha\cdot(\mathbf{T}\boldsymbol{\varepsilon}_\beta) &= \mathbf{E}^\alpha\cdot(\mathbf{B}^{\mathrm{T}}\mathbf{T}\mathbf{A})\mathbf{E}_\beta = B_i{}^\alpha T^i{}_j A^j{}_\beta
\end{aligned}\right\} \tag{2.4.37}$$

are respectively covariant, contravariant, left-covariant mixed and right-covariant mixed components of $\mathbf{T}$ on the embedded basis, where T_{ij}, T^{ij}, $T_i{}^j$, $T^i{}_j$ are the corresponding components on the basis $\{\mathbf{e}_i\}$. The generalization to higher-order tensors is straightforward. We note in passing that, as indicated in Section 2.4.1, $\mathbf{A}$ and $\mathbf{B}$ are associated with covariance and contravariance respectively in the induction from Eulerian to Lagrangean fields. In particular, the embedded metric tensor has covariant and contravariant components $\boldsymbol{\varepsilon}_\alpha\cdot\boldsymbol{\varepsilon}_\beta = \mathbf{E}_\alpha\cdot(\mathbf{A}^{\mathrm{T}}\mathbf{A})\mathbf{E}_\beta$ and $\boldsymbol{\varepsilon}^\alpha\cdot\boldsymbol{\varepsilon}^\beta = \mathbf{E}^\alpha\cdot(\mathbf{B}^{\mathrm{T}}\mathbf{B})\mathbf{E}^\beta$ respectively (recall the definition of a metric tensor given in Section 1.4.3).

REFERENCES

T.C. Doyle and J.L. Ericksen, Non-linear Elasticity, *Advances in Applied Mechanics*, Vol. 4, pp. 53–115 (1956).

R. Hill, On Constitutive Inequalities for Simple Materials—I, *Journal of the Mechanics and Physics of Solids*, Vol. 16, pp. 544–555 (1968).

R. Hill, Aspects of Invariance in Solid Mechanics, *Advances in Applied Mechanics*, Vol. 18, pp. 1–75 (1978).

R.S. Rivlin and D.W. Saunders, Large Elastic Deformations of Isotropic Materials VII: Experiments on the Deformation of Rubber, *Philosophical Transactions of the Royal Society of London*, Vol. A243, pp. 251–288 (1951).

A.J.M. Spencer, The Static Theory of Finite Elasticity, *Journal of the Institute of Mathematics and its Applications*, Vol. 6, pp. 164–200 (1970).

L.R.G. Treloar, *The Physics of Rubber Elasticity*, 3rd edition, Oxford University Press (1975).

C.A. Truesdell, *A First Course in Rational Continuum Mechanics*, Vol. I, Academic Press (1977).

C.A. Truesdell and W. Noll, The Non-linear Field Theories of Mechanics, *Handbuch der Physik*, Vol. III/3 (Ed. S. Flügge), Springer (1965).

C.-C. Wang and C.A. Truesdell, *Introduction to Rational Elasticity*, Noordhoff (1973).

CHAPTER 3

Balance Laws, Stress and Field Equations

3.1 MASS CONSERVATION

It is assumed that there is a function† m defined over the set of bodies such that, for an arbitrary body‡ B,

$$m(B) \geq 0. \tag{3.1.1}$$

Further, m is assumed to be *additive* in the sense that

$$m(B \cup B') = m(B) + m(B')$$

for arbitrary disjoint bodies B and B', where, in the notation of set theory, $B \cup B'$ denotes the (disjoint) union of B and B'. Finally, it is assumed that m is continuous§ with respect to the volume of the body B. From this it follows that $m(B) \to 0$ as the volume of B tends to zero.

The quantity $m(B)$ is called the *mass* of the body B. The above assumptions formalize in mathematical terms the intuitive notion of the physical nature of mass in the continuum context (the assumption of continuity of m excludes from consideration concentrated masses of the type familiar in classical particle mechanics). Since no reference is made to an observer in its definition, the mass $m(B)$ is intrinsic to the body B, independent of the motion of B, and therefore an objective scalar. Thus

$$\frac{\mathrm{d}}{\mathrm{d}t} m(B) = 0 \tag{3.1.2}$$

†In passing we remark that, in more abstract treatments, m is a *measure* defined over the measure space B (for arbitrary B) and, in continuum mechanics, the measurable sets for m are usually taken to be the Borel sets of B.

‡Equally, for an arbitrary sub-body of B. In order to economize on language, however, we take it as read that properties ascribed to 'body' apply also to 'sub-body'.

§Strictly, *absolutely* continuous in the sense of measure theory.

for each body B. In other words, $m(B)$ is independent of the configuration $\mathscr{B}$ occupied by B from the viewpoint of an arbitrary observer. This is a statement of the principle of *conservation of mass*.

The requirement of continuity ensures that for each configuration $\mathscr{B}$ of B there exists a scalar field ρ defined over $\mathscr{B}$ at time t such that

$$m(B) = \int_{\mathscr{B}} \rho(\mathbf{x}, t)\,\mathrm{d}v \tag{3.1.3}$$

for an arbitrary body B, where $\mathrm{d}v$ is the volume element appropriate to $\mathscr{B}$. We call ρ the mass *density* of the material of the body B in the configuration $\mathscr{B}$.

It is assumed that ρ is smooth and, because of (3.1.1) and (3.1.3), we have

$$\rho(\mathbf{x}, t) \geq 0 \qquad \mathbf{x} \in \mathscr{B}. \tag{3.1.4}$$

Since $m(B)$ is independent of the configuration $\mathscr{B}$ of B, equation (3.1.3) allows us to write

$$\int_{\mathscr{B}} \rho(\mathbf{x}, t)\,\mathrm{d}v = \int_{\mathscr{B}_0} \rho_0(\mathbf{X})\,\mathrm{d}V \tag{3.1.5}$$

for an arbitrary choice of reference configuration $\mathscr{B}_0$ for B, where ρ_0 is the mass density of B in the configuration $\mathscr{B}_0$ and $\mathrm{d}V$ is the volume element for $\mathscr{B}_0$, both independent of t.

From (3.1.5) applied to an arbitrary body, the continuity of ρ, and (2.2.14) it follows that densities ρ_0 and ρ are related by

$$\rho = J^{-1}\rho_0 \tag{3.1.6}$$

and hence, from (2.2.15),

$$\det \mathbf{A} \equiv J = \rho_0/\rho > 0, \tag{3.1.7}$$

where $\mathbf{A}$ is the gradient of the deformation from $\mathscr{B}_0$ to $\mathscr{B}$. Since $J > 0$ it follows that $\rho = 0$ if and only if $\rho_0 = 0$. Equation (3.1.6) is a local form of the principle of conservation of mass.

Since ρ is smooth, Lagrangean time differentiation of (3.1.6) followed by reference to (2.3.17) and (2.3.4) leads to

$$-\dot{\rho}/\rho = \dot{J}/J = \mathrm{tr}\,(\boldsymbol{\Gamma}) = \mathrm{div}\,\mathbf{v}, \tag{3.1.8}$$

where $\mathbf{v}(\mathbf{x}, t) = \dot{\boldsymbol{\chi}}(\mathbf{X}, t)$. This provides the usual dynamic local form of mass conservation, namely

$$\dot{\rho} + \rho \operatorname{div} \mathbf{v} = 0. \tag{3.1.9}$$

The corresponding global form is given by

$$\frac{\mathrm{d}}{\mathrm{d}t} \int_{\mathscr{B}} \rho(\mathbf{x}, t) \, \mathrm{d}v = 0 \tag{3.1.10}$$

in which, we emphasize, $\mathscr{B}$ changes with time (thus a distinction must be made between the Eulerian and Lagrangean time derivatives in carrying out differentiation under the integral; recall (3.1.5)).

A generalization of (3.1.10) is

$$\frac{\mathrm{d}}{\mathrm{d}t} \int_{\mathscr{B}} \rho f \, \mathrm{d}v = \int_{\mathscr{B}} \rho \dot{f} \, \mathrm{d}v, \tag{3.1.11}$$

where f is an arbitrary (scalar, vector or tensor) field defined over $\mathscr{B}$.

For an *isochoric* deformation $J = 1$ from (2.2.16) and hence

$$\rho = \rho_0 \tag{3.1.12}$$

for all deformations from $\mathscr{B}_0$. Then

$$\operatorname{tr}(\boldsymbol{\Gamma}) \equiv \operatorname{div} \mathbf{v} = 0 \tag{3.1.13}$$

or, equivalently,

$$\operatorname{tr}(\boldsymbol{\Sigma}) \equiv \dot{\lambda}_1/\lambda_1 + \dot{\lambda}_2/\lambda_2 + \dot{\lambda}_3/\lambda_3 = 0 \tag{3.1.14}$$

in view of (2.3.8) and (2.3.27).

Problem 3.1.1 Show that

$$\rho \dot{\mathbf{v}} = \frac{\partial}{\partial t}(\rho \mathbf{v}) + \operatorname{div}(\rho \mathbf{v} \otimes \mathbf{v}),$$

where ρ is the density in the current configuration $\mathscr{B}$ during the motion of a body B, $\mathbf{v}$ is the velocity of a material particle and $\partial/\partial t$ denotes the *Eulerian* time derivative.

Problem 3.1.2 Show that (3.1.10) may be expressed in the form

$$\int_{\mathscr{D}} \frac{\partial}{\partial t} \rho(\mathbf{x}, t) \, \mathrm{d}v = -\int_{\partial \mathscr{D}} \rho(\mathbf{x}, t) \mathbf{v}(\mathbf{x}, t) \cdot \mathbf{ds},$$

where $\mathscr{D}$ is the fixed (i.e. Eulerian) region of $\mathscr{E}$ occupied by the body at time t (i.e. $\mathscr{B}$ coincides with $\mathscr{D}$ at time t), and $\partial\mathscr{D}$ is its bounding surface.

Problem 3.1.3 Show that ρ and $\dot{\rho}$ are objective but that $\partial\rho/\partial t$ is not, where $\partial/\partial t$ denotes the Eulerian time derivative.

3.2 MOMENTUM BALANCE EQUATIONS

The *linear momentum* of the body B is defined as

$$\int_{\mathscr{B}} \rho(\mathbf{x}, t)\mathbf{v}(\mathbf{x}, t)\,\mathrm{d}v$$

in the motion described by (2.3.1), where $\mathbf{v}(\mathbf{x}, t) = \dot{\boldsymbol{\chi}}(\mathbf{X}, t)$ and $\mathscr{B}$ denotes the current configuration of the body. Note that this is an Eulerian integral which can be put in the Lagrangean form

$$\int_{\mathscr{B}_0} \rho_0(\mathbf{X}, t)\mathbf{v}\{\boldsymbol{\chi}(\mathbf{X}, t)\}\,\mathrm{d}V$$

by use of (2.3.1) and (3.1.6).

The *rotational momentum* (or moment of momentum) of B with respect to a point $\mathbf{x}_0$ of $\mathscr{E}$ (which need not be a point of $\mathscr{B}$) is defined as

$$\int_{\mathscr{B}} \rho(\mathbf{x}, t)(\mathbf{x} - \mathbf{x}_0) \wedge \mathbf{v}(\mathbf{x}, t)\,\mathrm{d}v.$$

Its value depends on the choice of $\mathbf{x}_0$. Note that neither the linear nor the rotational momentum is objective.

The reader is assumed to be familiar with the concept of force in the context of Newtonian mechanics and we do not enter a discussion of fundamentals here. For a full discussion of force in the continuum context see Truesdell (1977), for example.

In continuum mechanics we are concerned with the motion of deforming bodies and we regard the forces on a given body as arising from external influences (which we need not specify here) *independent of the observer or frame of reference.* We refer to such forces as *applied forces* in order to distinguish them from *inertial forces* which depend fundamentally on the choice of the observer and his frame of reference.

An applied force (or system of forces) acts on the current configuration $\mathscr{B}$ of the body B. It comprises a *body force*

$$\int_{\mathscr{B}} \rho(\mathbf{x}, t)\mathbf{b}(\mathbf{x}, t)\,\mathrm{d}v$$

and a *contact force*

$$\int_{\partial\mathscr{B}} \mathbf{t}(\mathbf{x}, \partial\mathscr{B})\, \mathrm{d}a,$$

where the *body-force density* $\mathbf{b}$ is a vector field defined over $\mathscr{B}$ and $\mathbf{t}$ is the *contact-force density*,† a vector field defined over any piecewise smooth oriented surface in $\mathscr{B}$ (including the boundary $\partial\mathscr{B}$ of $\mathscr{B}$). Note that we have indicated dependence of $\mathbf{t}$ on the boundary over which it acts. Other terminologies for $\mathbf{t}$ are *load* or *traction* or *stress vector* field per unit area of surface (in $\mathscr{B}$).

A body force affects each point of $\mathscr{B}$ directly (gravity is the most familiar example of a body force). A contact force, on the other hand, has a direct effect only on surface points but, of course, its influence is noticed by all points of the body in general by transmission across surfaces. (Typically a contact force consists of friction tangential to the surface and pressure normal to the surface.)

As we indicated above, we regard the notion of an applied force as independent of observer. This means that the force densities $\mathbf{b}$ and $\mathbf{t}$ (which are Eulerian vector fields) transform according to the rule (2.4.23) under an observer transformation, i.e. they are objective Eulerian vector fields. The body and contact forces themselves satisfy the same transformation rule except that they are independent of $\mathbf{x}$.

The resultant applied force on the body in the configuration $\mathscr{B}$ is

$$\int_{\mathscr{B}} \rho(\mathbf{x}, t)\mathbf{b}(\mathbf{x}, t)\, \mathrm{d}v + \int_{\partial\mathscr{B}} \mathbf{t}(\mathbf{x}, \partial\mathscr{B})\, \mathrm{d}a$$

and the resultant moment of the applied forces about a point $\mathbf{x}_0$ of $\mathscr{E}$ is

$$\int_{\mathscr{B}} \rho(\mathbf{x}, t)(\mathbf{x} - \mathbf{x}_0) \wedge \mathbf{b}(\mathbf{x}, t)\, \mathrm{d}v + \int_{\partial\mathscr{B}} (\mathbf{x} - \mathbf{x}_0) \wedge \mathbf{t}(\mathbf{x}, \partial\mathscr{B})\, \mathrm{d}a.$$

(In this book we omit from consideration body couples not arising from the applied body force.) Note that the moment of the applied forces, in contrast to the resultant applied force, is not objective because of the dependence on the choice of $\mathbf{x}_0$. The resultant moment is also called the resultant *torque* of the applied forces.

A fundamental assumption (essentially Newton's first law of motion) is that there exists an observer or, equivalently, a frame of reference of an observer (recall Section 2.1.1) for which the linear momentum of any body

†In general $\mathbf{t}$ depends on time t, but we omit its explicit dependence on t here to avoid congestion.

B in motion is independent of time provided that the resultant applied force acting on the configuration $\mathscr{B}$ of the body vanishes for all time. We refer to such an observer as an *inertial observer* and to a non-rotating, non-accelerating frame of reference of that observer as an *inertial frame*.

Given the existence of an inertial observer it follows that if the applied force is non-zero then the linear momentum changes in time. There is then a balance between the resultant applied force and the rate of change of linear momentum. Similarly, there is a balance between the resultant moment of the applied forces and the rate of change of the rotational momentum of the body. These two balances are independent and provide two fundamental principles (assumptions) of continuum mechanics known as *Euler's laws of motion*. The *balance of linear momentum* is written as

$$\int_{\mathscr{B}} \rho(\mathbf{x},t)\mathbf{b}(\mathbf{x},t)\,\mathrm{d}v + \int_{\partial\mathscr{B}} \mathbf{t}(\mathbf{x},\partial\mathscr{B})\,\mathrm{d}a = \frac{\mathrm{d}}{\mathrm{d}t}\int_{\mathscr{B}} \rho(\mathbf{x},t)\mathbf{v}(\mathbf{x},t)\,\mathrm{d}v$$

$$\equiv \int_{\mathscr{B}} \rho(\mathbf{x},t)\dot{\mathbf{v}}(\mathbf{x},t)\,\mathrm{d}v \qquad (3.2.1)$$

and the *balance of rotational momentum*

$$\int_{\mathscr{B}} \rho(\mathbf{x},t)(\mathbf{x}-\mathbf{x}_0)\wedge\mathbf{b}(\mathbf{x},t)\,\mathrm{d}v + \int_{\partial\mathscr{B}} (\mathbf{x}-\mathbf{x}_0)\wedge\mathbf{t}(\mathbf{x},\partial\mathscr{B})\,\mathrm{d}a$$

$$= \frac{\mathrm{d}}{\mathrm{d}t}\int_{\mathscr{B}} \rho(\mathbf{x},t)(\mathbf{x}-\mathbf{x}_0)\wedge\mathbf{v}(\mathbf{x},t)\,\mathrm{d}v$$

$$\equiv \int_{\mathscr{B}} \rho(\mathbf{x},t)(\mathbf{x}-\mathbf{x}_0)\wedge\dot{\mathbf{v}}(\mathbf{x},t)\,\mathrm{d}v. \qquad (3.2.2)$$

The final identity in (3.2.2) requires that $\mathbf{x}_0$ be a fixed point of $\mathscr{E}$ as far as the observer O is concerned. These two laws are not objective as they stand but can be put in objective form by incorporating inertial forces into the definition of $\mathbf{b}$ (see Wang and Truesdell, 1973 p. 118, for example).

3.3 THE CAUCHY STRESS TENSOR

3.3.1 Linear dependence of the stress vector on the surface normal

It is assumed that the stress vector $\mathbf{t}$ at a point $\mathbf{x}$ depends on the surface only through the (unit) normal $\mathbf{n}$ to the considered surface at $\mathbf{x}$, i.e. $\mathbf{t}$ has the same value for all surfaces through $\mathbf{x}$ which have normal in the direction $\mathbf{n}$

at $\mathbf{x}$. This is *Cauchy's fundamental postulate.* We now replace the notation $\mathbf{t}(\mathbf{x}, \partial\mathscr{B})$ by $\mathbf{t}(\mathbf{x}, \mathbf{n})$ to make the dependence on $\mathbf{n}$ explicit.

An immediate consequence is the result

$$\mathbf{t}(\mathbf{x}, -\mathbf{n}) = -\mathbf{t}(\mathbf{x}, \mathbf{n}). \tag{3.3.1}$$

Essentially this states that the contact force (per unit area) exerted by material on one side (side (1)) of a surface on the material on the other side (side (2)) is equal and opposite to the force exerted by the material on side (2) on the material of side (1). A proof of (3.3.1) is as follows.

Consider the volume formed by a circular cylinder centred at the point $\mathbf{x}$ with base area a having unit normal $\mathbf{n}$ parallel to the generators of the cylinder. Let h be the height of the cylinder. Application of the balance law (3.2.1) to this volume with the assumption that the body force and acceleration are suitably bounded leads to

$$\frac{1}{a}\int_{\mathscr{A}} [\mathbf{t}(\mathbf{x}, \mathbf{n}) + \mathbf{t}(\mathbf{x}, -\mathbf{n})]\, \mathrm{d}a = \mathbf{0}$$

in the limit as $h \rightarrow 0$, where the integration is over an end-face $\mathscr{A}$ of the cylinder (opposite end-faces have unit normals $\mathbf{n}$ and $-\mathbf{n}$ pointing out of the enclosed volume). Application of this equation to a cylinder of arbitrarily small cross-sectional area and use of the (assumed) continuity of $\mathbf{t}(\mathbf{x}, \mathbf{n})$ in $\mathbf{x}$ then yields (3.3.1).

We now state *Cauchy's theorem*: *provided it is continuous in* $\mathbf{x}$ *the stress vector* $\mathbf{t}(\mathbf{x}, \mathbf{n})$ *depends linearly on* $\mathbf{n}$, i.e. there exists a second-order tensor field $\mathbf{T}$ independent of $\mathbf{n}$ such that

$$\mathbf{t}(\mathbf{x}, \mathbf{n}) = \mathbf{T}(\mathbf{x})\mathbf{n} \tag{3.3.2}$$

for all $\mathbf{x}$ in $\mathscr{B}$ and for arbitrary unit vectors $\mathbf{n}$. The tensor $\mathbf{T}(\mathbf{x})$ is called the *Cauchy stress tensor*. It is also referred to as the *true* stress tensor. A standard proof of (3.3.2) now follows.

Consider a tetrahedron with three faces lying in the rectangular Cartesian coordinate planes through a point $\mathbf{x}$ and normal to the (orthonormal) basis vectors $\mathbf{e}_i$ so that $-\mathbf{e}_i$ $(i = 1, 2, 3)$ are the unit normals to these faces directed out of the enclosed volume. Let $\mathbf{n}$ be the unit (outward) normal to the arbitrarily oriented fourth face of the tetrahedron and suppose a is the area of this face. The other faces then have areas an_i, where $n_i = \mathbf{n} \cdot \mathbf{e}_i > 0$ $(i = 1, 2, 3)$.

Let $\mathbf{x}$ be a distance h from the fourth face. Then the tetrahedron has volume $\frac{1}{3}ha$. If h is small then

$$\int_{\mathscr{B}} \rho(\mathbf{x}, t)[\dot{\mathbf{v}}(\mathbf{x}, \mathbf{t}) - \mathbf{b}(\mathbf{x}, t)]\, \mathrm{d}v \simeq \tfrac{1}{3}\rho(\mathbf{x}, t)[\dot{\mathbf{v}}(\mathbf{x}, t) - \mathbf{b}(\mathbf{x}, t)]ha$$

provided $\dot{\mathbf{v}}$ and $\mathbf{b}$ are suitably bounded over the tetrahedron $\mathscr{B}$. This expression vanishes in the limit $h \to 0$. Application of the balance law (3.2.1) to the tetrahedron and use of this result yields

$$\frac{1}{a}\int_{\mathscr{A}} [\mathbf{t}(\mathbf{x}, \mathbf{n}) + \sum_{i=1}^{3} \mathbf{t}(\mathbf{x}, -\mathbf{e}_i)n_i]\,\mathrm{d}a = \mathbf{0}$$

in the limit $h \to 0$, the integration over the surface $\partial\mathscr{B}$ of the tetrahedron having been separated into integration over the four constituent faces. In the above integral $\mathscr{A}$ represents the fourth face of the tetrahedron.

By continuity of $\mathbf{t}(\mathbf{x}, \mathbf{n})$ in $\mathbf{x}$ for arbitrary $\mathbf{n}$ and use of (3.3.1) it follows from the above equation on taking the limit $a \to 0$ that

$$\mathbf{t}(\mathbf{x}, \mathbf{n}) = -\sum_{i=1}^{3} \mathbf{t}(\mathbf{x}, -\mathbf{e}_i)n_i = \sum_{i=1}^{3} \mathbf{t}(\mathbf{x}, \mathbf{e}_i)n_i. \tag{3.3.3}$$

Since the basis $\{\mathbf{e}_i\}$ is fixed, arbitrary and independent of $\mathbf{n}$, equation (3.3.3) shows that $\mathbf{t}(\mathbf{x}, \mathbf{n})$ is linearly dependent on the components n_i of the normal $\mathbf{n}$ and we have therefore established Cauchy's theorem.

In components (3.3.3) is written

$$t_i(\mathbf{x}, \mathbf{n}) = t_i(\mathbf{x}, \mathbf{e}_j)n_j \qquad (i = 1, 2, 3;\ \text{sum over } j).$$

By setting $t_i(\mathbf{x}, \mathbf{e}_j) = T_{ij}(\mathbf{x})$, we obtain

$$t_i(\mathbf{x}, \mathbf{n}) = T_{ij}(\mathbf{x})n_j \tag{3.3.4}$$

and on identification of $T_{ij}(\mathbf{x})$ with the (rectangular Cartesian) components of a second-order tensor $\mathbf{T}(\mathbf{x})$ we recover (3.3.2) in component form. (We recall that the components T_{ij} were discussed in Section 1.2.1 to help motivate the definition of a second-order tensor.)

It is left as an exercise for the reader to show that $\mathbf{T}$ is an objective second-order Eulerian tensor field.[†]

Equation (3.3.4) yields the following interpretation of the components of $\mathbf{T}(\mathbf{x})$: $T_{ij}(\mathbf{x})$ is the $\mathbf{e}_i$-component of the force (per unit area) at a point $\mathbf{x}$ on an element of surface in the current configuration whose unit normal is in the $\mathbf{e}_j$-direction. In particular, for $j = 2$, T_{12}, T_{22}, T_{32} are the components of force on a plane element of surface normal to $\mathbf{e}_2$ as illustrated in Fig. 3.1 (the argument $\mathbf{x}$ has been omitted).

Each component T_{ii} ($i = 1, 2, 3$; no summation) is called a *normal stress*

[†] It should be remembered that the time dependence of $\mathbf{t}$, and hence of $\mathbf{T}$, has been omitted from explicit consideration thus far.

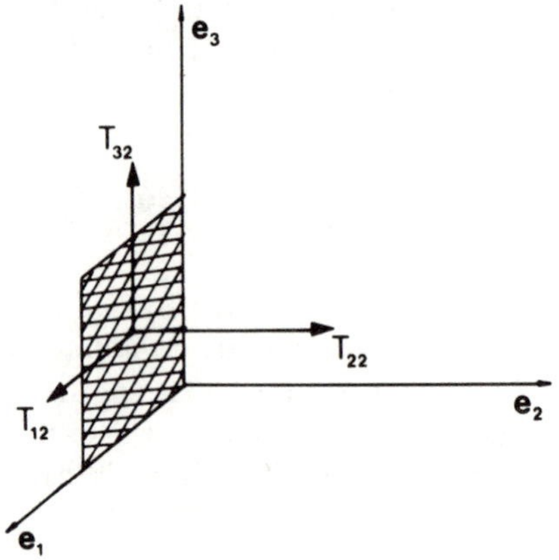

Fig. 3.1 Components T_{12}, T_{22}, T_{32} of the stress vector acting on an element of surface area normal to $\mathbf{e}_2$.

since it acts in the direction of the unit normal to a coordinate surface. The components T_{ij} $(i \neq j)$ act tangentially to coordinate surfaces and are called *shear stresses.* We show in Section 3.3.2 that not all the shear stresses are independent. In fact $T_{ij} = T_{ji}$, i.e. $\mathbf{T}$ is a symmetric second-order tensor.

In general, the *normal stress component* in a direction $\mathbf{n}$ is

$$\mathbf{n}\cdot\mathbf{t}(\mathbf{x},\mathbf{n}) \equiv \mathbf{n}\cdot[\mathbf{T}(\mathbf{x})\mathbf{n})] \tag{3.3.5}$$

and the *shear stress vector* is

$$\mathbf{t}(\mathbf{x},\mathbf{n}) - [\mathbf{n}\cdot\mathbf{t}(\mathbf{x},\mathbf{n})]\mathbf{n}. \tag{3.3.6}$$

3.3.2 Cauchy's laws of motion

In view of the result (3.3.2) the equation of linear momentum balance (3.2.1) becomes†

$$\int_{\mathscr{B}} \rho(\mathbf{x},t)\mathbf{b}(\mathbf{x},t)\,\mathrm{d}v + \int_{\partial\mathscr{B}} \mathbf{T}(\mathbf{x},t)\mathbf{n}\,\mathrm{d}a = \int_{\mathscr{B}} \rho(\mathbf{x},t)\dot{\mathbf{v}}(\mathbf{x},t)\,\mathrm{d}v$$

and application of the divergence theorem to the surface integral leads to

$$\int_{\mathscr{B}} [\rho(\mathbf{x},t)\mathbf{b}(\mathbf{x},t) + \operatorname{div}\mathbf{T}^{\mathrm{T}}(\mathbf{x},t) - \rho(\mathbf{x},t)\dot{\mathbf{v}}(\mathbf{x},t)]\,\mathrm{d}v = \mathbf{0}.$$

† The dependence of $\mathbf{T}$ on time is now included.

This is valid for arbitrary bodies $\mathscr{B}$. Hence, provided ρ, $\mathbf{b}$ and $\dot{\mathbf{v}}$ are continuous (already assumed in previous sections) and $\mathbf{T}$ is once continuously differentiable (now assumed) there follows the field equation

$$\operatorname{div}\mathbf{T}^{\mathrm{T}} + \rho\mathbf{b} = \rho\dot{\mathbf{v}}. \tag{3.3.7}$$

This is known as *Cauchy's first law of motion.*

Expressed in terms of rectangular Cartesian components (3.3.7) becomes

$$\frac{\partial T_{ij}}{\partial x_j} + \rho b_i = \rho \dot{v}_i. \tag{3.3.8}$$

Substitution of (3.3.7) into the equation of rotational momentum balance (3.2.2) gives

$$\int_{\mathscr{B}} (\mathbf{x}-\mathbf{x}_0)\wedge[\operatorname{div}\mathbf{T}^{\mathrm{T}}(\mathbf{x},t)]\,\mathrm{d}v = \int_{\partial\mathscr{B}} (\mathbf{x}-\mathbf{x}_0)\wedge[\mathbf{T}(\mathbf{x},t)\mathbf{n}]\,\mathrm{d}a.$$

In order to avoid using components at this stage we express this equation in an equivalent tensor form as follows. From (1.3.28) we deduce that $\mathbf{m}\wedge\mathbf{n}$ is the axial vector of the antisymmetric second-order tensor $\mathbf{n}\otimes\mathbf{m}-\mathbf{m}\otimes\mathbf{n}$, where $\mathbf{m}$ and $\mathbf{n}$ are arbitrary vectors. Hence the above equation may be written

$$\int_{\mathscr{B}} \{(\mathbf{x}-\mathbf{x}_0)\otimes[\operatorname{div}\mathbf{T}^{\mathrm{T}}(\mathbf{x},t)] - [\operatorname{div}\mathbf{T}^{\mathrm{T}}(\mathbf{x},t)]\otimes(\mathbf{x}-\mathbf{x}_0)\}\,\mathrm{d}v$$
$$= \int_{\partial\mathscr{B}} \{(\mathbf{x}-\mathbf{x}_0)\otimes[\mathbf{T}(\mathbf{x},t)\mathbf{n}] - [\mathbf{T}(\mathbf{x},t)\mathbf{n}]\otimes(\mathbf{x}-\mathbf{x}_0)\}\,\mathrm{d}a.$$

Application of the divergence theorem to the right-hand side of this equation and use of the result

$$\operatorname{div}[\mathbf{T}^{\mathrm{T}}(\mathbf{x},t)\otimes(\mathbf{x}-\mathbf{x}_0)] = [\operatorname{div}\mathbf{T}^{\mathrm{T}}(\mathbf{x},t)]\otimes(\mathbf{x}-\mathbf{x}_0) + \mathbf{T}(\mathbf{x},t)$$

(recall Problem 1.5.23) reduces the equation of rotational momentum balance to

$$\int_{\mathscr{B}} [\mathbf{T}(\mathbf{x},t) - \mathbf{T}^{\mathrm{T}}(\mathbf{x},t)]\,\mathrm{d}v = \mathbf{0}.$$

Since this must hold for arbitrary $\mathscr{B}$ it follows by continuity of $\mathbf{T}(\mathbf{x},t)$ that

$$\mathbf{T}^{\mathrm{T}} = \mathbf{T}, \tag{3.3.9}$$

i.e. the Cauchy stress tensor is symmetric. This is *Cauchy's second law of motion.* In rectangular Cartesian components it becomes

$$T_{ij} = T_{ji}. \tag{3.3.10}$$

We emphasize that once (3.3.7) is established the equation of balance of rotational momentum is equivalent simply to (3.3.9).

In summary, we collect together the field equations governing the motion of a continuum, namely the equation of mass conservation (3.1.9) and Cauchy's laws of motion (3.3.7) and (3.3.9):

$$\left.\begin{aligned} \dot{\rho} + \rho \operatorname{div} \mathbf{v} &= 0, \\ \mathbf{T}^{\mathrm{T}} &= \mathbf{T}, \\ \operatorname{div} \mathbf{T} + \rho \mathbf{b} &= \rho \dot{\mathbf{v}}. \end{aligned}\right\} \tag{3.3.11}$$

This set of equations is Eulerian in character since it involves Eulerian scalar, vector and tensor fields and $\mathbf{x}$ and $\mathbf{t}$ are regarded as independent variables. We therefore refer to the equations (3.3.11) as *Eulerian field equations.*

If the body B is *at rest* $\mathbf{v}(\mathbf{x}, t) = \mathbf{0}$ for all $\mathbf{x} \in \mathscr{B}$ and (3.3.7) becomes the *equilibrium equation*

$$\operatorname{div} \mathbf{T} + \rho \mathbf{b} = \mathbf{0}. \tag{3.3.12}$$

There is then no dependence on time and $(3.3.11)_1$ is replaced by (3.1.7). Of course, equation (3.3.12) does not imply that the body is at rest in general since a (non-zero) constant value of $\mathbf{v}(\mathbf{x}, t)$ is not excluded.

If, additionally, there are no body forces then (3.3.12) reduces further to

$$\operatorname{div} \mathbf{T} = \mathbf{0} \tag{3.3.13}$$

and the stress field $\mathbf{T}$ is then said to be *self-equilibrated.*

Once the symmetry $\mathbf{T}^{\mathrm{T}} = \mathbf{T}$ is given (3.3.11) provides four scalar equations. There are ten scalar quantities to be determined, however: the density ρ, the three components of $\mathbf{x}$ and the six independent components of $\mathbf{T}$. (It is assumed that the body-force density $\mathbf{b}$ and the mass density ρ_0 in the configuration $\mathscr{B}_0$ are known.) The set of equations is completed by specifying $\mathbf{T}$ as a function or functional of t, $\mathbf{x}$, $\mathbf{v} \equiv \dot{\mathbf{x}}$, $\operatorname{Grad} \mathbf{x}$ and, in general, higher derivatives of $\mathbf{x} = \boldsymbol{\chi}(\mathbf{X}, t)$ also. The equation resulting from this specification is called a *constitutive equation*, and its precise form depends on the nature of the material behaviour being modelled. A detailed discussion of constitutive theory is provided in Chapter 4 where attention is for the most part confined to elastic materials.

The differential equations themselves require supplementation by boundary conditions and (in general) initial conditions for the motion $\mathbf{x} = \boldsymbol{\chi}(\mathbf{X}, t)$ to be determinable (in principle) for a given type of material. Some standard boundary conditions are therefore described in Chapter 5 at the beginning of the discussion of boundary value problems. We remark in passing, however, that an initial-boundary-value problem so defined is not necessarily well-posed, nor is the uniqueness or even existence of a solution guaranteed in general. Thorough analysis of such questions is beyond the scope of this book.

Finally in this section we note that a consequence of the symmetry (3.3.9) is that the Cauchy stress tensor is expressible in the spectral form

$$\mathbf{T} = \sum_{i=1}^{3} t_i \mathbf{v}^{(i)} \otimes \mathbf{v}^{(i)}, \tag{3.3.14}$$

where t_i ($i = 1, 2, 3$) are the principal values of $\mathbf{T}$, i.e. the *principal Cauchy stresses*, and $\mathbf{v}^{(i)}$ ($i = 1, 2, 3$) are the principal directions of $\mathbf{T}$. These $\mathbf{v}^{(i)}$ do not in general coincide with the directions $\mathbf{v}^{(i)}$ of the Eulerian principal axes defined in Section 2.3.2. However, the two sets of directions do coincide for isotropic elastic solids (as we see in Chapter 4) and this is why the notation is duplicated here.

Problem 3.3.1 If $\mathbf{t}$ denotes the traction per unit area on a surface whose normal is in the direction $\mathbf{n}$, show that the square of the magnitude of the shear stress on that surface is

$$\mathbf{t} \cdot \mathbf{t} - (\mathbf{n} \cdot \mathbf{t})^2.$$

If n_i ($i = 1, 2, 3$) are the components of $\mathbf{n}$ relative to the principal axes of the Cauchy stress tensor, show that the above expression may be written

$$(t_2 - t_3)^2 n_2^2 n_3^2 + (t_3 - t_1)^2 n_3^2 n_1^2 + (t_1 - t_2)^2 n_1^2 n_2^2,$$

where t_i ($i = 1, 2, 3$) are the principal Cauchy stresses.

Show that the average of this over all possible directions $\mathbf{n}$ is

$$\tfrac{1}{15}\{(t_2 - t_3)^2 + (t_3 - t_1)^2 + (t_1 - t_2)^2\}$$

and deduce that this is expressible as

$$\tfrac{2}{15}\{[I_1(\mathbf{T})]^2 - 3I_2(\mathbf{T})\},$$

where $I_1(\mathbf{T})$ and $I_2(\mathbf{T})$ are the first two principal invariants of $\mathbf{T}$.

Problem 3.3.2 Show that if $\mathbf{n}\cdot\mathbf{t}$ is stationary with respect to $\mathbf{n}(\mathbf{n}\cdot\mathbf{n}=1)$ then $\mathbf{t}=(\mathbf{n}\cdot\mathbf{t})\mathbf{n}$ and find the stationary values of $\mathbf{n}\cdot\mathbf{t}$ in terms of the principal Cauchy stresses.

Find similarly the stationary values of

$$\mathbf{t}\cdot\mathbf{t}-(\mathbf{n}\cdot\mathbf{t})^2.$$

Problem 3.3.3 The stress at a point $\mathbf{x}$ is said to be *uniaxial* at $\mathbf{x}$ if there exists a unit vector $\mathbf{k}$ and a scalar T such that $\mathbf{T}=T\mathbf{k}\otimes\mathbf{k}$ at $\mathbf{x}$. If $\mathbf{T}$ is self-equilibrated and $\mathbf{k}$ is independent of $\mathbf{x}$ deduce that T is constant on any plane parallel to $\mathbf{k}$.

Show that the square of the magnitude of the shear stress on a surface whose unit normal is $\mathbf{n}$ at $\mathbf{x}$ is

$$T^2(\mathbf{k}\cdot\mathbf{n})^2[1-(\mathbf{k}\cdot\mathbf{n})^2],$$

and that this has a maximum value of $\frac{1}{4}T^2$ amongst all possible surfaces at $\mathbf{x}$ for fixed $\mathbf{k}$.

Problem 3.3.4 If a body B is in equilibrium in configuration $\mathscr{B}$ show that the average

$$\bar{\mathbf{T}}\equiv\frac{1}{v(\mathscr{B})}\int_{\mathscr{B}}\mathbf{T}\,\mathrm{d}v$$

of the Cauchy stress tensor is expressible as

$$\frac{1}{2v(\mathscr{B})}\int_{\mathscr{B}}\rho(\mathbf{x}\otimes\mathbf{b}+\mathbf{b}\otimes\mathbf{x})\,\mathrm{d}v+\frac{1}{2v(\mathscr{B})}\int_{\partial\mathscr{B}}(\mathbf{x}\otimes\mathbf{t}+\mathbf{t}\otimes\mathbf{x})\,\mathrm{d}a,$$

where $v(\mathscr{B})$ is the volume of $\mathscr{B}$ and the arguments have been omitted from the fields in the integrands.

If $\mathbf{b}=\mathbf{0}$ in $\mathscr{B}$ and $\mathbf{t}=T\mathbf{n}$ on $\partial\mathscr{B}$, where T is a constant, show that $\bar{\mathbf{T}}=T\mathbf{I}$, where $\mathbf{I}$ is the identity.

If $\mathbf{b}=\mathbf{0}$ in $\mathscr{B}$ and $\mathbf{t}=T(\mathbf{k}.\mathbf{n})\mathbf{k}$ on $\partial\mathscr{B}$, where $\mathbf{k}$ and T are constant, show that $\bar{\mathbf{T}}=T\mathbf{k}\otimes\mathbf{k}$.

3.4 THE NOMINAL STRESS TENSOR

3.4.1 Definition of nominal stress

On use of Nanson's formula (2.2.18) relating current and reference elements of surface area we see that the resultant contact force on the boundary $\partial\mathscr{B}$

of the current configuration $\mathscr{B}$ of the body B may be written

$$\int_{\partial\mathscr{B}} \mathbf{T}\mathbf{n}\,\mathrm{d}a = \int_{\partial\mathscr{B}_0} J\mathbf{T}\mathbf{B}\mathbf{N}\,\mathrm{d}A, \tag{3.4.1}$$

where $\mathbf{N}$ is the unit (outward) normal to the boundary $\partial\mathscr{B}_0$ of the reference configuration $\mathscr{B}_0$.

We introduce the notation $\mathbf{S}^{\mathrm{T}} = J\mathbf{T}\mathbf{B}$ so that the above may also be written

$$\int_{\partial\mathscr{B}_0} \mathbf{S}^{\mathrm{T}}\mathbf{N}\,\mathrm{d}A,$$

and, in view of the symmetry of $\mathbf{T}$, the tensor field $\mathbf{S}$ is given by

$$\mathbf{S} = J\mathbf{B}^{\mathrm{T}}\mathbf{T}, \tag{3.4.2}$$

where J and $\mathbf{B}$ are defined in Section 2.2.1.

Since $\mathbf{T}$ is Eulerian, $\mathbf{S}$ is clearly a two-point tensor field (in fact, an induced tensor field of $J\mathbf{T}$ and therefore objective). It is called the *nominal stress tensor* (field). Its transpose is sometimes referred to as the first Piola–Kirchhoff stress tensor (field) but we do not use this terminology here.

Locally, the load $\mathbf{dl} \equiv \mathbf{t}\,\mathrm{d}a$ on an element of surface $\mathrm{d}a$ in the configuration $\mathscr{B}$ may be written

$$\mathbf{dl} = \mathbf{T}\,\mathbf{da} = \mathbf{S}^{\mathrm{T}}\,\mathbf{dA} \tag{3.4.3}$$

and this provides a load-area interpretation for $\mathbf{S}$ parallel to the one for $\mathbf{T}$ given in Section 3.3.1: with respect to orthonormal bases $\{\mathbf{E}_\alpha\}$ and $\{\mathbf{e}_i\}$ in the reference and current configurations respectively the component $S_{\alpha i}$ of $\mathbf{S}$ is the $\mathbf{e}_i$-component of force (per unit *reference* area) on an element of surface in the current configuration whose normal was in the $\mathbf{E}_\alpha$-direction in the reference configuration. Note that $\mathbf{dl}$ is an Eulerian vector.

3.4.2. The Lagrangean field equations

The linear momentum balance equation (3.2.1) may now be written in terms of integrals over $\mathscr{B}_0$ and $\partial\mathscr{B}_0$:

$$\int_{\mathscr{B}_0} \rho_0(\mathbf{X})\mathbf{b}_0(\mathbf{X},t)\,\mathrm{d}V + \int_{\partial\mathscr{B}_0} \mathbf{S}^{\mathrm{T}}(\mathbf{X},t)\mathbf{N}\,\mathrm{d}A = \int_{\mathscr{B}_0} \rho_0(\mathbf{X})\ddot{\boldsymbol{\chi}}(\mathbf{X},t)\,\mathrm{d}V,$$

where $\mathbf{b}_0(\mathbf{X}, t) = \mathbf{b}\{\boldsymbol{\chi}(\mathbf{X}, t), t\}$. Application of the divergence theorem as in Section 3.3.2 then yields

$$\operatorname{Div} \mathbf{S} + \rho_0 \mathbf{b}_0 = \rho_0 \ddot{\boldsymbol{\chi}}. \tag{3.4.4}$$

In this equation the independent variables are $\mathbf{X}$ and t and we therefore refer to this as the *Lagrangean equation of motion* (although the vectors involved are actually Eulerian). With respect to rectangular Cartesian bases (3.4.4) has the form

$$\frac{\partial S_{\alpha i}}{\partial X_\alpha} + \rho_0 b_{0i} = \rho_0 \ddot{\chi}_i.$$

From (3.4.2) it follows that symmetry of $\mathbf{T}$ is equivalent to either of

$$\mathbf{AS} = \mathbf{S}^{\mathrm{T}}\mathbf{A}^{\mathrm{T}} \quad \text{or} \quad \mathbf{SB} = \mathbf{B}^{\mathrm{T}}\mathbf{S}^{\mathrm{T}}. \tag{3.4.5}$$

The set of Lagrangean field equations (3.4.4) and (3.4.5) is then completed by the addition of the mass conservation equation

$$\det \mathbf{A} = \rho_0/\rho. \tag{3.4.6}$$

The Lagrangean and Eulerian field equations are equivalent for any continuum (the equations may be modified to include couple stresses). Whether the Lagrangean or Eulerian specification is used, however, depends on the circumstances of the problem in question and also on the nature of the material being studied. It is most common in fluid dynamics, for example, for the Eulerian equations to be used but in solid mechanics a Lagrangean formulation is often appropriate since a reference configuration can be chosen to coincide with the initial geometry of the considered solid.

Problem 3.4.1 Derive equation (3.4.4) directly from (3.3.7) by using the definition (3.4.2) and the result (2.2.20).

Problem 3.4.2 Show that

$$\int_{\mathscr{B}_0} \mathbf{A}\, \mathrm{d}V = \int_{\partial\mathscr{B}_0} \mathbf{x} \otimes \mathbf{N}\, \mathrm{d}A,$$

$$\int_{\mathscr{B}_0} \mathbf{S}\, \mathrm{d}V = \int_{\partial\mathscr{B}_0} \mathbf{X} \otimes \mathbf{S}^{\mathrm{T}}\mathbf{N}\, \mathrm{d}A + \int_{\mathscr{B}_0} \rho_0 \mathbf{X} \otimes \mathbf{b}_0\, \mathrm{d}V,$$

and

$$\int_{\mathscr{B}_0} \mathbf{A}\mathbf{S}\,\mathrm{d}V = \int_{\partial\mathscr{B}_0} \mathbf{x}\otimes\mathbf{S}^{\mathrm{T}}\mathbf{N}\,\mathrm{d}A + \int_{\mathscr{B}_0} \rho_0\mathbf{x}\otimes\mathbf{b}_0\,\mathrm{d}V$$

for a body in equilibrium in a configuration $\mathscr{B}$, where $\mathbf{A}$ denotes the deformation gradient, $\mathscr{B}_0$ the reference configuration and the arguments have been omitted from the fields.

3.5 CONJUGATE STRESS ANALYSIS

3.5.1 Work rate and energy balance

On taking the scalar product of equation (3.3.7) with the velocity $\mathbf{v}$ and using (3.3.9) we obtain

$$(\operatorname{div}\mathbf{T})\cdot\mathbf{v} + \rho\mathbf{b}\cdot\mathbf{v} = \rho\mathbf{v}\cdot\dot{\mathbf{v}}$$

and this may be rearranged as

$$\operatorname{div}(\mathbf{T}\mathbf{v}) - \operatorname{tr}(\mathbf{T}\boldsymbol{\Gamma}) + \rho\mathbf{b}\cdot\mathbf{v} = \rho\mathbf{v}\cdot\dot{\mathbf{v}},$$

where $\boldsymbol{\Gamma} = \operatorname{grad}\mathbf{v}$.

Integration over the volume of the current configuration $\mathscr{B}$ followed by the use of the divergence theorem and the conservation of mass leads to

$$\int_{\mathscr{B}} \rho\mathbf{b}\cdot\mathbf{v}\,\mathrm{d}v + \int_{\partial\mathscr{B}} \mathbf{t}\cdot\mathbf{v}\,\mathrm{d}a = \frac{\mathrm{d}}{\mathrm{d}t}\int_{\mathscr{B}} \tfrac{1}{2}\rho\mathbf{v}\cdot\mathbf{v}\,\mathrm{d}v + \int_{\mathscr{B}} \operatorname{tr}(\mathbf{T}\boldsymbol{\Sigma})\,\mathrm{d}v \qquad (3.5.1)$$

since the symmetry of $\mathbf{T}$ implies

$$\operatorname{tr}(\mathbf{T}\boldsymbol{\Gamma}) = \operatorname{tr}(\mathbf{T}\boldsymbol{\Sigma}), \qquad (3.5.2)$$

where $\boldsymbol{\Sigma} = \frac{1}{2}(\boldsymbol{\Gamma} + \boldsymbol{\Gamma}^{\mathrm{T}})$ is the Eulerian strain-rate.

The term on the left-hand side of (3.5.1) represents the *rate of working of the applied forces* on the body in configuration $\mathscr{B}$. The term

$$\frac{\mathrm{d}}{\mathrm{d}t}\int_{\mathscr{B}} \tfrac{1}{2}\rho\mathbf{v}\cdot\mathbf{v}\,\mathrm{d}v$$

is the *rate of change of the kinetic energy* of the body, $\frac{1}{2}\rho\mathbf{v}\cdot\mathbf{v}$ being the kinetic energy density, and

$$\int_{\mathscr{B}} \operatorname{tr}(\mathbf{T}\boldsymbol{\Sigma})\,\mathrm{d}v \tag{3.5.3}$$

is the *rate of working of the stresses* on the body (also called the *stress power*). Equation (3.5.1) is the *mechanical energy balance equation.* In general energy is not conserved during the motion of a continuum and the stress power will incorporate both dissipative and conservative contributions. This means that the stress power term in (3.5.1) cannot be written as the rate of change of an integral over $\mathscr{B}$. The special case in which (3.5.1) represents conservation of energy arises when the considered material response is elastic and an elastic strain-energy function exists. This is discussed in Chapter 4.

From (3.4.2), (2.3.5) and (3.5.2) we obtain

$$\operatorname{tr}(\mathbf{S}\dot{\mathbf{A}}) = \operatorname{tr}(\mathbf{S}\boldsymbol{\Gamma}\mathbf{A}) = J\operatorname{tr}(\mathbf{T}\boldsymbol{\Sigma}) \tag{3.5.4}$$

and this clearly represents the stress power per unit volume of $\mathscr{B}_0$. Equation (3.5.1) may now be put in Lagrangean form

$$\int_{\mathscr{B}_0} \rho_0 \mathbf{b}_0 \cdot \dot{\chi}\,\mathrm{d}V + \int_{\partial\mathscr{B}_0} (\mathbf{S}^{\mathrm{T}}\mathbf{N})\cdot\dot{\chi}\,\mathrm{d}A = \frac{\mathrm{d}}{\mathrm{d}t}\int_{\mathscr{B}_0} \tfrac{1}{2}\rho_0 \dot{\chi}\cdot\dot{\chi}\,\mathrm{d}V + \int_{\mathscr{B}_0} \operatorname{tr}(\mathbf{S}\dot{\mathbf{A}})\,\mathrm{d}V \tag{3.5.5}$$

which is also derivable directly from (3.4.4) by the same method as that used to obtain (3.5.1).

Note that objectivity of J, $\mathbf{T}$ and $\boldsymbol{\Sigma}$ implies that the *stress power is objective.* However, although $\mathbf{S}$ and $\mathbf{A}$ are objective (two-point tensor fields), $\dot{\mathbf{A}}$ is not, but this fact does not affect the objectivity of (3.5.4).The reader may find it instructive to investigate this by considering what happens to the individual terms in $\operatorname{tr}(\mathbf{S}\dot{\mathbf{A}})$ under an observer transformation.

The 'weighted' Cauchy stress $J\mathbf{T}$ which occurs in (3.5.4) is often referred to as the *Kirchhoff stress.* Since this arises frequently in what follows it is convenient to introduce the notation

$$\hat{\mathbf{T}} = J\mathbf{T}. \tag{3.5.6}$$

3.5.2 Conjugate stress tensors

Let $\mathbf{E}^{(m)}$ denote a general Lagrangean strain tensor given by

$$\mathbf{E}^{(m)} = \begin{cases} \dfrac{1}{m}(\mathbf{U}^m - \mathbf{I}) & m \neq 0 \\ \ln \mathbf{U} & m = 0 \end{cases} \tag{3.5.7}$$

(recall (2.2.8)). Then, from equation (2.3.7), we have

$$\dot{\mathbf{E}}^{(2)} = \mathbf{A}^{\mathrm{T}}\boldsymbol{\Sigma}\mathbf{A} \tag{3.5.8}$$

and similarly

$$\dot{\mathbf{E}}^{(-2)} = \mathbf{B}^{\mathrm{T}}\boldsymbol{\Sigma}\mathbf{B} \tag{3.5.9}$$

in particular.

It follows immediately that the stress power density (3.5.4) may be written

$$\mathrm{tr}(\mathbf{S}\dot{\mathbf{A}}) = \mathrm{tr}(\hat{\mathbf{T}}\boldsymbol{\Sigma}) = \mathrm{tr}(\mathbf{T}^{(2)}\dot{\mathbf{E}}^{(2)}) = \mathrm{tr}(\mathbf{T}^{(-2)}\dot{\mathbf{E}}^{(-2)}), \tag{3.5.10}$$

where we have introduced the notation

$$\mathbf{T}^{(2)} = \mathbf{B}^{\mathrm{T}}\hat{\mathbf{T}}\mathbf{B}, \qquad \mathbf{T}^{(-2)} = \mathbf{A}^{\mathrm{T}}\hat{\mathbf{T}}\mathbf{A} \tag{3.5.11}$$

to represent the (symmetric) Lagrangean tensors induced from the Eulerian tensor $\hat{\mathbf{T}}$ and associated with the strain tensors $\mathbf{E}^{(2)}$ and $\mathbf{E}^{(-2)}$ respectively through (3.5.10). We refer to $\mathbf{T}^{(2)}$ and $\mathbf{T}^{(-2)}$ as *stress tensors* because of their intimate relationship with $\hat{\mathbf{T}}$. Their interpretations in terms of load and area are somewhat artificial, however, as we now see. From (3.4.2), (3.4.3) and the definitions (3.5.11), we obtain

$$\mathbf{B}^{\mathrm{T}}\mathbf{dl} = \mathbf{T}^{(2)}\mathrm{d}A, \qquad J\mathbf{A}^{\mathrm{T}}\mathbf{dl} = \mathbf{T}^{(-2)}\mathbf{B}^{\mathrm{T}}\mathbf{da} \tag{3.5.12}$$

and these involve the induced 'loads' $\mathbf{A}^{\mathrm{T}}\mathbf{dl}$ and $\mathbf{B}^{\mathrm{T}}\mathbf{dl}$ of $\mathbf{dl}$ and the induced 'area element' $\mathbf{B}^{\mathrm{T}}\mathbf{da}$.

More generally, we can associate with $\mathbf{E}^{(m)}$ a symmetric stress tensor $\mathbf{T}^{(m)}$ such that

$$\mathrm{tr}(\mathbf{T}^{(m)}\dot{\mathbf{E}}^{(m)}) \tag{3.5.13}$$

gives the stress power density independently of m. Since $\dot{\mathbf{E}}^{(2)}$ may also be written as $\frac{1}{2}(\mathbf{U}\dot{\mathbf{U}} + \dot{\mathbf{U}}\mathbf{U})$ we may rewrite $\mathrm{tr}(\mathbf{T}^{(2)}\dot{\mathbf{E}}^{(2)})$ as a linear form in $\dot{\mathbf{U}}$, namely

$$\mathrm{tr}\{\tfrac{1}{2}\mathbf{T}^{(2)}(\mathbf{U}\dot{\mathbf{U}} + \dot{\mathbf{U}}\mathbf{U})\} = \mathrm{tr}\{\tfrac{1}{2}(\mathbf{T}^{(2)}\mathbf{U} + \mathbf{U}\mathbf{T}^{(2)})\dot{\mathbf{U}}\}.$$

It follows that

$$\mathbf{T}^{(1)} = \tfrac{1}{2}(\mathbf{T}^{(2)}\mathbf{U} + \mathbf{U}\mathbf{T}^{(2)}) \tag{3.5.14}$$

and, by use of the polar decomposition theorem, (3.5.11) and (3.4.2), this may also be expressed

$$\mathbf{T}^{(1)} = \tfrac{1}{2}(\mathbf{SR} + \mathbf{R}^{\mathrm{T}}\mathbf{S}^{\mathrm{T}}) = \tfrac{1}{2}(\mathbf{U}^{-1}\mathbf{R}^{\mathrm{T}}\hat{\mathbf{T}}\mathbf{R} + \mathbf{R}^{\mathrm{T}}\hat{\mathbf{T}}\mathbf{R}\mathbf{U}^{-1}). \tag{3.5.15}$$

The strain tensor $\mathbf{E}^{(1)} = \mathbf{U} - \mathbf{I}$ associated with $\mathbf{T}^{(1)}$ is called the *right-stretch strain tensor*.

Pairs $(\mathbf{T}^{(m)}, \mathbf{E}^{(m)})$ of stress and strain tensors defined in this way are said to be *conjugate stress and strain tensors* (the notion of conjugacy in this context was introduced by Hill, 1968). Note that each of $\mathbf{E}^{(m)}$ and $\mathbf{T}^{(m)}$ is an objective Lagrangean tensor. Other objective conjugate pairs based either on more general Lagrangean strain tensors or on two-point deformation tensors may also be found. The pair $(\mathbf{S}, \mathbf{A})$ is an example of the latter kind and we shall consider the former kind shortly. First, however, we derive certain relations between $\mathbf{T}^{(m)}$ and $\mathbf{T}^{(1)}$, where m is now restricted to being a positive integer.

From (3.5.7) we obtain

$$\dot{\mathbf{E}}^{(m)} = \frac{1}{m}(\dot{\mathbf{U}}\mathbf{U}^{m-1} + \mathbf{U}\dot{\mathbf{U}}\mathbf{U}^{m-2} + \cdots + \mathbf{U}^{m-2}\dot{\mathbf{U}}\mathbf{U} + \mathbf{U}^{m-1}\dot{\mathbf{U}})$$

and substitution of this into the identity

$$\operatorname{tr}(\mathbf{T}^{(m)}\dot{\mathbf{E}}^{(m)}) = \operatorname{tr}(\mathbf{T}^{(1)}\dot{\mathbf{E}}^{(1)})$$

then yields

$$\operatorname{tr}\left\{\frac{1}{m}(\mathbf{U}^{m-1}\mathbf{T}^{(m)} + \mathbf{U}^{m-2}\mathbf{T}^{(m)}\mathbf{U} + \cdots + \mathbf{U}\mathbf{T}^{(m)}\mathbf{U}^{m-2} + \mathbf{T}^{(m)}\mathbf{U}^{m-1})\dot{\mathbf{U}}\right\} = \operatorname{tr}(\mathbf{T}^{(1)}\dot{\mathbf{U}}).$$

This holds for arbitrary symmetric $\dot{\mathbf{U}}$ and we therefore deduce

$$\frac{1}{m}(\mathbf{U}^{m-1}\mathbf{T}^{(m)} + \mathbf{U}^{m-2}\mathbf{T}^{(m)}\mathbf{U} + \cdots + \mathbf{U}\mathbf{T}^{(m)}\mathbf{U}^{m-2} + \mathbf{T}^{(m)}\mathbf{U}^{m-1}) = \mathbf{T}^{(1)}. \tag{3.5.16}$$

A corresponding formula for negative integers can be obtained by noting that

$$\dot{\mathbf{E}}^{(-m)} = \mathbf{U}^{-m}\dot{\mathbf{E}}^{(m)}\mathbf{U}^{-m}$$

but the case $m = 0$ is excluded.

In general it is not an easy matter to invert (3.5.16) to obtain $\mathbf{T}^{(m)}$ explicitly in terms of $\mathbf{T}^{(1)}$ and $\mathbf{U}$ in tensorial form. Connections between components of $\mathbf{T}^{(m)}$ and $\mathbf{T}^{(1)}$ on the Lagrangean principal axes are easier to obtain, however, as we see later. In the meantime we observe that the identity

$$\frac{1}{m}(\mathbf{U}^m\mathbf{T}^{(m)} - \mathbf{T}^{(m)}\mathbf{U}^m) = \mathbf{U}\mathbf{T}^{(1)} - \mathbf{T}^{(1)}\mathbf{U} \tag{3.5.17}$$

may be obtained from (3.5.16). Its derivation is left to the reader.

An instructive alternative route to the conjugate of $\mathbf{E}^{(1)}$ is as follows. Since, by use of the polar decomposition theorem $\dot{\mathbf{A}} = \mathbf{R}\dot{\mathbf{U}} + \dot{\mathbf{R}}\mathbf{U}$, we obtain

$$\mathrm{tr}(\mathbf{S}\dot{\mathbf{A}}) = \mathrm{tr}(\mathbf{S}\mathbf{R}\dot{\mathbf{U}}) + \mathrm{tr}(\mathbf{U}\mathbf{S}\dot{\mathbf{R}}) = \mathrm{tr}(\mathbf{S}\mathbf{R}\dot{\mathbf{U}}) = \mathrm{tr}(\mathbf{T}^{(1)}\dot{\mathbf{U}}),$$

the term $\mathrm{tr}(\mathbf{U}\mathbf{S}\dot{\mathbf{R}}) = \mathrm{tr}(\mathbf{A}\mathbf{S}\dot{\mathbf{R}}\mathbf{R}^{\mathrm{T}}) = \mathrm{tr}(\hat{\mathbf{T}}\dot{\mathbf{R}}\mathbf{R}^{\mathrm{T}})$ vanishing because of the symmetry of $\hat{\mathbf{T}}$ and antisymmetry of $\dot{\mathbf{R}}\mathbf{R}^{\mathrm{T}}$. Since antisymmetric coefficients of $\dot{\mathbf{U}}$ are indeterminate in the above identity the first equation in (3.5.15) follows. However, the tensor $\mathbf{SR}$ was used extensively (although not in this form) by Biot (1965) and its symmetric part $\mathbf{T}^{(1)}$ is often referred to as the Biot stress tensor (it is also called the Jaumann stress tensor by some writers).

Only recently has it been recognized that $\mathbf{T}^{(1)}$ is helpful in the understanding of certain fundamental problems in elasticity theory and it has not therefore been used extensively in the literature. We shall devote some attention to it in later chapters. By contrast $\mathbf{T}^{(2)}$, which is called the (second) Piola–Kirchhoff stress tensor, is much used in the literature. On the other hand, $\mathbf{T}^{(-2)}$ is rarely mentioned (and therefore, understandably, has no name).

The unsymmetrized form $\mathbf{SR}$ of Biot stress has the interpretation

$$\mathbf{R}^{\mathrm{T}}\mathbf{dl} = (\mathbf{SR})^{\mathrm{T}}\mathbf{dA}$$

in terms of load and area, but $\mathbf{T}^{(1)}$ itself has no such interpretation (nor, in general, does $\mathbf{T}^{(m)}(m \neq \pm 2)$). The (Lagrangean) load vector $\mathbf{R}^{\mathrm{T}}\mathbf{dl}$ differs from $\mathbf{dl}$ only through the orientation $\mathbf{R}$ induced by the deformation.

The Eulerian strain-rate $\mathbf{\Sigma}$ cannot in general be written as the (Lagrangean) time derivative of some strain tensor and therefore $\hat{\mathbf{T}}$ is not a conjugate stress tensor (in general). By the some token, neither is the Lagrangean stress tensor $\mathbf{R}^{\mathrm{T}}\hat{\mathbf{T}}\mathbf{R}$. These are just two examples of the general rule that a given stress tensor does not necessarily have associated with it a conjugate strain tensor. Equally, for a given deformation or strain tensor, there does not necessarily exist a stress tensor conjugate to it. We illustrate this by considering the Eulerian strain tensors $\mathbf{F}^{(m)}$ defined by

$$\mathbf{F}^{(m)} = \begin{cases} \dfrac{1}{m}(\mathbf{V}^m - \mathbf{I}) & m \neq 0 \\ \ln \mathbf{V} & m = 0. \end{cases} \tag{3.5.18}$$

Since $\mathbf{U} = \mathbf{R}^{\mathrm{T}}\mathbf{V}\mathbf{R}$, it follows that

$$\mathbf{E}^{(m)} = \mathbf{R}^{\mathrm{T}}\mathbf{F}^{(m)}\mathbf{R}$$

and hence, on differentiation of this and use of the antisymmetry of $\mathbf{R}^{\mathrm{T}}\dot{\mathbf{R}}$,

$$\dot{\mathbf{E}}^{(m)} = \mathbf{R}^{\mathrm{T}}\dot{\mathbf{F}}^{(m)}\mathbf{R} + \mathbf{E}^{(m)}\mathbf{R}^{\mathrm{T}}\dot{\mathbf{R}} - \mathbf{R}^{\mathrm{T}}\dot{\mathbf{R}}\mathbf{E}^{(m)}.$$

By substituting this into the definition of stress power given in (3.5.13), we obtain

$$\operatorname{tr}(\mathbf{T}^{(m)}\dot{\mathbf{E}}^{(m)}) = \operatorname{tr}(\mathbf{R}\mathbf{T}^{(m)}\mathbf{R}^{\mathrm{T}}\dot{\mathbf{F}}^{(m)}) + \operatorname{tr}\{(\mathbf{T}^{(m)}\mathbf{E}^{(m)} - \mathbf{E}^{(m)}\mathbf{T}^{(m)})\mathbf{R}^{\mathrm{T}}\dot{\mathbf{R}}\}$$

after some rearrangement of tensors. Since $\dot{\mathbf{F}}^{(m)}$ and $\dot{\mathbf{R}}$ are independent we see that $\mathbf{F}^{(m)}$ is conjugate to the stress tensor $\mathbf{R}\mathbf{T}^{(m)}\mathbf{R}^{\mathrm{T}}$ if and only if $\mathbf{T}^{(m)}\mathbf{E}^{(m)} - \mathbf{E}^{(m)}\mathbf{T}^{(m)}$ is a symmetric tensor and this means

$$\mathbf{T}^{(m)}\mathbf{E}^{(m)} = \mathbf{E}^{(m)}\mathbf{T}^{(m)} \tag{3.5.19}$$

because $\mathbf{T}^{(m)}\mathbf{E}^{(m)} - \mathbf{E}^{(m)}\mathbf{T}^{(m)}$ is antisymmetric. Similar problems arise with any symmetric strain tensor which is not coaxial with the Lagrangean principal axes since its rate of change is not a linear form in $\dot{\mathbf{U}}$.

Equation (3.5.19) states that $\mathbf{T}^{(m)}$ is a coaxial with $\mathbf{E}^{(m)}$ or, equivalently, $\mathbf{T}^{(m)}$ is coaxial with $\mathbf{U}$ for each m and hence $\hat{\mathbf{T}}$ is coaxial with $\mathbf{V}$. It then follows from (3.5.16) that

$$\mathbf{T}^{(m)} = \mathbf{T}^{(1)}\mathbf{U}^{-(m-1)} = \mathbf{U}^{-(m-1)}\mathbf{T}^{(1)}. \tag{3.5.20}$$

For this special case, equation (3.5.15) gives

$$\mathbf{T}^{(1)} = \mathbf{R}^{\mathrm{T}}\hat{\mathbf{T}}\mathbf{R}\mathbf{U}^{-1} = \mathbf{R}^{\mathrm{T}}\hat{\mathbf{T}}\mathbf{V}^{-1}\mathbf{R} = \mathbf{S}\mathbf{R} \tag{3.5.21}$$

with the help of (3.4.2) and the polar decomposition theorem. Thus

$$\mathbf{S} = \mathbf{T}^{(1)}\mathbf{R}^{\mathrm{T}} \tag{3.5.22}$$

and this is a polar decomposition for $\mathbf{S}$ analogous to that for $\mathbf{A}$ but $\mathbf{T}^{(1)}$, it should be emphasized, is not in general positive definite. This latter equation has important consequences in the context of elasticity theory.

Equation (3.5.20) includes $m = 0$ so that $\mathbf{T}^{(0)}$, the stress tensor conjugate to logarithmic strain $\ln \mathbf{U}$ is given by

$$\mathbf{T}^{(0)} = \mathbf{T}^{(1)}\mathbf{U} = \mathbf{R}^{\mathrm{T}}\hat{\mathbf{T}}\mathbf{R} \tag{3.5.23}$$

provided (3.5.19) holds. We confirm this result during the discussion which now follows.

For the general class of strain tensors

$$\mathbf{F}(\mathbf{U}) = \sum_{i=1}^{3} f(\lambda_i)\mathbf{u}^{(i)} \otimes \mathbf{u}^{(i)}$$

defined in Section 2.2.7, equation (2.3.25) gives

$$\dot{\mathbf{F}} = \sum_{i=1}^{3} f'(\lambda_i)\dot{\lambda}_i \mathbf{u}^{(i)} \otimes \mathbf{u}^{(i)} + \sum_{i \neq j} \Omega_{ij}^{(\mathrm{L})} \{f(\lambda_j) - f(\lambda_i)\} \mathbf{u}^{(i)} \otimes \mathbf{u}^{(j)}.$$

Identification of the coefficients of $\dot{\lambda}_i$ and $\Omega_{ij}^{(\mathrm{L})}$ in the equation which relates $\mathrm{tr}(\mathbf{T}^{(1)}\dot{\mathbf{U}})$ to the equivalent expression based on $\mathbf{F}(\mathbf{U})$ then shows that the normal and shear components of the stress tensor conjugate to $\mathbf{F}(\mathbf{U})$ on the Lagrangean principal axes are respectively

$$T_{ii}^{(1)}/f'(\lambda_i) \qquad (i = 1, 2, 3) \tag{3.5.24}$$

and

$$T_{ij}^{(1)}(\lambda_j - \lambda_i)/\{f(\lambda_j) - f(\lambda_i)\} \qquad (i \neq j), \tag{3.5.25}$$

where $T_{ij}^{(1)}$ $(i, j = 1, 2, 3)$ are the components of $\mathbf{T}^{(1)}$ on the same axes. Note particularly the case of logarithmic strain for which $f(\lambda_i) = \ln \lambda_i$.

If $\mathbf{T}^{(1)}$ is coaxial with $\mathbf{U}$ then so also is the stress conjugate to $\mathbf{F}(\mathbf{U})$. Then $T_{ij}^{(1)} = 0, i \neq j$, and $T_{ii}^{(1)}/f'(\lambda_i), i = 1, 2, 3$, are the principal conjugate stresses. In particular, $T_{ii}^{(0)} = \lambda_i T_{ii}^{(1)}$ and this is equivalent to the tensorial relation (3.5.23). That $\mathbf{R}^{\mathrm{T}}\hat{\mathbf{T}}\mathbf{R}$ is conjugate to $\mathbf{E}^{(0)}$ in this case is explained by the fact that the difference

$$\mathbf{R}^{\mathrm{T}}\boldsymbol{\Sigma}\mathbf{R} - \dot{\mathbf{E}}^{(0)} = \sum_{i \neq j} \Omega_{ij}^{(\mathrm{L})} \left\{ \frac{\lambda_j^2 - \lambda_i^2}{2\lambda_i\lambda_j} - \ln(\lambda_j/\lambda_i) \right\} \mathbf{u}^{(i)} \otimes \mathbf{u}^{(j)},$$

obtainable from (2.3.25) specialized to logarithmic strain together with (2.3.26), has only shear components. In general, however, $\mathbf{T}^{(0)} \neq \mathbf{R}^{\mathrm{T}}\hat{\mathbf{T}}\mathbf{R}$.

Towards an alternative to the above (component) derivation of the conjugate stress tensor of $\mathbf{F}(\mathbf{U})$ it is instructive to represent the Lagrangean time derivative of $\mathbf{F}(\mathbf{U})$ as the result of a linear transformation $\mathscr{L}^1$ on $\dot{\mathbf{U}}$. We write

$$\dot{\mathbf{F}} = \mathscr{L}^1 \dot{\mathbf{U}}, \tag{3.5.26}$$

where $\mathscr{L}^1$ (which depends on $\mathbf{U}$ in general) is a fourth-order Lagrangean tensor which may be represented symbolically as

$$\mathscr{L}^1 = \partial \mathbf{F}/\partial \mathbf{U}. \tag{3.5.27}$$

In orthonormal components

$$\dot{F}_{ij} = \mathscr{L}^1_{ijkl}\dot{U}_{kl}, \qquad \mathscr{L}^1_{ijkl} = \partial F_{ij}/\partial U_{kl}$$

and $\mathscr{L}^1$ has the symmetries

$$\mathscr{L}^1_{ijkl} = \mathscr{L}^1_{jikl} = \mathscr{L}^1_{ijlk} \tag{3.5.28}$$

induced by the symmetry of $\mathbf{F}$ and $\mathbf{U}$.

It is easy to deduce from the representations of $\dot{\mathbf{U}}$ and $\dot{\mathbf{F}}$ on the basis $\{\mathbf{u}^{(i)}\}$ that $\mathscr{L}^1$ may be decomposed as

$$\mathscr{L}^1 = \sum_{i=1}^{3} f'(\lambda_i)\mathbf{u}^{(i)}\otimes\mathbf{u}^{(i)}\otimes\mathbf{u}^{(i)}\otimes\mathbf{u}^{(i)} + \frac{1}{2}\sum_{i\neq j}\left\{\frac{f(\lambda_j)-f(\lambda_i)}{\lambda_j-\lambda_i}\right\}\times$$
$$\times(\mathbf{u}^{(i)}\otimes\mathbf{u}^{(j)}\otimes\mathbf{u}^{(i)}\otimes\mathbf{u}^{(j)} + \mathbf{u}^{(i)}\otimes\mathbf{u}^{(j)}\otimes\mathbf{u}^{(j)}\otimes\mathbf{u}^{(i)}). \tag{3.5.29}$$

Thus, the only non-zero components of $\mathscr{L}^1$ on the basis $\{\mathbf{u}^{(i)}\}$ are given by

$$\mathscr{L}^1_{iiii} = f'(\lambda_i) \qquad i = 1, 2, 3$$

$$\mathscr{L}^1_{ijij} = \begin{cases} \tfrac{1}{2}\{f(\lambda_j)-f(\lambda_i)\}/(\lambda_j-\lambda_i) & \lambda_j \neq \lambda_i, \quad i\neq j \\ \tfrac{1}{2}f'(\lambda_i) & \lambda_j = \lambda_i, \quad i \neq j. \end{cases}$$

From this follows the symmetry $\mathscr{L}^1_{ijkl} = \mathscr{L}^1_{klij}$ in addition to (3.5.28) and we write this symbolically as $\mathscr{L}^{1\mathrm{T}} = \mathscr{L}^1$.

By writing $\mathbf{T}^{\mathbf{F}}$ for the stress tensor conjugate to $\mathbf{F}(\mathbf{U})$, it follows that

$$\mathrm{tr}(\mathbf{T}^{(1)}\dot{\mathbf{U}}) = \mathrm{tr}\{\mathbf{T}^{\mathbf{F}}(\mathscr{L}^1\dot{\mathbf{U}})\} = \mathrm{tr}\{(\mathscr{L}^1\mathbf{T}^{\mathbf{F}})\dot{\mathbf{U}}\}$$

and hence $\mathbf{T}^{(1)} = \mathscr{L}^1\mathbf{T}^{\mathbf{F}}$. On introduction of the fourth-order tensor $\mathscr{M}^1$ such that

$$\mathscr{L}^1\mathscr{M}^1 = \mathscr{M}^1\mathscr{L}^1 = \mathscr{I}^1, \qquad \mathscr{L}^1_{ijkl}\mathscr{M}^1_{klpq} = \mathscr{M}^1_{ijkl}\mathscr{L}^1_{klpq} = \mathscr{I}^1_{ijpq}, \tag{3.5.30}$$

we obtain the concise formula

$$\mathbf{T}^{\mathbf{F}} = \mathscr{M}^1\mathbf{T}^{(1)}, \tag{3.5.31}$$

where $\mathscr{I}^1$ is the identity mapping on the six-dimensional space of symmetric second-order tensors and has the (orthonormal) component form

$$\mathscr{I}^1_{ijkl} = \tfrac{1}{2}(\delta_{ik}\delta_{jl} + \delta_{il}\delta_{jk}). \tag{3.5.32}$$

It is left to the reader to show that $\mathscr{M}^1$ is expressible as

$$\mathscr{M}^1 = \sum_{i=1}^{3} [f'(\lambda_i)]^{-1}\mathbf{u}^{(i)} \otimes \mathbf{u}^{(i)} \otimes \mathbf{u}^{(i)} \otimes \mathbf{u}^{(i)} + \frac{1}{2}\sum_{i \neq j}\left\{\frac{\lambda_j - \lambda_i}{f(\lambda_j) - f(\lambda_i)}\right\} \times$$
$$\times (\mathbf{u}^{(i)} \otimes \mathbf{u}^{(j)} \otimes \mathbf{u}^{(i)} \otimes \mathbf{u}^{(j)} + \mathbf{u}^{(i)} \otimes \mathbf{u}^{(j)} \otimes \mathbf{u}^{(j)} \otimes \mathbf{u}^{(i)}).$$

The formula (3.5.31) will prove useful in the next section where we consider stress rates. In that context a representation for the sixth-order tensor

$$\mathscr{L}^2 = \partial^2 \mathbf{F}/\partial \mathbf{U} \partial \mathbf{U}, \qquad \mathscr{L}^2_{ijklmn} = \partial^2 F_{ij}/\partial U_{kl}\, \partial U_{mn} \tag{3.5.33}$$

is also required and we now indicate briefly how this is obtained. We now regard $\mathscr{L}^2$ as a bilinear mapping from the product $\mathscr{S} \times \mathscr{S}$ to $\mathscr{S}$, where $\mathscr{S}$ denotes the space of symmetric second-order tensors, and write

$$\mathscr{L}^2[\mathbf{T}, \mathbf{T}'] = \mathscr{L}^2_{ijklmn} T_{kl} T'_{mn},$$
$$\mathscr{L}^2[\cdot, \mathbf{T}] \equiv \mathscr{L}^2_{ijklmn} T_{mn}$$

for $\mathbf{T}$ and $\mathbf{T}'$ in $\mathscr{S}$.

From (3.5.27) and (3.5.33) we obtain

$$\dot{\mathscr{L}}^1 = \mathscr{L}^2[\cdot, \dot{\mathbf{U}}]$$

Substitution of (3.5.29) into this followed by use of (2.3.18)–(2.3.23) and some algebraic manipulations then shows that the only non-zero components of $\mathscr{L}^2$ on the basis $\{\mathbf{u}^{(i)}\}$ are given by

$$\mathscr{L}^2_{iiiiii} = f''(\lambda_i) \qquad i = 1, 2, 3,$$

$$\mathscr{L}^2_{iiijij} = \mathscr{L}^2_{ijiiij} = \mathscr{L}^2_{ijijii} = \begin{cases} \dfrac{f(\lambda_j) - f(\lambda_i) - (\lambda_j - \lambda_i) f'(\lambda_i)}{4(\lambda_j - \lambda_i)^2} & \lambda_j \neq \lambda_i \\[2ex] \tfrac{1}{8} f''(\lambda_i) & \lambda_j = \lambda_i \end{cases}$$

for $i \neq j$, and

$$\mathscr{L}^2_{ijjkki} = \begin{cases} \dfrac{[f(\lambda_j) - f(\lambda_i)](\lambda_i + \lambda_j - 2\lambda_k) - [f(\lambda_i) + f(\lambda_j) - 2f(\lambda_k)](\lambda_j - \lambda_i)}{8(\lambda_j - \lambda_k)(\lambda_k - \lambda_i)(\lambda_i - \lambda_j)} \\ \qquad\qquad\qquad\qquad\qquad\qquad\qquad\qquad \lambda_i \neq \lambda_j \neq \lambda_k \neq \lambda_i \\ \mathscr{L}^2_{iiikik} \qquad \lambda_j = \lambda_i \neq \lambda_k \\ \tfrac{1}{8} f''(\lambda_i) \qquad \lambda_k = \lambda_j = \lambda_i \end{cases}$$

for $i \neq j \neq k \neq i$. We remark that the results for the special cases $\lambda_j = \lambda_i$ and $\lambda_k = \lambda_j = \lambda_i$ in the formulae for $\mathscr{L}^1_{ijij}$, $\mathscr{L}^2_{iiijij}$ and $\mathscr{L}^2_{ijikki}$ may be arrived at from the results for $\lambda_i \neq \lambda_j \neq \lambda_k \neq \lambda_i$ by an application of the Taylor expansion.

The situation in which the reference configuration is taken to coincide with the current configuration is of special interest. For then $\mathbf{F(U)} = \mathbf{0}$, $\lambda_i = 1 (i = 1, 2, 3)$ and from (3.5.29) it then follows that

$$\mathscr{L}^1 = \frac{1}{2} \sum_{i,j=1}^{3} (\mathbf{u}^{(i)} \otimes \mathbf{u}^{(j)} \otimes \mathbf{u}^{(i)} \otimes \mathbf{u}^{(j)} + \mathbf{u}^{(i)} \otimes \mathbf{u}^{(j)} \otimes \mathbf{u}^{(j)} \otimes \mathbf{u}^{(i)}) = \mathscr{I}^1,$$

and the components of $\mathscr{I}^1$ are given by (3.5.32). It is left as an exercise for the reader to find the corresponding expression for $\mathscr{L}^2$ but in component form we have

$$\begin{aligned}\mathscr{L}^2_{ijklmn} &= \tfrac{1}{8} f''(1)(\delta_{ik}\delta_{jm}\delta_{ln} + \delta_{ik}\delta_{jn}\delta_{lm} + \delta_{il}\delta_{jm}\delta_{kn} + \delta_{il}\delta_{jn}\delta_{km} \\ &\quad + \delta_{im}\delta_{jk}\delta_{ln} + \delta_{im}\delta_{jl}\delta_{kn} + \delta_{in}\delta_{jk}\delta_{lm} + \delta_{in}\delta_{jl}\delta_{km}) \\ &\equiv f''(1)\mathscr{I}^2_{ijklmn} \end{aligned} \tag{3.5.34}$$

which defines the notation $\mathscr{I}^2_{ijklmn}$

3.5.3 Stress rates

In continuum mechanics objective rates of stress have an important role to play in the description of material behaviour through constitutive relations and we therefore place on record here some connections between the rates of certain conjugate stress tensors for later reference.

On taking the Lagrangean time derivative of (3.5.31) and making use of (3.5.30), we obtain a relation between the rates $\dot{\mathbf{T}}^{\mathsf{F}}$ and $\dot{\mathbf{T}}^{(1)}$ of the stress tensors $\mathbf{T}^{\mathsf{F}}$ and $\mathbf{T}^{(1)}$ respectively, namely

$$\mathscr{L}^1 \dot{\mathbf{T}}^{\mathsf{F}} = \dot{\mathbf{T}}^{(1)} - \dot{\mathscr{L}}^1 \mathscr{M}^1 \mathbf{T}^{(1)}$$

which may also be written as

$$\mathscr{L}^1 \dot{\mathbf{T}}^{\mathsf{F}} = \dot{\mathbf{T}}^{(1)} - \mathscr{L}^2[\mathscr{M}^1 \mathbf{T}^{(1)}, \dot{\mathbf{U}}] \tag{3.5.35}$$

Since $\mathbf{T}^{(2)} = \mathbf{B}^{\mathrm{T}} \hat{\mathbf{T}} \mathbf{B}$, we also have

$$\dot{\mathbf{T}}^{(2)} = \mathbf{B}^{\mathrm{T}}(\dot{\hat{\mathbf{T}}} - \mathbf{\Gamma}\hat{\mathbf{T}} - \hat{\mathbf{T}}\mathbf{\Gamma}^{\mathrm{T}})\mathbf{B} \tag{3.5.36}$$

on use of (2.3.6) and similarly

$$\dot{\mathbf{T}}^{(-2)} = \mathbf{A}^{\mathrm{T}}(\dot{\hat{\mathbf{T}}} + \mathbf{\Gamma}^{\mathrm{T}}\hat{\mathbf{T}} + \hat{\mathbf{T}}\mathbf{\Gamma})\mathbf{A} \tag{3.5.37}$$

by direct calculation. From the definition (3.5.6), we also have

$$\dot{\hat{\mathbf{T}}} = J\dot{\mathbf{T}} + J\,\mathrm{tr}(\mathbf{\Sigma})\mathbf{T}. \tag{3.5.38}$$

When the reference configuration is taken to coincide with the current configuration the deformation gradient $\mathbf{A}$ becomes the identity, each rate of strain $\dot{\mathbf{F}}$ equals the Eulerian strain-rate $\mathbf{\Sigma}$ (this follows from (3.5.8) and (3.5.26) with $\mathscr{L}^1 = \mathscr{I}^1$) and hence each $\mathbf{T}^{\mathrm{F}}$ is equal to the Cauchy stress tensor $\mathbf{T}$. On use of the component form of $\mathscr{L}^2[\mathbf{T}, \mathbf{\Sigma}]$ with (3.5.34) it then follows from (3.5.35) that

$$\dot{\mathbf{T}}^{\mathrm{F}} = \dot{\mathbf{T}}^{(1)} - \tfrac{1}{2}f''(1)(\mathbf{\Sigma}\mathbf{T} + \mathbf{T}\mathbf{\Sigma}), \tag{3.5.39}$$

and from (3.5.36) that

$$\dot{\mathbf{T}}^{(2)} = \dot{\hat{\mathbf{T}}} - \mathbf{\Gamma}\mathbf{T} - \mathbf{T}\mathbf{\Gamma}^{\mathrm{T}} \equiv \dot{\mathbf{T}} + \mathrm{tr}(\mathbf{\Sigma})\mathbf{T} - \mathbf{\Gamma}\mathbf{T} - \mathbf{T}\mathbf{\Gamma}^{\mathrm{T}}. \tag{3.5.40}$$

For the class of strain tensors $\mathbf{F}(\mathbf{U}) = \mathbf{E}^{(m)}$, we have $f''(1) = m - 1$. Using this for $m = 2$ we eliminate $\dot{\mathbf{T}}^{(1)}$ between (3.5.39) and (3.5.40) to obtain

$$\dot{\mathbf{T}}^{\mathrm{F}} = \dot{\hat{\mathbf{T}}} - \mathbf{\Omega}\mathbf{T} + \mathbf{T}\mathbf{\Omega} - \tfrac{1}{2}[f''(1) + 1](\mathbf{\Sigma}\mathbf{T} + \mathbf{T}\mathbf{\Sigma}), \tag{3.5.41}$$

where $\mathbf{\Omega}$ is the body spin defined in (2.3.13). With the logarithmic strain measure $\mathbf{E}^{(0)}$ this reduces to

$$\dot{\mathbf{T}}^{(0)} = \dot{\hat{\mathbf{T}}} - \mathbf{\Omega}\mathbf{T} + \mathbf{T}\mathbf{\Omega}. \tag{3.5.42}$$

The notation

$$\frac{\mathscr{D}}{\mathscr{D}t}\hat{\mathbf{T}} \equiv \dot{\hat{\mathbf{T}}} - \mathbf{\Omega}\hat{\mathbf{T}} + \hat{\mathbf{T}}\mathbf{\Omega} \tag{3.5.43}$$

to represent the rate of change of $\hat{\mathbf{T}}$ relative to a basis rotating with the local body spin $\mathbf{\Omega}$ is in common use and this derivative is known as the *Jaumann* or *co-rotational* rate of the stress $\hat{\mathbf{T}}$. When evaluated with the reference configuration coinciding with the current configuration (3.5.43) becomes (3.5.42).

We note that each of the *conjugate* stress rates listed above is objective but $\dot{\hat{\mathbf{T}}}$ is not.

Higher rates of stress can be dealt with in a similar way on the basis of equation (3.5.35) but we shall not discuss them here (see Problem 3.5.4, however). A detailed account has been given by Ogden (1974).

Because of the importance of the nominal stress tensor in applications we now give a connection between its rate and the stress rates discussed above. Since, by (3.4.2), (3.5.6) and (3.5.11),

$$\mathbf{S} = \mathbf{T}^{(2)}\mathbf{A}^{\mathrm{T}} \tag{3.5.44}$$

we obtain

$$\dot{\mathbf{S}} = \dot{\mathbf{T}}^{(2)}\mathbf{A}^{\mathrm{T}} + \mathbf{T}^{(2)}\dot{\mathbf{A}}^{\mathrm{T}} = \mathbf{B}^{\mathrm{T}}(\dot{\hat{\mathbf{T}}} - \boldsymbol{\Gamma}\hat{\mathbf{T}}) \tag{3.5.45}$$

or, when reference and current configurations coincide,

$$\dot{\mathbf{S}} = \dot{\mathbf{T}}^{(2)} + \mathbf{T}\boldsymbol{\Gamma}^{\mathrm{T}}. \tag{3.5.46}$$

Note that this rate is not objective.

Problem 3.5.1 Show that

$$\mathbf{E}^{(m)}\mathbf{T}^{(m)} - \mathbf{T}^{(m)}\mathbf{E}^{(m)} = \tfrac{1}{2}(\mathbf{A}^{\mathrm{T}}\mathbf{S}^{\mathrm{T}} - \mathbf{S}\mathbf{A})$$

for all positive integers m. Hence find expressions for the shear components of $\mathbf{T}^{(m)}$ on the Lagrangean axes in terms of those of $\mathbf{T}^{(1)}$.

If $\mathbf{P}$ denotes the orthogonal tensor described in Section 2.3.2 so that

$$\bar{\mathbf{F}} \equiv \mathbf{P}^{\mathrm{T}}\mathbf{F}\mathbf{P} = \sum_{i=1}^{3} f(\lambda_i)\,\mathrm{E}_i \otimes \mathrm{E}_i,$$

where $\{\mathrm{E}_i\}$ is a rectangular Cartesian basis, show that

$$\mathrm{tr}(\mathbf{T}^{\mathsf{F}}\dot{\mathbf{F}}) = \mathrm{tr}(\mathbf{P}^{\mathrm{T}}\mathbf{T}^{\mathsf{F}}\mathbf{P}\dot{\bar{\mathbf{F}}}) + \mathrm{tr}\{(\mathbf{F}\mathbf{T}^{\mathsf{F}} - \mathbf{T}^{\mathsf{F}}\mathbf{F})\boldsymbol{\Omega}^{(\mathrm{L})}\},$$

where $\boldsymbol{\Omega}^{(\mathrm{L})} = \dot{\mathbf{P}}\mathbf{P}^{\mathrm{T}}$. Deduce that

$$\mathbf{F}\mathbf{T}^{\mathsf{F}} - \mathbf{T}^{\mathsf{F}}\mathbf{F} = \tfrac{1}{2}(\mathbf{A}^{\mathrm{T}}\mathbf{S}^{\mathrm{T}} - \mathbf{S}\mathbf{A}).$$

Problem 3.5.2 Show that

$$\mathbf{T}^{(-1)} = \mathbf{U}\,\mathbf{T}^{(1)}\mathbf{U}$$

and prove that $\mathbf{A}^{\mathrm{T}}\hat{\mathbf{T}}$ is the stress tensor conjugate to the deformation tensor $-\mathbf{B}$.

Problem 3.5.3 Show that in general there do not exist strain tensors conjugate to either $\mathbf{A}^{\mathrm{T}}\hat{\mathbf{T}}\mathbf{B}$ or $\mathbf{B}^{\mathrm{T}}\hat{\mathbf{T}}\mathbf{A}$. Give load–area interpretations of these stress tensors analogous to those for $\mathbf{T}^{(2)}$ and $\mathbf{T}^{(-2)}$ in (3.5.12).

If $\hat{\mathbf{T}}\mathbf{V} = \mathbf{V}\hat{\mathbf{T}}$, where $\mathbf{V}$ is the left stretch tensor, deduce that

$$\mathbf{A}^{\mathrm{T}}\hat{\mathbf{T}}\mathbf{B} = \mathbf{B}^{\mathrm{T}}\hat{\mathbf{T}}\mathbf{A} = \mathbf{R}^{\mathrm{T}}\hat{\mathbf{T}}\mathbf{R}$$

and that this is then conjugate to the logarithmic strain tensor $\mathbf{E}^{(0)}$.

Problem 3.5.4 If $\mathbf{F}(\mathbf{U})$ is the strain tensor defined in (2.2.85) show that

$$\ddot{\mathbf{F}} = \mathscr{L}^1\ddot{\mathbf{U}} + \mathscr{L}^2[\dot{\mathbf{U}}, \dot{\mathbf{U}}]$$

Show also that

$$\ddot{\mathbf{E}}^{(2)} = \tfrac{1}{2}(\mathbf{U}\ddot{\mathbf{U}} + 2\dot{\mathbf{U}}^2 + \ddot{\mathbf{U}}\mathbf{U}).$$

Deduce that when the reference configuration is taken as the current configuration

$$\ddot{\mathbf{F}} = \ddot{\mathbf{U}} + f''(1)\mathbf{\Sigma}^2$$

and hence

$$\ddot{\mathbf{E}}^{(0)} = \frac{\mathscr{D}\mathbf{\Sigma}}{\mathscr{D}t}.$$

Obtain the formula

$$\ddot{\mathbf{T}}^{\mathrm{F}} = \mathscr{M}^1\ddot{\mathbf{T}}^{(1)} - 2\mathscr{M}^1\dot{\mathscr{L}}^1\mathscr{M}^1\dot{\mathbf{T}}^{(1)} - \mathscr{M}^1\ddot{\mathscr{L}}^1\mathscr{M}^1\mathbf{T}^{(1)}$$
$$+ 2\mathscr{M}^1\dot{\mathscr{L}}^1\mathscr{M}^1\dot{\mathscr{L}}^1\mathscr{M}^1\mathbf{T}^{(1)}$$

and hence, by deriving an expression for $\mathscr{L}^1$, show that this can be expressed as

$$\ddot{\mathbf{T}}^{\mathrm{F}} = \ddot{\mathbf{T}}^{(0)} - f''(1)(\mathbf{\Sigma}\dot{\mathbf{T}}^{(0)} + \dot{\mathbf{T}}^{(0)}\mathbf{\Sigma}) - \tfrac{1}{2}f''(1)(\mathbf{T}\ddot{\mathbf{E}}^{(0)} + \ddot{\mathbf{E}}^{(0)}\mathbf{T})$$
$$+ \tfrac{1}{2}[f''(1)]^2(\mathbf{T}\mathbf{\Sigma}^2 + 2\mathbf{\Sigma}\mathbf{T}\mathbf{\Sigma} + \mathbf{\Sigma}^2\mathbf{T}) - \tfrac{1}{3}f'''(1)(\mathbf{T}\mathbf{\Sigma}^2 + \mathbf{\Sigma}\mathbf{T}\mathbf{\Sigma} + \mathbf{\Sigma}^2\mathbf{T}),$$

where

$$\dot{\mathbf{T}}^{(0)} = \frac{\mathscr{D}\hat{\mathbf{T}}}{\mathscr{D}t}$$

and

$$\ddot{\mathbf{T}}^{(0)} = \frac{\mathscr{D}^2\hat{\mathbf{T}}}{\mathscr{D}t^2} + \tfrac{1}{2}(\mathbf{T}\mathbf{\Sigma}^2 - 2\mathbf{\Sigma}\mathbf{T}\mathbf{\Sigma} + \mathbf{\Sigma}^2\mathbf{T}),$$

all rates being evaluated for coincident current and reference configurations.

Problem 3.5.5 Show that

$$\mathrm{tr}(\dot{\mathbf{T}}^{\mathrm{F}}\dot{\mathbf{F}}) = \mathrm{tr}(\dot{\mathbf{T}}^{(1)}\dot{\mathbf{E}}^{(1)}) - \mathrm{tr}\{\mathscr{L}^2[\dot{\mathbf{E}}^{(1)}, \dot{\mathbf{E}}^{(1)}](\mathscr{M}^1\mathbf{T}^{(1)})\}$$

and simplify for the case of coincident reference and current configurations.

Show also that

$$\mathrm{tr}(\dot{\mathbf{S}}\dot{\mathbf{A}}) = \mathrm{tr}(\dot{\mathbf{T}}^{(2)}\dot{\mathbf{E}}^{(2)}) + \mathrm{tr}(\mathbf{T}^{(2)}\dot{\mathbf{A}}^{\mathrm{T}}\dot{\mathbf{A}})$$

and hence

$$\mathrm{tr}(\dot{\mathbf{S}}\dot{\mathbf{A}}) = \mathrm{tr}(\dot{\mathbf{T}}^{(0)}\dot{\mathbf{E}}^{(0)}) - 2\,\mathrm{tr}(\mathbf{T}\boldsymbol{\Sigma}^2) + \mathrm{tr}(\mathbf{T}\boldsymbol{\Gamma}^{\mathrm{T}}\boldsymbol{\Gamma})$$

in the reference configuration.

Problem 3.5.6 Let $T^{\mathrm{F}}{}_{ij}$ denote the components of $\mathbf{T}^{\mathrm{F}}$ on the Lagrangean axes. Show that $\dot{\mathbf{T}}^{\mathrm{F}}$ has components

$$\dot{T}^{\mathrm{F}}{}_{ij} + \Omega^{(\mathrm{L})}{}_{ip}T^{\mathrm{F}}{}_{pj} + \Omega^{(\mathrm{L})}{}_{jp}T^{\mathrm{F}}{}_{pi}$$

(sum over p), where $\Omega^{(\mathrm{L})}{}_{ij}$ are the components of $\boldsymbol{\Omega}^{(\mathrm{L})}$ on the Lagrangean axes.

Deduce that if $\mathbf{T}^{\mathrm{F}}$ is coaxial with $\mathbf{F}$ then the normal and shear components of $\dot{\mathbf{T}}^{\mathrm{F}}$ are

$$\dot{T}^{\mathrm{F}}{}_{ii} \qquad i = 1, 2, 3$$

and

$$\Omega^{(\mathrm{L})}{}_{ij}(T^{\mathrm{F}}{}_{jj} - T^{\mathrm{F}}{}_{ii}) \qquad i \neq j$$

respectively.

REFERENCES

M.A. Biot, *Mechanics of Incremental Deformations*, John Wiley (1965).

R. Hill, On Constitutive Inequalities for Simple Materials, *Journal of the Mechanics and Physics of Solids*, Vol. 16, pp. 229–242 (1968).

R.W. Ogden, On Stress Rates in Solid Mechanics with Applications to Elasticity Theory, *Proceedings of the Cambridge Philosophical Society*, Vol. 75, pp. 303–319 (1974).

C.A. Truesdell, *A First Course in Rational Continuum Mechanics*, Vol. I, Academic Press (1977).

C.-C. Wang and C.A. Truesdell, *Introduction to Rational Elasticity*, Noordhoff (1973).

CHAPTER 4

Elasticity

4.1 CONSTITUTIVE LAWS FOR SIMPLE MATERIALS

4.1.1 General remarks on constitutive laws

A geometrical description of the motion of a continuous medium has been given in Chapter 2, while in Chapter 3 we have established a means of describing the stress in the material arising from this motion and from the forces which are the underlying cause of the motion. In particular, the equations of motion and mass conservation (four scalar equations in all) governing the stress and the motion have been derived. In general, however, as mentioned in Section 3.3.2, these equations are insufficient to determine the motion (which involves ten dependent variables; six components of stress, three components of the position vector, and the density of the material). Moreover, the description is incomplete physically because there is as yet no means of distinguishing between two different continuous media; for example, between a solid and a liquid or between two solids (such as rubber and steel). The (four) equations (3.1.9) and (3.3.7), with (3.3.9) assumed, or (3.4.4) and (3.4.6), with (3.4.5) assumed, as they stand are applicable to any continuum for which symmetry of Cauchy stress is appropriate (this restriction excludes from consideration materials with couple stresses). Provided the material is not subject to internal constraints, ten scalar equations are required for the determination of the ten dependent variables and to close the mathematical system.

The set of equations is completed by use of a *constitutive law* which describes the (macroscopic) nature of the material in question; different constitutive laws govern distinct types of (macroscopic) material behaviour. To be more precise a constitutive law describes an *ideal material*, and the predictions of this law should provide a close approximation to the actual (observed) behaviour of the real material that the constitutive law is intended to model. Here we are concerned with purely mechanical theories so that thermodynamic variables, such as temperature and entropy, in particular, are excluded.

If the Cauchy stress $\mathbf{T}$ at any point in the material at time t is known for any motion up to time t then the relationship between $\mathbf{T}$ and the history of the motion up to and including time t describes the stress response of the material to an arbitrary motion. This relationship is referred to as a *constitutive equation.* It provides the missing set of six scalar equations to close the mathematical system (symmetric $\mathbf{T}$ having six independent components). That a material can be characterized by such a knowledge of $\mathbf{T}$ is an *assumption*, commonly referred to as the 'principle of determinism for the stress' (Truesdell and Noll, 1965, for example). Some modification of this assumption is required for materials subject to internal constraints. Other assumptions are needed to enable further simplification to be made in the theory of constitutive equations. We emphasize that such assumptions should be based on physical reasoning and not on mathematical expediency although, of course, the latter may be consistent with the former.

An important assumption in continuum mechanics is that two observers in relative motion make equivalent (mathematical and physical) deductions about the macroscopic properties of a material under test. In other words, material properties are unaffected by a superposed rigid motion, and the relationship between the stress and the motion has the same form for all observers. This is the 'principle of material objectivity'. We adopt this assumption here. The mathematical consequences of material objectivity and other assumptions are examined in Section 4.1.2 prior to their application in the context of elasticity theory in later sections. Section 4.1.2, however, may be omitted by the reader who is prepared to accept the elastic constitutive laws at face value without regard to the background of their derivation.

In general the assumption of material objectivity imposes certain limitations on the form of a constitutive equation, but a considerable degree of generality is nevertheless still maintained. How best to select from among the available constitutive laws more specific ones for consideration in relation to real materials is a difficult problem, but a number of ways of tackling this in principle are now mentioned. For materials which have inherent symmetries in one (or more than one) configuration (certain perfect crystals, for example) further limitations on the constitutive laws are induced. The prime example is that of an *isotropic* material whose response has no preferred direction relative to certain configurations. Such physical symmetries give rise to mathematical restrictions on the constitutive equations. This is discussed in detail in Sections 4.2 and 4.3.

Mathematical restrictions also arise quite naturally in the solution of boundary-value problems. These restrictions usually take the form of inequalities which are related very much to inequalities associated with such questions as existence and uniqueness of solutions of the governing differential equations and boundary conditions. At the physical level further restrictions can be achieved on the basis of what is observed in experimental tests on particular materials. Comprehensive experimental data should enable more specialized forms of constitutive equation to be deduced. Furthermore, fundamental physical models, based on the underlying molecular structure, can be used to provide information about the macroscopic nature of the material by transmitting the microscopic models into macroscopic constitutive laws which are special cases of the constitutive equations developed within the general mathematical framework. This latter aspect of the subject is touched on only in passing in this book, however.

In later chapters and in relation specifically to elasticity, we approach the problem of delimiting constitutive equations from two extremities. First, from the purely mathematical viewpoint in relation to general forms of elastic constitutive law, our aim is to extract information (usually in the form of inequalities) from investigation of and solution of boundary-value problems, general and specific. Second, from the physical viewpoint, explicit constitutive equations derived from molecular models and experimental observation are considered. The two approaches are not inconsistent and they meet on the middle ground where empirical laws are used in the solution of boundary-value problems. This is the substance of Chapters 5 to 7.

4.1.2 Simple materials

Suppose that a motion of the body B recorded by an observer O is specified by the equation

$$\mathbf{x} = \boldsymbol{\chi}(X, t) \qquad X \in B \tag{4.1.1}$$

as in (2.1.7), and let $\mathbf{T}(\mathbf{x}, t)$ be the Cauchy stress at time t at the material point X which has position $\mathbf{x}$ in $\mathscr{E}$ at time t.

Let $\boldsymbol{\chi}^t$ denote the history of the motion $\boldsymbol{\chi}$ up to time t so that

$$\boldsymbol{\chi}^t(X, s) = \boldsymbol{\chi}(X, t-s) \qquad s \geq 0. \tag{4.1.2}$$

Then the assumption that $\mathbf{T}(\mathbf{x}, t)$ depends on the history of the motion of B is expressed in the form

$$\mathbf{T}(\mathbf{x}, t) = \mathbf{G}(\boldsymbol{\chi}^t; X, t), \tag{4.1.3}$$

where $\mathbf{G}$ is a *functional*† with respect to its first argument and a function with respect to each of the second and third arguments. Note that $\mathbf{G}$ depends explicitly on the material point X but that each $\boldsymbol{\chi}^t$ is a possible motion history for B as a whole and need not be restricted to X. This means that in general the history of the motion of B contributes to the stress at each constituent point of B at the current time t.

† This standard terminology is used to indicate that $\mathbf{G}$ is an operator with respect to s (for example, an integral operator) and, in general, depends on $\boldsymbol{\chi}^t(X, s)$ for all $s \geq 0$.

The corresponding motion

$$\mathbf{x}^* = \boldsymbol{\chi}^*(X, t^*) \qquad X \in B \tag{4.1.4}$$

and Cauchy stress $\mathbf{T}^*(\mathbf{x}^*, t^*)$ perceived by a second observer O^* are subject to the restrictions

$$\boldsymbol{\chi}^*(X, t^*) = \mathbf{Q}(t)\boldsymbol{\chi}(X, t) + \mathbf{c}(t) \tag{4.1.5}$$

and

$$\mathbf{T}^*(\mathbf{x}^*, t^*) = \mathbf{Q}(t)\mathbf{T}(\mathbf{x}, t)\mathbf{Q}(t)^{\mathrm{T}} \tag{4.1.6}$$

under the observer transformation

$$\mathbf{x}^* = \mathbf{Q}(t)\mathbf{x} + \mathbf{c}(t), \qquad t^* = t - a \tag{4.1.7}$$

given in (2.1.2). We recall that the Cauchy stress is an objective Eulerian tensor (Section 3.3.1).

The assumption of *material objectivity* implies that

$$\mathbf{T}^*\{\boldsymbol{\chi}^*(X, t^*), t^*\} = \mathbf{G}(\boldsymbol{\chi}^{*t^*}; X, t^*) \tag{4.1.8}$$

for all observer transformations should be coupled with (4.1.3). This is a mathematical statement of the assumption that material properties are independent of observer.

Further simplification of the very general constitutive equation (4.1.3) is achieved by making the assumption that constitutive laws are spatially local in character, as follows.

Suppose that $\boldsymbol{\chi}$ and $\bar{\boldsymbol{\chi}}$ are two motions such that for some neighbourhood $\mathcal{N}(X)$ of X in B

$$\bar{\boldsymbol{\chi}}^t(X', s) = \boldsymbol{\chi}^t(X', s) \tag{4.1.9}$$

for all $s \geq 0$, all $X' \in \mathcal{N}(X)$ and arbitrary $X \in B$. Then if

$$\mathbf{G}(\bar{\boldsymbol{\chi}}^t; X, t) = \mathbf{G}(\boldsymbol{\chi}^t; X, t) \tag{4.1.10}$$

for all motions $\boldsymbol{\chi}$ and $\bar{\boldsymbol{\chi}}$ satisfying (4.1.9) the history of the motion outside $\mathcal{N}(X)$ does not affect the material response at X, and whether or not $\boldsymbol{\chi}^t$ and $\bar{\boldsymbol{\chi}}^t$ coincide outside $\mathcal{N}(X)$ is irrelevant. This condition is normally taken as an *a priori* assumption of constitutive theory and is referred to as the 'principle of local action'. Here we do not impose this locality assumption dogmatically but we do examine its consequences while reserving the right to modify it as empirical evidence dictates (for the purpose of this book, however, such a modification will not be required). We note that the assumption excludes action at a distance from being present in a constitutive law and body forces, such as gravity, and inertial forces, in particular, do not affect the intrinsic properties of a material at the continuum level.

Consideration of the motion in the neighbourhood $\mathcal{N}(X)$ of X relative to an observer moving with X chosen as his origin (so that $\chi^*(X,t^*)=\mathbf{0}$, $\mathbf{Q}(t)=\mathbf{I}$ and $t^*=t$ in (4.1.5)) leads to

$$\chi^*(X',t)=\chi(X',t)-\chi(X,t)\equiv\chi_X(X',t) \tag{4.1.11}$$

for all t and all $X'\in\mathcal{N}(X)$, where we have introduced the notation χ_X for the *localization* of the motion in $\mathcal{N}(X)$. Correspondingly,

$$\chi^{*t}(X',s)=\chi_X^t(X',s) \qquad s\geq 0, X'\in\mathcal{N}(X) \tag{4.1.12}$$

defines the localization χ_X^t of the history of the motion χ^t.

Use of (4.1.3) and (4.1.8) with $\mathbf{Q}(t)=\mathbf{I}$ and $t^*=t$ shows that

$$\mathbf{T}\{\chi(X,t),t\}=\mathbf{T}^*\{\chi^*(X,t),t\}=\mathbf{G}(\chi_X^t;X,t) \tag{4.1.13}$$

and the constitutive equation (4.1.3) is therefore replaced by its localized form for each $X\in B$. The restrictions on the constitutive functional $\mathbf{G}$ occurring in (4.1.13) imposed by material objectivity may be obtained by use of (4.1.3), (4.1.5), (4.1.6) and (4.1.8) with the help of (4.1.2). This is left as an exercise for the reader.

Recalling from (4.1.12) that

$$\chi_X^t(X',s)=\chi^t(X',s)-\chi^t(X,s) \tag{4.1.14}$$

for all $s\geq 0$ and all $X'\in\mathcal{N}(X)$, we now select a reference configuration χ_0 so that $\mathbf{X}=\chi_0(X)$ and, as in Section 2.1.3, we identify $\mathbf{X}$ with X and, for convenience, adjust the notation so that (4.1.14) is replaced by

$$\chi_X^t(\mathbf{X}',s)=\chi^t(\mathbf{X}',s)-\chi^t(\mathbf{X},s). \tag{4.1.15}$$

We emphasize, however, that the motion histories occurring in (4.1.15) now depend on the choice of χ_0.

Because of the assumed regularity of the motion we may approximate (4.1.15) in $\mathcal{N}(X)$ as

$$\chi_X^t(\mathbf{X}',s)\simeq[\operatorname{Grad}\chi^t(\mathbf{X},s)](\mathbf{X}'-\mathbf{X})$$

or, equivalently,

$$\chi_X^t(\mathbf{X}', s) \simeq [\mathrm{Grad}\,\chi_X^t(\mathbf{X}, s)](\mathbf{X}' - \mathbf{X}) \tag{4.1.16}$$

in accordance with the convention of Section 1.5.1. Thus, the history of the motion in $\mathcal{N}(X)$ is governed by the history $\mathrm{Grad}\,\chi_X^t$ of the deformation gradient at X with respect to an arbitrary choice of reference configuration. And line elements of material defined by $\mathbf{X}' - \mathbf{X}$ in $\chi_0\{\mathcal{N}(X)\}$ transform according to (4.1.16) under the deformation history. *The motion is locally homogeneous in* $\mathcal{N}(X)$.

A material is said to be *simple* at X if its response to every deformation at X is determined uniquely by its response to deformations homogeneous in a neighbourhood of X. For such a material the deformation in a neighbourhood of X is determined by $\mathrm{Grad}\,\chi_X^t$ relative to the reference configuration χ_0 and the constitutive equation (4.1.13) reduces to

$$\mathbf{T}\{\chi(\mathbf{X}, t), t\} = \mathbf{G}_0(\mathrm{Grad}\,\chi_X^t; \mathbf{X}, t), \tag{4.1.17}$$

where the subscript 0 indicates dependence on the choice of reference configuration.

Equation (4.1.17) characterizes the constitutive law of a *simple material.* We emphasize, however, that the concept of a simple material is an idealization (a mathematical model) which is dependent on the assumptions of material objectivity and local action. Nevertheless, the constitutive law of a simple material embraces all the purely mechanical constitutive laws encountered in physics and engineering (including those of a Newtonian fluid and a Hookean elastic solid).

If (4.1.17) is specialized further so that the stress depends on $\mathrm{Grad}\,\chi_X$ at time t only, and not on its history, then the resulting simple material is said to be an *elastic material.* For an elastic material the response is independent of the rate at which deformation takes place and the change in stress between two configurations is independent of the time taken and, moreover, independent of the path of deformation joining the configurations. This is discussed in detail in Section 4.2.

It is an easy matter to show that for a simple material (4.1.17) the requirement of material objectivity (4.1.8) imposes the restriction

$$\mathbf{Q}(t)\mathbf{G}_0(\mathbf{A}^t; \mathbf{X}, t)\mathbf{Q}(\mathbf{t})^{\mathrm{T}} = \mathbf{G}_0(\mathbf{Q}^t\mathbf{A}^t; \mathbf{X}, t) \tag{4.1.18}$$

on the response functional $\mathbf{G}_0$, where $\mathbf{A}^t = \mathrm{Grad}\,\chi_X^t$ and $\mathbf{Q}^t(s) = \mathbf{Q}(t-s)$.

Finally in this section, we note that, since (4.1.17) holds for an arbitrary choice of reference configuration, we must have

$$\mathbf{G}_{0'}(\mathbf{A}'^t; \mathbf{X}', t) = \mathbf{G}_0(\mathbf{A}^t; \mathbf{X}, t) \tag{4.1.19}$$

for all (smooth) changes of reference configuration $\chi_0 \to \chi_0'$, say, where $\mathbf{X}' = \chi_0'(X)$, $\mathbf{A}'^t$ denotes the deformation gradient relative to χ_0' and the subscript $0'$ indicates dependence of the functional on χ_0'.

4.1.3 Material uniformity and homogeneity

A reference configuration χ_0 of a neighbourhood $\mathcal{N}(X)$ has uniform density if the scalar field ρ_0 (the mass density defined in Section 3.1) over $\chi_0\{\mathcal{N}(X)\}$ has the same value for each $\mathbf{X}$ in $\chi_0\{\mathcal{N}(X)\}$. We use the notation ρ_0 for this constant value.

Then, two material points X and $\bar{X}$ are said to be *materially isomorphic* if there exist reference configurations χ_0 and $\bar{\chi}_0$ of neighbourhoods $\mathcal{N}(X)$ and $\bar{\mathcal{N}}(\bar{X})$ respectively such that (a) χ_0 and $\bar{\chi}_0$ have the same uniform density $(\rho_0 = \bar{\rho}_0)$, and (b) the response functionals at X and $\bar{X}$ respectively coincide, i.e.

$$\mathbf{G}_0(\mathbf{A}^t; \mathbf{X}, t) = \mathbf{G}_{\bar{0}}(\mathbf{A}^t; \bar{\mathbf{X}}, t) \tag{4.1.20}$$

for all invertible second-order tensors $\mathbf{A}^t$, where $\mathbf{X} = \chi_0(X)$ and $\bar{\mathbf{X}} = \bar{\chi}_0(\bar{X})$. This means that the material in $\mathcal{N}(\bar{X})$ responds in the same way as that in $\mathcal{N}(X)$ to the same deformation gradient history for some choice of reference configurations. In other words, the mechanical properties of X and $\bar{X}$ are indistinguishable.

If all the material particles of a body B are mutually materially isomorphic then B is said to be *materially uniform.* For a materially uniform body B, a different reference configuration needs to be chosen for each point $X \in B$ in general in accordance with (4.1.20). However, if there exists a single reference configuration, χ_0 say, for the whole of B such that

$$\mathbf{G}_0(\mathbf{A}^t; \mathbf{X}, t) = \mathbf{G}_0(\mathbf{A}^t; \bar{\mathbf{X}}, t)$$

for *all* points X and $\bar{X}$ in B, where $\mathbf{X} = \chi_0(X), \bar{\mathbf{X}} = \chi_0(\bar{X})$, the response of each point of B is the same relative to χ_0. Then B is said to be a *homogeneous body.* (A laminate is an example of a materially uniform but inhomogeneous body.)

4.2 CAUCHY ELASTIC MATERIALS

4.2.1 The constitutive equation for a Cauchy elastic material

According to the definition given in Section 4.1.2 an elastic simple material is one at each material point of which the state of stress in the current configuration is determined solely by the state of deformation of this configuration relative to an arbitrary choice of reference configuration. Such a material is called a *Cauchy elastic material.* The Cauchy stress $\mathbf{T}$ does not depend on the path of deformation taken from the reference configuration, but the work done by the stress is in general dependent on the deformation path.

This means that, as distinct from Green elasticity (to be discussed in Section 4.3), Cauchy elasticity has a non-conservative structure (in other words, the stress is not derivable from a scalar potential function).

Since the stress response of a Cauchy elastic material is independent of the time taken to achieve a given deformation we may drop the explicit time dependence from the constitutive law and write (4.1.17) as

$$\mathbf{T}\{\chi(\mathbf{X},t),t\} = \mathbf{G}_0(\mathbf{A};\mathbf{X}), \tag{4.2.1}$$

where $\mathbf{A}(\mathbf{X},t) = \text{Grad}\,\chi(\mathbf{X},t)$ and $\mathbf{G}_0$ is now a *function* of $\mathbf{A}$, not a functional. Note that time dependence enters the right-hand side of (4.2.1) only through $\mathbf{A}$.

Although the elastic properties of the material depend in general on the point $\mathbf{X}$ considered, as indicated in (4.2.1), it is convenient for what follows to omit explicit reference to $\mathbf{X}$. To simplify the notation further, we rewrite (4.2.1) as

$$\mathbf{T} = \mathbf{G}(\mathbf{A}), \tag{4.2.2}$$

noting that we have dropped the subscript from $\mathbf{G}$ and the arguments of $\mathbf{T}$. It is important to remember, though, that $\mathbf{G}$ is dependent on the choice of reference configuration. We refer to $\mathbf{G}$ as the *reponse function of the Cauchy elastic material* relative to the chosen reference configuration (χ_0, say). Thus the mechanical properties of the material are characterized by the function $\mathbf{G}$ (which, strictly, is a mapping from the space of invertible second-order two-point tensors to the space of symmetric second-order Eulerian tensors).

If the reference configuration χ_0 is changed to χ_0' and $\mathbf{P}_0 = \text{Grad}\,\boldsymbol{\kappa}_0(\mathbf{X})$, where $\boldsymbol{\kappa}_0(\mathbf{X}) = \chi_0'\{\chi_0^{-1}(\mathbf{X})\}$ as in Section 2.2.8, then the deformation gradients $\mathbf{A}$, $\mathbf{A}'$ of the current configuration relative to χ_0 and χ_0' respectively are related by

$$\mathbf{A}' = \mathbf{A}\mathbf{P}_0^{-1} \tag{4.2.3}$$

as shown in equation (2.2.90). Let $\mathbf{G}'$ denote the response function of the material relative to χ_0'. Then the requirement that material properties are independent of the choice of reference configuration is expressible as

$$\mathbf{G}(\mathbf{A}) = \mathbf{G}'(\mathbf{A}') \tag{4.2.4}$$

in the notation adopted in (4.2.2). This is the specialization of (4.1.19) to the case of Cauchy elasticity.

Elimination of $\mathbf{A}'$ between (4.2.3) and (4.2.4) shows that

$$\mathbf{G}'(\mathbf{A}\mathbf{P}_0^{-1}) = \mathbf{G}(\mathbf{A}), \tag{4.2.5}$$

and, for any given $\mathbf{P}_0$, equation (4.2.5) must hold for arbitrary deformation gradient tensors $\mathbf{A}$. This is merely a rule governing the dependence of the response function on the reference configuration and does not impose any restriction on $\mathbf{G}$ itself. In contrast the assumption of material objectivity does impose restrictions on $\mathbf{G}$, as we now see.

Under an observer transformation the deformation gradient and Cauchy stress transform according to

$$\mathbf{A}^* = \mathbf{Q}\mathbf{A}, \tag{4.2.6}$$

$$\mathbf{T}^* = \mathbf{Q}\mathbf{T}\mathbf{Q}^{\mathrm{T}} \tag{4.2.7}$$

respectively, the arguments having been omitted from the tensors for simplicity. Material objectivity states that material properties are independent of observer. This means that the function $\mathbf{G}$ is the same for all observers and hence

$$\mathbf{T}^* = \mathbf{G}(\mathbf{A}^*) \tag{4.2.8}$$

should be coupled with (4.2.2), where $\mathbf{A}^*$ and $\mathbf{T}^*$ are given by (4.2.6) and (4.2.7), $\mathbf{Q}$ being an arbitrary proper orthogonal (Eulerian) tensor. It follows that

$$\mathbf{G}(\mathbf{Q}\mathbf{A}) = \mathbf{Q}\mathbf{G}(\mathbf{A})\mathbf{Q}^{\mathrm{T}} \tag{4.2.9}$$

for all proper orthogonal $\mathbf{Q}$ and arbitrary deformation gradients $\mathbf{A}$. This restriction on $\mathbf{G}$ is the specialization of (4.1.18) to the case of Cauchy elasticity. The reader should verify that (4.2.9) is invariant under a change of reference configuration.

4.2.2 Alternative forms of the constitutive equation

Since (4.2.9) holds for all proper orthogonal $\mathbf{Q}$, we may take $\mathbf{Q} = \mathbf{R}^{\mathrm{T}}$, where $\mathbf{R}$ arises in the polar decomposition $\mathbf{A} = \mathbf{R}\mathbf{U}$ of the deformation gradient. It follows from (4.2.2) and (4.2.9) that

$$\mathbf{T} = \mathbf{G}(\mathbf{A}) = \mathbf{R}\mathbf{G}(\mathbf{U})\mathbf{R}^{\mathrm{T}}. \tag{4.2.10}$$

Thus, for any given reference configuration, the response of a Cauchy elastic material to an arbitrary deformation gradient $\mathbf{A}$ is determined by a knowledge of its response to positive-definite, symmetric deformation gradients. In other words, $\mathbf{G}$ is fully determined by its restriction to a domain consisting of positive-definite symmetric second-order (Lagrangean) tensors. In essence, (4.2.10) states that intrinsic properties of a Cauchy elastic material are not influenced by the rotational part $\mathbf{R}$ of the deformation. This is emphasized by rewriting (4.2.10) in the Lagrangean form

$$\mathbf{R}^{\mathrm{T}}\mathbf{T}\mathbf{R} = \mathbf{G}(\mathbf{U}). \tag{4.2.11}$$

In terms of the nominal stress tensor $\mathbf{S}$ defined in Section 3.4.1, we have

$$\mathbf{S} = (\det \mathbf{A})\mathbf{A}^{-1}\mathbf{T} = (\det \mathbf{A})\mathbf{A}^{-1}\mathbf{G}(\mathbf{A}) \equiv \mathbf{H}(\mathbf{A}), \tag{4.2.12}$$

where we have introduced the response function $\mathbf{H}$. Symmetry of $\mathbf{T}$ is equivalent to the restriction

$$\mathbf{A}\mathbf{H}(\mathbf{A}) = [\mathbf{H}(\mathbf{A})]^{\mathrm{T}}\mathbf{A}^{\mathrm{T}}. \tag{4.2.13}$$

It is easily shown that material objectivity requires that

$$\mathbf{H}(\mathbf{Q}\mathbf{A}) = \mathbf{H}(\mathbf{A})\mathbf{Q}^{\mathrm{T}} \tag{4.2.14}$$

for all proper orthogonal $\mathbf{Q}$, this being equivalent to (4.2.9). The corresponding equivalent of (4.2.10) is

$$\mathbf{S} = \mathbf{H}(\mathbf{U})\mathbf{R}^{\mathrm{T}}, \tag{4.2.15}$$

where

$$\mathbf{H}(\mathbf{U}) = (\det \mathbf{U})\mathbf{U}^{-1}\mathbf{G}(\mathbf{U}),$$

and the symmetry requirement (4.2.13) can now be expressed as

$$\mathbf{U}\mathbf{H}(\mathbf{U}) = [\mathbf{H}(\mathbf{U})]^{\mathrm{T}}\mathbf{U}. \tag{4.2.16}$$

Thus, the constitutive equation (4.2.10) of a Cauchy elastic solid may be written equivalently as (4.2.12) and (4.2.15) subject to (4.2.16). Another alternative is provided by use of the Biot stress tensor $\mathbf{T}^{(1)}$ defined in (3.5.15). In view of (4.2.15), this is expressible as

$$\mathbf{T}^{(1)} = \tfrac{1}{2}(\mathbf{S}\mathbf{R} + \mathbf{R}^{\mathrm{T}}\mathbf{S}^{\mathrm{T}}) = \tfrac{1}{2}\{\mathbf{H}(\mathbf{U}) + [\mathbf{H}(\mathbf{U})]^{\mathrm{T}}\}. \tag{4.2.17}$$

Since $\mathbf{T}^{(1)}$ is conjugate to the strain tensor $\mathbf{E}^{(1)} = \mathbf{U} - \mathbf{I}$, we may rewrite (4.2.17) as a relationship

$$\mathbf{T}^{(1)} = \mathbf{G}^{(1)}(\mathbf{E}^{(1)}) \tag{4.2.18}$$

between the conjugate variables $\mathbf{T}^{(1)}$ and $\mathbf{E}^{(1)}$, wherein the function $\mathbf{G}^{(1)}$ is defined.

Similarly, the constitutive equation is expressible as

$$\mathbf{T}^{(2)} = \mathbf{G}^{(2)}(\mathbf{E}^{(2)}) \tag{4.2.19}$$

in terms of the conjugate pair $(\mathbf{T}^{(2)}, \mathbf{E}^{(2)})$, where $\mathbf{E}^{(2)} = \frac{1}{2}(\mathbf{U}^2 - \mathbf{I})$ is the Green strain tensor and

$$\mathbf{G}^{(2)}(\mathbf{E}^{(2)}) = \mathbf{H}(\mathbf{U})\mathbf{U}^{-1},$$

symmetry of $\mathbf{T}^{(2)}$ being ensured by (4.2.16).

More generally, because of the monotonicity ascribed to the strain tensor $\mathbf{F}(\mathbf{U})$ defined by equation (2.2.85) in Section 2.2.7, it follows that $\mathbf{U}$ is uniquely determined by a given value of $\mathbf{F}$. And since the stress tensor $\mathbf{T}^{\mathbf{F}}$ conjugate to $\mathbf{F}$ is expressible, through (3.5.31), in terms of $\mathbf{U}$ and $\mathbf{T}^{(1)}$, we deduce that (4.2.18) and (4.2.19) may be generalized to

$$\mathbf{T}^{\mathbf{F}} = \mathbf{G}^{\mathbf{F}}(\mathbf{F}), \tag{4.2.20}$$

where $\mathbf{G}^{\mathbf{F}}$ denotes the response function of a Cauchy elastic solid relative to the conjugate pair $(\mathbf{T}^{\mathbf{F}}, \mathbf{F})$.

Since $\mathbf{U}$ is unaffected by an observer transformation (recall Section 2.4.1) and, by the assumption of material objectivity applied in Section 4.2.1, $\mathbf{G}$ is the same for all observers (with the same reference configuration), it follows that $\mathbf{H}(\mathbf{U})$, like $\mathbf{G}(\mathbf{U})$, is automatically an objective Lagrangean tensor. Thus, material objectivity is embodied in (4.2.18) by virtue of the objective Lagrangean conjugate variables employed, given that the response function $\mathbf{G}^{(1)}$ is the same for all observers (this follows from material objectivity of $\mathbf{G}$). By the same token, for the forms (4.2.19) and (4.2.20) of the constitutive equation the assumption of material objectivity is built in.

Although the Cauchy stress tensor itself is independent of the reference configuration, as expressed in (4.2.4), the conjugate stress tensors discussed above are changed by a change of reference configuration. For illustration, we observe that, for the change of reference configuration described in Section 4.2.1, the stress tensors $\mathbf{S}$ and $\mathbf{T}^{(2)}$ change to

$$\mathbf{S}' = (\det \mathbf{P}_0)^{-1}\mathbf{P}_0\mathbf{S} \tag{4.2.21}$$

and

$$\mathbf{T}^{(2)\prime} = (\det \mathbf{P}_0)^{-1}\mathbf{P}_0\mathbf{T}^{(2)}\mathbf{P}_0^{\mathrm{T}} \tag{4.2.22}$$

respectively, where $\mathbf{P}_0$ is defined prior to (4.2.3). It is a simple matter to show that material objectivity is preserved under a change of reference configuration.

4.2.3 Material symmetry

The restrictions imposed by material objectivity described in Section 4.2.1 apply to all Cauchy elastic materials. Further restrictions, of a similar mathematical nature, can be found for specific elastic materials which display inherent symmetries in some reference configuration, as mentioned in Section 4.1.1. This type of restriction is now discussed in a general framework and in terms of the response function $\mathbf{G}$ appropriate to Cauchy stress.

In the notation of Section 4.2.1, we suppose that $\mathbf{A}$ and $\mathbf{A}'$ are the deformation gradients relative to the reference configurations χ_0 and χ_0' respectively, with $\mathbf{P}_0$ denoting the deformation gradient from χ_0 to χ_0'. From equation (4.2.5), we recall that

$$\mathbf{G}'(\mathbf{A}\mathbf{P}_0^{-1}) = \mathbf{G}(\mathbf{A}) \tag{4.2.23}$$

for all deformation gradients $\mathbf{A}$ and each given $\mathbf{P}_0$, where $\mathbf{G}'$ is the response function of the material relative to χ_0'.

Suppose, however, that the response of the material relative to χ_0' is indistinguishable from that relative to χ_0 in respect of any mechanical test for some given $\mathbf{P}_0$. This means that $\mathbf{G}' \equiv \mathbf{G}$ and (4.2.23) becomes

$$\mathbf{G}(\mathbf{A}\mathbf{P}_0^{-1}) = \mathbf{G}(\mathbf{A}) \tag{4.2.24}$$

for all deformation gradients $\mathbf{A}$ and the $\mathbf{P}_0$ in question.

The property (4.2.24) motivates the following characterization of material symmetry for Cauchy elastic materials. Consider the set $\mathscr{G}$ of invertible second-order tensors $\mathbf{K}$ for which

$$\mathbf{G}(\mathbf{A}\mathbf{K}) = \mathbf{G}(\mathbf{A}) \tag{4.2.25}$$

for all deformation gradients $\mathbf{A}$. Then $\mathscr{G}$ is said to characterize the symmetry of the elastic material in question relative to the reference configuration χ_0.

It is easy to see that $\mathscr{G}$ has the structure of a group, for if $\mathbf{K}, \bar{\mathbf{K}} \in \mathscr{G}$ then

$$\mathbf{G}(\mathbf{A}\mathbf{K}\bar{\mathbf{K}}) = \mathbf{G}(\mathbf{A}\mathbf{K}) = \mathbf{G}(\mathbf{A})$$

by (4.2.25) and hence $\mathbf{K}\bar{\mathbf{K}} \in \mathscr{G}$. Also, since $\mathbf{K}^{-1}$ exists,

$$\mathbf{G}(\mathbf{A}) = \mathbf{G}(\mathbf{A}\mathbf{K}^{-1}\mathbf{K}) = \mathbf{G}(\mathbf{A}\mathbf{K}^{-1}),$$

again by (4.2.25), and hence $\mathbf{K}^{-1} \in \mathscr{G}$. And finally, the identity $\mathbf{I} \in \mathscr{G}$ since $\mathbf{A}\mathbf{I} = \mathbf{A}$. Thus, the set $\mathscr{G}$ forms a group and it is called the *symmetry group of the material relative to the reference configuration* χ_0.

We now consider the effect on $\mathscr{G}$ of a change of reference configuration, from χ_0 to χ'_0 say, with $\mathbf{A}' = \mathbf{A}\mathbf{P}_0^{-1}$. If $\mathbf{G}'$ denotes the response function relative to χ'_0 then

$$\mathbf{G}'(\mathbf{A}') = \mathbf{G}(\mathbf{A}) = \mathbf{G}(\mathbf{A}'\mathbf{P}_0), \tag{4.2.26}$$

as in Section 4.2.1. But, on application of (4.2.25) and (4.2.26), we obtain

$$\mathbf{G}'(\mathbf{A}') = \mathbf{G}(\mathbf{A}\mathbf{K}) = \mathbf{G}(\mathbf{A}'\mathbf{P}_0\mathbf{K}\mathbf{P}_0^{-1}\mathbf{P}_0) = \mathbf{G}'(\mathbf{A}'\mathbf{P}_0\mathbf{K}\mathbf{P}_0^{-1}).$$

This shows that if $\mathbf{K} \in \mathscr{G}$ then $\mathbf{P}_0\mathbf{K}\mathbf{P}_0^{-1} \in \mathscr{G}'$, where $\mathscr{G}'$ is the symmetry group of the material relative to χ'_0. It is a simple matter to show that, like $\mathscr{G}$, $\mathscr{G}'$ has the structure of a group. In fact, $\mathscr{G}'$ is the group conjugate (in the sense of group theory) to $\mathscr{G}$ relative to $\mathbf{P}_0$, and we may represent this conjugacy in the form

$$\mathscr{G}' = \mathbf{P}_0\mathscr{G}\mathbf{P}_0^{-1}. \tag{4.2.27}$$

This is known as Noll's rule.

The following two special cases should be noted. First, if $\mathbf{P}_0 \in \mathscr{G}$ then $\mathscr{G}' = \mathscr{G}$, and second, if $\mathbf{P}_0$ is a pure dilatation (i.e. $\mathbf{P}_0 = P_0\mathbf{I}$, where P_0 is a positive scalar) then $\mathscr{G}' = \mathscr{G}$ also. Thus, in particular, *the symmetry group of a Cauchy elastic material is unaffected by a change of reference configuration corresponding to an arbitrary pure dilatation.*

At this point we emphasize that in (4.2.25) $\mathbf{G}$ is defined over the set of deformation gradients relative to the reference configuration χ_0, i.e. over the set of invertible second-order tensors with *positive* determinant. The restriction to positive determinant is consistent with the requirement imposed in Section 2.2.2 where we ruled out from consideration deformations which change the relative orientations of triads of line elements. It follows that the elements of $\mathscr{G}$ should be restricted similarly, so that

$$\det \mathbf{K} > 0 \qquad \mathbf{K} \in \mathscr{G}. \tag{4.2.28}$$

Here we differ slightly from the conventional account of this topic (see, for example, Truesdell and Noll, 1965, or Truesdell, 1977)† in which the restriction (4.2.28) is not imposed and configurations of different orientations are admitted (subject only to $\det \mathbf{K} \neq 0$). It is a simple matter to extend the domain of definition of $\mathbf{G}$ to include tensors with negative determinant, but we do not do this here. It should be remembered in this context that deformation gradients with negative determinant do not correspond to physically realistic deformations of the material. A $\mathbf{K}$ with $\det \mathbf{K} < 0$ may, nevertheless,

†Note, however, that Wang and Truesdell (1973) restrict attention to orientation preserving deformations.

correspond to a purely geometrical symmetry of the material in the reference configuration χ_0. For example, reflection of a cubic crystal in a plane through its centre and parallel to a pair of faces is such a symmetry operation, and, as far as mechanical testing is concerned, no distinction between the material properties before and after reflection can be detected. But, as indicated above, a reflection does not correspond to a deformation which can be achieved in a continuum. In this way we distinguish between deformational and geometrical symmetries in material properties.

Henceforth we impose the restriction (4.2.28) in respect of deformational material symmetry but the set $\mathscr{G}$ can be supplemented by $\{-\mathbf{K}:\mathbf{K}\in\mathscr{G}\}$ if geometrical symmetries are also to be considered, in which case the domain of definition of the response function $\mathbf{G}$ should be restricted to $\det \mathbf{A} \neq 0$ rather than $\det \mathbf{A} > 0$.

Given (4.2.28), we may write

$$\mathbf{K} = (\det \mathbf{K})^{1/3}\mathbf{K}^*$$

where $\det \mathbf{K}^* = 1$. But, since a pure dilatation (in this case $(\det \mathbf{K})^{1/3}\mathbf{I}$) does not change the symmetry group of a Cauchy elastic material, this group can be characterized by considering only those elements of $\mathscr{G}$ for which $\det \mathbf{K} = 1$. Indeed, it is appropriate to restrict $\mathscr{G}$ to elements with $\det \mathbf{K} = 1$. The symmetry group $\mathscr{G}$ then becomes a subgroup of the *unimodular group* $\mathscr{U} = \{\mathbf{K}:\det\mathbf{K} = \pm 1\}$, i.e. the group of all second-order tensors with $|\det \mathbf{K}| = 1$.

Note that the restriction to unimodular tensors is preserved under a change of reference configuration, for $\mathbf{K}\in\mathscr{U}$ implies $\mathbf{P}_0\mathbf{K}\mathbf{P}_0^{-1}\in\mathscr{U}$ for *arbitrary* invertible $\mathbf{P}_0$. Accordingly, a symmetry group $\mathscr{G}$ becomes a symmetry group $\mathscr{G}'$ under a change of reference configuration. If $\mathscr{U}^+ = \{\mathbf{K}:\det\mathbf{K} = 1\}$ then $\mathscr{G} = \mathscr{U}^+$ implies that $\mathscr{G}' = \mathscr{U}^+$ for all reference configurations, but if $\mathscr{G}$ is a strict subgroup of $\mathscr{U}^+$ then $\mathscr{G}' \neq \mathscr{G}$ in general. In accordance with the general decomposition of a tensor $\mathbf{K}$ with $\det \mathbf{K} = 1$ given in Theorem 2.2.3, we observe that elements of $\mathscr{U}^+$ can consist of rotations, simple shears, and extensions with lateral contractions, and combinations of such deformations.

To summarize, mechanical tests cannot distinguish the properties of a material in two distinct reference configurations which are equivalent in the sense that they are related by an element of the symmetry group $\mathscr{G}$. Note that the densities in two such configurations are equal since $\rho_0/(\det \mathbf{K}) = \rho_0$ (recall equation (3.1.7)). If purely geometrical symmetries (i.e. reflections) are to be allowed then $\mathscr{G}$ must be extended to include $\{-\mathbf{K}:\mathbf{K}\in\mathscr{G}\}$ as noted above and the domain of $\mathbf{G}$ is then the set of all invertible second-order tensors. For the purposes of this book, however, it suffices to restrict attention to $\mathscr{G}$. Characterization of $\mathscr{G}$ is simplified by use of the following theorem.

Theorem 4.2.1 If $\mathbf{G}$ is the response function of a Cauchy elastic material relative to the reference configuration χ_0 with symmetry group $\mathscr{G}$ then

$$\mathbf{G}(\mathbf{AK}) = \mathbf{G}(\mathbf{A}) \tag{4.2.29}$$

for all deformation gradients $\mathbf{A}$ and all $\mathbf{K} \in \mathscr{G}$ if and only if (4.2.29) holds for all generators of the group $\mathscr{G}$.

The proof of the theorem is straightforward and is left to the reader. Note that a finite set $\{\mathbf{K}_1, \ldots, \mathbf{K}_n\}$ of elements of $\mathscr{G}$ is called a generating set of $\mathscr{G}$ if every element $\mathbf{K}$ of $\mathscr{G}$ is expressible in the form

$$\mathbf{K} = \mathbf{K}_1^{p_1} \ldots \mathbf{K}_n^{p_n} \mathbf{K}_1^{q_1} \ldots \mathbf{K}_n^{q_n} \ldots \mathbf{K}_1^{r_1} \ldots \mathbf{K}_n^{r_n}$$

for some (positive or negative) integers $p_1, \ldots, r_n$ and no member of the set is expressible in terms of the others. Use of this theorem is illustrated in Section 4.2.5.

4.2.4 Undistorted configurations and isotropy

If a proper orthogonal tensor $\mathbf{Q}$ is an element of the symmetry group $\mathscr{G}$ then, by (4.2.27),

$$\mathbf{Q}' = \mathbf{P}_0 \mathbf{Q} \mathbf{P}_0^{-1}, \tag{4.2.30}$$

where $\mathbf{Q}'$ (not in general orthogonal) is the corresponding element of $\mathscr{G}'$. From the polar decomposition theorem, we have $\mathbf{P}_0 = \mathbf{RU}$, where $\mathbf{R}$ is proper orthogonal and $\mathbf{U}$ is positive definite and symmetric and substitution of this into (4.2.30) then gives $\mathbf{Q}'\mathbf{RU} = \mathbf{RUQ}$. If $\mathbf{Q}'$ is (proper) orthogonal then it follows from uniqueness of the polar decomposition that $\mathbf{Q}' = \mathbf{RQR}^{\mathrm{T}}$ and

$$\mathbf{U} = \mathbf{QUQ}^{\mathrm{T}}. \tag{4.2.31}$$

Thus, for each (proper) orthogonal $\mathbf{Q} \in \mathscr{G}$, the stretch tensor $\mathbf{U}$ relating the reference configurations χ_0 and χ_0' must comply with (4.2.31) if the corresponding element $\mathbf{Q}' \in \mathscr{G}'$ is also to be orthogonal. Equation (4.2.31) therefore restricts the relationship between reference configurations for which the symmetry groups of the material consist of (proper) orthogonal elements.

By setting $\mathbf{A} = \mathbf{I}$, the identity deformation, in (4.2.2) and using (4.2.29) with $\mathbf{K} = \mathbf{Q}$ together with (4.2.9), we obtain

$$\mathbf{G}(\mathbf{I}) = \mathbf{QG}(\mathbf{I})\mathbf{Q}^{\mathrm{T}}. \tag{4.2.32}$$

Hence, the Cauchy stress $\mathbf{G}(\mathbf{I})$ in the reference configuration χ_0 must satisfy (4.2.32) for each proper orthogonal $\mathbf{Q} \in \mathscr{G}$. This is a consequence of material objectivity and material symmetry in tandem. Observe that the restrictions placed on $\mathbf{U}$ and $\mathbf{G}(\mathbf{I})$ are identical.

If the symmetry group of an elastic material contains the *proper orthogonal group* (i.e. the group consisting of all proper orthogonal second-order tensors) for some reference configuration then (4.2.31) and (4.2.32) must hold for all proper orthogonal $\mathbf{Q}$. It follows that

$$\mathbf{G}(\mathbf{I}) = \alpha\mathbf{I}, \tag{4.2.33}$$

where α is a scalar, and similarly for $\mathbf{U}$, i.e. $\mathbf{G}(\mathbf{I})$ and $\mathbf{U}$ are second-order isotropic tensors in the sense of Section 1.2.5. Thus, the stress in the considered reference configuration is hydrostatic and the deformation relating two such reference configurations is a pure dilatation (possibly coupled with a rotation).

An *isotropic elastic material* is an elastic material whose symmetry group contains the proper orthogonal group for at least one reference configuration. In such a reference configuration the mechanical response of the material exhibits no preferred direction, and it is this property that characterizes isotropy.

A reference configuration of an isotropic elastic material relative to which the symmetry group does contain the proper orthogonal group is called an *undistorted configuration* of the material. It follows from the above discussion that undistorted configurations (of an isotropic elastic material) are related by a pure dilatation (coupled with a rotation in general). Thus, if $\boldsymbol{\chi}_0$ and $\boldsymbol{\chi}_0'$ are two such undistorted configurations with symmetry groups $\mathscr{G}$ and $\mathscr{G}'$ respectively then the appropriate specialization of (4.2.27) is

$$\mathscr{G}' = \mathbf{R}\mathscr{G}\mathbf{R}^{\mathrm{T}}, \tag{4.2.34}$$

where $\mathbf{P}_0 = \mathbf{R}\mathbf{U}$ and $\mathbf{U}$ is a pure dilatation. Undistorted configurations can also be defined for elastic materials which are not isotropic, as we record in Section 4.2.5.

From the above definition of an isotropic elastic material, we obtain

$$\mathbf{T} = \mathbf{G}(\mathbf{A}) = \mathbf{G}(\mathbf{V}) = \mathbf{G}(\mathbf{V}\mathbf{Q}^{\mathrm{T}}) \tag{4.2.35}$$

on use of the polar decomposition $\mathbf{A} = \mathbf{V}\mathbf{R}$ and by selecting $\mathbf{K} = \mathbf{R}^{\mathrm{T}}$ and $\mathbf{K} = \mathbf{R}^{\mathrm{T}}\mathbf{Q}^{\mathrm{T}}$ respectively in (4.2.29), where $\mathbf{Q}$ is an arbitrary proper orthogonal tensor. It follows from material objectivity (4.2.4) and (4.2.35) that

$$\mathbf{G}(\mathbf{Q}\mathbf{V}\mathbf{Q}^{\mathrm{T}}) = \mathbf{Q}\mathbf{G}(\mathbf{V})\mathbf{Q}^{\mathrm{T}} \tag{4.2.36}$$

for all proper orthogonal $\mathbf{Q}$ and arbitrary positive definite symmetric $\mathbf{V}$, $\mathbf{G}$ being the response function of the material relative to the undistorted configuration χ_0. Equation (4.2.36) characterizes isotropy of an elastic material, and more generally a second-order tensor function $\mathbf{G}$ of a symmetric second-order tensor satisfying (4.2.35) is said to be an isotropic function.

If the symmetry group $\mathscr{G}$ is equal to the proper orthogonal group the material is said to be an *isotropic elastic solid*. This is discussed in detail in Section 4.2.6. But if $\mathscr{G} = \mathscr{U}^+$, the proper unimodular group, the resulting constitutive law reduces to that of an (inviscid) compressible fluid, as we now see.

By writing

$$\mathbf{A}^* = J^{-1/3}\mathbf{A}$$

in accordance with (2.2.67), where $J = \det \mathbf{A}$ and $\det \mathbf{A}^* = 1$, and choosing $\mathbf{K} = \mathbf{A}^{*-1}$ in (4.2.29), we obtain

$$\mathbf{G}(\mathbf{A}) = \mathbf{G}(J^{1/3}\mathbf{I}).$$

Since $\mathbf{G}(\mathbf{A}) = \mathbf{G}(\mathbf{V})$ from (4.2.35), it follows from (4.2.36) that

$$\mathbf{Q}\mathbf{G}(J^{1/3}\mathbf{I})\mathbf{Q}^{\mathrm{T}} = \mathbf{G}(J^{1/3}\mathbf{I}),$$

for all proper orthogonal $\mathbf{Q}$. This means that $\mathbf{G}(J^{1/3}\mathbf{I})$ is a (second-order) isotropic tensor in the sense of Section 1.2.5 and hence

$$\mathbf{G}(J^{1/3}\mathbf{I}) = \alpha(J)\mathbf{I}, \tag{4.2.37}$$

where α is a scalar function of J.

But, by (3.1.6), we have $J = \rho_0/\rho$, where ρ_0 is the density of the material in an undistorted configuration (the same for all such configurations). Therefore, with ρ_0 fixed we regard $\alpha(J)$ as a function of the current density ρ and write $p(\rho) = -\alpha(J)$. Thus, the Cauchy stress is given by

$$\mathbf{T} = \mathbf{G}(\mathbf{A}) = -p(\rho)\mathbf{I} \tag{4.2.38}$$

in this special case, $p(\rho)$ being the pressure in the current configuration. Equation (4.2.38) characterizes the mechanical properties of an elastic fluid. It is an easy matter to show that every configuration of an elastic fluid is an undistorted configuration.

A *stress-free* undistorted configuration is called a *natural configuration*. Let χ_0 be such a configuration. Then $\mathbf{G}(\mathbf{I}) = \mathbf{0}$, where $\mathbf{G}$ is the response function of the material relative to χ_0. We now examine the relation between χ_0 and any other possible natural configurations.

Suppose χ_0' is an undistorted configuration relative to which the response function is $\mathbf{G}'$. The stress in that configuration is $\mathbf{G}'(\mathbf{I})$ and from equation (4.2.5) we obtain

$$\mathbf{G}'(\mathbf{I}) = \mathbf{G}(\mathbf{P}_0),$$

where $\mathbf{P}_0$, as before, is the gradient of the deformation from χ_0 to χ_0'. Since χ_0 and χ_0' are undistorted configurations we can write $\mathbf{P}_0 = J^{1/3}\mathbf{R}$, where $J = \det \mathbf{P}_0$ and $\mathbf{R}$ is proper orthogonal. Application of (4.2.9) with $\mathbf{Q} = \mathbf{R}$ and $\mathbf{A} = J^{1/3}\mathbf{I}$ followed by (4.2.36) and (4.2.37) now leads to

$$\mathbf{G}'(\mathbf{I}) = \mathbf{G}(J^{1/3}\mathbf{I}) = \alpha(J)\mathbf{I}.$$

Hence χ_0' is a natural configuration if and only if $\alpha(J) = 0$.

In order to make further progress, we consider the properties of $\alpha(J)$. Since $\alpha(1) = 0$ and

$$\mathbf{G}'(\mathbf{I}) = \mathbf{G}(\mathbf{R}) = \mathbf{G}(\mathbf{I}),$$

any configuration differing from χ_0 by a rotation $\mathbf{R}$ is also a natural configuration. Next, we note that $\alpha(J)$ is the hydrostatic stress in response to the pure dilatation $J^{1/3}\mathbf{I}$ from the stress-free configuration χ_0. Physically, it is reasonable to assume that an increase ($J > 1$) in volume requires hydrostatic tension while a decrease ($J < 1$) requires pressure, i.e.

$$\alpha(J) \gtreqless 0 \text{ according as } J \gtreqless 1. \tag{4.2.39}$$

Given (4.2.39) it follows that χ_0' is a natural configuration if and only if $J = 1$ and therefore the natural configuration χ_0 is unique to within an arbitrary rotation. Such a configuration is also referred to (ambiguously) as the undeformed stress-free configuration.

4.2.5 Anisotropic elastic solids

For a number of solid materials there exist reference configurations relative to which the symmetry groups of their response functions contain strict subgroups of the proper orthogonal group but not the proper orthogonal group itself. These are called *anisotropic* elastic solids. Reference configurations with the above property are called *undistorted configurations* by analogy with the definition given in the case of isotropy.

The prime example of an anisotropic elastic solid is a *transversely isotropic* elastic solid. In an undistorted configuration, such a solid has a *preferred direction*, $\mathbf{n}$ say, and the symmetry group consists of arbitrary rotations about $\mathbf{n}$ together with rotations which carry $\mathbf{n}$ to $-\mathbf{n}$. Undistorted configurations of a transversely isotropic elastic solid are related by the deformation gradient $\mathbf{P}_0 = \mathbf{RU}$, where $\mathbf{R}$ is an arbitrary rotation and $\mathbf{U}$ has the form

$$\mathbf{U} = \alpha\mathbf{I} + \beta\mathbf{n} \otimes \mathbf{n}, \tag{4.2.40}$$

α and β being scalars. The deformation defined by (4.2.40) is a combination of a pure dilatation and an extension along $\mathbf{n}$. The stress $\mathbf{G}(\mathbf{I})$ in an undistorted configuration also has the form (4.2.40), where $\mathbf{G}$ is the response function of the material relative to that configuration.

As indicated above a transversely isotropic elastic solid has a single preferred direction, namely $\mathbf{n}$. Other anisotropic elastic solids may have more than one such direction. In particular, anisotropic elastic solids which possess three preferred directions in an undistorted configuration have the symmetries of the so-called *crystal classes*.

The crystal classes derive from the regular structure of crystal lattices. A simple crystal lattice is an infinite periodic array of atoms defined by three linearly independent vectors, $\mathbf{a}_1, \mathbf{a}_2, \mathbf{a}_3$ say, such that the atoms in the lattice are located (to within an arbitrary choice of origin) at the points

$$l_{ij}\mathbf{a}_j \quad (i = 1, 2, 3; \text{ sum over } j = 1, 2, 3),$$

where the coefficients l_{ij} are integers. The basis $\{\mathbf{a}_i\}$ is said to generate the lattice.

Any symmetry at the atomic level can, under certain assumptions, be carried over to the macroscopic level for the lattice and such symmetry is then embodied in the continuum theory of such solids. The elastic properties of these crystal classes can be discussed by applying the theory of Section 4.2.3. with symmetry groups which reflect the symmetry implied by the lattice geometry. For a monatomic lattice, in particular, the atoms of the lattice are identical, and the theory of elasticity of a monatomic crystal lattice arises. The following example, in which we use the notation $\mathbf{R}_{\mathbf{n}}^{\phi}$ for a right-handed rotation about the direction of the unit vector $\mathbf{n}$ through an angle $\phi (0 < \phi < 2\pi)$, illustrates these general remarks.

A simple cubic lattice is generated by a set of three mutually orthogonal unit vectors, $\mathbf{i}, \mathbf{j}, \mathbf{k}$ say, these also being the preferred directions of the solid. The symmetry group of the solid has 24 elements and is generated by the three rotations[†]

$$\mathbf{R}_{\mathbf{i}}^{(1/2)\pi}, \ \mathbf{R}_{\mathbf{j}}^{(1/2)\pi}, \ \mathbf{R}_{\mathbf{k}}^{(1/2)\pi}.$$

[†]The standard classification of the crystal classes admits improper orthogonal tensors into the symmetry groups but here we restrict attention to proper orthogonal tensors in accordance with the discussion in Section 4.2.3.

It is an easy matter to show that, in an undistorted configuration,

$$\mathbf{G}(\mathbf{I}) = \alpha \mathbf{I},$$

where α is a scalar, and that undistorted configurations are related by the deformation gradient $\mathbf{P}_0 = \mathbf{RU}$, where $\mathbf{U}$ is a pure dilatation and $\mathbf{R}$ is an arbitrary rotation. Descriptions of this and other crystal classes can be found in Truesdell and Noll (1965), for example, and we do not pursue the details here.

Thus far we have restricted attention to (proper) orthogonal elements of the symmetry groups of anisotropic elastic solids, and, indeed, in most discussions of crystal elasticity it is only such elements which are considered. However, recent work, summarized by Ericksen (1977), Hill (1978) and Parry (1976), has shown that such considerations do not provide a complete picture of the symmetry properties of perfect crystals and that non-orthogonal transformations need to be admitted. We now discuss briefly this aspect of the theory which, we emphasize, is still at an early stage of development. According to Hill (1978) every primitive (Bravais) lattice can be regarded as a homogeneous deformation of a simple cubic lattice. It is therefore sufficient to confine the discussion to a cubic lattice.

Let $\{\mathbf{a}_i\}$ be an orthonormal basis for the lattice with the interatomic spacing defining the unit of distance. This is not, however, the only possible basis for the lattice since the vectors

$$k_{ij}\mathbf{a}_j \qquad (i = 1,2,3;\ \text{sum over } j = 1,\ 2,\ 3)$$

also form a generating basis for the lattice provided each coefficient k_{ij} is an integer and also†

$$\det(k_{ij}) = 1.$$

This basis transformation can be interpreted as a homogeneous, density preserving, deformation of the lattice corresponding to integral lattice spacing transformations of atoms leaving no lattice point unoccupied. Such a deformation carries the material into a new configuration which is indistinguishable from the original as far as experimental determination of the elastic properties of the material is concerned. For example, a simple shear of amount equal to a single lattice spacing in a lattice plane is such a deformation, as illustrated in Fig. 4.1. A general transformation of this type can be achieved by the combination of such simple shears and rotations. If

†For consistency with Section 4.2.3 we omit basis transformations for which $\det(k_{ij}) = -1$.

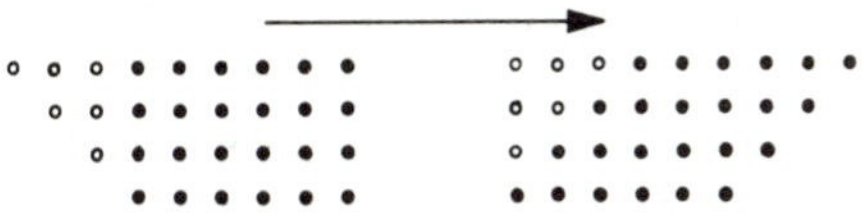

Fig. 4.1 Plane section of a cubic lattice (a) before, and (b) after, a simple shear by a single lattice spacing. Configurations (a) and (b) are indistinguishable locally.

the deformation in question has gradient $\mathbf{A}$, the basis vector $\mathbf{a}_i$ becomes $\mathbf{A}\mathbf{a}_i$ and $A_{ij}=k_{ji}$ are its components relative to the original basis.

The set of matrices $\mathsf{K}=(k_{ij})$ forms a group, the proper unimodular group in the integers. This group has just two generators, namely

$$\mathsf{K}_1=\begin{bmatrix}0&0&1\\1&0&0\\0&1&0\end{bmatrix},\qquad \mathsf{K}_2=\begin{bmatrix}1&1&0\\0&1&0\\0&0&1\end{bmatrix} \tag{4.2.41}$$

with $\mathsf{K}_1^3=\mathsf{I}$, where I represents the identity matrix. The matrix K_1 defines a rotation of $2\pi/3$ about a cube diagonal, while K_2 represents a simple shear deformation by a single lattice spacing in the plane of $\mathbf{a}_1$ and $\mathbf{a}_2$ (refer to Section 2.2.6 for a discussion of simple shear). For detailed discussion and references see Ericksen (1977).

We shall return to this topic briefly in Section 4.3 in the context of Green elasticity.

Problem 4.2.1 Derive equation (4.2.40) in respect of a transversely isotropic elastic solid.

Problem 4.2.2 The rhombic crystal class has mutually orthogonal preferred directions $\mathbf{i}, \mathbf{j}, \mathbf{k}$ and its (proper) orthogonal symmetry group is generated by $\mathbf{R}_\mathbf{i}^\pi$, $\mathbf{R}_\mathbf{j}^\pi$. Find all the elements of this group and show that undistorted configurations are related by a right stretch which is coaxial with $\mathbf{i}, \mathbf{j}, \mathbf{k}$ together with an arbitrary rotation.

Problem 4.2.3 For the cubic system with symmetry group generated by K_1 and K_2 as given in (4.2.41), show that the element $\mathsf{K}_3\mathsf{K}_2^{-1}\mathsf{K}_3$ of the group, where

$$\mathsf{K}_3=\mathsf{K}_1^{-2}\mathsf{K}_2^{-1}\mathsf{K}_1\mathsf{K}_2^{-1}\mathsf{K}_1^{-1}(\mathsf{K}_2\mathsf{K}_1)^2,$$

corresponds to rotation through $\pi/2$ about a cube edge.

Problem 4.2.4 The reference configuration χ_0 is an undistorted configuration of an isotropic elastic solid. Show that the solid possesses at least rhombic symmetry relative to an arbitrary reference configuration, χ_0' say.

4.2.6 Isotropic elastic solids

An isotropic elastic solid was defined in Section 4.2.4 as a material whose symmetry group relative to some reference configuration (an undistorted configuration) is equal to the proper orthogonal group. The response function $\mathbf{G}$ of the material relative to that configuration satisfies the identity

$$\mathbf{G}(\mathbf{Q}\mathbf{V}\mathbf{Q}^{\mathrm{T}}) = \mathbf{Q}\mathbf{G}(\mathbf{V})\mathbf{Q}^{\mathrm{T}} \tag{4.2.42}$$

for all proper orthogonal $\mathbf{Q}$ and arbitrary positive definite symmetric $\mathbf{V}$. More generally, a function $\mathbf{G}$ satisfying this requirement is referred to as a (second-order) isotropic tensor function defined over the space of positive definite symmetric second-order tensors. (Strictly, the restriction to positive definite $\mathbf{V}$ is not in general required but, in the context of elasticity, it is imposed and, moreover, the domain of definition of $\mathbf{G}$ may then be extended by use of (4.2.35) to the space of second-order tensors with positive determinant.) Examples of such functions are the strain tensors $\mathbf{F}(\mathbf{U})$ defined in Section 2.2.7, these being isotropic functions of the right stretch tensor $\mathbf{U}$.

Before proceeding to examine specific representations for $\mathbf{G}(\mathbf{V})$ it is first necessary to give a general definition of a scalar invariant of a second-order tensor. By way of motivation we recall from Sections 1.2.4 and 1.3.2 that $\mathrm{tr}\,\mathbf{V}$, $\mathrm{tr}(\mathbf{V}^2)$ and $\det\mathbf{V}$ are examples of scalar invariants of $\mathbf{V}$. This means that their values are unaffected if $\mathbf{V}$ is replaced by $\mathbf{Q}\mathbf{V}\mathbf{Q}^{\mathrm{T}}$, where $\mathbf{Q}$ is proper orthogonal. More specifically, if $\mathbf{V}$ denotes the left stretch tensor then $\mathrm{tr}\,\mathbf{V}$ is unaffected by an observer transformation which changes $\mathbf{V}$ according to (2.4.11).

Let ϕ be a scalar function defined over the space of (positive definite) symmetric tensors. Then its value $\phi(\mathbf{V})$ at $\mathbf{V}$ is said to be a *scalar invariant* of $\mathbf{V}$ if

$$\phi(\mathbf{Q}\mathbf{V}\mathbf{Q}^{\mathrm{T}}) = \phi(\mathbf{V}) \tag{4.2.43}$$

for all proper orthogonal $\mathbf{Q}$, and this definition embraces the examples cited above. Such a ϕ is also called an *isotropic scalar function*. The following theorem provides important representations for general isotropic scalar functions.

Theorem 4.2.2 Let ϕ be a scalar function defined over the space of symmetric (positive definite) second-order tensors. Then its value $\phi(\mathbf{V})$ at $\mathbf{V}$ is a scalar invariant of $\mathbf{V}$ if and only if there exists a function $\hat{\phi}$, defined on

$(0, \infty)^3$, such that

$$\phi(\mathbf{V}) = \hat{\phi}(\lambda_1, \lambda_2, \lambda_3), \tag{4.2.44}$$

where

$$\hat{\phi}(\lambda_1, \lambda_2, \lambda_3) = \hat{\phi}(\lambda_1, \lambda_3, \lambda_2) = \hat{\phi}(\lambda_3, \lambda_1, \lambda_2). \tag{4.2.45}$$

and $\lambda_1, \lambda_2, \lambda_3$ are the principal values of $\mathbf{V}$. Furthermore, the representation (4.2.44) together with (4.2.45) is equivalent to the representation

$$\phi(\mathbf{V}) = \Phi(I_1, I_2, I_3), \tag{4.2.46}$$

where Φ is also defined on $(0, \infty)^{3\dagger}$ and I_1, I_2, I_3 are the principal invariants of $\mathbf{V}$.

Proof (a) First we suppose that $\phi(\mathbf{V})$ is a scalar invariant of $\mathbf{V}$ so that (4.2.43) holds. By writing $\mathbf{V}$ in the spectral form

$$\mathbf{V} = \lambda_1 \mathbf{v}^{(1)} \otimes \mathbf{v}^{(1)} + \lambda_2 \mathbf{v}^{(2)} \otimes \mathbf{v}^{(2)} + \lambda_3 \mathbf{v}^{(3)} \otimes \mathbf{v}^{(3)}$$

we obtain

$$\mathbf{QVQ}^{\mathrm{T}} = \lambda_1 \mathbf{Qv}^{(1)} \otimes \mathbf{Qv}^{(1)} + \lambda_2 \mathbf{Qv}^{(2)} \otimes \mathbf{Qv}^{(2)} + \lambda_3 \mathbf{Qv}^{(3)} \otimes \mathbf{Qv}^{(3)}$$

It follows from (4.2.43) that

$$\phi(\lambda_1 \mathbf{Qv}^{(1)} \otimes \mathbf{Qv}^{(1)} + \lambda_2 \mathbf{Qv}^{(2)} \otimes \mathbf{Qv}^{(2)} + \lambda_3 \mathbf{Qv}^{(3)} \otimes \mathbf{Qv}^{(3)})$$

has the same value for all proper orthogonal $\mathbf{Q}$. Hence $\phi(\mathbf{V})$ is independent of the orientation of the principal directions $\mathbf{v}^{(1)}, \mathbf{v}^{(2)}, \mathbf{v}^{(3)}$ of $\mathbf{V}$ and must depend on $\mathbf{V}$ only through its principal values $\lambda_1, \lambda_2, \lambda_3$. This establishes (4.2.44).

To prove (4.2.45) we choose $\mathbf{Q}$ to be a rotation of $\pi/2$ about $\mathbf{v}^{(3)}$ so that $\mathbf{Qv}^{(1)} = \mathbf{v}^{(2)}, \mathbf{Qv}^{(2)} = -\mathbf{v}^{(1)}$ and $\mathbf{Qv}^{(3)} = \mathbf{v}^{(3)}$. It follows that

$$\phi(\mathbf{V}) = (\lambda_2 \mathbf{v}^{(1)} \otimes \mathbf{v}^{(1)} + \lambda_1 \mathbf{v}^{(2)} \otimes \mathbf{v}^{(2)} + \lambda_3 \mathbf{v}^{(3)} \otimes \mathbf{v}^{(3)})$$

and hence

$$\hat{\phi}(\lambda_1, \lambda_2, \lambda_3) = \hat{\phi}(\lambda_2, \lambda_1, \lambda_3).$$

The remainder of (4.2.45) follows in similar fashion.

† Subject to $I_1^2 \geq 3I_2$ and $I_2^2 \geq 3I_1 I_3$, with equality holding if and only if $\lambda_1 = \lambda_2 = \lambda_3$.

Next we recall from Section 1.3.2 that the principal values $\lambda_1, \lambda_2, \lambda_3$ are the solutions of the characteristic equation

$$\lambda^3 - I_1\lambda^2 + I_2\lambda - I_3 = 0.$$

This cubic equation has coefficients which are scalar invariants of $\mathbf{V}$. Its solutions are (in principle but in general not explicitly) expressible as functions of these coefficients and hence (4.2.44) may be replaced by (4.2.46). In passing we remark that this has shown that $\lambda_1, \lambda_2, \lambda_3$ are themselves scalar invariants of $\mathbf{V}$.

(b) Secondly, since I_1, I_2, I_3 and $\lambda_1, \lambda_2, \lambda_3$ are scalar invariants of $\mathbf{V}$ it follows immediately from either (4.2.44) or (4.2.46) that $\phi(\mathbf{V})$ is a scalar invariant of $\mathbf{V}$.

We remark that I_1, I_2, I_3 are symmetric functions of $\lambda_1, \lambda_2, \lambda_3$, i.e. they satisfy (4.2.45). Furthermore, in (4.2.46) I_1, I_2, I_3 may be replaced by any other set of three independent symmetric invariants. For example, whereas

$$I_1 = \lambda_1 + \lambda_2 + \lambda_3, I_2 = \lambda_2\lambda_3 + \lambda_3\lambda_1 + \lambda_1\lambda_2, I_3 = \lambda_1\lambda_2\lambda_3,$$

$\phi(\mathbf{V})$ may be regarded equally as a function of

$$\lambda_1 + \lambda_2 + \lambda_3, \lambda_1^2 + \lambda_2^2 + \lambda_3^2, \lambda_1^3 + \lambda_2^3 + \lambda_3^3.$$

During the above proof we obtained the subsidiary result that the principal values of $\mathbf{V}$ are scalar invariants of $\mathbf{V}$. This result may be recovered as a special case of the following theorem.

Theorem 4.2.3 If $\mathbf{G}$ is an isotropic tensor function defined in accordance with (4.2.42) then the principal values of $\mathbf{G}(\mathbf{V})$ are scalar invariants of $\mathbf{V}$.

Proof Let $g(\mathbf{V})$ be a principal value of $\mathbf{G}(\mathbf{V})$. Then, by (1.3.14), the characteristic equation for $\mathbf{G}(\mathbf{V})$ is

$$\det\{\mathbf{G}(\mathbf{V}) - g(\mathbf{V})\mathbf{I}\} = 0. \tag{4.2.47}$$

The corresponding principal value of $\mathbf{G}(\mathbf{QVQ}^{\mathrm{T}})$ is $g(\mathbf{QVQ}^{\mathrm{T}})$ and hence

$$\det\{\mathbf{G}(\mathbf{QVQ}^{\mathrm{T}}) - g(\mathbf{QVQ}^{\mathrm{T}})\mathbf{I}\} = 0.$$

By isotropy and orthogonality this may be rewritten as

$$\det\{\mathbf{QG}(\mathbf{V})\mathbf{Q}^{\mathrm{T}} - g(\mathbf{QVQ}^{\mathrm{T}})\mathbf{QIQ}^{\mathrm{T}}\} = 0$$

and, on use of (1.1.25), it reduces to

$$\det\{\mathbf{G}(\mathbf{V}) - g(\mathbf{Q}\mathbf{V}\mathbf{Q}^{\mathrm{T}})\mathbf{I}\} = 0.$$

Since this holds for all proper orthogonal $\mathbf{Q}$ we deduce that

$$g(\mathbf{Q}\mathbf{V}\mathbf{Q}^{\mathrm{T}}) = g(\mathbf{V})$$

for all proper orthogonal $\mathbf{Q}$, i.e. $g(\mathbf{V})$ is a scalar invariant of $\mathbf{V}$.

Another important consequence of isotropy is that $\mathbf{G}(\mathbf{V})$ is coaxial with $\mathbf{V}$, a result which is formalized in the following theorem.

Theorem 4.2.4 If $\mathbf{G}$ is an isotropic tensor function then $\mathbf{G}(\mathbf{V})$ is coaxial with $\mathbf{V}$, and hence

$$\mathbf{G}(\mathbf{V})\mathbf{V} = \mathbf{V}\mathbf{G}(\mathbf{V}). \tag{4.2.48}$$

Proof Let $\mathbf{v}$ be a unit eigenvector of $\mathbf{V}$ corresponding to eigenvalue λ and choose $\mathbf{Q}$ to be a rotation about $\mathbf{v}$ through an angle π. On specializing (1.3.31) and adapting (1.2.10), we obtain

$$\mathbf{Q} = 2\mathbf{v} \otimes \mathbf{v} - \mathbf{I}.$$

Use of the formula $\mathbf{V}\mathbf{v} = \lambda\mathbf{v}$ leads to

$$\mathbf{Q}\mathbf{V}\mathbf{Q}^{\mathrm{T}} = \mathbf{V}$$

and therefore (4.2.42) yields

$$\mathbf{Q}\mathbf{G}(\mathbf{V})\mathbf{Q}^{\mathrm{T}} = \mathbf{G}(\mathbf{V})$$

for this particular $\mathbf{Q}$. It follows that

$$\mathbf{Q}\mathbf{G}(\mathbf{V})\mathbf{v} = \mathbf{G}(\mathbf{V})\mathbf{Q}\mathbf{v} = \mathbf{G}(\mathbf{V})\mathbf{v}$$

since $\mathbf{Q}\mathbf{v} = \mathbf{v}$.

Thus $\mathbf{G}(\mathbf{V})\mathbf{v}$ is an eigenvector of $\mathbf{Q}$ (with unit eigenvalue). But, from the result of Problem 1.3.12, we know that an orthogonal tensor has only one real eigenvalue and, correspondingly, one real eigenvector. In the present case this eigenvector is $\mathbf{v}$ and it must follow that there exists some scalar, α say, such that

$$\mathbf{G}(\mathbf{V})\mathbf{v} = \alpha\mathbf{v}.$$

Consequently every eigenvector of $\mathbf{V}$ is also an eigenvector of $\mathbf{G}(\mathbf{V})$ and the theorem is proved.

By way of illustration, we consider the function defined by

$$\mathbf{G}(\mathbf{V}) = \phi_0\mathbf{I} + \phi_1\mathbf{V} + \cdots + \phi_N\mathbf{V}^N, \tag{4.2.49}$$

where $\phi_0, \phi_1, \ldots, \phi_N$ are scalar invariants of $\mathbf{V}$ and N is a positive integer, It follows immediately from (4.2.42) and (4.2.43) that this function is isotropic and, moreover, (4.2.48) clearly holds. The following theorem shows that the most general isotropic tensor function is expressible in the form (4.2.49) with $N = 2$.

Theorem 4.2.5 A symmetric second-order tensor-valued function $\mathbf{G}$ defined over the space of symmetric second-order tensors is isotropic if and only if it has the representation

$$\mathbf{G}(\mathbf{V}) = \phi_0\mathbf{I} + \phi_1\mathbf{V} + \phi_2\mathbf{V}^2, \tag{4.2.50}$$

where ϕ_0, ϕ_1, ϕ_2 are scalar invariants of $\mathbf{V}$.

Proof (a) If (4.2.50) holds with ϕ_0, ϕ_1, ϕ_2 scalar invariants of $\mathbf{V}$ then $\mathbf{G}(\mathbf{V})$ is clearly isotropic.

(b) If $\mathbf{G}(\mathbf{V})$ is isotropic we must show that it necessarily has the representation (4.2.50). From Theorems 4.2.3 and 4.2.4 we know that $\mathbf{G}(\mathbf{V})$ is coaxial with $\mathbf{V}$ and its principal values are scalar invariants of $\mathbf{V}$. Let $\lambda_1, \lambda_2, \lambda_3$ and g_1, g_2, g_3 be the principal values of $\mathbf{V}$ and $\mathbf{G}(\mathbf{V})$ respectively and consider the equations

$$\phi_0 + \phi_1\lambda_i + \phi_2\lambda_i^2 = g_i \qquad (i = 1, 2, 3) \tag{4.2.51}$$

for the three unknowns ϕ_0, ϕ_1, ϕ_2. Assuming that λ_i and g_i are given and that the λ_i are distinct it follows that ϕ_0, ϕ_1, ϕ_2 are determined uniquely in terms of λ_i and g_i which are themselves scalar invariants of $\mathbf{V}$. Thus ϕ_0, ϕ_1, ϕ_2 are scalar invariants of $\mathbf{V}$, and $\mathbf{G}(\mathbf{V})$ and $\mathbf{V}$ are coaxial; equation (4.2.50) follows. The case in which λ_i are not distinct is left to the reader.

For an isotropic elastic solid the Cauchy stress $\mathbf{T}$ is given by (4.2.35), namely

$$\mathbf{T} = \mathbf{G}(\mathbf{V}), \tag{4.2.52}$$

together with (4.2.50), and hence

$$\mathbf{TV} = \mathbf{VT}. \tag{4.2.53}$$

Thus, relative to an undistorted reference configuration, the Cauchy stress tensor is coaxial with the Eulerian strain ellipsoid and the properties attributed to coaxial stress and deformation tensors in Section 3.5.2 are then inherited.

Explicitly, we now have

$$\mathbf{T} = \mathbf{G}(\mathbf{V}) = \phi_0 \mathbf{I} + \phi_1 \mathbf{V} + \phi_2 \mathbf{V}^2, \tag{4.2.54}$$

where ϕ_0, ϕ_1, ϕ_2 are invariants of the left stretch tensor $\mathbf{V}$. According to Theorem 4.2.2 each of ϕ_0, ϕ_1, ϕ_2 may be written as a function of the principal invariants of $\mathbf{V}$, namely

$$\left.\begin{aligned} I_1(\mathbf{V}) &\equiv \lambda_1 + \lambda_2 + \lambda_3, \\ I_2(\mathbf{V}) &\equiv \lambda_2\lambda_3 + \lambda_3\lambda_1 + \lambda_1\lambda_2, \\ I_3(\mathbf{V}) &\equiv \lambda_1\lambda_2\lambda_3, \end{aligned}\right\} \tag{4.2.55}$$

these being symmetric in the principal stretches $\lambda_1, \lambda_2, \lambda_3$.

Equally, we may regard ϕ_0, ϕ_1, ϕ_2 as symmetric functions of $\lambda_1, \lambda_2, \lambda_3$ and write the principal components of (4.2.54) as

$$t_i = \phi_0 + \phi_1 \lambda_i + \phi_2 \lambda_i^2 \qquad (i = 1, 2, 3), \tag{4.2.56}$$

where t_i are the principal Cauchy stresses. It is then convenient to introduce the *scalar response function* g for an isotropic Cauchy elastic solid such that

$$t_i = g(\lambda_i, \lambda_j, \lambda_k), \tag{4.2.57}$$

where (i,j,k) is a permutation of $(1, 2, 3)$, g being defined on $(0, \infty)^3$.

Thus,

$$g(\lambda_i, \lambda_j, \lambda_k) = \phi_0 + \phi_1 \lambda_i + \phi_2 \lambda_i^2 \tag{4.2.58}$$

and this embodies the symmetry

$$g(\lambda_i, \lambda_j, \lambda_k) = g(\lambda_i, \lambda_k, \lambda_j) \tag{4.2.59}$$

implied by the symmetry of ϕ_0, ϕ_1, ϕ_2 as functions of $\lambda_1, \lambda_2, \lambda_3$. The single scalar response function g, together with coaxiality (4.2.53), provides a representation for the constitutive law of an isotropic Cauchy elastic solid that is an alternative to (4.2.54). Apart from the symmetry (4.2.59) and appropriate regularity, the function g is completely arbitrary. In later chapters, however, we shall see that g must be restricted further on both mathematical and physical grounds.

In practical terms isotropy may be interpreted in the following way. Suppose a cube of material is cut from a homogeneous specimen in an undistorted reference configuration χ_0 and inserted into a testing machine which applies stretches $\lambda_1, \lambda_2, \lambda_3$ parallel to the cube edges. The material is isotropic if the Cauchy stress is coaxial with the cube edges and the resulting principal stresses t_1, t_2, t_3 do not depend on the original orientation of the cube in χ_0. The response function g can then be determined empirically from measured values of λ_1, λ_2, λ_3 and t_1, t_2, t_3.

We have now seen just two of many possible representations for the constitutive law of an isotropic Cauchy elastic solid. Indeed, the deformation tensor $\mathbf{V}$ used as the basic independent variable in the above may be replaced by any one of an infinite number of suitable functions of $\mathbf{V}$, as we now illustrate.

Let $\mathbf{A}$ denote the deformation gradient and take $\mathbf{A}\mathbf{A}^{\mathrm{T}} = \mathbf{V}^2$ as the independent deformation variable. Then we may write

$$\mathbf{T} = \psi_0 \mathbf{I} + \psi_1 \mathbf{V}^2 + \psi_2 \mathbf{V}^4 \tag{4.2.60}$$

as an alternative to (4.2.54). Use of the Cayley–Hamilton theorem (1.3.19) enables other such alternatives to be obtained.

Problem 4.2.4 Compare (4.2.54) and (4.2.60) and use the Cayley–Hamilton theorem to express ψ_0, ψ_1, ψ_2 in terms of ϕ_0, ϕ_1, ϕ_2.

The representations considered above are based on the use of the Cauchy stress tensor, but the constitutive law of an isotropic Cauchy elastic solid may equally be expressed in terms of other stress tensors. In particular, we now examine the specialization to isotropy of the alternative forms of constitutive law discussed in Section 4.2.2.

First, we note that, by (2.2.36), (4.2.42) with $\mathbf{Q} = \mathbf{R}^{\mathrm{T}}$ and (4.2.53), we obtain

$$\mathbf{R}^{\mathrm{T}}\mathbf{T}\mathbf{R} = \mathbf{R}^{\mathrm{T}}\mathbf{G}(\mathbf{V})\mathbf{R} = \mathbf{G}(\mathbf{U}) \tag{4.2.61}$$

(recall (4.2.11)) together with

$$\mathbf{G}(\mathbf{U})\mathbf{U} = \mathbf{U}\mathbf{G}(\mathbf{U}). \tag{4.2.62}$$

In fact, apart from the factor $J = \det \mathbf{A}$, this is just the stress tensor $\mathbf{T}^{(0)}$ given in (3.5.23) and which is conjugate to logarithmic strain $\mathbf{E}^{(0)} = \ln \mathbf{U}$ in the case of isotropy. It follows that the constitutive law may be written

$$\mathbf{T}^{(0)} = \mathbf{G}^{(0)}(\mathbf{E}^{(0)}), \tag{4.2.63}$$

where

$$\mathbf{G}^{(0)}(\mathbf{E}^{(0)}) = J\mathbf{G}(\mathbf{U}),$$

subject to (4.2.62). It is an easy matter to show that $\mathbf{G}(\mathbf{U})$ is an isotropic function of $\mathbf{U}$, i.e. (4.2.42) holds with $\mathbf{V}$ replaced by $\mathbf{U}$. Hence $\mathbf{G}(\mathbf{U})$ or, equivalently, $\mathbf{G}^{(0)}(\mathbf{E}^{(0)})$ has representations analogous to (4.2.54), (4.2.57) and (4.2.60). The same applies to the constitutive laws considered below.

More generally, recalling (3.5.20), we have

$$\mathbf{T}^{(m)} = \mathbf{G}^{(m)}(\mathbf{E}^{(m)}) \tag{4.2.64}$$

with

$$\mathbf{G}^{(m)}(\mathbf{E}^{(m)})\mathbf{E}^{(m)} = \mathbf{E}^{(m)}\mathbf{G}^{(m)}(\mathbf{E}^{(m)}) \tag{4.2.65}$$

in respect of the conjugate variables $\mathbf{T}^{(m)}$ and $\mathbf{E}^{(m)}$, each $\mathbf{T}^{(m)}$ being coaxial with the Lagrangean principal axes.

In respect of nominal stress we have

$$\mathbf{S} = \mathbf{H}(\mathbf{A}) = \mathbf{H}(\mathbf{U})\mathbf{R}^{\mathrm{T}} \tag{4.2.66}$$

from (4.2.12). But since

$$\mathbf{S} = \mathbf{T}^{(2)}\mathbf{A}^{\mathrm{T}} = \mathbf{T}^{(2)}\mathbf{U}\mathbf{R}^{\mathrm{T}}$$

it follows by coaxiality of $\mathbf{T}^{(2)}$ and $\mathbf{U}$ that $\mathbf{SR}$ is symmetric. Hence $\mathbf{H}(\mathbf{U})$ is symmetric and therefore (4.2.17) and (4.2.64) with $m = 1$ give

$$\mathbf{T}^{(1)} = \mathbf{G}^{(1)}(\mathbf{E}^{(1)}) = \mathbf{H}(\mathbf{U}). \tag{4.2.67}$$

Finally, we obtain

$$\mathbf{S} = \mathbf{T}^{(1)}\mathbf{R}^{\mathrm{T}}. \tag{4.2.68}$$

This specializes (3.5.22) to the case of isotropic elasticity, and, we repeat, provides a polar decomposition for $\mathbf{S}$ analogous to that for $\mathbf{A}$. However, because $\mathbf{T}^{(1)}$ need not be positive definite, the decomposition (4.2.68) is not in general unique.

The reader is referred to Section 3.5.2 for further discussion of conjugate stress tensors. The consequences for elasticity which the remarks on coaxiality contained therein imply should be studied.

4.2.7 Internal constraints

A number of important applications of the theory arise when the deformation is subject to some form of local constraint. For example, the material may be constrained to be inextensible in some direction or the deformation may be volume preserving. Such constraints are idealizations as far as real materials are concerned but they can provide a good first approximation to actual material response. The associated mathematics is also simplified by the reduction in the number of independent deformation components.

Suppose the deformation is constrained locally according to the single scalar equation

$$C(\mathbf{A}) = 0, \tag{4.2.69}$$

where C is a function with sufficient regularity for our purposes.

The constraint holds good when the deformed material is subject to a rigid rotation, i.e. $C(\mathbf{A})$ is objective. Thus

$$C(\mathbf{QA}) = C(\mathbf{A})$$

for all proper orthogonal $\mathbf{Q}$, and the choice $\mathbf{Q} = \mathbf{R}^{\mathrm{T}}$ then leads to

$$C(\mathbf{U}) = 0, \tag{4.2.70}$$

where $\mathbf{U}$ is the right stetch tensor. Equally, the constraint may be expressed in terms of any other deformation or strain tensor.

Because of this constraint and the consequent interdependence of the components of $\mathbf{A}$ the stress in an elastic solid can no longer be specified as a function of the deformation alone, as we now show. First, we note that differentiation of (4.2.69) with respect to time gives

$$\dot{C} \equiv \operatorname{tr}\left(\frac{\partial C}{\partial \mathbf{A}}\dot{\mathbf{A}}\right) = 0, \tag{4.2.71}$$

where $\dot{\mathbf{A}}$ is the deformation rate introduced in Section 2.3.1 and $\partial C/\partial \mathbf{A}$ denotes the second-order tensor (the gradient of C with respect to $\mathbf{A}$) whose Cartesian components are defined by

$$\left(\frac{\partial C}{\partial \mathbf{A}}\right)_{\alpha i} = \frac{\partial C}{\partial A_{i\alpha}} \tag{4.2.72}$$

in terms of the corresponding components of $\mathbf{A}$ (recall Section 2.2.1). Equation (4.2.71) imposes a constraint on $\dot{\mathbf{A}}$ which is implied by that on $\mathbf{A}$.

In terms of nominal stress $\mathbf{S}$ the stress power defined in Section 3.5.1 is given by

$$\mathrm{tr}\,(\mathbf{S}\dot{\mathbf{A}})$$

per unit reference volume, and for an unconstrained elastic solid we have

$$\mathbf{S} = \mathbf{H}(\mathbf{A})$$

from (4.2.12). But, in view of (4.2.71), an arbitrary scalar multiple of $\partial C/\partial \mathbf{A}$ may be added to $\mathbf{S}$ without affecting the stress power. Thus, when the constraint (4.2.69) is given the above constitutive law must be replaced by†

$$\mathbf{S} = \mathbf{H}(\mathbf{A}) + q\frac{\partial C}{\partial \mathbf{A}}(\mathbf{A}). \qquad (4.2.73)$$

This is the (nominal) stress-deformation relation for an elastic solid subject to a single constraint, where q is an arbitrary scalar and $\mathbf{H}$ is the material response function. Thus, the stress is determined by the deformation to within an additive arbitrary stress for which the stress power vanishes during any motion compatible with the constraints. The scalar q has the role of a Lagrange multiplier.

Similarly, it follows from (4.2.70) that the Biot stress tensor is given by

$$\mathbf{T}^{(1)} = \mathbf{H}(\mathbf{U}) + q\frac{\partial C}{\partial \mathbf{U}}(\mathbf{U}), \qquad (4.2.74)$$

and corresponding equations can be written down in respect of other conjugate stress tensors.

Problem 4.2.5 Show that in terms of Cauchy stress the constitutive relation for a constrained elastic solid may be written

$$\mathbf{T} = \mathbf{G}(\mathbf{A}) + p\mathbf{A}\frac{\partial C}{\partial \mathbf{A}}(\mathbf{A}), \qquad (4.2.75)$$

where p is an arbitrary scalar.

For a material constrained by incompressibility, we have

$$C(\mathbf{A}) \equiv \det \mathbf{A} - 1 = 0$$

†This step is justified on the basis of variational arguments in Section 5.4.4.

and it follows from the result of Problem 2.2.4 that

$$\frac{\partial C}{\partial \mathbf{A}} = \mathbf{B}^{\mathrm{T}}.$$

Hence (4.2.73) becomes

$$\mathbf{S} = \mathbf{H}(\mathbf{A}) - p\mathbf{B}^{\mathrm{T}}, \tag{4.2.76}$$

where we have replaced q by $-p, p$ commonly being referred to as the *arbitrary hydrostatic pressure.* In respect of Cauchy stress we may write

$$\mathbf{T} = \mathbf{G}(\mathbf{A}) - p\mathbf{I}. \tag{4.2.77}$$

More particularly, for an isotropic elastic solid this yields the representation

$$\mathbf{T} = -p\mathbf{I} + \phi_1\mathbf{V} + \phi_2\mathbf{V}^2, \tag{4.2.78}$$

where ϕ_1, ϕ_2 are scalar invariants of $\mathbf{V}$, i.e. symmetric functions of $\lambda_1, \lambda_2, \lambda_3$ subject to $\lambda_1\lambda_2\lambda_3 = 1$. And in terms of principal components we have

$$t_i = g(\lambda_i, \lambda_j, \lambda_k) - p, \tag{4.2.79}$$

where g is a scalar response function satisfying (4.2.59) and

$$g(\lambda_i, \lambda_j, \lambda_k) = \phi_1\lambda_i + \phi_2\lambda_i^2. \tag{4.2.80}$$

As a second example we consider the inextensibility constraint discussed in Section 2.2.6. From equation (2.2.61) we recall that if $\mathbf{L}$ denotes the direction of inextensibility in the reference configuration then the constraint may be written as

$$C(\mathbf{A}) \equiv \mathbf{L}\cdot(\mathbf{A}^{\mathrm{T}}\mathbf{A}\mathbf{L}) - 1 = 0, \tag{4.2.81}$$

$\mathbf{L}$ being a unit vector.

It follows that

$$\frac{\partial C}{\partial \mathbf{A}} = 2\mathbf{L} \otimes \mathbf{A}\mathbf{L}$$

and hence

$$\mathbf{S} = \mathbf{H}(\mathbf{A}) + 2q\mathbf{L} \otimes \mathbf{A}\mathbf{L}.$$

Corresponding expressions for other stress tensors may also be written down but we do not give details here.

Problem 4.2.6 Obtain an expression analogous to (4.2.73) in respect of the conjugate stress and strain $(\mathbf{T}^{(2)}, \mathbf{E}^{(2)})$ by writing $C^{(2)}(\mathbf{E}^{(2)}) \equiv C(\mathbf{A})$.

Problem 4.2.7 If there are N distinct constraints

$$C_r(\mathbf{A}) = 0 \qquad r = 1, 2, \ldots, N$$

deduce that the appropriate generalization of (4.2.73) is

$$\mathbf{S} = \mathbf{H}(\mathbf{A}) + \sum_{r=1}^{N} q_r \frac{\partial C_r}{\partial \mathbf{A}}(\mathbf{A}),$$

where q_r $(r = 1, 2, \ldots, N)$ is an arbitarary scalar.

Show that for a material with two inextensible directions $\mathbf{L}_1, \mathbf{L}_2$ in some reference configuration the Cauchy stress tensor is expressible as

$$\mathbf{T} = \mathbf{G}(\mathbf{A}) + p_1 \mathbf{A}\mathbf{L}_1 \otimes \mathbf{A}\mathbf{L}_1 + p_2 \mathbf{A}\mathbf{L}_2 \otimes \mathbf{A}\mathbf{L}_2,$$

where p_1, p_2 are arbitrary scalars.

What is the maximum number of independent constraints for which a non-trivial deformation is possible?

Problem 4.2.8 For the non-trivial plane deformation of an incompressible elastic solid, show that there is at most one (in-plane) inextensible direction. Deduce that when there is one such direction the deformation is necessarily a simple shear and the in-plane components of Cauchy stress for an isotropic material may be written

$$T_{11} = -p + p' + \psi_1(1 + \gamma^2) + \psi_2(1 + 3\gamma^2 + \gamma^4),$$

$$T_{22} = -p + \psi_1 + \psi_2(1 + \gamma^2),$$

$$T_{12} = \psi_1 \gamma + \psi_2 \gamma(2 + \gamma^2),$$

where γ is the amount of shear, p and p' are arbitrary constants and ψ_1, ψ_2 are functions of γ.

4.2.8 Differentiation of a scalar function of a tensor

In Section 4.2.7 the need to differentiate a scalar function of a second-order tensor arose and a rule (4.2.72) was assigned for the explicit calculation of this differentiation in terms of Cartesian components. More generally, if $\mathbf{T}$

is a second-order tensor with components relative to a general curvilinear basis $\{\mathbf{e}_i\}$ and its reciprocal basis defined by the decomposition

$$\mathbf{T} = T^{ij}\mathbf{e}_i \otimes \mathbf{e}_j = T^i{}_j\mathbf{e}_i \otimes \mathbf{e}^j = T_i{}^j\mathbf{e}^i \otimes \mathbf{e}_j = T_{ij}\mathbf{e}^i \otimes \mathbf{e}^j,$$

we may write

$$\frac{\partial \phi}{\partial \mathbf{T}} = \frac{\partial \phi}{\partial T^{ji}}\mathbf{e}^i \otimes \mathbf{e}^j = \frac{\partial \phi}{\partial T^j{}_i}\mathbf{e}_i \otimes \mathbf{e}^j = \frac{\partial \phi}{\partial T_j{}^i}\mathbf{e}^i \otimes \mathbf{e}_j = \frac{\partial \phi}{\partial T_{ji}}\mathbf{e}_i \otimes \mathbf{e}_j. \tag{4.2.82}$$

We note that these formulae apply only if the basis is chosen independently of $\mathbf{T}$ so that changes in $\mathbf{T}$ are recorded as changes in its components and ϕ may be regarded as a function of the appropriate set of nine components of $\mathbf{T}$ given above.

This point is emphasized because in the case of a symmetric tensor $\mathbf{T}$ (with only six independent components) the principal directions of $\mathbf{T}$ form a suitable basis but they *are* dependent on $\mathbf{T}$. Thus, if t_1, t_2, t_3 denote the principal values of $\mathbf{T}$ and $\mathbf{e}^{(1)}, \mathbf{e}^{(2)}, \mathbf{e}^{(3)}$ the (orthonormal) principal directions then

$$\mathbf{T} = \sum_{i=1}^{3} t_i \mathbf{e}^{(i)} \otimes \mathbf{e}^{(i)}.$$

If follows that with respect to this basis the components of the differential $\mathrm{d}\mathbf{T}$ are given by

$$(\mathrm{d}\mathbf{T})_{ij} = \mathrm{d}t_i \delta_{ij} + (t_j - t_i)\mathbf{e}^{(i)} \cdot \mathrm{d}\mathbf{e}^{(j)}.$$

(Note that $(\mathrm{d}\mathbf{T})_{ij} \neq \mathrm{d}T_{ij}$, where $T_{ij} = t_i \delta_{ij}$.)

Thus, if ϕ is a function of $\mathbf{T}$ it may also be regarded as a function of t_1, t_2, t_3 and $\mathbf{e}^{(1)}, \mathbf{e}^{(2)}, \mathbf{e}^{(3)}$. In particular

$$\left(\frac{\partial \phi}{\partial \mathbf{T}}\right)_{ii} = \frac{\partial \phi}{\partial t_i}. \tag{4.2.83}$$

The shear components of $\partial \phi / \partial \mathbf{T}$ are not needed here and we do not make them explicit.

Important applications of the above arise when ϕ is an isotropic function of $\mathbf{T}$, i.e. depends only on t_1, t_2, t_3 and not on $\mathbf{e}^{(1)}, \mathbf{e}^{(2)}, \mathbf{e}^{(3)}$. Then we have

$$\frac{\partial \phi}{\partial \mathbf{T}} = \sum_{i=1}^{3} \frac{\partial \phi}{\partial t_i} \mathbf{e}^{(i)} \otimes \mathbf{e}^{(i)} \tag{4.2.84}$$

and

$$\mathbf{T}\frac{\partial \phi}{\partial \mathbf{T}} = \frac{\partial \phi}{\partial \mathbf{T}}\mathbf{T}, \tag{4.2.85}$$

i.e. $\partial\phi/\partial\mathbf{T}$ is coaxial with $\mathbf{T}$.

Problem 4.2.9 Let

$$\mathbf{U} = \sum_{i=1}^{3} \lambda_i \mathbf{u}^{(i)} \otimes \mathbf{u}^{(i)}$$

denote the right stretch tensor, where λ_i are the principal stretches and $\mathbf{u}^{(i)}$ the Lagrangean principal axes.

Show that

$$\frac{\partial \lambda_i}{\partial \mathbf{U}} = \mathbf{u}^{(i)} \otimes \mathbf{u}^{(i)}. \tag{4.2.86}$$

If $I_1(\mathbf{U}), I_2(\mathbf{U}), I_3(\mathbf{U})$ are the principal invariants of $\mathbf{U}$, deduce that

$$\frac{\partial I_1}{\partial \mathbf{U}}(\mathbf{U}) = \mathbf{I}, \frac{\partial I_2}{\partial \mathbf{U}}(\mathbf{U}) = I_1(\mathbf{U})\mathbf{I} - \mathbf{U}, \frac{\partial I_3}{\partial \mathbf{U}}(\mathbf{U}) = I_3(\mathbf{U})\mathbf{U}^{-1}, \tag{4.2.87}$$

where $\mathbf{I}$ is the identity tensor.

Problem 4.2.10 If T_{ij} denote Cartesian components of a second-order tensor $\mathbf{T}$ then, in general,

$$\frac{\partial T_{ij}}{\partial T_{kl}} = \delta_{ik}\delta_{jl},$$

but if $\mathbf{T}$ is symmetric this is replaced by the symmetrized form

$$\frac{\partial T_{ij}}{\partial T_{kl}} = \tfrac{1}{2}(\delta_{ik}\delta_{jl} + \delta_{il}\delta_{jk}). \tag{4.2.88}$$

Use this latter formula to give a proof of (4.2.87) without the help of (4.2.86).

Problem 4.2.11 Let $\mathbf{A}$ denote the deformation gradient. Show that

$$\frac{\partial}{\partial \mathbf{A}} I_1(\mathbf{A}^{\mathrm{T}}\mathbf{A}) = 2\mathbf{A}^{\mathrm{T}},$$

$$\frac{\partial}{\partial \mathbf{A}} I_2(\mathbf{A}^{\mathrm{T}}\mathbf{A}) = 2\{I_1(\mathbf{A}^{\mathrm{T}}\mathbf{A})\mathbf{I} - \mathbf{A}^{\mathrm{T}}\mathbf{A}\}\mathbf{A}^{\mathrm{T}},$$

$$\frac{\partial}{\partial \mathbf{A}} I_3(\mathbf{A}^{\mathrm{T}}\mathbf{A}) = 2I_3(\mathbf{A}^{\mathrm{T}}\mathbf{A})\mathbf{B}^{\mathrm{T}}.$$

Problem 4.2.12 Let ϕ be an objective function of the deformation gradient, i.e.

$$\phi(\mathbf{A}) = \phi(\mathbf{U}).$$

(a) If ϕ is isotropic show that $\phi(\mathbf{U}) = \phi(\mathbf{V})$ and

$$\frac{\partial \phi}{\partial \mathbf{A}} = \frac{\partial \phi}{\partial \mathbf{U}}\mathbf{R}^{\mathrm{T}} = \mathbf{R}^{\mathrm{T}}\frac{\partial \phi}{\partial \mathbf{V}}.$$

Confirm this result for $\phi(\mathbf{A}) = \det \mathbf{A}$.
(b) If ϕ is not isotropic prove that

$$\frac{\partial \phi}{\partial \mathbf{A}}\mathbf{R} - \frac{\partial \phi}{\partial \mathbf{U}}$$

is antisymmetric and for

$$\phi(\mathbf{A}) = \mathbf{L}\cdot(\mathbf{A}^{\mathrm{T}}\mathbf{A}\mathbf{L})$$

show that

$$\frac{\partial \phi}{\partial \mathbf{A}}\mathbf{R} - \frac{\partial \phi}{\partial \mathbf{U}} = \mathbf{L} \otimes \mathbf{U}\mathbf{L} - \mathbf{U}\mathbf{L} \otimes \mathbf{L}.$$

(c) If

$$\frac{\partial \phi}{\partial \mathbf{A}} = \frac{\partial \phi}{\partial \mathbf{U}}\mathbf{R}^{\mathrm{T}}$$

prove that ϕ is necessarily isotropic.

4.3 GREEN ELASTIC MATERIALS

4.3.1 The strain-energy function

From Sections 3.5.1 and 3.5.2 we recall that the stress power per unit volume may be written in terms of the nominal stress tensor $\mathbf{S}$ as

$$\mathrm{tr}(\mathbf{S}\dot{\mathbf{A}}) = \mathrm{tr}\{\mathbf{H}(\mathbf{A})\dot{\mathbf{A}}\} \tag{4.3.1}$$

for a Cauchy elastic solid. In general there does not exist a scalar function of $\mathbf{A}$, $W(\mathbf{A})$ say, such that

$$\dot{W} = \operatorname{tr}\{\mathbf{H}(\mathbf{A})\dot{\mathbf{A}}\}. \tag{4.3.2}$$

In other words, $\operatorname{tr}\{\mathbf{H}(\mathbf{A})\,\mathrm{d}\mathbf{A}\}$ is not an exact differential in general. However, if such a W exists then

$$\dot{W} = \operatorname{tr}\left\{\frac{\partial W}{\partial \mathbf{A}}(\mathbf{A})\dot{\mathbf{A}}\right\} \tag{4.3.3}$$

and, since the components of $\dot{\mathbf{A}}$ are independent for an unconstrained material, comparison of this with (4.3.2) shows that

$$\mathbf{H}(\mathbf{A}) = \frac{\partial W}{\partial \mathbf{A}}(\mathbf{A}). \tag{4.3.4}$$

When (4.3.2) holds the mechanical energy balance equation (3.5.5) may be rewritten as

$$\int_{\mathscr{B}_0} \rho_0 \mathbf{b}_0 \cdot \dot{\boldsymbol{\chi}}\,\mathrm{d}V + \int_{\partial\mathscr{B}_0} (\mathbf{S}^{\mathrm{T}}\mathbf{N})\cdot\dot{\boldsymbol{\chi}}\,\mathrm{d}A = \frac{\mathrm{d}}{\mathrm{d}t}\int_{\mathscr{B}_0} (\tfrac{1}{2}\rho_0\dot{\boldsymbol{\chi}}\cdot\dot{\boldsymbol{\chi}} + W)\,\mathrm{d}V \tag{4.3.5}$$

As before the left-hand side of (4.3.5) represents the total rate of working of the body and surface forces, but the right-hand side is now expressible as the rate of change of the total mechanical energy of the body. This mechanical energy consists of kinetic energy, with density $\frac{1}{2}\rho_0\dot{\boldsymbol{\chi}}\cdot\dot{\boldsymbol{\chi}}$ per unit reference volume, and potential energy W per unit reference volume.

Since W is purely elastic in origin it is referred to as the *elastic potential energy function.* It is a measure of the energy stored in the material as a result of deformation and is therefore sometimes called the elastic *stored energy function.* More commonly, the phrase *strain-energy function* is used to describe W and this is the terminology we adopt in this book. The rate $\dot{W}$ is simply the rate of change of strain energy density or, equivalently, the stress power per unit reference volume. The total elastic energy stored in the body is

$$\int_{\mathscr{B}_0} W\,\mathrm{d}V \equiv \int_{\mathscr{B}} WJ^{-1}\,\mathrm{d}v, \tag{4.3.6}$$

where $J(\equiv \det \mathbf{A})$ is the ratio of current to reference volume elements (as described in Section 2.2.2). Note that we have defined W per unit volume rather than per unit mass since this avoids the need for a multiplying factor of ρ_0 on the right-hand side of (4.3.4).

An elastic material for which a strain-energy function exists is called a *Green elastic* or *hyperelastic* material. The mechanical properties of such a material are then characterized by the strain-energy function W, and the (nominal) stress-deformation relation is simply

$$\mathbf{S} = \frac{\partial W}{\partial \mathbf{A}}(\mathbf{A}). \tag{4.3.7}$$

Since Green elasticity is a special case of Cauchy elasticity all the properties attributed to $\mathbf{H}(\mathbf{A})$ in Section 4.2 carry over to $\partial W(\mathbf{A})/\partial \mathbf{A}$. In what follows we shall see that these properties have implications for $W(\mathbf{A})$ itself. However, it is desirable, as far as possible, to make the current section self-contained in the sense that it does not draw heavily on the results of Section 4.2. This enables the theory of elasticity to be developed from the assumption that a strain-energy function exists without reference to Cauchy elasticity. In particular, the consequences for W of objectivity and material symmetry are established directly. It should be pointed out here that in the literature the term 'elasticity' is often used as a synonym for 'Green elasticity'. The reader should also be warned that the viability of a theory of elasticity in which there does not exist a strain-energy function has been called into question (on the basis of thermodynamic arguments). In this book, however, we make no claim either way and merely present both Cauchy elasticity and Green elasticity as parts of the mathematical development. Nevertheless, we remark that in applications Green elasticity has proved more useful than Cauchy elasticity.

Since, for a Green elastic material, the rate of working of the stresses on the body is given by (4.3.3), it follows that the work done on a path of deformation which takes the deformation gradient from some initial value, $\mathbf{A}_0$ say, to its current value $\mathbf{A}$ is

$$W(\mathbf{A}) - W(\mathbf{A}_0) = \int_{\mathbf{A}_0}^{\mathbf{A}} \mathrm{tr}\left\{\frac{\partial W}{\partial \mathbf{A}}(\mathbf{A})\,\mathrm{d}\mathbf{A}\right\}. \tag{4.3.8}$$

The integral is along a path in the space of deformation gradients.

Clearly, the change in energy in traversing the path is independent of the rate of deformation and, moreover, is path independent, i.e. depends only on the value of W at each of the end points just as in conventional potential theory. Furthermore, $\mathbf{S}$ and $\mathbf{A}$ are conjugate variables relative to the potential W in the classical sense since (4.3.7) holds.

If the starting point of the deformation path is the reference configuration then $\mathbf{A}_0$ is replaced by $\mathbf{I}$ in (4.3.8) but we note that in general $W(\mathbf{I}) \neq 0$.

However, it may sometimes be convenient to set $W(\mathbf{I})=0$ arbitrarily. This action is appropriate, for example, when a natural configuration (defined at the end of Section 4.2.4) is taken as the reference configuration (so that $\mathbf{S}=\mathbf{0}$ in that configuration). Note that the right-hand side of (4.3.5) is unaffected by the addition of an arbitrary constant to W.

With the help of results given in Section 3.5.2 we write the stress power as

$$\operatorname{tr}(\mathbf{S}\dot{\mathbf{A}})=\operatorname{tr}(\mathbf{T}^{(1)}\dot{\mathbf{U}}),$$

where, in respect of Cauchy elasticity, we have

$$\mathbf{S}=\mathbf{H}(\mathbf{A})=\mathbf{H}(\mathbf{U})\mathbf{R}^{\mathrm{T}}. \tag{4.3.9}$$

and

$$\mathbf{T}^{(1)}=\tfrac{1}{2}\{\mathbf{H}(\mathbf{U})+[\mathbf{H}(\mathbf{U})]^{\mathrm{T}}\}. \tag{4.3.10}$$

When this is specialized to Green elasticity with strain energy $W(\mathbf{A})$ the additional connection

$$\dot{W}=\operatorname{tr}(\mathbf{T}^{(1)}\dot{\mathbf{U}}) \tag{4.3.11}$$

results.

The objectivity of Cauchy elasticity embodied in (4.3.9) and (4.3.10) carries over to W because the right-hand side of (4.3.11) is independent of $\mathbf{R}$ and $\dot{\mathbf{R}}$ and therefore integration of (4.3.11) makes it clear that W is a function of $\mathbf{U}$ alone, i.e.

$$W(\mathbf{A})=W(\mathbf{U}) \tag{4.3.12}$$

is independent of $\mathbf{R}$ in the polar decomposition $\mathbf{A}=\mathbf{RU}$. Since (4.3.12) holds for arbitrary $\mathbf{A}$ and therefore arbitrary proper orthogonal $\mathbf{R}$ it is equivalent to the statement

$$W(\mathbf{QA})=W(\mathbf{A}) \tag{4.3.13}$$

for all proper orthogonal $\mathbf{Q}$ and arbitrary deformation gradients $\mathbf{A}$.

Thus, objectivity of the stress-deformation relation for a Cauchy elastic material implies that the strain-energy function of a Green elastic material satisfies (4.3.13). In other words W is an objective scalar function of $\mathbf{A}$. At this point we recall that it was shown in Section 3.5.1 that the stress power is objective. In the context of Green elasticity this means that $\dot{W}$ is objective and therefore, by integration, objectivity of W follows immediately.

Equation (4.3.13) has the physical interpretation that the elastic stored energy is unaffected by a superposed rigid-body motion after deformation. In fact, (4.3.13) with (4.3.7) may be taken as the starting point for the development of the theory without reference to Cauchy elasticity.

On taking $\mathbf{A}=\mathbf{I}$ in (4.3.13) we obtain, in particular,

$$W(\mathbf{Q})=W(\mathbf{I}) \tag{4.3.14}$$

for all proper orthogonal $\mathbf{Q}$. The corresponding result in respect of nominal stress is obtained from (4.2.14) as

$$\mathbf{H}(\mathbf{Q})=\mathbf{H}(\mathbf{I})\mathbf{Q}^{\mathrm{T}}. \tag{4.3.15}$$

For a natural configuration $\mathbf{H}(\mathbf{I})=\mathbf{0}$ and it follows that $\mathbf{H}(\mathbf{Q})=\mathbf{0}$ for all proper orthogonal $\mathbf{Q}$. In other words, when the stress is zero the deformation is not uniquely determined since a rotational deformation does not then change the value of the stress. Thus the response function $\mathbf{H}$ is not a one-to-one mapping. This problem does not arise with the response of functions associated with the Lagrangean strain tensors $\mathbf{E}^{(m)}$ since rotations are not involved (see (4.3.10), for example). Such response functions are one-to-one mappings at least for some neighbourhood of the unstrained natural configuration (this result is consistent with the classical theory of linear elasticity and we justify it in Chapter 6, but the wider truth, or otherwise, of this statement can be established only on an empirical basis).

By definition $\partial W(\mathbf{U})/\partial\mathbf{U}$ is symmetric. It follows from (4.3.10)–(4.3.12) that

$$\mathbf{T}^{(1)}=\frac{\partial W}{\partial\mathbf{U}}(\mathbf{U})=\tfrac{1}{2}\{\mathbf{H}(\mathbf{U})+[\mathbf{H}(\mathbf{U})]^{\mathrm{T}}\}. \tag{4.3.16}$$

Note, in particular, that since $\dot{\mathbf{U}}$ is symmetric only the symmetric part of $\mathbf{H}(\mathbf{U})$ is determinable from (4.3.11). More generally, in respect of the strain tensor $\mathbf{E}^{(m)}$, we have

$$\mathbf{T}^{(m)}=\frac{\partial W^{(m)}}{\partial\mathbf{E}^{(m)}}(\mathbf{E}^{(m)}), \tag{4.3.17}$$

where

$$W^{(m)}(\mathbf{E}^{(m)})=W(\mathbf{U}), \tag{4.3.18}$$

i.e. the value of the strain energy is independent of the strain tensor used in its representation. Equation (4.3.16) corresponds to $m=1$ in the above. We note that the right-hand side of (4.3.17) can also be written as $\mathbf{G}^{(m)}(\mathbf{E}^{(m)})$, where $\mathbf{G}^{(m)}$ is the stress response function used in (4.2.64).

Clearly, W is the potential function for each conjugate pair $(\mathbf{T}^{(m)}, \mathbf{E}^{(m)})$ in addition to $(\mathbf{S}, \mathbf{A})$. We recollect Section 3.5.2, however, and remark that in general W does not have the same interpretation in respect of either Cauchy stress $\mathbf{T}$ or Kirchhoff stress $\hat{\mathbf{T}}$ since neither has a conjugate strain. These stress tensors are derivable from W by means of

$$\hat{\mathbf{T}} \equiv J\mathbf{T} = \mathbf{A}\frac{\partial W}{\partial \mathbf{A}}(\mathbf{A}) = \mathbf{A}\frac{\partial W^{(2)}}{\partial \mathbf{E}^{(2)}}(\mathbf{E}^{(2)})\mathbf{A}^{\mathrm{T}}, \tag{4.3.19}$$

for example.

Problem 4.3.1 By writing $\mathbf{H}(\mathbf{A}) = \partial W(\mathbf{A})/\partial \mathbf{A}$, show that (4.3.13) implies (4.2.14).

4.3.2 Symmetry groups for hyperelastic materials

We recall that in Section 4.2.3 the symmetry group of a Cauchy elastic solid relative to a given reference configuration was defined as the group $\mathscr{G}$ of proper unimodular deformations satisfying

$$\mathbf{G}(\mathbf{AK}) = \mathbf{G}(\mathbf{A}) \tag{4.3.20}$$

for each $\mathbf{K}$ in $\mathscr{G}$ and arbitrary deformation gradients $\mathbf{A}$, where $\mathbf{G}$ is the (Cauchy stress) response function relative to the reference configuration in question.

Because of the connection (4.3.7) between nominal stress and deformation gradient for a Green elastic material it is best to work in terms of nominal stress for the purposes in this section. That equation (4.3.20) is equivalent to

$$\mathbf{H}(\mathbf{AK}) = \mathbf{K}^{-1}\mathbf{H}(\mathbf{A}) \tag{4.3.21}$$

is readily established by use of (3.4.2), where

$$\mathbf{H}(\mathbf{A}) = \frac{\partial W}{\partial \mathbf{A}}(\mathbf{A}), \tag{4.3.22}$$

W being the strain-energy function relative to the considered reference configuration. It follows from (4.3.21) that

$$\frac{\partial W}{\partial \mathbf{A}}(\mathbf{A}) = \mathbf{K}\frac{\partial W}{\partial \mathbf{A}'}(\mathbf{A}') = \frac{\partial W}{\partial \mathbf{A}}(\mathbf{AK}), \tag{4.3.23}$$

where $\mathbf{A}' = \mathbf{AK}$, since

$$\frac{\partial W}{\partial \mathbf{A}'}(\mathbf{A}') = \mathbf{K}^{-1}\frac{\partial W}{\partial \mathbf{A}}(\mathbf{A})$$

for fixed $\mathbf{K}$.

Integration of (4.3.23) leads to

$$W(\mathbf{A}) = W(\mathbf{AK}) + W(\mathbf{I}) - W(\mathbf{K}) \tag{4.3.24}$$

for each $\mathbf{K}$ in $\mathscr{G}$, the constant of integration being arrived at by setting $\mathbf{A} = \mathbf{I}$. In general, since not every Cauchy elastic material possesses a strain-energy function, $\mathscr{G}$ is not the symmetry group of a Green elastic material. The latter is defined as the group, $\mathscr{G}^*$ say, of proper unimodular deformations such that

$$W(\mathbf{AK}) = W(\mathbf{A}) \tag{4.3.25}$$

for each $\mathbf{K} \in \mathscr{G}^*$ and arbitrary deformation gradients $\mathbf{A}$. Comparison of (4.3.24) and (4.3.25) shows that $\mathscr{G}^* \subset \mathscr{G}$. In fact, that $\mathscr{G}^*$ is a *normal subgroup* of $\mathscr{G}$ is easily established (see Problem 4.3.2).

Physically, (4.3.25) means that the elastic properties of a Green elastic material relative to two reference configurations related by an element of $\mathscr{G}^*$ cannot be distinguished, and the stored energy of the material is unaffected by the implied change of reference configuration.

Problem 4.3.2 If $\mathbf{K} \in \mathscr{G}$ and $\mathbf{K}^* \in \mathscr{G}^*$ show that

$$W(\mathbf{K}) + W(\mathbf{K}^{-1}) = 2W(\mathbf{I})$$

and then

$$W(\mathbf{AKK^*K}^{-1}) = W(\mathbf{A}),$$

i.e. $\mathbf{KK^*K}^{-1} \in \mathscr{G}^*$. This means that $\mathscr{G}^*$ is a normal subgroup of $\mathscr{G}$.

Problem 4.3.3 Show that if $\mathscr{G}$ consists of only orthogonal elements then (4.3.24) and (4.3.25) are equivalent, i.e. $\mathscr{G}^* = \mathscr{G}$.

Problem 4.3.4 Suppose that (a) the reference configuration is changed, from χ_0 to χ_0' say, so that the configuration with gradient $\mathbf{A}$ relative to χ_0 has gradient $\mathbf{A}' = \mathbf{AP}_0^{-1}$ relative to χ_0' as in Section 2.2.8, (b) W and W' are the strain-energy functions relative to χ_0 and χ_0' respectively so that

$$W'(\mathbf{A}') = W'(\mathbf{A}'\mathbf{K}') + W'(\mathbf{I}) - W'(\mathbf{K}') \tag{4.3.26}$$

holds along with (4.3.24), where $\mathbf{K}' \in \mathscr{G}'$ (the symmetry group of the stress response function relative to χ_0').

Use the work interpretation of strain energy to show that

$$W'(\mathbf{A}') - W'(\mathbf{I}) = W(\mathbf{A}) - W(\mathbf{P}_0).$$

Prove that

$$W'(\mathbf{A}') - W'(\mathbf{I}) = W(\mathbf{A}'\mathbf{P}_0\mathbf{K}) - W(\mathbf{P}_0) + W(\mathbf{I}) - W(\mathbf{K}),$$

where $\mathbf{K}\in\mathscr{G}$, and then show that

$$W'(\mathbf{A}') = W'(\mathbf{A}'\mathbf{P}_0\mathbf{K}\mathbf{P}_0^{-1}) + W(\mathbf{I}) - W(\mathbf{K}).$$

Hence establish that (4.3.26) holds with $\mathbf{K}' = \mathbf{P}_0\mathbf{K}\mathbf{P}_0^{-1}$. This recovers Noll's result (4.2.27). Deduce that Noll's rule also holds in respect of (4.3.25).

Since $\mathbf{E}^{(2)} = \frac{1}{2}(\mathbf{A}^{\mathrm{T}}\mathbf{A} - \mathbf{I})$, the Green strain tensor, it follows from (4.3.12) and (4.3.18) that we may write

$$W(\mathbf{A}) = W^{(2)}(\mathbf{E}^{(2)}) = \hat{W}(\mathbf{A}^{\mathrm{T}}\mathbf{A}),$$

thereby defining the strain energy as a function $\hat{W}$ over the space of right Cauchy–Green deformation tensors $\mathbf{A}^{\mathrm{T}}\mathbf{A}$. Objectivity of $\hat{W}(\mathbf{A}^{\mathrm{T}}\mathbf{A})$ is ensured by that of $\mathbf{A}^{\mathrm{T}}\mathbf{A}$.

Equation (4.3.25) can now be replaced by its equivalent

$$\hat{W}(\mathbf{K}^{\mathrm{T}}\mathbf{A}^{\mathrm{T}}\mathbf{A}\mathbf{K}) = \hat{W}(\mathbf{A}^{\mathrm{T}}\mathbf{A}) \tag{4.3.27}$$

for $\mathbf{K}\in\mathscr{G}^*$ and arbitrary $\mathbf{A}$. Note that the restrictions on $\hat{W}$ imposed by (4.3.27) are unaffected if $\mathbf{K}$ is replaced by $-\mathbf{K}$. We now examine these restrictions in respect of certain crystals classes.

First, we consider the monoclinic crystal class whose orthogonal symmetry group has a single generator. We write this as $\mathbf{R}_{\mathbf{k}}^{\pi}$ and take $\mathbf{i}, \mathbf{j}, \mathbf{k}$ to be orthonormal basis vectors, as in Section 4.2.5. Let $\mathbf{A}^{\mathrm{T}}\mathbf{A} = \mathbf{C}$ and let $\mathbf{C}$ have components $C_{pq} = C_{qp}$ relative to $\mathbf{i}$, $\mathbf{j}$, $\mathbf{k}$. In components it is easy to show that on taking $\mathbf{K} = \mathbf{R}_{\mathbf{k}}^{\pi}$ equation (4.3.27) leads to

$$\begin{aligned}&\hat{W}(C_{11}, C_{22}, C_{33}, C_{23}, C_{31}, C_{12})\\ &\quad = \hat{W}(C_{11}, C_{22}, C_{33}, -C_{23}, -C_{31}, C_{12}).\end{aligned}$$

This invariance is embodied in the quantities

$$C_{11}, C_{22}, C_{33}, C_{12}, C_{31}^2, C_{23}^2, C_{23}C_{31}.$$

In particular, any polynomial in these seven terms satisfies the symmetry requirements. Moreover, an arbitrary strain energy for a monoclinic crystal can be approximated as closely as desired by a suitable polynomial of this kind. Detailed discussion of how $\hat{W}$ must depend on the components of $\mathbf{C}$

for each of the crystal classes can be found in Green and Adkins (1970) and we do not pursue the matter here except by way of the following problem.

Problem 4.3.5 If C_{pq} denote the components of $\mathbf{C} = \mathbf{A}^{\mathrm{T}}\mathbf{A}$ relative to the basis **i, j, k** show that (a) for a transversely isotropic material with preferred direction **k** the strain energy depends on $\mathbf{C}$ only through

$$I_1(\mathbf{C}), I_2(\mathbf{C}), I_3(\mathbf{C}), C_{33}, C_{31}^2 + C_{23}^2$$

and (b) for the rhombic crystal class defined in Problem 4.2.2

$$\hat{W}(C_{11}, C_{22}, C_{33}, C_{23}^2, C_{31}^2, C_{12}^2, C_{12}C_{23}C_{31})$$

has the required symmetry.

Next, recalling Section 4.2.5, we note that for a simple cubic crystal the symmetry group generated by $\mathbf{K}_1$ (a rotation of $2\pi/3$ about a cube diagonal) and $\mathbf{K}_2$ (a simple shear through a single lattice spacing in a lattice plane) imposes the invariance

$$\hat{W}(\mathbf{K}_1^{\mathrm{T}}\mathbf{A}^{\mathrm{T}}\mathbf{A}\mathbf{K}_1) = \hat{W}(\mathbf{K}_2^{\mathrm{T}}\mathbf{A}^{\mathrm{T}}\mathbf{A}\mathbf{K}_2) = \hat{W}(\mathbf{A}^{\mathrm{T}}\mathbf{A}), \tag{4.3.28}$$

from which (4.3.27) follows for all $\mathbf{K}$ in the group.

Returning to the general case, we observe that because of the invariance (4.3.27) $\hat{W}$ cannot be prescribed arbitrary on the space, $\mathscr{D}$ say, of right Cauchy–Green deformation tensors since if $\mathbf{C} \in \mathscr{D}$, then $\hat{W}(\mathbf{C}') = \hat{W}(\mathbf{C})$ for each $\mathbf{C}' \in \mathscr{D}$ of the form $\mathbf{C}' = \mathbf{K}^{\mathrm{T}}\mathbf{C}\mathbf{K}$, where $\mathbf{K}$ is an element of the symmetry group $\mathscr{G}^*$.

Clearly, $\mathscr{D}$ can be partitioned into disjoint subsets, $\mathscr{D}_I$ say, where I is contained in some labelling set, such that any two elements of $\mathscr{D}_I$, $\mathbf{C}$ and $\mathbf{C}'$ say, related according to $\mathbf{C}' = \mathbf{K}^{\mathrm{T}}\mathbf{C}\mathbf{K}$ for some $\mathbf{K} \in \mathscr{G}^*$. For each $\mathscr{D}_I$ a particular element, $\mathbf{C}_I$ say, can be singled out so that each element of $\mathscr{D}_I$ is expressible uniquely in the form $\mathbf{K}^{\mathrm{T}}\mathbf{C}_I\mathbf{K}$. The collection of elements $\mathbf{C}_I$ forms a subset $\mathscr{D}_0 \subset \mathscr{D}$. Since a given $\mathbf{C} \in \mathscr{D}$ belongs to one and only one subset $\mathscr{D}_I$ it follows that each $\mathbf{C}$ has the unique decomposition

$$\mathbf{C} = \mathbf{K}^{\mathrm{T}}\mathbf{C}_I\mathbf{K}, \tag{4.3.29}$$

where $\mathbf{C}_I \in \mathscr{D}_0$ and $\mathbf{K} \in \mathscr{G}^*$.

Because of (4.3.29) and the invariant of $\hat{W}$, values of $\hat{W}$ on $\mathscr{D}$ can be generated by assigning arbitrary values to $\hat{W}$ on $\mathscr{D}_0$. We therefore write

$$W(\mathbf{A}) = \hat{W}(\mathbf{C}) = w(\mathbf{C}_I), \tag{4.3.30}$$

where the function w may be prescribed arbitrarily on $\mathscr{D}_0$.

Through (4.3.30), objective strain-energy functions W can in principle be constructed by making specific choices of the function w. In practice, however, this is not a straightforward matter because the elements of the set $\mathscr{D}_0$ are not easy to determine explicitly and, moreover, $\widehat{W}$ and its first derivatives are required to be continuous across the boundary of $\mathscr{D}_0$ in $\mathscr{D}$. Even for the simple cubic case, for which (4.3.28) holds, the task is considerable, as Parry (1976) has demonstrated. Detailed discussion of such problems is beyond the scope of this book and the interested reader is referred to Parry's paper for further information.

4.3.3 Stress-deformation relations for constrained hyperelastic materials

From (4.2.73) we recall that, for a Cauchy elastic material subject to a single constraint $C(\mathbf{A}) = 0$, the nominal stress is given by

$$\mathbf{S} = \mathbf{H}(\mathbf{A}) + q\frac{\partial C}{\partial \mathbf{A}}(\mathbf{A}).$$

In view of the connection (4.3.4) it follows that a Green elastic material with strain energy $W(\mathbf{A})$ subject to the same constraint has nominal stress

$$\mathbf{S} = \frac{\partial W}{\partial \mathbf{A}}(\mathbf{A}) + q\frac{\partial C}{\partial \mathbf{A}}(\mathbf{A}). \tag{4.3.31}$$

Equivalently, because of the objectivity of W and C, the stress-deformation relation can be written

$$\mathbf{T}^{(1)} = \frac{\partial W}{\partial \mathbf{U}}(\mathbf{U}) + q\frac{\partial C}{\partial \mathbf{U}}(\mathbf{U}) \tag{4.3.32}$$

in terms of Biot stress (recall (4.3.16)). Also, since the Kirchhoff stress $\hat{\mathbf{T}}$ is given by $\hat{\mathbf{T}} = \mathbf{AS} = J\mathbf{T}$, where $\mathbf{T}$ is Cauchy stress and $J = \det \mathbf{A}$, we have the further alternative

$$\hat{\mathbf{T}} = \mathbf{A}\frac{\partial W}{\partial \mathbf{A}}(\mathbf{A}) + q\mathbf{A}\frac{\partial C}{\partial \mathbf{A}}(\mathbf{A}). \tag{4.3.33}$$

For future reference, we now single out the incompressibility constraint so that $C(\mathbf{A}) \equiv \det \mathbf{A} - 1 = 0$ and $\partial C(\mathbf{A})/\partial \mathbf{A} = \mathbf{B}^{\mathrm{T}}$. It follows from (4.3.31)–(4.3.33) that

$$\mathbf{S} = \frac{\partial W}{\partial \mathbf{A}}(\mathbf{A}) - p\mathbf{B}^{\mathrm{T}}, \tag{4.3.34}$$

$$\mathbf{T}^{(1)} = \frac{\partial W}{\partial \mathbf{U}}(\mathbf{U}) - p\mathbf{U}^{-1}, \tag{4.3.35}$$

$$\hat{\mathbf{T}} = \mathbf{T} = \mathbf{A}\frac{\partial W}{\partial \mathbf{A}}(\mathbf{A}) - p\mathbf{I}, \tag{4.3.36}$$

where p is the arbitrary hydrostatic pressure introduced in Section 4.2.7.

Of course, corresponding equations relative to other (conjugate) stress tensors may also be written down. The general case is left as an exercise for the reader.

Problem 4.3.6 A *dilatation group* consists of all transformations $\mathbf{H}$ such that $\mathbf{H}\mathbf{e}_i = \lambda_i \mathbf{e}_i$ ($i = 1, 2, 3$; no summation), where $\mathbf{e}_i$ are three fixed non-coplanar vectors and $\lambda_1\lambda_2\lambda_3 = 1$. Show that if a symmetry group is a dilatation group in one reference configuration then it is a dilatation group in all reference configurations. Show also that the $\mathbf{e}_i$ transform like material line elements under a change of reference configuration.

The symmetry group of a certain material is a dilatation group such that $\mathbf{e}_i$ are orthonormal. Obtain a decomposition $\mathbf{A} = \mathbf{A}'\mathbf{H}$ for the deformation gradient, where $\mathbf{H}$ is an element of the dilatation group and $\mathbf{A}'$ is such that, with respect to the basis $\{\mathbf{e}_i\}$, the diagonal components of $\mathbf{A}'^{\mathrm{T}}\mathbf{A}'$ are all equal.

Deduce that the strain-energy function of the material is expressible as a function of the four arguments $C_{12}\sqrt{C_{33}}$, $C_{23}\sqrt{C_{11}}$, $C_{13}\sqrt{C_{22}}$, $C_{11}C_{22}C_{33}$, where C_{ij} are the components of $\mathbf{C} = \mathbf{A}^{\mathrm{T}}\mathbf{A}$ on $\{\mathbf{e}_i\}$.

4.3.4 Stress-deformation relations for isotropic hyperelastic materials

We now turn to the important special case in which the symmetry group of the Green elastic material is equal to the (proper) orthogonal group. Thus

$$W(\mathbf{AQ}) = W(\mathbf{A})$$

holds for all (proper) orthogonal $\mathbf{Q}$ in addition to the objectivity requirement

$$W(\mathbf{QA}) = W(\mathbf{A}).$$

Combination of these two equations with appropriate choices of $\mathbf{Q}$ shows that

$$W(\mathbf{QVQ}^{\mathrm{T}}) = W(\mathbf{V}) \tag{4.3.37}$$

or, equivalently,

$$W(\mathbf{QUQ}^{\mathrm{T}}) = W(\mathbf{U})$$

for all (proper) orthogonal $\mathbf{Q}$ and arbitrary positive-definite symmetric $\mathbf{V}$(or $\mathbf{U}$), where $\mathbf{V}$ and $\mathbf{U}$ are the left and right stretch tensors respectively. (Note that the restrictions imposed on W are unaffected if $\mathbf{Q}$ is allowed to be improper orthogonal in (4.3.37).)

It follows from equation (4.3.37) that W is an isotropic scalar function of $\mathbf{V}$ (and hence of $\mathbf{U}$) in the sense discussed in Sections 4.2.6 and 4.2.8. In particular, we may regard W as a function of the principal invariants $I_1(\mathbf{V})$, $I_2(\mathbf{V})$, $I_3(\mathbf{V})$ or, equivalently, as a symmetric function of the principal stretches λ_1, λ_2, λ_3. In what follows we adopt the latter alternative and write

$$W(\mathbf{U}) \equiv W(\mathbf{V}) = W(\lambda_1, \lambda_2, \lambda_3) \tag{4.3.38}$$

for convenience in order to avoid introducing a separate notation to indicate the revised functional dependence. Of course, W must satisfy the symmetry requirements

$$W(\lambda_1, \lambda_2, \lambda_3) = W(\lambda_1, \lambda_3, \lambda_2) = W(\lambda_3, \lambda_1, \lambda_2) \tag{4.3.39}$$

subject to any constraint which may be imposed on the deformation subsequently.

Since Biot stress $\mathbf{T}^{(1)}$ is conjugate to the right stretch $\mathbf{U}$, use of the result (4.2.84) leads to

$$\mathbf{T}^{(1)} = \frac{\partial W}{\partial \mathbf{U}} = \sum_{i=1}^{3} \frac{\partial W}{\partial \lambda_i} \mathbf{u}^{(i)} \otimes \mathbf{u}^{(i)}, \tag{4.3.40}$$

where the unit vectors $\mathbf{u}^{(i)}$ define the Lagrangean principal axes. Hence the principal components of Biot stress are given simply by

$$t_i^{(1)} = \frac{\partial W}{\partial \lambda_i} \qquad i = 1, 2, 3. \tag{4.3.41}$$

Note that here and for much of the remainder of this section the arguments of W are omitted.

From (4.2.68) we have the result $\mathbf{S} = \mathbf{T}^{(1)}\mathbf{R}^{\mathrm{T}}$ for isotropic elasticity. Since $\hat{\mathbf{T}} = \mathbf{AS}$, and $\mathbf{U}$ is coaxial with $\mathbf{T}^{(1)}$, it follows from (4.3.40) that

$$\mathbf{R}^{\mathrm{T}}\hat{\mathbf{T}}\mathbf{R} = \sum_{i=1}^{3} \lambda_i \frac{\partial W}{\partial \lambda_i} \mathbf{u}^{(i)} \otimes \mathbf{u}^{(i)}. \tag{4.3.42}$$

Thus the principal components of Kirchhoff stress are

$$\hat{t}_i = \lambda_i \frac{\partial W}{\partial \lambda_i} \qquad i = 1, 2, 3, \tag{4.3.43}$$

$\hat{\mathbf{T}}$ being coaxial with the Eulerian principal axes $\mathbf{v}^{(i)} = \mathbf{R}\mathbf{u}^{(i)}$. Correspondingly, the principal components t_i of Cauchy stress are given by

$$Jt_i = \lambda_i \frac{\partial W}{\partial \lambda_i} \qquad i = 1, 2, 3, \tag{4.3.44}$$

where

$$J = \lambda_1 \lambda_2 \lambda_3. \tag{4.3.45}$$

Nominal stress itself has the representation

$$\mathbf{S} = \sum_{i=1}^{3} \frac{\partial W}{\partial \lambda_i} \mathbf{u}^{(i)} \otimes \mathbf{v}^{(i)}, \tag{4.3.46}$$

but in general this is not symmetric.

Each of the stress tensors $\mathbf{T}^{(m)}$ and conjugate strain tensors $\mathbf{E}^{(m)}$ is coaxial with $\mathbf{U}$ and their principal values are related by

$$t_i^{(m)} = \frac{\partial W}{\partial e_i^{(m)}} \qquad i = 1, 2, 3, \tag{4.3.47}$$

where $e_i^{(m)} = (\lambda_i^m - 1)/m$. This includes the special cases corresponding to $m = 1$ and $m = 0$ specified above.

For constrained isotropic Green elastic materials the stress-deformation relations derived above need modification in accordance with formulae listed in Section 4.3.3. We deal only with the incompressibility constraint here so that

$$\lambda_1 \lambda_2 \lambda_3 = 1 \tag{4.3.48}$$

and the principal Cauchy stresses t_i are given by

$$t_i = \lambda_i \frac{\partial W}{\partial \lambda_i} - p \qquad i = 1, 2, 3, \tag{4.3.49}$$

this replacing (4.3.44). The analogue of (4.3.47) in this case is

$$t_i^{(m)} = \frac{\partial W}{\partial e_i^{(m)}} - p\lambda_i^{-m} \qquad i = 1, 2, 3. \tag{4.3.50}$$

We note that the stress-deformation relation for an isotropic Green elastic material may be expressed concisely in the forms (4.3.44) and (4.3.49), in particular, for unconstrained and incompressible materials respectively, remembering that in each case the Cauchy stress tensor $\mathbf{T}$ is coaxial with the left stretch tensor $\mathbf{V}$.

Problem 4.3.7 If I_1, I_2, I_3 are the principal invariants of the left stretch tensor $\mathbf{V}$, let $W(I_1, I_2, I_3)$ denote the strain energy of an isotropic elastic solid. Obtain the representation

$$\mathbf{T} = \phi_0 \mathbf{I} + \phi_1 \mathbf{V} + \phi_2 \mathbf{V}^2$$

for the Cauchy stress tensor given in (4.2.54), expressing ϕ_0, ϕ_1, ϕ_2 in terms of I_1, I_2, I_3, $\partial W/\partial I_1$, $\partial W/\partial I_2$, $\partial W/\partial I_3$.

Deduce that

$$\frac{t_i - t_j}{\lambda_i - \lambda_j} = I_3^{-1} \frac{\partial W}{\partial I_1} + \lambda_i^{-1} \lambda_j^{-1} \frac{\partial W}{\partial I_2},$$

where t_i, λ_i are the principal components of $\mathbf{T}$, $\mathbf{V}$ respectively.

Derive corresponding results for an incompressible material.

Problem 4.3.8 A Green elastic material with strain energy $W(\lambda_1, \lambda_2, \lambda_3)$ is subject to the constraint

$$\lambda_1^2 L_1^2 + \lambda_2^2 L_2^2 + \lambda_3^2 L_3^2 = 1,$$

where L_1, L_2, L_3 are the components of unit vector $\mathbf{L}$ relative to the Lagrangean principal directions $\mathbf{u}^{(i)}$.

Show that

$$\mathbf{R}^{\mathrm{T}} \hat{\mathbf{T}} \mathbf{R} = \sum_{i=1}^{3} \lambda_i \frac{\partial W}{\partial \lambda_i} \mathbf{u}^{(i)} \otimes \mathbf{u}^{(i)} + \sum_{i,j=1}^{3} 2q L_i L_j \lambda_i \lambda_j \mathbf{u}^{(i)} \otimes \mathbf{u}^{(j)}.$$

Deduce that if $\mathbf{L} = \mathbf{u}^{(3)}$ is independent of position then the resulting deformation is equivalent to (unconstrained) plane strain normal to $\mathbf{u}^{(3)}$.

Deduce also that $\mathbf{R}^{\mathrm{T}} \hat{\mathbf{T}} \mathbf{R}$ is coaxial with the Lagrangean principal axes if and only if $\mathbf{L}$ lies along one of these axes.

Note that for constrained isotropic elastic solids coaxiality of conjugate stress and strain tensors is not maintained in general, as distinct from in the unconstrained situation.

Problem 4.3.9 Show that for a Green elastic material with strain energy $W(I_1, I_2, I_3)$, where I_1, I_2, I_3 are the principal invariants of $\mathbf{A}^{\mathrm{T}}\mathbf{A}$, the nominal stress tensor is given by

$$\mathbf{S} = 2 \frac{\partial W}{\partial I_1} \mathbf{A}^{\mathrm{T}} + 2 \frac{\partial W}{\partial I_2} \{ I_1 \mathbf{I} - \mathbf{A}^{\mathrm{T}} \mathbf{A} \} \mathbf{A}^{\mathrm{T}} + 2 I_3 \frac{\partial W}{\partial I_3} \mathbf{B}^{\mathrm{T}}.$$

Obtain the corresponding expression for an incompressible material.

4.3.5 Strain-energy functions for isotropic elastic materials

So far the only mathematical restrictions imposed on the form of strain-energy function are those arising from objectivity and material symmetry. In particular, for isotropic elastic solids there is no restriction on the strain-energy function W other than that it be a symmetric function of $\lambda_1, \lambda_2, \lambda_3$, although the conditions $W(1,1,1)=0$ and $\partial W(1,1,1)/\partial\lambda_i=0$ ($i=1,2,3$) are appropriate for a natural configuration. As mentioned in a more general context in Section 4.1.1, other mathematically founded restrictions arise naturally in the course of solution of boundary-value problems where, for example, the requirement of existence of solutions places some limit on the class of admissible strain-energy functions. Aspects of this are illustrated in Chapters 5 and 6.

At the physical level, the predicted mechanical behaviour of the material must conform with what is observed in experimental tests on the real material that the strain-energy function is intended to model, at least within the accessible experimental domain. Such conformity imposes further mathematical restrictions on W. This point is illustrated by means of some elementary examples in Section 4.4. Further illustrations are contained in Chapters 5–7, where specific forms of W are utilized.

As a prelude to the development in Chapters 5–7 we now present several general representations for the strain energy of both unconstrained and incompressible isotropic elastic materials. Specific examples of strain-energy functions are then identified for future reference.

Unconstrained materials In Section 4.3.4 we noted that W may be regarded as a function of either the principal stretches or of the principal invariants of $\mathbf{V}$. Equally, in accordance with the discussion of scalar invariants in Section 4.2.6, it may be treated as a function of any three independent invariants of the deformation. To be specific we consider the invariants in most common use in the literature, namely

$$\left.\begin{aligned} I_1 &= \lambda_1^2+\lambda_2^2+\lambda_3^2, \\ I_2 &= \lambda_2^2\lambda_3^2+\lambda_3^2\lambda_1^2+\lambda_1^2\lambda_2^2, \\ I_3 &= \lambda_1^2\lambda_2^2\lambda_3^2, \end{aligned}\right\} \tag{4.3.51}$$

the principal invariants of $\mathbf{A}\mathbf{A}^{\mathrm{T}}=\mathbf{V}^2$.

Subject to the regularity assumption that W is continuously differentiable infinitely many times with respect to I_1, I_2, I_3, we may write $W(I_1,I_2,I_3)$ as an infinite series in powers of I_1-3, I_2-3, I_3-1. Thus,

$$W(I_1,I_2,I_3)=\sum_{p,q,r=0}^{\infty} c_{pqr}(I_1-3)^p(I_2-3)^q(I_3-1)^r, \tag{4.3.52}$$

where p, q, r take values $0,1,2,\ldots$ and the coefficients c_{pqr} are independent of the deformation, on any closed region of the space of I_1, I_2, I_3 which includes the reference configuration $I_1 = I_2 = 3$, $I_3 = 1$. Moreover, an isotropic strain-energy function can be approximated as closely as desired by an expansion of this form containing a finite number of terms. Indeed, several special cases of (4.3.52) which involved only the first few terms have been used in the literature, but we do not give details here.

The requirement that the energy vanishes in the reference configuration is met provided $c_{000} = 0$, and it is easy to show that the reference configuration is stress free if and only if $c_{100} + 2c_{010} + c_{001} = 0$.

By setting $c_{pqr} = 0$ $(r = 1,2,\ldots)$ and $c_{pqr} = 0$ $(p,q = 1,2,3,\ldots)$ in (4.3.52), we obtain

$$W(I_1, I_2, I_3) = \sum_{p,q=0}^{\infty} c_{pq0}(I_1 - 3)^p(I_2 - 3)^q + \sum_{r=1}^{\infty} c_{00r}(I_3 - 1)^r. \tag{4.3.53}$$

With the latter sum written as $g(I_3)$ for convenience and only a small number of terms included in the first sum this form of W has some merit, as we demonstrate in later sections of this book. More specifically, by choosing $c_{pq0} = 0$ $(p = 2,3,\ldots,\ q = 1,2,\ldots)$ we obtain

$$W(I_1, I_2, I_3) = c_{100}(I_1 - 3) + g(I_3) \tag{4.3.54}$$

which is, of course, independent of I_2.

To a lesser extent expansions similar to (4.3.52), but based on the first two principal invariants of $\mathbf{A}^{*\mathrm{T}}\mathbf{A}^*$ together with $I_3(\mathbf{A}^{\mathrm{T}}\mathbf{A}) \equiv J^2$, where $\mathbf{A}^* = J^{-1/3}\mathbf{A}$, have been employed. These have the advantage of separating out the dependence of W on the distortional and dilatational parts of the deformation.

Turning now to the explicit representation of W in terms of λ_1, λ_2, λ_3, we write

$$W(\lambda_1, \lambda_2, \lambda_3) = \sum_{p,q,r=0}^{\infty} a_{pqr}\{[\lambda_1^p(\lambda_2^q + \lambda_3^q) + \lambda_2^p(\lambda_3^q + \lambda_1^q) + \lambda_3^p(\lambda_1^q + \lambda_2^q)](\lambda_1\lambda_2\lambda_3)^r - 6\}. \tag{4.3.55}$$

This automatically satisfies the requirement that W vanishes in the reference configuration, i.e. $W(1,1,1) = 0$, and we therefore set $a_{000} = 0$. It is also consistent with (4.3.39). Note that a more immediate analogue of (4.3.39) is obtained if λ_i is replaced by $\lambda_i - 1$ throughout (4.3.55) and the term -6 removed.

It is emphasized that in (4.3.52) and (4.3.55) we have used one notation W to represent two different functions. The values of those functions, however, are equal for any given deformation.

A particular case of (4.3.55) which is analogous to (4.3.53) is

$$W(\lambda_1, \lambda_2, \lambda_3) = \sum_{p,q=0}^{\infty} a_{pq0}[\lambda_1^p(\lambda_2^q + \lambda_3^q) + \lambda_2^p(\lambda_3^q + \lambda_1^q) + \lambda_3^p(\lambda_1^q + \lambda_2^q) - 6] + \sum_{r=1}^{\infty} 6a_{00r}[(\lambda_1\lambda_2\lambda_3)^r - 1], \tag{4.3.56}$$

and this may be reduced to

$$W(\lambda_1, \lambda_2, \lambda_3) = \sum_{p,r=0}^{\infty} 2a_{p0r}\{(\lambda_1^p + \lambda_2^p + \lambda_3^p)(\lambda_1\lambda_2\lambda_3)^r - 3\} \tag{4.3.57}$$

and then to

$$W(\lambda_1, \lambda_2, \lambda_3) = \sum_{p=1}^{\infty} 2a_{p00}(\lambda_1^p + \lambda_2^p + \lambda_3^p - 3) + g(J), \tag{4.3.58}$$

where $g(J)$ replaces $\sum_{r=1}^{\infty} 6a_{00r}(J^r - 1)$.

By setting $a_{p00} = 0$ for $p \neq 2$ we obtain

$$W(\lambda_1, \lambda_2, \lambda_3) = 2a_{200}(\lambda_1^2 + \lambda_2^2 + \lambda_3^2 - 3) + g(J), \tag{4.3.59}$$

and this is equivalent to (4.3.54). Note, however, that the arguments of g are different in (4.3.54) and (4.3.59).

Incompressible materials For an incompressible material $I_3 = 1$ and W depend on only two independent deformation invariants. In particular, the expansion (4.3.52) is replaced by

$$W(I_1, I_2) = \sum_{p,q=0}^{\infty} c_{pq}(I_1 - 3)^p(I_2 - 3)^q \tag{4.3.60}$$

with $c_{00} = 0$. From (4.3.51) with $I_3 = 1$ it is easy to show that

$$I_1^2 \geq 3I_2 \qquad I_2^2 \geq 3I_1, \tag{4.3.61}$$

the reference configuration (which corresponds to $I_1 = I_2 = 3$) being on the boundary of the domain (4.3.61).

We make note of the special case

$$W(I_1, I_2) = c_{10}(I_1 - 3) + c_{01}(I_2 - 3) \tag{4.3.62}$$

which is referred to as the *Mooney* or *Mooney–Rivlin* strain-energy function. If $c_{01} = 0$ this reduces to the *neo-Hookean* strain-energy function

$$W(I_1, I_2) = c_{10}(I_1 - 3). \tag{4.3.63}$$

The Mooney and neo-Hookean strain-energy functions have played an important part in the development of non-linear elasticity theory and its application.

Corresponding to (4.3.55) we have

$$W(\lambda_1, \lambda_2, \lambda_3) = \sum_{p,q=0}^{\infty} a_{pq}[\lambda_1^p(\lambda_2^q + \lambda_3^q) + \lambda_2^p(\lambda_3^q + \lambda_1^q) + \lambda_3^p(\lambda_1^q + \lambda_2^q) - 6] \tag{4.3.64}$$

with $a_{00} = 0$, subject to $\lambda_1\lambda_2\lambda_3 = 1$. Any two stretches, λ_1 and λ_2 say, are arbitrarily disposable subject to $\lambda_1 > 0$, $\lambda_2 > 0$ and the reference configuration corresponds to $\lambda_1 = \lambda_2 = 1$.

Since $\lambda_1\lambda_2\lambda_3 = 1$ it is not strictly necessary to restrict the integers p and q in (4.3.64) to be positive. This is particularly relevant in the special case of (4.3.64) which we write as

$$W(\lambda_1, \lambda_2, \lambda_3) = \sum_{p=-\infty}^{\infty} 2a_{p0}(\lambda_1^p + \lambda_2^p + \lambda_3^p - 3), \tag{4.3.65}$$

where, for $p < 0$, a_{p0} is defined as $a_{(-p)(-p)}$. Note that the choice $a_{p0} = 0$ for $p \neq \pm 2$ leads to the Mooney strain-energy function (4.3.62).

Equation (4.3.65) has the attractive feature of not involving products of the λ_i, except through $\lambda_1\lambda_2\lambda_3 = 1$. In fact, it is not necessary to restrict attention to exponents which are integers and we may replace (4.3.65) by

$$W(\lambda_1, \lambda_2, \lambda_3) = \sum_{p=1}^{N} \mu_p(\lambda_1^{\alpha_p} + \lambda_2^{\alpha_p} + \lambda_3^{\alpha_p} - 3)/\alpha_p, \tag{4.3.66}$$

where α_p $(p = 1, \ldots, N)$ is a real number and N is a positive integer (which may be taken as large or as small as desired). The same applies to (4.3.64) and to the corresponding unconstrained strain-energy function. Further discussion of this is contained in the review article by Ogden (1982). We remark in passing that strain-energy functions of the form (4.3.66) have led to very successful correlations with experimental stress-deformation data for rubberlike materials. The theoretical background for this is examined in Chapter 7.

The representations (4.3.52) and (4.3.55) are equivalent and completely general for unconstrained isotropic elastic materials. Indeed, it is only a matter of algebraic manipulation (albeit tedious) to express a_{pqr} in terms of c_{pqr} and conversely. If, however, the exponents in (4.3.55) are allowed to be real numbers (not just integers) then (4.3.55), although it remains equivalent to (4.3.52), cannot be rewritten explicitly in the form (4.3.52) in general. Similar comments apply in respect of (4.3.60) and (4.3.64) for incompressible materials.

In this section we have necessarily been selective and mentioned only two general representations for W. There are many other possibilities. Equally, only a few specific forms of W have been written down. Others will be introduced as the need arises in later sections, and further information about particular forms of W can be found in Treloar (1975) and Ogden (1982) where detailed references are provided.

Explicit stress-deformation relations for the strain-energy functions listed above can be obtained by use of the formulae given in Section 4.3.4. This exercise is left for the reader.

4.4 APPLICATION TO SIMPLE HOMOGENEOUS DEFORMATIONS

For definiteness we now consider an unconstrained material with strain-energy function

$$W = \tfrac{1}{2}\mu(\lambda_1^2 + \lambda_2^2 + \lambda_3^2 - 3 - 2\ln J) + \tfrac{1}{2}\mu'(J-1)^2 \tag{4.4.1}$$

and the corresponding incompressible material with

$$W = \tfrac{1}{2}\mu(\lambda_1^2 + \lambda_2^2 + \lambda_3^2 - 3), \qquad \lambda_1\lambda_2\lambda_3 = 1, \tag{4.4.2}$$

to which (4.4.1) reduces when $J = 1$. Note that (4.4.1) is a special case of (4.3.59), and (4.4.2) is identical to (4.3.63), μ and μ' being non-zero constants. In each case $W = 0$ in the reference configuration $\lambda_1 = \lambda_2 = \lambda_3 = 1$.

From (4.3.44) and (4.3.49) it follows that the principal components t_i of Cauchy stress are given by

$$Jt_i = \mu(\lambda_i^2 - 1) + \mu' J(J-1) \qquad i = 1, 2, 3 \tag{4.4.3}$$

and

$$t_i = \mu\lambda_i^2 - p \qquad i = 1, 2, 3 \tag{4.4.4}$$

for (4.4.1) and (4.4.2) respectively. In the reference configuration the unconstrained material is stress free while the incompressible material is subject to a hydrostatic stress $\mu - p$, where p is arbitrary.

The stress-deformation relations (4.4.3) and (4.4.4) are specialized in what follows to some of the homogeneous deformations discussed in Section 2.2.6.

Uniform extension with lateral contraction We consider a circular cylinder of material extended uniformly along its axis, this axis also being a principal axis of the deformation. Let λ_1 be the corresponding principal stretch. Then, by symmetry, the other two principal stretches are equal and may be written as $\lambda_1^{-1/2}J^{1/2}$. Because of isotropy the principal axes of Cauchy stress are along and normal to the cylinder axis. Thus, dealing first with the unconstrained material, we obtain from (4.4.3).

$$Jt_1 = \mu(\lambda_1^2 - 1) + \mu' J(J - 1)$$
$$Jt_2 = \mu(J\lambda_1^{-1} - 1) + \mu' J(J - 1)$$

with $t_3 = t_2$.

If $t_2 = 0$ the state of stress is referred to as *simple tension* and then

$$Jt_1 = \mu(\lambda_1^2 - 1) + \mu' J(J - 1) = \mu(\lambda_1^2 - J\lambda_1^{-1}). \tag{4.4.5}$$

This provides expressions for both J and t_1 in terms of λ_1.

Physical intuition dictates that the lateral stretch $\lambda_1^{-1/2}J^{1/2}$ decreases as λ_1 increases and that extension requires the application of tension ($t_1 > 0$). Equation (4.4.5) then implies that

$$\mu > 0. \tag{4.4.6}$$

It also follows that $J > 1$ if $\lambda_1 > 1$ provided

$$\mu' > 0. \tag{4.4.7}$$

(Recall that we have taken μ' to be non-zero.) In other words increase in volume accompanies extension. This makes physical sense since the hydrostatic part of the stress $\frac{1}{3}t_1$ is tensile, i.e. positive.

For the incompressible material equations (4.4.5) are replaced by

$$t_1 = \mu\lambda_1^2 - p = \mu(\lambda_1^2 - \lambda_1^{-1}) \tag{4.4.8}$$

and (4.4.6) holds. Note that p is now determined in terms of λ_1.

The stress-deformation relation (4.4.8) is shown in Fig. 4.2. Also plotted for comparison are the corresponding principal values $t_1^{(1)}, t_1^{(2)}$ of the stress

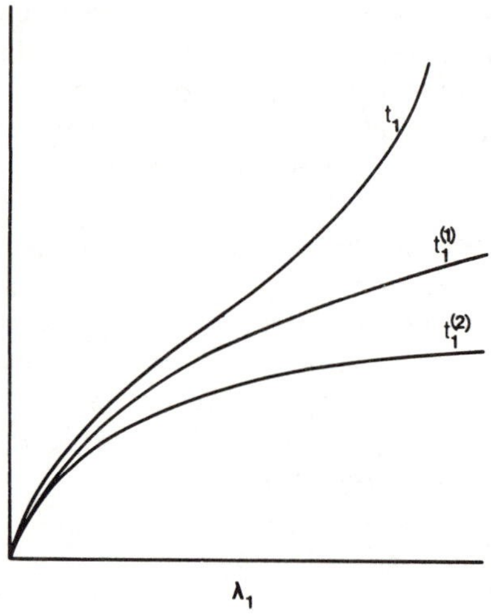

Fig. 4.2 Plot of the principal (tensile) stresses t_1, $t_1^{(1)}$, $t_1^{(2)}$ against the stretch λ_1 in simple tension for the incompressible material with strain-energy function (4.4.2).

tensors $\mathbf{T}^{(1)}, \mathbf{T}^{(2)}$. We recall that $t_1 = \lambda_1 t_1^{(1)} = \lambda_1^2 t_1^{(2)}$ and note that all three curves have gradient 3μ at $\lambda_1 = 1$.

For a compressible material $J \neq 1$ in general but the deformation can be made volume preserving by the application of lateral stress $t_2 = \mu(\lambda_1^{-1} - 1)$.

Pure dilatation For this deformation the principal stretches are all equal and the stress is purely hydrostatic. Thus, if λ_1 denotes the common value of the stretches then $J = \lambda_1^3$ and the principal Cauchy stress t_1 is given by

$$\lambda_1^3 t_1 = \mu(\lambda_1^2 - 1) + \mu' \lambda_1^3 (\lambda_1^3 - 1).$$

The inequalities (4.4.6) and (4.4.7) ensure that the volume of the material increases (respectively decreases) if t_1 is tensile (respectively compressive), as should be expected on physical grounds.

We note the growth conditions

$$t_1 \sim \mu' \lambda_1^3 \qquad \text{as} \quad \lambda_1 \to \infty,$$
$$t_1 \sim -\mu \lambda_1^{-3} \qquad \text{as} \quad \lambda_1 \to 0.$$

These have mathematical significance but caution should be exercised in their physical interpretation because suitably large deformations are unlikely to be achieved in practice. More generally, growth conditions of this kind play an important role in the discussion of, for example, questions of existence and uniqueness of the solution of boundary-value problems. We touch on this in Chapter 5. For an incompressible material $\lambda_1 = 1$ and t_1 is arbitrary.

Equibiaxial extension This deformation may be applied to a square sheet of material with $\lambda_2 = \lambda_1$ the stretches in the plane of the sheet and parallel to its edges. The stretch normal to the sheet is

$$\lambda_3 = J\lambda_1^{-2}.$$

The principal axes of the stress are also parallel to the edges of the sheet so that if no stress is applied normal to the sheet then

$$Jt_1 = \mu(\lambda_1^2 - 1) + \mu' J(J-1) = \mu(\lambda_1^2 - J^2\lambda_1^{-4}). \tag{4.4.9}$$

The resulting state of stress may be referred to as *equibiaxial tension*.

Implications of the inequalities (4.4.6) and (4.4.7), analogous to those for simple tension, are left for the reader to investigate.

For the incompressible material elimination of the hydrostatic pressure p yields

$$t_1 = \mu(\lambda_1^2 - \lambda_1^{-4})$$

and we plot t_1 as a function of λ_1 in Fig. 4.3. The stresses $t_1^{(1)}$ and $t_1^{(2)}$ have

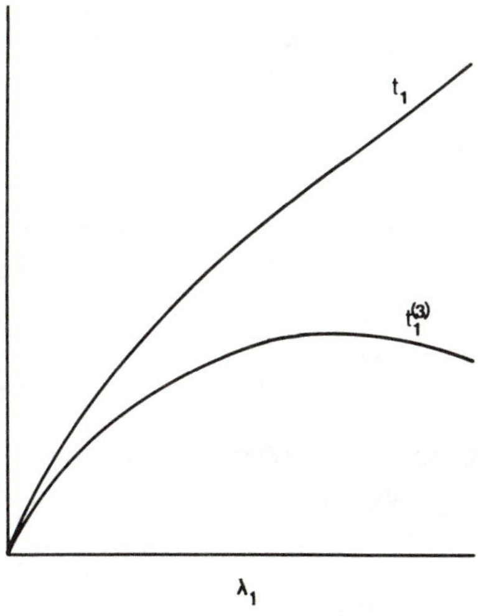

Fig. 4.3 Plot of the principal (tensile) stresses t_1, $t_1^{(3)}$ against the stretch λ_1 in equibiaxial tension for the incompressible material with strain-energy function (4.4.2).

similar characteristics to those shown in Fig. 4.2 for simple tension and are not therefore included in Fig. 4.3. We do, however, show $t_1^{(3)} = \lambda_1^{-3} t_1$ because it exhibits a maximum. This latter feature is relevant in the problem of inflation of a spherical balloon since (assuming spherical symmetry is maintained) the deformation in the skin of the balloon is approximately equibiaxial extension and the internal pressure is proportional to $t_1^{(3)}$. It is well known from experiments on meteorological balloons that the pressure achieves a maximum at some value of the radius. (Note, however, that the true stress increases monotonically.) Details of this problem are discussed in Chapter 5.

Simple shear For the simple shear deformation described in Section 2.2.6 the deformation gradient has the form

$$\mathbf{A} = \mathbf{I} + \gamma \mathbf{e}_1 \otimes \mathbf{e}_2,$$

where $\mathbf{e}_1$ and $\mathbf{e}_2$ are fixed orthonormal basis vectors in the plane of deformation. The Eulerian principal axes are given by

$$\left.\begin{aligned} \mathbf{v}^{(1)} &= \cos\theta\, \mathbf{e}_1 + \sin\theta\, \mathbf{e}_2, \\ \mathbf{v}^{(2)} &= -\sin\theta\, \mathbf{e}_1 + \cos\theta\, \mathbf{e}_2, \end{aligned}\right\} \tag{4.4.10}$$

where

$$\tan 2\theta = 2/\gamma \qquad 0 < \theta \leq \pi/4, \tag{4.4.11}$$

together with $\mathbf{v}^{(3)} = \mathbf{e}_3$ and

$$\gamma = \lambda_1 - \lambda_1^{-1}, \tag{4.4.12}$$

the principal stretches being $\lambda_1 (\geq 1)$, $\lambda_2 = \lambda_1^{-1}$, $\lambda_3 = 1$.

Because of the isotropy we may write the Cauchy stress tensor as

$$\mathbf{T} = \sum_{i=1}^{3} t_i \mathbf{v}^{(i)} \otimes \mathbf{v}^{(i)}. \tag{4.4.13}$$

If T_{ij} denote the components of $\mathbf{T}$ relative to the basis $\{\mathbf{e}_i\}$ then use of (4.4.10) in (4.4.13) leads, in particular, to the in-plane connections

$$\left.\begin{aligned} T_{11} &= t_1 \cos^2\theta + t_2 \sin^2\theta, \\ T_{12} &= (t_1 - t_2)\sin\theta\cos\theta, \\ T_{22} &= t_1 \sin^2\theta + t_2 \cos^2\theta. \end{aligned}\right\} \tag{4.4.14}$$

It is then easy to show from (4.4.11) and (4.4.12) that

$$T_{11} - T_{22} = \gamma T_{12}. \tag{4.4.15}$$

Because (4.4.15) is a relation between the stress components and the deformation and is independent of the choice of (isotropic elastic) constitutive law it is an example of what is referred to as a *universal relation*. Other examples will be seen in Chapter 5.

We now specialize to the constitutive laws (4.4.1) and (4.4.2) so that, for an unconstrained material,

$$t_1 = \mu(\lambda_1^2 - 1), \qquad t_2 = \mu(\lambda_1^{-2} - 1), \qquad t_3 = 0$$

from (4.4.3), while, for an incompressible material,

$$t_1 = \mu\lambda_1^2 - p, \qquad t_2 = \mu\lambda_1^{-2} - p, \qquad t_3 = \mu - p$$

from (4.4.4).

It follows that

$$T_{11} = \mu\gamma^2, \qquad T_{22} = 0, \qquad T_{12} = \mu\gamma$$

and

$$T_{11} = \mu\gamma^2 + \mu - p, \qquad T_{22} = \mu - p, \qquad T_{12} = \mu\gamma$$

respectively. Clearly, simple shear cannot be achieved by the application of shear stress T_{12} alone, and, in general, normal forces are required. On the surfaces normal to $\mathbf{e}_2$ the traction has components $(T_{12}, T_{22}, 0)$ and on the sloping faces of the deformed material the normal and shear components of traction are respectively $-\mu\gamma^2/(1+\gamma^2)$ and $\mu\gamma/(1+\gamma^2)$. (The unit normal to these faces is $\pm(\mathbf{e}_1 - \gamma\mathbf{e}_2)/(1+\gamma^2)^{1/2}$.)

If no normal forces were applied to the material its shape and volume would in general change under the action of the shear stress. The tendency to change volume on deformation in the absence of normal stresses is called the *Kelvin effect*, and the inequality $(T_{11} \neq T_{22})$ of the normal stresses is known as the *Poynting effect*. For an incompressible material there is no Kelvin effect but (unequal) normal stresses are still needed to achieve the required simple shear. For small deformations (where γ^2 can be neglected compared with γ) the Poynting effect can be ignored. Indeed, it does not appear in the classical linear theory of elastic shear.

Problem 4.4.1 For the simple shear of an isotropic elastic solid described above show that the normal and shear components of traction on the surfaces which in the reference configuration *were* normal to $\mathbf{e}_1$ are

$$(T_{11}+\gamma^2 T_{22}-2\gamma T_{12})/(1+\gamma^2), \qquad T_{12}/(1+\gamma^2).$$

Problem 4.4.2 Show that in the simple shear of the incompressible material with strain-energy function

$$W=\sum_{n=1}^{N}\mu_n(\lambda_1^n+\lambda_2^n+\lambda_3^n-3)/n$$

the shear stress T_{12} is given by

$$(\lambda_1+\lambda_1^{-1})T_{12}=\sum_{n=1}^{N}\mu_n(\lambda_1^n-\lambda_1^{-n}).$$

Deduce that for the Mooney strain-energy function T_{12} depends linearly on the amount of shear $\lambda_1-\lambda_1^{-1}$.

Problem 4.4.3 Show that for an incompressible isotropic Green elastic material which is inextensible in the fixed direction $\mathbf{L}$ the Cauchy stress tensor is expressible in the form

$$\mathbf{T}=\sum_{i=1}^{3}\lambda_i\frac{\partial W}{\partial\lambda_i}\mathbf{v}^{(i)}\otimes\mathbf{v}^{(i)}-p\mathbf{I}+\sum_{i,j,=1}^{3}2qL_iL_j\lambda_i\lambda_j\mathbf{v}^{(i)}\otimes\mathbf{v}^{(j)},$$

where the unit vectors $\mathbf{v}^{(i)}$ define the Eulerian principal axes, L_i are the components of $\mathbf{L}$ relative to those axes, and p,q are arbitrary.

For a plane deformation with $\mathbf{L}$ lying in the plane show that the deformation is necessarily a simple shear in the direction $\mathbf{L}$.

Prove that

$$T_{11}-T_{22}=\gamma T_{12}+2q,$$

where T_{ij} are the components of $\mathbf{T}$ with respect to basis vectors $\mathbf{L}=\mathbf{e}_1,\mathbf{e}_2$ in the plane of deformation and $\mathbf{e}_3$ normal to that plane.

If the material is bounded by plane faces normal to $\mathbf{e}_1$ and $\mathbf{e}_2$ in the reference configuration, find the components of traction on each of these faces after deformation.

Problem 4.4.4 If $\mathbf{C}=\mathbf{A}\mathbf{A}^{\mathrm{T}}$ and I_1,I_2,I_3 are the principal invariants of $\mathbf{C}$, show that

$$I_3^{1/2}\mathbf{T}=2\left(\frac{\partial W}{\partial I_1}+I_1\frac{\partial W}{\partial I_2}\right)\mathbf{C}-2\frac{\partial W}{\partial I_2}\mathbf{C}^2+2I_3\frac{\partial W}{\partial I_3}\mathbf{I}.$$

For the simple shear deformation defined above show that $I_1 = I_2$ and deduce that

$$T_{22} - T_{11} = 2\frac{\mathrm{d}\hat{W}}{\mathrm{d}I_1}\gamma^2 = \gamma T_{12},$$

where $\hat{W}(I_1) = W(I_1, I_1, 1)$.

REFERENCES

J.L. Ericksen, Special Topics in Elastostatics, *Advances in Applied Mechanics*, Vol. 17, pp. 189–244 (1977).

A.E. Green and J.E. Adkins, *Large Elastic Deformations*, 2nd edition, Oxford University Press (1970).

R. Hill, Aspects of Invariance in Solid Mechanics, *Advances in Applied Mechanics*, Vol. 18, pp. 1–75 (1978).

R.W. Ogden, Elastic Deformations of Rubberlike Solids, in *Mechanics of Solids*, the Rodney Hill 60th Anniversary Volume (Eds. H.G. Hopkins and M.J. Sewell), pp. 499–537, Pergamon (1982).

G.P. Parry, On the Elasticity of Monatomic Crystals, *Mathematical Proceedings of the Cambridge Philosophical Society*, Vol. 80, pp. 189–211 (1976).

L.R.G. Treloar, *The Physics of Rubber Elasticity*, 3rd edition, Oxford University Press (1975).

C.A. Truesdell, *A First Course in Rational Continuum Mechanics*, Vol. I, Academic Press (1977).

C.A. Truesdell and W. Noll, The Non-linear Field Theories of Mechanics, *Handbuch der Physik*, Vol. III/3 (Ed. S. Flügge), Springer (1965).

C.-C. Wang and C.A. Truesdell, *Introduction to Rational Elasticity*, Noordhoff (1973).

CHAPTER 5

Boundary-Value Problems

5.1 FORMULATION OF BOUNDARY-VALUE PROBLEMS

5.1.1 Equations of motion and equilibrium

In Section 3.3.2 we obtained the governing equation for the motion of a continuous medium in terms of the Cauchy stress tensor in the form

$$\operatorname{div} \mathbf{T} + \rho \mathbf{b} = \rho \dot{\mathbf{v}} \tag{5.1.1}$$

together with the mass conservation equation

$$\dot{\rho} + \rho \operatorname{div} \mathbf{v} = 0. \tag{5.1.2}$$

(Recall that the superposed dot denotes the material time derivative and that $\mathbf{T}$ is symmetric.)

For an elastic material, we now add to these Eulerian equations the constitutive equation

$$\mathbf{T} = \mathbf{G}(\mathbf{A}) \tag{5.1.3}$$

discussed in Section 4.2.1, where $\mathbf{A} = \operatorname{Grad} \boldsymbol{\chi}(\mathbf{X},t)$ is the deformation gradient relative to some given reference configuration. Essentially, subject to suitable prescribed initial and boundary conditions, the problem is to determine $\mathbf{x} = \boldsymbol{\chi}(\mathbf{X}, t)$ from equations (5.1.1)–(5.1.3). However, since $\mathbf{x}$ and t are the independent variables in (5.1.1) and (5.1.2), and, in particular, $\mathbf{x}$ cannot be known from the outset on parts of the current boundary of the material where it is not prescribed, it follows that the Eulerian formulation is not naturally suited to a discussion of elasticity problems. Nevertheless, in practice, an Eulerian approach can often yield useful information and should not be discarded out of hand.

For equilibrium problems (5.1.1) becomes

$$\operatorname{div} \mathbf{T} + \rho \mathbf{b} = \mathbf{0} \tag{5.1.4}$$

and (5.1.2) is replaced by its static equivalent

$$\rho = \rho_0 J^{-1} \equiv \rho_0 (\det \mathbf{A})^{-1}. \tag{5.1.5}$$

The elastic body is self-equilibrated if the body force **b** in (5.1.4) vanishes when **T** is given by (5.1.3).

We now turn to the Lagrangean equations which, with **X** and t as independent variables, are particularly suitable for non-linear elasticity problems. From Section 3.4.2 we have the equation of motion in the form

$$\operatorname{Div} \mathbf{S} + \rho_0 \mathbf{b}_0 = \rho_0 \ddot{\boldsymbol{\chi}} \tag{5.1.6}$$

with the nominal stress **S** subject to the restriction

$$\mathbf{AS} = \mathbf{S}^{\mathrm{T}} \mathbf{A}^{\mathrm{T}} \tag{5.1.7}$$

implied by symmetry of Cauchy stress.

For an elastic material, we have

$$\mathbf{S} = \mathbf{H}(\mathbf{A}) \tag{5.1.8}$$

from (4.2.12), and equation (5.1.6) then serves to determine $\boldsymbol{\chi}(\mathbf{X}, t)$. It is assumed that the reference density ρ_0 is known explicitly as a function of **X**. The body force density, on the other hand, depends on **X** only through $\mathbf{x} = \boldsymbol{\chi}(\mathbf{X}, t)$ since $\mathbf{b}_0(\mathbf{X}, t) = \mathbf{b}(\mathbf{x}, t)$ and, in general, is prescribed as a function of **x** (recall Section 3.2). The mass conservation equation (5.1.5) is not required for the determination of $\boldsymbol{\chi}(\mathbf{X}, t)$ and serves only to provide an expression for the density ρ in the current configuration.

In terms of rectangular Cartesian components and for a homogeneous material (5.1.8) allows (5.1.6) to be written

$$\mathscr{A}^1_{\alpha i \beta j} \frac{\partial^2 \chi_j}{\partial X_\alpha \partial X_\beta} + \rho_0 b_i = \rho_0 \ddot{\chi}_i, \tag{5.1.9}$$

where b_i depends on χ_k $(k = 1, 2, 3)$ and the coefficients $\mathscr{A}^1_{\alpha i \beta j}$, defined by

$$\mathscr{A}^1_{\alpha i \beta j} = \frac{\partial S_{\alpha i}}{\partial A_{j\beta}}, \tag{5.1.10}$$

depend non-linearly on the first derivatives $\partial \chi_k / \partial X_\gamma$ $(k, \gamma = 1, 2, 3)$ in general. Thus, (5.1.9) provides a set of three quasi-linear partial differential equations of the second order for χ_i $(i = 1, 2, 3)$ with X_1, X_2, X_3, t as independent variables.

The coefficients $\mathscr{A}^1_{\alpha i\beta j}$ are (rectangular Cartesian) components of a fourth-order (two-point) tensor $\boldsymbol{\mathscr{A}}^1$ which may be defined symbolically by

$$\boldsymbol{\mathscr{A}}^1 = \partial \mathbf{S}/\partial \mathbf{A}. \tag{5.1.11}$$

This is an example of a tensor of elastic moduli. More particularly, $\boldsymbol{\mathscr{A}}^1$ is the (fourth-order) tensor of first-order elastic moduli associated with the conjugate pair $(\mathbf{S}, \mathbf{A})$. This is mentioned only in passing at this point since a general discussion of elastic moduli tensors is reserved for Chapter 6. Clearly the precise form of $\boldsymbol{\mathscr{A}}^1$ will have an important influence in determining the nature of the solutions of (5.1.9), and indeed governs the very existence of such solutions for given boundary-value problems.

For a hyperelastic material with strain-energy function W per unit reference volume (5.1.8) specializes to

$$\mathbf{S} = \frac{\partial W}{\partial \mathbf{A}}(\mathbf{A}), \tag{5.1.12}$$

as in (4.3.7), and (5.1.10) and (5.1.11) are modified accordingly. Note, in particular, the symmetry

$$\mathscr{A}^1_{\alpha i\beta j} = \frac{\partial^2 W}{\partial A_{i\alpha}\partial A_{j\beta}} = \mathscr{A}^1_{\beta j\alpha i} \tag{5.1.13}$$

between pairs of indices (αi) and (βj).

If W is objective then, in particular, we have

$$W(\mathbf{A}) = W^{(2)}(\mathbf{E}^{(2)}),$$

where $\mathbf{E}^{(2)}$ is the Green strain tensor and $W^{(2)}$ is defined by (4.3.18). It follows immediately from the symmetry of $\mathbf{T}^{(2)} = \partial W^{(2)}(\mathbf{E}^{(2)})/\partial \mathbf{E}^{(2)}$ that

$$\mathbf{S} \equiv \frac{\partial W}{\partial \mathbf{A}}(\mathbf{A}) = \mathbf{T}^{(2)}\mathbf{A}^{\mathrm{T}}$$

and hence $\mathbf{AS}$ is symmetric. In other words, *the rotational balance equation* (5.1.7) *is a consequence of the assumption that W is objective*. Thus, if W is chosen to be objective from the outset the equation (5.1.7) follows automatically and need not be set down separately. On the other hand, (5.1.7) does not in general imply that W is objective. It suffices therefore to examine the differential equation (5.1.6) together with (5.1.12) for objective strain-energy functions.

In equilibrium, equation (5.1.6) is replaced by

$$\operatorname{Div} \mathbf{S} + \rho_0 \mathbf{b}_0 = \mathbf{0} \tag{5.1.14}$$

and this, with (5.1.12) inserted for objective W, is the *governing equation of non-linear elastostatics*. More explicitly, we may write this as

$$\operatorname{Div} \{\partial W(\mathbf{A})/\partial \mathbf{A}\} + \rho_0(\mathbf{X})\mathbf{b}\{\boldsymbol{\chi}(\mathbf{X})\} = \mathbf{0}, \tag{5.1.15}$$

where

$$\mathbf{A} = \operatorname{Grad} \boldsymbol{\chi}(\mathbf{X}) \tag{5.1.16}$$

and time dependence has been dropped.

We note that (5.1.6) and (5.1.14) are appropriate for either unconstrained or constrained materials, but that (5.1.9) and (5.1.15) are specific to unconstrained materials since the stress-deformation relations (5.1.8) and (5.1.12) respectively have been inserted. For a material with a single constraint

$$C(\mathbf{A}) = 0, \tag{5.1.17}$$

as discussed in Section 4.2.7, the constitutive equation (5.1.8) is replaced by

$$\mathbf{S} = \mathbf{H}(\mathbf{A}) + q \frac{\partial C}{\partial \mathbf{A}}(\mathbf{A}) \tag{5.1.18}$$

and this is inserted into the equation of motion (5.1.6). Equation (5.1.9) is then modified by the addition of the term

$$\frac{\partial}{\partial X_\alpha}\left(q \frac{\partial C}{\partial A_{i\alpha}}\right)$$

to its left-hand side and (5.1.15) adjusted by a similar addition of

$$\operatorname{Div}\left(q \frac{\partial C}{\partial \mathbf{A}}\right),$$

subject always to (5.1.17). The definition (5.1.11) is replaced by

$$\mathscr{A}^1 = \partial \mathbf{H}(\mathbf{A})/\partial \mathbf{A} \tag{5.1.19}$$

and its component form (5.1.10) follows suit.

For the incompressibility constraint det $\mathbf{A} = 1$, equation (5.1.18) becomes

$$\mathbf{S} = \mathbf{H}(\mathbf{A}) - p\mathbf{B}^{\mathrm{T}} \tag{5.1.20}$$

and we note this important special case for future reference. Since, by (2.2.20), Div $\mathbf{B}^{\mathrm{T}} = \mathbf{0}$ for an incompressible material, it follows on use of the result $\mathbf{B}$ Grad p = grad p that the equation of motion takes the form

$$\operatorname{Div}\mathbf{H}(\mathbf{A}) - \operatorname{grad} p + \rho_0 \mathbf{b}_0 = \rho_0 \ddot{\boldsymbol{\chi}},$$

with $\mathbf{H}(\mathbf{A}) = \partial W(\mathbf{A})/\partial \mathbf{A}$ for a hyperelastic material.

Thus far we have considered only the equations of motion and equilibrium together with rotational balance, objectivity and mass conservation. In order to describe a 'problem' we need further information. First, the region of 'space' to which the equations are to be applied must be specified. Specifically, this will be the physical space, $\mathscr{B}_0$ say, occupied by the elastic material (body) in some reference configuration (this being a subset of the Euclidean point space $\mathscr{E}$ in the sense of Section 2.1.2). Second, appropriate data need to be prescribed over the boundary $\partial\mathscr{B}_0$ of $\mathscr{B}_0$ (examples of such data are illustrated in Section 5.1.2). Third, for time-dependent problems, suitable initial data are required on $\mathscr{B}_0$. It is often sensible to take the initial position of the body (i.e. at time $t = 0$) as the reference configuration. The initial velocity of (each point of) $\mathscr{B}_0$ is one candidate for prescription.

In this book our attention will be confined largely to (quasi-) static problems so we do not discuss initial data in detail. Boundary conditions are examined in Section 5.1.2, but before moving on to this some further remarks are necessary. These are directed specifically to the equilibrium problem and for materials with a strain-energy function, but similar remarks are also applicable to Cauchy elastic materials and to time-dependent problems.

The equilibrium problem is defined by the governing differential equations and boundary conditions, with W assumed objective. We are required to determine the solution (or solutions) $\mathbf{x} = \boldsymbol{\chi}(\mathbf{X})$ of (5.1.14), with (5.1.12) for unconstrained materials and (5.1.18) plus (5.1.17) for materials subject to a single internal constraint, *if such a solution exists*. This raises the question of whether or not the problem is 'well-posed'. However, the classical notion of well-posedness (requiring existence and uniqueness of solution, and continuous dependence of the solution on the data), as applied to linear problems, is inappropriate for the equations of non-linear elasticity. In particular, uniqueness of solution cannot in general be expected; indeed, it is the examples of non-uniqueness (some of which will be illustrated later) which so characterize non-linear elasticity. (This should not, however, prevent us from investigating circumstances in which uniqueness might hold.)

The question of existence needs to be addressed, but existence within the class of twice continuously differentiable deformations χ (on which attention is concentrated in this book) cannot always be expected, and weak solutions of the governing equations should be catered for in a general treatment. In this book there is not space enough to include details of the functional analytic apparatus required for questioning existence and related matters. Full justice to this aspect of the theory can only be done in a separate monograph, and we do not pursue it here. For the reader interested in this rapidly developing area we refer to Ball (1977, 1982) and Hanyga (1984).

Of paramount consideration in such matters is the precise nature of the strain-energy function W and, more particularly, its effect, through the tensor $\mathscr{A}^1$ of elastic moduli, on the properties of the governing equations. Clearly, from the mathematical viewpoint, the choice of W must be governed by requirements such as existence. This raises an important question, which in general remains open: namely, what restrictions should be placed on W in order that $\mathscr{A}^1$ has a form which ensures, for example, existence of solution to certain boundary-value problems? We must also ask what restrictions ensure that the solutions thereby obtained lead to physically meaningful results. This important link between the mathematical results and their physical interpretation should not be forgotten. In part the physical interpretation relies on intuition, but much can also be learned from the results of experiments on actual elastic materials (or, more precisely, materials which can be modelled by an elastic constitutive law over some range of deformations). In particular, specific forms of W constructed on this basis can be used in the discussion of mathematical questions.

In Chapters 5 and 6 we are particularly concerned with obtaining explicit solutions to some specific boundary-value problems and finding restrictions on W thereby imposed. Existence of solutions (for these problems) is then established in a constructive manner.

Problem 5.1.1 If the strain-energy function W is objective and $\mathbf{T}^{(1)} = \partial W/\partial \mathbf{U}$ is the Biot stress tensor, show that

$$\mathbf{U}\mathbf{T}^{(1)} - \mathbf{T}^{(1)}\mathbf{U} = \mathbf{U}\boldsymbol{\Omega} + \boldsymbol{\Omega}\mathbf{U},$$

where $\boldsymbol{\Omega}$ is an antisymmetric tensor. Find expressions for the components of $\boldsymbol{\Omega}$ on the Lagrangean principal axes in terms of the principal stretches $\lambda_1, \lambda_2, \lambda_3$ and the corresponding components of $\mathbf{T}^{(1)}$. Deduce that $\boldsymbol{\Omega} = \mathbf{0}$ if and only if the material is isotropic relative to the chosen reference configuration. (The results of Problem 4.2.12 may be helpful.)

Problem 5.1.2 If $\mathbf{S}$ is a self-equilibrated nominal stress field, show that in terms of the stress $\mathbf{T}^{(2)}$ the equilibrium equation may be written as

$$\operatorname{Div} \mathbf{T}^{(2)} - (\mathbf{S}^{\mathrm{T}} \operatorname{Grad})\mathbf{B} = \mathbf{0}.$$

5.1.2 Boundary conditions

We consider a body which occupies the region $\mathscr{B}_0$, with boundary $\partial\mathscr{B}_0$, in some fixed reference configuration. Let $\mathscr{B}$ and $\partial\mathscr{B}$ be the corresponding region and boundary in the current configuration.

(a) The boundary condition of place This is the boundary condition in which the current position $\mathbf{x} = \boldsymbol{\chi}(\mathbf{X})$ is specified for all points $\mathbf{X}$ on $\partial\mathscr{B}_0$. Equivalently the displacement $\mathbf{u} = \mathbf{x} - \mathbf{X}$ may be specified on $\partial\mathscr{B}_0$ and the terminology *displacement boundary condition* is then appropriate. As an example, we consider a thick-walled spherical shell of material whose interior boundary is fixed while the outer boundary is twisted through a given angle about some specified direction through the centre of the shell. The corresponding two-dimensional geometry is depicted in Fig. 5.1. This geometry, in fact, can provide an example of non-uniqueness of solution for the problem with $\boldsymbol{\chi}(\mathbf{X}) = \mathbf{X}$ specified on both inner and outer boundaries. In addition to the trivial solution $\boldsymbol{\chi}(\mathbf{X}) = \mathbf{X}$ for all $\mathbf{X} \in \mathscr{B}_0$ there is a solution corresponding, for example, to rotation of the outer boundary through 2π while the inner boundary remains fixed.

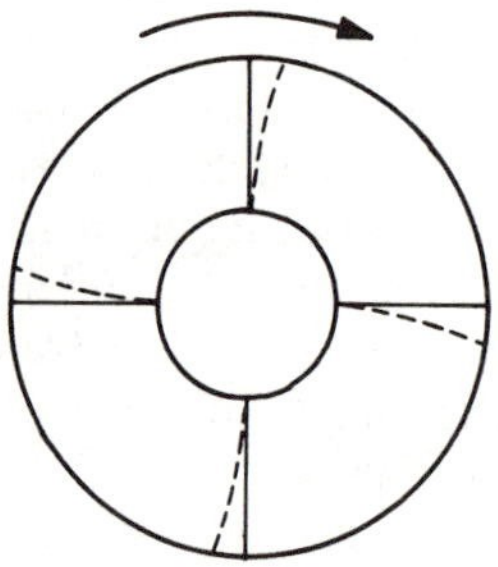

Fig. 5.1 Example of boundary conditions of place. Material in the form of a circular cylindrical annulus has its inner boundary fixed while the outer boundary is rotated through a prescribed angle at fixed radius. The broken curves are the deformed positions of initially radial material lines.

(b) The boundary condition of traction We recall from Section 3.4.1 that the load (or traction) on the current surface $\partial\mathscr{B}$ of a body is $\mathbf{S}^{\mathrm{T}}\mathbf{N}$ per unit area of $\partial\mathscr{B}_0$, where $\mathbf{N}$ is the unit outward normal to $\partial\mathscr{B}_0$. Since $\partial\mathscr{B}_0$ (as distinct from $\partial\mathscr{B}$) is known from the outset, and hence so is $\mathbf{N}$, a possible boundary condition is arrived at by specifying $\mathbf{S}^{\mathrm{T}}\mathbf{N}$ as a function of $\mathbf{X}$ on $\partial\mathscr{B}_0$. Thus, we write

$$\mathbf{S}^{\mathrm{T}}\mathbf{N} = \boldsymbol{\sigma}(\mathbf{X}) \tag{5.1.21}$$

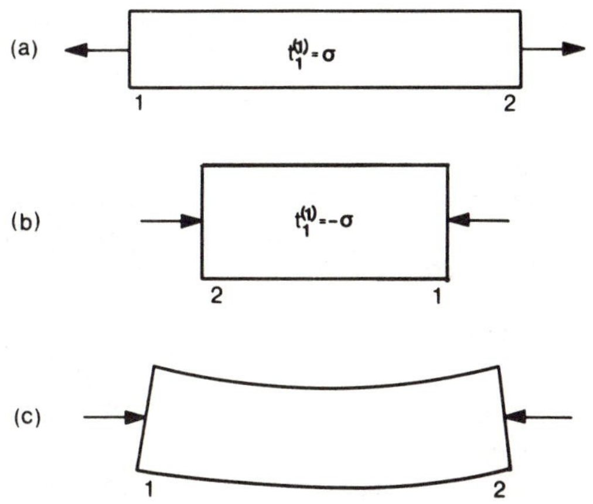

Fig. 5.2 Example of traction boundary condition, showing possible deformed configurations of a circular cylinder of uniform cross-section under uniaxial nominal traction of magnitude σ: (a) uniform extension, (b) uniform contraction, (c) buckled configuration. The ends of the cylinder are labelled 1 and 2.

say, where the vector $\boldsymbol{\sigma}(\mathbf{X})$ depends only on $\mathbf{X}$. The loading described by (5.1.21) is referred to as *dead loading*. In any dead-loading process which takes the body from $\mathscr{B}_0$ to $\mathscr{B}$ an increase in the magnitude of the load changes the deformation but, on the other hand, changes in the surface geometry resulting from the deformation do not have the effect of modifying that load. The importance of dead loading in applications is illustrated by the following example.

Consider a circular cylinder, of uniform cross-section and with plane ends normal to its axis, consisting of isotropic elastic material. We assume that the circular symmetry is maintained during deformation under the action of nominal traction of magnitude $\sigma(>0$, say) on the ends of the cylinder and parallel to the cylinder axis. The lateral surface is taken to be free of traction. Fig. 5.2(a) illustrates a possible deformed configuration corresponding to *extension* of the cylinder. But, as depicted in Fig. 5.2(b), it is possible for the cylinder to *contract* under the same nominal traction. This is now proved.

Let the unit vector $\mathbf{N}$ define the axis of the cylinder (in the reference configuration) so that $\pm\mathbf{N}$ are the normals to the end faces of the cylinder. Then, for this problem, (5.1.21) becomes

$$\mathbf{S}^{\mathrm{T}}\mathbf{N}=\sigma\mathbf{N}$$

on each end of the cylinder. Because the deformation, and hence the stress, is homogeneous it follows from the lateral boundary condition that

$$\mathbf{S}^{\mathrm{T}} = \sigma \mathbf{N} \otimes \mathbf{N} \tag{5.1.22}$$

throughout the cylinder.

But, since the material is isotropic, we have from equation (4.2.68) that $\mathbf{S} = \mathbf{T}^{(1)}\mathbf{R}^{\mathrm{T}}$, where $\mathbf{T}^{(1)}$ is the Biot stress tensor and $\mathbf{R}$ is the rotational part of the deformation. It therefore follows from (5.1.22) that

$$\mathbf{T}^{(1)} = \sigma \mathbf{R}^{\mathrm{T}}\mathbf{N} \otimes \mathbf{N},$$

and, since $\mathbf{T}^{(1)}$ is symmetric, we must have $\mathbf{RN} = \pm \mathbf{N}$. If arbitrary rotations about $\mathbf{N}$ are ignored, we deduce that either $\mathbf{R} = \mathbf{I}$, the identity, or $\mathbf{R} = 2\mathbf{M} \otimes \mathbf{M} - \mathbf{I}$, where $\mathbf{M}$ is an arbitrary unit vector satisfying $\mathbf{M}\cdot\mathbf{N} = 0$, this representing a rotation about $\mathbf{M}$ through π radians.

In the first case (simple tension) the principal Biot stress in the material is tensile and given by

$$\mathbf{N}\cdot(\mathbf{T}^{(1)}\mathbf{N}) = \sigma$$

and this is consistent with extension of the cylinder (Fig. 5.2(a)). In the second case (simple compression)

$$\mathbf{N}\cdot(\mathbf{T}^{(1)}\mathbf{N}) = -\sigma$$

and the Biot stress in the material is compressive, corresponding to contraction of the cylinder (Fig. 5.2(b)).

This provides an example of non-uniqueness of solution to the traction boundary-value problem and, at the same time, an example of non-uniqueness of the inversion of the stress-deformation relation $\mathbf{S} = \mathbf{H}(\mathbf{A})$. The deformation gradients corresponding to the two solutions are associated with very different stretch tensors $\mathbf{U}$ (one with $\mathbf{N}\cdot(\mathbf{UN}) > 1$, the other with $\mathbf{N}\cdot(\mathbf{UN}) < 1$) and are arrived at on two distinct paths of deformation from the reference configuration.

Another type of non-uniqueness arises if the cylinder buckles under compressive stress $-\sigma$ at some critical value of σ, and the cylinder bows (for example), as illustrated in Fig. 5.2(c). In this example, the deformation path (in the space of deformation gradients) is said to *bifurcate* at the point (the bifurcation point) corresponding to the critical value of σ. Beyond the bifurcation point two paths of deformation are possible—(b) and (c) in Fig. 5.2.

More generally than for the dead-loading boundary condition (5.1.21), the traction at a material point $\mathbf{X}$ may depend on the deformation $\boldsymbol{\chi}(\mathbf{X})$ and possibly also on the deformation in a neighbourhood of **X**, as measured through the deformation gradient Grad $\boldsymbol{\chi}(\mathbf{X})$. Thus, we may write

$$\mathbf{S}^{\mathrm{T}}\mathbf{N} = \boldsymbol{\sigma}(\mathbf{X}, \mathbf{x}, \mathbf{A}) \tag{5.1.23}$$

on $\partial\mathcal{B}_0$, where $\mathbf{x} = \boldsymbol{\chi}(\mathbf{X})$ and $\mathbf{A} = \mathrm{Grad}\,\boldsymbol{\chi}(\mathbf{X})$. In fact, $\boldsymbol{\sigma}$ may also be allowed to depend on higher derivatives of $\boldsymbol{\chi}(\mathbf{X})$ but for the purposes of this book (5.1.23) is sufficiently general; for a general discussion of traction boundary conditions, including (5.1.23), see Sewell (1967), who uses the terminology *configuration dependent* to describe such tractions.

The traction (5.1.23) is clearly sensitive to the deformation it produces and is modified continually as the deformation process proceeds. In particular, the traction adapts itself to the changing shape of the boundary $\partial\mathcal{B}$.

As a specific example, we consider loading by fluid pressure $P(\mathbf{x})$ per unit current area. The true traction is then given by

$$\mathbf{Tn} = -P\mathbf{n} \qquad \text{on} \quad \partial\mathcal{B}. \tag{5.1.24}$$

The corresponding nominal traction has the form

$$\mathbf{S}^{\mathrm{T}}\mathbf{N} = -PJ\mathbf{BN} \qquad \text{on} \quad \partial\mathcal{B}_0, \tag{5.1.25}$$

as can be seen by use of (2.2.18) and (3.4.2). Clearly (5.1.25) is a member of the class of tractions defined by (5.1.23) since $J = \det \mathbf{A}$, $\mathbf{B} = (\mathbf{A}^{\mathrm{T}})^{-1}$ and **N** is a function of **X** on the surface $\partial\mathcal{B}_0$. In many cases of practical interest P may be taken as a constant.

For the traction boundary-value problem in general, the boundary data must satisfy the integral forms of the balance equations, namely

$$\int_{\partial\mathcal{B}_0} \boldsymbol{\sigma}\,\mathrm{d}A + \int_{\mathcal{B}_0} \rho_0\mathbf{b}_0\,\mathrm{d}V = \mathbf{0} \tag{5.1.26}$$

and

$$\int_{\partial\mathcal{B}_0} (\mathbf{x} - \mathbf{x}_0) \wedge \boldsymbol{\sigma}\,\mathrm{d}A + \int_{\mathcal{B}_0} \rho_0(\mathbf{x} - \mathbf{x}_0) \wedge \mathbf{b}_0\,\mathrm{d}V = \mathbf{0}, \tag{5.1.27}$$

where $\mathbf{x}_0$ is an arbitrarily chosen origin. Since, in general, $\boldsymbol{\sigma}$ and $\mathbf{b}_0$ depend on the deformation one cannot be certain from the outset that these hold. Therefore they are to be regarded as compatibility conditions to be checked once the solution $\boldsymbol{\chi}(\mathbf{X})$ is found.

(c) Mixed boundary conditions A more general boundary-value problem than those described in (a) and (b) involves specification of the traction on part of the boundary, $\partial\mathscr{B}_0^\sigma$ say, and position on the remainder, $\partial\mathscr{B}_0^x$ say, where $\partial\mathscr{B}_0 = \partial\mathscr{B}_0^\sigma \cup \partial\mathscr{B}_0^x$. Consider, for example, a vertical column of material in the shape of a circular cylinder of uniform cross-section. On the foot of the column impose the boundary condition $\boldsymbol{\chi}(\mathbf{X}) = \mathbf{X}$ and let its lateral surface be free of traction. On the upper surface vertically downwards dead loading may be applied and this can lead to buckling, as shown in Fig. 5.3(a). Equally, a configuration-dependent loading which remains normal to the deforming (plane) upper surface of the column may be applied. A resulting buckled configuration is depicted in Fig. 5.3(b).

A second example, illustrated in Fig. 5.4, involves the squeezing of an

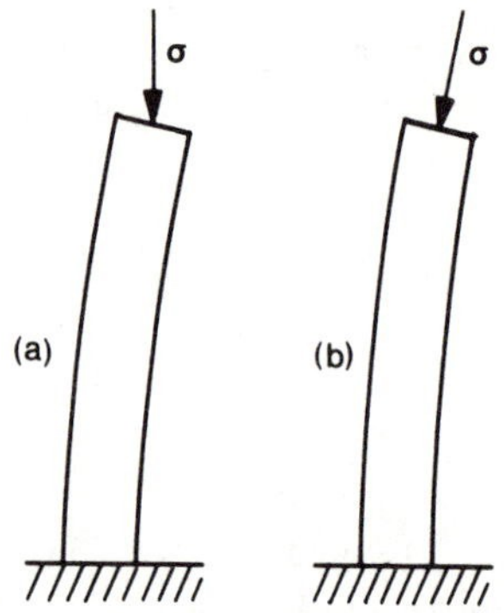

Fig. 5.3 Examples of mixed boundary conditions. Vertical column with fixed base under uniaxial compression, showing buckled configuration: (a) nominal traction, (b) configuration-dependent traction.

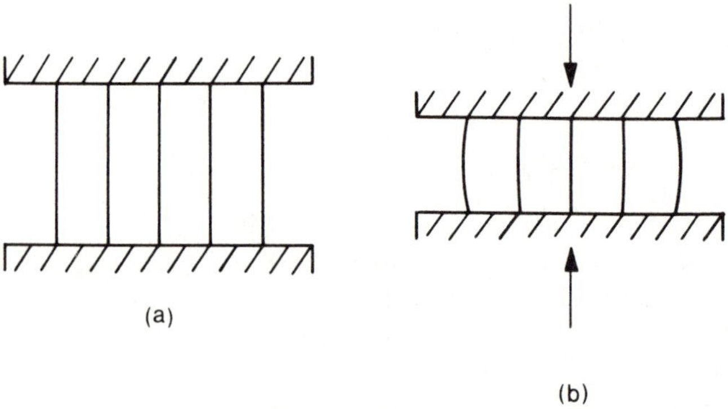

Fig. 5.4 Example of mixed boundary conditions. Rectangular block glued to parallel plates in compression with lateral faces traction free: (a) reference configuration, (b) current configuration.

initially rectangular block of material between two parallel rigid plates which are bonded to the upper and lower surfaces of the block. The displacement of the plates is specified and hence $\boldsymbol{\chi}(\mathbf{X})$ is prescribed on the upper and lower surfaces of the block. The lateral faces of the block are taken to be free of traction.

A third mixed boundary-value problem arises if the geometry shown in Fig. 5.1 is regarded as corresponding to a cross-section of a circular cylindrical tube. Then, the circular cylindrical geometry may be maintained by setting to zero the *axial* displacement along with the *shear* tractions on the (plane) ends of the tube. Vanishing of the shear tractions permits rotation of the tube ends normal to the axis of the cylinder. Non-uniqueness of solution to this problem, analogous to that in (a), is possible.

Finally, we remark that in many problems of practical interest only the resultant traction on a given surface may be specified and not its pointwise values. This eventuality will be catered for as the need arises.

Problem 5.1.3 Show that the rotational balance equation

$$\int_{\partial\mathscr{B}_0} (\mathbf{x}-\mathbf{x}_0) \wedge \mathbf{S}^{\mathrm{T}}\mathbf{N}\,\mathrm{d}A + \int_{\mathscr{B}_0} \rho_0(\mathbf{x}-\mathbf{x}_0) \wedge \mathbf{b}_0\,\mathrm{d}V = \mathbf{0}$$

holds if and only if $\mathbf{L}^{\mathrm{T}} = \mathbf{L}$, where

$$\mathbf{L} = \int_{\partial\mathscr{B}_0} (\mathbf{x}-\mathbf{x}_0) \otimes \mathbf{S}^{\mathrm{T}}\mathbf{N}\,\mathrm{d}A + \int_{\mathscr{B}_0} \rho_0(\mathbf{x}-\mathbf{x}_0) \otimes \mathbf{b}_0\,\mathrm{d}V.$$

If $\mathbf{L}^{\mathrm{T}} \neq \mathbf{L}$ show, by using the polar decomposition theorem for $\mathbf{L}$, that there exists a proper orthogonal tensor $\mathbf{Q}$, independent of $\mathbf{x}$, such that

$$\mathbf{QL} = \mathbf{L}^{\mathrm{T}}\mathbf{Q}^{\mathrm{T}}.$$

This means that if the deformed body is rotated by $\mathbf{Q}$ about $\mathbf{x}_0$ relative to the given system of surface and body forces then the resultant moment about $\mathbf{x}_0$ of these forces vanishes.

5.1.3 Restrictions on the deformation

In Section 5.1.1 we remarked on the fact that the strain-energy function and the resulting tensor $\mathscr{A}^1$ of elastic moduli should have forms which ensure existence of meaningful solutions to boundary-value problems. It is not enough, however, merely to show the existence of some deformation function $\boldsymbol{\chi}$ under given hypotheses on $\mathscr{A}^1$. Any 'solution' $\boldsymbol{\chi}$ must be consistent with certain (physically motivated) requirements. For example, we have the unilateral constraint on deformation gradients laid down in Section 2.2.2, namely

$$\det\{\text{Grad}\,\boldsymbol{\chi}(\mathbf{X})\} > 0. \tag{5.1.28}$$

This is a *local* condition which has to be checked once $\boldsymbol{\chi}$ is found.

For an incompressible material we have

$$\det\{\text{Grad}\,\boldsymbol{\chi}(\mathbf{X})\} = 1 \tag{5.1.29}$$

for every $\mathbf{X} \in \mathscr{B}_0$ and (5.1.28) is then automatically satisfied from the outset.

Another restriction which must be placed on $\boldsymbol{\chi}$ is the *global* analogue of (5.1.28), namely that $\boldsymbol{\chi}$ should be one-to-one on $\mathscr{B}_0$. Thus, for all points $\mathbf{X}$ and $\mathbf{X}'$ on $\mathscr{B}_0$,

$$\boldsymbol{\chi}(\mathbf{X}') = \boldsymbol{\chi}(\mathbf{X}) \qquad \text{if and only if } \mathbf{X}' = \mathbf{X}. \tag{5.1.30}$$

We note that, in general, neither (5.1.28) nor (5.1.30) implies the other although the two conditions are intimately related. Indeed, if $\boldsymbol{\chi}$ is *strictly convex*† on $\mathscr{B}_0$, i.e.

$$\{\boldsymbol{\chi}(\mathbf{X}') - \boldsymbol{\chi}(\mathbf{X})\} \cdot (\mathbf{X}' - \mathbf{X}) > 0 \tag{5.1.31}$$

for $\mathbf{X}' \neq \mathbf{X}$, then (5.1.30) clearly holds and, furthermore,

$$\det\{\text{Grad}\,\boldsymbol{\chi}(\mathbf{X})\} \geq 0$$

with strict equality holding except on a nowhere dense set in $\mathscr{B}_0$ (Bernstein and Toupin, 1962). However, (5.1.31) cannot hold in general, as can be seen by considering $\boldsymbol{\chi}$ to correspond to a rigid rotation which reverses the direction of $\mathbf{X}' - \mathbf{X}$.

The physical interpretation of (5.1.30) is that it excludes interpenetration of matter and mutual contact of different points of $\partial\mathscr{B}$.

Any one of (5.1.28)–(5.1.30) presents difficulties in the analysis of questions such as existence of solutions to boundary-value problems. However, when it comes to deriving solutions to problems of practical interest, by either analytical or numerical methods, it is not too difficult to check whether the resulting solutions do satisfy these restrictions. This is illustrated for particular problems in subsequent sections of this chapter, and, moreover, imposition of (5.1.29) from the outset is shown to lead to considerable simplification in some cases.

† *Strict convexity* is defined in Appendix 1.

5.2 PROBLEMS FOR UNCONSTRAINED MATERIALS

Some difficulties associated with boundary-value problems for non-linear elastic materials have been discussed briefly in Section 5.1. Because of the non-linearity, further difficulties are encountered when attempting to solve the equations for specific geometries and boundary data. Even if a simple form of strain-energy function is used, it is rarely possible to obtain explicit analytic solutions. The problems are perhaps most marked for unconstrained materials since the imposition of internal constraints can often alleviate the difficulties to a certain extent. We therefore devote this section to problems for unconstrained materials possessing an isotropic strain-energy function.

In the first instance we do not tackle boundary-value problems directly in that we do not seek explicit representations for solutions by methods analogous to those used in the linear theory of partial differential equations (e.g. Green's function methods). (On the whole this approach is not successful for non-linear problems.) What we do is adopt an *inverse procedure* whereby we *assume* an explicit form for the deformation (possibly suggested by the geometry in question) and calculate, through the constitutive law and governing equations, the resulting stress distribution and, in particular, the boundary tractions required for equilibrium to be maintained. However, this method does not always work because it is possible to choose a deformation which cannot be maintained by the application of surface tractions alone in respect of a general form of isotropic strain-energy function. A useful result in this context is provided by *Ericksen's Theorem*, a proof of which is given in Section 5.2.1. The theorem states that the *homogeneous deformations* are the only deformations (of an unconstrained isotropic elastic solid) which can be achieved by the application of surface tractions alone *independently of the form of strain-energy function.*

For unconstrained materials the inverse method is clearly of rather limited value. An alternative is the more general *semi-inverse method* in which the deformation is specified to within certain functions of position which are then to be found by application of the equilibrium equations. The nature of these functions depends critically on the choice of strain-energy function. Both the inverse and semi-inverse methods will be illustrated in the following sections.

In using the semi-inverse method we are faced with making a choice of strain-energy function W if any progress is to be made towards a solution of the considered problem. This forces us to address the question, raised in Section 5.1.1, of how to select suitable forms of W. In partial answer to this we make the following points. First, it may be possible to choose W in such a way that closed-form analytic solutions can be derived. Some examples of this method are provided in subsequent sections of this chapter, but we emphasize that such solutions are rare and the outcome of this approach, although of some mathematical merit, is not necessarily compatible with the physics of the problem. Second, it is sensible to use forms of W which have been constructed on the basis of experiments involving homogeneous deformations. In general these W are not simple enough to lead to analytical solutions, and numerical methods are needed in order to extract information of practical value. To a certain extent these two approaches meet on middle ground because the strain-energy functions for which closed-form solutions have been obtained so far are not devoid of physical interpretation, at least for some limited ranges of deformation.

Since non-homogeneous deformations depend on the form of W it is evident that, for a given W, not every deformation can be achieved if the material is subject only to boundary tractions. However, it should be possible in principle to produce an arbitrary deformation in every (unconstrained) elastic material† but, in general, this can only be done by the application of appropriate body forces. Essentially this can be regarded as a localized form of Ericksen's theorem since deformations in an elastic material are locally homogeneous. (Recall, however, that we have assumed that the deformation is twice continuously differentiable, and relaxation of this requirement may affect the above arguments.)

The semi-inverse method, although very useful, is not sufficiently general in scope for wide application in problems of practical interest, particularly for non-simple geometries. It is therefore necessary, in general, to use a direct method. In practice, each boundary-value problem raises its own difficulties and requires individually tailored techniques for its solution. These are illustrated in the following sections.

5.2.1 Ericksen's theorem

Let I_1, I_2, I_3 be the principal invariants of $\mathbf{A}^{\mathrm{T}}\mathbf{A}$ so that, from the results of Problem 4.2.11, we have

$$\frac{\partial I_1}{\partial \mathbf{A}} = 2\mathbf{A}^{\mathrm{T}}, \qquad \frac{\partial I_2}{\partial \mathbf{A}} = 2\{I_1\mathbf{I} - \mathbf{A}^{\mathrm{T}}\mathbf{A}\}\mathbf{A}^{\mathrm{T}}, \qquad \frac{\partial I_3}{\partial \mathbf{A}} = 2I_3\mathbf{B}^{\mathrm{T}}. \quad (5.2.1)$$

For an isotropic strain-energy function W, the nominal stress tensor $\mathbf{S}$ is given by

$$\mathbf{S} = \frac{\partial W}{\partial \mathbf{A}} = \sum_{i=1}^{3} \frac{\partial W}{\partial I_i}\frac{\partial I_i}{\partial \mathbf{A}}$$

with (5.2.1), and, in the absence of body forces the equilibrium equation

†Provided W is sufficiently well behaved.

$\operatorname{Div}\mathbf{S}=\mathbf{0}$ then leads to

$$\sum_{i=1}^{3}\frac{\partial W}{\partial I_i}\operatorname{Div}\left(\frac{\partial I_i}{\partial\mathbf{A}}\right)+\sum_{i=1}^{3}\sum_{j=1}^{3}\frac{\partial^2 W}{\partial I_i\partial I_j}(\operatorname{Grad} I_j)\frac{\partial I_i}{\partial\mathbf{A}}=\mathbf{0}.$$

If this is to hold for arbitrary W we must have

$$\operatorname{Div}\left(\frac{\partial I_i}{\partial\mathbf{A}}\right)=\mathbf{0}\qquad(i=1,2,3)\tag{5.2.2}$$

and

$$(\operatorname{Grad} I_j)\frac{\partial I_i}{\partial\mathbf{A}}+(\operatorname{Grad} I_i)\frac{\partial I_j}{\partial\mathbf{A}}=\mathbf{0}\qquad(i,j=1,2,3).\tag{5.2.3}$$

Since $\mathbf{A}$ and $I_1\mathbf{I}-\mathbf{A}^{\mathrm{T}}\mathbf{A}$ are non-singular the choice $i=j$ in (5.2.3) shows that

$$\operatorname{Grad} I_i=\mathbf{0}\qquad(i=1,2,3).\tag{5.2.4}$$

It then follows from the identity $\operatorname{Div}(J\mathbf{B}^{\mathrm{T}})=\mathbf{0}$ that for $i=3$ equation (5.2.2) is automatically satisfied. For $i=1$, we obtain

$$\operatorname{Div}\mathbf{A}^{\mathrm{T}}=\mathbf{0}\tag{5.2.5}$$

while for $i=2$ equation (5.2.2) then leads to

$$\operatorname{Div}(\mathbf{A}^{\mathrm{T}}\mathbf{A}\mathbf{A}^{\mathrm{T}})=\mathbf{0}.\tag{5.2.6}$$

In Cartesian components we write $A_{i\alpha}=x_{i,\alpha}\equiv\partial x_i/\partial X_\alpha$ and we note that $\operatorname{Grad} I_1=\mathbf{0}$ is expressible as

$$x_{i,\alpha}x_{i,\alpha\beta}=0\tag{5.2.7}$$

and (5.2.5) as

$$x_{i,\beta\beta}=0.\tag{5.2.8}$$

From (5.2.7) and (5.2.8) it follows that

$$x_{i,\alpha\beta}x_{i,\alpha\beta}=0,$$

and since the left-hand side of this equation represents a sum of squares we deduce that $x_{i,\alpha\beta}=0$. In tensor notation this is equivalent to

$$\text{Grad}\,\mathbf{A}=\mathbf{0}, \tag{5.2.9}$$

which means that the deformation is homogeneous. Note that we have used neither (5.2.6) nor (5.2.4) with $i=2$ in this proof but that these equations follow from the result (5.2.9).

The above establishes that *a deformation can be maintained* (by the application of surface tractions) *without body forces for arbitrary W if and only if it is homogeneous.* This is a statement of Ericksen's theorem. The required boundary tractions (which depend on W) may be calculated from the stress-deformation relation $\mathbf{S}=\partial W/\partial\mathbf{A}$ when the reference geometry of the material is specified.

Some examples of homogeneous deformations were studied in Section 4.4 for particular forms of W and we do not consider them further here. Attention is now turned to non-homogeneous deformations.

5.2.2 Spherically symmetric deformation of a spherical shell

We consider a thick-walled spherical shell whose reference geometry is defined by

$$A\le R\le B,\qquad 0\le\Theta\le\pi,\qquad 0\le\Phi\le 2\pi$$

in terms of spherical polar coordinates (R,Θ,Φ). As in Problem 2.2.16 we describe the current geometry by

$$a\le r\le b,\qquad \theta=\Theta,\quad \phi=\Phi,$$

in terms of spherical polar coordinates (r,θ,ϕ), and the deformation by

$$\mathbf{x}=f(R)\mathbf{X} \tag{5.2.10}$$

so that

$$r=f(R)R. \tag{5.2.11}$$

Since the material is unconstrained the function f is unknown and, given appropriate boundary conditions, the problem essentially is to find it. The deformation gradient $\mathbf{A}$ is given by

$$\mathbf{A}=f(R)\mathbf{I}+\frac{1}{R}f'(R)\mathbf{X}\otimes\mathbf{X}, \tag{5.2.12}$$

the symmetry of the geometry ensuring that $\mathbf{A}=\mathbf{U}$ (the right stretch tensor).

The principal stretches $\lambda_1, \lambda_2, \lambda_3$, corresponding to coordinate directions r, θ, ϕ respectively, are

$$\lambda_1 = Rf'(R) + f(R), \qquad \lambda_2 = \lambda_3 = f(R). \tag{5.2.13}$$

We assume that the reference configuration is stress free and that the material is isotropic relative to the reference configuration. From the spherical symmetry it follows that the nominal stress is given by $\mathbf{S} = \mathbf{T}^{(1)}$, where the Biot stress may be written

$$\mathbf{T}^{(1)} = t_1^{(1)}\mathbf{X}\otimes\mathbf{X}/R^2 + t_2^{(1)}(\mathbf{I} - \mathbf{X}\otimes\mathbf{X}/R^2) \tag{5.2.14}$$

in terms of its principal components $t_1^{(1)}$ and $t_2^{(1)} = t_3^{(1)}$.

On use of the formula (1.5.54) applied with R, Θ, Φ as independent variables, we obtain the equilibrium equation $\operatorname{Div}\mathbf{S} = \mathbf{0}$ in the form

$$\frac{\mathrm{d}t_1^{(1)}}{\mathrm{d}R} + \frac{2}{R}(t_1^{(1)} - t_2^{(1)}) = 0. \tag{5.2.15}$$

Equivalently, (5.2.15) may be written

$$\frac{\mathrm{d}t_1}{\mathrm{d}r} + \frac{2}{r}(t_1 - t_2) = 0 \tag{5.2.16}$$

in terms of the corresponding principal components of Cauchy stress. In what follows we find it convenient to use (5.2.15) rather than the more common (5.2.16).

For the strain-energy function $W(\lambda_1, \lambda_2, \lambda_3)$ we have

$$t_1^{(1)} = \frac{\partial W}{\partial \lambda_1}, \qquad t_2^{(1)} = \frac{\partial W}{\partial \lambda_2}, \tag{5.2.17}$$

(see (4.3.41)), evaluated for λ_1, λ_2, λ_3 given by (5.2.13). Substitution of (5.2.17) and (5.2.13) into (5.2.15) yields a non-linear second-order ordinary differential equation for $f(R)$. It may be written

$$\frac{\partial^2 W}{\partial \lambda_1^2} Rf''(R) + 2\left(\frac{\partial^2 W}{\partial \lambda_1^2} + \frac{\partial^2 W}{\partial \lambda_1 \partial \lambda_2}\right) f'(R) + \frac{2}{R}\left(\frac{\partial W}{\partial \lambda_1} - \frac{\partial W}{\partial \lambda_2}\right) = 0, \tag{5.2.18}$$

where each of the derivatives of W depends non-linearly on $f(R)$ and $f'(R)$. In order to solve this equation an explicit form of W needs to be chosen. In general it is not possible to obtain analytic solutions even for relatively simple forms of W, but we show below that for a particular form of W such a solution can be derived.

We consider the strain-energy function defined by

$$W = F(I_1) - \nu I_2 + \mu I_3, \tag{5.2.19}$$

where

$$\begin{aligned} I_1 &= \lambda_1 + \lambda_2 + \lambda_3, \\ I_2 &= \lambda_2\lambda_3 + \lambda_3\lambda_1 + \lambda_1\lambda_2, \\ I_3 &= \lambda_1\lambda_2\lambda_3 \end{aligned} \tag{5.2.20}$$

are the principal invariants of $\mathbf{U}$, μ and ν are constants, and F is a function whose properties need not be specified at this stage. The motivation for using (5.2.19) is that it is a generalization to three dimensions of the class of so-called 'harmonic' materials used in plane-strain theory. These lead to a number of explicit solutions for plane-strain problems, as we see in Section 5.2.6.

The principal Biot stresses are calculated as

$$\left.\begin{aligned} t_1^{(1)} &= F'(I_1) - \nu(\lambda_2 + \lambda_3) + \mu\lambda_2\lambda_3, \\ t_2^{(1)} &= F'(I_1) - \nu(\lambda_1 + \lambda_3) + \mu\lambda_3\lambda_1. \end{aligned}\right\} \tag{5.2.21}$$

Since the reference configuration ($\lambda_1 = \lambda_2 = \lambda_3 = 1$) is stress free we must have

$$\left.\begin{aligned} F(3) - 3\nu + \mu &= 0, \\ F'(3) - 2\nu + \mu &= 0 \end{aligned}\right\} \tag{5.2.22}$$

from (5.2.19) and (5.2.21) respectively.

Substitution of (5.2.21) into (5.2.15) leads to

$$\frac{\mathrm{d}}{\mathrm{d}R}\{F'(I_1) - \nu I_1\} + \nu\left\{\frac{\mathrm{d}\lambda_1}{\mathrm{d}R} + \frac{2}{R}(\lambda_1 - \lambda_2)\right\} = 0 \tag{5.2.23}$$

on use of (5.2.13). Note that μ does not appear in this equation. This can be expected from the form of (5.2.19) because of the kinematical identity $\mathrm{Div}\,(\partial I_3/\partial \mathbf{A}) = \mathbf{0}$.

Since $\lambda_3 = \lambda_2$ we also have the result

$$\frac{\mathrm{d}I_1}{\mathrm{d}R} = \frac{\mathrm{d}\lambda_1}{\mathrm{d}R} + \frac{2}{R}(\lambda_1 - \lambda_2) \tag{5.2.24}$$

so that (5.2.23) reduces to

$$F''(I_1)\frac{\mathrm{d}I_1}{\mathrm{d}R}=0. \tag{5.2.25}$$

Two possibilities arise: *either* (a)

$$F(I_1)=(2\nu-\mu)I_1-(3\nu-2\mu) \tag{5.2.26}$$

for consistency with (5.2.22), in which case no restriction is imposed on $f(R)$, i.e. the equilibrium equations are satisfied for arbitrary $f(R)$ in respect of the strain-energy function (5.2.19) with (5.2.26), *or* (b) $F(I_1)$ is arbitrary, subject to (5.2.22), and

$$\frac{\mathrm{d}I_1}{\mathrm{d}R}\equiv\frac{\mathrm{d}}{\mathrm{d}R}\left(R\frac{\mathrm{d}f}{\mathrm{d}R}+3f\right)=0. \tag{5.2.27}$$

The general solution of (5.2.27) is expressible as

$$f(R)=\tfrac{1}{3}\alpha+\beta/R^3, \tag{5.2.28}$$

where α and β are constants with $I_1=\alpha$. Since the principal stretches must be positive it follows that

$$\alpha>0, \tag{5.2.29}$$

while

$$\lambda_1=\tfrac{1}{3}\alpha-2\beta/R^3, \qquad \lambda_2=\tfrac{1}{3}\alpha+\beta/R^3. \tag{5.2.30}$$

If λ_1 and λ_2 are to be positive for $A\le R\le B$, the constant β must be restricted according to

$$-\tfrac{1}{3}\alpha A^3<\beta<\tfrac{1}{6}\alpha A^3. \tag{5.2.31}$$

It remains to calculate α and β from the boundary conditions on $R=A$ and $R=B$. To illustrate the procedure for obtaining α and β we first of all prescribe the inner radius a in the current configuration so that

$$\tfrac{1}{3}\alpha+\beta/A^3=a/A \tag{5.2.32}$$

is known. Secondly, we take the outer boundary $R=B$ to be free of traction so that, from (5.2.21),

$$t_1^{(1)}\equiv F'(\alpha)-2\nu\lambda+\mu\lambda^2=0, \tag{5.2.33}$$

where

$$\lambda = \tfrac{1}{3}\alpha + \beta/B^3. \tag{5.2.34}$$

In order to make further progress it is necessary to examine the signs of the constants μ and ν and the qualitative nature of $F'(I_1)$. To this end, we consider a pure shear deformation with $\lambda_3 = 1$ and $t_2^{(1)} = 0$, so that

$$t_1^{(1)} = F'(I_1) - \nu(\lambda_2 + 1) + \mu\lambda_2, \tag{5.2.35}$$

$$0 = F'(I_1) - \nu(\lambda_1 + 1) + \mu\lambda_1 \tag{5.2.36}$$

and hence, on elimination of $F'(I_1)$,

$$t_1^{(1)} = (\nu - \mu)(\lambda_1 - \lambda_2).$$

On physical grounds one can expect λ_1 to increase with $t_1^{(1)}$ and λ_2 to decrease correspondingly since $t_2^{(1)} = 0$. Thus $\lambda_1 - \lambda_2 > 0$ for $\lambda_1 > 1$. This indicates that $\nu - \mu > 0$. For definiteness we set

$$\nu > 0, \qquad \mu < 0 \tag{5.2.37}$$

with the possibility that either $\nu = 0$ or $\mu = 0$ (but not both).

It follows from (5.2.36) and (5.2.37) that $F'(I_1)$ increases monotonically with λ_1 in pure shear, and, since $I_1 = \lambda_1 + \lambda_2 + 1$, equation (5.2.36) determines λ_2 as a function of λ_1 *in principle*. For λ_2 to be determined uniquely by $\lambda_1, F'(I_1)$ must be a monotonic increasing function of I_1 and hence

$$F''(I_1) > 0 \tag{5.2.38}$$

for all $I_1 > 0$. Consideration of simple tension or pure dilatation, for example, also leads to the results (5.2.37) and (5.2.38).

From (5.2.31) and the fact that $A/B < 1$, we obtain the further inequalities

$$0 < F'(\alpha) < \nu\alpha - \tfrac{1}{4}\mu\alpha^2 \tag{5.2.39}$$

for all $\alpha > 0$. Clearly, it is possible to choose forms of $F(\alpha)$ which violate the right-hand inequality in (5.2.39) in particular for large enough α. In this sense there is a limitation on the range of deformations to which the strain-energy function (5.2.19) is applicable. Nevertheless the range may well be large enough to include the limited elastic response of some materials and (5.2.19), as a model of non-linear elastic behaviour, therefore has its attractions, not least the fact that it leads to the solution (5.2.28).

On the other hand it is also possible to choose $F(\alpha)$ so that the inequalities

(5.2.39) are satisfied for all $\alpha > 0$. A case in point is

$$F(I_1) = \tfrac{1}{3}\nu I_1^2 - \tfrac{1}{27}\mu I_1^3, \tag{5.2.40}$$

which is consistent with (5.2.22) and (5.2.38). This turns out to be a very interesting pathological form of $F(I_1)$. Indeed, substitution of (5.2.40) into (5.2.33) and elimination of β by means of (5.2.32) leads to the conclusion $\alpha = 3a/A$, and hence $\beta = 0$. The deformation in the spherical shell is therefore homogeneous and locally a pure dilatation. Since $I_1 = \alpha$ and $\lambda_1 = \frac{1}{3}\alpha$ it follows from (5.2.21) that $t_1^{(1)} = t_2^{(1)} = 0$ identically for $A \le R \le B$, *but no restriction is placed on* α. Clearly, the deformation is uniquely determined by the prescribed value of a/A since $r = aR/A$, and no stress is required to maintain this deformation. But, for the zero-traction boundary-value problem, any positive value of a/A is consistent with the boundary conditions and there exists a continuum of solutions.

We now consider a more realistic form of $F(I_1)$, namely

$$F(I_1) = \tfrac{1}{27}(2\nu - \mu)I_1^3 + \nu. \tag{5.2.41}$$

Use of this in (5.2.33) with the help of (5.2.32) leads to a unique value of α consistent with (5.2.29). This is given by

$$\tfrac{2}{3}\alpha = \frac{(\nu - \mu\eta a/A)(1-\eta) + [\nu^2(1-\eta)^2 + 2a\nu\eta(2\nu-\mu)/A - \mu a^2\eta^2(2\nu-\mu)/A^2]^{1/2}}{[2\nu - \mu\eta - \mu\eta(1-\eta)]}$$

where $\eta = A^3/B^3 < 1$.

The value of β is then obtained from (5.2.32) and the radial stress on $R = A$ is given by

$$t_1^{(1)} = \tfrac{1}{9}(2\nu - \mu)\alpha^2 - 2\nu a/A + \mu a^2/A^2.$$

It is also worth noting that, from (5.2.33), we also have

$$t_1^{(1)} = \left[2\nu - \mu\left(\lambda + \frac{a}{A}\right)\right]\left(\lambda - \frac{a}{A}\right)$$

and it follows that $t_1^{(1)} \gtreqless 0$ according as $\beta \lesseqgtr 0$. We note finally that (5.2.41) may violate (5.2.39) unless $\mu \le -8\nu/5$ so if this inequality is not satisfied the above solution is confined to values of α in the range

$$0 < \alpha < 36\nu/(8\nu + 5\mu).$$

The reader may find it instructive to examine the properties of $t_1^{(1)}$ as a function of a/A for a general value of η. Because of the cumbersome algebra involved in this exercise we do not provide details here, but the limiting case of a thin-walled shell ($\eta \to 1$) is now examined.

We write $(B-A)/A = \frac{1}{3}\varepsilon > 0$ and suppose that $\varepsilon \ll 1$. Then $\eta = 1 - \varepsilon + O(\varepsilon^2)$. This leads to

$$\tfrac{1}{3}\alpha = \left[\frac{(2\nu - \mu a/A)a/A}{(2\nu - \mu)}\right]^{1/2} + \frac{(\nu - \mu a/A)}{(2\nu - \mu)}\left[1 - \left\{\frac{(2\nu - \mu)a/A}{2\nu - \mu a/A}\right\}^{1/2}\right]\varepsilon + O(\varepsilon^2)$$

and hence

$$t_1^{(1)} = 2(\nu - \mu a/A)\left[\frac{(2\nu - \mu a/A)a/A}{(2\nu - \mu)}\right]^{1/2} \times \left[1 - \left\{\frac{(2\nu - \mu)a/A}{2\nu - \mu a/A}\right\}^{1/2}\right]\varepsilon + O(\varepsilon^2) \qquad (5.2.42)$$

It follows from the inequalities (5.2.37) that $t_1^{(1)} < 0$ when $a/A > 1$, i.e. an internal pressure is required to effect an expansion of the spherical shell.

The corresponding stress in the skin of the thin shell is

$$t_2^{(1)} = 3(\nu - \mu a/A)\left[a/A - \left\{\frac{(2\nu - \mu a/A)a/A}{(2\nu - \mu)}\right\}^{1/2}\right] + O(\varepsilon), \qquad (5.2.43)$$

calculated from (5.2.21) with $\lambda_2 = \lambda_3$ and $\lambda_1 = \alpha - 2a/A$, and $t_2^{(1)}$ is strictly positive for $a/A > 1$, i.e. the skin is in tension.

It follows from (5.2.42) and (5.2.43) that

$$-t_1^{(1)} = \tfrac{2}{3}\varepsilon t_2^{(1)}[1 + O(\varepsilon)]. \qquad (5.2.44)$$

In fact, this connection between $t_1^{(1)}$ and $t_2^{(1)}$ can be obtained directly by integration of the differential equation

$$\frac{\mathrm{d}}{\mathrm{d}R}(R^2 t_1^{(1)}) = 2R t_2^{(1)}$$

(a rearrangement of (5.2.15)) subject to $t_1^{(1)} = 0$ on $R = B$. In terms of Cauchy stress and the internal pressure P, equation (5.2.44) becomes

$$P \equiv -t_1 = \tfrac{2}{3}\varepsilon\lambda_1\lambda_2^{-1}t_2[1 + O(\varepsilon)], \qquad (5.2.45)$$

with λ_1, λ_2 and t_2 evaluated for $R = A$, say. This leads to the classical approximation $P = 2T/a$ for the pressure inside a thin spherical elastic shell, where $T \equiv (B - A)\lambda_1 t_2$ is the surface tension in the shell skin and a is the current radius.

In this section we have seen that a particular choice of strain-energy function can lead to explicit solutions of the governing differential equations. It is possible that choices of W other than (5.2.19), probably expressed in terms of invariants different from I_1, I_2, I_3, will also permit integration of the equations (5.2.18) in closed form, but we do not pursue the matter here. In general, however, it will be necessary to use forms of W which are not constructed artifically with the above objective and then the solution of (5.2.18) will require standard numerical methods. Such methods are not discussed here.

5.2.3 Extension and inflation of a circular cylindrical tube

A circular cylindrical tube has reference geometry described by

$$A \le R \le B, \qquad 0 \le \Theta \le 2\pi, \qquad 0 \le Z \le L,$$

where R, Θ, Z are cylindrical polar coordinates. It is deformed symmetrically in such a way that, in terms of current cylindrical polar coordinates r, θ, z,

$$r = Rf(R), \qquad \theta = \Theta, \qquad z = \lambda_3 Z, \tag{5.2.46}$$

where λ_3 is a constant and

$$a \le r \le b, \qquad 0 \le z \le l = \lambda_3 L.$$

In general $f(R)$ is unknown at the outset and hence the current internal and external radii a and b of the tube are unknown unless specified as part of the boundary data. (Note that for isochoric deformations $Rf(R)$ takes on the special form described in Section 2.2.6.)

The principal stretches corresponding to the R, Θ, Z directions are

$$\lambda_1 = Rf'(R) + f(R), \qquad \lambda_2 = f(R), \qquad \lambda_3 = \text{constant} \tag{5.2.47}$$

respectively, and the equilibrium equations reduce to

$$\frac{\mathrm{d}t_1^{(1)}}{\mathrm{d}R} + \frac{1}{R}(t_1^{(1)} - t_2^{(1)}) = 0 \tag{5.2.48}$$

analogously to (5.2.15). The principal stresses $t_1^{(1)}$ and $t_2^{(1)}$ are given by (5.2.17)

For the strain-energy function (5.2.19), equation (5.2.48) simplifies to

$$F''(I_1)\frac{\mathrm{d}}{\mathrm{d}R}(\lambda_1 + \lambda_2) = 0,$$

with

$$I_1 = \lambda_1 + \lambda_2 + \lambda_3$$

subject to (5.2.47). Provided $F(I_1)$ is not given by (5.2.26), the solution of this is

$$r = \tfrac{1}{2}(\alpha - \lambda_3)R + \beta/R, \tag{5.2.49}$$

with $I_1 = \alpha$ and β constants. The principal stretches are then

$$\lambda_1 = \tfrac{1}{2}(\alpha - \lambda_3) - \beta/R^2, \qquad \lambda_2 = \tfrac{1}{2}(\alpha - \lambda_3) + \beta/R^2, \tag{5.2.50}$$

and these are strictly positive for $A \leq R \leq B$ provided

$$-\tfrac{1}{2}(\alpha - \lambda_3) < \beta/A^2 < \tfrac{1}{2}(\alpha - \lambda_3). \tag{5.2.51}$$

Just as in Section 5.2.2 it remains to calculate α and β. If the axial stretch λ_3 is known then the resultant axial load N is obtained from

$$N = 2\pi \int_A^B t_3^{(1)} R\,dR$$

and, since

$$t_3^{(1)} = F'(\alpha) - \nu\alpha + \nu\lambda_3 + \mu\lambda_1\lambda_2,$$

it follows with the help of (5.2.50) that

$$N = \pi(B^2 - A^2)[F'(\alpha) - \nu\alpha + \nu\lambda_3 + \tfrac{1}{4}\mu(\alpha - \lambda_3)^2 - \mu\beta^2/A^2B^2]. \tag{5.2.52}$$

Alternatively, if N is prescribed, this equation determines λ_3 when α and β are known.

In order to obtain α and β we consider boundary conditions on the cylindrical surfaces $R = A, B$. For example, we may impose the conditions

$$t_1^{(1)} = -P\lambda_2\lambda_3 \quad \text{on} \quad R = A \tag{5.2.53}$$

$$t_1^{(1)} = 0 \quad \text{on } R = B \tag{5.2.54}$$

which state that the interior boundary is subject to pressure P per unit current area and the exterior boundary is traction free. Since

$$t_1^{(1)} = F'(\alpha) - \nu\alpha + \nu\lambda_1 + \mu\lambda_2\lambda_3$$

these lead to

$$F'(\alpha) - \nu\alpha + \tfrac{1}{2}(\nu + \mu\lambda_3)(\alpha - \lambda_3) - (\nu - \mu\lambda_3)\beta/B^2 = 0 \tag{5.2.55}$$

$$F'(\alpha) - \nu\alpha + \tfrac{1}{2}[\nu + (\mu + P)\lambda_3](\alpha - \lambda_3) - [\nu - (\mu + P)\lambda_3]\beta/A^2 = 0. \tag{5.2.56}$$

Hence β is obtained as a function of α and, in principle, α as a function of P. In practice, however, it is not easy to obtain α explicitly as a function of P. Moreover, the relationship between P and α is not necessarily monotonic and α may not be uniquely determined by P. This aspect of the theory is elaborated on in Section 5.3 in respect of incompressible materials for which algebraic manipulations are somewhat easier than those involved here.

For the special case $P = 0$ it follows from (5.2.55) and (5.2.56) that $\beta = 0$, $\lambda_1 = \lambda_2 = \text{constant}$, and

$$F'(\alpha) - \nu\alpha + \tfrac{1}{2}(\nu + \mu\lambda_3)(\alpha - \lambda_3) = 0.$$

This determines α for given λ_3. If $\lambda_3 = \lambda_1 = \lambda_2$ in this problem then $N = 0$ and the zero traction problem has a continuum of solutions for the form of F given by (5.2.40), a result analogous to that discussed in Section 5.2.2.

If, instead of imposing the boundary condition (5.2.53), we specify the internal radius as $a = \lambda_a A$ then λ_a, the value of λ_2 on $R = A$, is given by

$$\tfrac{1}{2}(\alpha - \lambda_3) + \beta/A^2 = \lambda_a, \tag{5.2.57}$$

this providing an explicit connection between α and β. Elimination of β between (5.2.55) and (5.2.57) leads to

$$\lambda_a = \tfrac{1}{2}(\alpha - \lambda_3) + \frac{B^2}{A^2(\nu - \mu\lambda_3)}[F'(\alpha) - \nu\alpha + \tfrac{1}{2}(\nu + \mu\lambda_3)(\alpha - \lambda_3)]$$

and hence

$$(\nu - \mu\lambda_3)\frac{A^2}{B^2}\frac{\mathrm{d}\lambda_a}{\mathrm{d}\alpha} = F''(\alpha) - \tfrac{1}{2}(\nu - \mu\lambda_3)(1 - A^2/B^2). \tag{5.2.58}$$

Bearing in mind the inequalities (5.2.37) and (5.2.38) we conclude that α is uniquely determined by λ_a provided

$$0 < F''(\alpha) - \tfrac{1}{2}(\nu - \mu\lambda_3)(1 - A^2/B^2) < \infty. \tag{5.2.59}$$

For the problem in question this strengthens (5.2.38), but notice that (5.2.59) depends on λ_3 and also, through the ratio A/B, on the geometry of the body.

Problem 5.2.1 Investigate the properties of $F(\alpha)$ and its derivatives which ensure that $\lambda_3 < 1$ when $N = 0$ and $\lambda_a > 1$, i.e. the tube shortens when inflated in the absence of end loading.

Problem 5.2.2 For a thin-walled tube obtain results analogous to those discussed in Section 5.2.2 for a spherical shell.

Problem 5.2.3 Use specific forms of F, consistent with (5.2.22) and (5.2.38), to obtain α explicitly as a function of λ_a or P in the problem considered above and investigate the monotonicity or otherwise of P as a function of λ_a.

5.2.4 Bending of a rectangular block into a sector of a circular tube

The geometry for this problem is described in Section 2.2.6 and we recall that the deformation is defined by

$$r = f(X_1), \qquad \theta = g(X_2), \qquad z = \lambda_3 X_3, \tag{5.2.60}$$

where r, θ, z are current cylindrical polar coordinates and X_1, X_2, X_3 are reference rectangular Cartesian coordinates with

$$-A \le X_1 \le A, \qquad -B \le X_2 \le B, \qquad -C \le X_3 \le C. \tag{5.2.61}$$

We assume that the deformation is symmetric about the X_1-axis so that

$$g(-X_2) = -g(X_2). \tag{5.2.62}$$

The deformation gradient may be written in the polar-decomposed form

$$\begin{aligned}\mathbf{A} = (\mathbf{e}_r \otimes \mathbf{E}_1 + \mathbf{e}_\theta \otimes \mathbf{E}_2 + \mathbf{e}_z \otimes \mathbf{E}_3)[f'(X_1)\mathbf{E}_1 \otimes \mathbf{E}_1 \\ + f(X_1)g'(X_2)\mathbf{E}_2 \otimes \mathbf{E}_2 + \lambda_3 \mathbf{E}_3 \otimes \mathbf{E}_3],\end{aligned} \tag{5.2.63}$$

the two factors of which correspond to the rotation $\mathbf{R}$ and right stretch $\mathbf{U}$ respectively.

Clearly, the Lagrangean principal axes coincide with the (fixed) Cartesian basis vectors $\mathbf{E}_1$, $\mathbf{E}_2$, $\mathbf{E}_3$ and the principal stretches are

$$\lambda_1 = f'(X_1), \qquad \lambda_2 = f(X_1)g'(X_2), \qquad \lambda_3 = \text{constant}. \tag{5.2.64}$$

It follows that the Biot stress is given by

$$\mathbf{T}^{(1)} = t_1^{(1)}\mathbf{E}_1 \otimes \mathbf{E}_1 + t_2^{(1)}\mathbf{E}_2 \otimes \mathbf{E}_2 + t_3^{(1)}\mathbf{E}_3 \otimes \mathbf{E}_3,$$

where

$$t_i^{(1)} = \frac{\partial W}{\partial \lambda_i} \qquad i = 1, 2, 3, \tag{5.2.65}$$

and the nominal stress by

$$\mathbf{S} = \mathbf{T}^{(1)}\mathbf{R}^{\mathrm{T}} = t_1^{(1)}\mathbf{E}_1 \otimes \mathbf{e}_r + t_2^{(1)}\mathbf{E}_2 \otimes \mathbf{e}_\theta + t_3^{(1)}\mathbf{E}_3 \otimes \mathbf{e}_z. \tag{5.2.66}$$

Since there is no dependence on X_3, we obtain from (5.2.66) the identity

$$\mathrm{Div}\,\mathbf{S} \equiv \frac{\partial}{\partial X_1}(t_1^{(1)}\mathbf{e}_r) + \frac{\partial}{\partial X_2}(t_2^{(1)}\mathbf{e}_\theta)$$

and because

$$\mathbf{e}_r = \cos\theta\mathbf{E}_1 + \sin\theta\mathbf{E}_2, \qquad \mathbf{e}_\theta = -\sin\theta\mathbf{E}_1 + \cos\theta\mathbf{E}_2,$$

with $\theta = g(X_2)$, the equilibrium equation $\mathrm{Div}\,\mathbf{S} = \mathbf{0}$ yields the two scalar equations

$$\frac{\partial t_1^{(1)}}{\partial X_1} - t_2^{(1)} g'(X_2) = 0, \tag{5.2.67}$$

$$\frac{\partial t_2^{(1)}}{\partial X_2} = 0, \tag{5.2.68}$$

Since λ_1 is independent of X_2 it follows from (5.2.68) that

$$\frac{\partial t_2^{(1)}}{\partial \lambda_2} f(X_1) g''(X_2) = 0.$$

With the assumption $f(X_1) \neq 0$ and $\partial t_2^{(1)}/\partial \lambda_2 \neq 0$, we deduce with the help of (5.2.62) that

$$g'(X_2) = \alpha, \qquad g(X_2) = \alpha X_2, \tag{5.2.69}$$

where α is a constant. Thus

$$\lambda_2 = \alpha f(X_1), \qquad \alpha\lambda_1 = \frac{\mathrm{d}\lambda_2}{\mathrm{d}X_1}. \tag{5.2.70}$$

Equation (5.2.67) now reduces to

$$\frac{dt_1^{(1)}}{dX_1} = \alpha t_2^{(1)} \tag{5.2.71}$$

and is independent of X_2.

The traction on a surface $r = \text{constant}$ is

$$2C\int_{-B}^{B} t_1^{(1)}\mathbf{e}_r\,dX_2.$$

By use of $\mathbf{e}_r = \cos\theta\mathbf{E}_1 + \sin\theta\mathbf{E}_2$ and the connection $\theta = \alpha X_2$ this reduces to

$$\frac{4}{\alpha}C\sin(\alpha B)t_1^{(1)}\mathbf{E}_1,$$

$t_1^{(1)}$ being independent of X_2. If this vanishes on both curved boundaries of the body then

$$t_1^{(1)} = 0 \qquad \text{on} \qquad X_1 = \pm A \tag{5.2.72}$$

(the possibility that $\sin\alpha B = 0$ may also be considered but we do not pursue this here).

The load on the faces $\theta = \pm\alpha B$ is purely normal to those faces and has magnitude given by

$$2C\int_{-A}^{A} t_2^{(1)}\,dX_1 = \frac{2C}{\alpha}[t_1^{(1)}]_{-A}^{A},$$

use having been made of (5.2.71). It follows that this load vanishes if the boundary conditions (5.2.72) are prescribed. The load on the faces $z = \pm\lambda_3 C$ is

$$2B\int_{-A}^{A} t_3^{(1)}\,dX_1 \tag{5.2.73}$$

and this can be calculated for given λ_3 once $f(X_1)$ has been determined.

Although the resultant load on each of $\theta = \pm\alpha B$ vanishes the value of $t_2^{(1)}$ is non-zero except possibly at isolated points. The moment of these stresses about the origin has magnitude

$$2C\int_{-A}^{A} rt_2^{(1)}\,dX_1 \tag{5.2.74}$$

and it is this which is required to hold the body in its current configuration. Note that this is independent of X_2.

In order to illustrate the theory we use the strain-energy function (5.2.19) once more. Substitution of the principal stresses (5.2.21) into (5.2.71) and use of (5.2.70) leads to

$$\frac{\mathrm{d}}{\mathrm{d}X_1}F'(I_1)-\alpha F'(I_1)=-\nu\alpha\lambda_3$$

which integrates to give

$$F'(I_1)=\nu\lambda_3+\beta \mathrm{e}^{\alpha X_1}, \tag{5.2.75}$$

where β is a constant. The right-hand side of (5.2.75) is a monotonic function of X_1 and therefore, in view of (5.2.38), equation (5.2.75) may be inverted to give I_1 uniquely as a function of X_1. We write

$$I_1=(F')^{-1}(\nu\lambda_3+\beta \mathrm{e}^{\alpha X_1})\equiv\Phi(X_1) \tag{5.2.76}$$

and therefore

$$f'(X_1)+\alpha f(X_1)+\lambda_3=\Phi(X_1),$$

integration of which gives

$$f(X_1)=\mathrm{e}^{-\alpha X_1}\int^{X_1}\{\Phi(X)-\lambda_3\}\mathrm{e}^{\alpha X}\,\mathrm{d}X+\gamma \mathrm{e}^{-\alpha X_1}, \tag{5.2.77}$$

where γ is a constant.

The constants α, β, γ are determined by boundary conditions on the surfaces $X_1=\pm A$ and $X_2=\pm B$, assuming that λ_3 is specified. Application of (5.2.72), for example, followed by use of (5.2.75) shows that

$$f(\pm A)=\beta \mathrm{e}^{\pm\alpha A}/\alpha(\nu-\mu\lambda_3),$$

which should be considered together with (5.2.76) and (5.2.77), while a third equation is provided by prescribing the moment (5.2.74). Alternatively, the values of $f(\pm A)$, $g(B)$ may be prescribed, subject to (5.2.62). In either case little progress can be made in respect of a general form of F and we leave it to the reader to obtain values of α, β, γ for suitably chosen simple forms of F.

Clearly, our semi-inverse approach to the traction boundary-value problem specified above leads to a well-defined solution. However, this does not mean that the assumed form of the solution (5.2.60) is the only possibility for this problem. In other words the solution may not be unique. Indeed, particularly if A/B is small one can expect buckled states of equilibrium to exist. This aspect of the theory will be examined in Chapter 6.

Problem 5.2.4 Investigate the inverse problem in which a sector of a circular tube is deformed into a rectangular block according to

$$x_1 = f(R), \qquad x_2 = g(\Theta), \qquad x_3 = \lambda_3 Z,$$

where R, Θ, Z are reference cylindrical polar coordinates and x_1, x_2, x_3 are current rectangular Cartesian coordinates.

5.2.5 Combined extension and torsion of a solid circular cylinder

We consider the deformation discussed in Section 2.2.6 but without the constraint of incompressibility so that (2.2.75) is replaced by

$$r = Rf(R), \qquad \theta = \Theta + \tau\lambda Z, \qquad z = \lambda Z, \tag{5.2.78}$$

the reference geometry of the cylinder being defined by

$$0 \leq R \leq A, \qquad 0 \leq \Theta \leq 2\pi, \qquad 0 \leq Z \leq L$$

in cylindrical polar coordinates.

By analogy with the calculation leading to (2.2.76), we find that the deformation gradient takes the form

$$\mathbf{A} = \lambda_1 \mathbf{e}_r \otimes \mathbf{E}_R + f(R)\mathbf{e}_\theta \otimes \mathbf{E}_\Theta + \lambda \mathbf{e}_z \otimes \mathbf{E}_Z + \lambda\tau r \mathbf{e}_\theta \otimes \mathbf{E}_Z, \tag{5.2.79}$$

and

$$\det \mathbf{A} = [Rf'(R) + f(R)]f(R)\lambda, \tag{5.2.80}$$

where

$$\lambda_1 = Rf'(R) + f(R). \tag{5.2.81}$$

It follows that the left Cauchy–Green deformation tensor is given by

$$\mathbf{A}^{\mathrm{T}}\mathbf{A} = \lambda_1^2 \mathbf{E}_R \otimes \mathbf{E}_R + \frac{r^2}{R^2}\mathbf{E}_\Theta \otimes \mathbf{E}_\Theta + \lambda^2(1 + \tau^2 r^2)\mathbf{E}_Z \otimes \mathbf{E}_Z$$

$$+ \frac{\lambda\tau r^2}{R}(\mathbf{E}_Z \otimes \mathbf{E}_\Theta + \mathbf{E}_\Theta \otimes \mathbf{E}_Z). \tag{5.2.82}$$

This equation defines the Lagrangean principal axes—one along $\mathbf{E}_R$, corresponding to principal stretch λ_1 given by (5.2.81). Let the remaining two principal axes be defined by

$$\mathbf{E}'_\Theta = \cos\psi \mathbf{E}_\Theta + \sin\psi \mathbf{E}_Z, \quad \mathbf{E}'_Z = -\sin\psi \mathbf{E}_\Theta + \cos\psi \mathbf{E}_Z \tag{5.2.83}$$

with principal stretches λ_2, λ_3 respectively, so that

$$\mathbf{A}^{\mathrm{T}}\mathbf{A} = \lambda_1^2 \mathbf{E}_R \otimes \mathbf{E}_R + \lambda_2^2 \mathbf{E}'_\Theta \otimes \mathbf{E}'_\Theta + \lambda_3^2 \mathbf{E}'_Z \otimes \mathbf{E}'_Z. \tag{5.2.84}$$

Comparison of (5.2.82) and (5.2.84) then shows that

$$\begin{aligned}
&\lambda_2^2 \cos^2\psi + \lambda_3^2 \sin^2\psi = r^2/R^2,\\
&\lambda_2^2 \sin^2\psi + \lambda_3^2 \cos^2\psi = \lambda^2 + \lambda^2\tau^2 r^2,\\
&\sin\psi\cos\psi(\lambda_2^2 - \lambda_3^2) = \lambda\tau r^2/R,
\end{aligned}$$

from which we deduce that

$$\tan 2\psi = \frac{2\lambda\tau r^2 R}{[r^2 - \lambda^2(1+\tau^2 r^2)R^2]} \tag{5.2.85}$$

and

$$\lambda_2^2 + \lambda_3^2 = \frac{r^2}{R^2} + \lambda^2 + \lambda^2\tau^2 r^2. \tag{5.2.86}$$

A second equation connecting λ_2 and λ_3 is obtained from (5.2.80) on noting (2.2.39). Thus

$$\lambda_2\lambda_3 = \lambda f(R). \tag{5.2.87}$$

Since

$$\mathbf{U}^{-1} = \lambda_1^{-1}\mathbf{E}_R \otimes \mathbf{E}_R + \lambda_2^{-1}\mathbf{E}'_\Theta \otimes \mathbf{E}'_\Theta + \lambda_3^{-1}\mathbf{E}'_Z \otimes \mathbf{E}'_Z$$

we obtain

$$\begin{aligned}
\mathbf{R} = \mathbf{A}\mathbf{U}^{-1} = \mathbf{e}_r \otimes \mathbf{E}_R &+ \lambda_2^{-1}\left[\left(\frac{r}{R}\cos\psi + \lambda\tau r\sin\psi\right)\mathbf{e}_\theta + \lambda\sin\psi\mathbf{e}_z\right]\otimes\mathbf{E}'_\Theta\\
&+ \lambda_3^{-1}\left[\left(\lambda\tau r\cos\psi - \frac{r}{R}\sin\psi\right)\mathbf{e}_\theta + \lambda\cos\psi\mathbf{e}_z\right]\otimes\mathbf{E}'_Z
\end{aligned}$$

and therefore, since the Biot stress tensor is given by

$$\mathbf{T}^{(1)} = t_1^{(1)}\mathbf{E}_R \otimes \mathbf{E}_R + t_2^{(1)}\mathbf{E}'_\Theta \otimes \mathbf{E}'_\Theta + t_3^{(1)}\mathbf{E}'_Z \otimes \mathbf{E}'_Z,$$

we write the nominal stress tensor in the form

$$\mathbf{S} = \mathbf{T}^{(1)}\mathbf{R}^{\mathrm{T}} = t_1^{(1)}\mathbf{E}_R \otimes \mathbf{e}_r + t_2^{(1)}\lambda_2^{-1}\mathbf{E}'_\Theta \otimes \left[\left(\frac{r}{R}\cos\psi + \lambda\tau r\sin\psi\right)\mathbf{e}_\theta + \lambda\sin\psi\mathbf{e}_z\right]$$
$$+ t_3^{(1)}\lambda_3^{-1}\mathbf{E}'_Z \otimes \left[\left(\lambda\tau r\cos\psi - \frac{r}{R}\sin\psi\right)\mathbf{e}_\theta + \lambda\cos\psi\mathbf{e}_z\right]. \quad (5.2.88)$$

It follows that on the end of the cylinder (with unit normal $\mathbf{E}_Z$) the traction $\mathbf{S}^{\mathrm{T}}\mathbf{E}_Z$ has non-zero components

$$S_{Z\theta} = \frac{r}{R}\sin\psi\cos\psi\left(\frac{t_2^{(1)}}{\lambda_2} - \frac{t_3^{(1)}}{\lambda_3}\right)$$
$$+ \lambda\tau r\left(\frac{t_2^{(1)}}{\lambda_2}\sin^2\psi + \frac{t_3^{(1)}}{\lambda_3}\cos^2\psi\right) \quad (5.2.89)$$

and

$$S_{Zz} = \lambda\left(\frac{t_2^{(1)}}{\lambda_2}\sin^2\psi + \frac{t_3^{(1)}}{\lambda_3}\cos^2\psi\right) \quad (5.2.90)$$

The axial force on the end of the cylinder is given by

$$N = 2\pi\int_0^A S_{Zz}R\,\mathrm{d}R$$

and the torsional couple by

$$M = 2\pi\int_0^A rS_{Z\theta}R\,\mathrm{d}R$$

about the cylinder axis.

Substitution of (5.2.89) and (5.2.90) into these formulae and elimination of ψ leads to

$$N = 2\pi\lambda\int_0^A\left\{\frac{\lambda_2 t_2^{(1)} - \lambda_3 t_3^{(1)}}{\lambda_2^2 - \lambda_3^2} - \frac{\lambda_2^2\lambda_3^2}{\lambda^2}\frac{((t_2^{(1)}/\lambda_2) - (t_3^{(1)}/\lambda_3))}{\lambda_2^2 - \lambda_3^2}\right\}R\,\mathrm{d}R \quad (5.2.91)$$

and

$$M = \frac{2\pi\tau}{\lambda}\int_0^A \lambda_2^2\lambda_3^2\frac{(\lambda_2 t_2^{(1)} - \lambda_3 t_3^{(1)})}{\lambda_2^2 - \lambda_3^2}R^3\,\mathrm{d}R. \quad (5.2.92)$$

It remains to calculate $f(R)$ from the equilibrium equation. Use of (1.5.52) in reference coordinates R, Θ, Z in respect of the two-point tensor $\mathbf{S}$ shows that the equilibrium equations reduce to

$$\frac{\mathrm{d}}{\mathrm{d}R}S_{Rr}+\frac{1}{R}S_{Rr}=\frac{1}{R}S_{\Theta\theta}+\lambda\tau S_{Z\theta}, \tag{5.2.93}$$

R being the only independent variable. Substitution of (5.2.88) into (5.2.93) leads to

$$\frac{\mathrm{d}}{\mathrm{d}R}(Rt_1^{(1)})=\frac{\lambda}{\lambda_2\lambda_3}\left\{\frac{\lambda_2^3t_2^{(1)}-\lambda_3^3t_3^{(1)}}{\lambda_2^2-\lambda_3^2}-\lambda^2\frac{(\lambda_2t_2^{(1)}-\lambda_3t_3^{(1)})}{\lambda_2^2-\lambda_3^2}\right\}, \tag{5.2.94}$$

ψ having been eliminated in favour of λ_2 and λ_3.

This form of the equation is convenient for the insertion of particular forms of constitutive law via (5.2.65), and we note that $\lambda_1, \lambda_2, \lambda_3$ are given in terms of $f(R)$ by (5.2.81), (5.2.86) and (5.2.87). A typical boundary condition is

$$t_1^{(1)}=0 \quad \text{on} \quad R=A. \tag{5.2.95}$$

We do not pursue this problem any further here since exact solutions of (5.2.94) are difficult to obtain and algebraically cumbersome, but the reader may find it instructive to attempt solution of (5.2.94) for very simple forms of strain-energy function along the lines of Sections 5.2.2 to 5.2.4.

Problem 5.2.5 The axial shear of a circular cylindrical tube is defined by

$$r=g(R), \qquad \theta=\Theta, \qquad z=Z+w(R)$$

in cylindrical polar coordinates, where

$$A\le R\le B, \qquad 0\le\Theta\le 2\pi, \qquad 0\le Z\le L,$$

with

$$g(A)=A, \qquad w(A)=0.$$

On the ends of the cylinder no traction is applied, and the deformation is achieved by prescribing $g(B)$ and $w(B)$ or, alternatively, by applying an axial shear traction of amount τ per unit length of the cylinder and, for example, imposing no radial traction on $R=B$.

Show that the deformation gradient has the form

$$\mathbf{A}=g'(R)\mathbf{e}_r\otimes\mathbf{E}_R+\frac{1}{R}g(R)\mathbf{e}_\theta\otimes\mathbf{E}_\Theta+w'(R)\mathbf{e}_z\otimes\mathbf{E}_R+\mathbf{e}_z\otimes\mathbf{E}_Z$$

and deduce that

$$\det \mathbf{A} = \frac{1}{R} g(R) g'(R).$$

Prove that the Lagrangean principal axes have directions

$$\cos\psi \mathbf{E}_R + \sin\psi \mathbf{E}_Z, \qquad \mathbf{E}_\Theta, \qquad -\sin\psi \mathbf{E}_R + \cos\psi \mathbf{E}_Z,$$

where

$$\lambda_1^2 \cos^2\psi + \lambda_3^2 \sin^2\psi = [w'(R)]^2 + [g'(R)]^2$$
$$\lambda_1^2 \sin^2\psi + \lambda_3^2 \cos^2\psi = 1$$
$$(\lambda_1^2 - \lambda_3^2) \sin\psi \cos\psi = w'(R)$$

and show that the principal stretches λ_1, λ_2, λ_3 satisfy

$$\lambda_1 \lambda_3 = g'(R), \qquad \lambda_2 = \frac{1}{R} g(R).$$

Derive the equilibrium equations in the forms

$$\frac{\mathrm{d}S_{Rr}}{\mathrm{d}R} + \frac{1}{R}(S_{Rr} - S_{\Theta\theta}) = 0, \tag{5.2.96}$$

$$\frac{\mathrm{d}}{\mathrm{d}R}(R S_{Rz}) = 0$$

and show that the latter equation integrates to give

$$\left(\frac{\lambda_1 t_1^{(1)} - \lambda_3 t_3^{(1)}}{\lambda_1^2 - \lambda_3^2}\right) w'(R) \equiv S_{Rz} = \tau/2\pi R, \tag{5.2.97}$$

where S_{Rr}, S_{Rz}, $S_{\Theta\theta}$, S_{Zr}, S_{Zz} are the components of nominal stress.

Show that

$$S_{Rr} = \lambda_1 \lambda_3 \left(\frac{\lambda_1 t_1^{(1)} - \lambda_3 t_3^{(1)}}{\lambda_1^2 - \lambda_3^2}\right) - \left(\frac{\lambda_3 t_1^{(1)} - \lambda_1 t_3^{(1)}}{\lambda_1^2 - \lambda_3^2}\right)$$

and obtain similar expressions for S_{Zr} and S_{Zz}.

Investigate the possibility of solving the coupled equations (5.2.96) and (5.2.97) for $w(R)$ and $g(R)$ in respect of specific choices of strain-energy function.

Problem 5.2.6 A solid cylinder is defined by

$$0 \leq R \leq A, \qquad 0 \leq \Theta \leq 2\pi, \qquad 0 \leq Z \leq L$$

in reference cylindrical polar coordinates. It is deformed by rotation about its axis with constant angular speed ω according to

$$r = Rf(R), \qquad \theta = \Theta + \omega t, \qquad z = \lambda Z,$$

where λ is a constant and t is time.

Show that the equation of motion is

$$\frac{\mathrm{d}t_1^{(1)}}{\mathrm{d}R} + \frac{1}{R}(t_1^{(1)} - t_2^{(1)}) = -\rho_0 \omega^2 r, \tag{5.2.98}$$

where ρ_0 is the reference density of the material (assumed independent of R), $t_i^{(1)} = \partial W/\partial \lambda_i, i = 1, 2, 3$, and the principal stretches $\lambda_1, \lambda_2, \lambda_3$ correspond to the R, Θ, Z coordinate directions respectively.

Prove that $f(R)$ cannot be a constant and investigate (5.2.98) for particular forms of W assuming that the surface $R = A$ is traction free and that the ends of the cylinder are subject to a (time-independent) axial tension.

5.2.6 Plane strain problems: complex variable methods

In this section we examine deformations which are constrained to take place normal to some fixed direction and independently of the coordinate in that direction. For definiteness we suppose that this direction is defined by the unit vector $\mathbf{E}_3$ and we choose rectangular Cartesian basis vectors $\{\mathbf{E}_i\}$ such that reference points are labelled by the position vector $\mathbf{X} = X_i\mathbf{E}_i$. If x_1, x_2, x_3 are the Cartesian coordinates of $\mathbf{X}$ in the deformed configuration, then the deformation defined by

$$\begin{aligned} x_3 &= X_3 \\ x_\alpha &= \chi_\alpha(X_1, X_2) \qquad \alpha = 1, 2 \end{aligned} \tag{5.2.99}$$

is called a *plane deformation* (or *plane strain*). Recall that plane *homogeneous* deformations were defined in Section 2.2.6. We note, in particular, that $\mathbf{E}_3$ is a principal axis of the deformation and corresponds to the principal stretch $\lambda_3 = 1$. It is an easy matter to extend the following discussion to allow for constant $\lambda_3 \neq 1$.

In vector notation we write $(5.2.99)_2$ as

$$\hat{\mathbf{x}} = \hat{\boldsymbol{\chi}}(\hat{\mathbf{X}}),$$

where ^ indicates the restriction of the quantity upon which it sits to the plane of $\mathbf{E}_1$ and $\mathbf{E}_2$. The deformation gradient is then

$$\mathbf{A} = \hat{\mathbf{A}} + \mathbf{E}_3 \otimes \mathbf{E}_3, \tag{5.2.100}$$

where

$$\hat{\mathbf{A}} = \partial\hat{\chi}(\hat{\mathbf{X}})/\partial\hat{\mathbf{X}},$$

and the polar decomposition of $\hat{\mathbf{A}}$ is

$$\hat{\mathbf{A}} = \hat{\mathbf{R}}\hat{\mathbf{U}}, \tag{5.2.101}$$

where $\hat{\mathbf{U}}$ is positive definite and symmetric and $\hat{\mathbf{R}}$ is a plane rotation.

Corresponding to (5.2.100) the nominal stress tensor is

$$\mathbf{S} = \hat{\mathbf{S}} + S_{33}\mathbf{E}_3 \otimes \mathbf{E}_3$$

and we note that in general $S_{33} \neq 0$ when $\lambda_3 = 1$. The strain-energy $W(\mathbf{A})$ is written $\hat{W}(\hat{\mathbf{A}})$ when $\lambda_3 = 1$ and we therefore have

$$\hat{\mathbf{S}} = \frac{\partial \hat{W}}{\partial \hat{\mathbf{A}}}(\hat{\mathbf{A}}), \tag{5.2.102}$$

with

$$S_{33} = t_3^{(1)} = \frac{\partial W}{\partial \lambda_3} \tag{5.2.103}$$

evaluated for $\lambda_3 = 1$.

For the in-plane Biot stress we have the representation

$$\hat{\mathbf{T}}^{(1)} = \frac{\partial \hat{W}}{\partial \hat{\mathbf{U}}}(\hat{\mathbf{U}}) \tag{5.2.104}$$

and, since attention is concentrated on isotropic materials, it follows from (4.2.68) that

$$\hat{\mathbf{S}} = \hat{\mathbf{T}}^{(1)}\hat{\mathbf{R}}^{\mathrm{T}}. \tag{5.2.105}$$

Let the angles $\theta_{\mathrm{L}}, \theta_{\mathrm{E}}$ respectively describe the orientation of the in-plane Lagrangean and Eulerian principal axes in the sense that

$$\mathbf{u}^{(1)} = \cos\theta_{\mathrm{L}}\mathbf{E}_1 + \sin\theta_{\mathrm{L}}\mathbf{E}_2, \qquad \mathbf{u}^{(2)} = -\sin\theta_{\mathrm{L}}\mathbf{E}_1 + \cos\theta_{\mathrm{L}}\mathbf{E}_2,$$
$$\mathbf{v}^{(1)} = \cos\theta_{\mathrm{E}}\mathbf{E}_1 + \sin\theta_{\mathrm{E}}\mathbf{E}_2, \qquad \mathbf{v}^{(2)} = -\sin\theta_{\mathrm{E}}\mathbf{E}_1 + \cos\theta_{\mathrm{E}}\mathbf{E}_2.$$

Then

$$\hat{\mathbf{T}}^{(1)} = t_1^{(1)}\mathbf{u}^{(1)} \otimes \mathbf{u}^{(1)} + t_2^{(1)}\mathbf{u}^{(2)} \otimes \mathbf{u}^{(2)}$$

and, from (5.2.105), it follows that

$$\hat{\mathbf{S}} = t_1^{(1)}\mathbf{u}^{(1)} \otimes \mathbf{v}^{(1)} + t_2^{(1)}\mathbf{u}^{(2)} \otimes \mathbf{v}^{(2)}$$

since $\hat{\mathbf{R}} = \mathbf{v}^{(1)} \otimes \mathbf{u}^{(1)} + \mathbf{v}^{(2)} \otimes \mathbf{u}^{(2)}$. Note that the Cartesian components of $\hat{\mathbf{R}}$ are represented by the rotation matrix

$$\begin{bmatrix} \cos(\theta_{\mathrm{L}} - \theta_{\mathrm{E}}) & \sin(\theta_{\mathrm{L}} - \theta_{\mathrm{E}}) \\ -\sin(\theta_{\mathrm{L}} - \theta_{\mathrm{E}}) & \cos(\theta_{\mathrm{L}} - \theta_{\mathrm{E}}) \end{bmatrix}$$

and

$$t_\alpha^{(1)} = \frac{\partial \hat{W}}{\partial \lambda_\alpha} \qquad \alpha = 1, 2.$$

Let $S_{\alpha\beta}$, $\alpha, \beta \in \{1, 2\}$, denote the Cartesian components of $\hat{\mathbf{S}}$. Then these may be put together as

$$\left.\begin{aligned} S_{11} + S_{22} + \mathrm{i}(S_{12} - S_{21}) &= (t_1^{(1)} + t_2^{(1)})\mathrm{e}^{\mathrm{i}(\theta_E - \theta_L)}, \\ S_{11} - S_{22} + \mathrm{i}(S_{12} + S_{21}) &= (t_1^{(1)} - t_2^{(1)})\mathrm{e}^{\mathrm{i}(\theta_E + \theta_L)}. \end{aligned}\right\} \tag{5.2.106}$$

In terms of $S_{\alpha\beta}$ the equilibrium equations are

$$\frac{\partial S_{11}}{\partial X_1} + \frac{\partial S_{21}}{\partial X_2} = 0,$$

$$\frac{\partial S_{12}}{\partial X_1} + \frac{\partial S_{22}}{\partial X_2} = 0$$

when there are no body forces. Respectively, these two equations imply the existence of *stress functions*, h_2, h_1 say, such that

$$\left.\begin{aligned} S_{11} &= \frac{\partial h_2}{\partial X_2}, & S_{21} &= -\frac{\partial h_2}{\partial X_1}, \\ S_{12} &= -\frac{\partial h_1}{\partial X_2}, & S_{22} &= \frac{\partial h_1}{\partial X_1}, \end{aligned}\right\} \tag{5.2.107}$$

where h_1 and h_2 are defined over that part of the (1, 2)-plane occupied by the material in the reference configuration $\mathscr{B}_0$. Let this domain be denoted by $\hat{\mathscr{B}}_0$ and its boundary by $\partial\hat{\mathscr{B}}_0$.

Introduction of the complex variable

$$Z = X_1 + \mathrm{i}X_2$$

and the complex stress function

$$h = h_1 + \mathrm{i}h_2,$$

depending on Z and $\bar{Z}$, into (5.2.107) then enables (5.2.106) to be written as

$$\left.\begin{aligned} \frac{\partial h}{\partial Z} &= \tfrac{1}{2}(t_1^{(1)} + t_2^{(1)})\,\mathrm{e}^{\mathrm{i}(\theta_E - \theta_L)}, \\ \frac{\partial h}{\partial \bar{Z}} &= -\tfrac{1}{2}(t_1^{(1)} - t_2^{(1)})\,\mathrm{e}^{\mathrm{i}(\theta_E + \theta_L)}, \end{aligned}\right\} \tag{5.2.108}$$

where $\bar{Z} = X_1 - \mathrm{i}X_2$ denotes the complex conjugate of Z.

By defining the complex function

$$z = x_1 + \mathrm{i}x_2$$

we obtain

$$\left.\begin{aligned} \frac{\partial z}{\partial Z} &= \tfrac{1}{2}\{A_{11} + A_{22} + \mathrm{i}(A_{21} - A_{12})\} \\ \frac{\partial z}{\partial \bar{Z}} &= \tfrac{1}{2}\{A_{11} - A_{22} + \mathrm{i}(A_{12} + A_{21})\}, \end{aligned}\right\} \tag{5.2.109}$$

where $A_{\alpha\beta}$, α, $\beta \in \{1, 2\}$, are the components of $\hat{\mathbf{A}}$.

On comparing the right-hand sides in (5.2.109) with the left-hand sides in (5.2.106) and noting the similarity of (5.2.105) and the transpose of (5.2.101), we deduce from (5.2.106) and (5.2.109) that

$$\left.\begin{aligned} \frac{\partial z}{\partial Z} &= \tfrac{1}{2}(\lambda_1 + \lambda_2)\,\mathrm{e}^{\mathrm{i}(\theta_E - \theta_L)}, \\ \frac{\partial z}{\partial \bar{Z}} &= \tfrac{1}{2}(\lambda_1 - \lambda_2)\,\mathrm{e}^{\mathrm{i}(\theta_E + \theta_L)}. \end{aligned}\right\} \tag{5.2.110}$$

Note that

$$\left.\begin{aligned} \left|\frac{\partial z}{\partial Z}\right| &= \tfrac{1}{2}(\lambda_1 + \lambda_2) \\ \left|\frac{\partial z}{\partial \bar{Z}}\right| &= \tfrac{1}{2}|\lambda_1 - \lambda_2| \end{aligned}\right\} \tag{5.2.111}$$

and λ_1 and λ_2 are therefore obtainable in terms of $|\partial z/\partial Z|$ and $|\partial z/\partial \bar{Z}|$. Also

$$\left.\begin{aligned} e^{i(\theta_E-\theta_L)} &= \frac{\partial z}{\partial Z}\bigg/\left|\frac{\partial z}{\partial Z}\right| \\ e^{i(\theta_E+\theta_L)} &= \frac{\partial z}{\partial \bar{Z}}\bigg/\left|\frac{\partial z}{\partial \bar{Z}}\right| \end{aligned}\right\} \tag{5.2.112}$$

and hence θ_E and θ_L are obtained. A knowledge of $\lambda_1, \lambda_2, \theta_E, \theta_L$ provides a complete description of the deformation.

Elimination of θ_E and θ_L between (5.2.108) and (5.2.110) gives

$$\frac{\partial h}{\partial Z} = \left(\frac{t_1^{(1)}+t_2^{(1)}}{\lambda_1+\lambda_2}\right)\frac{\partial z}{\partial Z}, \tag{5.2.113}$$

$$\frac{\partial h}{\partial \bar{Z}} = -\left(\frac{t_1^{(1)}-t_2^{(1)}}{\lambda_1-\lambda_2}\right)\frac{\partial z}{\partial \bar{Z}} \tag{5.2.114}$$

and then elimination of h leads to the differential equation

$$\frac{\partial}{\partial \bar{Z}}\left\{\left(\frac{t_1^{(1)}+t_2^{(1)}}{\lambda_1+\lambda_2}\right)\frac{\partial z}{\partial Z}\right\} + \frac{\partial}{\partial Z}\left\{\left(\frac{t_1^{(1)}-t_2^{(1)}}{\lambda_1-\lambda_2}\right)\frac{\partial z}{\partial \bar{Z}}\right\} = 0 \tag{5.2.115}$$

for the determination of z as a function of Z and $\bar{Z}$, bearing in mind that the coefficients in this equation depend non-linearly (in general) on the quantities (5.2.111).

The equations may also be set up in a dual fashion with h as the dependent variable and coefficients dependent on $|\partial h/\partial Z|$ and $|\partial h/\partial \bar{Z}|$ but we do not discuss the details here. In this connection reference may be made to Isherwood and Ogden (1977a) and Varley and Cumberbatch (1977, 1980).

With (5.2.115) holding in $\hat{\mathscr{B}}_0$ we need to prescribe boundary conditions on $\partial\hat{\mathscr{B}}_0$ in order to complete the boundary-value problem formulation. We assume that the complex deformation function is specified on all or part of $\partial\hat{\mathscr{B}}_0$ and the traction on the complementary part. It remains to express the traction in complex form. Let N_1, N_2 be the components of the unit outward normal to $\partial\hat{\mathscr{B}}_0$. Then, in terms of the arclength parameter, s say, which describes $\partial\hat{\mathscr{B}}_0$ in the anti-clockwise sense through $X_1(s)$, $X_2(s)$ we have $N_1 = \mathrm{d}X_2/\mathrm{d}s$, $N_2 = -\mathrm{d}X_1/\mathrm{d}s$. The components of traction are

$$t_1 = S_{11}N_1 + S_{21}N_2, \qquad t_2 = S_{12}N_1 + S_{22}N_2$$

and therefore, with the help of (5.2.107), we obtain

$$t_1 = \frac{\mathrm{d}h_2}{\mathrm{d}s}, \qquad t_2 = -\frac{\mathrm{d}h_1}{\mathrm{d}s} \quad \text{on } \partial\hat{\mathscr{B}}_0.$$

On introduction of the complex traction $t = t_1 + \mathrm{i}t_2$ these equations lead finally to

$$t = -\mathrm{i}\frac{\mathrm{d}h}{\mathrm{d}s} \quad \text{on } \partial\hat{\mathscr{B}}_0. \tag{5.2.116}$$

For the *dead-load* traction boundary condition (recall the definition in Section 5.1.2), t is prescribed as a function of s and, by integration of (5.2.116), this implies that h is known likewise to within an additive constant. Thus, effectively, specification of t is equivalent to specification of h. More generally, for a configuration-dependent boundary condition, t depends on the deformation on $\partial\hat{\mathscr{B}}_0$ and, specializing (5.1.23) to the current context, we have

$$t = \sigma\{\hat{\mathbf{X}}, \hat{\chi}(\hat{\mathbf{X}}), \hat{\mathbf{A}}(\hat{\mathbf{X}})\} \quad \text{on } \partial\hat{\mathscr{B}}_0, \tag{5.2.117}$$

where σ is a given (complex) function of its arguments. With t given by (5.2.117), equation (5.2.116) cannot be integrated explicitly to give h as a function of s on $\partial\hat{\mathscr{B}}_0$ since $\hat{\chi}$ is unknown at the outset.

In practice it turns out that more can be learned by using the governing equations in the separated form (5.2.113) and (5.2.114) rather than (5.2.115) and we now consider specialized forms of constitutive law which allow these equations to be integrated.

Discussion of harmonic strain-energy functions Clearly (5.2.113) can be integrated if

$$\frac{t_1^{(1)} + t_2^{(1)}}{\lambda_1 + \lambda_2} = 2\mu',$$

where μ' is a constant, and we therefore examine the consequences of this requirement, which we now write in the form

$$\frac{\partial\hat{W}}{\partial\lambda_1} + \frac{\partial\hat{W}}{\partial\lambda_2} = 2\mu'(\lambda_1 + \lambda_2). \tag{5.2.118}$$

Since λ_1 and λ_2 occur in the combinations $\lambda_1 \pm \lambda_2$ in equations (5.2.113) and (5.2.114) we introduce the notation

$$p = \lambda_1 + \lambda_2, \qquad q = |\lambda_1 - \lambda_2|, \tag{5.2.119}$$

recalling (5.2.111), these being symmetric invariants. Accordingly, we may regard $\hat{W}$ as a function of p and q and (5.2.118) becomes

$$\frac{\partial\hat{W}}{\partial p} = \mu' p.$$

Integration of this leads to

$$\hat{W}=\tfrac{1}{2}\mu' p^2+f(q), \tag{5.2.120}$$

where f is an arbitrary function. For definiteness we take $\lambda_1 \geq \lambda_2$ so that

$$t_1^{(1)}=\mu' p+f'(q), \qquad t_2^{(1)}=\mu' p-f'(q),$$

where $f' \equiv \mathrm{d}f/\mathrm{d}q$. Since $p=2$ and $q=0$ in the reference configuration, this configuration cannot be stress free unless $\mu'=0$, $f'(0)$, in which case the stress vanishes for any deformation with $\lambda_1=\lambda_2$. Because of this restriction we consider (5.2.120) no further here, but we remark that when $\mu' \neq 0$ there may be circumstances when it can prove useful. For further discussion see Isherwood and Ogden (1977b) and Varley and Cumberbatch (1980).

The reservation just described does not arise in respect of (5.2.114) for, on setting

$$t_1^{(1)}-t_2^{(1)}=2\mu(\lambda_1-\lambda_2),$$

where μ is a constant, we obtain

$$\hat{W}=f(p)+\tfrac{1}{2}\mu q^2, \tag{5.2.121}$$

f again being an arbitrary function (subject to sensible smoothness requirements).

It follows that

$$t_1^{(1)}+t_2^{(1)}=2f'(p) \tag{5.2.122}$$

and the reference configuration is stress free with $\hat{W}=0$ provided

$$f(2)=0, \qquad f'(2)=0. \tag{5.2.123}$$

The materials whose properties are described in terms of (5.2.121) are referred to as *harmonic materials*, and the mathematical properties of the governing equations resulting from the use of (5.2.121) were in essence discovered by John (1960). Our formulation is expressed in variables different from those used in John's original paper.

Clearly, in respect of the (plane) strain-energy function (5.2.121), equation (5.2.114) can be integrated to give an explicit connection between h and z, but, before we examine this in detail, some further discussion of the properties of (5.2.121) is desirable. Firstly, we remark that (5.2.121) is equivalent to the specialization of (5.2.19) to $\lambda_3=1$. It should be pointed out, however, that the constant μ has a different meaning in each case. The precise connection between $F(I_1) \equiv F(p+1)$ and $f(p)$ is left for the reader to examine.

The equivalent of (5.2.38) in respect of (5.2.121) is

$$f''(p)+\mu>0, \tag{5.2.124}$$

and it follows from (5.2.37) that the μ defined here satisfies

$$\mu>0. \tag{5.2.125}$$

The latter inequality is also consistent with the classical theory of infinitesimal isotropic elasticity for, when specialized to infinitesimal strains, μ in (5.2.121) has the role of the shear modulus. This aspect of the theory, which will be enlarged upon in Chapter 6, is brought out by consideration of the special case

$$f(p)=\tfrac{1}{2}(\lambda+\mu)(p-2)^2, \tag{5.2.126}$$

where λ is a constant (λ and μ are the classical Lamé moduli). This is consistent with (5.2.123) and, by (5.2.124), λ satisfies

$$\lambda+2\mu>0. \tag{5.2.127}$$

The material having the strain-energy function thus specified is sometimes referred to as the *semi-linear material.*

For consistency with the classical theory the $f(p)$ in (5.2.121) is subject to

$$f''(2)=\lambda+\mu \tag{5.2.128}$$

in addition to (5.2.123).

General solution of the differential equations In respect of (5.2.121), the differential equations (5.2.113) and (5.2.114) become

$$\frac{\partial h}{\partial Z}=\frac{2f'(p)}{p}\frac{\partial z}{\partial Z}, \tag{5.2.129}$$

$$\frac{\partial h}{\partial \bar{Z}}=-2\mu\frac{\partial z}{\partial \bar{Z}} \tag{5.2.130}$$

with

$$\tfrac{1}{2}p=\left|\frac{\partial z}{\partial Z}\right|. \tag{5.2.131}$$

Integration of (5.2.130) yields

$$h=-2\mu z+2\mu g(Z), \tag{5.2.132}$$

where g is an arbitrary function which ultimately is to be determined from the boundary conditions. Substitution of this into (5.2.129) leads to

$$g'(Z) = \{1 + f'(p)/\mu p\} \frac{\partial z}{\partial Z}. \tag{5.2.133}$$

For convenience we introduce the notation defined by

$$2\mu\Phi(p) = f'(p) + \mu p \tag{5.2.134}$$

so that, by (5.2.123) and (5.2.124),

$$\Phi(2) = 1 \tag{5.2.135}$$

and

$$\Phi'(p) > 0 \tag{5.2.136}$$

for all $p > 0$.

Equation (5.2.133) now becomes

$$g'(Z) = \frac{2\Phi(p)}{p} \frac{\partial z}{\partial Z} \tag{5.2.137}$$

and, on taking the modulus of this and using (5.2.131), we obtain

$$|g'(Z)| = |\Phi(p)|. \tag{5.2.138}$$

From the monotonicity of Φ implied by (5.2.136) and the assumed smoothness of Φ we deduce that Φ is non-singular for $0 < p < \infty$ and hence that $g(Z)$ is an analytic function. Further, there exists at most one value of p, p_0 say, for which $\Phi(p) = 0$. Since $|g'(Z)| = 0$ implies that $g'(Z) = 0$, $g'(Z)$ has zeros where $p = p_0$. But the equation $p = p_0$ defines a curve in $\hat{\mathscr{B}}_0$ and therefore $g'(Z)$ vanishes at a point Z in $\hat{\mathscr{B}}_0$ if and only if it vanishes on a curve through Z. However, from the properties of analytic functions, we know that $g'(Z)$ is either zero at isolated points in $\hat{\mathscr{B}}_0$ or vanishes identically on $\hat{\mathscr{B}}_0$. We conclude that $g(Z)$ is a constant on $\hat{\mathscr{B}}_0$ when $p = p_0$.

It follows that $\Phi(p)$ is one-signed on $\hat{\mathscr{B}}_0$, i.e.

$$\Phi(p) \gtrless 0 \text{ on } \hat{\mathscr{B}}_0 \text{ according as } p \gtrless p_0. \tag{5.2.139}$$

In the reference configuration we have $\Phi(2) = 1$ from (5.2.135); therefore $p_0 < 2$. Thus, in a neighbourhood of the reference configuration, $\Phi(p) > 0$

and (5.2.138) is replaced by

$$\Phi(p) = |g'(Z)|. \tag{5.2.140}$$

The sign of the right-hand side should be changed whenever p becomes less than p_0 in the course of the deformation process. However, for the problems to which this method has been applied it turns out that $p > p_0$. For our purposes, therefore, it suffices to consider (5.2.140).

Because of the monotonicity of $\Phi(p)$, equation (5.2.140) determines p uniquely in terms of $|g'(Z)|$ and we write the inversion of (5.2.140) formally as

$$p = \Phi^{-1}(|g'(Z)|), \tag{5.2.141}$$

where Φ^{-1} is the inverse function of Φ. Use of (5.2.140) and (5.2.141) in (5.2.137) now gives

$$\frac{\partial z}{\partial Z} = \frac{1}{2}\frac{\Phi^{-1}(|g'(Z)|)}{|g'(Z)|}g'(Z). \tag{5.2.142}$$

In principle this may be integrated to provide an integral representation for z, namely

$$z = \tfrac{1}{2}\int^{Z} \Psi(\zeta, \bar{Z})\,\mathrm{d}\zeta + \bar{k}(\bar{Z}), \tag{5.2.143}$$

where $k(Z)$ is an arbitrary analytic function (since, by assumption, the deformation, as described in (5.2.109), has no singularities in $\hat{\mathscr{B}}_0$) and

$$\Psi(\zeta, \bar{Z}) = \Phi^{-1}\{[g'(\zeta)]^{\frac{1}{2}}[\bar{g}'(\bar{Z})]^{\frac{1}{2}}\}[g'(\zeta)]^{\frac{1}{2}}[\bar{g}'(\bar{Z})]^{-\frac{1}{2}}, \tag{5.2.144}$$

with

$$\Psi(Z, \bar{Z}) = \frac{\Phi^{-1}(|g'(Z)|)g'(Z)}{|g'(Z)|}. \tag{5.2.145}$$

From (5.2.132) the stress function is now given by

$$h/2\mu = -\tfrac{1}{2}\int^{Z} \Psi(\zeta, \bar{Z})\,\mathrm{d}\zeta + g(Z) - \bar{k}(\bar{Z}). \tag{5.2.146}$$

It remains to match these representations for z and h with the boundary conditions on appropriate parts of $\partial\hat{\mathscr{B}}_0$ in order to determine $g(Z)$ and $k(Z)$ explicitly. In general this is a difficult procedure but there are a number of problems in which results can be obtained in a straightforward manner, and we now exemplify these.

Radial deformation of a circular annulus This is the special case of the problem discussed in Section 5.2.3 corresponding to $\lambda_3 = 1$. Attention is restricted to a cross-section of the cylinder whose reference geometry is defined by

$$A^2 \leq Z\bar{Z} \leq B^2.$$

We suppose that the exterior boundary $Z\bar{Z} = B^2$ is unstressed so that, from (5.2.116) we deduce that h is constant on this boundary. Without loss of generality we take the boundary condition to be

$$h = 0 \quad \text{on} \quad Z\bar{Z} = B^2. \tag{5.2.147}$$

It follows from (5.2.146) that

$$\bar{k}(\bar{Z}) = g(B^2/\bar{Z}) - \tfrac{1}{2}\int^{B^2/\bar{Z}} \Psi(\zeta, \bar{Z})\,\mathrm{d}\zeta. \tag{5.2.148}$$

This equation defines $\bar{k}(\bar{Z})$ for $Z\bar{Z} = B^2$. By analytic continuation the definition of $\bar{k}(\bar{Z})$ can be extended to $A \leq |Z| \leq B$ provided $g(Z)$ is analytic for

$$A \leq |Z| \leq B^2/A,$$

a domain which extends outside the annulus.

Formally, substitution of (5.2.148) into (5.2.143) gives

$$z = \tfrac{1}{2}\int_{B^2/\bar{Z}}^{Z} \Psi(\zeta, \bar{Z})\,\mathrm{d}\zeta + g(B^2/\bar{Z}). \tag{5.2.149}$$

On the inner boundary of the annulus we impose the condition

$$z = \frac{a}{A}Z \quad \text{on} \quad Z\bar{Z} = A^2, \tag{5.2.150}$$

where a is real, this corresponding to radially symmetric deformation of the boundary. Equations (5.2.149) and (5.2.150) together give

$$\tfrac{1}{2}\int_{B^2Z/A^2}^{Z} \Psi(\zeta, A^2/Z)\,\mathrm{d}\zeta = aZ/A - g(B^2Z/A^2). \tag{5.2.151}$$

This is an identity for $|Z| = A$, and, in view of the form of Ψ described by (5.2.144), we deduce that it is satisfied if $g(Z)$ has the form

$$g(Z) = \eta Z, \tag{5.2.152}$$

where η is a constant. It follows that p is a constant.

Substitution of (5.2.152) back into (5.2.151) and use of (5.2.141) leads to

$$\{\Phi(p) - \tfrac{1}{2}p(1 - A^2/B^2)\}\eta/|\eta| = Aa/B^2.$$

From this and the properties of $\Phi(p)$ we deduce that η is real and positive, and hence the above reduces to

$$\Phi(p) - \tfrac{1}{2}p(1 - A^2/B^2) = Aa/B^2. \tag{5.2.153}$$

Given a/A, p is uniquely determined provided the left-hand side of (5.2.153) is monotonic increasing with p, i.e.

$$\Phi'(p) > \tfrac{1}{2}(1 - A^2/B^2). \tag{5.2.154}$$

We note that this is more stringent than the inequality (5.2.136) imposed without reference to specific geometry.

With $g(Z)$ given by (5.2.152) for $|Z| = A$ extended analytically we find that the solution (5.2.149) simplifies to

$$z = Aa/\bar{Z} + \tfrac{1}{2}p(Z - A^2/\bar{Z}) \tag{5.2.155}$$

with p in principle determined from (5.2.153) in terms of a/A and A^2/B^2. Of course, p depends on the form of $\Phi(p)$; in particular, for the semi-linear material with $f(p)$ given by (5.2.126), p can be calculated explicitly. Details are left for the reader to work out. The result (5.2.155) should be compared with (5.2.49) specialized to $\lambda_3 = 1$.

The local volume ratio $J = \lambda_1\lambda_2$ is calculated from (5.2.155) through

$$J = \left|\frac{\partial z}{\partial Z}\right|^2 - \left|\frac{\partial z}{\partial \bar{Z}}\right|^2$$

as

$$J = \tfrac{1}{4}p^2 - \left(\tfrac{1}{2}p - \frac{a}{A}\right)^2 \frac{A^4}{R^4},$$

where $A \le |Z| = R \le B$, and the least value of J within the annulus is clearly

$$\tfrac{1}{4}p^2 - \left(\tfrac{1}{2}p - \frac{a}{A}\right)^2 = \left(p - \frac{a}{A}\right)\frac{a}{A}.$$

Since this must be positive we see from (5.2.153) that

$$\Phi(p) < \tfrac{1}{2}p(1 + A^2/B^2), \tag{5.2.156}$$

and we deduce from this and (5.2.154) that

$$\Phi(p) \sim \kappa p \quad \text{as} \quad p \to \infty, \tag{5.2.157}$$

where

$$0 < \tfrac{1}{2}(1 - A^2/B^2) < \kappa < \tfrac{1}{2}(1 + A^2/B^2) < 1. \tag{5.2.158}$$

Further comments on inequalities restricting the properties of $\Phi(p)$ and its asymptotic forms for $p \to 0$ and $p \to \infty$ are reserved for the next section.

From (5.2.132) with (5.2.152), (5.2.153) and (5.2.155) we now obtain the stress function in the form

$$h/2\mu = \{\Phi(p) - \tfrac{1}{2}p\}(Z - B^2/\bar{Z}),$$

and on the circle $|Z| = R$ this becomes

$$h/2\mu = \{\Phi(p) - \tfrac{1}{2}p\}(R - B^2/R)\,\mathrm{e}^{\mathrm{i}\theta}.$$

It follows from (5.2.116) that the (complex) traction on this circle is

$$t = -\frac{\mathrm{i}}{R}\frac{\mathrm{d}h}{\mathrm{d}\theta} = 2\mu\{\Phi(p) - \tfrac{1}{2}p\}(1 - B^2/R^2)\,\mathrm{e}^{\mathrm{i}\theta}$$

and this is purely radial. In particular, on $R = A$ the traction is

$$-t_1^{(1)} = (B^2/A^2 - 1)f'(p). \tag{5.2.159}$$

We deduce that for radial expansion (respectively contraction), corresponding to hydrostatic pressure (respectively tension) on $|Z| = A$, we have $f'(p) > 0$ (respectively < 0). In view of (5.2.123), we observe that the radial stress is a monotonic function of the internal radius provided

$$f''(p) > 0. \tag{5.2.160}$$

Then

$$f'(p) \gtreqless 0 \quad \text{according as } p \gtreqless 2. \tag{5.2.161}$$

Note that (5.2.136) is a consequence of (5.2.160) but the converse is not true in general.

Torsional shear of an annulus We next consider the problem in which the inner radius of the annulus is held fixed while the outer surface is twisted through an angle α at fixed radius. The boundary conditions are therefore

$$\left.\begin{aligned} z &= Z \quad \text{on} \quad Z\bar{Z} = A^2, \\ z &= Z e^{i\alpha} \quad \text{on} \quad Z\bar{Z} = B^2. \end{aligned}\right\} \tag{5.2.162}$$

Use of the boundary conditions with (5.2.143) leads to

$$\frac{1}{2}\int_{A^2Z/B^2}^{Z} \Psi(\zeta, B^2/Z)\,d\zeta = (e^{i\alpha} - A^2/B^2)Z,$$

the details being similar to those described in the previous problem. As before we deduce that this equation is satisfied if $g(Z)$ has the form (5.2.152), where now

$$\tfrac{1}{2}\eta = \frac{\Phi(p)}{p}(e^{i\alpha} - A^2/B^2)/(1 - A^2/B^2).$$

Since $\Phi(p) = |\eta|$ we obtain

$$\tfrac{1}{2}p = (1 + A^4/B^4 - 2A^2\cos\alpha/B^2)^{\frac{1}{2}}(1 - A^2/B^2)^{-1}, \tag{5.2.163}$$

while the solution of (5.2.142) becomes

$$z = (e^{i\alpha} - A^2/B^2)(1 - A^2/B^2)^{-1}(Z - A^2/\bar{Z}) + A^2/\bar{Z}. \tag{5.2.164}$$

Two features of the deformation are worth particular note here. First (5.2.164) is independent of the form of $\Phi(p)$; of course, the corresponding stress function h depends on $\Phi(p)$ through η. Second, p is bounded for all α according to

$$1 \le \tfrac{1}{2}p \le (1 + A^2/B^2)(1 - A^2/B^2)^{-1}$$

and monotonically increasing with α for $0 \le \alpha \le \pi$. Equally, the shear strain, as measured through

$$\tfrac{1}{2}q = 2(A^2/R^2)(1 - A^2/B^2)^{-1}|\sin\tfrac{1}{2}\alpha|$$

is bounded.

Physically, this does not seem reasonable since one would expect the shear strain to continue increasing with α beyond $\alpha = \pi$. Further, the local volume ratio J is positive at each point of the annulus provided

$$\cos\alpha > \tfrac{1}{2}(1 + A^2/B^2).$$

This inequality restricts α to a range of values $-\alpha_0 < \alpha < \alpha_0$, where $\cos\alpha_0 = \frac{1}{2}(1 + A^2/B^2)$ and $0 < \alpha_0 < \frac{1}{2}\pi$. A possible interpretation of the failure of the solution beyond a critical value of the boundary data is that the material with a constitutive law of the form (5.2.121) cannot support relatively large shear stresses. Calculation of the radial and shear stresses on a circle $|Z| = R$ is left to the reader. Note that there is no overall change in volume (i.e. area) of the annulus but that $J \not\equiv 1$ for $A \le R \le B$. However, it is easy to show that $J = 1$ for $R^2 = AB$.

Problem 5.2.7 Obtain the general solution of the governing equations in respect of the strain-energy function (5.2.120) and consider the possibility of boundary-value problems having solutions with $q = \text{constant}$.

Problem 5.2.8 Show that a (plane strain) *isochoric* deformation of the form

$$z = Ze^{i\alpha(R)},$$

where $\alpha(R)$ is a real function of $R = |Z|$, is not possible for any *compressible* isotropic elastic solid in the absence of body forces.

For further discussion of the application of complex variable methods to plane problems see Ogden and Isherwood (1978), Varley and Cumberbatch (1980) and Green and Adkins (1970). In particular, the method has been used for the solution of problems in which circular and elliptical holes are admitted into infinite bodies under uniform stress at infinity.

5.2.7 Growth conditions

Growth conditions for harmonic materials We have seen in Section 5.2.6 that a number of inequalities arise naturally in the construction of solutions to boundary-value problems in respect of harmonic materials with strain-energy function (5.2.121). In particular, the inequality $\Phi'(p) > 0$ for all $p > 0$ is required in general. For specific problems this inequality is strengthened; see (5.2.154), for example. These inequalities relate primarily to questions of existence and uniqueness, but physically-motivated requirements also lead to similar inequalities. For example, if the (nominal) pressure on the interior of an annulus is a monotonic increasing function of the radius then (5.2.160) follows or, equivalently,

$$\Phi'(p) > \tfrac{1}{2}. \tag{5.2.165}$$

Note that this implies (5.2.154).

Consider now a (plane) hydrostatic stress test with $\lambda_1 = \lambda_2 = \frac{1}{2}p$. The (nominal) hydrostatic stress is

$$t_1^{(1)} = t_2^{(1)} = 2\mu\Phi(p) - \mu p. \tag{5.2.166}$$

The physically-reasonable requirements that $t_1^{(1)} \to \infty$ as $p \to \infty$ and $t_1^{(1)} \to -\infty$ as $p \to 0$ impose like growth conditions on $f'(p)$ and these imply

$$\left.\begin{aligned} \Phi(p) &\to \infty \quad \text{as} \quad p \to \infty, \\ \Phi(p) &\to -\infty \quad \text{as} \quad p \to 0. \end{aligned}\right\} \tag{5.2.167}$$

In view of the monotonicity of $\Phi(p)$, we deduce that Φ is a one-to-one mapping of the interval $(0, \infty)$ *onto* $(-\infty, \infty)$. In particular, the existence of a $p_0 \in (0, \infty)$ such that $\Phi(p_0) = 0$ is assured. However, as we remarked in Section 5.2.6, the deformation achieved in particular problems may be such that $p > p_0$ so that $\Phi(p) > 0$ at all points of the body.

For the hydrostatic test under consideration the Cauchy stress is given by $t_2 = t_1 = t_1^{(1)}/\lambda_1$ and therefore, from (5.2.166), we have

$$t_1 = 4\mu\Phi(p)/p - 2\mu.$$

This increases monotonically with the deformation provided

$$p\Phi'(p) - \Phi(p) > 0. \tag{5.2.168}$$

In relation to the problem of the inflation of an annulus, the inequality (5.2.168) ensures that the inflating pressure ($-t_1 = -t_1^{(1)}/\lambda_2$, evaluated for $R = A$ from (5.2.159) increases monotonically with the internal radius a. In fact, it is not difficult to show that

$$-\frac{\mathrm{d}t_1}{\mathrm{d}p} = \mu(B^2 - A^2)\{p\Phi'(p) - \Phi(p)\}/a^2.$$

For some materials and for certain values of the ratio B/A, however, the inequality $-\mathrm{d}t_1/\mathrm{d}p > 0$ is not to be expected for all p. Further remarks on this are provided by Ogden and Isherwood (1978), and we return to the point in Section 5.3 in relation to incompressible materials.

In connection with (5.2.168), we observe that $\Phi(p)/p \to -\infty$ as $p \to 0$ is a consequence of (5.2.167), but for $p \to \infty$, it cannot be deduced that $\Phi(p)/p \to \infty$. Indeed, it follows from (5.2.156) that $\Phi(p)/p$ is finite at this upper limit. This is consistent with (5.2.168) if $p\Phi'(p) - \Phi(p) \to 0$ as $p \to \infty$. A number of other inequalities which put restrictions on $\Phi(p)$ and $\Phi'(p)$ have been examined by Knowles and Sternberg (1975) and Ogden and Isherwood (1978), but, for the most part, they are made redundant by the inequalities we have considered here.

General remarks on growth conditions As the torsional shear problem illustrates, application of the strain-energy function (5.2.121) is not valid for the whole range of values of $p (0 < p < \infty)$ in some problems and the solution may break down at a finite value of p with $J=0$. For a general plane strain problem with $\hat{W}$ regarded as a function of independent invariants p and J it is to be expected that.

$$\hat{W}\to\infty \quad \text{as} \quad J\to 0, \tag{5.2.169}$$

independently of p, i.e. an infinite amount of energy is required to reduce to zero the length of one side of a rectangle of material (in plane strain) with p fixed. We note, in particular, that this condition is not met in respect of (5.2.121) since $q^2=p^2-4J$ remains finite as $J\to 0$, and this fact may explain the breakdown of solution referred to above.

In three dimensions the obvious analogue of (5.2.169) is

$$W(\mathbf{A})\to\infty \quad \text{as} \quad \det\mathbf{A}\to 0 \tag{5.2.170}$$

(for compressible materials) and this need not be restricted to isotropic materials.

At the other end of the scale the behaviour of $W(\mathbf{A})$ as $\det\mathbf{A}\to\infty$ is crucial in determining whether or not singularities in the deformation (such as those engendered by the formation of holes) are possible. For example, the condition

$$\frac{W(\mathbf{A})}{(\det\mathbf{A})^3}\to\infty \quad \text{as} \quad \det\mathbf{A}\to\infty \tag{5.2.171}$$

excludes the possibility of fracture occurring, and the formation of holes (cavitation) at finite energy cannot be predicted if (5.2.171) holds. We refer to Ball (1977, 1982) for further discussion of these points, and recall that in this book we are concerned only with twice continuously differentiable deformations, thus excluding the type of singularity just mentioned.

In view of the detailed discussion of inequalities for harmonic materials and their evident dependence on geometry for particular problems coupled with some knowledge of the wide range of phenomena that can occur in elastic solids, it is clear that each boundary-value problem needs special attention *vis-à-vis* growth conditions on W. In particular, it is unwise to be dogmatic about what growth conditions should be imposed since physically sensible behaviour might thereby be overlooked.

5.3 PROBLEMS FOR MATERIALS WITH INTERNAL CONSTRAINTS

5.3.1 Preliminaries

We recall from Section 5.1.1 that for a (Green-) elastic material subject to a single constraint

$$C(\mathbf{A})\equiv C(\mathbf{U})=0 \tag{5.3.1}$$

the stress-deformation relation may be written

$$\mathbf{S} = \frac{\partial W}{\partial \mathbf{A}} + q\frac{\partial C}{\partial \mathbf{A}} \tag{5.3.2}$$

in terms of nominal stress, where q is an arbitrary scalar multiplier. Equally, in terms of Biot stress

$$\mathbf{T}^{(1)} = \frac{\partial W}{\partial \mathbf{U}} + q\frac{\partial C}{\partial \mathbf{U}}. \tag{5.3.3}$$

For an isotropic material we know that

$$\frac{\partial W}{\partial \mathbf{A}} = \frac{\partial W}{\partial \mathbf{U}}\mathbf{R}^{\mathrm{T}},$$

but, in general $C(\mathbf{U})$ is not an isotropic function of $\mathbf{U}$. Hence, we cannot make use of the connection $\mathbf{S} = \mathbf{T}^{(1)}\mathbf{R}^{\mathrm{T}}$ appropriate for unconstrained isotropic materials. In particular, this is the case with the inextensibility constraint

$$C(\mathbf{A}) \equiv \mathbf{L}\cdot(\mathbf{A}^{\mathrm{T}}\mathbf{A}\mathbf{L}) - 1 = 0 \tag{5.3.4}$$

discussed in Section 2.2.6 since

$$\frac{\partial C}{\partial \mathbf{A}} = 2\mathbf{L} \otimes \mathbf{A}\mathbf{L} \tag{5.3.5}$$

follows immediately while a short calculation yields

$$\frac{\partial C}{\partial \mathbf{U}} = \mathbf{L} \otimes \mathbf{U}\mathbf{L} + \mathbf{U}\mathbf{L} \otimes \mathbf{L}, \tag{5.3.6}$$

$\mathbf{L}$ being a unit Lagrangean vector. It follows that $\partial C/\partial \mathbf{A} = (\partial C/\partial \mathbf{U})\mathbf{R}^{\mathrm{T}}$ if and only if $\mathbf{L}$ coincides with a principal direction of $\mathbf{U}$.

For the incompressibility constraint

$$C(\mathbf{A}) \equiv \det \mathbf{A} - 1 = 0, \tag{5.3.7}$$

on the other hand, we have from (5.1.20) applied to Green-elastic materials that

$$\mathbf{S} = \frac{\partial W}{\partial \mathbf{A}} - p\mathbf{B}^{\mathrm{T}}, \tag{5.3.8}$$

recalling that $\mathbf{B}^{\mathrm{T}}$ is the inverse of $\mathbf{A}$. Similarly

$$\mathbf{T}^{(1)} = \frac{\partial W}{\partial \mathbf{U}} - p\mathbf{U}^{-1} \tag{5.3.9}$$

from (4.3.35), and the connection $\mathbf{S} = \mathbf{T}^{(1)}\mathbf{R}^{\mathrm{T}}$ does follow in this case.

For constrained materials there is no simple analogue of Ericksen's theorem (Section 5.2.1). Indeed, for incompressible isotropic elastic solids, for example, a number of non-homogeneous deformations can be maintained by surface tractions alone independently of the form of W. It is not as yet known, however, whether or not the catalogue of such deformations, referred to as *universal* or *controllable* deformations, is complete. Several of the deformations discussed in the following sections are of this type, but we do not emphasize this aspect of the theory. The reader is referred to Spencer (1970), for example, for a summary of the problem of classification of universal deformations. Since several homogeneous deformations (which are universal) were examined in Section 4.4, we concentrate attention on non-homogeneous deformations in what follows.

In the absence of body forces the equilibrium equations are put as

$$\operatorname{Div} \mathbf{S} = \mathbf{0}, \tag{5.3.10}$$

with $\mathbf{S}$ given by (5.3.2), when there is only one constraint (5.3.1), or (5.3.8) when the constraint is specialized to (5.3.7).

5.3.2 Spherically symmetric deformation of a spherical shell

The spherically symmetric deformation of a shell of *incompressible* material was defined in Problem 2.2.16 and may be written in the form

$$r = Rf(R)$$

in accord with the notation used in Section 5.2.2 for unconstrained materials, but with $f(R)$ now given by

$$f(R) = \left(1 + \frac{a^3 - A^3}{R^3}\right)^{1/3} = \lambda. \tag{5.3.11}$$

The principal stretches corresponding to the spherical polar coordinates r, θ, ϕ are respectively $\lambda^{-2}, \lambda, \lambda$ in the notation defined by (5.3.11). With the further notation

$$\lambda_a = a/A, \qquad \lambda_b = b/B, \tag{5.3.12}$$

we record the connections

$$(\lambda_a^3 - 1) = \left(\frac{R}{A}\right)^3 (\lambda^3 - 1) = \left(\frac{B}{A}\right)^3 (\lambda_b^3 - 1), \tag{5.3.13}$$

derivable from (5.3.11), and note that either

$$\lambda_a \geq \lambda \geq \lambda_b \geq 1 \tag{5.3.14}$$

or

$$\lambda_a \leq \lambda \leq \lambda_b \leq 1 \tag{5.3.15}$$

with equality if and only if $\lambda = 1$ for $A \leq R \leq B$.

As in Section 5.2.2 the radial equation of equilibrium has the form

$$\frac{\mathrm{d}t_1^{(1)}}{\mathrm{d}R} + \frac{2}{R}(t_1^{(1)} - t_2^{(1)}) = 0 \tag{5.3.16}$$

but now, from (5.3.9) specialized to isotropy and principal axes, we have

$$t_1^{(1)} = \frac{\partial W}{\partial \lambda_1} - p\lambda_1^{-1}, \qquad t_2^{(1)} = \frac{\partial W}{\partial \lambda_2} - p\lambda_2^{-1} \tag{5.3.17}$$

evaluated for $\lambda_1 = \lambda^{-2}$, $\lambda_2 = \lambda_3 = \lambda$.

From (5.3.11) we obtain

$$R\frac{\mathrm{d}\lambda}{\mathrm{d}R} = -(\lambda - \lambda^{-2})$$

so that, on replacing the independent variable R by λ, equation (5.3.16) becomes

$$\frac{\mathrm{d}t_1^{(1)}}{\mathrm{d}\lambda} = -2\frac{(t_1^{(1)} - t_2^{(1)})}{\lambda - \lambda^{-2}}. \tag{5.3.18}$$

For definiteness we assume that the boundary $R = B$ is free of traction while $R = A$ is subject to a pressure P per unit current area. It follows from the specialization of (5.1.25) to the present circumstances that the boundary conditions are

$$t_1^{(1)} = \begin{cases} -P\lambda_a^2 & \text{on} \quad R = A \\ 0 & \text{on} \quad R = B. \end{cases} \tag{5.3.19}$$

On introduction of the notation

$$\hat{W}(\lambda) = W(\lambda^{-2}, \lambda, \lambda) \tag{5.3.20}$$

we obtain

$$\tfrac{1}{2}\hat{W}'(\lambda) = t_2^{(1)} - \lambda^{-3} t_1^{(1)},$$

and integration of (5.3.18) with the help of (5.3.19) then yields

$$P = \int_{\lambda_b}^{\lambda_a} \frac{\hat{W}'(\lambda)\,\mathrm{d}\lambda}{\lambda^3 - 1} \tag{5.3.21}$$

after a little manipulation, where $\hat{W}' \equiv \mathrm{d}\hat{W}/\mathrm{d}\lambda$. We note, in particular, that (5.3.21) is valid for a general form of (incompressible) elastic strain-energy function and no analogous result was forthcoming in general for an unconstrained material (see Section 5.2.2).

Since, through (5.3.13), λ_b depends on λ_a, use of (5.3.13) after differentiation of (5.3.21) with respect to λ_a leads to

$$(\lambda_a - \lambda_a^{-2}) \frac{\mathrm{d}P}{\mathrm{d}\lambda_a} = \frac{\hat{W}'(\lambda_a)}{\lambda_a^2} - \frac{\hat{W}'(\lambda_b)}{\lambda_b^2}. \tag{5.3.22}$$

Clearly, the dependence of P on the (internal) radius $a = \lambda_a A$, as measured by λ_a, is governed by the properties of the function $\hat{W}'(\lambda)/\lambda^2$. This is brought out further in the case of a thin-walled shell, as we now show.

Let $(B - A)/A = \frac{1}{3}\varepsilon$, as in Section 5.2.2, and suppose $\varepsilon \ll 1$. Then, adapting the formula (5.2.44) to the present situation, we obtain

$$P = \tfrac{1}{3}\varepsilon \frac{\hat{W}'(\lambda)}{\lambda^2}, \tag{5.3.23}$$

neglecting terms of higher order in ε, where $\lambda = \lambda_a[1 + O(\varepsilon)]$ may take any value from λ_b to λ_a. Alternatively, (5.3.23) is obtainable directly from (5.3.21) on use of the mean-value theorem together with (5.3.13). It also follows (recall equation (5.2.44)) that $t_1^{(1)}$ is of order ε compared with $t_2^{(1)}$ so we have the further approximation

$$t_2^{(1)} = \tfrac{1}{2}\hat{W}'(\lambda). \tag{5.3.24}$$

The surface tension, T, in the wall of the thin shell is the traction per unit current length normal to a radial section through the wall, i.e. $(b - a)t_2$ apart from terms of order ε, where $b - a$ is the current thickness of the shell wall and t_2 is the in-surface component of Cauchy stress. Expressed in terms of Biot stress and reference geometry this is $(B-A)\lambda_1\lambda_2 t_2^{(1)}$ and hence, in terms of λ,

$$T=\tfrac{1}{6}A\varepsilon\frac{\hat{W}'(\lambda)}{\lambda}. \tag{5.3.25}$$

The corresponding Biot stress itself is given by (5.3.24) while the Cauchy stress is

$$t_2=\tfrac{1}{2}\lambda\hat{W}'(\lambda) \tag{5.3.26}$$

to the same order of approximation.

The comparative behaviour of $P, T, t_2^{(1)}$ and t_2 as functions of λ was alluded to in Section 4.4 in the context of equibiaxial deformations.

It is instructive to examine the properties of $\hat{W}'(\lambda)/\lambda^2$ and $\hat{W}'(\lambda)/\lambda$ in particular in respect of the model strain-energy function (4.3.66), for example. For the deformation considered here, we have

$$\hat{W}(\lambda)=\sum_{p=1}^{N}\frac{\mu_p}{\alpha_p}(2\lambda^{\alpha_p}+\lambda^{-2\alpha_p}-3)$$

so that

$$\lambda\hat{W}'(\lambda)=\sum_{p=1}^{N}2\mu_p(\lambda^{\alpha_p}-\lambda^{-2\alpha_p}).$$

It is easily deduced that the inequalities

$$\mu_p\alpha_p>0 \qquad (p=1,\ldots,N;\text{no summation}) \tag{5.3.27}$$

imply that t_2 is a monotonic increasing function of λ. The known elastic behaviour of real materials (to be discussed in Chapter 7) does not conflict with this property and we therefore adopt (5.3.27) here. Note, however, that in general they are sufficient for the stated property of t_2 but that this property does not in general necessitate the inequality $\mu_p\alpha_p>0$ for every $p\in\{1,2,\ldots,N\}$.

With (5.3.27) holding, the stronger inequalities

$$\alpha_p<-1 \quad \text{or} \quad \alpha_p>2 \qquad (p=1,\ldots,N) \tag{5.3.28}$$

are sufficient for $\hat{W}'(\lambda)/\lambda$ to be monotonic increasing in λ, while

$$\alpha_p<-\tfrac{3}{2} \quad \text{or} \quad \alpha_p>3 \qquad (p=1,\ldots,N) \tag{5.3.29}$$

are sufficient for $\hat{W}'(\lambda)/\lambda^2$ to be similarly monotonic.

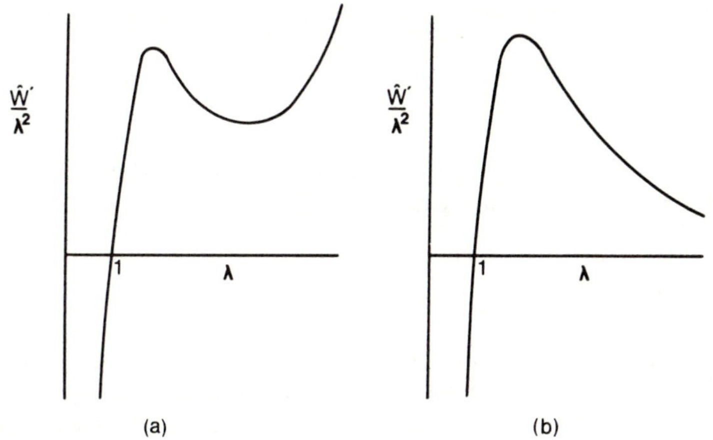

Fig. 5.5 Plot of $\hat{W}'(\lambda)/\lambda^2$ against λ for (a) the Mooney and (b) the neo-Hookean strain-energy functions.

In particular, for the Mooney strain-energy function (4.3.62), $\alpha_1 = 2$ and $\alpha_2 = -2$ so that each of (5.3.28) and (5.3.29) is violated. Nevertheless $\hat{W}'(\lambda)/\lambda$ is monotonic. On the other hand $\hat{W}'(\lambda)/\lambda^2$ has the characteristics illustrated in Fig. 5.5(a). The existence of a maximum and minimum should not be dismissed as physically unrealistic since this sort of behaviour has actually been observed in experiments on meteorological balloons. Moreover, the Mooney strain-energy function is widely accepted as a prototype model of the elastic behaviour of rubberlike materials. These points will be discussed further in Chapters 6 and 7.

The corresponding curve for the neo-Hookean strain-energy function (4.3.63) is shown in Fig. 5.5(b). We note that the decay of $\hat{W}'(\lambda)/\lambda^2$ for large λ is a reflection of the limited range of applicability of this model strain-energy function in respect of real materials (see Chapter 7 for detailed discussion of this point).

Problem 5.3.1 The spherically symmetric deformation of a spherical shell of isotropic elastic material which is *inextensible* in the radial direction is defined by

$$\lambda \equiv \frac{r}{R} = 1 + \frac{a - A}{R}$$

in the notation used in the above section.

Show that the principal components of Biot stress are given by

$$t_1^{(1)} = \frac{\partial W}{\partial \lambda_1} + 2q, \qquad t_2^{(1)} = t_3^{(1)} = \frac{\partial W}{\partial \lambda_2}$$

evaluated for $\lambda_1 = 1$, $\lambda_2 = \lambda_3 = \lambda$, where q is an arbitrary multiplier.

Show also that the analogue of (5.3.21) is

$$P = (1 - \lambda_a^{-1})^2 \int_{\lambda_b}^{\lambda_a} \frac{\hat{W}'(\lambda)\,\mathrm{d}\lambda}{(\lambda - 1)^3},$$

where $\hat{W}(\lambda) = W(1, \lambda, \lambda)$. Deduce that for thin-walled shells this reduces to the corresponding equation for incompressible materials, namely (5.3.23).

If $P > 0$ (and $\lambda > 1$) prove that the total volume of the shell material increases by an amount

$$4\pi(B - A)(a - A)(a + B)$$

during deformation from the reference configuration.

Problem 5.3.2 A circular cylinder of incompressible isotropic elastic material is deformed homogeneously by the action of axial loading and zero lateral tractions. By analogy with (5.3.20) the strain-energy function is defined by

$$\hat{W}(\lambda) = W(\lambda, \lambda^{-\frac{1}{2}}, \lambda^{-\frac{1}{2}}),$$

where $\lambda_1 = \lambda$ is the axial stretch.

Show that the axial Cauchy stress component is given by

$$t_1 = \lambda \hat{W}'(\lambda).$$

Deduce that, for the strain-energy function (4.3.66), this is a monotonic increasing function of λ if the inequalities (5.3.27) hold. Deduce further that a necessary condition for the corresponding Biot stress to possess a turning point for some $\lambda > 1$ is

$$-2 < \alpha_p < 1 \qquad \text{some } p \in \{1, \ldots, N\}.$$

5.3.3 Combined extension and inflation of a circular cylindrical tube

The incompressible specialization of the deformation examined in Section 5.2.3 is described by the equations

$$r^2 - a^2 = \lambda_z^{-1}(R^2 - A^2), \qquad \theta = \Theta, \qquad z = \lambda_z Z,$$

as in (2.2.71), and the principal stretches are

$$\lambda_1 = (\lambda\lambda_z)^{-1}, \qquad \lambda_2 = \frac{r}{R} \equiv \lambda, \qquad \lambda_3 = \lambda_z,$$

the notation λ being defined therein. Equations (2.2.73) should also be recalled.

We introduce the notation

$$\hat{W}(\lambda, \lambda_z) = W(\lambda_z^{-1}\lambda^{-1}, \lambda, \lambda_z) \tag{5.3.30}$$

and it follows that in terms of Biot stress

$$\left.\begin{aligned} \frac{\partial \hat{W}}{\partial \lambda} &= t_2^{(1)} - \lambda^{-2}\lambda_z^{-1}t_1^{(1)}, \\ \frac{\partial \hat{W}}{\partial \lambda_z} &= t_3^{(1)} - \lambda^{-1}\lambda_z^{-2}t_1^{(1)}. \end{aligned}\right\} \tag{5.3.31}$$

Integration of the equilibrium equation (5.2.48) and application of the boundary conditions (5.2.53) and (5.2.54) with the help of (2.2.73) yields

$$P = \int_{\lambda_b}^{\lambda_a} (\lambda^2\lambda_z - 1)^{-1} \frac{\partial \hat{W}}{\partial \lambda} \mathrm{d}\lambda, \tag{5.3.32}$$

the calculations being parallel to those leading to the result (5.3.21). Similarly the resultant axial loading, N, on the ends of the tube is expressible in the form

$$N = 2\pi A^2(\lambda_a^2\lambda_z - 1)\lambda_z \int_{\lambda_b}^{\lambda_a} (\lambda^2\lambda_z - 1)^{-2}\lambda t_3^{(1)} \mathrm{d}\lambda, \tag{5.3.33}$$

with

$$t_3^{(1)} = \frac{\partial \hat{W}}{\partial \lambda_z} - \lambda^{-1}\lambda_z^{-2} \int_{\lambda_b}^{\lambda} (\lambda^2\lambda_z - 1)^{-1} \frac{\partial \hat{W}}{\partial \lambda} \mathrm{d}\lambda. \tag{5.3.34}$$

The results

$$\lambda_a^{-1} \frac{\partial \hat{W}}{\partial \lambda_a}(\lambda_a, \lambda_z) - \lambda_b^{-1} \frac{\partial \hat{W}}{\partial \lambda_b}(\lambda_b, \lambda_z) = (\lambda_a\lambda_z - \lambda_a^{-1}) \frac{\partial P}{\partial \lambda_a} \tag{5.3.35}$$

and

$$\begin{aligned} (\lambda_a^2\lambda_z - 1)^{-1} &\frac{\partial \hat{W}}{\partial \lambda_z}(\lambda_a, \lambda_z) - (\lambda_b^2\lambda_z - 1)^{-1} \frac{\partial \hat{W}}{\partial \lambda_z}(\lambda_b, \lambda_z) \\ &= \lambda_a^{-1}\lambda_z^{-1} \frac{\partial N^*}{\partial \lambda_a} + (\lambda_a^2\lambda_z - 1)^{-1}(P\lambda_a^{-1}\lambda_z^{-2} - 2N^*), \end{aligned} \tag{5.3.36}$$

where $N^* = N/2\pi A^2$, follow from (5.3.32) and (5.3.33) with the help of (2.2.73). In the above λ_a and λ_z are taken as the independent variables.

For a thin-walled shell with $(B-A)/A = \varepsilon \ll 1$ the approximations

$$P = \varepsilon\lambda_z^{-1}\lambda^{-1}\frac{\partial\hat{W}}{\partial\lambda}, \tag{5.3.37}$$

$$N^* = \varepsilon\frac{\partial\hat{W}}{\partial\lambda_z} \tag{5.3.38}$$

may be deduced from (5.3.32) and (5.3.33) respectively, λ denoting any value from λ_b to λ_a. By writing

$$I_1 = \lambda + \lambda_z + \lambda^{-1}\lambda_z^{-1}, \quad I_2 = \lambda^{-1} + \lambda_z^{-1} + \lambda\lambda_z$$

and regarding W as a function of the independent symmetric invariants I_1 and I_2 it is easy to show from (5.3.37) and (5.3.38) that

$$P = 0 \quad \text{if and only if} \quad \lambda = \lambda_z^{-1/2}, \tag{5.3.39}$$

$$N = 0 \quad \text{if and only if} \quad \lambda = \lambda_z^{-2}. \tag{5.3.40}$$

More generally, for a thick-walled shell (5.3.39) also holds for $A \le R \le B$, the resulting deformation is homogeneous, and the results for simple tension described in Problem 5.3.2 are recovered. On the other hand (5.3.40) is not in general true for a thick-walled shell. Detailed calculations are left for the reader.

Problem 5.3.3 Show that for a thick-walled circular cylindrical tube made of incompressible isotropic elastic material with strain-energy function

$$W = \tfrac{1}{2}\mu_1(\lambda_1^2 + \lambda_2^2 + \lambda_3^2 - 3) - \tfrac{1}{2}\mu_2(\lambda_1^{-2} + \lambda_2^{-2} + \lambda_3^{-2} - 3),$$

the internal pressure P is given by

$$P = (\mu_1\lambda_z^{-2} - \mu_2)[\lambda_z \ln(\lambda_a/\lambda_b) - \tfrac{1}{2}(\lambda_a^{-2} - \lambda_b^{-2})].$$

Find the corresponding expression for the axial load N and, if $N = 0$, investigate the behaviour of λ_z as a function of λ_a.

If λ_z is fixed determine whether the equation $\mathrm{d}P/\mathrm{d}\lambda_a = 0$ has solutions $\lambda_a > 1$. Examine similarly the dependence of N on λ_z for fixed λ_a (remembering that λ_b depends on λ_z).

Problem 5.3.4 Show that the result (5.3.33) may be rewritten in the form

$$N/\pi A^2 = (\lambda_a^2\lambda_z - 1)\int_{\lambda_b}^{\lambda_a} (\lambda^2\lambda_z - 1)^{-2}\left(2\lambda_z\frac{\partial \hat{W}}{\partial \lambda_z} - \lambda\frac{\partial \hat{W}}{\partial \lambda}\right)\lambda\,\mathrm{d}\lambda + P\lambda_a^2.$$

Problem 5.3.5 If, for a thin-walled shell, we assume that $\partial N^*/\partial\lambda_z > 0$ prove that inflation is accompanied by shortening of the tube at fixed N^* provided $\partial^2\hat{W}/\partial\lambda\partial\lambda_z > 0$.

5.3.4 Flexure of a rectangular block

For compressible materials this problem was examined in Section 5.2.4. When the incompressibility constraint is imposed the deformation described by (5.2.60) with (5.2.69) simplifies to

$$r = \left(\beta + \frac{2X_1}{\alpha\lambda_3}\right)^{1/2}, \quad \theta = \alpha X_2, \quad z = \lambda_3 X_3, \tag{5.3.41}$$

as shown in (2.2.80) in different notation. From (5.2.64) it now follows that the principal stretches are

$$\lambda_1 = 1/\alpha\lambda_3 r, \qquad \lambda_2 = \alpha r, \qquad \lambda_3 \tag{5.3.42}$$

and we note that α, β and λ_3 are constants.

The principal Biot stresses in this case are

$$t_i^{(1)} = \frac{\partial W}{\partial \lambda_i} - p\lambda_i^{-1} \tag{5.3.43}$$

and, as in Section 5.2.4, the governing differential equation, which serves to determine p, is

$$\frac{\mathrm{d}t_1^{(1)}}{\mathrm{d}X_1} = \alpha t_2^{(1)}. \tag{5.3.44}$$

Regarding $W(\lambda_1, \lambda_2, \lambda_3)$ as a function of X_1 through (5.3.42) and (5.3.41), we find that

$$\begin{aligned}\frac{\mathrm{d}W}{\mathrm{d}X_1} &= -\lambda_1^2 r^{-1}\frac{\partial W}{\partial \lambda_1} + \lambda_1\lambda_2 r^{-1}\frac{\partial W}{\partial \lambda_2}\\ &= \lambda_1 r^{-1}(\lambda_2 t_2^{(1)} - \lambda_1 t_1^{(1)})\end{aligned}$$

and hence, with the help of (5.3.44),

$$\frac{\mathrm{d}W}{\mathrm{d}X_1}=\frac{\mathrm{d}}{\mathrm{d}X_1}(\lambda_1 t_1^{(1)}).$$

We deduce that

$$\lambda_1 t_1^{(1)}=W+\gamma, \tag{5.3.45}$$

where γ is a constant, or, equivalently,

$$p=\lambda_1\frac{\partial W}{\partial\lambda_1}-W-\gamma. \tag{5.3.46}$$

Application of the boundary condition (5.2.72) yields

$$W(\lambda_2^{-1}\lambda_3^{-1},\lambda_2,\lambda_3)+\gamma=0 \tag{5.3.47}$$

for

$$\lambda_2=\alpha\left(\beta\pm\frac{2A}{\alpha\lambda_3}\right)^{1/2} \tag{5.3.48}$$

and this provides two equations relating β and γ.

Assuming λ_3 is given the (normal) load, N, on the faces $X_3=\pm C$ of the block is obtained as follows. A short calculation based on (5.2.73) leads to

$$N=2B\lambda_3^{-1}\int_{-A}^{A}\left(\hat{W}-\lambda_1\frac{\partial\hat{W}}{\partial\lambda_1}+\gamma\right)\mathrm{d}X_1, \tag{5.3.49}$$

where

$$\hat{W}(\lambda_1,\lambda_2)=W(\lambda_1,\lambda_2,\lambda_1^{-1}\lambda_2^{-1}). \tag{5.3.50}$$

The moment, M, of the normal stresses $t_2^{(1)}$ on the faces $\theta=\pm\alpha B$ about the origin ($r=\theta=0$) is calculated as

$$M=-2C\alpha^{-1}\int_{r-}^{r+}t_1^{(1)}\,\mathrm{d}r$$

from (5.2.74) with (5.3.44), where

$$r_{\pm}=\left(\beta\pm\frac{2A}{\alpha\lambda_3}\right)^{1/2}.$$

Equivalently, by means of (5.3.45), this may be expressed in the form

$$M=2C\lambda_3^{-1}\alpha^{-2}\int_{\lambda_{1-}}^{\lambda_{1+}}\lambda_1^{-3}\tilde{W}\mathrm{d}\lambda_1, \tag{5.3.51}$$

where

$$\tilde{W}(\lambda_1,\lambda_3)=W(\lambda_1,\lambda_1^{-1}\lambda_3^{-1},\lambda_3) \tag{5.3.52}$$

and

$$\lambda_{1\pm}=\lambda_3^{-1}\alpha^{-1}\left(\beta\pm\frac{2A}{\alpha\lambda_3}\right)^{-1/2}.$$

Note that γ does not appear in (5.3.51).

Further discussion of this problem is provided in Green and Adkins (1970) and Green and Zerna (1968).

Problem 5.3.6 Evaluate M, N, β and γ explicitly in terms of λ_3 and α for the neo-Hookean strain-energy function

$$W=\tfrac{1}{2}\mu(\lambda_1^2+\lambda_2^2+\lambda_3^2-3).$$

Problem 5.3.7 Show that

$$\frac{\partial}{\partial\lambda_3}(N\lambda_3/2B)=\int_{-A}^{A}\lambda_1^2\frac{\partial^2\hat{W}}{\partial\lambda_1^2}\mathrm{d}X_1,$$

where $\hat{W}$ is given by (5.3.50).

Show also that

$$\alpha\frac{\partial}{\partial\alpha}(M\lambda_3\alpha^2/2C)=\tilde{W}^-/\lambda_{1-}^2-\tilde{W}^+/\lambda_{1+}^2+A\alpha\lambda_3(\tilde{W}^++\tilde{W}^-),$$

where $\tilde{W}$ is defined in (5.3.52) and $\tilde{W}^{\pm}$ indicates its evaluation for $\lambda_1=\lambda_{1\pm}$, λ_3 and α being independent.

5.3.5 Extension and torsion of a circular cylinder

We consider the deformation

$$r=\lambda^{-1/2}R,\qquad \theta=\Theta+\tau\lambda Z,\qquad z=\lambda Z \tag{5.3.53}$$

defined in (2.2.75) for a solid cylinder of incompressible material having radius A and length L in the reference configuration. As in Section 5.2.5 the Lagrangean principal axes are

$$\mathbf{E}_R,\quad \cos\psi\mathbf{E}_\Theta+\sin\psi\mathbf{E}_Z,\quad -\sin\psi\mathbf{E}_\Theta+\cos\psi\mathbf{E}_Z,$$

but, since (5.2.78) is now specialized to (5.3.53), equation (5.2.85) is replaced by

$$\tan 2\psi=2\tau R/(\lambda^{-1}-\lambda^2-\tau^2\lambda R^2). \tag{5.3.54}$$

The principal stretches λ_2 and λ_3 satisfy

$$\lambda_2\lambda_3=\lambda^{1/2},\qquad \lambda_2^2+\lambda_3^2=\lambda^{-1}+\lambda^2+\tau^2\lambda R^2 \tag{5.3.55}$$

and $\lambda_1=\lambda^{-1/2}$ is the principal stretch corresponding to $\mathbf{E}_R$.

Because of these connections, the expressions (5.2.91) and (5.2.92) for the axial load and torsional couple reduce to

$$N=2\pi\lambda\int_0^A\left\{\frac{\lambda_2t_2^{(1)}-\lambda_3t_3^{(1)}}{\lambda_2^2-\lambda_3^2}-\lambda^{-1}\frac{\lambda_2^{-1}t_2^{(1)}-\lambda_3^{-1}t_3^{(1)}}{\lambda_2^2-\lambda_3^2}\right\}R\,\mathrm{d}R \tag{5.3.56}$$

and

$$M=2\pi\tau\int_0^A\left(\frac{\lambda_2t_2^{(1)}-\lambda_3t_3^{(1)}}{\lambda_2^2-\lambda_3^2}\right)R^3\,\mathrm{d}R \tag{5.3.57}$$

respectively, the principal Biot stresses now being given by

$$t_i^{(1)}=\frac{\partial W}{\partial\lambda_i}-p\lambda_i^{-1}. \tag{5.3.58}$$

The equilibrium equation (5.2.94) becomes

$$\frac{\mathrm{d}}{\mathrm{d}R}(Rt_1^{(1)})=\lambda^{1/2}\left\{\frac{\lambda_2^3t_2^{(1)}-\lambda_3^3t_3^{(1)}}{\lambda_2^2-\lambda_3^2}-\lambda^2\left(\frac{\lambda_2t_2^{(1)}-\lambda_3t_3^{(1)}}{\lambda_2^2-\lambda_3^2}\right)\right\}, \tag{5.3.59}$$

and this enables p to be determined subject to the boundary condition

$$t_1^{(1)}=0\qquad \text{on } R=A. \tag{5.3.60}$$

By use of the result

$$\int_0^A t_1^{(1)} R\mathrm{d}R = -\int_0^A R\frac{\mathrm{d}}{\mathrm{d}R}(Rt_1^{(1)})\mathrm{d}R$$

together with (5.3.59) it is possible to rearrange (5.3.56) in the form

$$N = \tfrac{1}{2}\pi\lambda^{-1}\int_0^A \Big\{(\lambda_2 t_2^{(1)} + \lambda_3 t_3^{(1)} - 2\lambda_1 t_1^{(1)}) - 3\left(\frac{\lambda_2 t_2^{(1)} - \lambda_3 t_3^{(1)}}{\lambda_2^2 - \lambda_3^2}\right)(\lambda_2^2 + \lambda_3^2 - 2\lambda^2)\Big\} R\,\mathrm{d}R \quad (5.3.61)$$

so that p is eliminated. The details of the calculation are left as an exercise.

Furthermore, by introducing the notation

$$\hat{W}(\lambda_1, \lambda_2) = W(\lambda_1, \lambda_2, \lambda_1^{-1}\lambda_2^{-1}) \quad (5.3.62)$$

and changing the variable of integration from R to λ_2 in (5.3.57) and (5.3.61), we obtain

$$M = 2\pi\tau^{-3}\lambda^{-2}\int_\lambda^{\lambda_A}(\lambda_2^2 + \lambda\lambda_2^{-2} - \lambda^2 - \lambda^{-1})\frac{\partial \hat{W}}{\partial \lambda_2}\mathrm{d}\lambda_2 \quad (5.3.63)$$

and

$$N = \tfrac{1}{2}\pi\tau^{-2}\lambda^{-2}\int_\lambda^{\lambda_A}\Big\{\left(\lambda_2\frac{\partial \hat{W}}{\partial \lambda_2} - 2\lambda_1\frac{\partial \hat{W}}{\partial \lambda_1}\right)(\lambda_2^2 - \lambda\lambda_2^{-2}) - 3(\lambda_2^2 + \lambda\lambda_2^{-2} - 2\lambda^2)\lambda_2\frac{\partial \hat{W}}{\partial \lambda_2}\Big\}\lambda_2^{-1}\mathrm{d}\lambda_2, \quad (5.3.64)$$

where λ_A is the largest positive solution for λ_2 of (5.3.55) with $R = A$. Thus

$$2\lambda_A = [(\lambda + \lambda^{-1/2})^2 + \lambda\tau^2 A^2]^{1/2} + [(\lambda - \lambda^{-1/2})^2 + \lambda\tau^2 A^2]^{1/2}, \quad (5.3.65)$$

and, for fixed A and λ, λ_A is therefore a measure of the torsion.

It follows from (5.3.63) and (5.3.64) that

$$\frac{\partial}{\partial \lambda_A}(M\tau^3\lambda^3/2\pi) = \tau^2\lambda A^2\frac{\partial \hat{W}}{\partial \lambda_2} \quad (5.3.66)$$

and

$$\frac{\partial}{\partial\lambda_A}(2N\tau^2\lambda^2/\pi)=\left(\lambda_2\frac{\partial\hat{W}}{\partial\lambda_2}-2\lambda_1\frac{\partial\hat{W}}{\partial\lambda_1}\right)(\lambda_2-\lambda\lambda_2^{-3})$$
$$-3(\lambda_2^2+\lambda\lambda_2^{-2}-2\lambda^2)\frac{\partial\hat{W}}{\partial\lambda_2}, \tag{5.3.67}$$

where the right-hand sides are evaluated for $\lambda_2=\lambda_A$, $\lambda_1=\lambda^{-1/2}$ and the differentiations on the left-hand sides are carried out for fixed λ.

Problem 5.3.8 Calculate $\partial\lambda_A/\partial\tau$ for fixed λ and obtain expressions for $\partial\hat{W}/\partial\lambda_1$ and $\partial\hat{W}/\partial\lambda_2$ at $\lambda_2=\lambda_A$ in terms of $\partial M^*/\partial\tau$ and $\partial N^*/\partial\tau$, where $M^*=M\tau^3\lambda^2/2\pi$ and $N^*=2N\tau^2\lambda^2/\pi$.

By writing $\Phi(I_1,I_2)=\hat{W}(\lambda_1,\lambda_2)$, where $I_1=\lambda_1+\lambda_2+\lambda_1^{-1}\lambda_2^{-1}$, $I_2=\lambda_1^{-1}+\lambda_2^{-1}+\lambda_1\lambda_2$, deduce corresponding expressions for $\partial\Phi/\partial I_1$ and $\partial\Phi/\partial I_2$.

Problem 5.3.9 Evaluate M and N for the Mooney strain-energy function

$$W=\tfrac{1}{2}\mu_1(\lambda_1^2+\lambda_2^2+\lambda_3^3-3)-\tfrac{1}{2}\mu_2(\lambda_1^{-2}+\lambda_2^{-2}+\lambda_3^{-2}-3)$$

and hence, or otherwise, obtain $\partial M/\partial\tau$ and $\partial N/\partial\tau$. If $\mu_1>0$ and $\mu_2\leq 0$ deduce that $\partial M/\partial\tau>0$ and $\partial N/\partial\tau<0$ and interpret these results.

Problem 5.3.10 A circular cylindrical tube of incompressible isotropic elastic material has internal and external radii A and B and length L in some reference configuration. It is deformed according to

$$r^2=\lambda^{-1}(R^2-A^2)+a^2,\quad \theta=\Theta+\tau\lambda Z,\quad z=\lambda Z$$

and is subject to the boundary conditions

$$t_1^{(1)}=\begin{cases}-P & \text{on}\quad R=A\\ 0 & \text{on}\quad R=B\end{cases}$$

in the notation of Sections 5.3.3. and 5.3.5.

Obtain expressions for the resultant axial loading N and resultant torsional couple M on the ends of the tube and the pressure P on $R=A$ which are analogous to (5.3.57), and (5.3.61) and (5.3.32) respectively.

5.3.6 Shear of a circular cylindrical tube

From equation (2.2.81) the combined axial and torsional shear of a circular cylindrical tube of incompressible material is given by

$$r=R,\quad \theta=\Theta+\varphi(R),\quad z=Z+w(R)$$

in cylindrical polar coordinates, and the deformation gradient may be written

$$\mathbf{A} = \mathbf{e}_r \otimes \mathbf{E}_R + \mathbf{e}_\theta \otimes \mathbf{E}_\Theta + \mathbf{e}_z \otimes \mathbf{E}_Z + w'\mathbf{e}_z \otimes \mathbf{E}_R + R\varphi'\mathbf{e}_\theta \otimes \mathbf{E}_R, \tag{5.3.68}$$

where $w' \equiv \mathrm{d}w/\mathrm{d}r$, $\varphi' \equiv \mathrm{d}\varphi/\mathrm{d}r$. As indicated in Section 2.2.6 this corresponds locally to a simple shear of amount

$$v \equiv \lambda - \lambda^{-1} = (w'^2 + r^2\varphi'^2)^{1/2}, \tag{5.3.69}$$

where $\lambda \geq 1$ and the principal stretches $\lambda^{-1}, 1, \lambda$ are associated with the Eulerian axes

$$\left.\begin{aligned} \mathbf{v}^{(1)} &\equiv \cos\psi\mathbf{e}_r - \sin\psi(\sin\chi\mathbf{e}_\theta + \cos\chi\mathbf{e}_z),\\ \mathbf{v}^{(2)} &\equiv \cos\chi\mathbf{e}_\theta - \sin\chi\mathbf{e}_z,\\ \mathbf{v}^{(3)} &\equiv \sin\psi\mathbf{e}_r + \cos\psi(\sin\chi\mathbf{e}_\theta + \cos\chi\mathbf{e}_z) \end{aligned}\right\} \tag{5.3.70}$$

in that order. Furthermore, we recall the connections

$$\tan\chi = r\varphi'/w' \tag{5.3.71}$$

and

$$\tan 2\psi = 2/v. \tag{5.3.72}$$

For an isotropic elastic material the Cauchy stress $\mathbf{T}$ is coaxial with the Eulerian strain ellipsoid and may be written in the spectral form

$$\mathbf{T} = t_1\mathbf{v}^{(1)} \otimes \mathbf{v}^{(1)} + t_2\mathbf{v}^{(2)} \otimes \mathbf{v}^{(2)} + t_3\mathbf{v}^{(3)} \otimes \mathbf{v}^{(3)}.$$

It follows from (5.3.70) that the components of $\mathbf{T}$ relative to $\mathbf{e}_r, \mathbf{e}_\theta, \mathbf{e}_z$ are

$$\left.\begin{aligned} T_{rr} &= t_1\cos^2\psi + t_3\sin^2\psi,\\ T_{\theta\theta} &= (t_1\sin^2\psi + t_3\cos^2\psi)\sin^2\chi + t_2\cos^2\chi,\\ T_{zz} &= (t_1\sin^2\psi + t_3\cos^2\psi)\cos^2\chi + t_2\sin^2\chi,\\ T_{r\theta} &= (t_3 - t_1)\sin\psi\cos\psi\sin\chi,\\ T_{rz} &= (t_3 - t_1)\sin\psi\cos\psi\cos\chi,\\ T_{\theta z} &= (t_1\sin^2\psi + t_3\cos^2\psi - t_2)\sin\chi\cos\chi. \end{aligned}\right\} \tag{5.3.73}$$

We note, in particular, the connections

$$T_{r\theta} = T_{rz}\tan\chi, \tag{5.3.74}$$

$$2T_{\theta z} = (T_{zz} - T_{\theta\theta})\tan 2\chi, \tag{5.3.75}$$

and elimination of χ leads to the universal relation

$$T_{rz}T_{r\theta}(T_{zz} - T_{\theta\theta}) = T_{\theta z}(T_{rz}^2 - T_{r\theta}^2) \tag{5.3.76}$$

between the stress components.

The equilibrium equation $\operatorname{div}\mathbf{T} = \mathbf{0}$ yields

$$\left.\begin{aligned} \frac{\mathrm{d}T_{rr}}{\mathrm{d}r} + \frac{1}{r}(T_{rr} - T_{\theta\theta}) = 0,\\ \frac{\mathrm{d}}{\mathrm{d}r}(r^2 T_{r\theta}) = 0,\\ \frac{\mathrm{d}}{\mathrm{d}r}(rT_{rz}) = 0, \end{aligned}\right\} \tag{5.3.77}$$

and we note in passing that, since $r = R$ is the only independent variable, these equations are easily recast in terms of the components of nominal stress. In this regard we record the connection $\mathbf{T} = \mathbf{AS}$ appropriate to incompressible materials and observe, in particular, that

$$\left.\begin{aligned} &T_{rr} = S_{Rr}, \qquad T_{\theta\theta} = S_{\Theta\theta} + R\varphi' S_{R\theta},\\ &T_{r\theta} = S_{R\theta} = S_{\Theta r} + R\varphi' S_{Rr}, \quad T_{rz} = S_{Rz} = S_{Zr} + w' S_{Rr}. \end{aligned}\right\} \tag{5.3.78}$$

The second and third equations in (5.3.77) yield

$$T_{r\theta} = \sigma/2\pi r^2 \tag{5.3.79}$$

and

$$T_{rz} = \tau/2\pi r \tag{5.3.80}$$

respectively, where σ and τ are constants. In fact, τ represents the axial shear force on a cylinder $r = \text{constant}$ per unit length of the cylinder, while σ is the resultant moment of the azimuthal shear forces about the cylinder axis (also per unit length). It follows immediately that

$$T_{r\theta}/T_{rz} = \sigma/\tau r. \tag{5.3.81}$$

Since the right-hand side is independent of the constitutive law, equation (5.3.81) is another universal relation between stress components.

From (5.3.71) and (5.3.74) we now obtain

$$r\varphi'/w' = \sigma/\tau r \tag{5.3.82}$$

and therefore, with the formula $\sin 2\psi = 2/(\lambda + \lambda^{-1})$ derived from (5.3.72) and (5.3.69), we deduce that each of (5.3.79) and (5.3.80) leads to

$$\frac{t_3 - t_1}{\lambda + \lambda^{-1}} = \frac{(\sigma^2 + \tau^2 r^2)^{1/2}}{2\pi r^2}. \tag{5.3.83}$$

Essentially, this equation serves to determine λ in terms of r for given σ and τ.

For a hyperelastic material with strain-energy function $W(\lambda_1, \lambda_2, \lambda_3)$ we have

$$t_i = \lambda_i \frac{\partial W}{\partial \lambda_i} - p \qquad (i = 1, 2, 3)$$

evaluated for $\lambda_1 = \lambda^{-1}, \lambda_2 = 1, \lambda_3 = \lambda$ and, by writing

$$\hat{W}(\lambda) = W(\lambda^{-1}, 1, \lambda),$$

we therefore obtain

$$\lambda \hat{W}'(\lambda) = t_3 - t_1.$$

Insertion of this into the left-hand side of (5.3.83) renders it explicit in terms of λ. Thus

$$\frac{\lambda \hat{W}'(\lambda)}{\lambda + \lambda^{-1}} = \frac{(\sigma^2 + \tau^2 r^2)^{1/2}}{2\pi r^2}.$$

However, it turns out to be more convenient to use $v \equiv \lambda - \lambda^{-1}$ rather than λ as the deformation parameter here and the above equation then becomes

$$\tilde{W}'(v) = \frac{(\sigma^2 + \tau^2 r^2)^{1/2}}{2\pi r^2}, \tag{5.3.84}$$

where $\tilde{W}(v) \equiv \hat{W}(\lambda)$.

The right-hand side of (5.3.84) is a monotonic function of r, decreasing as r increases, and therefore v (and hence λ) is uniquely determined as a function of r if $\tilde{W}'(v)$ is a monotonic function of v. In fact, to make physical sense $\tilde{W}'(v)$ should be monotonic increasing with v. We assume that this is the case and write

$$v = \tilde{v}(r) \tag{5.3.85}$$

as the solution of (5.3.84).

It follows from (5.3.69) and (5.3.82) that

$$w'(r) = \frac{\tau r \tilde{v}(r)}{(\sigma^2 + \tau^2 r^2)^{1/2}}, \qquad r\varphi'(r) = \frac{\sigma \tilde{v}(r)}{(\sigma^2 + \tau^2 r^2)^{1/2}}. \tag{5.3.86}$$

Supposing that the inner surface $r = a$ of the tube is held fixed during deformation and that the outer surface $r = b$ is subjected to an axial displacement d and an azimuthal angular displacement δ, we have

$$\left.\begin{aligned} &w(a) = \varphi(a) = 0 \\ &w(b) = d, \quad \varphi(b) = \delta. \end{aligned}\right\} \tag{5.3.87}$$

Integration of (5.3.86) and use of these boundary conditions gives

$$d = \tau \int_a^b \frac{r\tilde{v}(r)\,\mathrm{d}r}{(\sigma^2 + \tau^2 r^2)^{1/2}}, \tag{5.3.88}$$

$$\delta = \sigma \int_a^b \frac{\tilde{v}(r)\,\mathrm{d}r}{r(\sigma^2 + \tau^2 r^2)^{1/2}}, \tag{5.3.89}$$

and these provide expressions for d and δ in terms of σ and τ.

The normal stress on a surface $r = \text{constant}$ is T_{rr} and this is obtained by integration of the first equation in (5.3.77). Thus

$$T_{rr} = T_{rr}(a) - \int_a^r (T_{rr} - T_{\theta\theta}) \frac{\mathrm{d}r}{r}. \tag{5.3.90}$$

From (5.3.73) we obtain

$$T_{rr} - T_{\theta\theta} = \tfrac{1}{2}(t_1 - t_3)\cos 2\psi(2 - \cos^2\chi) + \tfrac{1}{2}(t_1 + t_3 - 2t_2)\cos^2\chi \tag{5.3.91}$$

and, since the principal Cauchy stresses occur in this expression only through their differences, the hydrostatic pressure p does not appear in the integrand of (5.3.90). Hence the integrand is a known function of λ (or v) and therefore of r from (5.3.83), in principle if not explicitly. Essentially (5.3.90) serves to determine p through T_{rr}. Once T_{rr} has been evaluated $T_{\theta\theta}$ is obtained from

(5.3.91) and then T_{zz} from

$$T_{zz} - T_{\theta\theta} = \tfrac{1}{2}(t_3 - t_1)\cos 2\psi \cos 2\chi + \tfrac{1}{2}(t_1 + t_3 - 2t_2)\cos 2\chi. \tag{5.3.92}$$

On a surface $r = \text{constant}$, the traction components are T_{rr}, $T_{r\theta}$, T_{rz}. In view of (5.3.79) and (5.3.80), $T_{r\theta}$ and T_{rz} are determined from given values of σ and τ or, equivalently, δ and d. On the end faces of the tube the traction components are

$$S_{Zr} = T_{rz} - w'T_{rr},$$
$$S_{Z\theta} = T_{\theta z} - w'T_{r\theta},$$
$$S_{Zz} = T_{zz} - w'T_{rz},$$

and, since these are independent of z, we consider their resultants. By symmetry there is no resultant radial traction over the end faces. The resultant moment, M say, of the azimuthal shear stresses is

$$M = 2\pi \int_a^b S_{Z\theta} r^2 \, dr$$

and, with the help of (5.3.79) and (5.3.87), this may be rewritten as

$$M = 2\pi \int_a^b T_{\theta z} r^2 \, dr - \sigma d. \tag{5.3.93}$$

From (5.3.73) we obtain

$$T_{\theta z} = \tfrac{1}{4}(t_1 + t_3 - 2t_2)\sin 2\chi + \tfrac{1}{4}(t_3 - t_1)\cos 2\psi \sin 2\chi \tag{5.3.94}$$

and, since this is known as a function of the deformation, M cannot be specified independently of σ and τ.

The resultant axial load, N say, on the end faces is calculated similarly as

$$N = 2\pi \int_a^b T_{zz} r \, dr - \tau d. \tag{5.3.95}$$

Since $T_{zz} - T_{rr}$ is known, we may prescribe one of N, $T_{rr}(a)$ or $T_{rr}(b)$ in order to determine the constant of integration in (5.3.90). In particular, $N = 0$ would be an appropriate experimentally realizable boundary condition.

For the special case of pure torsional shear $\tau = d = 0$ and $\chi = \tfrac{1}{2}\pi$ and hence $M=0$ follows since $T_{\theta z}=0$ from (5.3.94). For pure axial shear $\sigma=\delta=\chi=0$ and again $M=0$. In the combined deformation, however, $M\neq 0$ in general.

In Problem 5.2.5 the axial shear deformation of an unconstrained material was considered. The results described in that problem, in particular equation (5.2.97), should be compared with the corresponding results given above specialized to the case $\sigma=0$.

To illustrate the above results we consider the Mooney strain-energy function

$$W=\tfrac{1}{2}\mu_1(\lambda_1^2+\lambda_2^2+\lambda_3^2-3)-\tfrac{1}{2}\mu_2(\lambda_1^{-2}+\lambda_2^{-2}+\lambda_3^{-2}-3),$$

where $\mu_1>0$, $\mu_2<0$. With $\lambda_1=\lambda^{-1}$, $\lambda_2=1$, $\lambda_3=\lambda$, this may be rewritten as

$$\tilde{W}=\tfrac{1}{2}(\mu_1-\mu_2)v^2. \tag{5.3.96}$$

It follows from (5.3.84) that

$$v=\frac{(\sigma^{*2}+\tau^{*2}r^2)^{1/2}}{2\pi r^2}, \tag{5.3.97}$$

where $\sigma^*=\sigma/(\mu_1-\mu_2)$, $\tau^*=\tau/(\mu_1-\mu_2)$, and hence, from (5.3.86) and (5.3.87), that

$$w(r)=\frac{\tau^*}{2\pi}\ln(r/a), \qquad \varphi(r)=\frac{\sigma^*}{4\pi a^2}\left(1-\frac{a^2}{r^2}\right).$$

It follows that

$$d=\frac{\tau^*}{2\pi}\ln(b/a), \qquad \delta=\frac{\sigma^*}{4\pi a^2}(1-a^2/b^2), \tag{5.3.98}$$

and these show that d and δ depend linearly on τ and σ respectively, there being no coupling between the azimuthal and axial shears in these equations.

A short calculation leads to

$$M=\mu_2\sigma^* d \tag{5.3.99}$$

and this shows that M is of order d/L times the total azimuthal shear force applied to the lateral boundary. Note, in particular, that $M=0$ for the neo-Hookean strain-energy function ($\mu_2=0$).

For the considered material, we have

$$T_{rr}-T_{\theta\theta}=\frac{(\mu_1-\mu_2)\sigma^{*2}}{4\pi^2 r^4}+\frac{\mu_1\tau^{*2}}{4\pi^2 r^2}$$

and substitution of this into (5.3.90) leads to

$$T_{rr}=T_{rr}(a)-\frac{(\mu_1-\mu_2)\sigma^{*2}}{16\pi^2a^4}\left(1-\frac{a^2}{r^4}\right)-\frac{\mu_1\tau^{*2}}{8\pi^2a^2}\left(1-\frac{a^2}{r^2}\right).$$

Use of this and the expression for $T_{zz}-T_{rr}$ in (5.3.95) then gives

$$N=\pi b^2T_{rr}(b)-\pi a^2T_{rr}(a)-\tfrac{1}{2}(\mu_1-2\mu_2)\tau^*d$$
$$-(2\mu_1-\mu_2)\sigma^{*2}\ln(b/a)/2\pi$$

which confirms that specification of $T_{rr}(a)$ is equivalent to that of N.

Finally, we note that for the Mooney material the stresses arising in the combined deformation are obtained by adding the stresses required for the constituent axial and torsional deformations except in the case of

$$T_{\theta z}=\mu_1\sigma^*\tau^*/4\pi^2r^3 \tag{5.3.100}$$

which vanishes if either $\sigma=0$ or $\tau=0$. Detailed calculations are left as an exercise.

Problem 5.3.11 Show that the unit Lagrangean principal axes associated with the Eulerian principal axes given in (5.3.70) are

$$\lambda(\cos\psi-\nu\sin\psi)\mathbf{E}_R-\lambda\sin\psi\sin\chi\mathbf{E}_\Theta-\lambda\sin\psi\cos\chi\mathbf{E}_Z,$$
$$\cos\chi\mathbf{E}_\Theta-\sin\chi\mathbf{E}_Z,$$
$$\lambda^{-1}(\sin\psi-\nu\cos\chi)\mathbf{E}_R+\lambda^{-1}\cos\psi\sin\chi\mathbf{E}_\Theta+\lambda^{-1}\cos\psi\cos\chi\mathbf{E}_Z.$$

Problem 5.3.12 Use equation (5.3.88) to show that if $\sigma=0$ then

$$\tau^2\frac{\mathrm{d}}{\mathrm{d}\tau}\left(\frac{d}{\tau}\right)=a\tilde{v}(a)-b\tilde{v}(b).$$

Deduce, using the assumed monotonicity properties of $\tilde{W}'(v)$, that

$$\tau\frac{\mathrm{d}(d)}{\mathrm{d}\tau}>a[\tilde{v}(a)-\tilde{v}(b)]>0.$$

For the Mooney material, show further that

$$\frac{\mathrm{d}(d)}{\mathrm{d}\tau}=\frac{d}{\tau}.$$

From equation (5.3.89) with $\tau = 0$ show that

$$2\sigma \frac{\mathrm{d}(\delta)}{\mathrm{d}\sigma} = \tilde{v}(a) - \tilde{v}(b).$$

Problem 5.3.13 The combined extension, inflation, axial and torsional shear of a circular cylindrical tube is defined by

$$r^2 = \lambda_z^{-1}(R^2 - A^2) + a^2, \qquad \theta = \Theta + \varphi(R), \qquad z = \lambda_z Z + w(R).$$

Find the principal stretches and the orientations of the Eulerian principal axes and use this information to generalize the results described in Section 5.3.6.

5.3.7 Rotation of a solid circular cylinder about its axis

We consider a solid circular cylinder of reference radius A and length L as defined in Problem 5.2.6. The cylinder rotates about its axis with constant angular speed ω. For an incompressible material the principal stretches are $\lambda^{-1/2}$, $\lambda^{-1/2}$, λ, corresponding respectively to the r, θ, z directions, and λ is a constant. Since $\lambda_1 = \lambda_2$, and hence $t_1^{(1)} = t_2^{(1)}$, the governing equation (5.2.98) simplifies to

$$\frac{\mathrm{d}t_1}{\mathrm{d}r} = -\rho\omega^2 r$$

where $t_1 = \lambda^{-1/2} t_1^{(1)}$ is the radial Cauchy stress and $\rho = \rho_0$ the (uniform) material density.

Taking the lateral surface $r = a \equiv \lambda^{-1/2} A$ to be stress free, we obtain

$$t_1 = \tfrac{1}{2}\rho\omega^2(a^2 - r^2). \tag{5.3.101}$$

We write the strain-energy function as

$$\hat{W}(\lambda) = W(\lambda^{-1/2}, \lambda^{-1/2}, \lambda)$$

so that

$$\lambda \hat{W}'(\lambda) = t_3 - t_1 \tag{5.3.102}$$

and this equation determines the axial stress t_3.

With the help of (5.3.101) and (5.3.102) we calculate the resultant axial load N as

$$N = 2\pi \int_0^a t_3 r \,\mathrm{d}r = \pi A^2\{\hat{W}'(\lambda) + \tfrac{1}{4}\rho\omega^2 A^2 \lambda^{-2}\}. \tag{5.3.103}$$

We consider three special cases:

(a) $\omega = 0$. This corresponds to simple tension with

$$N = \pi A^2 \hat{W}'(\lambda).$$

Assuming that extension (respectively contraction) accompanies tension (respectively compression), we deduce that

$$\hat{W}'(\lambda) \gtreqless 0 \quad \text{according as} \quad \lambda \gtreqless 1. \tag{5.3.104}$$

(b) $\lambda = 1$. If the length of the cylinder is fixed we have

$$N = \tfrac{1}{4}\pi\rho\omega^2 A^4,$$

i.e. tension is required.

(c) $N = 0$. For a freely rotating cylinder, we obtain

$$\lambda^2 \hat{W}'(\lambda) = -\tfrac{1}{4}\rho\omega^2 A^4. \tag{5.3.105}$$

In view of (5.3.104) this implies that *rotation is accompanied by shortening of the cylinder.*

If $\lambda^2 \hat{W}'(\lambda)$ is a monotonic function of λ then λ is uniquely determined by ω from (5.3.105) for given ρ and A. We now examine this function in respect of the strain energy (4.3.66), for which

$$\hat{W}(\lambda) = \sum_{p=1}^{N} \mu_p(2\lambda^{-(1/2)\alpha_p} + \lambda^{\alpha_p} - 3)/\alpha_p.$$

It follows that

$$\frac{d}{d\lambda}\{\lambda^2 \hat{W}'(\lambda)\} = \sum_{p=1}^{N} \mu_p\{(\alpha_p + 1)\lambda^{\alpha_p} + \tfrac{1}{2}(\alpha_p - 2)\lambda^{-(1/2)\alpha_p}\}$$

and a sufficient condition for this to be positive is

$$\alpha_p \leq -1 \quad \text{or} \quad \alpha_p \geq 2 \qquad \text{each } p \in \{1, \ldots, N\} \tag{5.3.106}$$

provided the material constants satisfy $\mu_p \alpha_p > 0$, as in (5.3.27). We note, in particular, that (5.3.106) holds for the Mooney material.

5.4 VARIATIONAL PRINCIPLES AND CONSERVATION LAWS

5.4.1 Virtual work and related principles

We recall from Section 5.1.1 that the equations which govern the deformation and stress fields $\boldsymbol{\chi}$ and $\mathbf{S}$ for a Green-elastic body in equilibrium are

$$\operatorname{Div}\mathbf{S} + \rho_0\mathbf{b} = \mathbf{0}, \tag{5.4.1}$$

$$\mathbf{S} = \frac{\partial W}{\partial \mathbf{A}}, \tag{5.4.2}$$

$$\mathbf{A} = \operatorname{Grad}\boldsymbol{\chi} \tag{5.4.3}$$

for each point of $\mathscr{B}_0$ (the position occupied by the body in some reference configuration), where $\mathbf{b}$ is a function of $\mathbf{x} = \boldsymbol{\chi}(\mathbf{X})$. In addition, we have the rotational balance equation

$$\mathbf{S}^{\mathrm{T}}\mathbf{A}^{\mathrm{T}} = \mathbf{A}\mathbf{S}, \tag{5.4.4}$$

which is a consequence of (5.4.2) for objective strain-energy functions W, as remarked in Section 5.1.1, and in this section we take W to be objective.

Let $\partial\mathscr{B}_0$ be the boundary of $\mathscr{B}_0$ and suppose that $\boldsymbol{\chi}$ is specified on the part $\partial\mathscr{B}_0^x$ of $\partial\mathscr{B}_0$ and that the traction, $\mathbf{S}^{\mathrm{T}}\mathbf{N}$ per unit reference area, is prescribed on $\partial\mathscr{B}_0^\sigma$, where $\partial\mathscr{B}_0^x$ and $\partial\mathscr{B}_0^\sigma$ are disjoint and $\partial\mathscr{B}_0^x \cup \partial\mathscr{B}_0^\sigma = \partial\mathscr{B}_0$. We write

$$\boldsymbol{\chi} = \boldsymbol{\xi} \quad \text{on} \quad \partial\mathscr{B}_0^x, \tag{5.4.5}$$

$$\mathbf{S}^{\mathrm{T}}\mathbf{N} = \boldsymbol{\sigma} \quad \text{on} \quad \partial\mathscr{B}_0^\sigma, \tag{5.4.6}$$

where $\boldsymbol{\xi}$ is a known function on $\partial\mathscr{B}_0^x$ and, as in Section 5.1.2, $\boldsymbol{\sigma}$ is in general a deformation-sensitive loading depending (in a known way) on $\boldsymbol{\chi}$ and $\mathbf{A}$ on $\partial\mathscr{B}_0^\sigma$.

A (twice-continuously differentiable†) vector field $\boldsymbol{\chi}$ defined on $\mathscr{B}_0$ and satisfying the boundary condition (5.4.5) is said to be a *kinematically admissible deformation.* Although such a field may be associated with a second-order tensor field $\mathbf{S}$ through (5.4.2) and (5.4.3), this $\mathbf{S}$ will not necessarily satisfy one or both of (5.4.1) and (5.4.6).

Dually, a (continuously differentiable†) second-order tensor field $\mathbf{S}$ defined on $\mathscr{B}_0$ and satisfying the equilibrium equation (5.4.1) and the boundary condition (5.4.6) with $\mathbf{b}$ and $\boldsymbol{\sigma}$ defined for any kinematically admissible deformation $\boldsymbol{\chi}$ is said to be a *statically admissible* (nominal) *stress.* Clearly, $\mathbf{S}$ depends on the choice of $\boldsymbol{\chi}$ in general, but $\boldsymbol{\chi}$ and $\mathbf{S}$ are not necessarily

†These restrictions may be relaxed in a more general treatment.

related through (5.4.2)–(5.4.4). When there are no body forces and $\boldsymbol{\sigma}$ corresponds to dead loading the definition of a statically admissible stress does not involve the choice of a kinematically admissible deformation.

Suppose that $\boldsymbol{\chi}^*$ and $\boldsymbol{\chi}'$ are two kinematically admissible deformations and that $\mathbf{S}'$ is a statically admissible stress corresponding to $\boldsymbol{\chi}'$ with body force $\mathbf{b}' \equiv \mathbf{b}(\boldsymbol{\chi}')$ and surface traction $\boldsymbol{\sigma}' \equiv \boldsymbol{\sigma}(\mathbf{X}, \boldsymbol{\chi}', \mathbf{A}')$, where $\mathbf{A}' = \operatorname{Grad} \boldsymbol{\chi}'$. Thus

$$\operatorname{Div} \mathbf{S}' + \rho_0 \mathbf{b}' = \mathbf{0} \quad \text{in } \mathcal{B}_0,$$
$$\mathbf{S}'^{\mathrm{T}}\mathbf{N} = \boldsymbol{\sigma}' \quad \text{on } \partial\mathcal{B}_0^{\sigma},$$

and it follows on use of the divergence theorem that

$$\int_{\partial\mathcal{B}_0^x} \mathbf{N}\cdot(\mathbf{S}'\boldsymbol{\xi})\,\mathrm{d}A + \int_{\partial\mathcal{B}_0^{\sigma}} \boldsymbol{\sigma}'\cdot\boldsymbol{\chi}^*\,\mathrm{d}A = \int_{\partial\mathcal{B}_0} \mathbf{N}\cdot(\mathbf{S}'\boldsymbol{\chi}^*)\,\mathrm{d}A \tag{5.4.7}$$
$$= \int_{\mathcal{B}_0} \operatorname{Div}(\mathbf{S}'\boldsymbol{\chi}^*)\,\mathrm{d}V = \int_{\mathcal{B}_0} \operatorname{tr}(\mathbf{A}^*\mathbf{S}')\,\mathrm{d}V - \int_{\mathcal{B}_0} \rho_0 \mathbf{b}'\cdot\boldsymbol{\chi}^*\,\mathrm{d}V,$$

where $\mathbf{A}^* = \operatorname{Grad} \boldsymbol{\chi}^*$. (This result does not rely on the existence of a strain-energy function.)

Firstly, if we take $\boldsymbol{\chi}^* = \boldsymbol{\chi}' = \boldsymbol{\chi}$ and $\mathbf{S}' = \mathbf{S}$ to be actual fields satisfying (5.4.1)–(5.4.6) then (5.4.7) becomes

$$\int_{\partial\mathcal{B}_0^x} \mathbf{N}\cdot(\mathbf{S}\boldsymbol{\xi})\,\mathrm{d}A + \int_{\partial\mathcal{B}_0^{\sigma}} \boldsymbol{\sigma}\cdot\boldsymbol{\chi}\,\mathrm{d}A = \int_{\mathcal{B}_0} \operatorname{tr}(\mathbf{A}\mathbf{S})\,\mathrm{d}V - \int_{\mathcal{B}_0} \rho_0 \mathbf{b}\cdot\boldsymbol{\chi}\,\mathrm{d}V. \tag{5.4.8}$$

Secondly, on setting $\mathbf{S}' = \mathbf{S}$ in (5.4.7) corresponding to the deformation $\boldsymbol{\chi}$, defining $\delta\boldsymbol{\chi}$ as the difference $\boldsymbol{\chi}^* - \boldsymbol{\chi}$ so that

$$\delta\boldsymbol{\chi} = \mathbf{0} \quad \text{on} \quad \partial\mathcal{B}_0^x, \tag{5.4.9}$$

substituting $\boldsymbol{\chi}^* = \boldsymbol{\chi} + \delta\boldsymbol{\chi}$ in (5.4.7), and subtracting (5.4.8) from the result, we obtain

$$\int_{\partial\mathcal{B}_0^{\sigma}} \boldsymbol{\sigma}\cdot\delta\boldsymbol{\chi}\,\mathrm{d}A + \int_{\mathcal{B}_0} \rho_0 \mathbf{b}\cdot\delta\boldsymbol{\chi}\,\mathrm{d}V = \int_{\mathcal{B}_0} \operatorname{tr}\{\mathbf{S}(\operatorname{Grad}\delta\boldsymbol{\chi})\}\,\mathrm{d}V. \tag{5.4.10}$$

We may interpret this equation as follows: the left-hand side of (5.4.10) represents the work done by the actual surface tractions and body forces in an imagined or *virtual* displacement $\delta\boldsymbol{\chi}$ from the current configuration, assuming that these tractions and body forces remain fixed during the displacement and that $\delta\boldsymbol{\chi}$ satisfies (5.4.9); the right-hand side represents the corresponding virtual increase in stress work at fixed $\mathbf{S}$.

Equation (5.4.10) is referred to as the *virtual displacement principle* or the *principle of virtual work*.

On introducing the notation $\delta\mathbf{A} \equiv \mathbf{A}^* - \mathbf{A} = \operatorname{Grad}\delta\boldsymbol{\chi}$, we obtain

$$\operatorname{tr}\{\mathbf{S}(\operatorname{Grad}\delta\boldsymbol{\chi})\} = \operatorname{tr}\left(\frac{\partial W}{\partial \mathbf{A}}\delta\mathbf{A}\right).$$

For infinitesimal virtual displacements $\delta\boldsymbol{\chi}$ we have the approximation

$$\delta W = \operatorname{tr}\left(\frac{\partial W}{\partial \mathbf{A}}\delta\mathbf{A}\right) \tag{5.4.11}$$

to the first order in $\delta\mathbf{A}$ and (5.4.10) may therefore be approximated by

$$\int_{\partial\mathscr{B}_0^\sigma} \boldsymbol{\sigma}\cdot\delta\boldsymbol{\chi}\,\mathrm{d}A + \int_{\mathscr{B}_0} \rho_0\mathbf{b}\cdot\delta\boldsymbol{\chi}\,\mathrm{d}V = \int_{\mathscr{B}_0} \delta W\,\mathrm{d}V, \tag{5.4.12}$$

where δW is the change in elastic stored energy (density) due to an infinitesimal displacement $\delta\boldsymbol{\chi}$ from the current configuration. A parallel is now apparent between the principle of virtual work and the energy balance equation (4.3.5) specialized to the static context.

Let $\boldsymbol{\chi}$ and $\mathbf{S}$ be deformation and stress fields satisfying (5.4.1)–(5.4.6) and suppose $\mathbf{S}'$ is a statically admissible stress field associated with $\boldsymbol{\chi}$, i.e. $\mathbf{S}'$ satisfies (5.4.1) and (5.4.6) but is not necessarily related to $\boldsymbol{\chi}$ through (5.4.2)–(5.4.4). Then the difference $\delta\mathbf{S} = \mathbf{S}' - \mathbf{S}$ is such that

$$\operatorname{Div}\delta\mathbf{S} = \mathbf{0} \quad \text{in } \mathscr{B}_0, \tag{5.4.13}$$

$$\delta\mathbf{S}^{\mathrm{T}}\mathbf{N} = \mathbf{0} \quad \text{on } \partial\mathscr{B}_0^\sigma. \tag{5.4.14}$$

Application of (5.4.7) to $\mathbf{S}$ and $\mathbf{S}'$ in turn with $\boldsymbol{\chi}^* = \boldsymbol{\chi}' = \boldsymbol{\chi}$ and subtraction of the results yields

$$\int_{\partial\mathscr{B}_0^x} \mathbf{N}\cdot(\delta\mathbf{S}\boldsymbol{\xi})\,\mathrm{d}A = \int_{\mathscr{B}_0} \operatorname{tr}(\mathbf{A}\delta\mathbf{S})\,\mathrm{d}V \tag{5.4.15}$$

for all virtual stresses $\delta\mathbf{S}$ satisfying (5.4.13) and (5.4.14), irrespective of the interpretations of $\mathbf{S}$ and $\mathbf{S}'$ given above, $\mathbf{A}$ being the actual deformation gradient.

We refer to (5.4.15) as the *virtual stress principle*. It is the complement of

(5.4.10) in the sense that, for actual deformation and stress fields χ and $\mathbf{S}$, equation (5.4.10) holds for arbitrary virtual displacements $\delta\chi$ subject to (5.4.9) while (5.4.15) holds for arbitrary virtual stresses $\delta\mathbf{S}$ satisfying (5.4.13) and (5.4.14). Thus, (5.4.15) may also be called the *principle of virtual complementary work* (with $\mathrm{tr}(\mathbf{A}\delta\mathbf{S})$ complementary to $\mathrm{tr}(\mathbf{S}\delta\mathbf{A})$ in the sense that for infinitesimal virtual displacements and stresses their sum is a perfect differential $\mathrm{tr}\{\delta(\mathbf{AS})\}$).

By adding (5.4.10) and (5.4.15), we obtain

$$\int_{\partial\mathscr{B}_0^x} \mathbf{N}\cdot(\delta\mathbf{S}\boldsymbol{\xi})\,\mathrm{d}A + \int_{\partial\mathscr{B}_0^\sigma} \boldsymbol{\sigma}\cdot\delta\chi\,\mathrm{d}A + \int_{\mathscr{B}_0} \rho_0\mathbf{b}\cdot\delta\chi\,\mathrm{d}V$$
$$= \int_{\mathscr{B}_0} \mathrm{tr}(\mathbf{A}\delta\mathbf{S} + \mathbf{S}\delta\mathbf{A})\,\mathrm{d}V \qquad (5.4.16)$$

for *independent* $\delta\chi$ and $\delta\mathbf{S}$ satisfying (5.4.9), (5.4.13) and (5.4.14). This may be regarded as the 'virtual' counterpart of (5.4.8).

Returning now to the principle of virtual work, we rewrite (5.4.10) in the form

$$\int_{\mathscr{B}_0} \mathrm{tr}(\mathbf{S}\,\delta\mathbf{A})\,\mathrm{d}V - \int_{\mathscr{B}_0} \rho_0\mathbf{b}\cdot\delta\chi\,\mathrm{d}V - \int_{\partial\mathscr{B}_0^\sigma} \boldsymbol{\sigma}\cdot\delta\chi\,\mathrm{d}A = 0, \qquad (5.4.17)$$

where $\mathbf{S} = \partial W/\partial\mathbf{A}$, $\mathbf{A} = \mathrm{Grad}\,\chi$ and χ satisfies (5.4.5). Since

$$\mathrm{tr}(\mathbf{S}\,\delta\mathbf{A}) = \mathrm{Div}(\mathbf{S}\,\delta\chi) - (\mathrm{Div}\,\mathbf{S})\cdot\delta\chi,$$

equation (5.4.17) may be rearranged as

$$\int_{\partial\mathscr{B}_0^\sigma} (\mathbf{S}^{\mathrm{T}}\mathbf{N} - \boldsymbol{\sigma})\cdot\delta\chi\,\mathrm{d}A - \int_{\mathscr{B}_0} (\mathrm{Div}\,\mathbf{S} + \rho_0\mathbf{b})\cdot\delta\chi\,\mathrm{d}V = 0. \qquad (5.4.18)$$

Clearly (5.4.18) holds if $\mathbf{S}$ satisfies (5.4.1) and (5.4.6). Conversely, if the left-hand side of (5.4.18) vanishes for arbitrary $\delta\chi$ subject to (5.4.9), then (5.4.1) and (5.4.6) follow by the calculus of variations, $\delta\chi$ being interpreted as a variation in χ.

A true variational principle, however, can be obtained if and only if the left-hand side of (5.4.17) is expressible as the variation of some functional of χ, this functional being made stationary within the considered class of kinematically admissible deformations by any deformation satisfying (5.4.1)–(5.4.6). Such a functional exists only for certain forms of body force $\mathbf{b}$ and boundary data $\boldsymbol{\sigma}$, and the details are now discussed. We remark that derivation of a complementary variational principle on the basis of (5.4.15) raises certain difficulties which will be examined in Section 5.4.3.

5.4.2 The principle of stationary potential energy

For (infinitesimal) variations $\delta\boldsymbol{\chi}$ in $\boldsymbol{\chi}$ and corresponding variations $\delta\mathbf{A} \equiv \operatorname{Grad}\delta\boldsymbol{\chi}$ in $\mathbf{A}$, the variation δW in $W(\mathbf{A})$ is given by (5.4.11) and (5.4.17) therefore becomes

$$\int_{\mathscr{B}_0} \delta W \,\mathrm{d}V - \int_{\mathscr{B}_0} \rho_0 \mathbf{b}\cdot\delta\boldsymbol{\chi}\,\mathrm{d}V - \int_{\partial\mathscr{B}_0^\sigma} \boldsymbol{\sigma}\cdot\delta\boldsymbol{\chi}\,\mathrm{d}A = 0. \tag{5.4.19}$$

We require first of all that $\mathbf{b}\cdot\delta\boldsymbol{\chi}$ is expressible in the form δ(scalar function of $\boldsymbol{\chi}$). This means that $\mathbf{b}$ is a conservative force, i.e.

$$\mathbf{b} = -\operatorname{grad}\phi, \tag{5.4.20}$$

where ϕ is a scalar function of $\boldsymbol{\chi}$, and $\mathbf{b}\cdot\delta\boldsymbol{\chi} = -\delta\phi$. Thus (5.4.19) may be written

$$\delta\int_{\mathscr{B}_0} (W + \rho_0\phi)\mathrm{d}V - \int_{\partial\mathscr{B}_0^\sigma} \boldsymbol{\sigma}\cdot\delta\boldsymbol{\chi}\,\mathrm{d}A = 0. \tag{5.4.21}$$

It remains to consider the form of the surface traction $\boldsymbol{\sigma}$. Essentially two distinct cases arise. Firstly, if $\boldsymbol{\sigma}$ is independent of $\operatorname{Grad}\boldsymbol{\chi}$† and depends on the deformation only through $\boldsymbol{\chi}$ then, in parallel with (5.4.20), we have

$$\boldsymbol{\sigma} = -\operatorname{grad}\psi, \tag{5.4.22}$$

where ψ is a scalar function of $\boldsymbol{\chi}$ and $\mathbf{X}$ independently, the gradient being with respect to $\boldsymbol{\chi}$. Equation (5.4.21) now becomes

$$\delta\left\{\int_{\mathscr{B}_0} (W + \rho_0\phi)\mathrm{d}V + \int_{\partial\mathscr{B}_0^\sigma} \psi\,\mathrm{d}A\right\} = 0 \tag{5.4.23}$$

or, equivalently, (5.4.18) with (5.4.20), (5.4.22), (5.4.2) and (5.4.3).

For a dead-loading surface traction, $\boldsymbol{\sigma}$ depends only on $\mathbf{X}$, $\psi = -\boldsymbol{\sigma}\cdot\boldsymbol{\chi}$, and (5.4.23) reduces to

$$\delta\left\{\int_{\mathscr{B}_0} (W + \rho_0\phi)\mathrm{d}V - \int_{\partial\mathscr{B}_0^\sigma} \boldsymbol{\sigma}\cdot\boldsymbol{\chi}\,\mathrm{d}A\right\} = 0. \tag{5.4.24}$$

In the second case we write

$$\boldsymbol{\sigma} = \boldsymbol{\Sigma}^{\mathrm{T}}\mathbf{N},$$

†Under certain conditions a *surface potential* ψ can be constructed when $\boldsymbol{\sigma}$ depends on $\operatorname{Grad}\boldsymbol{\chi}$ and higher derivatives of $\boldsymbol{\chi}$. See Sewell (1967) for details.

where $\boldsymbol{\Sigma}$ is a known function of $\boldsymbol{\chi}$ and $\mathbf{A} = \mathrm{Grad}\,\boldsymbol{\chi}$ and its dependence on $\mathbf{X}$ is prescribed on $\partial\mathscr{B}_0^\sigma$. By (5.4.9) and the divergence theorem, we have

$$\int_{\partial\mathscr{B}_0^\sigma} \boldsymbol{\sigma}\cdot\delta\boldsymbol{\chi}\,\mathrm{d}A = \int_{\partial\mathscr{B}_0} \mathbf{N}\cdot(\boldsymbol{\Sigma}\,\delta\boldsymbol{\chi})\,\mathrm{d}A = \int_{\mathscr{B}_0} \{(\mathrm{Div}\,\boldsymbol{\Sigma})\cdot\delta\boldsymbol{\chi} + \mathrm{tr}(\boldsymbol{\Sigma}\,\delta\mathbf{A})\}\,\mathrm{d}V.$$

If there exists a scalar function φ of $\mathbf{X}$, $\boldsymbol{\chi}$ and $\mathbf{A}$ such that

$$\delta\varphi = (\mathrm{Div}\,\boldsymbol{\Sigma})\cdot\delta\boldsymbol{\chi} + \mathrm{tr}(\boldsymbol{\Sigma}\,\delta\mathbf{A}) \tag{5.4.25}$$

for variations $\delta\boldsymbol{\chi}$ and $\delta\mathbf{A}$ in $\boldsymbol{\chi}$ and $\mathbf{A}$ at fixed $\mathbf{X}$, then (5.4.21) gives

$$\delta\int_{\mathscr{B}_0} (W + \rho_0\phi - \varphi)\,\mathrm{d}V = 0. \tag{5.4.26}$$

From (5.4.25) we have

$$\boldsymbol{\Sigma} = \frac{\partial\varphi}{\partial\mathbf{A}}, \qquad \mathrm{Div}\,\boldsymbol{\Sigma} = \mathrm{grad}\,\varphi, \tag{5.4.27}$$

the gradient being with respect to $\boldsymbol{\chi}$ at fixed $\mathbf{X}$ and $\mathbf{A}$. As an example we consider

$$\boldsymbol{\Sigma} = -PJ\mathbf{B}^{\mathrm{T}}, \tag{5.4.28}$$

where P is a constant, $\mathbf{B}^{\mathrm{T}}$ is the inverse of $\mathbf{A}$ and $J = \det\mathbf{A}$. This corresponds to constant pressure loading on $\partial\mathscr{B}_0^\sigma$, as described in (5.1.25). It follows that Div $\boldsymbol{\Sigma} = \mathbf{0}$ and $\varphi = -PJ$ is independent of $\boldsymbol{\chi}$ (recall that $\partial J/\partial\mathbf{A} = J\mathbf{B}^{\mathrm{T}}$). Equation (5.4.26) now becomes

$$\delta\int_{\mathscr{B}_0} (W + \rho_0\phi + PJ)\,\mathrm{d}V = 0. \tag{5.4.29}$$

Let $E\{\boldsymbol{\chi}\}$ denote the functional of $\boldsymbol{\chi}$ which occurs in either (5.4.23) or (5.4.26), or in their specializations (5.4.24) or (5.4.29). We note that $E\{\boldsymbol{\chi}\}$ has the explicit forms

$$\int_{\mathscr{B}_0} \{W[\mathrm{Grad}\,\boldsymbol{\chi}(\mathbf{X})] + \rho_0(\mathbf{X}))\phi[\boldsymbol{\chi}(\mathbf{X})]\}\,\mathrm{d}V + \int_{\partial\mathscr{B}_0^\sigma} \psi[\mathbf{X}, \boldsymbol{\chi}(\mathbf{X})]\,\mathrm{d}A \tag{5.4.30}$$

and

$$\int_{\mathscr{B}_0} \{W[\operatorname{Grad}\boldsymbol{\chi}(\mathbf{X})] + \rho_0(\mathbf{X})\phi[\boldsymbol{\chi}(\mathbf{X})] - \varphi[\mathbf{X}, \boldsymbol{\chi}(\mathbf{X}), \operatorname{Grad}\boldsymbol{\chi}(\mathbf{X})]\}\,\mathrm{d}V \tag{5.4.31}$$

in respect of (5.4.23) and (5.4.26) respectively.

In either case, using the method which leads to (5.4.18) from (5.4.17), we obtain

$$\delta E = \int_{\partial\mathscr{B}_0^\sigma} (\mathbf{S}^\mathrm{T}\mathbf{N} - \boldsymbol{\sigma})\cdot\delta\boldsymbol{\chi}\,\mathrm{d}A - \int_{\mathscr{B}_0} (\operatorname{Div}\mathbf{S} + \rho_0\mathbf{b})\cdot\delta\boldsymbol{\chi}\,\mathrm{d}V, \tag{5.4.32}$$

where $\mathbf{S} = \partial W/\partial\mathbf{A}$. The variational principle is stated as follows: *the deformation* $\boldsymbol{\chi}$ *is a solution of the given boundary-value problem if and only if the variation* δE *of* $E\{\boldsymbol{\chi}\}$ *vanishes for all variations* $\delta\boldsymbol{\chi}$ *with* $\delta\boldsymbol{\chi} = \mathbf{0}$ *on* $\partial\mathscr{B}_0^x$. In other words, the actual deformation renders E stationary in the considered class of kinematically admissible deformations. It is appropriate to refer to $E\{\boldsymbol{\chi}\}$ as the (potential) *energy functional*, and to the variational principle as the *principle of stationary potential energy*.

In the above, attention was restricted implicitly to homogeneous materials, but it is a simple matter to extend the results so as to include explicit dependence of W on $\mathbf{X}$.

Remark If the term

$$-\int_{\partial\mathscr{B}_0^x} (\boldsymbol{\chi} - \boldsymbol{\xi})\cdot(\mathbf{S}^\mathrm{T}\mathbf{N})\,\mathrm{d}A$$

is added to the functional $E\{\boldsymbol{\chi}\}$, where $\mathbf{S} = \partial W/\partial\mathbf{A}$ and $\mathbf{A} = \operatorname{Grad}\boldsymbol{\chi}$, then the class of admissible deformations need not be restricted to those satisfying the boundary condition (5.4.5). The variation δE is then given by (5.4.32) supplemented by the integral

$$-\int_{\partial\mathscr{B}_0^x} (\boldsymbol{\chi} - \boldsymbol{\xi})\cdot(\delta\mathbf{S}^\mathrm{T}\mathbf{N})\,\mathrm{d}A.$$

This applies in respect of either (5.4.30) or (5.4.31), but in the former case the restriction (5.4.9) may also be relaxed. The statement of the variational principle should be modified accordingly.

5.4.3 Complementary and mixed variational principles

We now consider the possibility of converting the virtual stress principle into a variational principle. From (5.4.15), we have

$$\int_{\partial\mathscr{B}_0^x} \mathbf{N}\cdot(\delta\mathbf{S}\boldsymbol{\xi})\,\mathrm{d}A - \int_{\mathscr{B}_0} \operatorname{tr}(\mathbf{A}\,\delta\mathbf{S})\,\mathrm{d}V = 0. \tag{5.4.33}$$

First, we note that if $\mathbf{S}$ and $\mathbf{A} = \operatorname{Grad}\boldsymbol{\chi}$ are actual stress and deformation fields then

$$\operatorname{tr}(\mathbf{A}\,\delta\mathbf{S}) = \delta\{\operatorname{tr}(\mathbf{AS}) - W(\mathbf{A})\}, \tag{5.4.34}$$

and (5.4.33) may therefore be written in the variational form

$$\delta\left\{\int_{\partial\mathscr{B}_0^x} \mathbf{N}\cdot(\mathbf{S}\boldsymbol{\xi})\,\mathrm{d}A - \int_{\mathscr{B}_0} [\operatorname{tr}(\mathbf{AS}) - W(\mathbf{A})]\,\mathrm{d}V\right\} = 0. \tag{5.4.35}$$

This involves no restriction to conservative body forces and boundary tractions. However, as it stands, the functional in (5.4.35) depends on both $\mathbf{S}$ and, through $\mathbf{A}$, $\boldsymbol{\chi}$. On the other hand, the energy functional, as displayed in (5.4.30) or (5.4.31), depends only on $\boldsymbol{\chi}$, and in order to arrive at a truly complementary variational principle it is therefore necessary to construct a functional of $\mathbf{S}$ alone. *Formally*, this can be done as follows.

We introduce the *complementary energy density* $W_c(\mathbf{S}, \mathbf{A})$ defined by

$$W_c(\mathbf{S}, \mathbf{A}) = \operatorname{tr}(\mathbf{AS}) - W(\mathbf{A}). \tag{5.4.36}$$

If the stress-deformation relation

$$\mathbf{S} = \frac{\partial W}{\partial \mathbf{A}} \tag{5.4.37}$$

is invertible then $\mathbf{A}$ may be regarded as a function of $\mathbf{S}$, W_c is defined likewise as a function of $\mathbf{S}$ by the Legendre transformation (5.4.36), and the inverse relationship is written

$$\mathbf{A} = \frac{\partial W_c}{\partial \mathbf{S}}. \tag{5.4.38}$$

In general, however, the inverse of (5.4.37) *is not uniquely defined, either globally or locally*, and it is a difficult matter to determine which branch of the inversion is appropriate for the problem at hand. (This lack of uniqueness, intimated at the end of Section 4.2.6, will be discussed in detail in Chapter 6.) Nevertheless, it is instructive to complete the formal analysis on the assumption that W_c can be made explicit as a function of $\mathbf{S}$; we replace the notation in (5.4.36) by $W_c(\mathbf{S})$ for this purpose.

The *complementary energy functional* is defined as

$$E_c\{\mathbf{S}\} = \int_{\partial\mathscr{B}_0^x} \mathbf{N}\cdot(\mathbf{S}\boldsymbol{\xi})\,\mathrm{d}A - \int_{\mathscr{B}_0} W_c(\mathbf{S})\,\mathrm{d}V \tag{5.4.39}$$

for statically admissible stress fields corresponding to some fixed choice $\boldsymbol{\chi}$ of kinematically admissible deformation in $\mathbf{b}$ and $\boldsymbol{\sigma}$. Then

$$\delta E_c = \int_{\partial\mathscr{B}_0^x} \mathbf{N}\cdot(\delta\mathbf{S}\boldsymbol{\xi})\,\mathrm{d}A - \int_{\mathscr{B}_0} \operatorname{tr}(\mathbf{A}\,\delta\mathbf{S})\,\mathrm{d}V, \tag{5.4.40}$$

where $\mathbf{A} = \partial W_c/\partial\mathbf{S}$.

In general, $\mathbf{A}$ thus defined is not necessarily the gradient of a deformation field satisfying the boundary condition (5.4.5). However, if the variations $\delta\mathbf{S}$ are subject to (5.4.13) and (5.4.14), it follows that δE_c can be rearranged as

$$\delta E_c = \int_{\mathscr{B}_0} \operatorname{tr}\{(\operatorname{Grad}\boldsymbol{\chi} - \mathbf{A})\,\delta\mathbf{S}\}\,\mathrm{d}V. \tag{5.4.41}$$

Hence $\delta E_c = 0$ if and only if $\mathbf{A} = \operatorname{Grad}\boldsymbol{\chi}$. The differential 'constraint' $\operatorname{Div}\delta\mathbf{S} = \mathbf{0}$ on the variations $\delta\mathbf{S}$ does not affect the conclusion.

Expressed otherwise, *$E_c\{\mathbf{S}\}$ is stationary within the class of statically admissible stress fields if and only if $\mathbf{S}$ corresponds to a solution of the boundary-value problem.* (The solution need not in general be unique.) We refer to this as the *principle of stationary complementary energy*. Strictly, since $\mathbf{S}$ depends on the choice of $\boldsymbol{\chi}$ in $\mathbf{b}$ and $\boldsymbol{\sigma}$, this is not a truly complementary principle unless the body force is constant and there is dead loading on $\partial\mathscr{B}_0^\sigma$. However, $\boldsymbol{\chi}$ plays a passive role as far as variations are concerned here.

If, by contrast, $\boldsymbol{\chi}$ is allowed to be active independently of $\mathbf{S}$ then the need to invert the stress-deformation relation (5.4.37) can be circumvented. This is now illustrated by a return to the functional in (5.4.35), which we write as

$$E_c\{\mathbf{S},\boldsymbol{\chi}\} = \int_{\partial\mathscr{B}_0^x} \mathbf{N}\cdot(\mathbf{S}\boldsymbol{\xi})\,\mathrm{d}A - \int_{\mathscr{B}_0} \{\operatorname{tr}(\mathbf{AS}) - W(\mathbf{A})\}\,\mathrm{d}V, \tag{5.4.42}$$

where $\mathbf{A} = \operatorname{Grad}\boldsymbol{\chi}$. We consider kinematically admissible deformation fields $\boldsymbol{\chi}$ and statically admissible stress fields $\mathbf{S}$ corresponding to some deformation field $\boldsymbol{\chi}'$ (which need not satisfy the boundary condition (5.4.5)). There is no restriction on the forms of $\mathbf{b}$ and $\boldsymbol{\sigma}$.

For independent variations $\delta\boldsymbol{\chi}$ and $\delta\mathbf{S}$ satisfying (5.4.9), (5.4.13) and

(5.4.14), we have

$$\delta E_c = -\int_{\mathscr{B}_0} \operatorname{tr}\{(\mathbf{S} - \partial W/\partial \mathbf{A})\delta \mathbf{A}\}\,\mathrm{d}V + \int_{\mathscr{B}_0} \operatorname{tr}\{(\operatorname{Grad}\boldsymbol{\chi}' - \operatorname{Grad}\boldsymbol{\chi})\delta \mathbf{S}\}\,\mathrm{d}V + \int_{\partial\mathscr{B}_0^x} (\boldsymbol{\xi} - \boldsymbol{\chi}')\cdot(\delta \mathbf{S}^{\mathrm{T}}\mathbf{N})\,\mathrm{d}A. \tag{5.4.43}$$

Since $\boldsymbol{\chi} = \boldsymbol{\xi}$ on $\partial\mathscr{B}_0^x$ it follows that $\mathbf{S}$ and $\boldsymbol{\chi} = \boldsymbol{\chi}'$ correspond to a solution of the given boundary-value problem if and only if $E_c\{\mathbf{S}, \boldsymbol{\chi}\}$ is stationary in the considered class of stress and deformation fields.

Problem 5.4.1 Show that the combined virtual displacement and stress principle (5.4.16) can be written in the variational form $\delta(E - E_c) = 0$, where $E\{\boldsymbol{\chi}\}$ is given by (5.4.30) or (5.4.31) and $E_c\{\mathbf{S}\}$ by (5.4.39). Show further that

$$\delta(E - E_c) = -\int_{\mathscr{B}_0} (\operatorname{Div}\mathbf{S}_\chi + \rho_0 \mathbf{b})\cdot\delta\boldsymbol{\chi}\,\mathrm{d}V + \int_{\partial\mathscr{B}_0^\sigma} (\mathbf{S}^{\mathrm{T}}\mathbf{N} - \boldsymbol{\sigma})\cdot\delta\boldsymbol{\chi}\,\mathrm{d}A + \int_{\mathscr{B}_0} \operatorname{tr}\{(\mathbf{A} - \operatorname{Grad}\boldsymbol{\chi})\,\delta \mathbf{S}\}\,\mathrm{d}V + \int_{\partial\mathscr{B}_0^x} (\boldsymbol{\chi} - \boldsymbol{\xi})\cdot(\delta \mathbf{S}^{\mathrm{T}}\mathbf{N})\,\mathrm{d}A,$$

where $\mathbf{S}_\chi = \partial W/\partial(\operatorname{Grad}\boldsymbol{\chi})$ and $\mathbf{A} = \partial W_c/\partial \mathbf{S}$, with the variations $\delta\boldsymbol{\chi}$, $\delta\mathbf{S}$ subject to (5.4.9), (5.4.13) and (5.4.14).

Problem 5.4.2 For the functionals defined in Problem 5.4.1 show that if $\boldsymbol{\chi}$ and $\mathbf{S}$ correspond to an actual solution of the boundary-value problem, then

$$E\{\boldsymbol{\chi}\} - E_c\{\mathbf{S}\} = \int_{\mathscr{B}_0} \rho_0(\phi + \boldsymbol{\chi}\cdot\mathbf{b})\,\mathrm{d}V + \int_{\partial\mathscr{B}_0^\sigma} (\psi + \boldsymbol{\chi}\cdot\boldsymbol{\sigma})\,\mathrm{d}A$$

in respect of (5.4.30). Deduce that the right-hand side vanishes if and only if the loading is dead and the body force constant.

Problem 5.4.3 If $\mathbf{S}$ is a statically admissible stress field corresponding to the deformation $\boldsymbol{\chi}$ in $\mathbf{b}$ and $\boldsymbol{\sigma}$ and $\mathbf{A}$ is a second-order tensor field, show that the variation of the functional

$$E_c\{\mathbf{S},\mathbf{A}\} \equiv \int_{\partial\mathcal{B}_0^x} \mathbf{N}\cdot(\mathbf{S}\boldsymbol{\xi})\,\mathrm{d}A - \int_{\mathcal{B}_0} \{\mathrm{tr}\,(\mathbf{AS}) - W(\mathbf{A})\}\,\mathrm{d}V$$

is given by

$$\delta E_c = \int_{\partial\mathcal{B}_0^x} (\boldsymbol{\xi}-\boldsymbol{\chi})\cdot(\delta\mathbf{S}^{\mathrm{T}}\mathbf{N})\,\mathrm{d}A - \int_{\mathcal{B}_0} \mathrm{tr}\left\{\left(\mathbf{S}-\frac{\partial W}{\partial \mathbf{A}}\right)\delta\mathbf{A}\right\}\mathrm{d}V + \int_{\mathcal{B}_0} \mathrm{tr}\,\{(\mathrm{Grad}\,\boldsymbol{\chi}-\mathbf{A})\,\delta\mathbf{S}\}\,\mathrm{d}V$$

for all variations $\delta\mathbf{S}$ subject to (5.4.13) and (5.4.14) and all variations $\delta\mathbf{A}$. Compare this result with (5.4.43).

Problem 5.4.4 Show that the variation of the functional

$$E_c\{\mathbf{S},\boldsymbol{\chi}\} = \int_{\partial\mathcal{B}_0^x} \mathbf{N}\cdot(\mathbf{S}\boldsymbol{\xi})\,\mathrm{d}A - \int_{\mathcal{B}_0} W_c(\mathbf{S})\,\mathrm{d}V - \int_{\mathcal{B}_0} \boldsymbol{\chi}\cdot(\mathrm{Div}\,\mathbf{S})\,\mathrm{d}V + \int_{\partial\mathcal{B}_0^\sigma} \mathbf{N}\cdot(\mathbf{S}\boldsymbol{\chi})\mathrm{d}A + \int_{\partial\mathcal{B}_0} \rho_0\phi(\boldsymbol{\chi})\mathrm{d}V + \int_{\partial\mathcal{B}_0^\sigma} \psi(\boldsymbol{\chi})\mathrm{d}A$$

is expressible as

$$\delta E_c = -\int_{\mathcal{B}_0} (\mathrm{Div}\,\mathbf{S} + \rho_0\mathbf{b})\cdot\delta\boldsymbol{\chi}\,\mathrm{d}V + \int_{\mathcal{B}_0} \mathrm{tr}\{(\mathrm{Grad}\,\boldsymbol{\chi}-\mathbf{A})\delta\mathbf{S}\}\,\mathrm{d}V + \int_{\partial\mathcal{B}_0^x} (\boldsymbol{\xi}-\boldsymbol{\chi})\cdot(\delta\mathbf{S}^{\mathrm{T}}\mathbf{N})\mathrm{d}A + \int_{\partial\mathcal{B}_0^\sigma} (\mathbf{S}^{\mathrm{T}}\mathbf{N}-\boldsymbol{\sigma})\cdot\delta\boldsymbol{\chi}\,\mathrm{d}A,$$

where $\mathbf{A} = \partial W_c/\partial\mathbf{S}$. *Note that there is no requirement for admissible* $\boldsymbol{\chi}$ *and* $\mathbf{S}$ *to be kinematically admissible and statically admissible respectively in the sense of* Section 5.4.1. Moreover, the variations $\delta\boldsymbol{\chi}$ and $\delta\mathbf{S}$ are not restricted to those satisfying (5.4.9), (5.4.13) and (5.4.14).

Show that, for dead loading on $\partial\mathcal{B}_0^\sigma$ and constant body force, $E_c\{\mathbf{S},\boldsymbol{\chi}\}$ reduces to $E_c\{\mathbf{S}\}$ given by (5.4.39) if $\mathbf{S}$, $\boldsymbol{\chi}$ correspond to a solution of the boundary-value problem.

Problem 5.4.5 The functional in Problem 5.4.4 may be rewritten as a functional of three independent fields $\boldsymbol{\chi}$, $\mathbf{S}$, $\mathbf{A}$, where $\boldsymbol{\chi}$ and $\mathbf{S}$ are as before but $\mathbf{A}$ is now a second-order tensor (which need not be the gradient of a deformation field):

$$E_c\{\mathbf{S},\chi,\mathbf{A}\}=\int_{\mathscr{B}_0}\{W(\mathbf{A})-\operatorname{tr}(\mathbf{AS})\}\,\mathrm{d}V+\int_{\partial\mathscr{B}_0^x}\mathrm{N}\cdot(\mathbf{S}\xi)\,\mathrm{d}A$$
$$-\int_{\mathscr{B}_0}\chi\cdot(\operatorname{Div}\mathbf{S})\,\mathrm{d}V+\int_{\partial\mathscr{B}_0^\sigma}\mathrm{N}\cdot(\mathbf{S}\chi)\,\mathrm{d}A$$
$$+\int_{\mathscr{B}_0}\rho_0\phi(\chi)\,\mathrm{d}V+\int_{\partial\mathscr{B}_0^\sigma}\psi(\chi)\,\mathrm{d}A.$$

This avoids the need to invert the stress-deformation relation in order to find $W_c(\mathbf{S})$.

Show that

$$\delta E_c=\int_{\mathscr{B}_0}\operatorname{tr}\left\{\left(\frac{\partial W}{\partial\mathbf{A}}-\mathbf{S}\right)\delta\mathbf{A}\right\}\mathrm{d}V-\int_{\mathscr{B}_0}(\operatorname{Div}\mathbf{S}+\rho_0\mathbf{b})\cdot\delta\chi\,\mathrm{d}V$$
$$+\int_{\mathscr{B}_0}\operatorname{tr}\{(\operatorname{Grad}\chi-\mathbf{A})\,\delta\mathbf{S}\}\,\mathrm{d}V+\int_{\partial\mathscr{B}_0^x}(\xi-\chi)\cdot(\delta\mathbf{S}^{\mathrm{T}}\mathrm{N})\,\mathrm{d}A$$
$$+\int_{\partial\mathscr{B}_0^\sigma}(\mathbf{S}^{\mathrm{T}}\mathrm{N}-\sigma)\cdot\delta\chi\,\mathrm{d}A,$$

with no restrictions on admissible χ, $\mathbf{S}$ and $\mathbf{A}$ and the variations $\delta\chi,\delta\mathbf{S}$ and $\delta\mathbf{A}$ other than appropriate differentiability.

Deduce that χ, $\mathbf{S}$ and $\mathbf{A}$, with $\mathbf{A}=\operatorname{Grad}\chi$, correspond to a solution of the boundary-value problem if and only if $E_c\{\mathbf{S},\chi,\mathbf{A}\}$ is stationary within the considered class of fields (χ twice continuously differentiable and $\mathbf{S}$ and $\mathbf{A}$ once continuously differentiable), all variations $\delta\chi$, $\delta\mathbf{S}$, $\delta\mathbf{A}$ being independent.

An alternative way to formulate variational principles based on the use of complementary energy is to adopt a stress tensor other than $\mathbf{S}$ and thereby avoid the need to invert (5.4.37). With this in mind we recall from (3.5.15) that the Biot stress tensor is defined by

$$\mathbf{T}^{(1)}=\tfrac{1}{2}(\mathbf{SR}+\mathbf{R}^{\mathrm{T}}\mathbf{S}^{\mathrm{T}}),\tag{5.4.44}$$

where $\mathbf{R}$ is the rotation tensor arising from the polar decomposition $\mathbf{A}=\mathbf{RU}$. For a Green-elastic material we have

$$\mathbf{T}^{(1)}=\frac{\partial W}{\partial\mathbf{U}}\tag{5.4.45}$$

for objective W.

The difficulty associated with the inversion of (5.4.37) does not arise in respect of (5.4.45). In general (5.4.45) is invertible in some bounded domain in $\mathbf{U}$-space provided certain (physically motivated) conditions are satisfied (these will be examined in Chapter 6). In the present discussion we assume for convenience that (5.4.45) is invertible (locally and globally) in the space of symmetric (positive definite) second-order tensors. This enables us to define uniquely a complementary energy density $W_c(\mathbf{T}^{(1)})$ through the Legendre transformation

$$W(\mathbf{U}) + W_c(\mathbf{T}^{(1)}) = \mathrm{tr}\,(\mathbf{T}^{(1)}\mathbf{U}). \tag{5.4.46}$$

It follows that

$$\mathbf{U} = \frac{\partial W_c}{\partial \mathbf{T}^{(1)}} \tag{5.4.47}$$

and $\mathbf{U}$ is defined uniquely for given $\mathbf{T}^{(1)}$.

We note in passing that, because of the symmetry of $\mathbf{U}$, it follows from (5.4.44) that

$$\mathrm{tr}\,(\mathbf{T}^{(1)}\mathbf{U}) = \mathrm{tr}\,(\mathbf{SA}) \tag{5.4.48}$$

and therefore $W_c(\mathbf{T}^{(1)})$ is numerically equal to (the appropriate branch of) $W_c(\mathbf{S})$. Hence the duplication of notation (which is ambiguous in the case of $\mathbf{S}$).

The complementary energy functional (5.4.39) may now be replaced by

$$E_c\{\mathbf{S}, \mathbf{R}\} = \int_{\partial \mathscr{B}_0^x} \mathrm{N}\cdot(\mathbf{S}\boldsymbol{\xi})\,\mathrm{d}A - \int_{\mathscr{B}_0} W_c(\mathbf{T}^{(1)})\,\mathrm{d}V, \tag{5.4.49}$$

where $\mathbf{S}$ is statically admissible, as for (5.4.39), $\mathbf{R}$ is a rotation, $\mathbf{T}^{(1)}$ is defined by (5.4.44) and the function W_c defined by (5.4.46).

On taking the variation of (5.4.49), we obtain

$$\delta E_c = \int_{\partial \mathscr{B}_0^x} (\boldsymbol{\xi} - \boldsymbol{\chi})\cdot(\delta\mathbf{S}^{\mathrm{T}}\mathrm{N})\,\mathrm{d}A + \int_{\mathscr{B}_0} \mathrm{tr}\,\{(\mathrm{Grad}\,\boldsymbol{\chi} - \mathbf{A})\,\delta\mathbf{S}\}\,\mathrm{d}V$$
$$- \int_{\mathscr{B}_0} \mathrm{tr}\,(\mathbf{AS\Omega})\,\mathrm{d}V,$$

where $\mathbf{A} = \mathbf{RU} = \mathbf{R}(\partial W_c/\partial \mathbf{T}^{(1)})$ and $\boldsymbol{\Omega} \equiv (\delta\mathbf{R})\mathbf{R}^{\mathrm{T}}$ is antisymmetric. The variations $\delta\mathbf{S}$ are subject to (5.4.13) and (5.4.14) while $\delta\mathbf{R}$ is unrestricted. Note that $\mathrm{tr}\,(\mathbf{AS\Omega})$ vanishes for all antisymmetric $\boldsymbol{\Omega}$ if and only if $\mathbf{AS}$ is symmetric, i.e. the rotational balance equation is satisfied. If, as is assumed implicitly, W is objective then the final integral in δE_c is redundant.

Conceivably, the constitutive law of an elastic material could be defined by W_c without reference to W. As far as (5.4.49) is concerned this would avoid any algebraic complications involved in the inversion of (5.4.45), and this approach therefore has a distinct advantage from the practical point of view. Objectivity is assured since $\mathbf{T}^{(1)}$ itself is objective. We do not pursue the matter here.

Problem 5.4.6 Consider the functional

$$E\{\chi, \mathbf{S}, \mathbf{R}\} = \int_{\mathscr{B}_0} \{\mathrm{tr}(\mathbf{S}\,\mathrm{Grad}\,\chi) - W_c(\mathbf{T}^{(1)}) + \rho_0\phi(\chi)\}\,\mathrm{d}V + \int_{\partial\mathscr{B}_0^\sigma} \psi(\chi)\,\mathrm{d}A,$$

where $\mathbf{R}$ is a rotation, $\mathbf{S}$ a second-order tensor, and χ a kinematically-admissible deformation. (Note that this reduces to $E\{\chi\}$ for an actual solution of the boundary-value problem.)

If $\delta\chi = \mathbf{0}$ on $\partial\mathscr{B}_0^x$ and there is no restriction on the variations $\delta\mathbf{S}$ and $\delta\mathbf{R}$, show that

$$\delta E = -\int_{\mathscr{B}_0} (\mathrm{Div}\,\mathbf{S} + \rho_0\mathbf{b})\cdot\delta\chi\,\mathrm{d}V + \int_{\mathscr{B}_0} \mathrm{tr}\{(\mathrm{Grad}\,\chi - \mathbf{A})\,\delta\mathbf{S}\}\,\mathrm{d}V$$

$$+ \int_{\partial\mathscr{B}_0^\sigma} (\mathbf{S}^{\mathrm{T}}\mathrm{N} - \sigma)\cdot\delta\chi\,\mathrm{d}A - \int_{\mathscr{B}_0} \mathrm{tr}(\mathbf{AS\Omega})\,\mathrm{d}V,$$

where $\mathbf{A} = \mathbf{R}(\partial W_c/\partial\mathbf{T}^{(1)})$ and $\mathbf{\Omega} = (\delta\mathbf{R})\mathbf{R}^{\mathrm{T}}$.

Show also that δE is unaffected if the integral

$$\int_{\partial\mathscr{B}_0^\sigma} \psi(\chi)\,\mathrm{d}A$$

in $E\{\chi, \mathbf{S}, \mathbf{R}\}$ is replaced by

$$-\int_{\mathscr{B}_0} \varphi(\chi, \mathrm{Grad}\,\chi)\,\mathrm{d}V,$$

where φ is defined in Section 5.4.2.

The above result may be generalized so that χ need not satisfy the boundary condition (5.4.5). This is done by adding the term

$$\int_{\partial\mathscr{B}_0^x} (\boldsymbol{\xi} - \boldsymbol{\chi})\cdot(\mathbf{S}^{\mathrm{T}}\mathbf{N})\,\mathrm{d}A$$

to the functional so that δE is supplemented by

$$\int_{\partial\mathscr{B}_0^x} (\boldsymbol{\xi} - \boldsymbol{\chi})\cdot(\delta\mathbf{S}^{\mathrm{T}}\mathbf{N})\,\mathrm{d}A.$$

Compare these results with those in Problem 5.4.4.

Many possible modifications of the above variational principles are available. For example, into the various functionals may be admitted fields $\mathbf{S}$ and $\boldsymbol{\chi}$ which satisfy some, all or none of (5.4.1), (5.4.5) and (5.4.6) provided the functionals are adjusted accordingly. This, as we have seen, has consequences for the variations $\delta\boldsymbol{\chi}$ and $\delta\mathbf{S}$.

We now turn to the use of the stress tensor $\mathbf{T}^{(2)}$ defined in Section 3.5.2 as the conjugate of the Green strain tensor

$$\mathbf{E}^{(2)} = \tfrac{1}{2}(\mathbf{A}^{\mathrm{T}}\mathbf{A} - \mathbf{I}),$$

where $\mathbf{A}^{\mathrm{T}}\mathbf{A}$ is the right Cauchy–Green deformation tensor. In the notation of Section 4.3.2, the strain-energy function has the form

$$W(\mathbf{U}) = W^{(2)}(\mathbf{E}^{(2)}) = \hat{W}(\mathbf{A}^{\mathrm{T}}\mathbf{A}),$$

and

$$\mathbf{T}^{(2)} = \frac{\partial W^{(2)}}{\partial \mathbf{E}^{(2)}} = 2\frac{\partial \hat{W}}{\partial(\mathbf{A}^{\mathrm{T}}\mathbf{A})}. \tag{5.4.50}$$

Rather than employing $\mathbf{E}^{(2)}$ in what follows we use $\tfrac{1}{2}\mathbf{A}^{\mathrm{T}}\mathbf{A}$ since it involves slightly less algebra.

Relative to $\mathbf{T}^{(2)}$ and $\tfrac{1}{2}\mathbf{A}^{\mathrm{T}}\mathbf{A}$ we define the complementary energy $\hat{W}_{\mathrm{c}}(\mathbf{T}^{(2)})$ by

$$\hat{W}_{\mathrm{c}}(\mathbf{T}^{(2)}) = \tfrac{1}{2}\mathrm{tr}\,(\mathbf{T}^{(2)}\mathbf{A}^{\mathrm{T}}\mathbf{A}) - \hat{W}(\mathbf{A}^{\mathrm{T}}\mathbf{A}). \tag{5.4.51}$$

In doing this we have assumed that (5.4.50) is invertible on the same basis as (5.4.45). It is left as an exercise to show that

$$W_{\mathrm{c}}(\mathbf{T}^{(1)}) = \hat{W}_{\mathrm{c}}(\mathbf{T}^{(2)}) + \tfrac{1}{2}\mathrm{tr}\,(\mathbf{T}^{(2)}\mathbf{A}^{\mathrm{T}}\mathbf{A}). \tag{5.4.52}$$

On insertion of (5.4.52) into the functional in Problem 5.4.6 and removal of the dependence on $\mathbf{R}$, we obtain a new functional, namely

$$\hat{E}\{\chi, \mathbf{T}^{(2)}\} = \int_{\mathscr{B}_0} \{\tfrac{1}{2}\mathrm{tr}(\mathbf{S}\mathbf{A}) - \hat{W}_c(\mathbf{T}^{(2)}) + \rho_0\phi(\chi)\}\,\mathrm{d}V + \int_{\partial\mathscr{B}_0^\sigma} \psi(\chi)\,\mathrm{d}A + \int_{\partial\mathscr{B}_0^x} (\boldsymbol{\xi} - \chi)\cdot(\mathbf{S}^{\mathrm{T}}\mathbf{N})\,\mathrm{d}A, \qquad (5.4.53)$$

where $\mathbf{A} = \mathrm{Grad}\,\chi$ and $\mathbf{S} = \mathbf{T}^{(2)}\mathbf{A}^{\mathrm{T}}$. It is not required that χ and $\mathbf{S}$ satisfy (5.4.1), (5.4.5) and (5.4.6), but $\mathbf{T}^{(2)}$ must be symmetric.

Problem 5.4.7 Show that

$$\delta\hat{E} = -\int_{\mathscr{B}_0} (\mathrm{Div}\,\mathbf{S} + \rho_0\mathbf{b})\cdot\delta\chi\,\mathrm{d}V + \int_{\partial\mathscr{B}_0^\sigma} (\mathbf{S}^{\mathrm{T}}\mathbf{N} - \boldsymbol{\sigma})\cdot\delta\chi\,\mathrm{d}A + \int_{\partial\mathscr{B}_0^x} (\boldsymbol{\xi} - \chi)\cdot(\delta\mathbf{S}^{\mathrm{T}}\mathbf{N})\,\mathrm{d}A + \int_{\mathscr{B}_0} \mathrm{tr}\left\{\left(\tfrac{1}{2}\mathbf{A}^{\mathrm{T}}\mathbf{A} - \frac{\partial\hat{W}}{\partial\mathbf{T}^{(2)}}\right)\delta\mathbf{T}^{(2)}\right\}\mathrm{d}V$$

for all variations $\delta\chi$ and $\delta\mathbf{T}^{(2)}$ with $\delta\mathbf{S} = \delta(\mathbf{T}^{(2)}\mathbf{A}^{\mathrm{T}})$.

For further discussion of complementary variational principles relating to inversion of (5.4.37) see Ogden (1975, 1977), while a recent survey of variational principles in non-linear elasticity has been given by Guo Zhong-Heng (1980). Applications are detailed in the books by Washizu (1975) and Oden (1972), the latter being concerned with the numerical solution of boundary-value problems by use of finite-element methods. For variational principles involving discontinuous fields (not admitted here) we refer to Nemat-Nasser (1974).

The stationary principles discussed here cannot in general be strengthened to extremum principles, but circumstances in which the transition to extremum principles can be made are outlined in Chapter 6.

5.4.4 Variational principles with constraints

Suppose the material is such that the deformation field χ is subject to the internal constraint

$$C(\mathrm{Grad}\,\chi) = 0 \qquad (5.4.54)$$

for each point of $\mathscr{B}_0$, as in Section 4.2.7. Since the components of $\mathrm{Grad}\,\delta\chi$ are not then all independent, the energy functional in Section 5.4.2 must be supplemented by the term

$$\int_{\mathscr{B}_0} qC(\operatorname{Grad}\boldsymbol{\chi})\,\mathrm{d}V,$$

where q (which depends on $\mathbf{X}$ in general) is a Lagrange multiplier.

For example, starting with the functional (5.4.30), we obtain

$$E\{\boldsymbol{\chi}\} = \int_{\mathscr{B}_0} \{W(\operatorname{Grad}\boldsymbol{\chi}) + qC(\operatorname{Grad}\boldsymbol{\chi}) + \rho_0\phi(\boldsymbol{\chi})\}\,\mathrm{d}V + \int_{\partial\mathscr{B}_0^\sigma} \psi(\boldsymbol{\chi})\,\mathrm{d}A \tag{5.4.55}$$

and hence

$$\delta E = \int_{\mathscr{B}_0} \left\{\operatorname{Div}\left(\frac{\partial W}{\partial \mathbf{A}} + q\frac{\partial C}{\partial \mathbf{A}}\right) + \rho_0\mathbf{b}\right\}\cdot\delta\boldsymbol{\chi}\,\mathrm{d}V + \int_{\partial\mathscr{B}_0^\sigma} \left\{\left(\frac{\partial W}{\partial \mathbf{A}} + q\frac{\partial C}{\partial \mathbf{A}}\right)^{\mathrm{T}}\mathbf{N} - \boldsymbol{\sigma}\right\}\cdot\delta\boldsymbol{\chi}\,\mathrm{d}A. \tag{5.4.56}$$

Thus, $\mathbf{S}$ in (5.4.32) is replaced by $\partial W/\partial\mathbf{A} + q\partial C/\partial\mathbf{A}$ in (5.4.56). Admissible $\boldsymbol{\chi}$ in (5.4.55) are subject to (5.4.5) and (5.4.54).

Similar adjustments should be made for the variational principles in Section 5.4.3. Note that the Lagrange multiplier q is of a different nature from those arising in Sections 5.4.2 and 5.4.3 in respect of the 'constraints' (5.4.1), (5.4.5) and (5.4.6) for which the fields $\boldsymbol{\chi}$ and $\mathbf{S}^{\mathrm{T}}\mathbf{N}$ are, in effect, Lagrange multipliers.

5.4.5 Conservation laws and the energy momentum tensor

In integral form the equilibrium equation may be written

$$\int_{\partial\mathscr{B}_0} \mathbf{S}^{\mathrm{T}}\mathbf{N}\,\mathrm{d}A + \int_{\mathscr{B}_0} \rho_0\mathbf{b}\,\mathrm{d}V = \mathbf{0}, \tag{5.4.57}$$

the corresponding rotational balance equation being

$$\int_{\partial\mathscr{B}_0} (\boldsymbol{\chi}\otimes\mathbf{S}^{\mathrm{T}}\mathbf{N} - \mathbf{S}^{\mathrm{T}}\mathbf{N}\otimes\boldsymbol{\chi})\,\mathrm{d}A + \int_{\mathscr{B}_0} \rho_0(\boldsymbol{\chi}\otimes\mathbf{b} - \mathbf{b}\otimes\boldsymbol{\chi})\,\mathrm{d}V = \mathbf{0}. \tag{5.4.58}$$

These basic equations are valid for an arbitrary body (or sub-body) $\mathscr{B}_0$ with boundary $\partial\mathscr{B}_0$.

Further integral identities, similar in structure to the above and requiring the local balance equations (5.4.1) and (5.4.4) for their derivation, are now examined.

For a *homogeneous* material, use of (5.4.1) shows that

$$\mathrm{Div}(W\mathbf{I}-\mathbf{SA})=\rho_0\mathbf{A}^{\mathrm{T}}\mathbf{b}, \tag{5.4.59}$$

where $\mathbf{A}=\mathrm{Grad}\,\boldsymbol{\chi}$, and this yields the identity

$$\int_{\partial\mathscr{B}_0}\mathbf{M}\mathbf{N}\,\mathrm{d}A-\int_{\mathscr{B}_0}\rho_0\mathbf{A}^{\mathrm{T}}\mathbf{b}\,\mathrm{d}V=\mathbf{0}, \tag{5.4.60}$$

where the notation

$$\mathbf{M}=W\mathbf{I}-\mathbf{A}^{\mathrm{T}}\mathbf{S}^{\mathrm{T}} \tag{5.4.61}$$

has been introduced for convenience.

Next, use of (5.4.59) with the notation (5.4.61) shows that

$$\int_{\partial\mathscr{B}_0}\mathbf{M}\mathbf{N}\otimes\mathbf{X}\,\mathrm{d}A-\int_{\mathscr{B}_0}\rho_0\mathbf{A}^{\mathrm{T}}\mathbf{b}\otimes\mathbf{X}\,\mathrm{d}V=\int_{\mathscr{B}_0}\mathbf{M}\,\mathrm{d}V \tag{5.4.62}$$

and, on taking the trace of this equation, we obtain

$$\int_{\partial\mathscr{B}_0}(\mathbf{M}\mathbf{N})\cdot\mathbf{X}\,\mathrm{d}A-\int_{\mathscr{B}_0}\rho_0(\mathbf{A}^{\mathrm{T}}\mathbf{b})\cdot\mathbf{X}\,\mathrm{d}V=\int_{\mathscr{B}_0}\{3W-\mathrm{tr}(\mathbf{SA})\}\,\mathrm{d}V$$

or, equivalently,

$$\int_{\partial\mathscr{B}_0}\{W\mathbf{N}\cdot\mathbf{X}-(\mathbf{A}^{\mathrm{T}}\mathbf{S}^{\mathrm{T}}\mathbf{N})\cdot\mathbf{X}+(\mathbf{S}^{\mathrm{T}}\mathbf{N})\cdot\boldsymbol{\chi}\}\,\mathrm{d}A$$

$$=\int_{\mathscr{B}_0}\{3W+\rho_0(\mathbf{A}^{\mathrm{T}}\mathbf{b})\cdot\mathbf{X}-\rho_0\mathbf{b}\cdot\boldsymbol{\chi}\}\,\mathrm{d}V. \tag{5.4.63}$$

In the case of an isotropic elastic material further information can be obtained, for then $\mathbf{SA}=\mathbf{T}^{(1)}\mathbf{U}$ is symmetric (so that $\mathbf{M}$ is symmetric) and (5.4.62) leads to the identity

$$\int_{\partial\mathscr{B}_0}(\mathbf{M}\mathbf{N}\otimes\mathbf{X}-\mathbf{X}\otimes\mathbf{M}\mathbf{N})\,\mathrm{d}A$$

$$-\int_{\mathscr{B}_0}\rho_0(\mathbf{A}^{\mathrm{T}}\mathbf{b}\otimes\mathbf{X}-\mathbf{X}\otimes\mathbf{A}^{\mathrm{T}}\mathbf{b})\,\mathrm{d}V=\mathbf{0}. \tag{5.4.64}$$

The structures of (5.4.60) and (5.4.64) should be compared with those of (5.4.57) and (5.4.58) respectively.

We now return to the energy functional $E\{\boldsymbol{\chi}\}$ introduced in Section 5.4.2. The variation of either (5.4.30) or (5.4.31) may be put as

$$\delta E = \int_{\mathscr{B}_0} (\delta W - \rho_0 \mathbf{b} \cdot \delta\boldsymbol{\chi})\,\mathrm{d}V - \int_{\partial\mathscr{B}_0^\sigma} \boldsymbol{\sigma} \cdot \delta\boldsymbol{\chi}\,\mathrm{d}A \tag{5.4.65}$$

and we take

$$\delta\boldsymbol{\chi} = \mathbf{0} \quad \text{on} \quad \partial\mathscr{B}_0^x. \tag{5.4.66}$$

Noting that $J^{-1}W$ is the strain energy per unit *current* volume, we obtain

$$\delta W = J\delta(J^{-1}W) + W \operatorname{div} \delta\boldsymbol{\chi},$$

where we have used the 'incremental' continuity equation $J \operatorname{div} \delta\boldsymbol{\chi} = \delta J$ (recall equation (2.3.17)). Since $\delta(J^{-1}W) = \delta\boldsymbol{\chi} \cdot \operatorname{grad}(J^{-1}W)$, it follows that $\delta W = J \operatorname{div}(J^{-1}W\delta\boldsymbol{\chi})$, and, by use of the divergence theorem applied in the current configuration followed by application of Nanson's formula (2.2.18), we arrive at

$$\delta E = \int_{\partial\mathscr{B}_0^\sigma} (\mathbf{BMN}) \cdot \delta\boldsymbol{\chi}\,\mathrm{d}A - \int_{\mathscr{B}_0} \rho_0 \mathbf{b} \cdot \delta\boldsymbol{\chi}\,\mathrm{d}V, \tag{5.4.67}$$

where $\mathbf{B} = (\mathbf{A}^{\mathrm{T}})^{-1}$. The boundary conditions (5.4.66) and $\mathbf{S}^{\mathrm{T}}\mathbf{N} = \boldsymbol{\sigma}$ on $\partial\mathscr{B}_0^\sigma$ have also been used. Of course, $\delta E = 0$ for an actual deformation, but the expression (5.4.67) has an alternative interpretation that attaches physical significance to $\mathbf{M}$, as we see in what follows.

Suppose $\mathscr{B}_0$ is a subset of some body in its reference configuration and let $\partial\mathscr{B}_0$ be the boundary of $\mathscr{B}_0$. When the body is deformed, $\mathscr{B}_0$ and $\partial\mathscr{B}_0$ become $\mathscr{B}$ and $\partial\mathscr{B}$ respectively. The total energy associated with the deformation of $\mathscr{B}_0$ is

$$E\{\boldsymbol{\chi}\} = \int_{\mathscr{B}_0} W(\operatorname{Grad} \boldsymbol{\chi})\,\mathrm{d}V + \int_{\partial\mathscr{B}_0} \psi(\boldsymbol{\chi})\,\mathrm{d}A, \tag{5.4.68}$$

where the potential energy ψ of the surface tractions on $\partial\mathscr{B}_0$ is taken to depend only on $\boldsymbol{\chi}$ for simplicity and body forces are set to zero.

Similarly, the total energy of a second subset $\mathscr{B}_0'$ (with boundary $\partial\mathscr{B}_0'$) is

$$E'\{\boldsymbol{\chi}\} = \int_{\mathscr{B}_0'} W(\operatorname{Grad} \boldsymbol{\chi})\,\mathrm{d}V + \int_{\partial\mathscr{B}_0'} \psi(\boldsymbol{\chi})\,\mathrm{d}A. \tag{5.4.69}$$

If the position of $\mathscr{B}_0'$ differs from that of $\mathscr{B}_0$ by a uniform infinitesimal displacement $\delta\boldsymbol{\xi}$ then, apart from an error of order $|\delta\boldsymbol{\xi}|^2$, $E'\{\chi\}$ can be calculated in terms of integrals over $\mathscr{B}_0$ and $\partial\mathscr{B}_0$ by replacing $\mathbf{X}$ by $\mathbf{X}-\delta\boldsymbol{\xi}$ in the integrands of (5.4.68). It follows that the energy difference is given by

$$E'\{\chi\}-E\{\chi\}=-\int_{\mathscr{B}_0}\delta\boldsymbol{\xi}\cdot\operatorname{Grad}W\,\mathrm{d}V-\int_{\partial\mathscr{B}_0}\delta\boldsymbol{\xi}\cdot\operatorname{Grad}\psi\,\mathrm{d}A.$$

Using the fact that $\delta\boldsymbol{\xi}$ is uniform and the connections $\operatorname{Grad}\psi=\mathbf{A}^{\mathrm{T}}\operatorname{grad}\psi=-\mathbf{A}^{\mathrm{T}}\boldsymbol{\sigma}=-\mathbf{A}^{\mathrm{T}}\mathbf{S}^{\mathrm{T}}\mathbf{N}$, where $\mathbf{N}$ is the outward unit normal to $\partial\mathscr{B}_0$, together with the divergence theorem, we obtain

$$E'\{\chi\}-E\{\chi\}=-\delta\boldsymbol{\xi}\cdot\int_{\partial\mathscr{B}_0}\mathbf{M}\mathbf{N}\,\mathrm{d}A. \tag{5.4.70}$$

For $\mathbf{b}=\mathbf{0}$ this is precisely the negative of δE in (5.4.67) if we take $\delta\chi=\mathbf{A}\,\delta\boldsymbol{\xi}$.

Now, according to (5.4.60), we have

$$\int_{\partial\mathscr{B}_0}\mathbf{M}\mathbf{N}\,\mathrm{d}A=\mathbf{0}$$

(since $\mathbf{b}=\mathbf{0}$) provided $\mathscr{B}_0$ is simply connected and contains no singularities of $\mathbf{M}$. If, however, $\mathscr{B}_0$ encloses a single internal boundary $\mathscr{S}$ of the considered body then the above integral does not vanish (in general) and, furthermore, we have

$$\mathbf{F}\equiv\int_{\mathscr{S}}\mathbf{M}\mathbf{N}\,\mathrm{d}A=\int_{\partial\mathscr{B}_0}\mathbf{M}\mathbf{N}\,\mathrm{d}A \tag{5.4.71}$$

for *any* closed surface $\partial\mathscr{B}_0$ in the material enclosing $\mathscr{S}$ when there is no singularity and no other internal surface in the region between $\partial\mathscr{B}_0$ and $\mathscr{S}$.

In these circumstances (5.4.70) may be written

$$\delta E=-\mathbf{F}\cdot\delta\boldsymbol{\xi},$$

where $\mathbf{F}$ is defined by (5.4.71). This δE represents the *decrease* in the total energy of the body associated with a change in reference geometry corresponding to a displacement $\delta\boldsymbol{\xi}$ of the internal surface. (The energy outside the original $\partial\mathscr{B}_0$ is unchanged to order $|\delta\boldsymbol{\xi}|$.) This means that $\mathbf{F}$ may be interpreted as an 'effective' force on the cavity $\mathscr{S}$ due to the surrounding deformation field. The force per unit area of $\mathscr{S}$ is $\mathbf{M}\mathbf{N}$ and this provides a physical interpretation of $\mathbf{M}$, the latter being referred to as the *energy momentum tensor*. In view of (5.4.71) we see that $\mathbf{F}$ can be calculated for any surface $\partial\mathscr{B}_0$ surrounding $\mathscr{S}$ and consistent with the restrictions stated above.

Equally, $\mathscr{S}$ may be taken as a material surface enclosing a singularity in the form of a defect (such as a dislocation). Then $\mathbf{F}$ can be regarded as the force on the defect in the elastic field. It was in this context that the energy momentum tensor was originally introduced by Eshelby (1951). For more recent discussions we refer to Eshelby (1975) and Chadwick (1975).

In the sense that the value of the integral on the right of (5.4.71) is unaffected by a 'deformation' of $\partial\mathscr{B}_0$ within the material (provided $\partial\mathscr{B}_0$ does not pass through a singularity in the process), the integral is referred to as a *surface-independent integral.* Its two-dimensional counterpart, which has useful applications in fracture mechanics, is known as a *path-independent integral.* In the same sense, for an isotropic material,

$$\int_{\partial\mathscr{B}_0} \{\mathbf{X}\otimes\mathbf{M}\mathbf{N} - \mathbf{M}\mathbf{N}\otimes\mathbf{X}\}\,\mathrm{d}A$$

is surface independent and vanishes when (5.4.71) does (recall (5.4.64) for $\mathbf{b}=\mathbf{0}$).

The term *conservation law* is used for an identity such as

$$\int_{\partial\mathscr{B}_0} \mathbf{M}\mathbf{N}\,\mathrm{d}A = \mathbf{0},$$

appropriate to the situation when $\partial\mathscr{B}_0$ encloses neither singularity nor internal boundary. Other examples of conservation laws are (5.4.57), (5.4.58) and (5.4.64) with $\mathbf{b}=\mathbf{0}$, and these are the only ones of this type available in non-linear elasticity. Equation (5.4.63) is not included since it involves the volume integral of W. For detailed discussion of conservation laws in elasticity and their applications see Knowles and Sternberg (1972) and Green (1973).

REFERENCES

J.M. Ball, Convexity Conditions and Existence Theorems in Non-linear Elasticity, *Archive for Rational Mechanics and Analysis*, Vol. 63, pp. 337–403 (1977).

J.M. Ball, Discontinuous Equilibrium Solutions and Cavitation in Non-linear Elasticity, *Philosophical Transactions of the Royal Society of London Series A*, Vol. 306, pp. 557–611 (1982).

B. Bernstein and R.A. Toupin, Some Properties of the Hessian Matrix of a Strictly Convex Function, *Journal für reine und angewandte Mathematik*, Vol, 210, pp. 65–72 (1962).

P. Chadwick, Applications of an Energy-Momentum Tensor in Non-linear Elastostatics, *Journal of Elasticity*, Vol. 5, pp. 249–258 (1975).

J.D. Eshelby, The Force on an Elastic Singularity, *Philosophical Transactions of the Royal Society of London, Series A*, Vol. 244, pp. 87–112 (1951).

J.D. Eshelby, The Elastic Energy Momentum Tensor, *Journal of Elasticity*, Vol. 5, pp. 321–335 (1975).

A.E. Green, On Some General Formulae in Finite Elastostatics, *Archive for Rational Mechanics and Analysis*, Vol. 50, pp. 73–80 (1973).

A.E. Green and J.E. Adkins, *Large Elastic Deformations*, 2nd edition, Oxford University Press (1970).

A.E. Green and W. Zerna, *Theoretical Elasticity*, 2nd edition, Oxford University Press (1968).

Guo Zhong-Heng, The Unified Theory of Variational Principles in Non-linear Elasticity, *Archives of Mechanics*, Vol. 32, pp. 577–596 (1980).

A. Hanyga, *Mathematical Theory of Non-linear Elasticity*, Ellis Horwood (1984).

D.A. Isherwood and R.W. Ogden, Finite Plane Strain Problems for Compressible Elastic Solids: General Solution and Volume Changes, *Rheologica Acta*, Vol. 16, pp. 113–122 (1977a).

D.A. Isherwood and R.W. Ogden, Towards the Solution of Finite Plane Strain Problems for Compressible Elastic Solids, *International Journal of Solids and Structures*, Vol. 13, pp. 105–123 (1977b).

F. John, Plane Strain Problems for a Perfectly Elastic Material of Harmonic Type, *Communications on Pure and Applied Mathematics*, Vol. 13, pp. 239–296 (1960).

J.K. Knowles and E. Sternberg, On a Class of Conservation Laws in Linearized and Finite Elastostatics, *Archive for Rational Mechanics and Analysis*, Vol. 44, pp. 187–211 (1972).

J.K. Knowles and E. Sternberg, On the Singularity Induced by Certain Mixed Boundary Conditions in Linearized and Non-linear Elastostatics, *International Journal of Solids and Structures*, Vol. 11, pp. 1173–1201 (1975).

S. Nemat-Nasser, General Variational Principles in Non-linear and Linear Elasticity with Applications, *Mechanics Today* (Ed. S. Nemat-Nasser), Vol. 1, pp. 214–261, Pergamon Press (1974).

J.T. Oden, *Finite Elements of Non-linear Continua*, McGraw-Hill (1972).

R.W. Ogden, A Note on Variational Theorems in Non-linear Elastostatics, *Mathematical Proceedings of the Cambridge Philosophical Society*, Vol. 77, pp. 609–615 (1975).

R.W. Ogden, Inequalities Associated with the Inversion of Elastic Stress-Deformation Relations and their Implications, *Mathematical Proceedings of the Cambridge Philosophical Society*, Vol. 81, pp. 313–324 (1977).

R.W. Ogden and D.A. Isherwood, Solution of some Finite Plane Strain Problems for Compressible Elastic Solids, *Quarterly Journal of Mechanics and Applied Mathematics*, Vol. 31, pp. 219–249 (1978).

M.J. Sewell, On Configuration-Dependent Loading, *Archive for Rational Mechanics and Analysis*, Vol. 23, pp. 327–351 (1967).

A.J.M. Spencer, The Static Theory of Finite Elasticity, *Journal of the Institute of Mathematics and Its Applications*, Vol. 6, pp. 164–200 (1970).

E. Varley and E. Cumberbatch, The Finite Deformation of an Elastic Material Surrounding an Elliptical Hole, in *Finite Elasticity*, Applied Mechanics Symposia Series Vol. 27 (Ed. R.S. Rivlin), American Society of Mechanical Engineers, pp. 41–64 (1977).

E. Varley and E. Cumberbatch, Finite Deformation of Elastic Materials Surrounding Cylindrical Holes, *Journal of Elasticity*, Vol. 10, pp. 1–65 (1980).

K. Washizu, *Variational Methods in Elasticity and Plasticity*, 2nd edition, Pergamon Press (1975).

CHAPTER 6

Incremental Elastic Deformations

6.1 INCREMENTAL CONSTITUTIVE RELATIONS

6.1.1 Deformation increments

We recall from Section 2.2 that the deformation of a body relative to a given reference configuration (in which the body occupies the region $\mathscr{B}_0$) is denoted by $\boldsymbol{\chi}$, so that

$$\mathbf{x} = \boldsymbol{\chi}(\mathbf{X})$$

is the current position of the material particle which is at $\mathbf{X}$ in $\mathscr{B}_0$. Time dependence is not admitted here.

Suppose that the deformation is changed to $\boldsymbol{\chi}'$ and let

$$\mathbf{x}' = \boldsymbol{\chi}'(\mathbf{X}) \qquad \mathbf{X} \in \mathscr{B}_0.$$

The displacement of a material particle due to this change is described by

$$\mathbf{x}' - \mathbf{x} = \boldsymbol{\chi}'(\mathbf{X}) - \boldsymbol{\chi}(\mathbf{X}).$$

We rewrite this as

$$\delta\mathbf{x} = \delta\boldsymbol{\chi}(\mathbf{X}),$$

where the operator δ is defined by $\delta\boldsymbol{\chi} = \boldsymbol{\chi}' - \boldsymbol{\chi}$.

If the displacement $\delta\mathbf{x}$ is 'small' for each $\mathbf{X} \in \mathscr{B}_0$ so that terms of order $|\delta\mathbf{x}|^2$ are negligible† in comparison with those of order $|\delta\mathbf{x}|$ then we refer to $\delta\boldsymbol{\chi}$ as an *incremental deformation* from the configuration described by $\boldsymbol{\chi}$.

Corresponding to an incremental deformation, the deformation gradient

† We do not quantify precisely the meaning of *small* here.

$\mathbf{A} = \operatorname{Grad} \boldsymbol{\chi}(\mathbf{X})$ changes according to

$$\operatorname{Grad} \delta\boldsymbol{\chi}(\mathbf{X}) = \delta \operatorname{Grad} \boldsymbol{\chi}(\mathbf{X}), \tag{6.1.1}$$

i.e. the operators δ and Grad commute. Note that this formula is *exact* and is valid even if $\delta\boldsymbol{\chi}(\mathbf{X})$ is not incremental in the sense defined above.

By contrast, the increment δJ in $J = \det \mathbf{A}$, given by

$$\delta J = J \operatorname{tr}\{(\delta\mathbf{A})\mathbf{B}^{\mathrm{T}}\}, \tag{6.1.2}$$

is not exact since terms of second and higher orders in $\delta\mathbf{A}$ have been neglected. It is left as an exercise for the reader to show that the second-order terms omitted from the right-hand side of (6.1.2) are

$$\tfrac{1}{2}J[\operatorname{tr}\{(\delta\mathbf{A})\mathbf{B}^{\mathrm{T}}\}]^2 - \tfrac{1}{2}J \operatorname{tr}\{\mathbf{B}^{\mathrm{T}}(\delta\mathbf{A})\mathbf{B}^{\mathrm{T}}(\delta\mathbf{A})\}.$$

More generally, for any (sufficiently regular) scalar function ϕ, defined on the space of second-order tensors and taking the value $\phi(\mathbf{A})$ for the deformation gradient $\mathbf{A}$, we have the linear approximation

$$\delta\phi = \operatorname{tr}\{(\partial\phi/\partial\mathbf{A})\delta\mathbf{A}\}, \tag{6.1.3}$$

where the derivative $\partial\phi/\partial\mathbf{A}$ is defined in accordance with the convention described in Section 4.2.8.

Of course, the right-hand side of (6.1.3) is merely the first term in the Taylor series for $\delta\phi$. The second term may be written

$$\tfrac{1}{2}\operatorname{tr}\{((\partial^2\phi/\partial\mathbf{A}^2)\delta\mathbf{A})\delta\mathbf{A}\},$$

where $\partial^2\phi/\partial\mathbf{A}^2$ is a fourth-order tensor with Cartesian components defined by

$$(\partial^2\phi/\partial\mathbf{A}^2)_{jilk} = \partial^2\phi/\partial A_{ij}\partial A_{kl}.$$

We also define the product $(\partial^2\phi/\partial\mathbf{A}^2)\delta\mathbf{A}$ as the second-order tensor with Cartesian components

$$(\partial^2\phi/\partial A_{ij}\partial A_{kl})\delta A_{kl}.$$

Higher-order derivatives of ϕ and their products with $\delta\mathbf{A}$ are defined similarly. For example, in the convention introduced following equation (3.5.33) in Section 3.5.2, we have

$$\left.\begin{aligned}(\partial^3\phi/\partial\mathbf{A}^3)[\cdot, \delta\mathbf{A}] &\equiv (\partial^3\phi/\partial A_{ij}\partial A_{kl}\partial A_{mn})\delta A_{mn},\\ (\partial^3\phi/\partial\mathbf{A}^3)[\delta\mathbf{A}, \delta\mathbf{A}] &\equiv (\partial^3\phi/\partial A_{ij}\partial A_{kl}\partial A_{mn})\delta A_{mn}\delta A_{kl}.\end{aligned}\right\} \tag{6.1.4}$$

Tensor functions of $\mathbf{A}$ are treated in a similar manner. Here we consider a second-order tensor function $\mathbf{F}$. Its increment at $\mathbf{A}$ may be written formally as a Taylor series

$$\delta\mathbf{F} = \mathscr{L}^1\delta\mathbf{A} + \tfrac{1}{2}\mathscr{L}^2[\delta\mathbf{A}, \delta\mathbf{A}] + \ldots, \tag{6.1.5}$$

where

$$\mathscr{L}^1 = \partial\mathbf{F}/\partial\mathbf{A}, \qquad \mathscr{L}^2 = \partial^2\mathbf{F}/\partial\mathbf{A}^2, \ldots . \tag{6.1.6}$$

are respectively fourth-, sixth-, ... order tensors with Cartesian components

$$\mathscr{L}^1_{ijkl} = \partial F_{ij}/\partial A_{kl}, \qquad \mathscr{L}^2_{ijklmn} = \partial^2 F_{ij}/\partial A_{kl}\partial A_{mn}, \ldots . \tag{6.1.7}$$

The above formalities serve essentially to introduce the convention we adopt for the derivatives of functions of second-order tensors as a prelude to consideration of stress increments in Section 6.1.2. Note the similarity between deformation rates (as discussed in Section 2.3.1) and deformation increments. We observe, however, that, whereas rate equations are exact, the (linearized) incremental equations are approximations. Thus, in order to obtain incremental formulae the rates (of stress, strain etc.) occurring in rate equations may be replaced by the corresponding increments.

For example, the rate $\dot{\mathbf{E}}$ of the Green strain tensor

$$\mathbf{E} = \tfrac{1}{2}(\mathbf{A}^{\mathrm{T}}\mathbf{A} - \mathbf{I})$$

is given by

$$\dot{\mathbf{E}} = \tfrac{1}{2}(\mathbf{A}^{\mathrm{T}}\dot{\mathbf{A}} + \dot{\mathbf{A}}^{\mathrm{T}}\mathbf{A}).$$

The corresponding incremental equation is

$$\delta\mathbf{E} = \tfrac{1}{2}\{\mathbf{A}^{\mathrm{T}}(\delta\mathbf{A}) + (\delta\mathbf{A}^{\mathrm{T}})\mathbf{A}\}, \tag{6.1.8}$$

the error in this approximation being $\tfrac{1}{2}(\delta\mathbf{A}^{\mathrm{T}})(\delta\mathbf{A})$.

Let $\delta\mathbf{R}$ and $\delta\mathbf{U}$ denote the increments in $\mathbf{R}$ and $\mathbf{U}$ arising in the increment of the polar decomposition $\mathbf{A} = \mathbf{RU}$. Then

$$\delta\mathbf{A} = (\delta\mathbf{R})\mathbf{U} + \mathbf{R}(\delta\mathbf{U}) \tag{6.1.9}$$

and

$$\delta\mathbf{E} = \tfrac{1}{2}\{\mathbf{U}(\delta\mathbf{U}) + (\delta\mathbf{U})\mathbf{U}\} \tag{6.1.10}$$

correct to the first order.

6.1.2 Stress increments and elastic moduli

Consider first the stress-deformation relation in the form

$$\mathbf{S} = \mathbf{H}(\mathbf{A}),$$

where $\mathbf{S}$ is the nominal stress tensor. Then, in accordance with (6.1.5), the nominal stress increment, linearized in $\delta\mathbf{A}$, may be written

$$\delta\mathbf{S} = \mathscr{A}^1 \delta\mathbf{A}, \tag{6.1.11}$$

where we have introduced the notation

$$\mathscr{A}^1 = \partial\mathbf{S}/\partial\mathbf{A}. \tag{6.1.12}$$

We refer to $\mathscr{A}^1$ as the *tensor of first-order elastic moduli* associated with the conjugate pair $(\mathbf{S}, \mathbf{A})$. It is a fourth-order tensor. More generally, we define

$$\mathscr{A}^\nu = \partial^\nu\mathbf{S}/\partial\mathbf{A}^\nu \tag{6.1.13}$$

as the tensor of νth-order† elastic moduli for the same conjugate pair. It is a tensor of order $2\nu + 2$.

In Cartesian components, we have

$$\left.\begin{aligned} \mathscr{A}^1_{\alpha i\beta j} &= \partial S_{\alpha i}/\partial A_{j\beta}, \\ \mathscr{A}^2_{\alpha i\beta j\gamma k} &= \partial^2 S_{\alpha i}/\partial A_{j\beta}\partial A_{k\gamma} = \mathscr{A}^2_{\alpha i\gamma k\beta j} \end{aligned}\right\} \tag{6.1.14}$$

and similarly for higher-order moduli. For a Green-elastic material

$$\mathbf{S} = \partial W/\partial\mathbf{A}, \qquad S_{\alpha i} = \partial W/\partial A_{i\alpha}$$

and the moduli then have the pairwise symmetries

$$\left.\begin{aligned} \mathscr{A}^1_{\alpha i\beta j} &= \mathscr{A}^1_{\beta j\alpha i}, \\ \mathscr{A}^2_{\alpha i\beta j\gamma k} &= \mathscr{A}^2_{\alpha i\gamma k\beta j} = \mathscr{A}^2_{\gamma k\alpha i\beta j}, \end{aligned}\right\} \tag{6.1.15}$$

and similarly for $\mathscr{A}^\nu$.

The tensor $\mathscr{A}^1$ is particularly important since, as we have seen in Section 5.1.1, it arises in the governing differential equations and its properties have crucial influence on the nature of the solutions of these equations. We shall return to a discussion of $\mathscr{A}^1$ in Section 6.2.

Just as we have associated the moduli $\mathscr{A}^\nu$ with $(\mathbf{S}, \mathbf{A})$, we may likewise

†Sometimes in the literature referred to as the tensor of $(\nu + 1)$th-order moduli.

define moduli based on the conjugate stress and deformation pairs considered in Section 3.5. To illustrate this we examine the class of strain tensors $\mathbf{E}^{(m)}$ defined in (3.5.7) and the associated conjugate stresses $\mathbf{T}^{(m)}$ introduced in Section 3.5.2. The results are easily generalized to other conjugate pairs.

We write

$$\left.\begin{aligned}\mathscr{L}^{(m)1} &= \partial\mathbf{T}^{(m)}/\partial\mathbf{E}^{(m)},\\ \mathscr{L}^{(m)2} &= \partial^2\mathbf{T}^{(m)}/\partial\mathbf{E}^{(m)2}\end{aligned}\right\} \tag{6.1.16}$$

respectively for the tensors of first- and second-order elastic moduli associated with $(\mathbf{T}^{(m)}, \mathbf{E}^{(m)})$. Similarly, $\mathscr{L}^{(m)\nu}$ may be defined $(\nu = 1, 2, \ldots)$. Recall that m may take any real value.

In components (6.1.16) are

$$\left.\begin{aligned}&\mathscr{L}^{(m)1}_{\alpha\beta\gamma\delta} = \mathscr{L}^{(m)1}_{\beta\alpha\gamma\delta} = \mathscr{L}^{(m)1}_{\alpha\beta\delta\gamma} = \partial T^{(m)}_{\alpha\beta}/\partial E^{(m)}_{\gamma\delta},\\ &\mathscr{L}^{(m)2}_{\alpha\beta\gamma\delta\varepsilon\kappa} = \mathscr{L}^{(m)2}_{\beta\alpha\gamma\delta\varepsilon\kappa} = \mathscr{L}^{(m)2}_{\alpha\beta\delta\gamma\varepsilon\kappa} = \mathscr{L}^{(m)2}_{\alpha\beta\gamma\delta\kappa\varepsilon} = \mathscr{L}^{(m)2}_{\alpha\beta\varepsilon\kappa\gamma\delta} = \partial^2 T^{(m)}_{\alpha\beta}/\partial E^{(m)}_{\gamma\delta}\partial E^{(m)}_{\varepsilon\kappa},\end{aligned}\right\} \tag{6.1.17}$$

the symmetries induced by those of $\mathbf{E}^{(m)}$ and $\mathbf{T}^{(m)}$ being noted. For a Green-elastic material $\mathbf{T}^{(m)} = \partial W/\partial\mathbf{E}^{(m)}$, and we then have the further pairwise symmetries

$$\left.\begin{aligned}\mathscr{L}^{(m)1}_{\alpha\beta\gamma\delta} &= \mathscr{L}^{(m)1}_{\gamma\delta\alpha\beta},\\ \mathscr{L}^{(m)2}_{\alpha\beta\gamma\delta\varepsilon\kappa} &= \mathscr{L}^{(m)2}_{\gamma\delta\alpha\beta\varepsilon\kappa}.\end{aligned}\right\} \tag{6.1.18}$$

Connections between moduli corresponding to different conjugate variables are now obtained. From the results of Section 3.5.2 it follows that $\mathbf{T}^{(m)}$ and $\mathbf{T}^{(1)}$ are related by

$$\mathbf{T}^{(1)} = \mathscr{L}^1\mathbf{T}^{(m)}, \tag{6.1.19}$$

where $\mathscr{L}^1$ is the fourth-order tensor defined on the right-hand side of (3.5.29) specialized to $f(\lambda_i) = (\lambda_i^m - 1)/m$.

On differentiation of (6.1.19) with respect to $\mathbf{E}^{(1)} \equiv \mathbf{U} - \mathbf{I}$ we obtain

$$\mathscr{L}^{(1)1} = \mathscr{L}^2\mathbf{T}^{(m)} + \mathscr{L}^1\mathscr{L}^{(m)1}\mathscr{L}^1, \tag{6.1.20}$$

where $\mathscr{L}^2 = \partial\mathscr{L}^1/\partial\mathbf{U}$, the symmetries of $\mathscr{L}^2$ having been used. The component form of (6.1.20) is

$$\mathscr{L}^{(1)1}_{\alpha\beta\gamma\delta} = \mathscr{L}^2_{\alpha\beta\gamma\delta\varepsilon\kappa}T^{(m)}_{\varepsilon\kappa} + \mathscr{L}^1_{\alpha\beta\varepsilon\kappa}\mathscr{L}^{(m)1}_{\varepsilon\kappa\lambda\mu}\mathscr{L}^1_{\lambda\mu\gamma\delta}.$$

The corresponding component form of the connection between the second-order moduli $\mathscr{L}^{(1)2}$ and $\mathscr{L}^{(m)2}$ is

$$\mathscr{L}^{(1)2}_{\alpha\beta\gamma\delta\varepsilon\kappa} = \mathscr{L}^{3}_{\alpha\beta\gamma\delta\varepsilon\kappa\lambda\mu} T^{(m)}_{\lambda\mu} + \mathscr{L}^{2}_{\alpha\beta\gamma\delta\lambda\mu} \mathscr{L}^{(m)1}_{\lambda\mu\nu\rho} \mathscr{L}^{1}_{\nu\rho\varepsilon\kappa} + \\ + \mathscr{L}^{2}_{\alpha\beta\varepsilon\kappa\lambda\mu} \mathscr{L}^{(m)1}_{\lambda\mu\nu\rho} \mathscr{L}^{1}_{\nu\rho\gamma\delta} + \mathscr{L}^{1}_{\alpha\beta\lambda\mu} \mathscr{L}^{(m)1}_{\lambda\mu\nu\rho} \mathscr{L}^{2}_{\nu\rho\gamma\delta\varepsilon\kappa} + \\ + \mathscr{L}^{1}_{\alpha\beta\lambda\mu} \mathscr{L}^{(m)1}_{\lambda\mu\nu\rho\sigma\tau} \mathscr{L}^{1}_{\nu\rho\gamma\delta} \mathscr{L}^{1}_{\sigma\tau\varepsilon\kappa},$$

where $\mathscr{L}^3 = \partial \mathscr{L}^2/\partial \mathbf{U}$. Connections between higher-order moduli are obtained in a similar manner.

The moduli $\mathscr{L}^{(m)\nu}$ may be linked to $\mathscr{A}^{\nu}$ by using (6.1.20) and its generalization to higher orders together with the close connection between $\mathscr{L}^{(2)\nu}$ and $\mathscr{A}^{\nu}$ generated from

$$\mathbf{S} = \mathbf{T}^{(2)}\mathbf{A}^{\mathrm{T}}. \tag{6.1.21}$$

Differentiation of this with respect to $\mathbf{A}$ in component form gives

$$\mathscr{A}^{1}_{\alpha i\beta j} = \mathscr{L}^{(2)1}_{\alpha\gamma\beta\varepsilon} A_{i\gamma} A_{j\varepsilon} + T^{(2)}_{\alpha\beta}\delta_{ij}, \tag{6.1.22}$$

use having been made of the result

$$\frac{\partial E^{(2)}_{\lambda\mu}}{\partial A_{j\beta}} = \tfrac{1}{2}(\delta_{\lambda\beta} A_{j\mu} + \delta_{\mu\beta} A_{j\lambda}).$$

A further differentiation leads to

$$\mathscr{A}^{2}_{\alpha i\beta j\gamma\kappa} = \mathscr{L}^{(2)2}_{\alpha\lambda\beta\mu\gamma\nu} A_{i\lambda} A_{j\mu} A_{k\nu} + \mathscr{L}^{(2)1}_{\alpha\beta\gamma\lambda} A_{k\lambda}\delta_{ij} \\ + \mathscr{L}^{(2)1}_{\alpha\gamma\beta\lambda} A_{j\lambda}\delta_{ik} + \mathscr{L}^{(2)1}_{\alpha\lambda\beta\gamma} A_{i\lambda}\delta_{jk}, \tag{6.1.23}$$

and similarly for higher-order moduli. For the corresponding tensor representations of (6.1.22) and (6.1.23) see Chadwick and Ogden (1971a).

6.1.3 Instantaneous moduli

The tensors of elastic moduli considered in Section 6.1.2 relate to an arbitrary reference configuration. Suppose we now consider a fixed reference configuration (which may be taken as a natural configuration if appropriate) and let $\mathbf{A}$ be the deformation gradient which relates this to the current configuration. We then refer to the moduli as *fixed-reference moduli*.

Equally, we may take the reference configuration to coincide with the current configuration at any stage of the deformation. The resulting elastic moduli are called *instantaneous moduli*. We now relate these to the fixed-reference moduli.

Firstly, we note that, since line elements of material transform according to $\mathbf{dx} = \mathbf{A}\,\mathbf{dX}$, we have the incremental connection $\delta(\mathbf{dx}) = (\delta\mathbf{A})\mathbf{dX}$. If the current configuration is now chosen as the reference configuration then the right-hand side of this equation becomes $(\delta\mathbf{A}_0)\,\mathbf{dx}$, where $\delta\mathbf{A}_0$ is the value of $\delta\mathbf{A}$ in this configuration. Since $\mathbf{dx}$, and hence $\delta(\mathbf{dx})$, is independent of the reference configuration, we obtain the connection

$$\delta\mathbf{A} = (\delta\mathbf{A}_0)\mathbf{A}. \tag{6.1.24}$$

In what follows the subscript zero indicates evaluation in the current configuration of the quantity to which it is attached. For example,

$$\delta\mathbf{E}_0^{(2)} = \tfrac{1}{2}(\delta\mathbf{A}_0^{\mathrm{T}} + \delta\mathbf{A}_0) \tag{6.1.25}$$

and

$$\delta\mathbf{E}^{(2)} = \mathbf{A}^{\mathrm{T}}(\delta\mathbf{E}_0^{(2)})\mathbf{A}, \tag{6.1.26}$$

these being analogous to (2.3.8) and (2.3.7) respectively.

The load $\mathbf{S}^{\mathrm{T}}\mathbf{N}\,\mathrm{d}A$ on a reference area element $\mathbf{N}\,\mathrm{d}A$ is also independent of the choice of reference configuration. Thus, since the current area element is given by $J\mathbf{B}\mathbf{N}\,\mathrm{d}A$ (see equation (2.2.18)) we obtain

$$\delta\mathbf{S}_0 = J^{-1}\mathbf{A}\delta\mathbf{S}, \tag{6.1.27}$$

where $\delta\mathbf{S}$ is the increment in nominal stress evaluated for the fixed reference configuration while $\delta\mathbf{S}_0$ is its value relative to the current configuration.

On substituting

$$\delta\mathbf{S}_0 = \boldsymbol{\mathscr{A}}_0^1\delta\mathbf{A}_0, \qquad \delta\mathbf{S} = \boldsymbol{\mathscr{A}}^1\delta\mathbf{A} \tag{6.1.28}$$

into (6.1.27) and making use of (6.1.24) we obtain, in component form,

$$\mathscr{A}^1_{0ijkl} = J^{-1}A_{i\alpha}A_{k\beta}\mathscr{A}^1_{\alpha j\beta l}. \tag{6.1.29}$$

For the second-order moduli

$$\mathscr{A}^2_{0ijklmn} = J^{-1}A_{i\alpha}A_{k\beta}A_{m\gamma}\mathscr{A}^2_{\alpha j\beta l\gamma n}, \tag{6.1.30}$$

and likewise for higher-order moduli.

Problem 6.1.1 Show that the first- and second-order fixed-reference and instantaneous moduli associated with the conjugate pair $(\mathbf{T}^{(2)},\mathbf{E}^{(2)})$ have rectangular Cartesian components satisfying

$$\mathscr{L}^{(2)1}_{0ijkl} = J^{-1}A_{i\alpha}A_{j\beta}A_{k\gamma}A_{l\lambda}\mathscr{L}^{(2)1}_{\alpha\beta\gamma\lambda}, \tag{6.1.31}$$

$$\mathscr{L}^{(2)2}_{0ijklmn} = J^{-1}A_{i\alpha}A_{j\beta}A_{k\gamma}A_{l\lambda}A_{m\mu}A_{n\nu}\mathscr{L}^{(2)2}_{\alpha\beta\gamma\lambda\mu\nu} \tag{6.1.32}$$

Problem 6.1.2 Obtain the connections

$$\mathscr{A}^1_{0ijkl} = \mathscr{L}^{(2)1}_{0ijkl} + T_{ik}\delta_{jl}, \tag{6.1.33}$$

$$\mathscr{L}^{(2)1}_{0ijkl} = \mathscr{L}^{(m)1}_{0ijkl} + \tfrac{1}{4}(m-2)(\delta_{ik}T_{jl} + \delta_{il}T_{jk} + \delta_{jk}T_{il} + \delta_{jl}T_{ik}), \tag{6.1.34}$$

where T_{ij} are the components of Cauchy stress. Derive corresponding expressions for the second-order moduli.

The formulae relating the various moduli given in this section are valid for constrained as well as for unconstrained materials. We note, however, that the moduli for constrained materials are defined relative to that part of the stress which does not include Lagrange multipliers. For example, we have from (4.2.73), for a material with a single constraint, the stress-deformation relation

$$\mathbf{S} = \mathbf{H}(\mathbf{A}) + q\frac{\partial C(\mathbf{A})}{\partial \mathbf{A}} \tag{6.1.35}$$

relative to the conjugate pair $(\mathbf{S}, \mathbf{A})$.

We write

$$\boldsymbol{\mathscr{A}}^1 = \partial \mathbf{H}(\mathbf{A})/\partial \mathbf{A}, \tag{6.1.36}$$

not $\partial \mathbf{S}/\partial \mathbf{A}$, for the tensor of first-order elastic moduli, and similarly for higher-order moduli. Thus,

$$\delta \mathbf{S} = \boldsymbol{\mathscr{A}}^1 \delta \mathbf{A} + \delta q\, \partial C/\partial \mathbf{A} + q(\partial^2 C/\partial \mathbf{A}^2)\delta \mathbf{A}, \tag{6.1.37}$$

and, in particular, for an incompressible material this becomes

$$\delta \mathbf{S} = \boldsymbol{\mathscr{A}}^1 \delta \mathbf{A} + \delta q \mathbf{B}^{\mathrm{T}} - q\mathbf{B}^{\mathrm{T}}(\delta \mathbf{A})\mathbf{B}^{\mathrm{T}} \tag{6.1.38}$$

since $\mathrm{tr}(\mathbf{B}^{\mathrm{T}}\delta \mathbf{A}) = 0$ follows from (6.1.2).

Finally, when the reference configuration is taken as the current configuration, equation (6.1.38) reduces to

$$\delta \mathbf{S}_0 = \boldsymbol{\mathscr{A}}^1_0 \delta \mathbf{A}_0 + \delta q \mathbf{I} - q\delta \mathbf{A}_0. \tag{6.1.39}$$

Parallel results may be obtained for other conjugate pairs, but we do not give details here.

In the following section we specialize the results of Section 6.1.3 to the case of isotropic materials in order to obtain explicit formulae for certain elastic moduli in terms of the principal stresses and stretches.

6.1.4 Elastic moduli for isotropic materials

We recall from Section 4.2.6 that for an isotropic elastic material the conjugate stress and strain tensors $(\mathbf{T}^{(m)}, \mathbf{E}^{(m)})$ are coaxial with the Lagrangean principal axes $\{\mathbf{u}^{(i)}\}$. Thus, in spectral form,

$$\mathbf{E}^{(m)} = \sum_{i=1}^{3} e_i^{(m)} \mathbf{u}^{(i)} \otimes \mathbf{u}^{(i)}, \quad \mathbf{T}^{(m)} = \sum_{i=1}^{3} t_i^{(m)} \mathbf{u}^{(i)} \otimes \mathbf{u}^{(i)}, \tag{6.1.40}$$

where $e_i^{(m)}$ and $t_i^{(m)}$ $(i = 1, 2, 3)$ are the principal values of $\mathbf{E}^{(m)}$ and $\mathbf{T}^{(m)}$ respectively.

From the results of Section 2.3.2 we may write the strain rate† $\dot{\mathbf{E}}^{(m)}$ as

$$\dot{\mathbf{E}}^{(m)} = \sum_{i=1}^{3} \dot{e}_i^{(m)} \mathbf{u}^{(i)} \otimes \mathbf{u}^{(i)} + \sum_{i \neq j} \Omega_{ij}^{(\mathrm{L})} (e_j^{(m)} - e_i^{(m)}) \mathbf{u}^{(i)} \otimes \mathbf{u}^{(j)} \tag{6.1.41}$$

and similarly for $\dot{\mathbf{T}}^{(m)}$.

Since $t_i^{(m)}$, i = 1, 2, 3, depends only on $e_j^{(m)}$, $j = 1, 2, 3$, we have the following expression for $\dot{\mathbf{T}}^{(m)}$:

$$\dot{\mathbf{T}}^{(m)} = \sum_{i,j=1}^{3} \frac{\partial t_i^{(m)}}{\partial e_j^{(m)}} \dot{e}_j^{(m)} \mathbf{u}^{(i)} \otimes \mathbf{u}^{(i)} + \sum_{i \neq j} \Omega_{ij}^{(\mathrm{L})} (e_j^{(m)} - e_i^{(m)}) \left(\frac{t_j^{(m)} - t_i^{(m)}}{e_j^{(m)} - e_i^{(m)}} \right) \mathbf{u}^{(i)} \otimes \mathbf{u}^{(j)}. \tag{6.1.42}$$

In view of (6.1.41) this may be represented in the form

$$\dot{\mathbf{T}}^{(m)} = \mathscr{L}^{(m)1} \dot{\mathbf{E}}^{(m)} \tag{6.1.43}$$

with $\mathscr{L}^{(m)1}$ defined by $(6.1.16)_1$.

On comparing (6.1.42) and (6.1.43) we deduce that $\mathscr{L}^{(m)1}$ may be represented in the form

$$\mathscr{L}^{(m)1} = \sum_{i,j=1}^{3} \frac{\partial t_i^{(m)}}{\partial e_j^{(m)}} \mathbf{u}^{(i)} \otimes \mathbf{u}^{(i)} \otimes \mathbf{u}^{(j)} \otimes \mathbf{u}^{(j)} + \tfrac{1}{2} \sum_{i \neq j} \left(\frac{t_j^{(m)} - t_i^{(m)}}{e_j^{(m)} - e_i^{(m)}} \right) (\mathbf{u}^{(i)} \otimes \mathbf{u}^{(j)} \otimes \mathbf{u}^{(i)} \otimes \mathbf{u}^{(j)} + \mathbf{u}^{(i)} \otimes \mathbf{u}^{(j)} \otimes \mathbf{u}^{(j)} \otimes \mathbf{u}^{(i)}), \tag{6.1.44}$$

† For notational convenience we use rates rather than increments in this section.

this being a generalization of the result (3.5.29). Thus, on the Lagrangean principal axes, the non-zero components of $\mathscr{L}^{(m)1}$ are

$$\left.\begin{aligned} \mathscr{L}^{(m)1}_{iijj} &= \frac{\partial t_i^{(m)}}{\partial e_j^{(m)}}, \\ \mathscr{L}^{(m)1}_{ijij} &= \frac{1}{2}\left(\frac{t_i^{(m)} - t_j^{(m)}}{e_i^{(m)} - e_j^{(m)}}\right) \qquad i \neq j, \end{aligned}\right\} \tag{6.1.45}$$

$(i,j = 1,2,3$; no summation over repeated indices) subject to the symmetries

$$\mathscr{L}^{(m)1}_{ijkl} = \mathscr{L}^{(m)1}_{jikl} = \mathscr{L}^{(m)1}_{ijlk} \tag{6.1.46}$$

and, for a Green-elastic material,

$$\mathscr{L}^{(m)1}_{ijkl} = \mathscr{L}^{(m)1}_{klij}. \tag{6.1.47}$$

In the above and for the remainder of this section we follow closely the work of Chadwick and Ogden (1971b) with some differences of notation.

The tensor $\mathscr{L}^{(m)2}$ of second-order moduli is obtained in a similar way by evaluating the rate of (6.1.44) and equating coefficients of $\dot{e}_i^{(m)}, \Omega_{ij}^{(\mathrm{L})}$ $(i,j = 1,2,3)$ in the equation

$$\dot{\mathscr{L}}^{(m)1} = \mathscr{L}^{(m)2}[\cdot, \dot{\mathbf{E}}^{(m)}]. \tag{6.1.48}$$

Referred to the Lagrangean principal axes, the resulting non-zero components of $\mathscr{L}^{(m)2}$ are found to be

$$\left.\begin{aligned} \mathscr{L}^{(m)2}_{iijjkk} &= \frac{\partial}{\partial e_k^{(m)}} \mathscr{L}^{(m)1}_{iijj}, \\ \mathscr{L}^{(m)2}_{ijijkk} &= \frac{\partial}{\partial e_k^{(m)}} \mathscr{L}^{(m)1}_{ijij} \qquad i \neq j, \\ \mathscr{L}^{(m)2}_{iijkjk} &= \frac{1}{2}\frac{\mathscr{L}^{(m)1}_{iijj} - \mathscr{L}^{(m)1}_{iikk}}{e_j^{(m)} - e_k^{(m)}} - \mathscr{L}^{(m)1}_{jkjk}\frac{(\delta_{ij} - \delta_{ik})}{e_j^{(m)} - e_k^{(m)}} \qquad j \neq k, \\ \mathscr{L}^{(m)2}_{ijjkki} &= \frac{1}{4}\frac{\mathscr{L}^{(m)1}_{ijij} - \mathscr{L}^{(m)1}_{kiki}}{e_j^{(m)} - e_k^{(m)}} + \frac{1}{4}\frac{\mathscr{L}^{(m)1}_{jkjk} - \mathscr{L}^{(m)1}_{ijij}}{e_k^{(m)} - e_i^{(m)}} \qquad i \neq j \neq k \neq i, \end{aligned}\right\} \tag{6.1.49}$$

$(i,j,k \in \{1,2,3\}$; no summation on repeated indices) subject to the symmetries

$$\mathscr{L}^{(m)2}_{ijklmn} = \mathscr{L}^{(m)2}_{jiklmn} = \mathscr{L}^{(m)2}_{ijlkmn} = \mathscr{L}^{(m)2}_{ijklnm}. \tag{6.1.50}$$

For a Green-elastic material we have the additional symmetry

$$\mathscr{L}^{(m)2}_{ijklmn} = \mathscr{L}^{(m)2}_{klijmn}. \tag{6.1.51}$$

Full details of the above derivations are given in Chadwick and Ogden (1971b).

In has been assumed implicitly in the above that $e_1^{(m)}, e_2^{(m)}, e_3^{(m)}$ are distinct. The corresponding results when two or all three are equal are obtained by formally taking the limits as $e_i^{(m)}$ approaches $e_j^{(m)}$. For example,

$$\frac{t_i^{(m)} - t_j^{(m)}}{e_i^{(m)} - e_j^{(m)}} \quad \text{becomes} \quad \frac{\partial}{\partial e_i^{(m)}}(t_i^{(m)} - t_j^{(m)})$$

with $e_j^{(m)}$ being set equal to $e_i^{(m)}$ after differentiation. In what follows we assume that the constitutive law has regularity sufficient to ensure the existence of such limits as are required. We consider separately the cases of two and three equal principal strains.

(a) *Suppose* $e_i^{(m)} = e_j^{(m)} \neq e_k^{(m)}$, *where* (i,j,k) *is a cyclic permutation of* $(1,2,3)$.

Then, from the resulting symmetries in the constitutive relations (replace t_i and λ_i by $t_i^{(m)}$ and $e_i^{(m)}$ respectively in (4.2.57) and (4.2.59)), we deduce that

$$t_i^{(m)} = t_j^{(m)}, \quad \frac{\partial t_i^{(m)}}{\partial e_i^{(m)}} = \frac{\partial t_j^{(m)}}{\partial e_j^{(m)}}, \quad \frac{\partial t_i^{(m)}}{\partial e_j^{(m)}} = \frac{\partial t_j^{(m)}}{\partial e_i^{(m)}}, \quad \frac{\partial t_k^{(m)}}{\partial e_i^{(m)}} = \frac{\partial t_k^{(m)}}{\partial e_j^{(m)}}. \tag{6.1.52}$$

Further differentiations yield the connections

$$\left.\begin{aligned}
&\frac{\partial^2 t_i^{(m)}}{\partial e_i^{(m)2}} + \frac{\partial^2 t_i^{(m)}}{\partial e_i^{(m)}\partial e_j^{(m)}} = \frac{\partial^2 t_j^{(m)}}{\partial e_i^{(m)}\partial e_j^{(m)}} + \frac{\partial^2 t_j^{(m)}}{\partial e_j^{(m)2}},\\
&\frac{\partial^2 t_i^{(m)}}{\partial e_i^{(m)}\partial e_j^{(m)}} + \frac{\partial^2 t_i^{(m)}}{\partial e_j^{(m)2}} = \frac{\partial^2 t_j^{(m)}}{\partial e_i^{(m)2}} + \frac{\partial^2 t_j^{(m)}}{\partial e_i^{(m)}\partial e_j^{(m)}},\\
&\frac{\partial^2 t_k^{(m)}}{\partial e_i^{(m)2}} = \frac{\partial^2 t_k^{(m)}}{\partial e_j^{(m)2}}, \qquad \frac{\partial^2 t_i^{(m)}}{\partial e_i^{(m)}\partial e_k^{(m)}} = \frac{\partial^2 t_j^{(m)}}{\partial e_j^{(m)}\partial e_k^{(m)}},\\
&\frac{\partial^2 t_i^{(m)}}{\partial e_j^{(m)}\partial e_k^{(m)}} = \frac{\partial^2 t_j^{(m)}}{\partial e_i^{(m)}\partial e_k^{(m)}}, \qquad \frac{\partial^2 t_k^{(m)}}{\partial e_i^{(m)}\partial e_k^{(m)}} = \frac{\partial^2 t_k^{(m)}}{\partial e_j^{(m)}\partial e_k^{(m)}},
\end{aligned}\right\} \tag{6.1.53}$$

and similarly for higher derivatives.

The expressions for the components of $\mathscr{L}^{(m)1}$ given in (6.1.45) are unchanged except that $\mathscr{L}^{(m)1}_{ijij}$ is now given by

$$\mathscr{L}^{(m)1}_{ijij} = \tfrac{1}{2}\frac{\partial}{\partial e_i^{(m)}}(t_i^{(m)} - t_j^{(m)}) \tag{6.1.54}$$

subject to (6.1.52).

The corresponding adjustments in the components of $\mathscr{L}^{(m)2}$ are as follows. In (6.1.49) the expression for $\mathscr{L}^{(m)2}_{ijijkk}$ is replaced by

$$\left.\begin{aligned} \mathscr{L}^{(m)}_{ijijii} &= \frac{1}{4}\frac{\partial^2}{\partial e_i^{(m)2}}(t_i^{(m)} - t_j^{(m)}), \\ \mathscr{L}^{(m)2}_{ijijkk} &= \frac{1}{2}\frac{\partial^2}{\partial e_i^{(m)}\partial e_k^{(m)}}(t_i^{(m)} - t_j^{(m)}), \end{aligned}\right\} \tag{6.1.55}$$

and that for $\mathscr{L}^{(m)2}_{iijkjk}$ by

$$\left.\begin{aligned} \mathscr{L}^{(m)2}_{iiijij} &= \frac{1}{4}\left(\frac{\partial^2 t_i^{(m)}}{\partial e_i^{(m)2}} + \frac{\partial^2 t_j^{(m)}}{\partial e_i^{(m)2}} - 2\frac{\partial^2 t_i^{(m)}}{\partial e_i^{(m)}\partial e_j^{(m)}}\right), \\ \mathscr{L}^{(m)2}_{kkijij} &= \frac{1}{2}\left(\frac{\partial^2 t_k^{(m)}}{\partial e_i^{(m)2}} - \frac{\partial^2 t_k^{(m)}}{\partial e_i^{(m)}\partial e_j^{(m)}}\right), \end{aligned}\right\} \tag{6.1.56}$$

recalling that now we have the restriction $i \neq j \neq k \neq i$. This restriction is not required in $(6.1.49)_1$ which, along with $(6.1.49)_4$, remains valid.

(b) *Suppose* $e_1^{(m)} = e_2^{(m)} = e_3^{(m)}$. *Then the deformation corresponds to a state of pure dilatation.*

Connections between the first and second derivatives additional to those given in (6.1.52) and (6.1.53) then arise. Indeed, only two independent components of $\mathscr{L}^{(m)1}$ remain, while $\mathscr{L}^{(m)2}$ retains four independent components (three in the case of a Green-elastic material) as detailed below.

Since $t_1^{(m)} = t_2^{(m)} = t_3^{(m)}$ it follows that $\partial t_i^{(m)}/\partial e_i^{(m)}$ is independent of i while $\partial t_i^{(m)}/\partial e_j^{(m)}$ has the same value for any pair of indices $i, j \neq i$ $(i, j \in \{1, 2, 3\})$. We therefore introduce the notation

$$\left.\begin{aligned} \beta_1^{(m)} &= \frac{\partial t_i^{(m)}}{\partial e_j^{(m)}} \qquad i \neq j, \\ \beta_1^{(m)} + 2\beta_2^{(m)} &= \frac{\partial t_i^{(m)}}{\partial e_i^{(m)}} \qquad i = 1, 2, 3;\ \text{no summation.} \end{aligned}\right\} \tag{6.1.57}$$

Similarly, we introduce the following notations defined in terms of the independent second derivatives of $t_1^{(m)}, t_2^{(m)}, t_3^{(m)}$ evaluated for $e_1^{(m)} = e_2^{(m)} = e_3^{(m)}$:

$$\left.\begin{aligned}
\gamma_1^{(m)} &= \frac{\partial^2 t_i^{(m)}}{\partial e_j^{(m)} \partial e_k^{(m)}} && (i,j,k) \text{ permutation of } (1, 2, 3),\\
\gamma_1^{(m)} + 2\gamma_2^{(m)} &= \frac{\partial^2 t_i^{(m)}}{\partial e_j^{(m)2}} && i \neq j,\\
\gamma_1^{(m)} + 2\gamma_2^{(m)\prime} &= \frac{\partial^2 t_i^{(m)}}{\partial e_i^{(m)} \partial e_j^{(m)}} && i \neq j,\\
\gamma_1^{(m)} + 2\gamma_2^{(m)} + 4\gamma_2^{(m)\prime} + 8\gamma_3^{(m)} &= \frac{\partial^2 t_i^{(m)}}{\partial e_i^{(m)2}} && i = 1,2,3; \text{ no summation.}
\end{aligned}\right\} \tag{6.1.58}$$

No additional changes are needed in the expressions for the components of $\mathscr{L}^{(m)1}$ and $\mathscr{L}^{(m)2}$ except in the case of $(6.1.49)_4$ which now reduces to

$$\begin{aligned}
\mathscr{L}_{ijjkki}^{(m)2} &= \frac{1}{8}\left(\frac{\partial^2 t_i^{(m)}}{\partial e_i^{(m)2}} - \frac{\partial^2 t_i^{(m)}}{\partial e_j^{(m)2}} + 2\frac{\partial^2 t_i^{(m)}}{\partial e_j^{(m)} \partial e_k^{(m)}} - 2\frac{\partial^2 t_i^{(m)}}{\partial e_i^{(m)} \partial e_j^{(m)}}\right)\\
&= \gamma_3^{(m)}
\end{aligned} \tag{6.1.59}$$

for $i \neq j \neq k \neq i$.

The components of $\mathscr{L}^{(m)1}$ and $\mathscr{L}^{(m)2}$ may now be put together compactly in the forms

$$\mathscr{L}_{ijkl}^{(m)1} = \beta_1^{(m)} \delta_{ij}\delta_{kl} + 2\beta_2^{(m)} \mathscr{I}_{ijkl}^{1}, \tag{6.1.60}$$

$$\begin{aligned}
\mathscr{L}_{ijklmn}^{(m)2} &= \gamma_1^{(m)} \delta_{ij}\delta_{kl}\delta_{mn} + 2\gamma_2^{(m)} \delta_{ij} \mathscr{I}_{klmn}^{1} +\\
&\quad + 2\gamma_2^{(m)\prime} (\delta_{kl} \mathscr{I}_{mnij}^{1} + \delta_{mn} \mathscr{I}_{ijkl}^{1}) + 8\gamma_3^{(m)} \mathscr{I}_{ijklmn}^{2},
\end{aligned} \tag{6.1.61}$$

where $\mathscr{I}_{ijkl}^{1}$ is defined by (3.5.32) and

$$\mathscr{I}_{ijklmn}^{2} = \tfrac{1}{4}(\delta_{ik} \mathscr{I}_{jlmn}^{1} + \delta_{jl} \mathscr{I}_{ikmn}^{1} + \delta_{il} \mathscr{I}_{jkmn}^{1} + \delta_{jk} \mathscr{I}_{ilmn}^{1}), \tag{6.1.62}$$

the latter being an alternative expression for the coefficient of $f''(1)$ in (3.5.34).

Note that the *scalar* elastic moduli $\beta_1^{(m)}, \beta_2^{(m)}, \gamma_1^{(m)}, \gamma_2^{(m)}, \gamma_2^{(m)\prime}, \gamma_3^{(m)}$ depend on the value of $e_1^{(m)} = e_2^{(m)} = e_3^{(m)}$ and that for a Green-elastic material

$$\gamma_2^{(m)\prime} = \gamma_2^{(m)}. \tag{6.1.63}$$

Problem 6.1.3 By considering the definitions (6.1.57) for different values of

m show that

$$\beta_1^{(m)} = (1 + me^{(m)})^{-2}\beta_1^{(0)},$$
$$\beta_2^{(m)} = (1 + me^{(m)})^{-2}\beta_2^{(0)} - \tfrac{1}{2}m(1 + me^{(m)})^{-1}t^{(m)},$$

where $e^{(m)}$ and $t^{(m)}$ are the common values of $e_1^{(m)} = e_2^{(m)} = e_3^{(m)}$ and $t_1^{(m)} = t_2^{(m)} = t_3^{(m)}$ respectively.

Find corresponding connections between $\gamma_1^{(m)}, \gamma_2^{(m)}, \gamma_2^{(m)\prime}, \gamma_3^{(m)}$ and $\gamma_1^{(0)}, \gamma_2^{(0)}, \gamma_2^{(0)\prime}, \gamma_3^{(0)}$.

Problem 6.1.4 Show that on specialization to appropriate principal axes the formulae (6.1.31) and (6.1.32) may be rewritten as

$$\mathscr{L}^{(2)1}_{0ijkl} = J^{-1}\lambda_i\lambda_j\lambda_k\lambda_l\mathscr{L}^{(2)1}_{ijkl},$$
$$\mathscr{L}^{(2)2}_{0ijklmn} = J^{-1}\lambda_i\lambda_j\lambda_k\lambda_l\lambda_m\lambda_n\mathscr{L}^{(2)2}_{ijklmn},$$

where $\lambda_1, \lambda_2, \lambda_3$ are the principal stretches and $J = \lambda_1\lambda_2\lambda_3$.

Use the connection $t_i^{(2)} = J\lambda_i^{-2}t_i$, where t_1, t_2, t_3 are the principal Cauchy stresses, and the above formulae together with (6.1.45) and (6.1.49) specialized to $m = 2$ to establish the expressions

$$\mathscr{L}^{(2)1}_{0iijj} = \lambda_j\frac{\partial t_i}{\partial \lambda_j} + (1 - 2\delta_{ij})t_i,$$

$$\mathscr{L}^{(2)1}_{0ijij} = \frac{\lambda_j^2 t_i - \lambda_i^2 t_j}{\lambda_i^2 - \lambda_j^2} \qquad i \neq j,$$

$$\mathscr{L}^{(2)2}_{0iijjkk} = \lambda_j\frac{\partial}{\partial \lambda_j}\mathscr{L}^{(2)1}_{0iikk} + (1 - 2\delta_{ij} - 2\delta_{jk})\mathscr{L}^{(2)1}_{0iikk},$$

$$\mathscr{L}^{(2)2}_{0ijijkk} = \lambda_k\frac{\partial}{\partial \lambda_k}\mathscr{L}^{(2)1}_{0ijij} + (1 - 2\delta_{ik} - 2\delta_{jk})\mathscr{L}^{(2)1}_{0ijij} \qquad i \neq j,$$

$$\mathscr{L}^{(2)2}_{0iijkjk} = \frac{\lambda_k^2\mathscr{L}^{(2)1}_{0iijj} - \lambda_j^2\mathscr{L}^{(2)1}_{0iikk} - 2\lambda_i^2\mathscr{L}^{(2)1}_{0jkjk}(\delta_{ij} - \delta_{ik})}{\lambda_j^2 - \lambda_k^2} \qquad j \neq k,$$

$$\mathscr{L}^{(2)2}_{0ijjkki} = \frac{1}{2}\frac{\lambda_k^2\mathscr{L}^{(2)1}_{0ijij} - \lambda_j^2\mathscr{L}^{(2)1}_{0kiki}}{\lambda_j^2 - \lambda_k^2}$$
$$+ \frac{1}{2}\frac{\lambda_i^2\mathscr{L}^{(2)1}_{0jkjk} - \lambda_k^2\mathscr{L}^{(2)1}_{0ijij}}{\lambda_k^2 - \lambda_i^2} \qquad i \neq j \neq k \neq i,$$

for $\lambda_1 \neq \lambda_2 \neq \lambda_3 \neq \lambda_1$.

Problem 6.1.5 For a pure dilatation with $\lambda_i = \lambda, t_i = t$ show that

$$\mathscr{L}^{(2)1}_{0ijkl} = \mu_1 \delta_{ij}\delta_{kl} + 2\mu_2 \mathscr{I}^1_{ijkl},$$

$$\mathscr{L}^{(2)2}_{0ijklmn} = \nu_1 \delta_{ij}\delta_{kl}\delta_{mn} + 2\nu_2 \delta_{ij}\mathscr{I}^1_{klmn} + 2\nu_2'(\delta_{kl}\mathscr{I}^1_{mnij} + \delta_{mn}\mathscr{I}^1_{ijkl}) + 8\nu_3 \mathscr{I}^2_{ijklmn},$$

where $\mu_1 = \lambda\beta_1^{(2)}$, $\mu_2 = \lambda\beta_2^{(2)}$, $\nu_1 = \lambda^3\gamma_1^{(2)}$, $\nu_2 = \lambda^3\gamma_2^{(2)}$, $\nu_2' = \lambda^3\gamma_2^{(2)\prime}$, $\nu_3 = \lambda^3\gamma_3^{(2)}$.

Show that these scalar moduli are expressible as

$$\mu_1 = \lambda\frac{\partial t_i}{\partial \lambda_j} + t, \qquad \mu_2 = \tfrac{1}{2}\lambda\left(\frac{\partial t_i}{\partial \lambda_i} - \frac{\partial t_i}{\partial \lambda_j}\right) - t,$$

$$\nu_1 = \lambda^2 \frac{\partial^2 t_i}{\partial \lambda_j \partial \lambda_k} + 2\mu_1 - t,$$

$$\nu_2 = \tfrac{1}{2}\lambda^2\left(\frac{\partial^2 t_i}{\partial \lambda_j^2} - \frac{\partial^2 t_i}{\partial \lambda_j \partial \lambda_k}\right) - \tfrac{1}{2}(\mu_1 + t),$$

$$\nu_2' = \tfrac{1}{2}\lambda^2\left(\frac{\partial^2 t_i}{\partial \lambda_i \partial \lambda_j} - \frac{\partial^2 t_i}{\partial \lambda_j \partial \lambda_k}\right) - \mu_1 + \mu_2 + t,$$

$$\nu_3 = \tfrac{1}{8}\lambda^2\left(\frac{\partial^2 t_i}{\partial \lambda_i^2} - \frac{\partial^2 t_i}{\partial \lambda_j^2} + 2\frac{\partial^2 t_i}{\partial \lambda_j \partial \lambda_k} - 2\frac{\partial^2 t_i}{\partial \lambda_j \partial \lambda_k}\right) - \tfrac{1}{4}(5\mu_2 + t),$$

where $i \neq j \neq k \neq i$. Confirm that if the material is hyperelastic then $\nu_2' = \nu_2$.

Problem 6.1.6 Deduce from (6.1.22) and (6.1.23) that for an isotropic material the components of $\mathscr{A}^1$, $\mathscr{A}^2$, $\mathscr{L}^{(2)1}$ and $\mathscr{L}^{(2)2}$ on (Eulerian for current indices, Lagrangean for reference indices) principal axes are related according to

$$\mathscr{A}^1_{ijkl} = \lambda_j\lambda_l \mathscr{L}^{(2)1}_{ijkl} + t_i^{(2)}\delta_{ik}\delta_{jl},$$

$$\mathscr{A}^2_{ijklmn} = \lambda_j\lambda_l\lambda_n \mathscr{L}^{(2)2}_{ijklmn} + \lambda_j \mathscr{L}^{(2)1}_{ijkm}\delta_{ln} + \lambda_l \mathscr{L}^{(2)1}_{imkl}\delta_{jn} + \lambda_n \mathscr{L}^{(2)1}_{ikmn}\delta_{jl}.$$

(we do not distinguish reference indices by Greek letters here).

Show that the corresponding instantaneous moduli satisfy

$$\mathscr{A}^1_{0ijkl} = \mathscr{L}^{(2)1}_{0ijkl} + t_i\delta_{ik}\delta_{jl},$$

$$\mathscr{A}^2_{0ijklmn} = \mathscr{L}^{(2)2}_{0ijklmn} + \mathscr{L}^{(2)1}_{0ikmn}\delta_{jl} + \mathscr{L}^{(2)1}_{0imkl}\delta_{jn} + \mathscr{L}^{(2)1}_{0ijkm}\delta_{ln}.$$

For the case of pure dilatation use the results of Problem 6.1.5 to show that

$$\mathscr{A}^1_{0ijkl} = \mu_1\delta_{ij}\delta_{kl} + (\mu_2 + t)\delta_{ik}\delta_{jl} + \mu_2\delta_{il}\delta_{jk}$$

and write down an analogous expression for $\mathscr{A}^2_{0ijklmn}$.

Problem 6.1.7 Show that

$$\mathscr{A}^1_{0iijj} = \lambda_j \frac{\partial t_i}{\partial \lambda_j} + (1 - \delta_{ij})t_i,$$

$$\mathscr{A}^1_{0ijij} = \left(\frac{t_i - t_j}{\lambda_i^2 - \lambda_j^2}\right)\lambda_i^2 \qquad i \neq j,$$

$$\mathscr{A}^1_{0ijji} = \mathscr{A}^1_{0jiij} = \mathscr{A}^1_{0ijij} - t_i \qquad i \neq j,$$

and write down corresponding explicit expressions for the components of $\boldsymbol{\mathscr{A}}^2_0$.

6.1.5. Elastic moduli for incompressible isotropic materials

Since the incompressibility constraint is important in applications we now give separate details of certain elastic moduli for incompressible isotropic elastic solids.

We recall that the elastic moduli associated with the conjugate pair $(\mathbf{S}, \mathbf{A})$ for a constrained material were defined on the basis of equations (6.1.35) and (6.1.36), the moduli associated with $(\mathbf{T}^{(m)}, \mathbf{E}^{(m)})$ being defined in a similar manner. The incompressibility constraint.

$$C(\mathbf{A}) \equiv \det \mathbf{A} - 1 = 0 \tag{6.1.64}$$

does not alter the expressions for the components of $\mathscr{L}^{(m)1}$ and $\mathscr{L}^{(m)2}$ given by (6.1.45) and (6.1.49), but we note that (6.1.64) implies

$$(1 + me_1^{(m)})(1 + me_2^{(m)})(1 + me_3^{(m)}) = 1. \tag{6.1.65}$$

The corresponding results for $\boldsymbol{\mathscr{A}}^1$ and $\boldsymbol{\mathscr{A}}^2$ generated from formulae given in Problem 6.1.6 are likewise unaltered provided $t_i^{(2)}$ does not include the 'constraint stress'.

Since the formula $t_i^{(m)} = \lambda_i^{-m} J t_i$ is replaced by $t_i^{(m)} = \lambda_i^{-m} t_i$ for incompressible materials, the transition from (6.1.45) and (6.1.49) to the corresponding components of the instantaneous moduli is slightly modified.

In particular, the results of Problem 6.1.4 are replaced by

$$\left.\begin{aligned} \mathscr{L}^{(2)1}_{0iijj} &= \lambda_j \frac{\partial t_i}{\partial \lambda_j} - 2\delta_{ij} t_i, \\ \mathscr{L}^{(2)1}_{0ijij} &= \frac{\lambda_j^2 t_i - \lambda_i^2 t_j}{\lambda_i^2 - \lambda_j^2} \qquad i \neq j, \end{aligned}\right\} \tag{6.1.66}$$

$$\left.\begin{aligned} \mathscr{L}^{(2)2}_{0iijjkk} &= \lambda_j \frac{\partial}{\partial \lambda_j} \mathscr{L}^{(2)1}_{0iikk} - 2(\delta_{ij} + \delta_{jk}) \mathscr{L}^{(2)1}_{0iikk}, \\ \mathscr{L}^{(2)2}_{0ijijkk} &= \lambda_k \frac{\partial}{\partial \lambda_k} \mathscr{L}^{(2)1}_{0ijij} - 2(\delta_{ik} + \delta_{jk}) \mathscr{L}^{(2)1}_{0ijij} \qquad i \neq j, \\ \mathscr{L}^{(2)2}_{0iijkjk} &= \frac{\lambda_k^2 \mathscr{L}^{(2)1}_{0iijj} - \lambda_j^2 \mathscr{L}^{(2)1}_{0iikk} - 2\lambda_i^2(\delta_{ij} - \delta_{ik}) \mathscr{L}^{(2)1}_{0jkjk}}{\lambda_j^2 - \lambda_k^2} \quad j \neq k, \\ \mathscr{L}^{(2)2}_{0ijjkki} &= \frac{1}{2} \frac{\lambda_k^2 \mathscr{L}^{(2)1}_{0ijij} - \lambda_j^2 \mathscr{L}^{(2)1}_{0kiki}}{\lambda_j^2 - \lambda_k^2} + \frac{1}{2} \frac{\lambda_i^2 \mathscr{L}^{(2)1}_{0jkjk} - \lambda_k^2 \mathscr{L}^{(2)1}_{0ijij}}{\lambda_k^2 - \lambda_i^2} \\ &\qquad i \neq j \neq k \neq i, \end{aligned}\right\} \tag{6.1.67}$$

where $\lambda_1 \lambda_2 \lambda_3 = 1$, $\lambda_1 \neq \lambda_2 \neq \lambda_3 \neq \lambda_1$ and t_i does not include the arbitrary hydrostatic part of the Cauchy stress.

Problem 6.1.8 For an incompressible isotropic Green-elastic material show that

$$\mathscr{A}^1_{0iijj} = \lambda_i \lambda_j \frac{\partial^2 W}{\partial \lambda_i \partial \lambda_j},$$

$$\mathscr{A}^1_{0ijij} = \frac{\left(\lambda_i \dfrac{\partial W}{\partial \lambda_i} - \lambda_j \dfrac{\partial W}{\partial \lambda_j}\right) \lambda_i^2}{(\lambda_i^2 - \lambda_j^2)} \qquad i \neq j,$$

$$\mathscr{A}^1_{0ijji} = \mathscr{A}^1_{0jiij} = \mathscr{A}^1_{0ijij} - \lambda_i \frac{\partial W}{\partial \lambda_i} \qquad i \neq j,$$

and use the results of Problem 6.1.6 together with (6.1.67) to obtain corresponding expressions for the components of $\mathscr{A}^2_0$.

Show further that if $\lambda_i = \lambda_j$, $i \neq j$, then

$$\mathscr{A}^1_{0ijij} = \tfrac{1}{2}\left(\mathscr{A}^1_{0iiii} - \mathscr{A}^1_{0iijj} + \lambda_i \frac{\partial W}{\partial \lambda_i}\right).$$

Problem 6.1.9 Show that for an incompressible elastic material the reference configuration ($\lambda_1 = \lambda_2 = \lambda_3 = 1$) is the only configuration corresponding to a pure dilatational deformation (it may be subject to an arbitrary hydrostatic stress).

By expressing the incremental incompressibility constraint in the form $\mathrm{tr}(\delta \mathbf{E}_0^{(2)}) = 0$† deduce that the terms involving $\mu_1, \nu_1, \nu_2, \nu_2'$ in the expressions for $\boldsymbol{\mathscr{L}}_0^{(2)1}$ and $\boldsymbol{\mathscr{L}}_0^{(2)2}$ given in Problem 6.1.5 either do not contribute to the expansion

$$\boldsymbol{\mathscr{L}}_0^{(2)1}\,\delta\mathbf{E}_0^{(2)} + \tfrac{1}{2}\boldsymbol{\mathscr{L}}_0^{(2)2}[\delta\mathbf{E}_0^{(2)}, \delta\mathbf{E}_0^{(2)}]$$

or contribute only a hydrostatic component (which may be absorbed into the hydrostatic constraint stress increment in $\delta\mathbf{T}_0^{(2)}$).

Conclude that for an incompressible isotropic material the elastic moduli may be reduced to

$$\mathscr{L}^{(2)1}_{0ijkl} = 2\mu \mathscr{I}^1_{ijkl},$$
$$\mathscr{L}^{(2)2}_{0ijklmn} = 8\nu \mathscr{I}^2_{ijklmn}$$

relative to the (unique) undistorted configuration, where μ and ν are constants.

Note that only one material constant is associated with each of the first- and second-order moduli. In fact, for a Green-elastic material the third- and fourth-order moduli also involve only one constant each. Moreover, to the fourth order, an incompressible isotropic elastic material is necessarily Green-elastic. For details see Ogden (1974a).

6.1.6 Linear and second-order elasticity

Consider now the incremental stress-deformation relation in the form

$$\delta\mathbf{S} = \boldsymbol{\mathscr{A}}^1\delta\mathbf{A} + \tfrac{1}{2}\boldsymbol{\mathscr{A}}^2[\delta\mathbf{A}, \delta\mathbf{A}] \qquad (6.1.68)$$

up to terms of the second order in $\delta\mathbf{A}$, with the fixed reference configuration taken as a natural configuration (i.e. one in which the stress vanishes).

†Note that this equation is not correct to second order in $\delta\mathbf{E}_0^{(2)}$.

Equivalently, we may write

$$\delta \mathbf{T}^{(m)} = \mathscr{L}^{(m)1}\delta \mathbf{E}^{(m)} + \tfrac{1}{2}\mathscr{L}^{(m)2}[\delta \mathbf{E}^{(m)}, \delta \mathbf{E}^{(m)}] \tag{6.1.69}$$

in terms of the conjugate variables $(\mathbf{T}^{(m)}, \mathbf{E}^{(m)})$, also to the second order.

We note that the increments of the different strain measures are connected through

$$\delta \mathbf{E}^{(m)} = \mathscr{L}^{1}\delta \mathbf{U} + \tfrac{1}{2}\mathscr{L}^{2}[\delta \mathbf{U}, \delta \mathbf{U}] \tag{6.1.70}$$

where $\mathscr{L}^1$ and $\mathscr{L}^2$ are as defined in Section 6.1.2 and $\delta \mathbf{E}^{(1)} = \delta \mathbf{U}$. We also have the connection

$$\delta \mathbf{E}^{(2)} = \tfrac{1}{2}\{\mathbf{A}^{\mathrm{T}}\delta \mathbf{A} + (\delta \mathbf{A}^{\mathrm{T}})\mathbf{A} + (\delta \mathbf{A}^{\mathrm{T}})(\delta \mathbf{A})\}. \tag{6.1.71}$$

In order to analyse the elastic behaviour of the material in a neighbourhood of the chosen natural configuration we take the current configuration to coincide with the natural configuration (and hence also the reference configuration). It follows that the moduli tensors $\mathscr{A}^1, \mathscr{A}^2, \mathscr{L}^{(m)1}, \mathscr{L}^{(m)2}$ (to which we do not attach zero subscripts here) are then *material constants* at any point of the considered material body. In other words they are independent of the deformation. Thus, to the second-order in the deformation, the elastic response of the material near a natural configuration is governed by stress-deformation relations of the form (6.1.68) or (6.1.69), $\mathscr{A}^1$, $\mathscr{A}^2$, $\mathscr{L}^{(m)1}, \mathscr{L}^{(m)2}$ being constant tensors.

When evaluated in the reference configuration, $\mathscr{L}^1$ and $\mathscr{L}^2$ are given by

$$\mathscr{L}^1 = \mathscr{I}^1, \qquad \mathscr{L}^2 = (m-1)\mathscr{I}^2,$$

where $\mathscr{I}^1$ and $\mathscr{I}^2$ are the tensors with Cartesian components defined by (3.5.32) and (6.1.62) respectively. It follows that (6.1.70) and (6.1.71) reduce to

$$\left.\begin{aligned} \delta \mathbf{E}^{(m)} &= \delta \mathbf{U} + \tfrac{1}{2}(m-1)(\delta \mathbf{U})^2, \\ \delta \mathbf{E}^{(2)} &= \tfrac{1}{2}\{\delta \mathbf{A} + \delta \mathbf{A}^{\mathrm{T}} + (\delta \mathbf{A}^{\mathrm{T}})(\delta \mathbf{A})\}. \end{aligned}\right\} \tag{6.1.72}$$

In particular, we deduce from (6.1.72) that, *to the first order, we need not distinguish between the increments of different strain measures.*

Furthermore, from (6.1.33) and (6.1.34) specialized to the natural configuration, we have

$$\mathscr{L}^{(m)1} = \mathscr{A}^1 \tag{6.1.73}$$

for each m. Thus, in a natural configuration, the tensors of first-order moduli associated with different pairs of conjugate variables are identical. In particular, $\mathcal{A}^1$ adopts the symmetries of $\mathcal{L}^{(m)1}$ described in (6.1.46).

It now follows that to the first order in $\delta \mathbf{A}$

$$\delta \mathbf{T}^{(m)} = \mathcal{L}^{(m)1}\delta \mathbf{E}^{(m)} = \mathcal{A}^1 \delta \mathbf{A} = \delta \mathbf{S},$$

i.e. the stress increments associated with different conjugate variables coincide in a natural configuration. The same is not true if second-order terms are included, as is clear from (6.1.23) specialized so that $A_{i\alpha} = \delta_{i\alpha}$. We shall consider the second-order theory presently, but, for the moment, we concentrate on the *first-order (or linear) theory of elasticity.*

In order to discuss the linear theory it is convenient to replace the increment $\delta \mathbf{A}$ by the displacement gradient

$$\mathbf{D} = \mathbf{A} - \mathbf{I} = \operatorname{Grad} \mathbf{u},$$

where $\mathbf{u}$ is the displacement vector defined in Section 2.2.3. Since

$$\mathbf{D} = \operatorname{Grad} \mathbf{u} = (\operatorname{grad} \mathbf{u})(\mathbf{I} + \mathbf{D})$$

it follows that in the linear approximation $\mathbf{D}$ may be regarded as $\operatorname{grad} \mathbf{u}$. In other words, we need not distinguish between $\mathbf{X}$ and $\mathbf{x}$ as the independent variable for $\mathbf{u}$. With this in mind we introduce the *linear* (or infinitesimal) *strain tensor*.

$$\mathbf{E} = \tfrac{1}{2}(\mathbf{D} + \mathbf{D}^{\mathrm{T}}), \qquad E_{ij} = \tfrac{1}{2}\left(\frac{\partial u_i}{\partial x_j} + \frac{\partial u_j}{\partial x_i}\right) \tag{6.1.74}$$

(to which each $\mathbf{E}^{(m)}$ reduces when linearized in $\mathbf{D}$).

The *linear elastic constitutive law* may now be written as

$$\mathbf{T} = \mathcal{A}\mathbf{E}, \qquad T_{ij} = \mathcal{A}_{ijkl}E_{kl}, \tag{6.1.75}$$

where (the Cauchy stress) $\mathbf{T}$ is now referred to as the *linear stress tensor* and $\mathcal{A}$ is the common value of $\mathcal{L}^{(m)1}$ and $\mathcal{A}^1$, carrying the symmetries

$$\mathcal{A}_{ijkl} = \mathcal{A}_{jikl} = \mathcal{A}_{ijlk}. \tag{6.1.76}$$

For a Green-elastic material the additional symmetry

$$\mathcal{A}_{ijkl} = \mathcal{A}_{klij} \tag{6.1.77}$$

holds, and the strain-energy function is given by

$$W = \tfrac{1}{2}\operatorname{tr}\{(\mathcal{A}\mathbf{E})\mathbf{E}\} \equiv \tfrac{1}{2}\mathcal{A}_{ijkl}E_{ij}E_{kl}. \tag{6.1.78}$$

We note that (6.1.75) is referred to as (the generalized) Hooke's Law in the context of the classical (or infinitesimal) theory of elasticity. It should be emphasized, however, that its status here is strictly that of a linear approximation to the general theory of elasticity.

If the material is isotropic then, in accordance with the results of Problems 6.1.5 and 6.1.6, we have

$$\mathscr{A}_{ijkl} = \lambda\delta_{ij}\delta_{kl} + 2\mu\mathscr{I}^1_{ijkl}, \tag{6.1.79}$$

where λ and μ are constants (the classical Lamé moduli). Note that $\mathscr{A}$ is a fourth-order isotropic tensor according to the definition given in Section 1.2.5. From (6.1.75) and (6.1.79) we now deduce the stress-strain relation

$$\mathbf{T} = 2\mu\mathbf{E} + \lambda(\operatorname{tr}\mathbf{E})\mathbf{I}, \tag{6.1.80}$$

while (6.1.78) becomes

$$W = \mu\operatorname{tr}(\mathbf{E}^2) + \tfrac{1}{2}\lambda(\operatorname{tr}\mathbf{E})^2. \tag{6.1.81}$$

Note that (6.1.80) may be written as

$$\mathbf{T} = 2\mu\mathbf{E}^* + \kappa(\operatorname{tr}\mathbf{E})\mathbf{I}, \tag{6.1.82}$$

where

$$\mathbf{E}^* = \mathbf{E} - \tfrac{1}{3}(\operatorname{tr}\mathbf{E})\mathbf{I} \tag{6.1.83}$$

is the *distortional* part of the strain, $\operatorname{tr}\mathbf{E}$ is the *dilatational* part of the strain,

$$\kappa = \lambda + \tfrac{2}{3}\mu$$

is the *bulk modulus*, and μ is the *shear modulus*. Respectively, $\operatorname{tr}\mathbf{E}$ and $\mathbf{E}^*$ are the linearizations of $\det\mathbf{A} - 1$ and the symmetric part of $J^{-1/3}\mathbf{A} - \mathbf{I}$.

For physical reasons (summarized by Truesdell and Noll (1965)) the classical inequalities

$$\mu > 0, \qquad \kappa > 0 \tag{6.1.84}$$

are imposed.

If the material is incompressible then $\operatorname{tr}\mathbf{E} = 0$ and (6.1.81) and (6.1.80) are replaced by

$$W = \mu\operatorname{tr}(\mathbf{E}^2), \tag{6.1.85}$$

$$\mathbf{T} = 2\mu\mathbf{E} + q\mathbf{I} \tag{6.1.86}$$

respectively, where q is an arbitrary scalar function of $\mathbf{x}$.

For the general theory of isotropic elasticity to be consistent with the classical theory in the linear approximation the strain-energy function $W(\lambda_1, \lambda_2, \lambda_3)$ must satisfy

$$\left.\begin{aligned} &W(1,1,1) = 0,\\ &\frac{\partial W}{\partial \lambda_i}(1,1,1) = 0 \quad i = 1,2,3,\\ &\frac{\partial^2 W}{\partial \lambda_i \partial \lambda_j}(1,1,1) = \lambda + 2\mu\delta_{ij} \quad i,j \in \{1,2,3\} \end{aligned}\right\} \tag{6.1.87}$$

in respect of an unconstrained material, and, for an incompressible material,

$$\left.\begin{aligned} &W(1,1,1) = 0,\\ &\frac{\partial W}{\partial \lambda_i}(1,1,1) = \frac{\partial W}{\partial \lambda_j}(1,1,1) \qquad i,j \in \{1,2,3\},\\ &\frac{\partial^2 W}{\partial \lambda_i^2}(1,1,1) = \frac{\partial^2 W}{\partial \lambda_j^2}(1,1,1) \qquad i,j \in \{1,2,3\},\\ &\frac{\partial^2 W}{\partial \lambda_i \partial \lambda_j}(1,1,1) \qquad \text{independent of } i, j \neq i\\ &\frac{\partial^2 W}{\partial \lambda_i^2}(1,1,1) - \frac{\partial^2 W}{\partial \lambda_i \partial \lambda_j}(1,1,1) + \frac{\partial W}{\partial \lambda_i}(1,1,1) = 2\mu \quad i \neq j \end{aligned}\right\} \tag{6.1.88}$$

Analogous restrictions apply in the case of anisotropic materials.

Turning now to the second-order theory we note that the nominal stress tensor is given by

$$\mathbf{S} = \mathscr{A}^1\mathbf{D} + \tfrac{1}{2}\mathscr{A}^2[\mathbf{D}, \mathbf{D}] \tag{6.1.89}$$

while from (6.1.69) specialized to $m = 2$ we have

$$\mathbf{T}^{(2)} = \mathscr{L}^{(2)1}\mathbf{E}^{(2)} + \tfrac{1}{2}\mathscr{L}^{(2)2}[\mathbf{E}^{(2)}, \mathbf{E}^{(2)}]. \tag{6.1.90}$$

Since

$$\mathbf{E}^{(2)} = \tfrac{1}{2}(\mathbf{D} + \mathbf{D}^{\mathrm{T}} + \mathbf{D}^{\mathrm{T}}\mathbf{D})$$

it follows from the symmetry of $\mathscr{L}^{(2)1}$ that $\mathbf{T}^{(2)}$ may also be written

$$\mathbf{T}^{(2)} = \mathscr{L}^{(2)1}\mathbf{D} + \tfrac{1}{2}\mathscr{L}^{(2)1}(\mathbf{D}^{\mathrm{T}}\mathbf{D}) + \tfrac{1}{2}\mathscr{L}^{(2)2}[\mathbf{D}, \mathbf{D}]. \qquad (6.1.91)$$

The connection $\mathbf{S} = \mathbf{T}^{(2)}\mathbf{A}^{\mathrm{T}} = \mathbf{T}^{(2)} + \mathbf{T}^{(2)}\mathbf{D}^{\mathrm{T}}$ shows that

$$\mathbf{S} - \mathbf{T}^{(2)} = (\mathscr{L}^{(2)1}\mathbf{D})\mathbf{D}^{\mathrm{T}} \qquad (6.1.92)$$

correct to the second order in $\mathbf{D}$. Similar equations may be written down in respect of other stress tensors. Either (6.1.89) or (6.1.90) may be regarded as the *constitutive law of second-order elasticity* but, in view of (6.1.92), it is important to distinguish between different stress tensors and between different measures of deformation (as distinct from in the linear approximation). In the above $\mathbf{D}$ is interpreted as Grad **u** and can only be replaced by grad **u** if an appropriate adjustment is made in the second-order terms in (6.1.89) and (6.1.91).

From the results of Problem 6.1.5 we deduce that for an isotropic Green-elastic material

$$\left.\begin{aligned} \mathscr{L}^{(2)1}_{ijkl} &= \lambda\delta_{ij}\delta_{kl} + 2\mu\mathscr{I}^{2}_{ijkl}, \\ \mathscr{L}^{(2)2}_{ijklmn} &= \nu_1\delta_{ij}\delta_{kl}\delta_{mn} + 2\nu_2(\delta_{ij}\mathscr{I}^{1}_{klmn} + \delta_{kl}\mathscr{I}^{1}_{mnij} \\ &\quad + \delta_{mn}\mathscr{I}^{1}_{ijkl}) + 8\nu_3\mathscr{I}^{2}_{ijklmn}, \end{aligned}\right\} \qquad (6.1.93)$$

where λ, μ, ν_1, ν_2, ν_3 are constants.

For incompressible materials the above are replaced by

$$\left.\begin{aligned} \mathscr{L}^{(2)1}_{ijkl} &= 2\mu\mathscr{I}^{1}_{ijkl}, \\ \mathscr{L}^{(2)2}_{ijklmn} &= 8\nu\mathscr{I}^{2}_{ijklmn}, \end{aligned}\right\} \qquad (6.1.94)$$

and the strain-energy function is given by

$$W = \mu\,\mathrm{tr}(\mathbf{E}^{(2)2}) + \tfrac{4}{3}\nu\,\mathrm{tr}(\mathbf{E}^{(2)3}). \qquad (6.1.95)$$

It is interesting to note in passing that the Mooney strain-energy function, defined by equation (4.3.62), contains two material constants and embraces the second-order theory described by (6.1.95).

Problem 6.1.10 Write down the strain-energy function corresponding to (6.1.93) in terms of $\mathbf{D}$ and obtain an explicit expression for $\mathbf{S}$ which involves the constants λ, μ, ν_1, ν_2, ν_3.

Connections between sets of second-order constants used by different authors are listed in Truesdell and Noll (1965). This reference contains more detailed discussion of second-order elasticity and, in particular, its application in the solution of boundary-value problems. For a modern authoritative account of the linear theory of elasticity we refer to Gurtin (1973).

6.2 STRUCTURE AND PROPERTIES OF THE INCREMENTAL EQUATIONS

6.2.1 Incremental boundary-value problems

As a prelude to the discussion of incremental boundary-value problems it is convenient here to recall the ingredients of the mixed boundary-value problem formulated in Sections 5.1.1 and 5.1.2.

For a body whose reference configuration is $\mathscr{B}_0$ we write the boundary conditions as

$$\mathbf{x} = \boldsymbol{\xi}(\mathbf{X}) \qquad \text{on } \partial\mathscr{B}_0^x, \tag{6.2.1}$$

$$\mathbf{S}^{\mathrm{T}}\mathbf{N} = \boldsymbol{\sigma}(\mathbf{X}, \mathbf{x}, \mathbf{A}) \qquad \text{on } \partial\mathscr{B}_0^\sigma, \tag{6.2.2}$$

where $\boldsymbol{\xi}$ and $\boldsymbol{\sigma}$ are prescribed functions of their arguments, $\mathbf{x} = \boldsymbol{\chi}(\mathbf{X})$ for $\mathbf{X} \in \mathscr{B}_0$ defines the deformation, and $\mathbf{A} = \operatorname{Grad} \boldsymbol{\chi}(\mathbf{X})$. The nominal stress $\mathbf{S}$ is given by an appropriate form of elastic constitutive law; in particular,

$$\mathbf{S} = \frac{\partial W}{\partial \mathbf{A}} \tag{6.2.3}$$

for an unconstrained Green-elastic material, and this is modified to

$$\mathbf{S} = \frac{\partial W}{\partial \mathbf{A}} + q\mathbf{B}^{\mathrm{T}} \tag{6.2.4}$$

for an incompressible material, where q is a Lagrange multiplier. The equilibrium and rotational balance equations are

$$\operatorname{Div} \mathbf{S} + \rho_0 \mathbf{b} = \mathbf{0} \tag{6.2.5}$$

and

$$\mathbf{AS} = \mathbf{S}^{\mathrm{T}}\mathbf{A}^{\mathrm{T}} \tag{6.2.6}$$

respectively, the latter being a consequence of objectivity of the constitutive law.

Suppose that a solution $\boldsymbol{\chi}$ *for this problem is known* (it need not be unique). We now pose the problem of finding the incremental deformation $\delta\boldsymbol{\chi}$ resulting from incremental changes in the boundary conditions.

Let the boundary conditions (6.2.1) and (6.2.2) be subjected to the increments

$$\delta \mathbf{x} = \delta \boldsymbol{\xi}(\mathbf{X}) \qquad \text{on } \partial \mathscr{B}_0^x, \tag{6.2.7}$$

$$\delta \mathbf{S}^{\mathrm{T}} \mathrm{N} = \delta \boldsymbol{\sigma}(\mathbf{X}, \mathbf{x}, \mathbf{A}) \qquad \text{on } \partial \mathscr{B}_0^\sigma, \tag{6.2.8}$$

where $\delta \mathbf{S}$ is the increment in nominal stress. Note, in particular, that the increment $\delta \boldsymbol{\sigma}$ in the nominal traction depends in general on the increments $\delta \mathbf{x} \equiv \delta \boldsymbol{\chi}(\mathbf{X})$ and $\delta \mathbf{A} \equiv \mathrm{Grad}\, \delta \chi(\mathbf{X})$ in addition to the increments in any loading parameters included in $\boldsymbol{\sigma}$. For example, in the case of pressure loading, $\boldsymbol{\sigma} = -JP\mathbf{B}\mathrm{N}$ and

$$\delta \boldsymbol{\sigma} = -\delta P J \mathbf{B}\mathrm{N} - JP \operatorname{tr}(\mathbf{B}^{\mathrm{T}} \delta \mathbf{A}) \mathbf{B}\mathrm{N} + JP \mathbf{B} \delta \mathbf{A}^{\mathrm{T}} \mathbf{B}\mathrm{N} \tag{6.2.9}$$

to the first order in $\delta \mathbf{A}$, where δP is the prescribed increment in P.

The incremental counterpart of (6.2.5) is

$$\mathrm{Div}\, \delta \mathbf{S} + \rho_0 \delta \mathbf{b} = \mathbf{0}, \tag{6.2.10}$$

where, *to the first order in* $\delta \mathbf{A}$, we have

$$\delta \mathbf{S} = \boldsymbol{\mathscr{A}}^1 \delta \mathbf{A} \tag{6.2.11}$$

for an unconstrained material and

$$\delta \mathbf{S} = \boldsymbol{\mathscr{A}}^1 \delta \mathbf{A} + \delta q \mathbf{B}^{\mathrm{T}} - q \mathbf{B}^{\mathrm{T}} \delta \mathbf{A} \mathbf{B}^{\mathrm{T}} \tag{6.2.12}$$

for an incompressible material (from (6.1.11) and (6.1.38) respectively). Equation (6.2.12) is accompanied by the incompressibility constraint $\delta(\det \mathbf{A}) = 0$ which, to the first order in $\delta \mathbf{A}$, becomes

$$\operatorname{tr}(\mathbf{B}^{\mathrm{T}} \delta \mathbf{A}) = 0. \tag{6.2.13}$$

To the same order, the incremental version of (6.2.6) is

$$\delta \mathbf{A} \mathbf{S} + \mathbf{A} \delta \mathbf{S} = \delta \mathbf{S}^{\mathrm{T}} \mathbf{A}^{\mathrm{T}} + \mathbf{S}^{\mathrm{T}} \delta \mathbf{A}^{\mathrm{T}}. \tag{6.2.14}$$

In respect of (6.2.11) the Cartesian component form of (6.2.10) is

$$\frac{\partial}{\partial X_\alpha} \left(\mathscr{A}^1_{\alpha i \beta j} \frac{\partial \delta \chi_j}{\partial X_\beta} \right) + \rho_0 \delta b_i = 0 \tag{6.2.15}$$

or, equivalently,

$$\mathscr{A}^1_{\alpha i\beta j}\frac{\partial^2\delta\chi_j}{\partial X_\alpha \partial X_\beta}+\mathscr{A}^2_{\alpha i\beta j\gamma k}\frac{\partial^2\chi_k}{\partial X_\gamma \partial X_\alpha}\frac{\partial\delta\chi_j}{\partial X_\beta}+\rho_0\delta b_i=0. \tag{6.2.16}$$

If the deformation $\boldsymbol{\chi}$ is homogeneous then $\boldsymbol{\mathscr{A}}^1$ is independent of $\mathbf{X}$ and (6.2.15) reduces to

$$\mathscr{A}^1_{\alpha i\beta j}\frac{\partial^2\delta\chi_j}{\partial X_\alpha \partial X_\beta}+\rho_0\delta b_i=0. \tag{6.2.17}$$

For an incompressible material insertion of (6.2.12) into (6.2.10) and use of (6.2.13) leads to

$$\mathrm{Div}(\boldsymbol{\mathscr{A}}^1\delta\mathbf{A})+\mathbf{B}\,\mathrm{Grad}\,\delta q-\mathbf{B}\delta\mathbf{A}^{\mathrm{T}}\mathbf{B}\,\mathrm{Grad}\,q+\rho_0\delta\mathbf{b}=\mathbf{0}. \tag{6.2.18}$$

Its Cartesian component form simplifies to

$$\mathscr{A}^1_{\alpha i\beta j}\frac{\partial^2\delta\chi_j}{\partial X_\alpha \partial X_\beta}+\frac{\partial\delta q}{\partial x_i}+\rho_0\delta b_i=0 \tag{6.2.19}$$

when the underlying deformation is homogeneous.

When the reference configuration is chosen to coincide with the current configuration equations (6.2.10) and (6.2.14) respectively become

$$\mathrm{div}\,\delta\mathbf{S}_0+\rho\delta\mathbf{b}=\mathbf{0}, \tag{6.2.20}$$

$$\delta\mathbf{A}_0\mathbf{T}+\delta\mathbf{S}_0=\delta\mathbf{S}_0^{\mathrm{T}}+\mathbf{T}\delta\mathbf{A}_0^{\mathrm{T}}, \tag{6.2.21}$$

where ρ is the current density and $\mathbf{T}$ is Cauchy stress. The incremental constitutive laws (6.2.11) and (6.2.12) are replaced by

$$\delta\mathbf{S}_0=\boldsymbol{\mathscr{A}}_0^1\delta\mathbf{A}_0, \tag{6.2.22}$$

and

$$\delta\mathbf{S}_0=\boldsymbol{\mathscr{A}}_0^1\delta\mathbf{A}_0+\delta q\mathbf{I}-q\delta\mathbf{A}_0, \tag{6.2.23}$$

with

$$\mathrm{tr}(\delta\mathbf{A}_0)=0 \tag{6.2.24}$$

in the latter case, where

$$\delta\mathbf{A}_0=\mathrm{grad}\,\delta\boldsymbol{\chi}. \tag{6.2.25}$$

We note also that the boundary condition (6.2.9) becomes

$$\delta \mathbf{S}_0^{\mathrm{T}}\mathbf{n} \equiv \delta \boldsymbol{\sigma}_0 = -\delta P\mathbf{n} - P\operatorname{tr}(\delta \mathbf{A}_0)\mathbf{n} + P\delta \mathbf{A}_0^{\mathrm{T}}\mathbf{n}, \tag{6.2.26}$$

where $\mathbf{n}$ is the unit normal to the current boundary $\partial\mathscr{B}$ of the body, the middle term on the right-hand side vanishing for an incompressible material.

In the above formulation of the incremental mixed boundary-value problem we have assumed that second- and higher order terms in $\delta\chi$ and its derivatives may be neglected. We are thus dealing with the *linearized theory of incremental deformations superposed on a finite deformation.* This linearization is to be taken as implicit throughout this chapter unless stated otherwise.

As mentioned in Section 5.1 the tensor of moduli $\mathscr{A}^1$ plays an important role in the determination of the nature of the solutions of boundary-value problems. This is equally true in respect of incremental boundary-value problems with $\mathscr{A}^1$ (or $\mathscr{A}_0^1$) then depending on the known state of deformation χ. We note, in particular, that (6.2.17) has precisely the same structure as the equilibrium form of equation (5.1.9). In the following sections we examine some relevant properties of $\mathscr{A}^1$ and also of the stress-deformation relation $\mathbf{S} = \mathbf{H}(\mathbf{A})$ before proceeding to investigate the solution of specific boundary-value problems in Section 6.3. An important first step, taken in Sections 6.2.2 and 6.2.3, is to establish sufficient conditions for uniqueness of solution to certain boundary-value problems from both local (i.e. incremental) and global standpoints.

6.2.2 Uniqueness: global considerations

In order to provide some perspective for the discussion, in Section 6.2.3, of uniqueness of solution in respect of the incremental boundary-value problem formulated in Section 6.2.1 we consider here the question of uniqueness for the global boundary-value problem (6.2.1)–(6.2.6). For the sake of continuity, examination of this latter question has been postponed from its more natural place in Chapter 5, where examples of non-uniqueness were cited.

Let χ, χ' be two possible solutions of (6.2.1)–(6.2.6) with corresponding deformation gradients $\mathbf{A}, \mathbf{A}'$, nominal stresses $\mathbf{S}, \mathbf{S}'$ and body forces $\mathbf{b}, \mathbf{b}'$ in $\mathscr{B}_0$ and surface tractions $\boldsymbol{\sigma}, \boldsymbol{\sigma}'$ on $\partial\mathscr{B}_0^\sigma$. Then, by use of the divergence theorem together with (6.2.1), (6.2.2) and (6.2.5), we obtain

$$\int_{\mathscr{B}_0} \operatorname{tr}\{(\mathbf{S}' - \mathbf{S})(\mathbf{A}' - \mathbf{A})\}\,\mathrm{d}V$$
$$= \int_{\partial\mathscr{B}_0^\sigma} (\boldsymbol{\sigma}' - \boldsymbol{\sigma})\cdot(\chi' - \chi)\,\mathrm{d}A + \int_{\mathscr{B}_0} \rho_0(\mathbf{b}' - \mathbf{b})\cdot(\chi' - \chi)\,\mathrm{d}V. \tag{6.2.27}$$

If, in particular, the body force is independent of χ and there is dead loading on

$\partial\mathscr{B}_0^\sigma$ then $\mathbf{b}'=\mathbf{b}$, $\boldsymbol{\sigma}'=\boldsymbol{\sigma}$ and it follows from (6.2.27) that

$$\int_{\mathscr{B}_0} \operatorname{tr}\{(\mathbf{S}'-\mathbf{S})(\mathbf{A}'-\mathbf{A})\}\mathrm{d}V=0. \tag{6.2.28}$$

Thus, two distinct solutions χ and χ' necessarily satisfy (6.2.27) or (6.2.28), as appropriate. For much of this section we concentrate attention on the dead-loading problem since it enables the restrictions on the stress required for uniqueness to be made explicit.

Suppose, now, that there exists† a solution χ to the considered problem. Then this solution is unique if there is no kinematically admissible deformation field χ' for which (6.2.28) holds, i.e.

$$\int_{\mathscr{B}_0} \operatorname{tr}\{(\mathbf{S}'-\mathbf{S})(\mathbf{A}'-\mathbf{A})\}\mathrm{d}V\neq 0 \tag{6.2.29}$$

for all $\chi'\not\equiv\chi$ satisfying (6.2.1). On the other hand, non-uniqueness is not guaranteed if (6.2.28) holds for some χ' because the associated nominal stress $\mathbf{S}'$ need not satisfy (6.2.2) and (6.2.5).

Clearly, a sufficient condition for uniqueness is that

$$\int_{\mathscr{B}_0} \operatorname{tr}\{(\mathbf{S}'-\mathbf{S})(\mathbf{A}'-\mathbf{A})\}\mathrm{d}V>0 \tag{6.2.30}$$

for all deformation fields χ, $\chi'\not\equiv\chi$ satisfying (6.2.1). Equally, a valid sufficient condition for uniqueness is provided by (6.2.30) with the inequality reversed; however, any χ which makes the integral in (6.2.30) negative for some χ' in a neighbourhood of χ corresponds to an unstable configuration, as we explain shortly, and it is therefore of no practical interest to consider this option.

A stronger inequality, which implies (6.2.30), is

$$\operatorname{tr}\{(\mathbf{S}'-\mathbf{S})(\mathbf{A}'-\mathbf{A})\}>0 \tag{6.2.31}$$

for all pairs $\mathbf{A}$, $\mathbf{A}'\neq\mathbf{A}$ (not necessarily the gradients of deformation fields). With $\mathbf{S}=\mathbf{H}(\mathbf{A})$, the inequality (6.2.31) states that $\mathbf{H}$ is a *strictly convex function*.‡

† To establish existence of solutions to boundary-value problems in non-linear elasticity is difficult and requires much apparatus from functional analysis. Only a handful of results have thus far been proved and discussion of these is beyond the scope of this book. Nevertheless, reference will be made to selected papers dealing with existence theory at appropriate points in the remaining sections of this chapter.

‡ For formal definitions of strict convexity of scalar-, vector- and tensor-valued functions see Appendix 1.

In view of our earlier remarks to the effect that uniqueness is not in general to be expected, it is clear that inequalities such as (6.2.30) and (6.2.31) are unduly restrictive on the elastic constitutive law. Nevertheless, it is instructive to characterize that part of $\mathbf{A}$-space over which (6.2.31) holds. This is part of the objective of the present section.

For a Green-elastic material with strain energy function W we consider, instead of (6.2.31), the inequality

$$W(\mathbf{A}') - W(\mathbf{A}) - \operatorname{tr}\{\mathbf{S}(\mathbf{A}' - \mathbf{A})\} > 0 \tag{6.2.32}$$

for all $\mathbf{A}$, $\mathbf{A}' \neq \mathbf{A}$. This states that W is a strictly convex (scalar) function. By reversing the roles of $\mathbf{A}$ and $\mathbf{A}'$ in (6.2.32) and adding the result to (6.2.32), it is easy to see that (6.2.32) implies (6.2.31). The converse is not true in general, however (see Appendix 1).

In the case of dead-load surface tractions and body forces integration of (6.2.32) over $\mathscr{B}_0$ and application of the divergence theorem with the boundary conditions on $\partial\mathscr{B}_0$ leads to

$$\int_{\mathscr{B}_0} \{W(\mathbf{A}') - \rho_0 \mathbf{b}\cdot\boldsymbol{\chi}'\}\mathrm{d}V - \int_{\partial\mathscr{B}_0^\sigma} \boldsymbol{\sigma}\cdot\boldsymbol{\chi}'\mathrm{d}A$$
$$> \int_{\mathscr{B}_0} \{W(\mathbf{A}) - \rho_0 \mathbf{b}\cdot\boldsymbol{\chi}\}\mathrm{d}V - \int_{\partial\mathscr{B}_0^\sigma} \boldsymbol{\sigma}\cdot\boldsymbol{\chi}\mathrm{d}A. \tag{6.2.33}$$

The inequality (6.2.33) may also be written as

$$E\{\boldsymbol{\chi}'\} > E\{\boldsymbol{\chi}\}, \tag{6.2.34}$$

where $E\{\cdot\}$ is the energy functional introduced in Section 5.4. If, given that $\boldsymbol{\chi}$ is a solution, this holds for all kinematically admissible $\boldsymbol{\chi}' \not\equiv \boldsymbol{\chi}$ then uniqueness is guaranteed.

To see an immediate consequence of (6.2.32) we take $\mathbf{A}' = \mathbf{Q}\mathbf{A}$, where $\mathbf{Q}$ is an arbitrary proper orthogonal tensor. Because of the objectivity of W and the connection $(\det \mathbf{A})\mathbf{T} = \mathbf{A}\mathbf{S}$, where $\mathbf{T}$ is Cauchy stress, (6.2.32) reduces to $\operatorname{tr}\{\mathbf{T}(\mathbf{I} - \mathbf{Q})\} > 0$. When decomposed on the principal axes of $\mathbf{T}$ in terms of the principal stresses t_1, t_2, t_3, this becomes

$$t_1(1 - Q_{11}) + t_2(1 - Q_{22}) + t_3(1 - Q_{33}) > 0.$$

Finally, by taking $\mathbf{Q}$ to be a rotation about each of the three principal axes in turn, we obtain the restrictions

$$t_2 + t_3 > 0, \qquad t_3 + t_1 > 0, \qquad t_1 + t_2 > 0 \tag{6.2.35}$$

on the principal stresses. We shall return to these inequalities and their interpretation in Section 6.2.3.

We now turn to the stability interpretation of (6.2.33) for which purpose it is rewritten as

$$\int_{\mathscr{B}_0} \{W(\mathbf{A}') - W(\mathbf{A})\}\,\mathrm{d}V$$
$$> \int_{\partial\mathscr{B}_0^\sigma} \boldsymbol{\sigma}\cdot(\chi' - \chi)\,\mathrm{d}A + \int_{\mathscr{B}_0} \rho_0\mathbf{b}\cdot(\chi' - \chi)\,\mathrm{d}V. \qquad (6.2.36)$$

Let χ correspond to an actual solution and let χ' be an arbitrary kinematically admissible deformation.Then (6.2.36) states that the increase in the strain energy in moving from the equilibrium configuration χ to χ' exceeds the work done by the prescribed tractions and body forces. It follows from the classical criterion of stability that the equilibrium configuration is *stable* (for a more detailed discussion of elastic stability we refer to Hill, 1957). Expressed in the form (6.2.34), (6.2.36) shows that for a stable equilibrium configuration the energy functional is minimized within the class of kinematically admissible configurations.

In fact, χ is stable if the weaker inequality

$$E\{\chi'\} \geq E\{\chi\} \qquad (6.2.37)$$

holds for all kinematically admissible χ', with the possibility of equality holding for some $\chi' \not\equiv \chi$. It follows that stability does not in general imply uniqueness and, conversely, from (6.2.29), that a unique equilibrium configuration need not be stable. We remark, however, that it is not of practical interest to study uniqueness of unstable equilibrium configurations.

The inequality (6.2.37) is a *global sufficient condition for stability of* χ. It is too restrictive to be regarded as a stability criterion, i.e. as necessary and sufficient for stability of χ, for elastic bodies. If, however, χ' is confined to some neighbourhood of χ then (6.2.37) is a suitable *local stability criterion*, often referred to as the *infinitesimal stability criterion* (see Truesdell and Noll, 1965, for details). This is the criterion which we discuss in Section 6.2.3 in the context of incremental deformations.

In Section 5.4 we commented that in general the principle of stationary potential energy cannot be strengthened to an extremum principle, i.e. (6.2.37) does not hold in general for an equilibrium solution χ. The above discussion makes it clear that (6.2.37) may be violated even if χ is stable in the local sense. Nevertheless, if χ is locally stable then (6.2.37) can be regarded as a restricted form of extremum principle.

In order to provide mathematical insight into the possibility of global non-uniqueness exemplified in Section 5.1.2 it suffices to illustrate that the stress-deformation relation

$$\mathbf{S} = \mathbf{H}(\mathbf{A}) \tag{6.2.38}$$

is not in general uniquely invertible. In particular, we are concerned with isotropic materials, for which, we recall from (4.2.68), we have

$$\mathbf{S} = \mathbf{T}^{(1)}\mathbf{R}^{\mathrm{T}} \tag{6.2.39}$$

in analogy with the polar decomposition $\mathbf{A} = \mathbf{R}\mathbf{U}$, where

$$\mathbf{T}^{(1)} = \mathbf{H}(\mathbf{U}) \tag{6.2.40}$$

is the Biot stress tensor (automatically symmetric by objectivity).

That two possible states of deformation could correspond to a given nominal stress $\mathbf{S}$ was shown in Section 5.1.2 for the simple tension/compression problem. Here we examine the general case, showing that (at least) *four* $\mathbf{A}$'s correspond to a given $\mathbf{S}$ and, exceptionally, certain $\mathbf{S}$'s are associated with *infinitely many* $\mathbf{A}$'s.

Given $\mathbf{S}$, we apply the polar decomposition (6.2.39), where $\mathbf{R}$ is proper orthogonal (identifiable as the rotational part of the deformation) and $\mathbf{T}^{(1)}$ is symmetric (identifiable as Biot stress and associated through (6.2.40) with the right stretch tensor $\mathbf{U}$). Unlike the polar decomposition of the deformation gradient, however, (6.2.39) is not unique since $\mathbf{T}^{(1)}$ is neither positive nor negative definite in general. In what follows we catalogue the set of such decompositions. The results depend on whether or not $\mathbf{S}\mathbf{S}^{\mathrm{T}}$ has distinct principal values and on the dimension of the null space of $\mathbf{S}$ (regarded as a linear mapping).

Case 1: $\mathbf{S}$ *is non-singular*

Since $\mathbf{S}$ is non-singular, $\mathbf{S}\mathbf{S}^{\mathrm{T}}$ has positive principal values, $\sigma_1, \sigma_2, \sigma_3$, say. Also, from (6.2.39), we have

$$\mathbf{T}^{(1)2} = \mathbf{S}\mathbf{S}^{\mathrm{T}} \tag{6.2.41}$$

so it follows that the principal Biot stresses are given by

$$t_i^{(1)} = \pm\sqrt{\sigma_i} \qquad i = 1, 2, 3. \tag{6.2.42}$$

There are eight possible combinations of signs for $t_1^{(1)}, t_2^{(1)}, t_3^{(1)}$ contained in (6.2.42)—one for each octant of the three-dimensional $t_i^{(1)}$-space. However, from (6.2.39) we also have

$$t_1^{(1)}t_2^{(1)}t_3^{(1)} = \det \mathbf{T}^{(1)} = \det \mathbf{S} \tag{6.2.43}$$

so that four each of these combinations correspond to each sign of det $\mathbf{S}$.

Three distinct situations now arise and we treat them separately:

(a) $\sigma_i \neq \sigma_j$ $(i \neq j)$

If $\sigma_1, \sigma_2, \sigma_3$ are distinct then $\mathbf{S}\mathbf{S}^{\mathrm{T}}$ defines uniquely the principal axes of $\mathbf{T}^{(1)2}$, hence of $\mathbf{T}^{(1)}$ (which is coaxial with $\mathbf{T}^{(1)2}$) and therefore of the Lagrangean strain ellipsoid (since the material is isotropic and $\mathbf{T}^{(1)}$ is coaxial with $\mathbf{U}$). Let $\{\mathbf{u}^{(i)}\}$ denote these (orthonormal) axes. Then

$$\mathbf{S}\mathbf{S}^{\mathrm{T}} = \sum_{i=1}^{3} \sigma_i \mathbf{u}^{(i)} \otimes \mathbf{u}^{(i)} \tag{6.2.44}$$

and consequently

$$\mathbf{T}^{(1)} = \sum_{i=1}^{3} t_i^{(1)} \mathbf{u}^{(i)} \otimes \mathbf{u}^{(i)}. \tag{6.2.45}$$

If det $\mathbf{S} > 0$ then the four possible sign combinations for $t_i^{(1)}$ are

$$\left.\begin{array}{llll}
\text{(i)} & t_1^{(1)} = \sqrt{\sigma_1}, & t_2^{(1)} = \sqrt{\sigma_2}, & t_3^{(1)} = \sqrt{\sigma_3}, \\
\text{(ii)} & t_1^{(1)} = \sqrt{\sigma_1}, & t_2^{(1)} = -\sqrt{\sigma_2}, & t_3^{(1)} = -\sqrt{\sigma_3}, \\
\text{(iii)} & t_1^{(1)} = -\sqrt{\sigma_1}, & t_2^{(1)} = \sqrt{\sigma_2}, & t_3^{(1)} = -\sqrt{\sigma_3}, \\
\text{(iv)} & t_1^{(1)} = -\sqrt{\sigma_1}, & t_2^{(1)} = -\sqrt{\sigma_2}, & t_3^{(1)} = \sqrt{\sigma_3}.
\end{array}\right\} \tag{6.2.46}$$

By taking (6.2.39) to correspond to (i) above the remaining three polar decompositions are expressible in the form

$$\mathbf{S} = \mathbf{T}^{(1)\prime}\mathbf{R}^{\prime\mathrm{T}}, \tag{6.2.47}$$

where

$$\mathbf{T}^{(1)\prime} = \mathbf{T}^{(1)}\mathbf{Q}_i^{\pi} = \mathbf{Q}_i^{\pi}\mathbf{T}^{(1)}, \qquad \mathbf{R}' = \mathbf{R}\mathbf{Q}_i^{\pi} \tag{6.2.48}$$

and $\mathbf{Q}_i^{\pi}$ is a rotation through π about $\mathbf{u}^{(i)}$ $(i = 1, 2, 3)$.

Similarly, if det $\mathbf{S} < 0$, the possible signs for $t_i^{(1)}$ are

$$\left.\begin{array}{llll}
\text{(v)} & t_1^{(1)} = -\sqrt{\sigma_1}, & t_2^{(1)} = \sqrt{\sigma_2}, & t_3^{(1)} = \sqrt{\sigma_3}, \\
\text{(vi)} & t_1^{(1)} = -\sqrt{\sigma_1}, & t_2^{(1)} = -\sqrt{\sigma_2}, & t_3^{(1)} = -\sqrt{\sigma_3}, \\
\text{(vii)} & t_1^{(1)} = \sqrt{\sigma_1}, & t_2^{(1)} = \sqrt{\sigma_2}, & t_3^{(1)} = -\sqrt{\sigma_3}, \\
\text{(viii)} & t_1^{(1)} = \sqrt{\sigma_1}, & t_2^{(1)} = -\sqrt{\sigma_2}, & t_3^{(1)} = \sqrt{\sigma_3}.
\end{array}\right\} \tag{6.2.49}$$

In either case there are just *four* possible polar decompositions for $\mathbf{S}$ of the form (6.2.39).

(b) $\sigma_i = \sigma_j \neq \sigma_k$ (i, j, k is a permutation of 1, 2, 3)

For definiteness we take $\sigma_1 = \sigma_2 \neq \sigma_3$. In this degenerate case the Lagrangean axes are no longer uniquely defined by $\mathbf{SS}^\mathrm{T}$. Indeed, $\mathbf{u}^{(3)}$ is uniquely defined (to within sign), but $\mathbf{u}^{(1)}$ and $\mathbf{u}^{(2)}$ are arbitrarily disposable subject to orthonormality and right-handedness of $\{\mathbf{u}^{(i)}\}$. Thus, we write

$$\mathbf{SS}^\mathrm{T} = \sigma_1(\mathbf{I} - \mathbf{u}^{(3)} \otimes \mathbf{u}^{(3)}) + \sigma_3 \mathbf{u}^{(3)} \otimes \mathbf{u}^{(3)}. \tag{6.2.50}$$

The possible signs for $t_i^{(1)}$ are, with σ_2 replaced by σ_1, again given by (6.2.46) and (6.2.49) for $\det \mathbf{S} > 0$ and $\det \mathbf{S} < 0$ respectively.

As in (a) we may write $\mathbf{T}^{(1)}$ in the form (6.2.45) but now $\mathbf{u}^{(1)}$ and $\mathbf{u}^{(2)}$ are determined only to within an arbitrary rotation about $\mathbf{u}^{(3)}$. Possible polar decompositions for $\mathbf{S}$ are again generated by (6.2.47) with (6.2.48) subject to the orientation of the principal axes of $\mathbf{T}^{(1)}$ being undetermined to the extent mentioned above. Because of this indeterminancy there is no essential distinction between (ii) and (iii) or between (v) and (viii) in (a) with $\sigma_2 = \sigma_1$.

(c) $\sigma_i = \sigma_j$ ($i \neq j$)

Let $\sigma_i = \sigma$ ($i = 1, 2, 3$). Then we have

$$\mathbf{SS}^\mathrm{T} = \sigma \mathbf{I}. \tag{6.2.51}$$

The signs of $t_i^{(1)}$ are once more distributed according to (6.2.46) or (6.2.49) appropriately specialized, and $\mathbf{T}^{(1)}$ is again expressible in the form (6.2.45). Now, however, the orientation of the orthonormal Lagrangean axes $\{\mathbf{u}^{(i)}\}$ is completely arbitrary. Polar decompositions for $\mathbf{S}$ are generated from (6.2.47) and (6.2.48) subject to this arbitrariness. With $\sigma_i = \sigma$ ($i = 1, 2, 3$), (ii), (iii) and (iv) in (6.2.46) are then indistinguishable, as are (v), (vii) and (viii) in (6.2.49).

Case 2: $\mathbf{S}$ *is singular* If $\mathbf{S}$ is singular, at least one of $\sigma_1, \sigma_2, \sigma_3$ must vanish. We deal separately with the cases in which one, two or all three of $\sigma_1, \sigma_2, \sigma_3$ are zero.

(a) $\sigma_1 \neq 0$, $\sigma_2 \neq 0$, $\sigma_3 = 0$

The results are obtained on specializing Cases 1(a), (b) above by putting $\sigma_3 = 0$.

(b) $\sigma_1 = \sigma_2 = 0$, $\sigma_3 \neq 0$

This is a special case of 1(b) which leaves the two possibilities

$$\mathbf{T}^{(1)} = \pm \sqrt{\sigma_3}\, \mathbf{u}^{(3)} \otimes \mathbf{u}^{(3)}, \tag{6.2.52}$$

with the Lagrangean axes (and also $\mathbf{R}$) defined to within an arbitrary rotation about $\mathbf{u}^{(3)}$ (see the example discussed in Section 5.1.2).

(c) $\sigma_1 = \sigma_2 = \sigma_3 = 0$
Here $\mathbf{T}^{(1)} = \mathbf{0}$, the orientation of the Lagrangean axes is arbitrary, and $\mathbf{R}$ is an arbitrary rotation.

The above exhausts the possible polar decompositions of $\mathbf{S}$ and characterizes the associated Biot stress $\mathbf{T}^{(1)}$ and rotation $\mathbf{R}$. In order to determine $\mathbf{A}$ it remains to find $\mathbf{U}$ from (6.2.40).

Suppose that $\mathbf{H}$ is globally one-to-one on the space of symmetric tensors. Then $\mathbf{U}$ is defined uniquely by $\mathbf{T}^{(1)}$ through the inverse of (6.2.40). It follows that, for a given $\mathbf{S}$, the number of solutions $\mathbf{A} = \mathbf{RU}$ of (6.2.38) is equal to the number of polar decompositions described in Cases 1 and 2, as appropriate. In particular, in Case 1(a) there are precisely four distinct inversions of (6.2.38), all having the same Lagrangean axes. Note that the orientation of the Lagrangean axes is arbitrary whenever $\mathbf{S}$ corresponds to a point on a plane $t_i^{(1)} \pm t_j^{(1)} = 0$ $(i \neq j)$ in $t_i^{(1)}$-space (the intersection of such planes is the stress-free origin).

Unique invertibility of (6.2.40) is a consequence of strict convexity of $\mathbf{H}$, which requires that

$$\operatorname{tr}\{[\mathbf{H}(\mathbf{U}') - \mathbf{H}(\mathbf{U})](\mathbf{U}' - \mathbf{U})\} > 0 \tag{6.2.53}$$

for all symmetric $\mathbf{U}$ and $\mathbf{U}' \neq \mathbf{U}$. There are good physical reasons for supposing that the inequality (6.2.53) holds for real elastic materials, at least for some bounded domain which encloses the stress-free origin $\mathbf{U} = \mathbf{I}$. In particular, for consistency with the inequalities (6.1.84) of infinitesimal elasticity, strict convexity of $\mathbf{H}$ is required in some neighbourhood of $\mathbf{U} = \mathbf{I}$. If, however, $\mathbf{H}$ is not uniquely invertible globally then the count of inversions of (6.2.38) needs modifying according to the number of $\mathbf{U}$'s associated with each $\mathbf{T}^{(1)}$ through (6.2.40).

Since we are dealing with isotropic materials, and the principal axes of $\mathbf{U}$ are therefore the same as those of $\mathbf{T}^{(1)}$, the inverse of (6.2.40) is determined by the inverse of the relation between $t_1^{(1)}$, $t_2^{(1)}$, $t_3^{(1)}$ and the principal stretches $\lambda_1, \lambda_2, \lambda_3$. That this relation is one-to-one follows from (6.2.53) can be seen by taking $\mathbf{U}'$, with principal values λ_i' and corresponding principal stresses $t_i^{(1)'}$, to be coaxial with $\mathbf{U}$, in which case (6.2.53) specializes to

$$\sum_{i=1}^{3} (t_i^{(1)'} - t_i^{(1)})(\lambda_i' - \lambda_i) > 0, \tag{6.2.54}$$

where $\lambda_i' \neq \lambda_i$ for at least one $i \in \{1, 2, 3\}$.

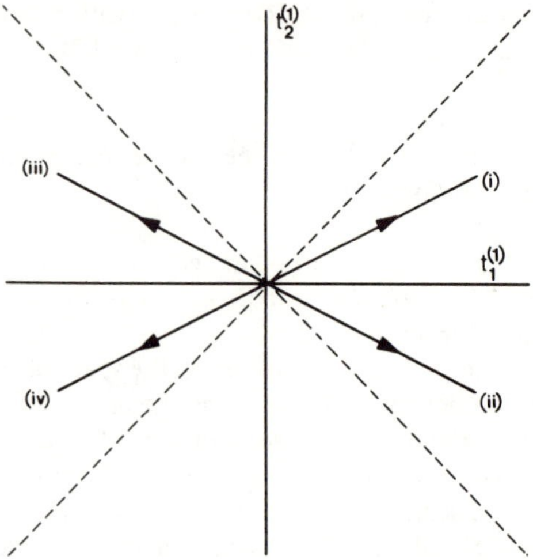

Fig. 6.1 Straight-line loading paths in the $(t_1^{(1)}, t_2^{(1)})$-plane of principal Biot stress space, labelled (i)–(iv) to comply with (6.2.46).

We now illustrate diagrammatically the inversion of (6.2.38), for which purpose we assume that (6.2.54) holds. We take $\sigma_3 = 0$ so that the possible $t_i^{(1)}$'s are given by (6.2.46) with $\sigma_3 = 0$. For definiteness we take σ_1 and σ_2 to be increasing from zero along a path of proportional loading. The resulting stress paths in $t_i^{(1)}$-space are straight lines in the plane $t_3^{(1)} = 0$. These are shown in Fig. 6.1, labelled (i)–(iv) in accordance with (6.2.46). (Of course, other stress paths may change type among (i)–(iv) by crossing an axis $t_1^{(1)} = 0$ or $t_2^{(1)} = 0$.) Except at the origin the paths shown correspond to values of σ_i for which the Lagrangean axes are determined uniquely by **S**, but other paths may meet with or partially coincide with the singular lines $t_1^{(1)} \pm t_2^{(1)} = 0$ (the broken lines in Fig. 6.1) on which the Lagrangean axes are disposable. We note in passing that, of the four paths, only (i) corresponds to $t_i^{(1)} + t_j^{(1)} > 0$ for $i, j \in \{1, 2, 3\}$ and, as we shall see in Section 6.2.3, is the only stable path (these inequalities must be distinguished from (6.2.35) in general).

The corresponding deformation paths are typified in Fig. 6.2 as plots in the (λ_1, λ_2)-plane (although, in general, λ_3 varies also). In general these paths are not straight lines. The traces of the lines $t_1^{(1)} \pm t_2^{(1)} = 0$ are the broken curves, and we note, in particular, that $t_1^{(1)} = t_2^{(1)}$ implies $\lambda_1 = \lambda_2$ is a consequence of (6.2.54). A proof of this latter result is embodied in Appendix 1.

In the analysis of the inversion of (6.2.38) given above **S** may be regarded as the nominal stress in a body homogeneously deformed by the action of

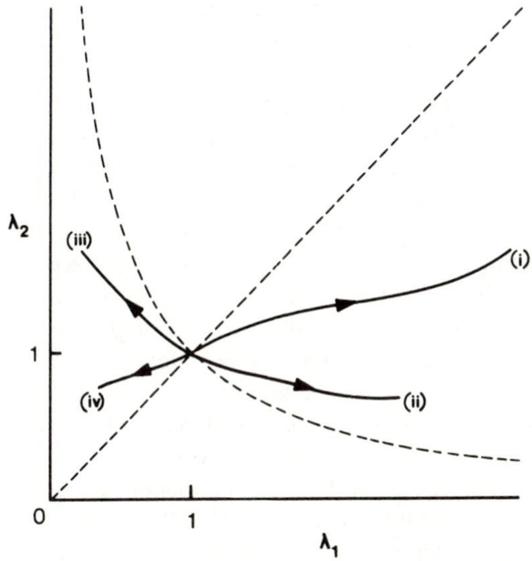

Fig. 6.2 Deformation paths in the (λ_1, λ_2)-plane corresponding to the stress paths shown in Fig. 6.1.

dead-load tractions on its boundary and zero body forces, with rotational balance assured by symmetry of $\mathbf{T}^{(1)}$. A general dead-load traction problem can be treated in a similar manner, as we now show.

Under body force density $\mathbf{b}$ (independent of the deformation) in $\mathscr{B}_0$ and dead-load $\boldsymbol{\sigma}$ on $\partial\mathscr{B}_0$ the global equilibrium equation is

$$\int_{\mathscr{B}_0} \rho_0 \mathbf{b}\,\mathrm{d}V + \int_{\partial\mathscr{B}_0} \boldsymbol{\sigma}\,\mathrm{d}A = \mathbf{0} \tag{6.2.55}$$

which, for this problem, is merely a condition on the data. For rotational balance a solution $\boldsymbol{\chi}$ is required to satisfy the equation

$$\mathbf{L}^{\mathrm{T}} = \mathbf{L}, \tag{6.2.56}$$

where

$$\mathbf{L} = \int_{\mathscr{B}_0} \rho_0 \boldsymbol{\chi} \otimes \mathbf{b}\,\mathrm{d}V + \int_{\partial\mathscr{B}_0} \boldsymbol{\chi} \otimes \boldsymbol{\sigma}\,\mathrm{d}A \tag{6.2.57}$$

is, in the terminology of Truesdell and Noll (1965), the *astatic load tensor* (relative to the origin).

In general, $\mathbf{L}$ may be decomposed as

$$\mathbf{L} = \mathbf{MQ}, \tag{6.2.58}$$

where $\mathbf{M}$ is symmetric and $\mathbf{Q}$ is proper orthogonal. This polar decomposition has identical properties to those of (6.2.39).

It follows from (6.2.58) that for a given deformation χ and given loads $\boldsymbol{\sigma}$ and $\mathbf{b}$ equilibrium may be achieved by a suitable rotation of the loads relative to the body (i.e. $\mathbf{b} \to \mathbf{Q}\mathbf{b}$, $\boldsymbol{\sigma} \to \mathbf{Q}\boldsymbol{\sigma}$). If m_1, m_2, m_3 are the principal values of $\mathbf{M}$ and $m_i + m_j = 0$ for some pair of indices $i, j \neq i$ then the equilibrated loads $\mathbf{Q}\boldsymbol{\sigma}$ and $\mathbf{Q}\mathbf{b}$ are said to have an *axis of equilibrium* (equilibrium is maintained if the loads are subject to an arbitrary rotation about that axis). The axis in question is normal to the plane of the eigenvectors corresponding to m_i and m_j. Clearly, classification of the equilibrium loads for a given deformation follows the pattern established for the polar decomposition (6.2.39).

The particular case in which $\chi(\mathbf{X}) = \mathbf{X}$ for all $\mathbf{X} \in \mathscr{B}_0$ is of special significance, for a knowledge of the load types for which (6.2.56) then holds enables the number of solutions *near* the trivial solutions to be found. (The *trivial solutions* to the boundary-value problem are the rigid rotations, corresponding to $\boldsymbol{\sigma} = \mathbf{b} = \mathbf{0}$.) A detailed analysis of such solutions has been provided recently in Chillingworth *et al.* (1982) and we refer to this paper for discussion of earlier relevant literature. We note, in particular, that Chillingworth *et al.* establish existence of solutions provided that at least two of m_1, m_2, m_3 are non-zero.

Thus far we have restricted attention to the details of the dead-load traction problem with constant body force. For more general traction boundary conditions and body forces a sufficient condition for uniqueness is obtained by replacing equality in (6.2.27) by strict inequality for all kinematically admissible χ and χ'. In general, however, explicit information as to the extent of non-uniqueness in such problems, along the lines of that extracted in the dead-load problem, has not been forthcoming. The pressure loading problem is exceptional in this respect, as we now show.

Suppose that the traction boundary condition is given in the form (6.2.2) with

$$\boldsymbol{\sigma} = -PJ\mathbf{B}\mathbf{N},$$

where P is a prescribed *constant*, and that the body force is constant. Then, uniqueness is guaranteed if

$$\int_{\mathscr{B}_0} \operatorname{tr}\{(\mathbf{S}' - \mathbf{S})(\mathbf{A}' - \mathbf{A})\}\,\mathrm{d}V$$

$$> -P \int_{\partial\mathscr{B}_0^{\sigma}} (J'\mathbf{B}'\mathbf{N} - J\mathbf{B}\mathbf{N})\cdot(\chi' - \chi)\,\mathrm{d}A$$

for all kinematically admissible χ and χ'. By use of (6.2.1), the divergence theorem, and the identity $\operatorname{Div}(J\mathbf{B}^{\mathrm{T}}) = \mathbf{0}$ applied to each of χ and χ', this inequality may be put in the form

$$\int_{\mathscr{B}_0} \mathrm{tr}\,\{(\mathbf{S}' + PJ'\mathbf{B}'^{\mathrm{T}} - \mathbf{S} - PJ\mathbf{B}^{\mathrm{T}})(\mathbf{A}' - \mathbf{A})\}\,\mathrm{d}V > 0. \tag{6.2.59}$$

The inequality (6.2.59) has the same structure as (6.2.30), with $\mathbf{S}$ replaced by $\mathbf{S} + PJ\mathbf{B}^{\mathrm{T}}$. Analysis of the inversion of (6.2.38) then carries over to the present problem under this replacement.

Finally in this section we comment on the inversion of the (nominal) stress-deformation relation for an incompressible material, i.e. of

$$\mathbf{S} = \mathbf{H}(\mathbf{A}) - p\mathbf{B}^{\mathrm{T}} \tag{6.2.60}$$

subject to $\det \mathbf{A} = 1$. The polar decomposition of $\mathbf{S}$ is not affected by imposition of the constraint, but the inversion of

$$\mathbf{T}^{(1)} = \mathbf{H}(\mathbf{U}) - p\mathbf{U}^{-1}, \tag{6.2.61}$$

with $\det \mathbf{U} = 1$, involves determination of the hydrostatic pressure p. Such an inversion will be exemplified in Section 6.3.2.

6.2.3 Incremental uniqueness and stability

We examine next the question of uniqueness in respect of the incremental boundary-value problem formulated in Section 6.2.1, attention being restricted to the dead-loading problem in the first instance.

Suppose that $\delta\boldsymbol{\chi}$ and $\delta\boldsymbol{\chi}'$ are two possible solutions, with corresponding nominal stresses $\delta\mathbf{S}$, $\delta\mathbf{S}'$ respectively. Let their differences be denoted by

$$\Delta(\delta\boldsymbol{\chi}) = \delta\boldsymbol{\chi}' - \delta\boldsymbol{\chi}, \qquad \Delta(\delta\mathbf{S}) = \delta\mathbf{S}' - \delta\mathbf{S} \tag{6.2.62}$$

and similarly for differences of other quantities associated with $\delta\boldsymbol{\chi}$ and $\delta\boldsymbol{\chi}'$.

It follows from (6.2.7), (6.2.8) and (6.2.10) that for dead-load surface tractions and body forces

$$\Delta(\delta\boldsymbol{\chi}) = \mathbf{0} \quad \text{on} \quad \partial\mathscr{B}_0^x, \tag{6.2.63}$$

$$\Delta(\delta\mathbf{S}^{\mathrm{T}})\mathbf{N} = \mathbf{0} \quad \text{on} \quad \partial\mathscr{B}_0^\sigma, \tag{6.2.64}$$

$$\mathrm{Div}\,\Delta(\delta\mathbf{S}) = \mathbf{0} \quad \text{in} \quad \mathscr{B}_0. \tag{6.2.65}$$

Use of the divergence theorem then leads to

$$\int_{\mathscr{B}_0} \mathrm{tr}\,\{\Delta(\delta\mathbf{S})\Delta(\delta\mathbf{A})\}\,\mathrm{d}V = 0 \tag{6.2.66}$$

which, on replacing $\delta\mathbf{S}$ by its linear approximation (6.2.11), may be written

$$\int_{\mathcal{B}_0} \operatorname{tr}\{[\mathscr{A}^1\Delta(\delta\mathbf{A})]\Delta(\delta\mathbf{A})\}\mathrm{d}V = 0.$$

By analogy with (6.2.30) the inequality

$$\int_{\mathcal{B}_0} \operatorname{tr}\{[\mathscr{A}^1\Delta(\delta\mathbf{A})]\Delta(\delta\mathbf{A})\}\mathrm{d}V > 0 \qquad (6.2.67)$$

for all admissible fields $\Delta(\delta\chi)$ satisfying (6.2.63), with $\Delta(\delta\mathbf{A}) = \operatorname{Grad}\Delta(\delta\chi) \not\equiv \mathbf{0}$, suffices for uniqueness of solution of the incremental boundary-value problem.

Following the pattern of Section 6.2.2, we consider next the inequality

$$\operatorname{tr}\{\Delta(\delta\mathbf{S})\Delta(\delta\mathbf{A})\} > 0 \qquad (6.2.68)$$

or, equivalently,

$$\operatorname{tr}\{[\mathscr{A}^1\Delta(\delta\mathbf{A})]\Delta(\delta\mathbf{A})\} > 0 \qquad (6.2.69)$$

for all fields $\Delta(\delta\mathbf{A}) \neq \mathbf{0}$. It follows from (6.2.68) that if $\Delta(\delta\mathbf{A}) \neq \mathbf{0}$ then $\Delta(\delta\mathbf{S}) \neq \mathbf{0}$, and conversely, i.e. the relation between $\delta\mathbf{S}$ and $\delta\mathbf{A}$ is one-to-one. In other words $\mathbf{S} = \mathbf{H}(\mathbf{A})$ is strictly convex *locally* (at $\mathbf{A}$).

The inequality (6.2.69) states that $\mathscr{A}^1$ is positive definite. However, as for the global problem, uniqueness is not to be expected in general, and (6.2.69) may fail for certain underlying deformations $\mathbf{A}$. Later in this section we characterize the region of $\mathbf{A}$-space over which (6.2.69) holds for isotropic materials and we also determine the set of $\mathbf{A}$-values for which $\mathscr{A}^1$ is singular.

In Section 6.2.2 we discussed the stability interpretation of the sufficient condition for uniqueness and we indicated that this interpretation carries over to the incremental problem. We now examine (6.2.37) for incremental deformations from χ in respect of the dead-load mixed boundary-value problem with χ prescribed on $\partial\mathcal{B}_0^x$. We set $\delta\chi = \chi' - \chi$. Then, for χ' kinematically admissible, it follows that $\delta\chi = \mathbf{0}$ on $\partial\mathcal{B}_0^x$.

A short calculation, using the Taylor expansion

$$W(\mathbf{A}') = W(\mathbf{A}) + \operatorname{tr}(\mathbf{S}\,\delta\mathbf{A}) + \tfrac{1}{2}\operatorname{tr}\{(\mathscr{A}^1\,\delta\mathbf{A})\,\delta\mathbf{A}\} + \ldots, \qquad (6.2.70)$$

the divergence theorem and the boundary conditions, shows that, to the second-order in $\delta\mathbf{A}$,

$$E\{\chi'\} - E\{\chi\} = \tfrac{1}{2}\int_{\mathcal{B}_0} \operatorname{tr}\{(\mathscr{A}^1\,\delta\mathbf{A})\,\delta\mathbf{A}\}\,\mathrm{d}V. \qquad (6.2.71)$$

Thus, the criterion for local (i.e. incremental or infinitesimal) stability of the deformation $\boldsymbol{\chi}$ may be expressed as

$$\int_{\mathscr{B}_0} \operatorname{tr}\{(\mathscr{A}^1\,\delta\mathbf{A})\,\delta\mathbf{A}\}\,\mathrm{d}V \geq 0 \tag{6.2.72}$$

for all $\delta\boldsymbol{\chi}$ which vanish on $\partial\mathscr{B}_0^x$, where $\mathscr{A}^1$ is evaluated at $\mathbf{A}$ and $\delta\mathbf{A} = \operatorname{Grad}\delta\boldsymbol{\chi}$. Equivalently, (6.2.72) may be written

$$\int_{\mathscr{B}_0} \operatorname{tr}(\delta\mathbf{S}\,\delta\mathbf{A})\,\mathrm{d}V \geq 0. \tag{6.2.73}$$

Clearly, $E\{\cdot\}$ is rendered a local minimum by an incrementally stable configuration within the considered class of deformations $\boldsymbol{\chi}'$.

If equality holds in (6.2.72) for some $\delta\mathbf{A} \not\equiv \mathbf{0}$ then the incremental stability is said to be *neutral.* If the integral (6.2.72) is negative for some $\delta\mathbf{A}$ the configuration in question in said to be *unstable.*

Since the admissible vector fields, $\Delta(\delta\boldsymbol{\chi})$ in (6.2.67) and $\delta\boldsymbol{\chi}$ in (6.2.72), satisfy the same condition on $\partial\mathscr{B}_0^x$ it follows that the restrictions placed on $\mathscr{A}^1$ by (6.2.67) and the strict form of (6.2.72), i.e.

$$\int_{\mathscr{B}_0} \operatorname{tr}\{(\mathscr{A}^1\,\delta\mathbf{A})\,\delta\mathbf{A}\}\,\mathrm{d}V > 0, \tag{6.2.74}$$

are identical. However, because equality is permitted in (6.2.72), incremental stability does not in general imply incremental uniqueness. Although the inequality (6.2.67) implies incremental stability, it is not necessary for incremental uniqueness and therefore incremental uniqueness does not in general entail incremental stability. These remarks parallel those made in Section 6.2.2 with regard to global problems.

It is of interest to note here that if a configuration $\boldsymbol{\chi}$ is unique in the incremental sense of (6.2.67) it may fail to be unique in the global sense. Thus, although (6.2.30) may hold for $\boldsymbol{\chi}'$ in some neighbourhood of $\boldsymbol{\chi}$, there may exist some kinematically admissible $\boldsymbol{\chi}'$, outside that neighbourhood, for which (6.2.28) holds. This follows because (6.2.67) cannot hold for every configuration.

With this in mind we now examine in detail the inversion of

$$\delta\mathbf{S} = \mathscr{A}^1\,\delta\mathbf{A} \tag{6.2.75}$$

together with the consequences of the inequality (6.2.69) or, equivalently,

$$\operatorname{tr}\{(\mathscr{A}^1\delta\mathbf{A})\,\delta\mathbf{A}\} > 0 \tag{6.2.76}$$

for $\delta\mathbf{A} \neq \mathbf{0}$ in respect of an *isotropic* material. For this task the Biot stress $\mathbf{T}^{(1)}$ is a useful vehicle, just as it is for the analysis of the inversion of (6.2.38).

First we study the singularity of $\mathscr{A}^1$. Thus, we require to find the set of $\mathbf{A}$'s for which $\mathscr{A}^1\delta\mathbf{A} = \mathbf{0}$ for some non-zero $\delta\mathbf{A}$. Taking the increment of (6.2.39) we obtain

$$\delta\mathbf{S} = (\delta\mathbf{T}^{(1)})\mathbf{R}^{\mathrm{T}} + \mathbf{T}^{(1)}\,\delta\mathbf{R}^{\mathrm{T}}$$

which, on setting $\delta\mathbf{S} = \mathbf{0}$ and making use of the orthogonality of $\mathbf{R}$ and the symmetry of $\delta\mathbf{T}^{(1)}$, leads to

$$\delta\mathbf{T}^{(1)} = \mathbf{T}^{(1)}\mathbf{R}^{\mathrm{T}}\,\delta\mathbf{R} = -\,\mathbf{R}^{\mathrm{T}}(\delta\mathbf{R})\mathbf{T}^{(1)}. \tag{6.2.77}$$

Because of the symmetry of $\delta\mathbf{T}^{(1)}$ rotational balance is automatic. Note that the increment $\delta\mathbf{A}$ may be expanded as

$$\delta\mathbf{A} = (\delta\mathbf{R})\mathbf{U} + \mathbf{R}\,\delta\mathbf{U}. \tag{6.2.78}$$

We now adapt the notation $\boldsymbol{\Omega}^{(\mathrm{L})}$ of Section 2.3.2 by writing

$$\boldsymbol{\Omega}^{(\mathrm{L})} = \sum_{i=1}^{3} \delta\mathbf{u}^{(i)} \otimes \mathbf{u}^{(i)}$$

for the incremental rotation of the Lagrangean axes. Then, referred to the underlying Lagrangean axes $\mathbf{u}^{(i)}$, $\delta\mathbf{T}^{(1)}$ has normal components

$$(\delta\mathbf{T}^{(1)})_{ii} = \delta t_i^{(1)}$$

and shear components

$$(\delta\mathbf{T}^{(1)})_{ij} = \Omega_{ij}^{(\mathrm{L})}(t_j^{(1)} - t_i^{(1)}) \qquad i \neq j.$$

These follow from the results of Sections 3.5.2 and 3.5.3.

Let $\Omega_{ij}^{(\mathrm{R})}$ denote the components of $\boldsymbol{\Omega}^{(\mathrm{R})} \equiv \mathbf{R}^{\mathrm{T}}\,\delta\mathbf{R}$ on the same axes. Then, decomposition of (6.2.77) on these axes leads to the equations

$$\delta t_i^{(1)} = 0 \tag{6.2.79}$$

and

$$\Omega_{ij}^{(\mathrm{L})}(t_i^{(1)} - t_j^{(1)}) = \Omega_{ij}^{(\mathrm{R})} t_i^{(1)} = -\,\Omega_{ij}^{(\mathrm{R})} t_j^{(1)} \qquad i \neq j. \tag{6.2.80}$$

In full (6.2.79) may be written

$$\sum_{j=1}^{3} (\partial t_i^{(1)}/\partial \lambda_j)\,\delta\lambda_j = 0 \qquad i = 1, 2, 3, \tag{6.2.81}$$

while (6.2.80) yields the two non-trivial possibilities

$$t_i^{(1)} + t_j^{(1)} = 0 \text{ with } \Omega_{ij}^{(\mathrm{L})} = -\tfrac{1}{2}\Omega_{ij}^{(\mathrm{R})} \neq 0 \quad i \neq j, \tag{6.2.82}$$

$$t_i^{(1)} - t_j^{(1)} = 0 \text{ with } \Omega_{ij}^{(\mathrm{R})} = 0, \Omega_{ij}^{(\mathrm{L})} \neq 0 \quad i \neq j. \tag{6.2.83}$$

The singularities of $\boldsymbol{\mathscr{A}}^1$ are therefore of three types, characterized by the following equations:

$$\left.\begin{array}{lll} \text{(a)} & \det(\partial t_i^{(1)}/\partial \lambda_j) = 0, & \\ \text{(b)} & t_i^{(1)} + t_j^{(1)} = 0 & i \neq j, \\ \text{(c)} & t_i^{(1)} - t_j^{(1)} = 0 & i \neq j, \lambda_i \neq \lambda_j. \end{array}\right\} \tag{6.2.84}$$

The case in which $\lambda_i = \lambda_j \, (i \neq j)$ requires special attention because of the additional symmetries it imposes, so we shall consider this separately. First, however, it is useful to note that, from the results of Section 6.1.4, the components of $\boldsymbol{\mathscr{A}}^1$ on principal axes are calculated as

$$\mathscr{A}^1_{iijj} = \partial t_i^{(1)}/\partial \lambda_j, \tag{6.2.85}$$

$$\mathscr{A}^1_{ijij} = \frac{\lambda_i t_i^{(1)} - \lambda_j t_j^{(1)}}{\lambda_i^2 - \lambda_j^2} \qquad i \neq j, \lambda_i \neq \lambda_j, \tag{6.2.86}$$

$$\mathscr{A}^1_{ijji} = \frac{\lambda_j t_i^{(1)} - \lambda_i t_j^{(1)}}{\lambda_i^2 - \lambda_j^2} \qquad i \neq j, \lambda_i \neq \lambda_j. \tag{6.2.87}$$

Strictly, (b) and (c) in (6.2.84) correspond to vanishing of

$$\mathscr{A}^1_{ijij} - \mathscr{A}^1_{ijji} \equiv \frac{t_i^{(1)} + t_j^{(1)}}{\lambda_i + \lambda_j} \qquad i \neq j \tag{6.2.88}$$

and

$$\mathscr{A}^1_{ijij} + \mathscr{A}^1_{ijji} \equiv \frac{t_i^{(1)} - t_j^{(1)}}{\lambda_i - \lambda_j} \qquad i \neq j \tag{6.2.89}$$

respectively, and the latter of these makes it clear why $\lambda_i = \lambda_j \; (i \neq j)$ needs separate treatment.

The solutions of $\boldsymbol{\mathscr{A}}^1 \delta \mathbf{A} = \mathbf{0}$ for $\delta \mathbf{A}$ in each case are given as follows:

$$\left.\begin{array}{lll} \text{(a)} & \mathbf{R}^{\mathrm{T}} \delta \mathbf{A} = \sum_{i=1}^{3} \delta \lambda_i \mathbf{u}^{(i)} \otimes \mathbf{u}^{(i)}, & \\ \text{(b)} & \mathbf{R}^{\mathrm{T}} \delta \mathbf{A} = -\Omega_{ij}^{(\mathrm{L})}(\lambda_i + \lambda_j)(\mathbf{u}^{(i)} \otimes \mathbf{u}^{(j)} - \mathbf{u}^{(j)} \otimes \mathbf{u}^{(i)}) & i \neq j, \\ \text{(c)} & \mathbf{R}^{\mathrm{T}} \delta \mathbf{A} = \Omega_{ij}^{(\mathrm{L})}(\lambda_j - \lambda_i)(\mathbf{u}^{(i)} \otimes \mathbf{u}^{(j)} + \mathbf{u}^{(j)} \otimes \mathbf{u}^{(i)}) & i \neq j. \end{array}\right\} \tag{6.2.90}$$

Each of $\delta \lambda_i$ and $\Omega_{ij}^{(\mathrm{L})}$ contains an arbitrary scalar factor.

In (a) the incremental deformation is *coxial with the underlying deformation*, $\mathbf{\Omega}^{(L)} = \mathbf{\Omega}^{(R)} = \mathbf{0}$, and $\delta\lambda_i$ is subject to (6.2.81).

In (b) there are three independent incremental deformations, corresponding to *shearing in each of the principal planes*, $\delta\lambda_i = 0$ $(i = 1, 2, 3)$, and $\mathbf{\Omega}^{(L)} = -\frac{1}{2}\mathbf{\Omega}^{(R)}$. It follows immediately from (2.3.28) that

$$\mathbf{R}^{\mathrm{T}}\mathbf{\Omega}^{(E)}\mathbf{R} = -\mathbf{\Omega}^{(L)}$$

so there is a change in the relative orientations of the Eulerian and Lagrangean axes coupled with the absolute rotation of the Lagrangean axes governed by $\mathbf{\Omega}^{(L)}$.

The results for (c) are similar to those for (b) except that $\mathbf{\Omega}^{(R)} = \mathbf{0}$ and

$$\mathbf{R}^{\mathrm{T}}\mathbf{\Omega}^{(E)}\mathbf{R} = \mathbf{\Omega}^{(L)},$$

so that the relative orientation of the Eulerian and Lagrangean axes is maintained during the shearing.

Problem 6.2.1 The body spin $\mathbf{\Omega}$ and the Eulerian strain-rate $\mathbf{\Sigma}$ are given by (2.3.16) and (2.3.15) respectively. By adapting these formulae to the present (incremental) circumstances and making use of the connection (2.3.27) show that the components of $\mathbf{\Omega}$ and $\mathbf{\Sigma}$ on the Eulerian axes are related according to

$$\Omega_{ij} = \frac{\lambda_i + \lambda_j}{\lambda_i - \lambda_j}\Sigma_{ij} \qquad i \neq j$$

for (b) and

$$\Omega_{ij} = \frac{\lambda_i - \lambda_j}{\lambda_i + \lambda_j}\Sigma_{ij} \qquad i \neq j$$

for (c).

Deduce that in terms of the principal components of Cauchy stress either of these becomes

$$\Omega_{ij} = \frac{t_i - t_j}{t_i + t_j}\Sigma_{ij} \qquad i \neq j.$$

Equations (6.2.84) define surfaces in stress or deformation space and, in particular, may be represented as surfaces in λ_i- or $t_i^{(1)}$-space. We note that the planes $t_i^{(1)} \pm t_j^{(1)} = 0$ in $t_i^{(1)}$-space occurring in (b) and (c) correspond precisely to those found in Section 6.2.2, on which the orientation of the Lagrangean axes has certain arbitrariness. The present results confirm this from the local viewpoint. However, at points not on these singular surfaces,

$\delta \mathbf{A}$ is uniquely determined by $\delta \mathbf{S}$.[†] This local result contrasts with that given in Section 6.2.2 for the global situation.

Where two (or more) of the singular surfaces of $\boldsymbol{\mathscr{A}}^1$ meet, two (or more) of the incremental deformations (6.2.90) are available. This is exemplified by the case $\lambda_i = \lambda_j$ $(i \neq j)$, which we now examine.

Suppose $\lambda_i = \lambda_j \neq \lambda_k$, where $i \neq j \neq k \neq i$. Then, from (6.1.52) specialized to $m = 1$ and with the notation (6.2.85) we have, in addition to $t_i^{(1)} = t_j^{(1)}$,

$$\left.\begin{aligned} \mathscr{A}^1_{iiii} &= \mathscr{A}^1_{jjjj}, \quad \mathscr{A}^1_{kkii} = \mathscr{A}^1_{kkjj}, \\ \mathscr{A}^1_{iijj} &= \mathscr{A}^1_{jjii}, \quad \mathscr{A}^1_{iikk} = \mathscr{A}^1_{jjkk}. \end{aligned}\right\} \tag{6.2.91}$$

These include the symmetries appropriate to Green elasticity.

In this limiting case (6.2.86) and (6.2.87) become

$$\left.\begin{aligned} \mathscr{A}^1_{ijij} &= \tfrac{1}{2}(\mathscr{A}^1_{iiii} - \mathscr{A}^1_{iijj} + t_i^{(1)}/\lambda_i) \qquad i \neq j, \\ \mathscr{A}^1_{ijji} &= \tfrac{1}{2}(\mathscr{A}^1_{iiii} - \mathscr{A}^1_{iijj} - t_i^{(1)}/\lambda_i) \qquad i \neq j. \end{aligned}\right\} \tag{6.2.92}$$

The determinant in (a) of (6.2.84) now factorizes to give

$$(\mathscr{A}^1_{iiii} - \mathscr{A}^1_{iijj})\{(\mathscr{A}^1_{iiii} + \mathscr{A}^1_{iijj})\mathscr{A}^1_{kkkk} - 2(\mathscr{A}^1_{iikk})^2\} = 0,$$

each factor yielding coaxial incremental deformations. First,

$$\mathscr{A}^1_{iiii} - \mathscr{A}^1_{iijj} = 0 \tag{6.2.93}$$

gives

$$\delta\lambda_i : \delta\lambda_j : \delta\lambda_k = 1 : -1 : 0 \tag{6.2.94}$$

which represents a pure shear in the $(\mathbf{u}^{(i)}, \mathbf{u}^{(j)})$-plane destroying the underlying axi-symmetry. Second, the equation

$$(\mathscr{A}^1_{iiii} + \mathscr{A}^1_{iijj})\mathscr{A}^1_{kkkk} - 2(\mathscr{A}^1_{iikk})^2 = 0 \tag{6.2.95}$$

gives

$$\delta\lambda_i : \delta\lambda_j : \delta\lambda_k = 1 : 1 : -2\mathscr{A}^1_{iikk}/\mathscr{A}^1_{kkkk}, \tag{6.2.96}$$

and this deformation maintains axi-symmetry.

By use of (6.2.89) we see that (c) in (6.2.84) may be rewritten as

$$\mathscr{A}^1_{ijij} + \mathscr{A}^1_{ijji} = 0 \qquad i \neq j.$$

[†] Further discussion of these points is provided in Section 6.2.8.

Equations (6.2.92) show that when $\lambda_i = \lambda_j$ this becomes (6.2.93) and hence the two types of singularity coincide. The associated incremental deformation, however, differs from that in (6.2.94). It is given by (c) in (6.2.90) which, with use of (2.3.27) followed by specialization to $\lambda_i = \lambda_j$, may be rewritten in the form

$$\mathbf{R}^{\mathrm{T}}(\delta\mathbf{A})\mathbf{A}^{-1}\mathbf{R} = \Sigma_{ij}(\mathbf{u}^{(i)} \otimes \mathbf{u}^{(j)} + \mathbf{u}^{(j)} \otimes \mathbf{u}^{(i)}) \qquad i \neq j,$$

where Σ_{ij} is arbitrary.

Equation (b) in (6.2.84) is not relevant for $\lambda_i = \lambda_j$ unless $t_i^{(1)} = t_j^{(1)} = 0$ in which case $\Omega_{ij}^{(\mathrm{R})}$ and $\Omega_{ij}^{(\mathrm{L})}$ are arbitrary.

Finally we examine the pure dilatation $\lambda_1 = \lambda_2 = \lambda_3$. In this case the determinant of (6.2.85) simplifies to give

$$(\mathscr{A}^1_{iiii} - \mathscr{A}^1_{iijj})^2(\mathscr{A}^1_{iiii} + 2\mathscr{A}^1_{iijj}) = 0 \tag{6.2.97}$$

for any pair $i, j \in \{1, 2, 3\}$.

The first factor corresponds to an isochoric, coaxial incremental deformation such that

$$\delta\lambda_1 + \delta\lambda_2 + \delta\lambda_3 = 0. \tag{6.2.98}$$

The second factor is associated with the pure dilatation

$$\delta\lambda_1 = \delta\lambda_2 = \delta\lambda_3. \tag{6.2.99}$$

Cases (b) and (c) are as discussed for $\lambda_i = \lambda_j$ $(i \neq j)$.

The singularities of $\mathscr{L}^{(1)1} = \partial\mathbf{T}^{(1)}/\partial\mathbf{U}$ are obtained from those of $\mathscr{A}^1$ by imposing the restrictions $\mathbf{R} = \mathbf{I}, \mathbf{\Omega}^{(\mathrm{R})} = \mathbf{0}$ in the above analysis. Thus, $\mathscr{L}^{(1)1}$ has the singularities of $\mathscr{A}^1$ with the exception of those of type (b) in (6.2.84). It follows that if $\mathscr{L}^{(1)1}$ is non-singular for all $\mathbf{U}$, i.e. $\mathbf{T}^{(1)} = \mathbf{H}(\mathbf{U})$ is locally invertible for all $\mathbf{U}$, then the only singularities of $\mathscr{A}^1$ are of type (b).

We turn now to the inequality (6.2.76). By referring to principal axes it is easy to see that (6.2.76) is expressible in the form

$$\begin{aligned}\operatorname{tr}\{(\mathscr{A}^1\delta\mathbf{A})\delta\mathbf{A}\} \equiv {} & \sum_{i,j=1}^{3} \frac{\partial t_i^{(1)}}{\partial\lambda_j}\delta\lambda_i\delta\lambda_j \\ & + \sum_{i\neq j}(\Omega_{ij}^{(\mathrm{L})} + \tfrac{1}{2}\Omega_{ij}^{(\mathrm{R})})^2(t_i^{(1)} - t_j^{(1)})(\lambda_i - \lambda_j) \\ & + \tfrac{1}{4}\sum_{i\neq j}(\Omega_{ij}^{(\mathrm{R})})^2(t_i^{(1)} + t_j^{(1)})(\lambda_i + \lambda_j) > 0 \end{aligned} \tag{6.2.100}$$

for $\delta\mathbf{A} \neq \mathbf{0}$.

Clearly, necessary and sufficient conditions for (6.2.100) are

$$
\begin{aligned}
&\text{(i)} && \text{The Jacobian matrix } (\partial t_i^{(1)}/\partial \lambda_j) \text{ is positive definite,} \\
&\text{(ii)} && (t_i^{(1)} - t_j^{(1)})/(\lambda_i - \lambda_j) > 0 \quad i \neq j \\
&\text{(iii)} && t_i^{(1)} + t_j^{(1)} > 0 \quad i \neq j
\end{aligned}
\tag{6.2.101}
$$

together, with (ii) incorporating the case $\lambda_i = \lambda_j$ by an appropriate limiting procedure.

It follows from (iii) and the ordering $t_i^{(1)} > t_j^{(1)}$ with $\lambda_i > \lambda_j$ implied by (ii) that the inequalities $t_i + t_j > 0$ mentioned in (6.2.35) are necessary for (6.2.100) but not in general sufficient for (iii).

Note that of the four $\mathbf{T}^{(1)}$'s described by either (6.2.46) or (6.2.49) only one is consistent with (iii) for all pairs of indices $i, j \neq i$, and therefore only one branch of the global inversion of $\mathbf{S} = \mathbf{H}(\mathbf{A})$ is incrementally stable (as illustrated in Fig. 6.1).

Problem 6.2.2 Show that when $\boldsymbol{\Omega}^{(\mathrm{R})} = \mathbf{0}$ the inequality (6.2.100) may be used to deduce that

$$
\begin{aligned}
\operatorname{tr}\{(\mathscr{L}^{(1)1}\delta\mathbf{U})\delta\mathbf{U}\} \equiv \sum_{i,j=1}^{3} \frac{\partial t_i^{(1)}}{\partial \lambda_j}\delta\lambda_i\delta\lambda_j \\
+ \sum_{i \neq j} (\Omega_{ij}^{(\mathrm{L})})^2 (t_i^{(1)} - t_j^{(1)})(\lambda_i - \lambda_j) > 0
\end{aligned}
\tag{6.2.102}
$$

for $\delta\mathbf{U} \neq \mathbf{0}$.

If (6.2.102) holds at each $\mathbf{U}$ in some convex domain show that (ii) in (6.2.101) is a consequence of (i) in that domain (see Appendix 1).

Use the results of Section 6.1.6 to show that (6.2.102) holds for $\mathbf{U} = \mathbf{I}$.

Finally in this section we write down uniqueness and stability conditions in respect of the boundary-value problem with traction boundary condition (6.2.8) and equilibrium equation (6.2.10). The boundary condition (6.2.64) is replaced by

$$\Delta(\delta\mathbf{S}^{\mathrm{T}})\mathbf{N} = \Delta(\delta\boldsymbol{\sigma}) \quad \text{on} \quad \partial\mathscr{B}_0^{\sigma}$$

and equation (6.2.65) by

$$\operatorname{Div}\Delta(\delta\mathbf{S}) + \rho_0\Delta(\delta\mathbf{b}) = \mathbf{0} \quad \text{in } \mathscr{B}_0.$$

Then, by use of the divergence theorem, the equilibrium equations and the boundary conditions, we obtain

$$\int_{\mathscr{B}_0} \operatorname{tr}\{\Delta(\delta\mathbf{S})\Delta(\delta\mathbf{A})\}\,\mathrm{d}V = \int_{\partial\mathscr{B}_0^\sigma} \Delta(\delta\boldsymbol{\sigma})\cdot\Delta(\delta\boldsymbol{\chi})\,\mathrm{d}A + \int_{\mathscr{B}_0} \rho_0\Delta(\delta\mathbf{b})\cdot\Delta(\delta\boldsymbol{\chi})\,\mathrm{d}V$$

if there are two distinct solutions $\delta\boldsymbol{\chi}, \delta\boldsymbol{\chi}'$ whose difference is $\Delta(\delta\boldsymbol{\chi})$. It follows, generalizing (6.2.67), that a sufficient condition for incremental uniqueness is

$$\int_{\mathscr{B}_0} \operatorname{tr}\{\Delta(\delta\mathbf{S})\Delta(\delta\mathbf{A})\}\,\mathrm{d}V - \int_{\partial\mathscr{B}_0^\sigma} \Delta(\delta\boldsymbol{\sigma})\cdot\Delta(\delta\boldsymbol{\chi})\,\mathrm{d}A - \int_{\mathscr{B}_0} \rho_0\Delta(\delta\mathbf{b})\cdot\Delta(\delta\boldsymbol{\chi})\,\mathrm{d}V > 0 \qquad (6.2.103)$$

for all admissible fields $\Delta(\delta\boldsymbol{\chi}) \not\equiv \mathbf{0}$ satisfying (6.2.63),

The corresponding incremental stability inequality is

$$\int_{\mathscr{B}_0} \operatorname{tr}(\delta\mathbf{S}\delta\mathbf{A})\,\mathrm{d}V - \int_{\partial\mathscr{B}_0^\sigma} \delta\boldsymbol{\sigma}\cdot\delta\boldsymbol{\chi}\,\mathrm{d}A - \int_{\mathscr{B}_0} \rho_0\delta\mathbf{b}\cdot\delta\boldsymbol{\chi}\,\mathrm{d}V \geq 0 \qquad (6.2.104)$$

for all $\delta\boldsymbol{\chi}$ satisfying $\delta\boldsymbol{\chi} = \mathbf{0}$ on $\partial\mathscr{B}_0^x$. Thus, in any kinematically admissible incremental deformation $\delta\boldsymbol{\chi}$ from the current equilibrium configuration the elastic energy stored exceeds the work done on the body by the boundary tractions and body forces (by half the quantity on the left-hand side of (6.2.104)) in accordance with the classical criterion of stability. However, in general, (6.2.104) is merely a sufficient condition for stability. It is also necessary if the loading is conservative. The reader interested in more general aspects of stability theory is referred to Knops and Wilkes (1973). For a summary of the relationship between uniqueness and stability see Gurtin (1982).

Problem 6.2.3 Examine the inequalities (6.2.103) and (6.2.104) when (a) $\delta\boldsymbol{\sigma}$ is given by (6.2.9), and (b) the material is incompressible.

6.2.4 Variational aspects of incremental problems

For the purposes of this section we adopt the notation $\dot{\boldsymbol{\chi}}, \dot{\mathbf{S}}, \ldots$ for increments in $\boldsymbol{\chi}, \mathbf{S}, \ldots$ in place of $\delta\boldsymbol{\chi}, \delta\mathbf{S}, \ldots$ used so far in this chapter. The prefix δ is then free to represent a variation in the quantity on which it operates, as in

Section 5.4. Thus, in what follows, $\delta\dot{\chi}$, $\delta\dot{\mathbf{S}}, \ldots$ are to be interpreted as variations in the increments $\dot{\chi}$, $\dot{\mathbf{S}}, \ldots$.

Suppose the (arbitrary) incremental deformations $\dot{\chi}$, $\dot{\chi}'$ are associated through the constitutive law with nominal stress increments $\dot{\mathbf{S}}$, $\dot{\mathbf{S}}'$ respectively at the same underlying state of deformation. Thus

$$\dot{\mathbf{S}} = \mathscr{A}^1 \dot{\mathbf{A}}, \qquad \dot{\mathbf{S}}' = \mathscr{A}^1 \dot{\mathbf{A}}', \tag{6.2.105}$$

and, for a Green-elastic material, we deduce from the symmetry of $\mathscr{A}^1$ the reciprocal relation

$$\operatorname{tr}(\dot{\mathbf{S}}'\dot{\mathbf{A}}) = \operatorname{tr}(\dot{\mathbf{S}}\dot{\mathbf{A}}'). \tag{6.2.106}$$

Let $\dot{\mathbf{b}}$, $\dot{\mathbf{b}}'$ be the body-force increments associated with $\dot{\chi}$ and $\dot{\chi}'$ respectively, and $\dot{\boldsymbol{\sigma}}$, $\dot{\boldsymbol{\sigma}}'$ similarly the surface traction increments. We take $\dot{\boldsymbol{\xi}}$, $\dot{\boldsymbol{\xi}}'$ to be the prescribed incremental displacements on $\partial\mathscr{B}_0^x$. By integration of (6.2.106) over $\mathscr{B}_0$, use of the divergence theorem, equilibrium equations and boundary conditions, we obtain

$$\begin{aligned}\int_{\mathscr{B}_0} \rho_0 \dot{\chi}\cdot\dot{\mathbf{b}}'\,\mathrm{d}V + \int_{\partial\mathscr{B}_0^\sigma} \dot{\chi}\cdot\dot{\boldsymbol{\sigma}}'\,\mathrm{d}A + \int_{\partial\mathscr{B}_0^x} \mathbf{N}\cdot(\dot{\mathbf{S}}'\dot{\boldsymbol{\xi}})\,\mathrm{d}A \\ = \int_{\mathscr{B}_0} \rho_0 \dot{\chi}'\cdot\dot{\mathbf{b}}\,\mathrm{d}V + \int_{\partial\mathscr{B}_0^\sigma} \dot{\chi}'\cdot\dot{\boldsymbol{\sigma}}\,\mathrm{d}A + \int_{\partial\mathscr{B}_0^x} \mathbf{N}\cdot(\dot{\mathbf{S}}\dot{\boldsymbol{\xi}}')\,\mathrm{d}A\end{aligned} \tag{6.2.107}$$

for all equilibrium solutions $\dot{\chi}$, $\dot{\chi}'$ for the given body, $\partial\mathscr{B}_0^x$ and $\partial\mathscr{B}_0^\sigma$ being the same for each set of data.

Equation (6.2.107) is referred to as *Betti's reciprocal theorem*. Its derivation relies on the self-adjointness of the governing differential equations embodied in the symmetry of $\mathscr{A}^1$. The classical Betti theorem of infinitesimal elasticity is included in (6.2.107) as the special case in which $\mathscr{A}^1$ is evaluated for $\mathbf{A} = \mathbf{I}$. We note that there is no global counterpart of Betti's theorem since $\operatorname{tr}(\mathbf{S}'\mathbf{A}) \neq \operatorname{tr}(\mathbf{S}\mathbf{A}')$ in general, in contrast with (6.2.106).

By following the procedure that led to (6.2.107) with $\dot{\chi}' = \dot{\chi}$ we obtain the identity

$$\int_{\mathscr{B}_0} \operatorname{tr}(\dot{\mathbf{S}}\dot{\mathbf{A}})\,\mathrm{d}V = \int_{\mathscr{B}_0} \rho_0 \dot{\chi}\cdot\dot{\mathbf{b}}\,\mathrm{d}V + \int_{\partial\mathscr{B}_0^\sigma} \dot{\chi}\cdot\dot{\boldsymbol{\sigma}}\,\mathrm{d}A + \int_{\partial\mathscr{B}_0^x} \mathbf{N}\cdot(\dot{\mathbf{S}}\dot{\boldsymbol{\xi}})\,\mathrm{d}A. \tag{6.2.108}$$

This is the incremental analogue of the global equation (5.4.8).

Now take $\dot{\chi}'$ to be the variation $\delta\dot{\chi}$ in $\dot{\chi}$ with $\dot{\boldsymbol{\sigma}}' = \delta\dot{\boldsymbol{\sigma}}$, $\dot{\mathbf{b}}' = \delta\dot{\mathbf{b}}$ and $\dot{\boldsymbol{\xi}}' = \mathbf{0}$.

On substitution into (6.2.107) we arrive at

$$\begin{aligned}\int_{\mathscr{B}_0} \operatorname{tr}(\dot{\mathbf{A}}\,\delta\dot{\mathbf{S}})\,\mathrm{d}V &= \int_{\mathscr{B}_0} \rho_0\dot{\boldsymbol{\chi}}\cdot\delta\dot{\mathbf{b}}\,\mathrm{d}V + \int_{\partial\mathscr{B}_0^\sigma} \dot{\boldsymbol{\chi}}\cdot\delta\dot{\boldsymbol{\sigma}}\,\mathrm{d}A \\ &\quad + \int_{\partial\mathscr{B}_0^x} \mathbf{N}\cdot(\delta\dot{\mathbf{S}}\,\dot{\boldsymbol{\xi}})\,\mathrm{d}A \\ &= \int_{\mathscr{B}_0} \rho_0\dot{\mathbf{b}}\cdot\delta\dot{\boldsymbol{\chi}}\,\mathrm{d}V + \int_{\partial\mathscr{B}_0^\sigma} \dot{\boldsymbol{\sigma}}\cdot\delta\dot{\boldsymbol{\chi}}\,\mathrm{d}A \\ &= \int_{\mathscr{B}_0} \operatorname{tr}(\dot{\mathbf{S}}\,\delta\dot{\mathbf{A}})\,\mathrm{d}V. \end{aligned} \qquad (6.2.109)$$

The right-hand equation in (6.2.109) is the incremental analogue of the virtual work principle (5.4.10), while the left-hand equation, with $\delta\dot{\mathbf{b}} = \delta\dot{\boldsymbol{\sigma}} = \mathbf{0}$, is the incremental counterpart of the virtual stress principle (5.4.15).

By making use of the self-adjoint property of the equations we now write (6.2.109) in the form

$$\tfrac{1}{2}\delta \int_{\mathscr{B}_0} \operatorname{tr}(\dot{\mathbf{S}}\dot{\mathbf{A}})\,\mathrm{d}V = \int_{\partial\mathscr{B}_0^\sigma} \dot{\boldsymbol{\sigma}}\cdot\delta\dot{\boldsymbol{\chi}}\,\mathrm{d}A + \int_{\mathscr{B}_0} \rho_0\dot{\mathbf{b}}\cdot\delta\dot{\boldsymbol{\chi}}\,\mathrm{d}V. \qquad (6.2.110)$$

In general the incremental loading is not conservative so that (6.2.110) cannot be recast as a variational principle. We now examine the circumstances which lead to such a principle.

Suppose $\dot{\boldsymbol{\sigma}}$ is expressible in the form

$$\dot{\boldsymbol{\sigma}} = \dot{\boldsymbol{\sigma}}^{\mathrm{c}} + \dot{\boldsymbol{\sigma}}^{\mathrm{s}}, \qquad (6.2.111)$$

where $\dot{\boldsymbol{\sigma}}^{\mathrm{c}}$ is the controllable part of $\dot{\boldsymbol{\sigma}}$, independent of the incremental deformation, and $\dot{\boldsymbol{\sigma}}^{\mathrm{s}}$ is the deformation sensitive part, depending on $\dot{\boldsymbol{\chi}}$, $\dot{\mathbf{A}}$ and vanishing for $\dot{\boldsymbol{\chi}} = \mathbf{0}$ (thus linear and homogeneous in $\dot{\boldsymbol{\chi}}$, $\dot{\mathbf{A}}$ if $\dot{\boldsymbol{\sigma}}$ is derived from $\boldsymbol{\sigma}$ in (6.2.2)). For example, for the pressure loading boundary condition use of (6.2.9) shows that

$$\left.\begin{aligned}\dot{\boldsymbol{\sigma}}^{\mathrm{c}} &= -\dot{P}J\mathbf{B}\mathbf{N}, \\ \dot{\boldsymbol{\sigma}}^{\mathrm{s}} &= JP\{\operatorname{tr}(\mathbf{B}^{\mathrm{T}}\dot{\mathbf{A}})\mathbf{I} - \mathbf{B}\dot{\mathbf{A}}^{\mathrm{T}}\}\mathbf{B}\mathbf{N},\end{aligned}\right\} \qquad (6.2.112)$$

where $\dot{P}$ is disposable.

If the data on $\partial\mathscr{B}_0^\sigma$ satisfy

$$\int_{\partial\mathscr{B}_0^\sigma} (\dot{\boldsymbol{\sigma}}^{\mathrm{s}}\cdot\dot{\boldsymbol{\chi}}' - \dot{\boldsymbol{\chi}}\cdot\dot{\boldsymbol{\sigma}}'^{\mathrm{s}})\,\mathrm{d}A = 0$$

for all $\dot{\chi}$, $\dot{\chi}'$ whose difference $\Delta\dot{\chi} = \dot{\chi}' - \dot{\chi}$ vanishes on $\partial\mathscr{B}_0^x$ they are said to be *self-adjoint*. An alternative way of writing this is

$$\int_{\partial\mathscr{B}_0^\sigma} (\dot{\boldsymbol{\sigma}}^{\mathrm{s}}\cdot\Delta\dot{\chi} - \dot{\chi}\cdot\Delta\dot{\boldsymbol{\sigma}}^{\mathrm{s}})\,\mathrm{d}A = 0 \tag{6.2.113}$$

or equivalently, if $\dot{\boldsymbol{\sigma}}^{\mathrm{s}}$ is linear and homogeneous in $\dot{\chi}$ and $\dot{\mathbf{A}}$, Δ may be replaced by δ in (6.2.113). In the latter case it follows that

$$\int_{\partial\mathscr{B}_0^\sigma} \dot{\boldsymbol{\sigma}}^{\mathrm{s}}\cdot\delta\dot{\chi}\,\mathrm{d}A = \tfrac{1}{2}\delta\int_{\partial\mathscr{B}_0^\sigma} \dot{\chi}\cdot\dot{\boldsymbol{\sigma}}^{\mathrm{s}}\,\mathrm{d}A \tag{6.2.114}$$

and (6.2.110) becomes

$$\delta\left\{\tfrac{1}{2}\int_{\mathscr{B}_0} \mathrm{tr}\,(\dot{\mathbf{S}}\dot{\mathbf{A}})\,\mathrm{d}V - \int_{\partial\mathscr{B}_0^\sigma} (\dot{\boldsymbol{\sigma}}^{\mathrm{c}} + \tfrac{1}{2}\dot{\boldsymbol{\sigma}}^{\mathrm{s}})\cdot\dot{\chi}\,\mathrm{d}A\right\} = \int_{\mathscr{B}_0} \rho_0\dot{\mathbf{b}}\cdot\delta\dot{\chi}\,\mathrm{d}V.$$

With $\dot{\mathbf{b}}$ treated likewise we obtain finally

$$\delta\left\{\tfrac{1}{2}\int_{\mathscr{B}_0} \mathrm{tr}\,(\dot{\mathbf{S}}\dot{\mathbf{A}})\,\mathrm{d}V - \int_{\partial\mathscr{B}_0^\sigma} (\dot{\boldsymbol{\sigma}}^{\mathrm{c}} + \tfrac{1}{2}\dot{\boldsymbol{\sigma}}^{\mathrm{s}})\cdot\dot{\chi}\,\mathrm{d}A - \int_{\mathscr{B}_0} \rho_0(\dot{\mathbf{b}}^{\mathrm{c}} + \tfrac{1}{2}\dot{\mathbf{b}}^{\mathrm{s}})\cdot\dot{\chi}\,\mathrm{d}V\right\} = 0. \tag{6.2.115}$$

Thus, an actual solution (unique or not) of the considered incremental boundary-value problem makes stationary the functional

$$\tfrac{1}{2}\int_{\mathscr{B}_0} \mathrm{tr}\,\{(\mathscr{A}^1\dot{\mathbf{A}})\dot{\mathbf{A}}\}\,\mathrm{d}V - \int_{\partial\mathscr{B}_0^\sigma} (\dot{\boldsymbol{\sigma}}^{\mathrm{c}} + \tfrac{1}{2}\dot{\boldsymbol{\sigma}}^{\mathrm{s}})\cdot\dot{\chi}\,\mathrm{d}A - \int_{\mathscr{B}_0} \rho_0(\dot{\mathbf{b}}^{\mathrm{c}} + \tfrac{1}{2}\dot{\mathbf{b}}^{\mathrm{s}})\cdot\dot{\chi}\,\mathrm{d}V \tag{6.2.116}$$

within the class of kinematically admissible incremental deformations $\dot{\chi}$, where $\dot{\boldsymbol{\sigma}}^{\mathrm{s}}$ and $\dot{\mathbf{b}}^{\mathrm{s}}$ depend on $\dot{\chi}$ and $\dot{\mathbf{A}} = \mathrm{Grad}\,\dot{\chi}$ in general. The converse is easily established by rearranging (6.2.115) in the form

$$-\int_{\mathscr{B}_0} (\mathrm{Div}\,\dot{\mathbf{S}} + \rho_0\dot{\mathbf{b}})\cdot\delta\dot{\chi}\,\mathrm{d}V + \int_{\partial\mathscr{B}_0^\sigma} (\dot{\mathbf{S}}^{\mathrm{T}}\mathbf{N} - \dot{\boldsymbol{\sigma}})\cdot\delta\dot{\chi}\,\mathrm{d}A = 0. \tag{6.2.117}$$

Compare (6.2.115) and (6.2.117) with (5.4.23) and (5.4.32) respectively.

Problem 6.2.4 If $\dot{\boldsymbol{\sigma}}^{\mathrm{s}}$ is given by (6.2.112) show that the integral in (6.2.113) may be written as

$$\int_{\partial\mathscr{B}_0^\sigma} P\mathbf{N}\cdot\{\mathrm{Curl}\,(\dot{\boldsymbol{\chi}}\wedge\delta\dot{\boldsymbol{\chi}})\}\,\mathrm{d}A.$$

Deduce that if P is uniform on $\partial\mathscr{B}_0^\sigma$ and $\delta\dot{\boldsymbol{\chi}}$ vanishes on the perimeter of $\partial\mathscr{B}_0^\sigma$ then $\dot{\boldsymbol{\sigma}}^{\mathrm{s}}$ is self-adjoint.

Problem 6.2.5 Derive a 'complementary' variational principle based on the left-hand equation in (6.2.109), and analogous to (5.4.35) when $\delta\dot{\mathbf{b}}\equiv\delta\dot{\boldsymbol{\sigma}}\equiv\mathbf{0}$.

In the notation of the present section the sufficient condition for uniqueness (6.2.103) becomes

$$\int_{\mathscr{B}_0}\mathrm{tr}\,(\Delta\dot{\mathbf{S}}\,\Delta\dot{\mathbf{A}})\,\mathrm{d}V-\int_{\partial\mathscr{B}_0^\sigma}\Delta\dot{\boldsymbol{\sigma}}\cdot\Delta\dot{\boldsymbol{\chi}}\,\mathrm{d}A-\int_{\mathscr{B}_0}\rho_0\,\Delta\dot{\mathbf{b}}\cdot\Delta\dot{\boldsymbol{\chi}}\,\mathrm{d}V>0 \tag{6.2.118}$$

for all admissible $\Delta\dot{\boldsymbol{\chi}}\not\equiv\mathbf{0}$ with $\Delta\dot{\boldsymbol{\chi}}=\mathbf{0}$ on $\partial\mathscr{B}_0^x$, where $\Delta\dot{\boldsymbol{\chi}}=\dot{\boldsymbol{\chi}}'-\dot{\boldsymbol{\chi}}$. This inequality holds, in particular, if $\dot{\boldsymbol{\chi}}$ is the unique solution of the incremental problem. With $\dot{\boldsymbol{\chi}}$ thus interpreted (6.2.118) may be written in the form

$$\tfrac{1}{2}\int_{\mathscr{B}_0}\Delta\{\mathrm{tr}\,(\dot{\mathbf{S}}\dot{\mathbf{A}})\}\,\mathrm{d}V-\int_{\partial\mathscr{B}_0^\sigma}(\dot{\boldsymbol{\sigma}}+\tfrac{1}{2}\Delta\dot{\boldsymbol{\sigma}})\cdot\Delta\dot{\boldsymbol{\chi}}\,\mathrm{d}A$$
$$-\int_{\mathscr{B}_0}\rho_0(\dot{\mathbf{b}}+\tfrac{1}{2}\Delta\dot{\mathbf{b}})\cdot\Delta\dot{\boldsymbol{\chi}}\,\mathrm{d}V>0,$$

use having been made of the divergence theorem, the equilibrium equations and boundary conditions. By applying the decomposition (6.2.111) to $\dot{\boldsymbol{\sigma}}$, and similarly to $\dot{\mathbf{b}}$, we make the further rearrangement

$$\Delta\left\{\tfrac{1}{2}\int_{\mathscr{B}_0}\mathrm{tr}\,(\dot{\mathbf{S}}\dot{\mathbf{A}})\,\mathrm{d}V-\int_{\partial\mathscr{B}_0^\sigma}(\dot{\boldsymbol{\sigma}}+\tfrac{1}{2}\dot{\boldsymbol{\sigma}}^{\mathrm{s}})\cdot\dot{\boldsymbol{\chi}}\,\mathrm{d}A\right.$$
$$\left.-\int_{\mathscr{B}_0}(\mathbf{b}^{\mathrm{c}}+\tfrac{1}{2}\dot{\mathbf{b}}^{\mathrm{s}})\cdot\dot{\boldsymbol{\chi}}\,\mathrm{d}V\right\}>\tfrac{1}{2}\int_{\partial\mathscr{B}_0^\sigma}(\dot{\boldsymbol{\sigma}}^{\mathrm{s}}\cdot\Delta\dot{\boldsymbol{\chi}}-\dot{\boldsymbol{\chi}}\cdot\Delta\dot{\boldsymbol{\sigma}}^{\mathrm{s}})\mathrm{d}A$$
$$+\tfrac{1}{2}\int_{\mathscr{B}_0}\rho_0(\dot{\mathbf{b}}^{\mathrm{s}}\cdot\Delta\dot{\boldsymbol{\chi}}-\dot{\boldsymbol{\chi}}\cdot\Delta\dot{\mathbf{b}}^{\mathrm{s}})\,\mathrm{d}V. \tag{6.2.119}$$

If the data are self-adjoint then, according to (6.2.113) and a similar equation for the body force, the right-hand side of (6.2.119) vanishes. The inequality then remaining shows that, within the class of admissible incremental deformations, the functional (6.2.116) is minimized by the unique solution $\dot{\boldsymbol{\chi}}$. The variational principle (6.2.115) can therefore be strengthened to a minimum principle at configurations for which the incremental uniqueness condition (6.2.118) holds. Such configurations are incrementally stable since, for self-adjoint data linearly dependent on $\dot{\boldsymbol{\chi}}$ and $\dot{\mathbf{A}}$, (6.2.118) implies (6.2.104). For further discussion of these matters, in a wider context, see Hill (1978).

It is left as an exercise for the reader to construct a complementary maximum principle associated with the variational principle alluded to in Problem 6.2.5. In this connection we refer to Shield and Fosdick (1966).

Problem 6.2.6 Show how the variational and extremum properties derived above are modified for an incompressible material.

6.2.5 Bifurcation analysis: dead-load tractions

Suppose the dead-load mixed boundary-value problem with data $\boldsymbol{\sigma}$, $\mathbf{b}$, $\boldsymbol{\xi}$ has solution $\boldsymbol{\chi}$. We now examine the inhomogeneous incremental boundary-value problem corresponding to increments $\dot{\boldsymbol{\sigma}}$, $\dot{\mathbf{b}}$, $\dot{\boldsymbol{\xi}}$ in the above data, the notation for increments introduced in Section 6.2.4 being retained. Thus

$$\mathrm{Div}\,(\mathscr{A}^1\dot{\mathbf{A}}) + \rho_0\dot{\mathbf{b}} = \mathbf{0} \quad \text{in} \quad \mathscr{B}_0, \tag{6.2.120}$$

$$\dot{\mathbf{S}}^{\mathrm{T}}\mathrm{N} = \dot{\boldsymbol{\sigma}} \quad \text{on} \quad \partial\mathscr{B}_0^\sigma, \tag{6.2.121}$$

$$\dot{\boldsymbol{\chi}} = \dot{\boldsymbol{\xi}} \quad \text{on} \quad \partial\mathscr{B}_0^x, \tag{6.2.122}$$

where $\dot{\boldsymbol{\sigma}}$ and $\dot{\mathbf{b}}$ are independent of the incremental deformation $\dot{\boldsymbol{\chi}}$.

From the results of Section 6.2.3 the solution of this boundary-value problem is unique if

$$\int_{\mathscr{B}_0} \mathrm{tr}\,\{(\mathscr{A}^1\,\Delta\dot{\mathbf{A}})\,\Delta\dot{\mathbf{A}}\}\,\mathrm{d}V > 0 \tag{6.2.123}$$

for all $\Delta\dot{\boldsymbol{\chi}} \not\equiv \mathbf{0}$ satisfying $\Delta\dot{\boldsymbol{\chi}} = \mathbf{0}$ on $\partial\mathscr{B}_0^x$ where $\Delta\dot{\mathbf{A}} = \mathrm{Grad}\,\Delta\dot{\boldsymbol{\chi}}$.

Equivalently, (6.2.123) may be written

$$\int_{\mathscr{B}_0} \mathrm{tr}\{(\mathscr{A}^1\dot{\mathbf{A}})\dot{\mathbf{A}}\}\,\mathrm{d}V > 0 \tag{6.2.124}$$

for all $\dot{\boldsymbol{\chi}} \not\equiv \mathbf{0}$ satisfying $\dot{\boldsymbol{\chi}} = \mathbf{0}$ on $\partial\mathscr{B}_0^x$. If this inequality holds then the underlying configuration $\boldsymbol{\chi}$ is (incrementally) stable.

When $\dot{\boldsymbol{\sigma}} = \dot{\mathbf{b}} = \dot{\boldsymbol{\xi}} = \mathbf{0}$ the equations and boundary conditions (6.2.120)–(6.2.122) are homogeneous and $\dot{\boldsymbol{\chi}} \equiv \mathbf{0}$ in $\mathscr{B}_0$ is always a solution, so if (6.2.124) holds it is the only solution. This means that the underlying dead-load problem has a unique solution (χ itself) in some neighbourhood of χ. For certain configurations, however, there may exist a $\dot{\chi} \not\equiv \mathbf{0}$ satisfying the homogeneous problem (6.2.120)–(6.2.122), i.e. $\mathscr{A}^1$ is such that

$$\int_{\mathscr{B}_0} \mathrm{tr}\,\{(\mathscr{A}^1\dot{\mathbf{A}})\dot{\mathbf{A}}\}\,\mathrm{d}V = 0, \tag{6.2.125}$$

and the solution of the incremental boundary-value problem is then definitely not unique. A configuration in which uniqueness fails is called here an *eigenconfiguration*,[†] or *e-configuration* for short. A solution $\dot{\chi} \not\equiv \mathbf{0}$ of the homogeneous problem in an e-configuration is called an *eigenmode*, or *e-mode*.

The inequality (6.2.124) ensures that the considered configuration is not an e-configuration. Because (6.2.124) therefore excludes e-modes its left-hand side is referred to as the *exclusion functional* (for the dead-load problem considered).

If there does exist an e-mode $\dot{\chi}$ then

$$\int_{\mathscr{B}_0} \mathrm{tr}\,\{(\mathscr{A}^1\dot{\mathbf{A}})\dot{\mathbf{A}}\}\mathrm{d}V = \int_{\partial\mathscr{B}_0^\sigma} (\dot{\mathbf{S}}^{\mathrm{T}}\mathbf{N})\cdot\dot{\chi}\,\mathrm{d}A - \int_{\mathscr{B}_0} (\mathrm{Div}\,\dot{\mathbf{S}})\cdot\dot{\chi}\,\mathrm{d}V = 0, \tag{6.2.126}$$

confirming (6.2.125). The variational counterpart of this is

$$\int_{\mathscr{B}_0} \mathrm{tr}\,\{(\mathscr{A}^1\dot{\mathbf{A}})\delta\dot{\mathbf{A}}\}\mathrm{d}V = \int_{\partial\mathscr{B}_0^\sigma} (\dot{\mathbf{S}}^{\mathrm{T}}\mathbf{N})\cdot\delta\dot{\chi}\,\mathrm{d}A - \int_{\mathscr{B}_0} (\mathrm{Div}\,\dot{\mathbf{S}})\cdot\delta\dot{\chi}\,\mathrm{d}V = 0, \tag{6.2.127}$$

and we note that

$$\int_{\mathscr{B}_0} \mathrm{tr}\,\{(\mathscr{A}^1\dot{\mathbf{A}})\,\delta\dot{\mathbf{A}}\}\,\mathrm{d}V = \tfrac{1}{2}\delta\int_{\mathscr{B}_0} \mathrm{tr}\,\{(\mathscr{A}^1\dot{\mathbf{A}})\dot{\mathbf{A}}\}\,\mathrm{d}V \tag{6.2.128}$$

since attention is confined to Green-elastic materials. Thus, an e-mode makes the exclusion functional stationary within the class of admissible (i.e. continuously differentiable variations $\delta\dot{\chi}$ that satisfy $\delta\dot{\chi} = \mathbf{0}$ on $\partial\mathscr{B}_0^x$. Conversely, any admissible incremental deformation $\dot{\chi} \not\equiv \mathbf{0}$ that makes the exclusion functional stationary is an e-mode. These results are special cases of those outlined in Section 6.2.4.

[†]The term *eigenstate* is commonly used in the literature (see Hill, 1978).

An e-mode defines an *adjacent equilibrium configuration* (i.e. $\chi + \dot{\chi}$ is near χ) under the given dead loads. (Because of the homogeneity of the incremental problem an e-mode has arbitrary magnitude, but it should be remembered that the equations are valid only if $\dot{\chi}$ is 'incremental' in the sense that higher-order terms have been neglected. The magnitude of an e-mode must therefore be restricted accordingly.) Thus, when in an e-configuration, a body remains in equilibrium if subjected to an eigenmodal deformation. Moreover, the energy functional

$$E\{\chi\} = \int_{\mathscr{B}_0} \{W(\mathbf{A}) - \rho_0 \mathbf{b}\cdot\chi\}\,\mathrm{d}V - \int_{\partial\mathscr{B}_0^{\sigma}} \boldsymbol{\sigma}\cdot\chi\,\mathrm{d}A \tag{6.2.129}$$

is stationary not only for the underlying configuration χ but also for each equilibrium configuration adjacent to χ. This can be seen by noting (6.2.125)–(6.2.128) and recalling from (6.2.71) that

$$E\{\chi + \dot{\chi}\} - E\{\chi\} = \tfrac{1}{2} \int_{\mathscr{B}_0} \mathrm{tr}\,\{(\mathscr{A}^1 \dot{\mathbf{A}})\dot{\mathbf{A}}\}\,\mathrm{d}V. \tag{6.2.130}$$

Now consider a deformation path generated by some monotonically increasing *loading parameter* (contained in $\boldsymbol{\sigma}$, for example). Suppose that along the path the exclusion functional is positive definite (i.e. (6.2.124) holds for all admissible $\dot{\chi}$) up to some critical value of the parameter in question, beyond which the exclusion functional is indefinite. By continuity it follows that the set of configurations where the exclusion functional is positive definite is bounded by configurations where it is semi-definite, i.e. where

$$\int_{\mathscr{B}_0} \mathrm{tr}\,\{(\mathscr{A}^1 \dot{\mathbf{A}})\dot{\mathbf{A}}\}\,\mathrm{d}V \geq 0 \tag{6.2.131}$$

for all admissible $\dot{\chi}$, with equality holding for some $\dot{\chi} \not\equiv \mathbf{0}$. Such configurations are necessarily e-configurations (but see the reference to a pathological situation in Hill, 1978, p. 60) and a $\dot{\chi} \not\equiv \mathbf{0}$ causing equality in (6.2.131) is an e-mode.†

This contrasts with the situation for e-configurations where the exclusion functional is indefinite, for then an admissible $\dot{\chi}$ may make the exclusion functional vanish but not be an e-mode (i.e. need not make the functional stationary). Such e-configurations are incrementally unstable and therefore

†Note the analogy with linear algebra: if $\mathbf{A}$ denotes a *symmetric* linear mapping of a finite-dimensional vector space into itself and $(\mathbf{A}\mathbf{v})\cdot\mathbf{v} \geq 0$ for all vectors $\mathbf{v}$ with equality holding for some vector $\mathbf{v} = \mathbf{u} \neq \mathbf{0}$ then $\mathbf{A}\mathbf{u} = \mathbf{0}$, i.e. $\mathbf{u}$ is an eigenvector of $\mathbf{A}$ corresponding to zero eigenvalue.

of no practical interest. Eigenconfigurations with the semi-definiteness property (6.2.131) are called *primary* e-configurations since they are the first e-configurations to be reached on a stable path of deformation of the type considered above.

Since (6.2.123) and (6.2.124) are equivalent *the criticality of the exclusion functional is independent of the inhomogeneous data* $\dot{\boldsymbol{\sigma}}$, $\dot{\mathbf{b}}$, $\dot{\boldsymbol{\xi}}$. It follows that the solution of the (homogeneous or inhomogeneous) incremental boundary-value problem (6.2.120)–(6.2.122) is unique at any stage along the considered stable path up to a primary e-configuration, at which uniqueness fails and the solution path *bifurcates*. The critical configuration in question, $\boldsymbol{\chi}_c$ say, is called a *bifurcation point* on the path.

In order to emphasize the close connection between the homogeneous and inhomogeneous incremental problems we suppose that $\dot{\boldsymbol{\chi}}$ is an e-mode† at $\boldsymbol{\chi}_c$ so that both $\boldsymbol{\chi}_c$ and $\boldsymbol{\chi}_c + \dot{\boldsymbol{\chi}}$ are solutions of the underlying dead-load problem corresponding to critical data $\boldsymbol{\sigma}_c$, $\mathbf{b}_c$, $\boldsymbol{\xi}_c$. The inhomogeneous problem at $\boldsymbol{\chi}_c$, with $\dot{\boldsymbol{\sigma}} \neq \mathbf{0}$, $\dot{\mathbf{b}} = \dot{\boldsymbol{\xi}} = \mathbf{0}$ for example, corresponds to $\boldsymbol{\sigma}$ being taken beyond its critical value $\boldsymbol{\sigma}_c$ to $\boldsymbol{\sigma}_c + \dot{\boldsymbol{\sigma}}$. Let $\dot{\boldsymbol{\mu}}$ be a solution of this problem. Then, since the difference of two solutions satisfies the homogeneous problem at $\boldsymbol{\chi}_c$, $\dot{\boldsymbol{\mu}} + \dot{\boldsymbol{\chi}}$ is also a solution ($\dot{\boldsymbol{\chi}}$ containing an arbitrary scalar factor). Note, however, that $\dot{\boldsymbol{\sigma}}$ is constrained by

$$\int_{\partial\mathscr{B}_0^\sigma} \dot{\boldsymbol{\sigma}}\cdot\dot{\boldsymbol{\chi}}\,\mathrm{d}A = 0 \tag{6.2.132}$$

at $\boldsymbol{\chi}_c$ (when $\dot{\mathbf{b}} = \dot{\boldsymbol{\xi}} = \mathbf{0}$, as assumed). This follows from the reciprocal theorem (6.2.107) with $\dot{\boldsymbol{\chi}}' = \dot{\boldsymbol{\mu}}$ and $\dot{\boldsymbol{\chi}}$ the e-mode, or, equivalently, by direct calculation using the divergence theorem, the equilibrium equations, the boundary conditions and the symmetry of $\mathscr{A}^1$. Thus, the incremental loading in an e-configuration is orthogonal (in the generalized sense of (6.2.132)) to each e-mode for that configuration. (Equation (6.2.132) may be generalized to the situation where $\dot{\mathbf{b}} \neq \mathbf{0}$, $\dot{\boldsymbol{\xi}} \neq \mathbf{0}$ by further use of (6.2.107).)

By contrast with (6.2.132), since $\dot{\boldsymbol{\mu}}$ is not an e-mode, we have

$$\int_{\partial\mathscr{B}_0^\sigma} \dot{\boldsymbol{\sigma}}\cdot\dot{\boldsymbol{\mu}}\,\mathrm{d}A \equiv \int_{\mathscr{B}_0} \mathrm{tr}\{[\mathscr{A}^1(\mathrm{Grad}\,\dot{\boldsymbol{\mu}})](\mathrm{Grad}\,\dot{\boldsymbol{\mu}})\}\,\mathrm{d}V > 0$$

at $\boldsymbol{\chi}_c$.

If (6.2.132) holds for some $\dot{\boldsymbol{\sigma}} \not\equiv \mathbf{0}$ on $\partial\mathscr{B}_0^\sigma$ then $\dot{\boldsymbol{\mu}}$ exists and bifurcation occurs. Recall, however, that we are dealing with the linear approximation near $\boldsymbol{\chi}_c$ so that (6.2.132) need only vanish to second order. Non-vanishing

†Generally only one independent e-mode exists in a given e-configuration but, exceptionally, there may be more than one, as the analysis for $\lambda_i = \lambda_j (i \neq j)$ in Section 6.2.3 indicates.

third and higher-order contributions to the integral in (6.2.132) are permitted. To the first order $\dot{\sigma}$ has no component in the direction $\dot{\chi}$ (in the sense of (6.2.132)) and the considered loading parameter therefore has a turning point with respect to the e-mode direction at χ_c. Any first-order load $\dot{\sigma}$ which violates (6.2.132) cannot be supported in equilibrium by the material. A satisfactory discussion of these points can only be given by involving second- (and possible higher-) order terms in the analysis of the boundary-value problem in a neighbourhood of χ_c, but this is beyond the scope of this book. Indeed, no results along these lines are available at present except when χ_c is homogeneous under all-round dead load (with the nominal stress correspondingly uniform). The linearized case for this latter problem has been dealt with in Section 6.2.3, the analysis being simplified by its restriction to a single material point. Insight into the general problem can be gained by an extension of this pointwise branching analysis to second and higher orders, and a brief account of this is therefore given in Section 6.2.8.

It is worth noting at this point that the magnitude of the e-mode, arbitrary in the linear approximation, is generally fixed by the second-order terms, as also are the initial curvatures of the two deformation paths emanating from χ_c. The tangents to these paths at χ_c are $\dot{\mu}$ and $\dot{\mu} + \dot{\chi}$. The branching behaviour is depicted in Fig. 6.3 which shows the stable path of deformation up to χ_c and the two solution branches through χ_c of the global problem.

We now turn to the problem, alluded to above, in which the underlying deformation and material properties are homogeneous. By taking the incremental deformation to have a uniform gradient we deduce that positive definiteness of $\mathscr{A}^1$ is a consequence of the exclusion condition (6.2.124). For all-round dead load (6.2.124) is equivalent to

$$\operatorname{tr}\{(\mathscr{A}^1\dot{\mathbf{A}})\dot{\mathbf{A}}\} > 0 \tag{6.2.133}$$

for all gradients $\dot{\mathbf{A}} \neq \mathbf{0}$. And in a primary e-configuration

$$\operatorname{tr}\{(\mathscr{A}^1\dot{\mathbf{A}})\dot{\mathbf{A}}\} \geq 0 \tag{6.2.134}$$

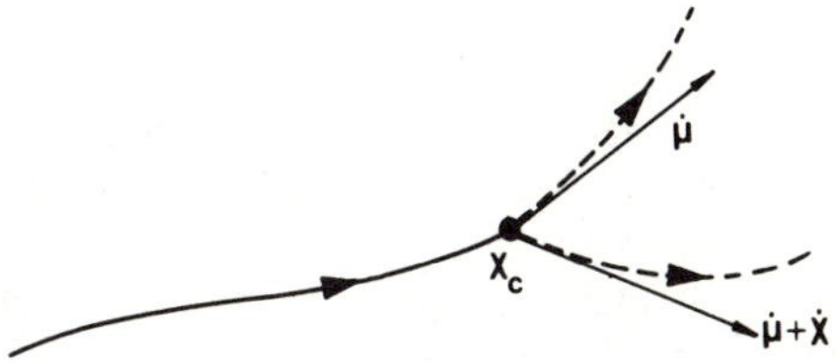

Fig. 6.3 Stable solution path (continuous curve) bifurcates at the e-configuration χ_c into two branches (broken curves) with tangents $\dot{\mu}$ and $\dot{\mu} + \dot{\chi}$ at χ_c, where $\dot{\chi}$ is an e-mode at χ_c.

for all $\dot{\mathbf{A}}$, with equality for some $\dot{\mathbf{A}} \equiv \text{Grad}\,\dot{\chi} \neq \mathbf{0}$ necessarily holding at each point of the body. The e-mode $\dot{\chi}$ is subject to the stationarity property

$$\dot{\mathbf{S}} \equiv \mathscr{A}^1 \dot{\mathbf{A}} = \mathbf{0} \tag{6.2.135}$$

in a primary e-configuration.

Clearly, any incremental deformation with a *uniform* gradient satisfying (6.2.135) is an e-mode (not necessarily primary). On the other hand, a primary e-mode necessarily either has a uniform gradient or has the form

$$\dot{\chi} = f(\mathbf{N}\cdot\mathbf{X})\mathbf{m}, \tag{6.2.136}$$

where $\mathbf{m}$ and $\mathbf{N}$ are fixed, but arbitrary, vectors and f is a continuous and (at least) piecewise continuously differentiable function. The proof of this result is left as an exercise with the remark that only the magnitude of $\dot{\mathbf{A}}$ is permitted to depend on $\mathbf{X}$ if (6.2.135) is to be satisfied at each point of the body.

With (6.2.136), the exclusion condition (6.2.124) becomes

$$\text{tr}\{[\mathscr{A}^1(\mathbf{m}\otimes\mathbf{N})](\mathbf{m}\otimes\mathbf{N})\} > 0$$

for arbitrary non-zero $\mathbf{m}$ and $\mathbf{N}$, while, on substitution of (6.2.136) into the equilibrium equation $\text{Div}\,\dot{\mathbf{S}} = \mathbf{0}$, we obtain

$$[\mathscr{A}^1(\mathbf{m}\otimes\mathbf{N})]^{\mathrm{T}}\mathbf{N}f''(\mathbf{N}\cdot\mathbf{X}) = \mathbf{0}.$$

It follows that either the deformation is homogeneous, i.e. $f''(\mathbf{N}\cdot\mathbf{X}) = 0$, or $[\mathscr{A}^1(\mathbf{m}\otimes\mathbf{N})]^{\mathrm{T}}\mathbf{N} = \mathbf{0}$. In particular, a deformation of the type (6.2.136) may be associated with discontinuous deformation gradients of the kind linked to *shear band* formation, for example, by taking f' to be piecewise constant and the discontinuity to be across a plane with normal $\mathbf{N}$ such that $\mathbf{m}\cdot(\mathbf{BN}) = 0$.

In general, however, inhomogeneous e-modes that are not primary are not of this special type and remain characterized by (6.2.126) and (6.2.127) jointly. Problems which involve such e-modes will be considered in Section 6.3.

For an isotropic material e-configurations associated with homogeneous e-modes have been catalogued in Section 6.2.3 in the context of solving (6.2.135), this amounting to finding the singularities of $\mathscr{A}^1$.

We record here that e-configurations form (*eigen*-) surfaces in $\mathbf{A}$-space characterized by one or more of equations (6.2.84), (6.2.95) or (6.2.97), as appropriate. The corresponding e-modes are (6.2.90), (6.2.96) and (6.2.98) with (6.2.99) respectively. Which is primary along a given deformation path depends on the properties of the considered constitutive law. The region of stability in **A**-space is the open set where inequalities (6.2.101) hold jointly, bounded by primary e-surfaces. In the linearized theory primary e-configurations are neutrally stable but may turn out to be unstable if assessed on the basis of a higher-order theory (see Section 6.2.8).

Problem 6.2.7 Show that in an e-configuration the rotational balance equations may be written in the form

$$(\mathbf{\Sigma}+\mathbf{\Omega})\mathbf{T}=\mathbf{T}(\mathbf{\Sigma}-\mathbf{\Omega}), \tag{6.2.137}$$

where $\mathbf{T}$ is the Cauchy stress tensor and $\mathbf{\Sigma}$ and $\mathbf{\Omega}$ are as defined in Problem 6.2.1 (Section 6.2.3). Deduce that when referred to the principal axes of $\mathbf{T}$ this can be rearranged as the final equation in Problem 6.2.1.

By use of (6.1.22) show that equation (6.2.135) is expressible as

$$(\mathscr{L}^{(2)1}\dot{\mathbf{E}}^{(2)})\mathbf{A}^{\mathrm{T}}+\mathbf{T}^{(2)}\dot{\mathbf{A}}^{\mathrm{T}}=\mathbf{0},$$

or as

$$\mathscr{L}_0^{(2)1}\mathbf{\Sigma}+\mathbf{T}(\mathbf{\Sigma}-\mathbf{\Omega})=\mathbf{0} \tag{6.2.138}$$

if the reference configuration is chosen to coincide with the current configuration.

Eliminate $\mathbf{\Omega}$ between (6.2.137) and (6.2.138) to show that for an isotropic material the component form of (6.2.138) becomes

$$\begin{cases} \displaystyle\sum_{j=1}^{3} \mathscr{L}_{0iijj}^{(2)1}\Sigma_{jj}+t_i\Sigma_{ii}=0 & i=1,2,3;\ \text{no summation}, \\ \left(\mathscr{L}_{0ijij}^{(2)1}+\dfrac{t_it_j}{t_i+t_j}\right)\Sigma_{ij}=0 & i\neq j;\ \text{no summation}. \end{cases}$$

Reconcile these equations with those given in (6.2.84), and note, in particular, that the latter may be rewritten as

$$\frac{(t_i^{(1)}-t_j^{(1)})(t_i^{(1)}+t_j^{(1)})}{(\lambda_i-\lambda_j)(t_i+t_j)}\Sigma_{ij}=0 \qquad i\neq j.$$

Deduce that the singularity $t_i+t_j=0$ $(i\neq j)$ is ruled out by this equation unless $t_i=t_j=0$ in which case an arbitrary rotation Ω_{ij} is permissible, along with arbitrary Σ_{ij} for that (i,j) pair.

Show that the exclusion condition (6.2.133) correspondingly becomes

$$\operatorname{tr}\{(\mathscr{L}_0^{(2)1}\mathbf{\Sigma})\mathbf{\Sigma}+\mathbf{T}(\mathbf{\Sigma}-\mathbf{\Omega})(\mathbf{\Sigma}+\mathbf{\Omega})\}>0$$

for arbitrary symmetric $\boldsymbol{\Sigma}$ and antisymmetric $\boldsymbol{\Omega}$ (not both zero), and deduce that $t_i + t_j > 0$ $(i \neq j)$ irrespective of the material symmetry. By considering isotropic materials prove that a configuration where $t_i + t_j = 0$ $(i \neq j)$ is unstable unless $t_i = t_j = 0$. Show that when this latter condition holds the exclusion function is positive provided $t_k > 0$ $(k \neq i, j)$ except for $\boldsymbol{\Sigma} = \mathbf{0}$, $\Omega_{ik} = \Omega_{jk} = 0$, $\Omega_{ij} \neq 0$ (an arbitrary rotation about the t_k-axis), in which case the function vanishes.

Problem 6.2.8 For a sphere of isotropic elastic material under all-round dead-load tension $t_i^{(1)} = t^{(1)} > 0$ with corresponding stretches $\lambda_i = \lambda$ show that the exclusion condition may be written

$$(\alpha - \beta)\operatorname{tr}(\boldsymbol{\Sigma}^2) + \beta(\operatorname{tr}\boldsymbol{\Sigma})^2 - \lambda^{-1}t^{(1)}\operatorname{tr}(\boldsymbol{\Omega}^2) > 0,$$

where $\alpha = \partial t_i^{(1)}/\partial\lambda_i$ $(i = 1, 2, 3)$ and $\beta = \partial t_i^{(1)}/\partial\lambda_j$ $(i \neq j)$ evaluated for $\lambda_i = \lambda$. Prove that e-configurations are characterized by $\alpha = \beta$ or $\mathrm{d}t^{(1)}/\mathrm{d}\lambda = 0$ and describe the e-modes.

Problem 6.2.9 Show that for an incompressible material the functional in (6.2.124) is replaced by

$$\int_{\mathscr{B}_0} \operatorname{tr}\{(\mathscr{A}^1\dot{\mathbf{A}})\dot{\mathbf{A}} + p\mathbf{B}^{\mathrm{T}}\dot{\mathbf{A}}\mathbf{B}^{\mathrm{T}}\dot{\mathbf{A}}\}\,\mathrm{d}V,$$

where $\operatorname{tr}(\dot{\mathbf{A}}\mathbf{B}^{\mathrm{T}}) = 0$.

Use this to deduce that the exclusion condition in Problem 6.2.8 becomes

$$(\alpha - \beta + \gamma - t)\operatorname{tr}(\boldsymbol{\Sigma}^2) - t\operatorname{tr}(\boldsymbol{\Omega}^2) > 0,$$

where

$$\alpha = \frac{\partial^2 W}{\partial\lambda_i^2}, \qquad \gamma = \frac{\partial W}{\partial\lambda_i} \qquad (i = 1, 2, 3)$$

and

$$\beta = \frac{\partial^2 W}{\partial\lambda_i\partial\lambda_j} \qquad (i \neq j)$$

are evaluated for $\lambda_1 = \lambda_2 = \lambda_3 = 1$. With the help of equation (6.1.88) deduce further that

$$0 < t < 2\mu.$$

Describe the primary e-configurations and e-modes.

6.2.6 Bifurcation analysis: non-adjoint and self-adjoint data

We now extend the analysis given in Section 6.2.5 to the situation where the traction and body force are as described in Section 6.2.4. First, we recall from (6.2.118) that the inequality

$$\int_{\mathscr{B}_0} \operatorname{tr}(\Delta\dot{\mathbf{S}}\Delta\dot{\mathbf{A}})\,\mathrm{d}V - \int_{\partial\mathscr{B}_0^\sigma} \Delta\dot{\boldsymbol{\sigma}}\cdot\Delta\dot{\boldsymbol{\chi}}\,\mathrm{d}A - \int_{\mathscr{B}_0} \rho_0\Delta\dot{\mathbf{b}}\cdot\Delta\dot{\boldsymbol{\chi}}\,\mathrm{d}V > 0$$

for all admissible $\Delta\dot{\boldsymbol{\chi}} \not\equiv \mathbf{0}$ satisfying $\Delta\dot{\boldsymbol{\chi}} = \mathbf{0}$ on $\partial\mathscr{B}_0^x$ is sufficient for uniqueness of the incremental boundary-value problem. Just as in Section 6.2.5, uniqueness is assured when the corresponding homogeneous problem ($\dot{\boldsymbol{\sigma}}^c = \dot{\mathbf{b}}^c = \mathbf{0}$) does not admit e-modes, i.e. non-trivial solutions of (6.2.120)–(6.2.122) with deformation-sensitive data $\dot{\mathbf{b}} = \dot{\mathbf{b}}^s$, $\dot{\boldsymbol{\sigma}} = \dot{\boldsymbol{\sigma}}^s$ and $\dot{\boldsymbol{\xi}} = \mathbf{0}$. The exclusion condition for this problem is

$$\int_{\mathscr{B}_0} \operatorname{tr}(\dot{\mathbf{S}}\dot{\mathbf{A}})\,\mathrm{d}V - \int_{\partial\mathscr{B}_0^\sigma} \dot{\boldsymbol{\sigma}}^s\cdot\dot{\boldsymbol{\chi}}\,\mathrm{d}A - \int_{\mathscr{B}_0} \rho_0\dot{\mathbf{b}}^s\cdot\dot{\boldsymbol{\chi}}\,\mathrm{d}V > 0 \tag{6.2.139}$$

for all admissible fields $\dot{\boldsymbol{\chi}} \not\equiv \mathbf{0}$ satisfying $\dot{\boldsymbol{\chi}} = \mathbf{0}$ on $\partial\mathscr{B}_0^x$. The equivalence of the above two inequalities can be seen by noting that $\Delta\dot{\boldsymbol{\sigma}} = \Delta\dot{\boldsymbol{\sigma}}^s$, $\Delta\dot{\mathbf{b}} = \Delta\dot{\mathbf{b}}^s$ for any $\dot{\boldsymbol{\sigma}}^c$, $\dot{\mathbf{b}}^c$.

The exclusion functional in (6.2.139) vanishes for an e-mode and the supporting e-configuration is primary if (6.2.139) holds for all admissible $\dot{\boldsymbol{\chi}}$ other than e-modes. By specializing (6.2.110) we see that e-modes also satisfy the variational equation

$$\tfrac{1}{2}\delta\int_{\mathscr{B}_0} \operatorname{tr}(\dot{\mathbf{S}}\dot{\mathbf{A}})\,\mathrm{d}V = \int_{\partial\mathscr{B}_0^\sigma} \dot{\boldsymbol{\sigma}}^s\cdot\delta\dot{\boldsymbol{\chi}}\,\mathrm{d}A + \int_{\mathscr{B}_0} \rho_0\dot{\mathbf{b}}^s\cdot\delta\dot{\boldsymbol{\chi}}\,\mathrm{d}V\cdot$$

for all admissible variations $\delta\dot{\boldsymbol{\chi}}$ satisfying $\delta\dot{\boldsymbol{\chi}} = \mathbf{0}$ on $\partial\mathscr{B}_0^x$. However, the exclusion functional itself is not stationary in an e-mode unless the deformation-sensitive data are self-adjoint† in accordance with the definition given in Section 6.2.4. For such data both the exclusion functional and its first variation vanish in an e-mode.

Problem 6.2.10 Show that with $\mathbf{b} = \mathbf{0}$ and uniform pressure P on $\partial\mathscr{B}_0^\sigma$ the exclusion condition (6.2.139) becomes

$$\int_{\mathscr{B}_0} \{\operatorname{tr}(\dot{\mathbf{S}}\dot{\mathbf{A}}) + JP(\operatorname{tr}\dot{\mathbf{A}}\mathbf{B}^{\mathrm{T}})^2 - JP\operatorname{tr}(\dot{\mathbf{A}}\mathbf{B}^{\mathrm{T}}\dot{\mathbf{A}}\mathbf{B}^{\mathrm{T}})\}\,\mathrm{d}V > 0, \tag{6.2.140}$$

where $J = \det\mathbf{A}$.

†In linear algebra an analogous problem is, in the notation of the footnote on p. 381, to solve $\mathbf{A}\mathbf{w} = \mathbf{0}$ subject to $(\mathbf{A}\mathbf{v})\cdot\mathbf{v} \geq 0$ for all $\mathbf{v}$ without the restriction to symmetric $\mathbf{A}$. The vector $\mathbf{u}$ making the quadratic form stationary satisfies $(\mathbf{A} + \mathbf{A}^{\mathrm{T}})\mathbf{u} = \mathbf{0}$, so that, in general, $\mathbf{u} \neq \mathbf{w}$.

Adapt the results and notations of Problem 6.2.8 to show that for a sphere of homogeneous isotropic elastic material under uniform hydrostatic *tension* T the exclusion condition reduces to

$$(\alpha-\beta+\lambda T)\operatorname{tr}(\mathbf{\Sigma}^2)+(\beta-\lambda T)(\operatorname{tr}\mathbf{\Sigma})^2>0.$$

Deduce that arbitrary isochoric e-modes are possible where $\alpha-\beta+\lambda T=0$, and pure dilatational modes where $\mathrm{d}T/\mathrm{d}\lambda=0$.

For the strain-energy function

$$W=\sum_{n=1}^{N}\mu_n(\alpha_1^{\alpha_n}+\lambda_2^{\alpha_n}+\lambda_3^{\alpha_n}-3-\alpha_n\ln J)/\alpha_n+\tfrac{1}{2}\kappa(J-1)^2,$$

where $\mu_n\alpha_n>0\,(n=1,2,\ldots,N)$ and $\kappa>0$, show that $\alpha-\beta+\lambda T>0$ for all $\lambda>0$ but that $\mathrm{d}T/\mathrm{d}\lambda=0$ is possible only under certain restrictions on α_n. Compare these results with the corresponding results for the dead-load data in Problem 6.2.8.

Problem 6.2.11 Show that for an incompressible material (6.2.140) is replaced by

$$\int_{\mathscr{B}_0}\operatorname{tr}\{\dot{\mathbf{S}}\dot{\mathbf{A}}+(p-P)\dot{\mathbf{A}}\mathbf{B}^{\mathrm{T}}\dot{\mathbf{A}}\mathbf{B}^{\mathrm{T}}\}\,\mathrm{d}V>0. \tag{6.2.141}$$

For the hydrostatic tension problem considered in Problem 6.2.10, deduce that the exclusion condition simplifies to $\mu\operatorname{tr}(\mathbf{\Sigma}^2)>0$ independently of T, where μ is the shear modulus defined by (6.1.88). Compare the results of Problem 6.2.9.

Remark For the problem of a sphere under tension, discontinuous solutions associated with the formation of cavities have been examined by Ball (1982). Such 'bifurcations' are not of the type considered here because they are not solutions of the linearized equations.

Problem 6.2.12 Show that when the deformation-sensitive data are self-adjoint the inhomogeneous data $\dot{\boldsymbol{\sigma}}^{\mathrm{c}},\dot{\mathbf{b}}^{\mathrm{c}}$ prescribed in an e-configuration are constrained by

$$\int_{\partial\mathscr{B}_0^{\sigma}}\dot{\boldsymbol{\sigma}}^{\mathrm{c}}\cdot\dot{\boldsymbol{\chi}}\,\mathrm{d}A+\int_{\mathscr{B}_0}\rho_0\dot{\mathbf{b}}^{\mathrm{c}}\cdot\dot{\boldsymbol{\chi}}\,\mathrm{d}V=0$$

for each e-mode $\dot{\boldsymbol{\chi}}$.

6.2.7 The strong ellipticity condition

With $\dot{\mathbf{A}}$ taken to be proportional to the rank-one tensor $\mathbf{m} \otimes \mathbf{N}$, where $\mathbf{m}$ is an Eulerian and $\mathbf{N}$ a Lagrangean vector, the inequality (6.2.133) specializes to

$$\operatorname{tr}\{[\mathscr{A}^1(\mathbf{m} \otimes \mathbf{N})](\mathbf{m} \otimes \mathbf{N})\} > 0. \tag{6.2.142}$$

If (6.2.142) holds for all $\mathbf{m} \otimes \mathbf{N} \neq \mathbf{0}$ it is called the *strong ellipticity condition.* In Cartesian components (6.2.142) takes the form

$$\mathscr{A}^1_{\alpha i \beta j} N_\alpha N_\beta m_i m_j > 0.$$

Since not every second-order tensor $\dot{\mathbf{A}}$ is of rank one it is clear that (6.2.142) does not in general imply (6.2.133). However, when the underlying deformation and the material are homogeneous (so that $\mathscr{A}^1$ is independent of $\mathbf{X}$) the strong ellipticity condition guarantees uniqueness of solution to the boundary-value problem of *place*, i.e. (6.2.142) implies (6.2.124) in these special circumstances, but not (6.2.133) pointwise. For further discussion of this point, together with references, see Truesdell and Noll (1965, Section 68).

The constitutive relation $\mathbf{S} = \mathbf{H}(\mathbf{A})$ is said to be *strongly elliptic* if $\mathscr{A}^1 \equiv \partial \mathbf{S}/\partial \mathbf{A}$ satisfies the strong ellipticity condition for all $\mathbf{A}$. In the case of a Green-elastic material the strain-energy function itself is said to be strongly elliptic. When the constitutive law is strongly elliptic the equilibrium equations (for either the global or the incremental problem) form a *strongly elliptic system* in accordance with the usual terminology of the theory of partial differential equations. Strictly, strong ellipticity requires only that the left-hand side of (6.2.142) be *non-zero* for all non-zero $\mathbf{m} \otimes \mathbf{N}$, but, since configurations where the left-hand side of (6.2.142) is negative for some $\mathbf{m} \otimes \mathbf{N}$ are unstable, we adopt (6.2.142) as the definition of strong ellipticity here.

On introduction of the second-order tensor $\mathbf{Q}(\mathbf{N})$, depending on $\mathbf{N}$, with Cartesian components defined by

$$Q_{ij}(\mathbf{N}) = \mathscr{A}^1_{\alpha i \beta j} N_\alpha N_\beta, \tag{6.2.143}$$

we see that (6.2.142) may be rewritten as

$$[\mathbf{Q}(\mathbf{N})\mathbf{m}] \cdot \mathbf{m} > 0. \tag{6.2.144}$$

This asserts that $\mathbf{Q}(\mathbf{N})$ is *positive definite* for each $\mathbf{N} \neq \mathbf{0}$. Because of its connection with the propagation of infinitesimal plane waves (to be discussed in Section 6.4) $\mathbf{Q}(\mathbf{N})$ is called the *acoustic tensor*, $\mathbf{N}$ then representing the direction of propagation of the wave. For a Green-elastic material $\mathbf{Q}(\mathbf{N})$ is symmetric.

The weaker inequality

$$[\mathbf{Q}(\mathbf{N})\mathbf{m}]\cdot\mathbf{m} \equiv \mathrm{tr}\{[\mathscr{A}^1(\mathbf{m}\otimes\mathbf{N})](\mathbf{m}\otimes\mathbf{N})\} \geq 0 \tag{6.2.145}$$

for all $\mathbf{m}\otimes\mathbf{N}$, known as the *Hadamard* (or *Legendre–Hadamard*) *condition*, follows from the incremental stability inequality (6.2.134). In fact, (6.2.145) necessarily follows from (6.2.131) provided that the class of admissible $\dot{\chi}$ in (6.2.131) is sufficiently large (for details see Truesdell and Noll, 1965, Section 68). On the other hand, when $\mathscr{A}^1$ is independent of $\mathbf{X}$ (6.2.145) ensures that (6.2.131) holds for the boundary-value problem of place, in parallel with the implication from (6.2.142) to (6.2.124).

Consider a path of deformation along which the strong ellipticity condition holds up to some critical configuration at which it fails in the sense that (6.2.145) then holds with equality for some $\mathbf{m}\otimes\mathbf{N}\neq\mathbf{0}$. In this critical configuration $\mathbf{Q}(\mathbf{N})$ is positive definite for all $\mathbf{N}\neq\mathbf{0}$ except for that (or those) $\mathbf{N}$ for which equality holds in (6.2.145). For each such $\mathbf{N}$ there is an $\mathbf{m}\neq\mathbf{0}$ such that

$$\mathbf{Q}(\mathbf{N})\mathbf{m} \equiv [\mathscr{A}^1(\mathbf{m}\otimes\mathbf{N})]^{\mathrm{T}}\mathbf{N} = \mathbf{0}, \tag{6.2.146}$$

the symmetry of $\mathbf{Q}(\mathbf{N})$ appropriate to Green-elasticity have been used. Thus, in configurations where strong ellipticity just fails in the above sense there exists an $\mathbf{N}$ such that $\mathbf{Q}(\mathbf{N})$ is singular, and therefore $\det\mathbf{Q}(\mathbf{N}) = 0$. Note, however, that it does *not* follow from (6.2.146) that $\mathscr{A}^1(\mathbf{m}\otimes\mathbf{N}) = \mathbf{0}$. This can easily be confirmed for isotropic materials since none of the e-modes (6.2.90) associated with the singularities of $\mathscr{A}^1$ are of rank one. Thus, $\mathscr{A}^1(\mathbf{m}\otimes\mathbf{N}) = \mathbf{0}$ implies $\mathbf{m}\otimes\mathbf{N} = \mathbf{0}$.

Since the left-hand side of (6.2.145) may be rearranged as a quadratic form in $\mathbf{N}$ we deduce that (6.2.146) is coupled with $[\mathscr{A}^1(\mathbf{m}\otimes\mathbf{N})]\mathbf{m} = \mathbf{0}$. For an isotropic material these two equations are equivalent, as may be deduced from the expression given in (6.2.150) below. Indeed, the consequences of the inequality (6.2.150) with regard to material properties are unaffected if the roles of $\mathbf{m}$ and $\mathbf{N}$ are reversed.

A physical interpretation of (6.2.146) is as follows. Suppose that an incremental deformation, given by $\dot{\chi} = \mathbf{m}f(\mathbf{N}\cdot\mathbf{X})$, is superposed on a homogeneous deformation in which (6.2.146) holds for some $\mathbf{m}\neq\mathbf{0}$, $\mathbf{N}\neq\mathbf{0}$ and (6.2.145) for all $\mathbf{m}\otimes\mathbf{N}$, where f is a suitably smooth function. Then the associated traction increment on a plane surface with reference normal $\mathbf{N}$ is calculated as

$$\dot{\mathbf{S}}^{\mathrm{T}}\mathbf{N} = \mathbf{Q}(\mathbf{N})\mathbf{m}f'(\mathbf{N}\cdot\mathbf{X}).$$

It follows from (6.2.146) that the nominal traction is stationary with respect to such an incremental deformation in the considered configuration. For a detailed discussion of this interpretation we refer to Hill (1979, Section 4).

With (6.2.146) in mind we record here that if $\mathbf{Q}(\mathbf{N})$ is merely non-singular, i.e.

$$\det \mathbf{Q}(\mathbf{N}) \neq 0, \tag{6.2.147}$$

and not necessarily positive definite then the equilibrium equations are said to be *elliptic*. Correspondingly, the constitutive law is elliptic in any configuration where (6.2.147) holds. Clearly, strong ellipticity implies ellipticity with $\det \mathbf{Q}(\mathbf{N}) > 0$, but the converse is not in general true (although for particular constitutive laws ellipticity and strong ellipticity may be equivalent).

Such ellipticity has been examined by Knowles and Sternberg (1975, 1977), who also studied the implications of loss of ellipticity ($\det \mathbf{Q}(\mathbf{N}) = 0$) with regard to the emergence of discontinuous solutions (near a crack tip, for example). A case in point is the shear-band solution touched on at the end of Section 6.2.5. For further references to this and related work see Knowles (1982). By contrast, certain types of singular solution are not precluded by strong ellipticity (Ball, 1982).

Of course, classification of the equilibrium equations into hyperbolic, parabolic and other categories according to the properties of $\mathscr{A}^1$ is also possible. Indeed for plane-strain problems a detailed classification has been given by Hill (1979) in relation to a wide class of incremental stress-deformation relations which includes elasticity as a special case.

For an isotropic material the components of $\mathscr{A}^1$ on principal axes are given by (6.2.85)–(6.2.87), so we deduce from (6.2.143) that the corresponding components of $\mathbf{Q}(\mathbf{N})$ are

$$Q_{ii}(\mathbf{N}) = \frac{\partial t_i^{(1)}}{\partial \lambda_i} N_i^2 + \left(\frac{\lambda_i t_i^{(1)} - \lambda_j t_j^{(1)}}{\lambda_i^2 - \lambda_j^2}\right) N_j^2 + \left(\frac{\lambda_i t_i^{(1)} - \lambda_k t_k^{(1)}}{\lambda_i^2 - \lambda_k^2}\right) N_k^2, \tag{6.2.148}$$

where (i, j, k) is a cyclic permutation of (1, 2, 3), and

$$Q_{ij}(\mathbf{N}) = \left(\frac{\partial t_i^{(1)}}{\partial \lambda_j} + \frac{\lambda_j t_i^{(1)} - \lambda_i t_j^{(1)}}{\lambda_i^2 - \lambda_j^2}\right) N_i N_j \quad i \neq j, \tag{6.2.149}$$

N_1, N_2, N_3 being the components of $\mathbf{N}$.

After a little algebra it is easy to see that (6.2.144) may be arranged as

$$[\mathbf{Q}(\mathbf{N})\mathbf{m}]\cdot\mathbf{m} \equiv \sum_{i,j=1}^{3} \frac{\partial t_i^{(1)}}{\partial \lambda_j} N_i N_j m_i m_j + \frac{1}{4}\sum_{i\neq j}\left\{\left(\frac{t_i^{(1)}+t_j^{(1)}}{\lambda_i+\lambda_j}\right)(N_i m_j - N_j m_i)^2 + \left(\frac{t_i^{(1)}-t_j^{(1)}}{\lambda_i-\lambda_j}\right)(N_i m_j + N_j m_i)^2\right\} > 0. \tag{6.2.150}$$

On reference to (6.2.101) we find confirmation that the exclusion condition implies strong ellipticity. On the other hand, no one of the three conditions in (6.2.101) is necessary for strong ellipticity in general.

For a Green-elastic material, necessary and sufficient conditions for strong ellipticity are

$$\begin{aligned} &Q_{ii}(\mathbf{N}) > 0 \qquad \text{some } i\in\{1,2,3\},\\ &Q_{ii}(\mathbf{N})Q_{jj}(\mathbf{N}) - Q_{ij}(\mathbf{N})^2 > 0 \qquad \text{some } j\neq i\in\{1,2,3\},\\ &\det \mathbf{Q}(\mathbf{N}) > 0 \end{aligned} \tag{6.2.151}$$

for all $\mathbf{N}\neq\mathbf{0}$. In view of (6.2.148) the first of these is seen immediately to be equivalent to

$$\frac{\partial t_i^{(1)}}{\partial \lambda_i} > 0 \qquad \text{some } i\in\{1,2,3\} \tag{6.2.152}$$

together with

$$\frac{\lambda_i t_i^{(1)} - \lambda_j t_j^{(1)}}{\lambda_i^2 - \lambda_j^2} > 0 \qquad \text{each } j\neq i\in\{1,2,3\}. \tag{6.2.153}$$

The second inequality in (6.2.151), when combined with (6.2.152) and (6.2.153) and reduced to a quadratic in N_1^2, N_2^2, yields

$$\frac{\partial t_i^{(1)}}{\partial \lambda_i}\frac{\partial t_j^{(1)}}{\partial \lambda_j} + \left(\frac{\lambda_i t_i^{(1)} - \lambda_j t_j^{(1)}}{\lambda_i^2 - \lambda_j^2}\right)^2 - \left(\frac{\partial t_i^{(1)}}{\partial \lambda_j} + \frac{\lambda_j t_i^{(1)} - \lambda_i t_j^{(1)}}{\lambda_i^2 - \lambda_j^2}\right)^2 + 2\left(\frac{\lambda_i t_i^{(1)} - \lambda_j t_j^{(1)}}{\lambda_i^2 - \lambda_j^2}\right)\left(\frac{\partial t_i^{(1)}}{\partial \lambda_i}\frac{\partial t_j^{(1)}}{\partial \lambda_j}\right)^{1/2} > 0 \quad j\neq i. \tag{6.2.154}$$

So far it has not been possible to obtain corresponding necessary and sufficient conditions on the material properties for the third inequality in

(6.2.151) when $N_i \neq 0$ for each $i \in \{1, 2, 3\}$.[†] Note, however, that when expanded as a cubic form in N_1^2, N_2^2, N_3^2, $\det \mathbf{Q}(\mathbf{N})$ may be written compactly as

$$\det \mathbf{Q}(\mathbf{N}) \equiv \sum_{i,j,k=1}^{3} Q_{ijk} N_i^2 N_j^2 N_k^2,$$

where

$$\begin{aligned}
Q_{iii} &= \alpha_{ii}\beta_{ij}\beta_{ik}, \\
Q_{iij} &= Q_{iji} = Q_{jii} = \tfrac{1}{3}\{\alpha_{ii}\beta_{ij}\beta_{jk} + \beta_{ik}[\alpha_{ii}\alpha_{jj} + \beta_{ij}^2 - (\alpha_{ij} + \gamma_{ij})^2]\}, \\
Q_{ijk} &= \tfrac{1}{6}\{\alpha_{11}\alpha_{22}\alpha_{33} + \alpha_{11}[\beta_{23}^2 - (\alpha_{23} + \gamma_{23})^2] \\
&\quad + \alpha_{22}[\beta_{13}^2 - (\alpha_{13} + \gamma_{13})^2] + \alpha_{33}[\beta_{12}^2 - (\alpha_{12} + \gamma_{12})^2] \\
&\quad + 2\beta_{12}\beta_{23}\beta_{31} + 2(\alpha_{12} + \gamma_{12})(\alpha_{23} + \gamma_{23})(\alpha_{31} + \gamma_{31})\}
\end{aligned}$$

for $i \neq j \neq k \neq i$ and

$$\begin{aligned}
\alpha_{ij} &= \alpha_{ji} = \frac{\partial t_i^{(1)}}{\partial \lambda_j}, \\
\beta_{ij} &= \beta_{ji} = \frac{\lambda_i t_i^{(1)} - \lambda_j t_j^{(1)}}{\lambda_i^2 - \lambda_j^2} \quad i \neq j, \\
\gamma_{ij} &= \gamma_{ji} = \frac{\lambda_j t_i^{(1)} - \lambda_i t_j^{(1)}}{\lambda_i^2 - \lambda_j^2} \quad i \neq j.
\end{aligned}$$

The inequalities (6.2.152) and (6.2.153) arise in connection with the propagation of waves in a principal direction of strain (Truesdell and Noll, 1965) but the appearance of (6.2.154) in the literature is relatively recent (Knowles and Sternberg, 1977; Sawyers and Rivlin, 1978; Hill, 1979). It seems to have been obtained first in the author's thesis (1970b) along with the above expansion for $\det \mathbf{Q}(\mathbf{N})$, but in different notation. For an incompressible material corresponding conditions were given by Sawyers and Rivlin (1977). The reader should confirm that for such a material the components of $\mathbf{Q}(\mathbf{N})$ are still given by (6.2.148) and (6.2.149) but with the arbitrary hydrostatic part of the stress omitted and $\mathbf{m}$ and $\mathbf{N}$ subject to $\mathbf{m} \cdot (\mathbf{B}\mathbf{N}) = 0$, where $\mathbf{B}$ is the inverse deformation gradient.

Let $\mathbf{Q}_0(\mathbf{n})$ be defined by

$$[\mathbf{Q}_0(\mathbf{n})\mathbf{m}] \cdot \mathbf{m} = \operatorname{tr}\{[\mathscr{A}_0^1(\mathbf{m} \otimes \mathbf{n})](\mathbf{m} \otimes \mathbf{n})\}, \tag{6.2.155}$$

[†] But see the recent paper by Zee and Sternberg (1982).

where $\mathscr{A}_0^1$ is the tensor of instantaneous moduli introduced in Section 6.1.3. By taking $\mathbf{n} = \mathbf{BN}$ and making use of (6.1.29) it is easy to deduce that

$$\mathbf{Q}_0(\mathbf{n}) = J^{-1}\mathbf{Q}(\mathbf{N}). \tag{6.2.156}$$

Clearly, $\mathbf{Q}_0(\mathbf{n})$ may be interpreted as the acoustic tensor for the situation in which the reference and current configurations coincide.

When both the reference and current configurations coincide with the natural configuration for an isotropic material with Lamé moduli λ, μ it follows, on use of (6.1.79), that

$$[\mathbf{Q}_0(\mathbf{n})\mathbf{m}]\cdot\mathbf{m} = \mu(\mathbf{n}\cdot\mathbf{n})(\mathbf{m}\cdot\mathbf{m}) + (\lambda+\mu)(\mathbf{m}\cdot\mathbf{n})^2$$

in that configuration. Necessary and sufficient conditions for the strong ellipticity condition to hold there are therefore simply

$$\mu > 0, \qquad \lambda + 2\mu > 0. \tag{6.2.157}$$

The corresponding result for an incompressible material is deduced from the above by imposing the constraint $\mathbf{m}\cdot\mathbf{n} = 0$.

The inequalities (6.2.157) are compatible with the classical inequalities (6.1.84). However, although (6.1.84) implies (6.2.157) the latter does not entail $\kappa \equiv \lambda + \frac{2}{3}\mu > 0$. Thus, the strong ellipticity condition does not ensure that the bulk modulus is positive. This fact is reinforced by the results of Section 5.2.2, for, by use of (6.2.150), the strain-energy function (5.2.19) with (5.2.22) is seen to be strongly elliptic if and only if

$$F''(I_1) > 0, \quad F'(I_1) > \nu\lambda_i \quad i = 1, 2, 3. \tag{6.2.158}$$

In particular, the $F(I_1)$ given by (5.2.40) yields $\kappa = 0$ but also satisfies the first inequality in (6.2.158) for all $I_1 > 0$ and the second for a wide range of deformations including all pure dilatations. The anomalous results revealed by this strain-energy function can be attributed to the vanishing of the bulk modulus.

On the one hand the strong ellipticity condition is not strong enough to ensure 'physically reasonable response' in the sense of $\kappa > 0$, while on the other hand it is too strong to permit discontinuous solutions of the 'shear band' type. Its status with regard to the conditions it imposes on material properties is therefore uncertain. The situation is confused further by consideration of elastic crystals such as those discussed in Section 4.2.5. Ericksen (1977, Section IVF) has shown that the Hadamard condition is incompatible with certain shear deformations of such crystals in the sense that $[\mathbf{Q}(\mathbf{N})\mathbf{m}]\cdot\mathbf{m} < 0$ must hold for some $\mathbf{m}\otimes\mathbf{N}$ at some points of the material if these deformations are to be admissible. The situation is somewhat clearer in relation to the solution of boundary-value problems because of the intimate connection of the strong ellipticity condition with uniqueness and stability criteria. Strong ellipticity also plays a central role in existence theory although, in general, it is not on its own sufficient to establish existence of solution to a given boundary-value problem. In fact, inequalities which imply the Hadamard condition (rather than strong ellipticity) have been used to prove existence of solution for a wide class of boundary value problems (Ball, 1977).

Problem 6.2.13 Show that the components of $\mathbf{Q}_0(\mathbf{n})$ on the Eulerian principal axes have the forms

$$Q_{0ii} = \lambda_i \frac{\partial t_i}{\partial \lambda_i} n_i^2 + \frac{t_i - t_j}{\lambda_i^2 - \lambda_j^2} \lambda_j^2 n_j^2 + \frac{t_i - t_k}{\lambda_i^2 - \lambda_k^2} \lambda_k^2 n_k^2,$$

$$Q_{0ij} = \left(\lambda_j \frac{\partial t_i}{\partial \lambda_j} + \frac{t_i - t_j}{\lambda_i^2 - \lambda_j^2} \lambda_i^2 \right) n_i n_j,$$

where $i \neq j \neq k \neq i$, when expressed in terms of the principal Cauchy stresses.

Evaluate $[\mathbf{Q}_0(\mathbf{n})\mathbf{m}]\cdot\mathbf{m}$ for the strain-energy function

$$W = \tfrac{1}{2}\alpha I_1 + \tfrac{1}{2}\beta I_2 + f(I_3),$$

where

$$I_1 = \lambda_1^2 + \lambda_2^2 + \lambda_3^2,$$

$$I_2 = \lambda_2^2\lambda_3^2 + \lambda_3^2\lambda_1^2 + \lambda_1^2\lambda_2^2,$$

$$I_3 = \lambda_1^2\lambda_2^2\lambda_3^2,$$

and obtain necessary and sufficient conditions for strong ellipticity of W.

Problem 6.2.14 Suppose

$$\mathbf{Q}_0(\mathbf{n})\mathbf{m} = \alpha\mathbf{l} + \beta\mathbf{m} + \gamma\mathbf{n},$$

where $\{\mathbf{l}, \mathbf{m}, \mathbf{n}\}$ is orthonormal, and deduce that this may be rewritten as

$$[\mathbf{Q}_0(\mathbf{n}) - \mathbf{n} \otimes \mathbf{Q}_0(\mathbf{n})\mathbf{n} - \mathbf{l} \otimes \mathbf{Q}_0(\mathbf{n})\mathbf{l}]\mathbf{m} = \beta\mathbf{m},$$

where $\beta = [\mathbf{Q}_0(\mathbf{n})\mathbf{m}]\cdot\mathbf{m}$.

If $\alpha = 0$ prove that

$$\beta^2 - \beta \operatorname{tr} \mathbf{Q}_0^*(\mathbf{n}) + \tfrac{1}{2}\{[\operatorname{tr} \mathbf{Q}_0^*(\mathbf{n})]^2 - \operatorname{tr}[\mathbf{Q}_0^*(\mathbf{n})]^2\} = 0,$$

where $\mathbf{Q}_0^*(\mathbf{n}) = \mathbf{Q}_0(\mathbf{n}) - \mathbf{n} \otimes \mathbf{Q}_0(\mathbf{n})\mathbf{n}$, and deduce that $\beta > 0$ if and only if

$$\begin{aligned}
&\operatorname{tr} \mathbf{Q}_0(\mathbf{n}) - [\mathbf{Q}_0(\mathbf{n})\mathbf{n}] \cdot \mathbf{n} > 0, \\
&2 \operatorname{tr} [\mathbf{Q}_0(\mathbf{n})]^2 - 4[\mathbf{Q}_0(\mathbf{n})\mathbf{n}] \cdot [\mathbf{Q}_0(\mathbf{n})\mathbf{n}] + \{[\mathbf{Q}_0(\mathbf{n})\mathbf{n}] \cdot \mathbf{n}\}^2 \\
&\quad > [\operatorname{tr} \mathbf{Q}_0(\mathbf{n})]^2 - 2[\operatorname{tr} \mathbf{Q}_0(\mathbf{n})][\mathbf{Q}_0(\mathbf{n})\mathbf{n}] \cdot \mathbf{n} + \{[\mathbf{Q}_0(\mathbf{n})\mathbf{n}] \cdot \mathbf{n}\}^2 \\
&\quad > \operatorname{tr} [\mathbf{Q}_0(\mathbf{n})]^2 - 2[\mathbf{Q}_0(\mathbf{n})\mathbf{n}] \cdot [\mathbf{Q}_0(\mathbf{n})\mathbf{n}] + \{[\mathbf{Q}_0(\mathbf{n})\mathbf{n}] \cdot \mathbf{n}\}^2
\end{aligned}$$

for arbitrary unit vectors $\mathbf{n}$.

Calculate β for the strain-energy function

$$W = \tfrac{1}{2}\mu_1(\lambda_1^2 + \lambda_2^2 + \lambda_3^2 - 3) - \tfrac{1}{2}\mu_2(\lambda_1^{-2} + \lambda_2^{-2} + \lambda_3^{-2} - 3)$$

subject to $\lambda_1\lambda_2\lambda_3 = 1$. If $\mu_1 > 0$ and $\mu_2 \leq 0$ prove that $\beta > 0$. Are these conditions necessary for $\beta > 0$?

6.2.8 Constitutive branching and constitutive inequalities

The inequality

$$\operatorname{tr}(\dot{\mathbf{S}}\dot{\mathbf{A}}) \equiv \operatorname{tr}\{(\mathscr{A}^1\dot{\mathbf{A}})\dot{\mathbf{A}}\} > 0 \tag{6.2.159}$$

for all $\dot{\mathbf{A}} \neq \mathbf{0}$ arose in Section 6.2.3 in connection with considerations of uniqueness and stability for the incremental mixed boundary-value problem with prescribed tractions of the dead-load type. In particular, we established that (6.2.159) cannot hold in every configuration, but, for configurations where it does hold uniqueness and stability are guaranteed.

Of special significance are primary e-configurations, in which

$$\operatorname{tr}\{(\mathscr{A}^1\dot{\mathbf{A}})\dot{\mathbf{A}}\} \geq 0 \tag{6.2.160}$$

for all $\dot{\mathbf{A}}$ with equality holding for some $\dot{\mathbf{A}} \neq \mathbf{0}$. Such configurations, reached along a stable path of deformation, signal failure of incremental uniqueness. In Section 6.2.3 we referred to such configurations as being neutrally stable. In this connection we recall from (6.2.71) that

$$\tfrac{1}{2}\int_{\mathscr{B}_0} \operatorname{tr}\{(\mathscr{A}^1\dot{\mathbf{A}})\dot{\mathbf{A}}\}\,\mathrm{d}V$$

represents the difference between the total energy functionals for the configurations $\chi + \dot{\chi}$ and χ, *correct to the second order in* $\dot{\mathbf{A}}$. For a proper assessment of the stability of primary e-configurations, however, it is necessary to include higher-order terms in $\dot{\mathbf{A}}$. To this end we note that the integrand in the above functional should be corrected to

$$\dot{W} - \operatorname{tr}(\mathbf{S}\dot{\mathbf{A}}) = \tfrac{1}{2}\operatorname{tr}\{(\mathscr{A}^1\dot{\mathbf{A}})\dot{\mathbf{A}}\} + \tfrac{1}{6}\operatorname{tr}\{\mathscr{A}^2[\dot{\mathbf{A}}, \dot{\mathbf{A}}]\dot{\mathbf{A}}\} + \cdots \quad (6.2.161)$$

The corresponding expansion for $\dot{\mathbf{S}}$ is

$$\dot{\mathbf{S}} = \mathscr{A}^1\dot{\mathbf{A}} + \tfrac{1}{2}\mathscr{A}^2[\dot{\mathbf{A}}, \dot{\mathbf{A}}] + \cdots \quad (6.2.162)$$

In what follows we examine the stability inequality $\dot{W} - \operatorname{tr}(\mathbf{S}\dot{\mathbf{A}}) > 0$ in respect of (6.2.161) for sufficiently small $\dot{\mathbf{A}}$ in the neighbourhood of an e-configuration, together with the inversion of (6.2.162).

Geometrical considerations We recall from Section 6.2.5 that e-configurations are characterized by the singularities of $\mathscr{A}^1$. Thus, e-surfaces in nine-dimensional $\mathbf{A}$-space are described by the equation

$$\det \mathscr{A}^1 = 0, \quad (6.2.163)$$

the specialization of which for isotropic materials was detailed in Sections 6.2.3 and 6.2.5. Let $\dot{\mathbf{A}}_e$ denote an e-mode† so that, on an e-surface,

$$\mathscr{A}^1\dot{\mathbf{A}}_e = \mathbf{0}. \quad (6.2.164)$$

The local geometry of an e-surface is determined from knowledge of the higher-order moduli $\mathscr{A}^2, \mathscr{A}^3, \ldots$ on the surface. For our purposes, however, it suffices to obtain a representation for the normal to a primary e-surface. Moreover, we confine attention to the primary e-surface bounding the stable region of $\mathbf{A}$-space where (6.2.159) holds and which contains the natural configuration $\mathbf{A} = \mathbf{I}$. This region is depicted in Fig. 6.4.

Let $\delta_s\mathbf{A}$ denote the derivative of $\mathbf{A}$ in an e-surface. Then, application of the operator δ_s to equation (6.2.164) yields

$$\mathscr{A}^2[\delta_s\mathbf{A}, \dot{\mathbf{A}}_e] + \mathscr{A}^1\,\delta_s\dot{\mathbf{A}}_e = \mathbf{0}, \quad (6.2.165)$$

it being assumed that $\dot{\mathbf{A}}_e$ varies smoothly over the e-surface. Taking the scalar product (i.e. the trace) of (6.2.165) with $\dot{\mathbf{A}}_e$ and making use of (6.2.164), we obtain

$$\operatorname{tr}\{\mathscr{A}^2[\dot{\mathbf{A}}_e, \dot{\mathbf{A}}_e]\delta_s\mathbf{A}\} = 0 \quad (6.2.166)$$

† Strictly, $\dot{\mathbf{A}}_e$ is the gradient of an e-mode, but it is convenient to adopt the more concise terminology here.

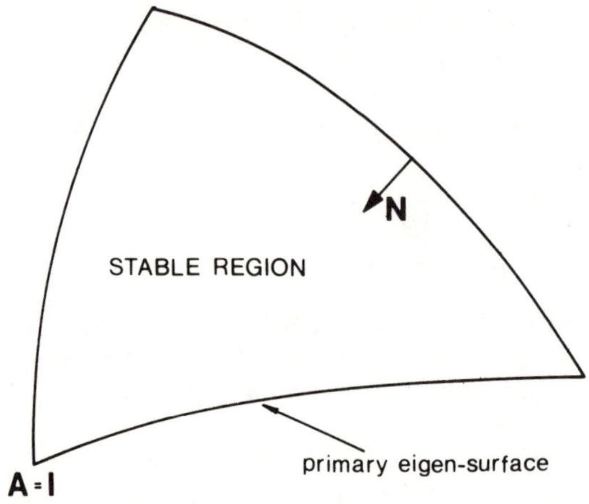

Fig. 6.4 Connected stable region in $\mathbf{A}$-space bounded by primary e-surfaces with inward normal $\mathbf{N}$.

for arbitrary tangents $\delta_s \mathbf{A}$ to the e-surface. We deduce that the transpose of $\mathscr{A}^2[\dot{\mathbf{A}}_e, \dot{\mathbf{A}}_e]$, regarded as a vector in $\mathbf{A}$-space, is normal to the e-surface and, accordingly, we introduce the notation

$$\mathbf{N}^{\mathrm{T}} = \mathscr{A}^2[\dot{\mathbf{A}}_e, \dot{\mathbf{A}}_e]. \tag{6.2.167}$$

In fact, $\mathbf{N}$ is the *inward* normal to the considered primary e-surface. This can be seen by considering an arbitrary increment $\dot{\mathbf{A}}$ pointing *into* the stable region from a point on the surface and observing that

$$\mathrm{tr}(\mathbf{N}^{\mathrm{T}}\dot{\mathbf{A}}) = \mathrm{tr}\{\mathscr{A}^2[\dot{\mathbf{A}}_e, \dot{\mathbf{A}}_e]\dot{\mathbf{A}}\} = \mathrm{tr}\{[(\mathscr{A}^1 + \mathscr{A}^2[\cdot, \dot{\mathbf{A}}])\dot{\mathbf{A}}_e]\dot{\mathbf{A}}_e\}.$$

Since $\mathscr{A}^1 + \mathscr{A}^2[\cdot, \dot{\mathbf{A}}]$, being $\mathscr{A}^1$ evaluated at $\mathbf{A} + \dot{\mathbf{A}}$ (just inside the stable region), is positive definite, it follows that $\mathrm{tr}(\mathbf{N}^{\mathrm{T}}\dot{\mathbf{A}}) > 0$.

The constitutive relation $\mathbf{S} = \mathbf{H}(\mathbf{A})$ maps an e-surface in $\mathbf{A}$-space onto a surface in $\mathbf{S}$-space. Following the terminology of Hill (1982), we call a surface of this kind a *trace surface*.

Consider an arbitrary $\dot{\mathbf{A}}$ orthogonal to the e-mode $\dot{\mathbf{A}}_e$, so that

$$\mathrm{tr}(\dot{\mathbf{A}}^{\mathrm{T}}\dot{\mathbf{A}}_e) = 0. \tag{6.2.168}$$

Then, in view of (6.2.164), each line element $\dot{\mathbf{A}} + \lambda\dot{\mathbf{A}}_e$ in $\mathbf{A}$-space, where λ is an arbitrary scalar, maps onto $\dot{\mathbf{S}} = \mathscr{A}^1\dot{\mathbf{A}}$ in the linear approximation. Moreover,

$$\mathrm{tr}(\dot{\mathbf{S}}\dot{\mathbf{A}}_e) = 0, \tag{6.2.169}$$

so that when $\dot{\mathbf{A}}$ is tangential to the e-surface and $\dot{\mathbf{S}}$ is therefore tangential to the trace surface we deduce that $\dot{\mathbf{A}}_e^T$, regarded as a vector in $\mathbf{S}$-space, is normal to the trace surface provided $\dot{\mathbf{A}}_e$ is *not* tangential to the e-surface. The restriction of $\mathbf{S} = \mathbf{H}(\mathbf{A})$ to an e-surface is then locally one-to-one and the trace surface is an eight-dimensional manifold. Exceptionally, when

$$\mathrm{tr}(\mathbf{N}^T\dot{\mathbf{A}}_e) = 0, \tag{6.2.170}$$

the constraints (6.2.168) and (6.2.170) reduce the trace of the e-surface to a seven-dimensional manifold.

In either case, by (6.2.162) and (6.2.164), we have

$$\mathrm{tr}(\dot{\mathbf{S}}\dot{\mathbf{A}}_e) = \tfrac{1}{2}\mathrm{tr}\{\mathscr{A}^2[\dot{\mathbf{A}}, \dot{\mathbf{A}}]\dot{\mathbf{A}}_e\} + \cdots, \tag{6.2.171}$$

which, being second order in $\dot{\mathbf{A}}$, confirms the approximation (6.2.169). The normality of $\dot{\mathbf{A}}_e^T$ to the trace surface is therefore correct only to the first order in $\dot{\mathbf{A}}$. This normal is unique when $\dot{\mathbf{A}}_e$ is not tangential to the e-surface, but if (6.2.170) holds there are two independent normals to the degenerate trace surface, as we now see.

The equations

$$\mathscr{A}^1\mathbf{M} = \mathbf{N}, \qquad \mathrm{tr}(\mathbf{M}^T\dot{\mathbf{A}}_e) = 0 \tag{6.2.172}$$

define a unique vector $\mathbf{M}$ orthogonal to $\dot{\mathbf{A}}_e$ in $\mathbf{A}$-space. It follows from (6.2.162) and the above definition that

$$\mathrm{tr}(\dot{\mathbf{S}}\mathbf{M}) = \tfrac{1}{2}\mathrm{tr}\{\mathscr{A}^2[\dot{\mathbf{A}}, \dot{\mathbf{A}}]\mathbf{M}\} + \cdots \tag{6.2.173}$$

for any $\dot{\mathbf{A}}$ orthogonal to $\mathbf{N}$, and therefore $\mathbf{M}^T$, to the same approximation as is $\dot{\mathbf{A}}_e^T$, is also normal to the trace manifold.

Constitutive branching In Section 6.2.3 we examined the non-uniqueness of the inverse of the incremental constitutive relation $\dot{\mathbf{S}} = \mathscr{A}^1\dot{\mathbf{A}}$, specifically in relation to isotropic materials. More generally, for a given $\dot{\mathbf{S}}$, the equations

$$\dot{\mathbf{S}} = \mathscr{A}^1\dot{\mathbf{A}}, \qquad \mathrm{tr}(\dot{\mathbf{A}}^T\dot{\mathbf{A}}_e) = 0 \tag{6.2.174}$$

define $\dot{\mathbf{A}}$ uniquely at points of the e-surface for which there is only one independent e-mode $\dot{\mathbf{A}}_e$. This analysis, of course, is valid only in the linear approximation, and, consequently, an $\dot{\mathbf{S}}$ prescribed in an e-configuration is constrained by (6.2.169). Subject to this constraint and (6.2.174) the linearized constitutive relation associates with a given $\dot{\mathbf{S}}$ the incremental deformation gradient $\dot{\mathbf{A}} + \lambda\dot{\mathbf{A}}_e$ for arbitrary λ, as we indicated following equation (6.2.168).

We now investigate how this analysis is modified by the inclusion of higher-order terms. Suppose that the material is subject to an arbitrary, smooth path of loading described by means of a single monotonically-increasing parameter, $\tau(\geq 0)$ say. In particular, on a path leading from a point on the trace surface we may write

$$\dot{\mathbf{S}} = \tau\dot{\mathbf{S}}^{(1)} + \tfrac{1}{2}\tau^2\dot{\mathbf{S}}^{(2)} + \cdots \tag{6.2.175}$$

for sufficiently small τ, where

$$\dot{\mathbf{S}}^{(n)} = \frac{\mathrm{d}^n}{\mathrm{d}\tau^n}\dot{\mathbf{S}}(\tau = 0) \qquad n = 1, 2, \ldots .$$

Correspondingly

$$\dot{\mathbf{A}} = \tau\dot{\mathbf{A}}^{(1)} + \tfrac{1}{2}\tau^2\dot{\mathbf{A}}^{(2)} + \cdots, \tag{6.2.176}$$

with

$$\dot{\mathbf{A}}^{(n)} = \frac{\mathrm{d}^n}{\mathrm{d}\tau^n}\dot{\mathbf{A}}(\tau = 0) \qquad n = 1, 2, \ldots .$$

After substitution of (6.2.175) and (6.2.176) into (6.2.162) equating coefficients of τ, $\tau^2, \ldots$ yields

$$\left.\begin{aligned} \dot{\mathbf{S}}^{(1)} &= \mathscr{A}^1\dot{\mathbf{A}}^{(1)}, \\ \dot{\mathbf{S}}^{(2)} &= \mathscr{A}^1\dot{\mathbf{A}}^{(2)} + \mathscr{A}^2[\dot{\mathbf{A}}^{(1)}, \dot{\mathbf{A}}^{(1)}], \\ \dot{\mathbf{S}}^{(3)} &= \mathscr{A}^1\dot{\mathbf{A}}^{(3)} + 3\mathscr{A}^2[\dot{\mathbf{A}}^{(1)}, \dot{\mathbf{A}}^{(2)}] \\ &\quad + \mathscr{A}^3[\dot{\mathbf{A}}^{(1)}, \dot{\mathbf{A}}^{(1)}, \dot{\mathbf{A}}^{(1)}], \end{aligned}\right\} \tag{6.2.177}$$

and similarly for higher-order terms; note that

$$\mathrm{tr}(\dot{\mathbf{S}}^{(1)}\dot{\mathbf{A}}_{\mathrm{e}}) = 0. \tag{6.2.178}$$

In what follows we consider those points of the primary e-surface for which there exists only one independent e-mode, $\dot{\mathbf{A}}_{\mathrm{e}}$, and we treat separately the cases $\mathrm{tr}(\mathbf{N}^{\mathrm{T}}\dot{\mathbf{A}}_{\mathrm{e}}) \neq 0$ and $\mathrm{tr}(\mathbf{N}^{\mathrm{T}}\dot{\mathbf{A}}_{\mathrm{e}}) = 0$, where $\mathbf{N}$ is given by (6.2.167). The analysis below needs modification for points at which there exist more than one e-mode, but we do not include this generalization here.

Case 1: $\mathrm{tr}(\mathbf{N}^{\mathrm{T}}\dot{\mathbf{A}}_{\mathrm{e}}) \neq 0$

First we note that by taking $\dot{\mathbf{A}}^{(1)} = \pm\dot{\mathbf{A}}_{\mathrm{e}}$ in (6.2.161) we obtain

$$\dot{W} - \mathrm{tr}(\mathbf{S}\dot{\mathbf{A}}) = \pm\tfrac{1}{6}\mathrm{tr}\{\mathscr{A}^2[\dot{\mathbf{A}}_{\mathrm{e}}, \dot{\mathbf{A}}_{\mathrm{e}}]\dot{\mathbf{A}}_{\mathrm{e}}\}\tau^3 + \cdots, \tag{6.2.179}$$

remembering that $\dot{\mathbf{A}}_e$ contains an arbitrary scalar multiplier. For sufficiently small τ the right-hand side is negative for one of the two signs since, by assumption, we have

$$\mathrm{tr}\{\mathscr{A}^2[\dot{\mathbf{A}}_e, \dot{\mathbf{A}}_e]\dot{\mathbf{A}}_e\} \neq 0. \tag{6.2.180}$$

This means that primary e-configurations of the considered type, although neutrally stable when assessed to order τ^2, must be regarded as *unstable* when terms of order τ^3 are admitted into the classical stability criterion.

We now consider incremental deformation gradient paths whose initial tangent $\dot{\mathbf{A}}^{(1)}$ is given by

$$\dot{\mathbf{A}}^{(1)} = \dot{\mathbf{A}}_0^{(1)} + \lambda \dot{\mathbf{A}}_e \tag{6.2.181}$$

with $\dot{\mathbf{A}}_0^{(1)} \neq \mathbf{0}$ satisfying

$$\mathrm{tr}(\dot{\mathbf{A}}_e^{\mathrm{T}} \dot{\mathbf{A}}_0^{(1)}) = 0, \tag{6.2.182}$$

λ being arbitrary. It follows that

$$\dot{\mathbf{S}}^{(1)} = \mathscr{A}^1 \dot{\mathbf{A}}_0^{(1)} \tag{6.2.183}$$

subject to (6.2.178). Thus, the loading paths corresponding to (6.2.181) have initial tangent $\dot{\mathbf{S}}^{(1)}$ independent of λ, and $\mathscr{A}^1$, regarded as a linear mapping locally, is one-to-one when restricted to the sub-space† locally orthogonal to $\dot{\mathbf{A}}_e$.

From the second equation in (6.2.177) coupled with (6.2.181) we obtain

$$\begin{aligned}\dot{\mathbf{S}}^{(2)} = \mathscr{A}^1 \dot{\mathbf{A}}^{(2)} + \mathscr{A}^2[\dot{\mathbf{A}}_0^{(1)}, \dot{\mathbf{A}}_0^{(1)}] + 2\lambda \mathscr{A}^2[\dot{\mathbf{A}}_0^{(1)}, \dot{\mathbf{A}}_e] \\ + \lambda^2 \mathscr{A}^2[\dot{\mathbf{A}}_e, \dot{\mathbf{A}}_e],\end{aligned} \tag{6.2.184}$$

and hence, by use of (6.2.164) and (6.2.167),

$$\begin{aligned}\mathrm{tr}(\dot{\mathbf{S}}^{(2)}\dot{\mathbf{A}}_e) = \lambda^2 \mathrm{tr}(\mathbf{N}^{\mathrm{T}}\dot{\mathbf{A}}_e) + 2\lambda\, \mathrm{tr}(\mathbf{N}\dot{\mathbf{A}}_0^{(1)}) \\ + \mathrm{tr}\{\mathscr{A}^2[\dot{\mathbf{A}}_0^{(1)}, \dot{\mathbf{A}}_0^{(1)}]\dot{\mathbf{A}}_e\}.\end{aligned} \tag{6.2.185}$$

Since the coefficients on the right-hand side are known from the first-order equations, (6.2.185) provides a quadratic equation from which two values of λ can be obtained when $\dot{\mathbf{S}}^{(2)}$ is specified. These values of λ are real and distinct

†Strictly, the domain of $\mathscr{A}^1$ is, in the terminology alluded to in Section 2.2.1, the *tangent space* at the point of **A**-space in question, and the restricted domain is the orthogonal complement of $\dot{\mathbf{A}}_e$ in that space.

provided

$$\operatorname{tr}(\dot{\mathbf{S}}^{(2)}\dot{\mathbf{A}}_e)\operatorname{tr}(\mathbf{N}^T\dot{\mathbf{A}}_e) > \operatorname{tr}\{\mathscr{A}^2[\dot{\mathbf{A}}_0^{(1)}, \dot{\mathbf{A}}_0^{(1)}]\dot{\mathbf{A}}_e\}\operatorname{tr}(\mathbf{N}^T\dot{\mathbf{A}}_e) - \{\operatorname{tr}(\mathbf{N}^T\dot{\mathbf{A}}_0^{(1)})\}^2. \tag{6.2.186}$$

Thus, in addition to the constraint (6.2.178) on $\dot{\mathbf{S}}^{(1)}$, we now have the constraint (6.2.186) on $\dot{\mathbf{S}}^{(2)}$. If $\dot{\mathbf{A}}^{(1)}$, $\dot{\mathbf{A}}^{(2)}$ are prescribed then $\dot{\mathbf{S}}^{(1)}$, $\dot{\mathbf{S}}^{(2)}$ are determined from (6.2.183) and (6.2.184) respectively, but, on the other hand, prescribed $\dot{\mathbf{S}}^{(1)}$, $\dot{\mathbf{S}}^{(2)}$ must be consistent with these constraints for $\dot{\mathbf{A}}^{(1)}$, $\dot{\mathbf{A}}^{(2)}$ to be found. Stresses $\dot{\mathbf{S}}^{(1)}$, $\dot{\mathbf{S}}^{(2)}$ satisfying (6.2.178) and (6.2.186) in a primary e-configuration are said to be *constitutively admissible* or *supportable.* Those violating either constraint are not supportable.

Clearly, when supportable $\dot{\mathbf{S}}^{(1)}$ is prescribed $\dot{\mathbf{A}}_0^{(1)}$ orthogonal to $\dot{\mathbf{A}}_e$ is uniquely determined from (6.2.183), but λ in (6.2.181) is arbitrary. When, in addition, supportable $\dot{\mathbf{S}}^{(2)}$ is specified λ is fixed by $\operatorname{tr}(\dot{\mathbf{S}}^{(2)}\dot{\mathbf{A}}_e)$. And since (6.2.185) yields two possible λ's the prescribed loading path is associated with two distinct deformation paths emanating from the point of the e-surface considered. Such local non-unique inversion of the constitutive law is called *constitutive branching.*

Once λ is found, that part of $\dot{\mathbf{A}}^{(2)}$ orthogonal to $\dot{\mathbf{A}}_e$ is, for each λ, uniquely determined by (6.2.184), and we write

$$\dot{\mathbf{A}}^{(2)} = \dot{\mathbf{A}}_0^{(2)} + \mu^{(2)}\dot{\mathbf{A}}_e, \tag{6.2.187}$$

where $\mu^{(2)}$ is arbitrary. Since

$$\begin{aligned}\operatorname{tr}(\dot{\mathbf{S}}^{(3)}\dot{\mathbf{A}}_e) = 3\operatorname{tr}\{\mathscr{A}^2[\dot{\mathbf{A}}_0^{(1)}, \dot{\mathbf{A}}_0^{(2)}]\dot{\mathbf{A}}_e\} + 3\lambda\operatorname{tr}(\mathbf{N}^T\dot{\mathbf{A}}_0^{(2)}) \\ + 3\mu^{(2)}\operatorname{tr}(\mathbf{N}^T\dot{\mathbf{A}}^{(1)}) \\ + \operatorname{tr}\{\mathscr{A}^3[\dot{\mathbf{A}}^{(1)}, \dot{\mathbf{A}}^{(1)}, \dot{\mathbf{A}}^{(1)}]\dot{\mathbf{A}}_e\},\end{aligned} \tag{6.2.188}$$

$\mu^{(2)}$ is determined uniquely for each λ when $\dot{\mathbf{S}}^{(3)}$ is prescribed provided $\operatorname{tr}(\mathbf{N}^T\dot{\mathbf{A}}^{(1)}) \neq 0$. (We do not consider details of the situation in which this proviso is not met.) This procedure is easily extended to higher-order terms by writing

$$\dot{\mathbf{A}}^{(n)} = \dot{\mathbf{A}}_0^{(n)} + \mu^{(n)}\dot{\mathbf{A}}_e \qquad n \geq 2,$$

where $\dot{\mathbf{A}}_0^{(n)}$ is orthogonal to $\dot{\mathbf{A}}_e$. With $\dot{\mathbf{A}}_0^{(n)}$ determined by prescribed $\dot{\mathbf{S}}^{(n)}$, $\mu^{(n)}$ is uniquely determined on each deformation path since it occurs linearly in $\operatorname{tr}(\dot{\mathbf{S}}^{(n+1)}\dot{\mathbf{A}}_e)$.

From (6.2.181) and (6.2.185) it is easy to show that the mean of the initial tangents to the two deformation paths is

$$\dot{\mathbf{A}}_0^{(1)} - \{\operatorname{tr}(\mathbf{N}^T\dot{\mathbf{A}}_0^{(1)})/\operatorname{tr}(\mathbf{N}^T\dot{\mathbf{A}}_e)\}\dot{\mathbf{A}}_e$$

and this, being orthogonal to $\mathbf{N}$, is tangential to the e-surface. It follows that on a primary e-surface one path heads into, and the other out of, the stable region. This is to be expected, of course, if the stable region is convex since (6.2.159) then implies that the stress-deformation relation is globally one-to-one on this region.

Finally, we consider the case in which $\dot{\mathbf{A}}_0^{(1)} = \mathbf{0}$ and we take

$$\dot{\mathbf{A}}^{(1)} = \pm \dot{\mathbf{A}}_e \tag{6.2.189}$$

with $\operatorname{tr}(\mathbf{N}^T\dot{\mathbf{A}}_e) > 0$ and the magnitude of $\dot{\mathbf{A}}_e$ arbitrary. Then $\dot{\mathbf{S}}^{(1)} = \mathbf{0}$,

$$\dot{\mathbf{S}}^{(2)} = \mathscr{A}^1\dot{\mathbf{A}}^{(2)} + \mathbf{N} \tag{6.2.190}$$

and

$$\operatorname{tr}(\dot{\mathbf{S}}^{(2)}\dot{\mathbf{A}}_e) = \operatorname{tr}(\mathbf{N}^T\dot{\mathbf{A}}_e) = \operatorname{tr}\{\mathscr{A}^2[\dot{\mathbf{A}}_e, \dot{\mathbf{A}}_e]\dot{\mathbf{A}}_e\} > 0. \tag{6.2.191}$$

Note, in particular, that $\dot{\mathbf{S}}^{(2)}$ is independent of the sign prefixing $\dot{\mathbf{A}}_e$ while $\dot{\mathbf{S}}^{(1)} = \mathbf{0}$ ensures that the nominal stress is stationary in the considered e-configuration.

Suppose now that the loading is prescribed, with a component along $\dot{\mathbf{A}}_e$. For this to be supportable it follows from (6.2.178) that $\dot{\mathbf{S}}^{(1)} = \mathbf{0}$ and the inequality (6.2.191) must hold. Consequently, $\dot{\mathbf{A}}_0^{(1)} = \mathbf{0}$ and $\dot{\mathbf{A}}^{(1)} = \pm \dot{\mathbf{A}}_e$, the magnitude of $\dot{\mathbf{A}}_e$ being determined from (6.2.191). The given loading path is therefore associated with two distinct deformation paths whose initial tangents are in opposite directions. This can be regarded as a limiting case of constitutive branching.

The second-order deformation gradient $\dot{\mathbf{A}}^{(2)}$ is determined uniquely on each branch by (6.2.190) together with

$$\operatorname{tr}(\dot{\mathbf{S}}^{(3)}\dot{\mathbf{A}}_e) = \pm 3\operatorname{tr}(\mathbf{N}^T\dot{\mathbf{A}}^{(2)}) \pm \operatorname{tr}\{\mathscr{A}^3[\dot{\mathbf{A}}_e, \dot{\mathbf{A}}_e, \dot{\mathbf{A}}_e]\dot{\mathbf{A}}_e\}, \tag{6.2.192}$$

the signs being in accordance with (6.2.189), $\dot{\mathbf{S}}^{(3)}$ being known from the geometry of the loading path. These results can be obtained directly by setting $\dot{\mathbf{A}}_0^{(1)} = \mathbf{0}$, $\lambda = \pm 1$ in the general formulae given above.

Case 2: $\operatorname{tr}(\mathbf{N}^T\dot{\mathbf{A}}_e) = 0$

In this case the e-mode is tangential to the e-surface. With $\dot{\mathbf{A}}^{(1)} = \pm \dot{\mathbf{A}}_e$ equation (6.2.161) now becomes

$$\dot{W} - \operatorname{tr}(\mathbf{S}\dot{\mathbf{A}}) = \tfrac{1}{24}\operatorname{tr}\{\mathscr{A}^3[\dot{\mathbf{A}}_e, \dot{\mathbf{A}}_e, \dot{\mathbf{A}}_e]\dot{\mathbf{A}}_e\}\tau^4 + \cdots \tag{6.2.193}$$

and the stability of the considered primary e-configuration is therefore governed by the sign of this fourth-order term.

On a deformation path whose initial tangent has the form (6.2.181) subject to (6.2.182) we obtain

$$\dot{\mathbf{S}}^{(1)} = \mathscr{A}^1 \dot{\mathbf{A}}_0^{(1)}, \qquad \mathrm{tr}(\dot{\mathbf{S}}^{(1)} \dot{\mathbf{A}}_e) = 0 \tag{6.2.194}$$

as in Case 1, while (6.2.185) reduces to

$$\mathrm{tr}(\dot{\mathbf{S}}^{(2)} \dot{\mathbf{A}}_e) = \mathrm{tr}\{\mathscr{A}^2[\dot{\mathbf{A}}_0^{(1)}, \dot{\mathbf{A}}_0^{(1)}]\dot{\mathbf{A}}_e\} + 2\lambda\, \mathrm{tr}(\mathbf{N}^{\mathrm{T}} \dot{\mathbf{A}}_0^{(1)}) \tag{6.2.195}$$

and (6.2.184) remains intact.

Suppose that a loading path with $\dot{\mathbf{S}}^{(1)} \neq \mathbf{0}$, subject to the orthogonality condition in (6.2.194), is prescribed. Then λ is determined uniquely by (6.2.195) provided $\mathrm{tr}(\mathbf{N}^{\mathrm{T}} \dot{\mathbf{A}}_0^{(1)}) \neq 0$. It follows from (6.2.181) – (6.2.183) that the initial tangent to the associated deformation path is given uniquely by

$$\dot{\mathbf{A}}^{(1)} = \dot{\mathbf{A}}_0^{(1)} + [\mathrm{tr}(\dot{\mathbf{S}}^{(2)} \dot{\mathbf{A}}_e) - \mathrm{tr}\{\mathscr{A}^2[\dot{\mathbf{A}}_0^{(1)}, \dot{\mathbf{A}}_0^{(1)}]\dot{\mathbf{A}}_e\}]\dot{\mathbf{A}}_e / 2\, \mathrm{tr}(\mathbf{N}^{\mathrm{T}} \dot{\mathbf{A}}_0^{(1)}) \tag{6.2.196}$$

and *no branching occurs*.

If, however, $\mathrm{tr}(\mathbf{N}^{\mathrm{T}} \dot{\mathbf{A}}_0^{(1)}) = 0$ then from (6.2.172) and (6.2.194) we obtain

$$\mathrm{tr}(\dot{\mathbf{S}}^{(1)} \mathbf{M}) = 0.$$

Thus $\dot{\mathbf{S}}^{(1)}$ is orthogonal to both independent normals to the trace manifold. The associated deformation is tangential to the e-surface. Indeed, subject to $\mathrm{tr}(\mathbf{N}^{\mathrm{T}} \dot{\mathbf{A}}^{(1)}) = 0$, $\dot{\mathbf{A}}^{(1)}$ is arbitrary in direction since λ is indeterminate. For prescribed $\dot{\mathbf{S}}^{(2)}$ to be supportable in this special case it must satisfy

$$\mathrm{tr}(\dot{\mathbf{S}}^{(2)} \dot{\mathbf{A}}_e) = \mathrm{tr}\{\mathscr{A}^2[\dot{\mathbf{A}}_0^{(1)}, \dot{\mathbf{A}}_0^{(1)}]\dot{\mathbf{A}}_e\}.$$

Next we examine deformation paths for which $\dot{\mathbf{A}}_0^{(1)} = \mathbf{0}$ and

$$\dot{\mathbf{A}}^{(1)} = \lambda \dot{\mathbf{A}}_e, \tag{6.2.197}$$

so that

$$\dot{\mathbf{S}}^{(1)} = \mathbf{0} \tag{6.2.198}$$

and, from (6.2.190) and (6.2.172),

$$\dot{\mathbf{S}}^{(2)} = \mathscr{A}^1 \dot{\mathbf{A}}^{(2)} + \lambda^2 \mathbf{N} = \mathscr{A}^1(\dot{\mathbf{A}}^{(2)} + \lambda^2 \mathbf{M}) \tag{6.2.199}$$

which implies

$$\mathrm{tr}(\dot{\mathbf{S}}^{(2)} \dot{\mathbf{A}}_e) = 0. \tag{6.2.200}$$

Assume first that $\dot{\mathbf{A}}^{(2)} + \lambda^2\mathbf{M} \neq \mu^{(2)}\dot{\mathbf{A}}_e$ for any scalar $\mu^{(2)}$, so that $\dot{\mathbf{S}}^{(2)} \neq \mathbf{0}$.

If, conversely, a loading path is prescribed subject to (6.2.198), (6.2.200) and $\dot{\mathbf{S}}^{(2)} \neq \mathbf{0}$ then the associated loading path is given by (6.2.197) and

$$\dot{\mathbf{A}}^{(2)} = \dot{\mathbf{A}}_0^{(2)} + \mu^{(2)}\dot{\mathbf{A}}_e, \qquad \mathrm{tr}\,(\dot{\mathbf{A}}_e^{\mathrm{T}}\dot{\mathbf{A}}_0^{(2)}) = 0 \tag{6.2.201}$$

with $\dot{\mathbf{A}}_0^{(2)} \neq -\lambda^2\mathbf{M}$. For each λ, $\dot{\mathbf{A}}_0^{(2)}$ is uniquely determined by (6.2.199) while λ itself is obtained from

$$\begin{aligned}\mathrm{tr}\,(\dot{\mathbf{S}}^{(3)}\dot{\mathbf{A}}_e) = \lambda^3\,\mathrm{tr}\,\{\mathscr{A}^3[\dot{\mathbf{A}}_e, \dot{\mathbf{A}}_e, \dot{\mathbf{A}}_e]\dot{\mathbf{A}}_e\}\\ -3\lambda^3\,\mathrm{tr}\,(\mathbf{M}^{\mathrm{T}}\mathbf{N}) + 3\lambda\,\mathrm{tr}\,(\dot{\mathbf{S}}^{(2)}\mathbf{M}),\end{aligned} \tag{6.2.202}$$

the specialization of (6.2.188) to the present situation. The multiplier $\mu^{(2)}$ may be found by use of $\mathrm{tr}\,(\dot{\mathbf{S}}^{(4)}\dot{\mathbf{A}}_e)$, and similarly for higher-order terms.

For convenience we write the cubic (6.2.202) in the form

$$\alpha\lambda^3 + 3\beta\lambda - \gamma = 0,$$

where

$$\begin{aligned}\alpha &= \mathrm{tr}\,\{\mathscr{A}^3[\dot{\mathbf{A}}_e, \dot{\mathbf{A}}_e, \dot{\mathbf{A}}_e]\dot{\mathbf{A}}_e\} - 3\,\mathrm{tr}\,(\mathbf{M}^{\mathrm{T}}\mathbf{N}),\\ \beta &= \mathrm{tr}\,(\dot{\mathbf{S}}^{(2)}\mathbf{M}) = \mathrm{tr}\,(\mathbf{N}^{\mathrm{T}}\dot{\mathbf{A}}_0^{(2)}),\\ \gamma &= \mathrm{tr}\,(\dot{\mathbf{S}}^{(3)}\dot{\mathbf{A}}_e).\end{aligned}$$

If $\alpha = 0$ then there is only one solution for λ and no branching occurs. Necessary and sufficient conditions for there to exist more than one real solution are

$$\begin{aligned}&\alpha \neq 0, \qquad \beta/\alpha < 0,\\ &2(\beta/\alpha)(-\beta/\alpha)^{1/2} \leq \gamma/\alpha \leq -2(\beta/\alpha)(-\beta/\alpha)^{1/2}.\end{aligned} \tag{6.2.203}$$

If strict inequality holds in (6.2.203) then there are two positive and one negative solutions for λ if $\gamma/\alpha < 0$ and two negative and one positive if $\gamma/\alpha > 0$. The case $\gamma = 0$ yields one zero, one positive and one negative solution, the latter two having the same magnitude; this corresponds to the prescribed loading being orthogonal to $\dot{\mathbf{A}}_e$. If the left-hand or right-hand equality holds in (6.2.203) then there are only two real values of λ, one positive and one negative, and they differ in magnitude. Constitutive branching with *three* deformation paths emanating from the considered e-configuration is therefore possible, each value of λ yielding a different value of $\mu^{(2)}$ in general.

Finally, we consider deformation paths for which

$$\dot{\mathbf{A}}^{(1)} = \lambda\dot{\mathbf{A}}_e, \qquad \dot{\mathbf{A}}^{(2)} = -\lambda^2\mathbf{M} + \mu^{(2)}\dot{\mathbf{A}}_e \tag{6.2.204}$$

so that

$$\dot{\mathbf{S}}^{(1)} = \dot{\mathbf{S}}^{(2)} = \mathbf{0}, \tag{6.2.205}$$

$$\dot{\mathbf{S}}^{(3)} = \mathscr{A}^1 \dot{\mathbf{A}}^{(2)} - 3\lambda^3 \mathscr{A}^2[\mathbf{M}, \dot{\mathbf{A}}_e] + 3\lambda\mu^{(2)}\mathbf{N} + \lambda^3 \mathscr{A}^3[\dot{\mathbf{A}}_e, \dot{\mathbf{A}}_e, \dot{\mathbf{A}}_e] \tag{6.2.206}$$

and

$$\operatorname{tr}(\dot{\mathbf{S}}^{(3)}\dot{\mathbf{A}}_e) = \lambda^3 \operatorname{tr}\{\mathscr{A}^3[\dot{\mathbf{A}}_e, \dot{\mathbf{A}}_e, \dot{\mathbf{A}}_e]\dot{\mathbf{A}}_e\} - 3\lambda^3 \operatorname{tr}(\mathbf{M}^{\mathrm{T}}\mathbf{N}). \tag{6.2.207}$$

If the loading path is prescribed with (6.2.205) and $\operatorname{tr}(\dot{\mathbf{S}}^{(3)}\dot{\mathbf{A}}_e) \neq 0$ then the resulting deformation path is given by (6.2.204) with λ and $\mu^{(2)}$ uniquely determined by (6.2.207) and $\operatorname{tr}(\dot{\mathbf{S}}^{(4)}\dot{\mathbf{A}}_e)$ respectively, and similarly for higher-order terms. Thus, no branching occurs.

Apart from notational differences and the fact that the present work is set in a nine- (rather than six-) dimensional space, the branching analysis described above follows closely that given by Hill (1982). Hill's results may be recovered by restricting $\mathbf{A}$, $\mathbf{S}$, $\dot{\mathbf{A}}$, $\dot{\mathbf{S}}$ to being symmetric, the interpretation of these variables then needing appropriate modification; specifically, the conjugate pair $(\mathbf{S}, \mathbf{A})$ may be replaced by $(\mathbf{T}^{(1)}, \mathbf{U})$, or by any other pair of conjugate variables of the type discussed in Section 3.5, and the space dimension reduced by three.

For isotropic materials, however, the results for $(\mathbf{T}^{(1)}, \mathbf{U})$ may simply be picked out from those given for $(\mathbf{S}, \mathbf{A})$ by setting $\mathbf{R} = \mathbf{I}$, $\dot{\mathbf{R}} = \mathbf{0}$. This procedure was illustrated in Section 6.2.3 in respect of the inequality $\operatorname{tr}(\dot{\mathbf{S}}\dot{\mathbf{A}}) > 0$ and its specialization to $\operatorname{tr}(\dot{\mathbf{T}}^{(1)}\dot{\mathbf{U}}) > 0$, and carries over to the specialization of

$$\dot{\mathbf{S}} \equiv \mathscr{A}^1 \dot{\mathbf{A}} = \mathbf{0}$$

to

$$\dot{\mathbf{T}}^{(1)} \equiv \mathscr{L}^{(1)1} \dot{\mathbf{U}} = \mathbf{0}. \tag{6.2.208}$$

In line with the terminology used in respect of $(\mathbf{S}, \mathbf{A})$ we refer to points in $\mathbf{U}$-space where (6.2.208) holds for some $\dot{\mathbf{U}} \neq \mathbf{0}$ as e-configurations, and the associated increments $\dot{\mathbf{U}}$ as e-modes, *relative to* $(\mathbf{T}^{(1)}, \mathbf{U})$. Once results for $(\mathbf{T}^{(1)}, \mathbf{U})$ have been obtained, corresponding results can be read off for the conjugate pair $(\mathbf{T}^{(m)}, \mathbf{E}^{(m)})$ by making the change of variables $(\mathbf{T}^{(1)}, \mathbf{U}) \to (\mathbf{T}^{(m)}, \mathbf{E}^{(m)})$. Branching analysis for isotropic materials relative to such conjugate variables has been examined by Hill (1982) and we refer to that paper for details. Specialization of the present analysis in nine-dimensional $\mathbf{A}$-space to the case of isotropy, based on the results in Sections 6.2.3 and 6.2.5, is left as an exercise for the reader.

This direct specialization is not in general valid for non-isotropic materials since, when $\mathbf{R} = \mathbf{I}$ and $\dot{\mathbf{R}} = \mathbf{0}$, it follows from (4.3.10) that $\dot{\mathbf{T}}^{(1)} =$

$\frac{1}{2}(\dot{\mathbf{S}}+\dot{\mathbf{S}}^{\mathrm{T}})$. Thus, although $\dot{\mathbf{S}}=\mathbf{0}$ implies $\dot{\mathbf{T}}^{(1)}=\mathbf{0}$, there may be non-zero $\dot{\mathbf{U}}$ which make $\dot{\mathbf{T}}^{(1)}=\mathbf{0}$ and $\dot{\mathbf{S}}$ antisymmetric.

Consider now the inequality

$$\mathrm{tr}\,(\dot{\mathbf{T}}^{(1)}\dot{\mathbf{U}}) \equiv \mathrm{tr}\{(\mathscr{L}^{(1)1}\dot{\mathbf{U}})\dot{\mathbf{U}}\} > 0 \qquad (6.2.209)$$

for all $\dot{\mathbf{U}} \neq \mathbf{0}$. From the results of Section 6.1.6 we deduce that for an isotropic material in its natural configuration (6.2.209) is equivalent to

$$\mu > 0, \qquad \kappa > 0, \qquad (6.2.210)$$

where μ and κ are the classical shear and bulk moduli. If (6.2.210) are accepted then (6.2.209) holds not only in the natural configuration but also, by continuity, in some connected domain *enclosing* the natural configuration. (This contrasts with the situation in respect of the inequality (6.2.159) since the natural configuration lies *on the boundary* of the stable region where (6.2.159) holds.)

In fact, for any conjugate pair $(\mathbf{T}^{(m)}, \mathbf{E}^{(m)})$, the inequality†

$$\mathrm{tr}(\dot{\mathbf{T}}^{(m)}\dot{\mathbf{E}}^{(m)}) \equiv \mathrm{tr}\{(\mathscr{L}^{(m)1}\dot{\mathbf{E}}^{(m)})\dot{\mathbf{E}}^{(m)}\} > 0 \qquad (6.2.211)$$

for all $\dot{\mathbf{E}}^{(m)} \neq \mathbf{0}$ also holds for some connected domain enclosing the natural configuration since, by the results of Sections 6.1.3 and 6.1.4 the tensor $\mathscr{L}^{(m)1}$ of first-order moduli is independent of m when evaluated in that configuration. This result extends to non-isotropic materials if (6.2.210) is replaced by positive-definiteness of the tensor $\mathscr{A}$ of linear elastic moduli in Section 6.1.6.

The extent of the domain (in $\mathbf{U}$-space, for example) for which (6.2.211) holds clearly depends on the choice of conjugate variables, and may be assessed on physical grounds (by reference to experiment) or mathematical grounds (by reference to its implications for the solution of boundary-value problems). This is the concern of the remainder of this section.

Constitutive inequalities An inequality, such as (6.2.211), which imposes restrictions on the constitutive law, and hence on the nature of the elastic response, whether for all deformations or for some subset of deformations, is called a *constitutive inequality*. Of particular interest are *objective* constitutive inequalities, which, unlike (6.2.159), are indifferent to superposed rigid rotations. Note that (6.2.211) is objective.

† An interpretation of this inequality, paralleling the stability interpretation of (6.2.159), is as follows. Under (6.2.211) the material may be regarded as *intrinsically stable* under tractions which 'follow' the material, i.e. rotate with the local rotation $\mathbf{R}$. For example, whereas $\mathbf{S}^{\mathrm{T}}\mathbf{N}=\boldsymbol{\sigma}(\mathbf{X},\mathbf{A})$ describes nominal loading the equation $\mathbf{T}^{(1)}\mathbf{N}=\boldsymbol{\sigma}(\mathbf{X},\mathbf{U})$ describes the corresponding $(m=1)$ follower tractions for an isotropic material $(\mathbf{S}=\mathbf{T}^{(1)}\mathbf{R}^{\mathrm{T}})$.

The search for a *universal constitutive inequality*, holding for the whole of $\mathbf{U}$-space and all elastic materials, has occupied much attention in the literature in recent years, but, as yet, no generally acceptable inequality has been forthcoming. Many constitutive inequalities, general and particular, have been proposed for consideration in this context, and we refer to Truesdell and Noll (1965), Wang and Truesdell (1973), Hill (1970) and Ogden (1970a) for detailed accounts. In the present section we merely examine the implications of a representative class of constitutive inequalities, with attention confined to isotropic materials.

First, either directly from formulae given in Section 3.5 or by means of the change of variables $(\mathbf{T}^{(1)}, \mathbf{U}) \rightarrow (\mathbf{T}^{(m)}, \mathbf{E}^{(m)})$ in (6.2.102), the inequality (6.2.211) yields

$$\operatorname{tr}(\dot{\mathbf{T}}^{(m)}\dot{\mathbf{E}}^{(m)}) \equiv \sum_{i,j=1}^{3} \frac{\partial t_i^{(m)}}{\partial e_j^{(m)}} \dot{e}_i^{(m)} \dot{e}_j^{(m)} + \sum_{i \neq j} (t_i^{(m)} - t_j^{(m)})(e_i^{(m)} - e_j^{(m)})(\Omega_{ij}^{(\mathrm{L})})^2 > 0. \qquad (6.2.212)$$

Necessary and sufficient conditions for (6.2.212) are

$$\left.\begin{array}{l} \text{(i) matrix } \quad (\partial t_i^{(m)}/\partial e_j^{(m)}) \text{ is positive definite,} \\ \text{(ii) } (t_i^{(m)} - t_j^{(m)})/(e_i^{(m)} - e_j^{(m)}) > 0 \quad i \neq j \end{array}\right\} \qquad (6.2.213)$$

jointly, in parallel with (i) and (ii) in (6.2.101).

Evidence from macroscopic mechanical tests on elastic solids is consistent with the predictions of (6.2.212) for a range of values of m including $m = 0$ and $m = 1$. In particular, for rubberlike solids, experimental data have been modelled by strain-energy functions satisfying (6.2.212) with $m = 0$ (Ogden, 1972a, b). (We refer to Ogden (1982) for a recent review of the elastic properties of rubberlike solids.) Comparison of the consequences of (6.2.212) for some specific material models has been provided by Ogden (1970a) following the fundamental work of Hill (1970) who assessed the inequalities on physical grounds. Further discussion is contained in Wang and Truesdell (1973). The relative strengths of the intrinsic stability criteria (6.2.212) for different m have been examined by Parry (1978).

For full details the reader should consult the above-mentioned references. Here, it suffices to draw attention to the following points. Firstly, for *incompressible* materials (6.2.212) fails because of the occurrence of the arbitrary hydrostatic pressure p (recall equation (4.3.50)), *except in the case* $m = 0$. But, if p is excluded and (6.2.212) regarded as a restriction on the strain-energy function (subject to the incremental incompressibility constraint) the validity of the inequality can be extended to a range of values of m. On the other hand, (6.2.212) with $m=0$ fails for an elastic *fluid* (4.2.38) since it requires the shear modulus to be positive. Secondly, with $m=1$ the inequality has an important role to play because of its close connection with (6.2.159), elicited in Section 6.2.3. Thirdly, since experimental tests only cover bounded domains in **A**-space, represented as domains in $(\lambda_1, \lambda_2, \lambda_3)$-space for isotropic materials, the question of whether, for some m, (6.2.212) holds for all deformations (at least for compressible solids) remains open. Moreover, the applicability of elasticity theory outside such bounded domains is itself questionable because, for example, there may exist yield surfaces beyond which permanent deformation occurs.

To illustrate that (6.2.212) may fail within the elastic regime for some m while holding for other values of m one need only refer to Fig. 4.3. This shows that, in an equibiaxial tension test, $\mathrm{d}t_1^{(m)}/\mathrm{d}\lambda_1 > 0$ when $\lambda_1 > 1$ for $m = 0, 1, 2$ but fails for $m = 3$ beyond some critical value of λ_1. The behaviour of $\partial t_1^{(m)}/\partial \lambda_1$, or equivalently $\partial t_1^{(m)}/\partial e_1^{(m)}$, mirrors that of $\mathrm{d}t_1^{(m)}/\mathrm{d}\lambda_1$. Although Fig. 4.3 is based on an artificial form of strain-energy function, the behaviour it displays is typical of that found for rubberlike materials in equibiaxial tension (Ogden, 1982).

The above discussion is based on the restriction to deformations that are twice continuously differentiable. It does not account for the possibility of deformations with weaker regularity, such as are associated with shearband formation. Thus, if, for example, failure of ellipticity is permitted then the domain of validity of (6.2.212), in particular for $m=1$, may be diminished since (6.2.212) implies the strong ellipticity condition (6.2.150) in the regime $t_i^{(1)} + t_j^{(1)} > 0$, $i \neq j$. The possibility of the existence of locally infinite deformations associated with the formation of cavities (Ball, 1982) also deserves consideration in the present context because their admission would allow extension of the elastic domain to the whole of $(\lambda_1, \lambda_2, \lambda_3)$-space.

Clearly, in assessing constitutive inequalities a number of factors, mathematical and physical, need to be accounted for. It is premature to assert that certain inequalities should always hold when they may have implications for as yet unexplored situations and may, in particular, rule out plausible physical phenomena. Nevertheless, examination of the consequences of such inequalities is a valuable exercise. These are the considerations which guide the approach to constitutive inequalities in this book.

We remark here that alongside constitutive inequalities such as (6.2.212) must be considered inequalities which arise naturally in the course of solution of boundary-value problems. Such inequalities, examples of which can be found in Sections 5.2 and 5.3, depend, in general, on the geometry of the problem in question and on the prescribed data.

In (6.2.211) the reference configuration is taken as the natural configuration relative to which the material is isotropic. If a different reference configuration is chosen then the consequences of (6.2.211) differ from (6.2.213).

To illustrate the point we take the reference and current configurations to coincide so that (6.2.211) becomes

$$\operatorname{tr}\{(\mathscr{L}_0^{(m)1}\boldsymbol{\Sigma})\boldsymbol{\Sigma}\} > 0, \tag{6.2.214}$$

where $\boldsymbol{\Sigma}$ is the Eulerian strain-rate and $\mathscr{L}_0^{(m)1}$ is the tensor of first-order *instantaneous* moduli associated with $(\mathbf{T}^{(m)}, \mathbf{E}^{(m)})$.

Reference to the Eulerian principal axes and use of the component form of $\mathscr{L}_0^{(m)1}$ obtainable from Section 6.1.3 shows that (6.2.214) may be rearranged as

$$J^{-1}\sum_{i,j=1}^{3}\frac{\partial t_i^{(m)}}{\partial e_j^{(m)}}\dot{e}_i^{(m)}\dot{e}_j^{(m)} + \sum_{i\neq j}\left\{\left(\frac{t_i - t_j}{\lambda_i^2 - \lambda_j^2}\right)(\lambda_i^2 + \lambda_j^2) - \tfrac{1}{2}m(t_i + t_j)\right\}\Sigma_{ij}^2 > 0, \tag{6.2.215}$$

where t_i are the components of Cauchy stress and $J = \lambda_1\lambda_2\lambda_3$. Necessary and sufficient conditions for (6.2.215) are

$$\left.\begin{array}{ll} \text{(i) matrix } (\partial t_i^{(m)}/\partial e_j^{(m)}) \text{ is positive definite} & \\ \text{(ii) } \left(\dfrac{t_i - t_j}{\lambda_i^2 - \lambda_j^2}\right)(\lambda_i^2 + \lambda_j^2) - \tfrac{1}{2}m(t_i + t_j) > 0 & i \neq j. \end{array}\right\} \tag{6.2.216}$$

From Appendix 1 we know that (ii) in (6.2.213) is a consequence of (i) for any convex domain in $(e_1^{(m)}, e_2^{(m)}, e_3^{(m)})$-space (in particular, for the whole space). It follows that on such a domain (6.2.215) implies (6.2.212). In fact, (6.2.214) is stronger than (6.2.211) since in general (ii) in (6.2.213) is not equivalent to (ii) in (6.2.216). It is left for the reader to establish the equivalence of (6.2.212) and (6.2.215) for $m = 0$ and $m = \pm 2$.

Note that the appearance of the term $t_i + t_j$ in (6.2.215) admits the possibility of this inequality failing at sufficiently large hydrostatic tension or compression (depending on the sign of m) except when $m = 0$. This applies in particular for $m = 1$ in which case (6.2.215) becomes the (strengthened) Coleman–Noll inequality, details of which can be found in Truesdell and Noll (1965).

The reservations expressed earlier about imposition of (6.2.211) for any m apply *a fortiori* to (6.2.214). They also apply to the global analogue of (6.2.211), namely

$$\operatorname{tr}\{[\mathbf{G}^{(m)}(\mathbf{E}^{(m)\prime}) - \mathbf{G}^{(m)}(\mathbf{E}^{(m)})](\mathbf{E}^{(m)\prime} - \mathbf{E}^{(m)})\} > 0 \tag{6.2.217}$$

for $\mathbf{E}^{(m)\prime} \neq \mathbf{E}^{(m)}$, where $\mathbf{T}^{(m)} = \mathbf{G}^{(m)}(\mathbf{E}^{(m)})$ generalizes (4.2.19). We do not discuss such inequalities here but refer to Appendix 1 where connections between local and global inequalities such as (6.2.211) and (6.2.217) are examined.

6.3 SOLUTION OF INCREMENTAL BOUNDARY-VALUE PROBLEMS

For ease of reference the equations and boundary conditions required in this section are summarized here. Firstly, from Section 6.2.1, when the reference configuration is chosen to coincide with the current configuration the equilibrium equation is

$$\operatorname{div} \dot{\mathbf{S}}_0 + \rho\dot{\mathbf{b}} = \mathbf{0}, \tag{6.3.1}$$

the dot notation for increments being retained.

The incremental constitutive law is

$$\dot{\mathbf{S}}_0 = \mathscr{A}_0^1 \dot{\mathbf{A}}_0 \tag{6.3.2}$$

for an unconstrained material, and

$$\dot{\mathbf{S}}_0 = \mathscr{A}_0^1 \dot{\mathbf{A}}_0 + \dot{q}\mathbf{I} - q\dot{\mathbf{A}}_0 \tag{6.3.3}$$

for an incompressible material, where

$$\dot{\mathbf{A}}_0 = \operatorname{grad} \dot{\chi}. \tag{6.3.4}$$

For an incompressible material the incremental incompressibility condition is

$$\operatorname{tr}(\dot{\mathbf{A}}_0) = 0. \tag{6.3.5}$$

Rotational balance is guaranteed by objectivity, i.e. $\dot{\mathbf{S}}_0 + \dot{\mathbf{A}}_0\mathbf{T}$ is symmetric, where $\mathbf{T}$ is the current Cauchy stress tensor.

The incremental boundary conditions are

$$\dot{\chi} = \dot{\xi} \quad \text{on } \partial\mathscr{B}^x, \tag{6.3.6}$$

$$\dot{\mathbf{S}}_0^{\mathrm{T}}\mathbf{n} = \dot{\boldsymbol{\sigma}}_0 \quad \text{on } \partial\mathscr{B}^\sigma, \tag{6.3.7}$$

where $\mathbf{n}$ is the unit outward normal to the current boundary $\partial\mathscr{B}$. In particular, when pressure is prescribed on $\partial\mathscr{B}^\sigma$ we have

$$\dot{\boldsymbol{\sigma}}_0 = -\dot{P}\mathbf{n} - P\operatorname{tr}(\dot{\mathbf{A}}_0)\mathbf{n} + P\dot{\mathbf{A}}_0^{\mathrm{T}}\mathbf{n}, \tag{6.3.8}$$

the second term on the right-hand side vanishing for an incompressible material.

At this point it is convenient to record the Cartesian component form of the equilibrium equations and to introduce the notation

$$\mathbf{v} = \dot{\boldsymbol{\chi}}. \tag{6.3.9}$$

After substitution of the constitutive equation into (6.3.1) we obtain

$$\mathscr{A}^1_{0ijkl}\frac{\partial^2 v_l}{\partial x_i \partial x_k} + \frac{\partial}{\partial x_i}(\mathscr{A}^1_{0ijkl})\frac{\partial v_l}{\partial x_k} + \rho\dot{b}_j = 0 \quad j = 1, 2, 3 \tag{6.3.10}$$

for an unconstrained material, and

$$\begin{aligned}\mathscr{A}^1_{0ijkl}\frac{\partial^2 v_l}{\partial x_i \partial x_k} + \frac{\partial}{\partial x_i}(\mathscr{A}^1_{0ijkl})\frac{\partial v_l}{\partial x_k} + \frac{\partial \dot{q}}{\partial x_k} \\ - \frac{\partial v_i}{\partial x_j}\frac{\partial q}{\partial x_i} + \rho\dot{b}_j = 0 \qquad j = 1, 2, 3\end{aligned} \tag{6.3.11}$$

coupled with

$$\frac{\partial v_i}{\partial x_i} = 0 \tag{6.3.12}$$

for an incompresible material. And when the underlying deformation is homogeneous (6.3.10) and (6.3.11) reduce to

$$\mathscr{A}^1_{0ijkl}\frac{\partial^2 v_l}{\partial x_i \partial x_k} + \rho\dot{b}_j = 0 \qquad j = 1, 2, 3 \tag{6.3.13}$$

and

$$\mathscr{A}^1_{0ijkl}\frac{\partial^2 v_l}{\partial x_i \partial x_k} + \frac{\partial \dot{q}}{\partial x_j} + \rho\dot{b}_j = 0 \qquad j = 1, 2, 3 \tag{6.3.14}$$

respectively. For component forms of (6.3.1) in other coordinate systems we shall draw on the results of Section 1.5.4 as the need arises.

For an isotropic Green-elastic material the components of $\mathscr{A}^1_0$ on the Eulerian principal axes are (see Problem 6.1.7) expressible as

$$\left.\begin{aligned}&\mathscr{A}^1_{0iijj} = J^{-1}\lambda_i\lambda_j\frac{\partial^2 W}{\partial\lambda_i\partial\lambda_j},\\ &\mathscr{A}^1_{0ijij} = J^{-1}\left(\lambda_i\frac{\partial W}{\partial\lambda_i} - \lambda_j\frac{\partial W}{\partial\lambda_j}\right)\lambda_i^2/(\lambda_i^2 - \lambda_j^2) \qquad i \neq j,\\ &\mathscr{A}^1_{0ijji} = \mathscr{A}^1_{0jiij} = \mathscr{A}^1_{0ijij} - J^{-1}\lambda_i\frac{\partial W}{\partial\lambda_i} \qquad i \neq j,\end{aligned}\right\} \tag{6.3.15}$$

where J is $\lambda_1\lambda_2\lambda_3$ (respectively 1) for an unconstrained (respectively incompressible) material. When $\lambda_i = \lambda_j$, $i \neq j$, we have the connection

$$\mathscr{A}^1_{0ijij} = \tfrac{1}{2}\left\{\mathscr{A}^1_{0iiii} - \mathscr{A}^1_{0iijj} + J^{-1}\lambda_i\frac{\partial W}{\partial \lambda_i}\right\} \tag{6.3.16}$$

and the number of independent components of $\boldsymbol{\mathscr{A}}^1_0$ is also reduced in accordance with

$$\left.\begin{aligned} \mathscr{A}^1_{0iiii} &= \mathscr{A}^1_{0jjjj}, \quad & \mathscr{A}^1_{0iikk} &= \mathscr{A}^1_{0jjkk},\\ \mathscr{A}^1_{0ikik} &= \mathscr{A}^1_{0jkjk}, \quad & \mathscr{A}^1_{0ikki} &= \mathscr{A}^1_{0jkkj}, \end{aligned}\right\} \tag{6.3.17}$$

where $k \neq i \neq j \neq k$. Further reductions occur when all three principal stretches are equal.

In Sections 6.3.1–6.3.3 attention is restricted to problems in which the underlying deformation is homogeneous, while one problem involving inhomogeneous deformation is examined in Section 6.3.4.

We concentrate on the details of a central example in each case and, in particular, we examine the conditions under which uniqueness fails and bifurcation occurs. Throughout this section we make use of the linearized equations, but it should be remembered from the discussion in Section 6.2.8 that a proper branching analysis can be conducted only if higher-order terms are retained. In respect of boundary-value problems, such an analysis is not yet available (see, however, Section 6.3.2).

The intention is not to provide an exhaustive account of problems whose solutions are known but to describe a representative selection of such problems together with methods for their solution. As in Chapter 5, only isotropic materials are considered.

6.3.1 Bifurcation of a pre-strained rectangular block

We consider a rectangular block of isotropic elastic material which has dimensions $2L_1$, $2L_2$, $2L_3$ when unstressed. Suppose it occupies the region

$$-L_i \le X_i \le L_i \qquad i = 1, 2, 3,$$

where X_1, X_2, X_3 are (Lagrangean) coordinates relative to a rectangular Cartesian basis defined by the edges of the block and with origin at the centre of the block.

The block is deformed uniformly so that the current coordinates x_1, x_2, x_3 of the material point X_1, X_2, X_3 are given by

$$x_i = \lambda_i X_i \qquad i = 1, 2, 3,$$

where $\lambda_1, \lambda_2, \lambda_3$ are the principal stretches of the deformation (which is a pure strain). The current geometry of the block is defined by

$$-l_i \equiv -\lambda_i L_i \leq x_i \leq \lambda_i L_i \equiv l_i \qquad i=1,2,3.$$

The Eulerian (and Lagrangean) principal axes coincide with the rectangular axes, and the principal Cauchy stresses are given by

$$t_i = J^{-1}\lambda_i \frac{\partial W}{\partial \lambda_i} \qquad i=1,2,3 \tag{6.3.18}$$

and

$$t_i = \lambda_i \frac{\partial W}{\partial \lambda_i} + q \qquad i=1,2,3 \tag{6.3.19}$$

for unconstrained and incompressible materials respectively.

First we discuss the general case in which no two of $\lambda_1, \lambda_2, \lambda_3$ are equal, leaving aside the special cases in which two or all three of $\lambda_1, \lambda_2, \lambda_3$ are equal for separate consideration at a later stage.

The equations governing superposed incremental deformations are obtained from (6.3.13) or (6.3.14), the only non-vanishing components of $\mathscr{A}_0^1$ being given by (6.3.15). With $\partial v_i/\partial x_j, \partial^2 v_i/\partial x_j \partial x_k, \ldots$ written as $v_{i,j}, v_{i,jk}, \ldots$ and body forces omitted, equation (6.3.13) yields

$$\left.\begin{aligned}
&\mathscr{A}_{01111}^1 v_{1,11} + \mathscr{A}_{02121}^1 v_{1,22} + \mathscr{A}_{03131}^1 v_{1,33} \\
&\quad + (\mathscr{A}_{01122}^1 + \mathscr{A}_{02112}^1) v_{2,12} + (\mathscr{A}_{01133}^1 + \mathscr{A}_{03113}^1) v_{3,13} = 0, \\
&(\mathscr{A}_{01122}^1 + \mathscr{A}_{02112}^1) v_{1,12} + \mathscr{A}_{01212}^1 v_{2,11} + \mathscr{A}_{02222}^1 v_{2,22} \\
&\quad + \mathscr{A}_{03232}^1 v_{2,33} + (\mathscr{A}_{02233}^1 + \mathscr{A}_{02332}^1) v_{3,23} = 0, \\
&(\mathscr{A}_{01133}^1 + \mathscr{A}_{03113}^1) v_{1,13} + (\mathscr{A}_{02233}^1 + \mathscr{A}_{02332}^1) v_{2,23} \\
&\quad + \mathscr{A}_{01313}^1 v_{3,11} + \mathscr{A}_{02323}^1 v_{3,22} + \mathscr{A}_{03333}^1 v_{3,33} = 0.
\end{aligned}\right\} \tag{6.3.20}$$

The corresponding equations for an incompressible material are obtained from (6.3.20) by adding $\partial \dot{q}/\partial x_1$, $\partial \dot{q}/\partial x_2$, $\partial \dot{q}/\partial x_3$ respectively to the left-hand sides of these equations and imposing the constraint

$$v_{1,1} + v_{2,2} + v_{3,3} = 0. \tag{6.3.21}$$

For an unconstrained material under *all-round dead-load tractions* bifurcation can occur where $\mathscr{A}_0^1$ is singular in the sense described in Section 6.2.5, but for other boundary conditions the bifurcation analysis is not so straightforward, as we see below.

In general the algebra involved in matching the solution of (6.3.20) to given boundary conditions is cumbersome and little can be learned without the use of specific forms of strain-energy function and numerical calculations. Accordingly, in order to avoid lengthy algebra and to illustrate the type of results obtainable we restrict attention to *plane incremental deformations*. In particular, we take

$$v_3 = 0, \quad v_\alpha = v_\alpha(x_1, x_2) \qquad \alpha = 1, 2, \tag{6.3.22}$$

and equations (6.3.20) reduce to

$$\left.\begin{aligned} &\mathscr{A}^1_{01111} v_{1,11} + \mathscr{A}^1_{02121} v_{1,22} + (\mathscr{A}^1_{01122} + \mathscr{A}^1_{02112}) v_{2,12} = 0, \\ &(\mathscr{A}^1_{01122} + \mathscr{A}^1_{02112}) v_{1,12} + \mathscr{A}^1_{01212} v_{2,11} + \mathscr{A}^1_{02222} v_{2,22} = 0. \end{aligned}\right\} \tag{6.3.23}$$

For incompressible materials $\dot{q}$ is independent of x_3 and

$$\left.\begin{aligned} &\mathscr{A}^1_{01111} v_{1,11} + \mathscr{A}^1_{02121} v_{1,22} + (\mathscr{A}^1_{01122} + \mathscr{A}^1_{02112}) v_{2,12} + \dot{q}_{,1} = 0, \\ &(\mathscr{A}^1_{01122} + \mathscr{A}^1_{02112}) v_{1,12} + \mathscr{A}^1_{01212} v_{2,11} + \mathscr{A}^1_{02222} v_{2,22} + \dot{q}_{,2} = 0, \end{aligned}\right\} \tag{6.3.24}$$

coupled with

$$v_{1,1} + v_{2,2} = 0. \tag{6.3.25}$$

We deal first with unconstrained materials and return to equations (6.3.24) and (6.3.25) later.

Analysis for unconstrained materials Elimination of either v_1 or v_2 from (6.3.23) leads to

$$\begin{aligned} &\mathscr{A}^1_{01111}\mathscr{A}^1_{01212} v_{\alpha,1111} + \{\mathscr{A}^1_{01111}\mathscr{A}^1_{02222} + \mathscr{A}^1_{01212}\mathscr{A}^1_{02121} \\ &- (\mathscr{A}^1_{01122} + \mathscr{A}^1_{02112})^2\} v_{\alpha,1122} \\ &+ \mathscr{A}^1_{02222}\mathscr{A}^1_{02121} v_{\alpha,2222} = 0 \qquad \alpha = 1, 2. \end{aligned} \tag{6.3.26}$$

Equations (6.3.23) and (6.3.26) can be satisfied by fields of the form

$$v_\alpha = m_\alpha f(p_1 x_1 + p_2 x_2) \qquad \alpha = 1, 2, \tag{6.3.27}$$

where m_α, p_α are constants and f is an arbitrary twice-differentiable function. Provided the second derivative of f does not vanish identically, the ratio p_1/p_2 is determined by substitution of (6.3.27) into (6.3.26), i.e. from the quartic equation

$$\mathscr{A}^1_{01111}\mathscr{A}^1_{01212}p_1^4+\{\mathscr{A}^1_{01111}\mathscr{A}^1_{02222}+\mathscr{A}^1_{01212}\mathscr{A}^1_{02121}-(\mathscr{A}^1_{01122}+\mathscr{A}^1_{02112})^2\}p_1^2p_2^2+\mathscr{A}^1_{02222}\mathscr{A}^1_{02121}p_2^4=0. \quad (6.3.28)$$

For each of the four values of p_1/p_2 obtained from (6.3.28) the corresponding ratio m_1/m_2 is given by either of the two equivalent equations

$$\left.\begin{aligned}(\mathscr{A}^1_{01111}p_1^2+\mathscr{A}^1_{02121}p_2^2)m_1+(\mathscr{A}^1_{01122}+\mathscr{A}^1_{02112})p_1p_2m_2&=0,\\(\mathscr{A}^1_{01122}+\mathscr{A}^1_{02112})p_1p_2m_1+(\mathscr{A}^1_{01212}p_1^2+\mathscr{A}^1_{02222}p_2^2)m_2&=0.\end{aligned}\right\} \quad (6.3.29)$$

In respect of a general class of materials having incrementally linear constitutive laws plane problems governed by equations of the form (6.3.23) have been analysed in detail by Hill (1979), and results for isotropic elastic solids can be picked out as special cases of his general results. Hill gave general solutions of (6.3.23) based on (6.3.27) for each of the elliptic, parabolic, hyperbolic and transitional regimes of the equations. These correspond respectively to those parts of $(\lambda_1, \lambda_2, \lambda_3)$-space for which (6.3.28) yields no real solutions, two real and two purely imaginary solutions, four real and distinct solutions, and pairwise equal solutions for p_1/p_2.

Here we are interested in the primary stable domain in $(\lambda_1, \lambda_2, \lambda_3)$-space, for which the equations are strongly elliptic (and therefore elliptic), and also the boundary of this domain, where strong and ordinary ellipticity jointly fail.

From the results of Section 6.2.7 we note that necessary and sufficient conditions for strong ellipticity of (6.3.23), or equivalently of (6.3.29), are expressible as

$$\mathscr{A}^1_{01111}>0, \qquad \mathscr{A}^1_{02222}>0, \qquad \mathscr{A}^1_{01212}>0, \qquad \mathscr{A}^1_{02121}>0 \quad (6.3.30)$$

(the third and fourth of which are equivalent), and

$$\mathscr{A}^1_{01111}\mathscr{A}^1_{02222}+\mathscr{A}^1_{01212}\mathscr{A}^1_{02121}-(\mathscr{A}^1_{01122}+\mathscr{A}^1_{02112})^2 > -2(\mathscr{A}^1_{01111}\mathscr{A}^1_{02222}\mathscr{A}^1_{01212}\mathscr{A}^1_{02121})^{1/2}. \quad (6.3.31)$$

For ease of subsequent manipulation we introduce the notation

$$\left.\begin{aligned}a&=\mathscr{A}^1_{01111}\mathscr{A}^1_{01212}, \qquad c=\mathscr{A}^1_{02222}\mathscr{A}^1_{02121},\\2b&=\mathscr{A}^1_{01111}\mathscr{A}^1_{02222}+\mathscr{A}^1_{01212}\mathscr{A}^1_{02121}\\&\quad-(\mathscr{A}^1_{01122}+\mathscr{A}^1_{02112})^2.\end{aligned}\right\} \quad (6.3.32)$$

It follows that strong ellipticity implies

$$a > 0, \qquad c > 0, \qquad b > -\sqrt{ac} \tag{6.3.33}$$

(but not conversely), and (6.3.28) becomes

$$ap_1^4 + 2bp_1^2p_2^2 + cp_2^4 = 0, \tag{6.3.34}$$

The roots of (6.3.34) are $\pm p_1/p_2$, where

$$p_1^2/p_2^2 = (-b \pm \sqrt{b^2 - ac})/a, \tag{6.3.35}$$

and the form of the general solution of (6.3.23) depends on the sign of $b^2 - ac$ in each of the regimes. First, we consider the possibilities arising in the *strongly elliptic regime.*

Case 1: $b > \sqrt{ac}$ (with $a > 0$, $c > 0$) The roots of (6.3.34) are all pure imaginary, $\pm p_1/p_2$ and $\pm p_1^*/p_2^*$ say, and the general *real* solution may be written as

$$\left.\begin{aligned} v_1 &= m_1\{f(p_1x_1 + p_2x_2) + \bar{f}(p_1x_1 - p_2x_2)\} \\ &\quad + m_1^*\{g(p_1^*x_1 + p_2^*x_2) + \bar{g}(p_1^*x_1 - p_2^*x_2)\}, \\ v_2 &= m_2\{f(p_1x_1 + p_2x_2) - \bar{f}(p_1x_1 - p_2x_2)\} \\ &\quad + m_2^*\{g(p_1^*x_1 + p_2^*x_2) - \bar{g}(p_1^*x_1 - p_2^*x_2)\}. \end{aligned}\right\} \tag{6.3.36}$$

where we have taken m_1, m_1^*, p_1, p_1^* to be real and m_2, m_2^*, p_2, p_2^* to be purely imaginary. An overbar denotes complex conjugate and f and g are arbitrary holomorphic functions of their arguments.

Case 2: $b = \sqrt{ac}$ (with $a > 0$, $c > 0$) There are two pairs of equal pure imaginary roots, $\pm p_1/p_2$ say, and the general real solution of (6.3.23) is

$$\left.\begin{aligned} v_1 &= m_1\{f(p_1x_1 + p_2x_2) + \bar{f}(p_1x_1 - p_2x_2) \\ &\quad + (p_1x_1 - p_2x_2)g'(p_1x_1 + p_2x_2) \\ &\quad + (p_1x_1 + p_2x_2)\bar{g}'(p_1x_1 - p_2x_2)\} \\ &\quad - m_1k\{g(p_1x_1 + p_2x_2) + \bar{g}(p_1x_1 - p_2x_2)\}, \\ v_2 &= m_2\{f(p_1x_1 + p_2x_2) - \bar{f}(p_1x_1 - p_2x_2) \\ &\quad - (p_1x_1 - p_2x_2)g'(p_1x_1 + p_2x_2) \\ &\quad - (p_1x_1 + p_2x_2)\bar{g}'(p_1x_1 - p_2x_2)\} \\ &\quad + m_2k\{g(p_1x_1 + p_2x_2) - \bar{g}(p_1x_1 - p_2x_2)\}, \end{aligned}\right\} \tag{6.3.37}$$

where m_1, p_1 are real, p_2, m_2 are pure imaginary, a prime denotes the derivative with respect to the indicated argument, and

$$k = (b + \mathscr{A}^1_{01212}\mathscr{A}^1_{02121})/(b - \mathscr{A}^1_{01212}\mathscr{A}^1_{02121}). \tag{6.3.38}$$

Case 3: $b^2 < ac$ (with $a > 0,\ c > 0$) Equation (6.3.34) has two pairs of complex conjugate roots, $\pm p_1/p_2$ and $\pm \bar{p}_1/\bar{p}_2$, and the real solution for v_1 and v_2 is

$$\left.\begin{aligned} v_1 &= m_1\{f(p_1x_1 + p_2x_2) + g(p_1x_1 - p_2x_2)\} \\ &\quad + \bar{m}_1\{\bar{f}(\bar{p}_1x_1 + \bar{p}_2x_2) + \bar{g}(\bar{p}_1x_1 - \bar{p}_2x_2)\}, \\ v_2 &= m_2\{f(p_1x_1 + p_2x_2) - g(p_1x_1 - p_2x_2)\} \\ &\quad + \bar{m}_2\{\bar{f}(\bar{p}_1x_1 + \bar{p}_2x_2) - \bar{g}(\bar{p}_1x_1 - \bar{p}_2x_2)\}. \end{aligned}\right\} \tag{6.3.39}$$

On the boundary of the strongly elliptic domain the following situations are possible.

Case 4: $b = -\sqrt{ac}$ (with $a > 0,\ c > 0$) The roots of (6.3.34) are pairwise equal and real, and the solution is given by (6.3.37).

Case 5: $a = 0$ (with $c > 0, b > 0$) One pair of values of p_2/p_1 is zero and the other roots are pure imaginary with $p_1^2/p_2^2 = -c/2b$. The general real solution for v_1 and v_2 is then

$$\left.\begin{aligned} v_1 &= m_1\{f(p_1x_1 + p_2x_2) + \bar{f}(p_1x_1 - p_2x_2)\} \\ &\quad + \phi(x_1) - \mathscr{A}^1_{01212}x_2\psi'(x_1), \\ v_2 &= m_2\{f(p_1x_1 + p_2x_2) - \bar{f}(p_1x_1 - p_2x_2)\} \\ &\quad + (\mathscr{A}^1_{01122} + \mathscr{A}^1_{02112})\psi(x_1), \end{aligned}\right\} \tag{6.3.40}$$

where ϕ is an arbitrary function and ψ is arbitrary and twice differentiable. The results for $c = 0$, $a > 0$, $b > 0$ are obtained in a similar way.

Case 6: $a = c = 0$ (with $b > 0$) If this corresponds to $\mathscr{A}^1_{01212} = \mathscr{A}^1_{02121} = 0$ then

$$\left.\begin{aligned} v_1 &= \mathscr{A}^1_{02222}x_1\phi'(x_2) - (\mathscr{A}^1_{01122} - t_1)\psi(x_1) + \chi_2(x_2), \\ v_2 &= \mathscr{A}^1_{01111}x_2\psi'(x_1) - (\mathscr{A}^1_{01122} - t_1)\phi(x_2) + \chi_1(x_1) \end{aligned}\right\} \tag{6.3.41}$$

while if $\mathscr{A}^1_{01111} = \mathscr{A}^1_{02222} = 0$ the solution is

$$\left.\begin{aligned} v_1 &= \mathscr{A}^1_{01212} x_2 \psi'(x_1) - (\mathscr{A}^1_{01122} + \mathscr{A}^1_{02112})\phi(x_2) + \chi_1(x_1), \\ v_2 &= \mathscr{A}^1_{02121} x_1 \phi'(x_2) - (\mathscr{A}'_{01122} + \mathscr{A}^1_{02112})\psi(x_1) + \chi_2(x_2), \end{aligned}\right\} \tag{6.3.42}$$

where χ_1, χ_2 are arbitrary and ϕ, ψ arbitrary twice differentiable functions.

Case 7: $a = b = 0$ (with $c > 0$) A straightforward calculation shows that the solution may be represented in the form

$$\begin{aligned} v_1 = &-\tfrac{1}{6}\mathscr{A}^1_{01212}\mathscr{A}^1_{02121} x_2^3 \phi''(x_1) \\ &- (\mathscr{A}^1_{01122} + \mathscr{A}^1_{02112})\{\tfrac{1}{2}x_2^2\psi'(x_1) + x_2\chi'(x_1)\} \\ &- \mathscr{A}^1_{02121}\mathscr{A}^1_{02222} x_2 \phi(x_1) + \kappa(x_1), \\ v_2 = &\ \mathscr{A}^1_{02121}\{\tfrac{1}{2}(\mathscr{A}^1_{01122} + \mathscr{A}^1_{02112})x_2^2\phi'(x_1) + x_2\psi(x_1) + \chi(x_1)\}, \end{aligned} \tag{6.3.43}$$

where ϕ, ψ, χ, κ are arbitrary functions possessing appropriate differentiability.

Case 8: $a = b = c = 0$

$$\left.\begin{aligned} &\text{If } \mathscr{A}^1_{01111} = \mathscr{A}^1_{02222} = 0,\ \mathscr{A}^1_{01212} > 0 \text{ then} \\ &\qquad v_1 = Ax_2 + \chi_1(x_1) + \frac{\partial\phi}{\partial x_1}(x_1, x_2), \\ &\qquad v_2 = -l^{-1}Ax_1 + \chi_2(x_2) + l\frac{\partial\phi}{\partial x_2}(x_1, x_2), \end{aligned}\right\} \tag{6.3.44}$$

where A is a constant,

$$\begin{aligned} l &= -\mathscr{A}^1_{02121}/(\mathscr{A}^1_{01122} + \mathscr{A}^1_{02112}) \\ &= -(\mathscr{A}^1_{01122} + \mathscr{A}^1_{02112})/\mathscr{A}^1_{01212}, \end{aligned} \tag{6.3.45}$$

and χ_1, χ_2, ϕ are arbitrary functions, ϕ being differentiable.

If, instead, $\mathscr{A}^1_{01212} = \mathscr{A}^1_{02121} = 0$, $\mathscr{A}^1_{01111} > 0$, $\mathscr{A}^1_{02222} > 0$ then (6.3.44) is replaced by

$$\left.\begin{aligned} v_1 &= Ax_1 + \chi_2(x_2) + \frac{\partial\phi}{\partial x_2}(x_1, x_2), \\ v_2 &= -m^{-1}Ax_2 + \chi_1(x_1) + m\frac{\partial\phi}{\partial x_1}(x_1, x_2), \end{aligned}\right\} \tag{6.3.46}$$

where

$$m = -\mathscr{A}^1_{01111}/(\mathscr{A}^1_{01122}+\mathscr{A}^1_{02112})$$
$$= -(\mathscr{A}^1_{01122}+\mathscr{A}^1_{02112})/\mathscr{A}^1_{02222}. \tag{6.3.47}$$

Finally, we note that if $\mathscr{A}^1_{01111}=\mathscr{A}^1_{01212}=0$, and hence $\mathscr{A}^1_{01122}+\mathscr{A}^1_{02112}=0$, and $\mathscr{A}^1_{02222}>0$ then no restriction is placed on v_1, and v_2 has the form

$$v_2 = x_2\phi(x_1)+\chi(x_1), \tag{6.3.48}$$

where ϕ and χ are arbitrary functions. A similar result is obtained when the roles of $\mathscr{A}^1_{01111}$ and $\mathscr{A}^1_{02222}$ are reversed.

At this point we observe from (6.3.7) and (6.3.2) that the nominal traction increment on a plane with current unit normal $\mathbf{n}$ is

$$\dot{\mathbf{S}}_0^{\mathrm{T}}\mathbf{n} \equiv \mathscr{A}^1_{0ijkl}v_{l,k}n_i,$$

and by taking $n_i=\delta_{i1},\delta_{i2},\delta_{i3}$ in turn we obtain the incremental tractions on planes normal to the three coordinate axes. These respectively, have components

$$\left.\begin{aligned}\dot{S}_{01j}&\equiv(\mathscr{A}^1_{01111}v_{1,1}+\mathscr{A}^1_{01122}v_{2,2},\mathscr{A}^1_{01212}v_{2,1}+\mathscr{A}^1_{01221}v_{1,2},0),\\ \dot{S}_{02j}&\equiv(\mathscr{A}^1_{02121}v_{1,2}+\mathscr{A}^1_{02112}v_{2,1},\mathscr{A}^1_{01122}v_{1,1}+\mathscr{A}^1_{02222}v_{2,2},0),\\ \dot{S}_{03j}&\equiv(0,0,\mathscr{A}^1_{01133}v_{1,1}+\mathscr{A}^1_{02233}v_{2,2})\end{aligned}\right\} \tag{6.3.49}$$

$(j=1,2,3)$, the last of which is the normal traction required to maintain $v_3=0$ on the plane faces $x_3=\pm l_3$.

Whereas the ratios p_1/p_2 are determined from (6.3.34), the values of p_1 (respectively p_2) are obtained from homogeneous boundary conditions on $x_1=\pm l_1$ (respectively $x_2=\pm l_2$). The boundary conditions on $x_2=\pm l_2$ (respectively $x_1=\pm l_1$) then lead to compatibility (or bifurcation) equations. These equations yield curves of critical values of (λ_1,λ_2) in (λ_1,λ_2)-space for which bifurcation can occur in respect of each value of p_1 at fixed λ_3. The existence of such curves depends on the regime type which itself is dependent on the form of elastic constitutive law. We recall, in particular, that since strong ellipticity does not in general imply the exclusion condition (6.2.133) bifurcation may occur in the strongly elliptic regime. On the other hand, although failure of strong ellipticity implies failure of (6.2.133), bifurcation will not necessarily occur where strong ellipticity fails (Section 6.2.7).

In order to illustrate the theory described above we now consider a specific boundary-value problem in some detail. For this purpose we choose the arbitrary functions arising in the solutions given in Cases 1–8 to be exponential, converted to trigonometric or hyperbolic function form as the boundary conditions and regime type dictate.

For definiteness we impose the boundary conditions

$$v_1 = 0 \quad \text{on} \quad x_1 = \pm l_1 \tag{6.3.50}$$

(i.e. no displacement normal to the faces $x_1 = \pm l_1$ of the block), together with $\dot{S}_{012} = 0$ on $x_1 = \pm l_1$ (i.e. no shear traction on these faces). Since (6.3.50) entails $v_{1,2} = 0$ on $x_1 = \pm l_1$, it follows from (6.3.49) that the latter reduces to

$$\mathscr{A}^1_{01212} v_{2,1} = 0 \quad \text{on} \quad x_1 = \pm l_1. \tag{6.3.51}$$

We take the faces $x_2 = \pm l_2$ to be traction free, so that

$$\left.\begin{aligned} \mathscr{A}^1_{02121} v_{1,2} + \mathscr{A}^1_{02112} v_{2,1} = 0 \quad \text{on} \quad x_2 = \pm l_2, \\ \mathscr{A}^1_{01122} v_{1,1} + \mathscr{A}^1_{02222} v_{2,2} = 0 \quad \text{on} \quad x_2 = \pm l_2. \end{aligned}\right\} \tag{6.3.52}$$

Equations (6.3.50) and (6.3.51) are satisfied by solutions of the form

$$v_1 = -u_1(x_2)\sin p_1 x_1, \qquad v_2 = u_2(x_2)\cos p_1 x_1 \tag{6.3.53}$$

with

$$p_1 = n\pi/l_1 \qquad n = 1, 2, \ldots, \tag{6.3.54}$$

and also by

$$v_1 = u_1(x_2)\cos p_1 x_1, \qquad v_2 = u_2(x_2)\sin p_1 x_1 \tag{6.3.55}$$

with

$$p_1 = (n - \tfrac{1}{2})\pi/l_1 \qquad n = 1, 2, \ldots. \tag{6.3.56}$$

The incremental deformation (6.3.53) is symmetric in x_1, while (6.3.55) is antisymmetric.

For consistency with functional dependence on the argument $p_1 x_1 + p_2 x_2$, as in (6.3.27), u_1 and u_2 are required to be trigonometric functions with arguments $p_2 x_2$ so that the ratios p_2/p_1 satisfy (6.3.34). To proceed further it is necessary to consider Cases 1–8 separately.

First, we introduce the notation

$$\alpha^2 = (b + \sqrt{b^2 - ac})/c, \qquad \beta^2 = (b - \sqrt{b^2 - ac})/c \tag{6.3.57}$$

so that the roots of (6.3.34) are

$$p_2/p_1 = \pm i\alpha, \pm i\beta. \tag{6.3.58}$$

Case 1: α, β real and distinct The functions u_1 and u_2 appearing in (6.3.53) and (6.3.55) have the forms

$$\left.\begin{aligned} u_1 &= A_1 \cosh \alpha p_1 x_2 + B_1 \sinh \alpha p_1 x_2 \\ &\quad + C_1 \cosh \beta p_1 x_2 + D_1 \sinh \beta p_1 x_2, \\ u_2 &= A_2 \sinh \alpha p_1 x_2 + B_2 \cosh \alpha p_1 x_2 \\ &\quad + C_2 \sinh \beta p_1 x_2 + D_2 \cosh \beta p_1 x_2, \end{aligned}\right\} \tag{6.3.59}$$

where $A_1, B_1, \ldots, D_2$ are constants and p_1 is given by either (6.3.54) or (6.3.56).

To ensure that (6.3.59) satisfy the equilibrium equation we must have the connections

$$\left.\begin{aligned} (\mathscr{A}^1_{01111} - \alpha^2 \mathscr{A}^1_{02121})\begin{bmatrix} A_1 \\ B_1 \end{bmatrix} &= \alpha(\mathscr{A}^1_{01122} + \mathscr{A}^1_{02112})\begin{bmatrix} A_2 \\ B_2 \end{bmatrix}, \\ (\mathscr{A}^1_{01111} - \beta^2 \mathscr{A}^1_{02121})\begin{bmatrix} C_1 \\ D_1 \end{bmatrix} &= \beta(\mathscr{A}^1_{01122} + \mathscr{A}^1_{02112})\begin{bmatrix} C_2 \\ D_2 \end{bmatrix}, \end{aligned}\right\} \tag{6.3.60}$$

where

$$\left.\begin{aligned} a - 2b\alpha^2 + c\alpha^4 = 0, \\ a - 2b\beta^2 + c\beta^4 = 0, \end{aligned}\right\} \tag{6.3.61}$$

four independent constants remaining.

Case 2: α real and $\beta = \alpha$ The solution (6.3.59) is replaced by

$$\left.\begin{aligned} u_1 &= A_1 \cosh \alpha p_1 x_2 + B_1 \sinh \alpha p_1 x_2 + C_1 p_1 x_2 \sinh \alpha p_1 x_2 \\ &\quad + D_1 \alpha p_1 x_2 \cosh \alpha p_1 x_2, \\ u_2 &= A_2 \sinh \alpha p_1 x_2 + B_2 \cosh \alpha p_1 x_2 + C_2 \alpha p_1 x_2 \cosh \alpha p_1 x_2 \\ &\quad + D_2 \alpha p_1 x_2 \sinh \alpha p_1 x_2, \end{aligned}\right\} \tag{6.3.62}$$

where

$$(\mathscr{A}^1_{01111} - \alpha^2 \mathscr{A}^1_{02121})\begin{bmatrix} C_1 \\ D_1 \end{bmatrix} = \alpha(\mathscr{A}^1_{01122} + \mathscr{A}^1_{02112})\begin{bmatrix} C_2 \\ D_2 \end{bmatrix}, \tag{6.3.63}$$

$$(\mathscr{A}^1_{01111}-\alpha^2\mathscr{A}^1_{02121})\begin{bmatrix}A_1\\B_1\end{bmatrix}-\alpha(\mathscr{A}^1_{01122}+\mathscr{A}^1_{02112})\begin{bmatrix}A_2\\B_2\end{bmatrix}$$

$$=(\mathscr{A}^1_{01111}+\alpha^2\mathscr{A}^1_{02121})\begin{bmatrix}C_1\\D_1\end{bmatrix} \tag{6.3.64}$$

and

$$\alpha^2=b/c=a/b. \tag{6.3.65}$$

It is left to the reader to relate the solution derived here to the general representation (6.3.37); note that the constant k defined by (6.3.38) may be rewritten as

$$(\mathscr{A}^1_{01111}+\alpha^2\mathscr{A}^1_{02121})/(\mathscr{A}^1_{01111}-\alpha^2\mathscr{A}^1_{02121}).$$

Case 3: α, β complex conjugates The solution may again be represented in the form (6.3.59), but, by writing $\alpha=\gamma+\mathrm{i}\delta$, $\beta=\gamma-\mathrm{i}\delta$, where

$$\gamma=\left(\frac{b+\sqrt{ac}}{2c}\right)^{1/2},\qquad \delta=\left(\frac{\sqrt{ac}-b}{2c}\right)^{1/2}, \tag{6.3.66}$$

it can be recast in real hyperbolic/trigonometric function form. It is convenient for our purposes to leave the solution in the complex form (6.3.59), bearing in mind that the ratios of the constants given in (6.3.60) are also complex.

Case 4: α pure imaginary and $\beta=\alpha$ The functions u_1 and u_2 are as given in (6.3.62), with the hyperbolic functions converted to trigonometric form if necessary.

Case 5: α real and $\beta=0$ In accordance with (6.3.40), we obtain

$$\left.\begin{aligned}u_1&=A_1\cosh\alpha p_1x_2+B_1\sinh\alpha p_1x_2+C_1-D_1p_1\mathscr{A}^1_{01212}x_2,\\ u_2&=A_2\sinh\alpha p_1x_2+B_2\cosh\alpha p_1x_2+D_1(\mathscr{A}^1_{01122}+\mathscr{A}^1_{02112}),\end{aligned}\right\} \tag{6.3.67}$$

with

$$-\alpha\mathscr{A}^1_{02121}\begin{bmatrix}A_1\\B_1\end{bmatrix}=(\mathscr{A}^1_{01122}+\mathscr{A}^1_{02112})\begin{bmatrix}A_2\\B_2\end{bmatrix} \tag{6.3.68}$$

and

$$\alpha^2=2b/c. \tag{6.3.69}$$

Case 6: $\alpha^{-1}=\beta=0$ Use of the boundary conditions (6.3.50) and (6.3.51) enables (6.3.41) and (6.3.42) to be simplified without the need to assume a trigonometric form for the x_1-dependence. First we deduce from (6.3.50) that (6.3.41) yields

$$\pm \mathscr{A}^1_{02222} l_1 \phi'(x_2) - (\mathscr{A}^1_{01122} - t_1)\psi(\pm l_1) + \chi_2(x_2) = 0$$

while (6.3.51) is automatically satisfied. Hence

$$\phi'(x_2) = \text{constant}, \qquad \chi_2(x_2) = \text{constant}. \tag{6.3.70}$$

Secondly, in respect of (6.3.42) the boundary conditions give

$$\mathscr{A}^1_{01212} x_2 \psi'(\pm l_1) - (\mathscr{A}^1_{01122} + \mathscr{A}^1_{02112})\phi(x_2) + \chi_1(\pm l_1) = 0,$$
$$\mathscr{A}^1_{02121} \phi'(x_2) - (\mathscr{A}^1_{01122} + \mathscr{A}^1_{02112})\psi'(\pm l_1) = 0,$$

from which we deduce

$$\phi'(x_2) = \text{constant}, \ \psi'(l_1) = \psi'(-l_1), \chi_1(l_1) = \chi_1(-l_1). \tag{6.3.71}$$

Case 7: $\alpha=\beta=0$ The general solution (6.3.43) satisfies the boundary conditions (6.3.50) and (6.3.51) provided

$$\kappa(\pm l_1) = \phi(\pm l_1) = \chi'(\pm l_1) = \psi'(\pm l_1) = \phi''(\pm l_1) = 0. \tag{6.3.72}$$

Case 8: α and β indeterminate For the three possibilities (6.3.44), (6.3.46) and (6.3.48) the boundary conditions

$$A = 0, \quad \frac{\partial \phi}{\partial x_1}(\pm l_1, x_2) + \chi_1(\pm l_1) = 0, \tag{6.3.73}$$

$$\frac{\partial \phi}{\partial x_2}(\pm l_1, x_2) + \chi_2(x_2) \pm A l_1 = 0, \tag{6.3.74}$$

and

$$v_1(\pm l_1, x_2) = 0 \tag{6.3.75}$$

respectively are obtained on use of (6.3.50) and (6.3.51).

It now remains to ensure that the various solutions given above satisfy the second set of boundary conditions, namely (6.3.52). First, we note that, in respect of Cases 1–5, (6.3.52) becomes

$$\left.\begin{aligned} &\mathscr{A}^1_{02121}u_1' + p_1\mathscr{A}^1_{02112}u_2 = 0 \quad \text{on } x_2 = \pm l_2,\\ &p_1\mathscr{A}^1_{01122}u_1 - \mathscr{A}^1_{02222}u_2' = 0 \quad \text{on } x_2 = \pm l_2. \end{aligned}\right. \tag{6.3.76}$$

For Case 1, substitution of (6.3.59) into (6.3.76) followed by some rearrangement of the resulting equations leads to

$$\left.\begin{aligned} &(\mathscr{A}^1_{02121}A_1\alpha + \mathscr{A}^1_{02112}A_2)\sinh\alpha p_1 l_2\\ &\quad + (\mathscr{A}^1_{02121}C_1\beta + \mathscr{A}^1_{02112}C_2)\sinh\beta p_1 l_2 = 0,\\ &(\mathscr{A}^1_{01122}A_1 - \mathscr{A}^1_{02222}A_2\alpha)\cosh\alpha p_1 l_2\\ &\quad + (\mathscr{A}^1_{01122}C_1 - \mathscr{A}^1_{02222}C_2\beta)\cosh\beta p_1 l_2 = 0, \end{aligned}\right\} \tag{6.3.77}$$

$$\left.\begin{aligned} &(\mathscr{A}^1_{02121}B_1\alpha + \mathscr{A}^1_{02112}B_2)\cosh\alpha p_1 l_2\\ &\quad + (\mathscr{A}^1_{02121}D_1\beta + \mathscr{A}^1_{02112}D_2)\cosh\beta p_1 l_2 = 0,\\ &(\mathscr{A}^1_{01122}B_1 - \mathscr{A}^1_{02222}B_2\alpha)\sinh\alpha p_1 l_2\\ &\quad + (\mathscr{A}^1_{01122}D_1 - \mathscr{A}^1_{02222}D_2\beta)\sinh\beta p_1 l_2 = 0. \end{aligned}\right\} \tag{6.3.78}$$

In view of the connections (6.3.60) we see that the two pairs of equations (6.3.77) and (6.3.78) are decoupled and may therefore be examined independently.

Equations (6.3.77) yield non-trivial solutions for A_1 and C_1 provided, after use of (6.3.60), the determinant of coefficients vanishes. After some algebraic manipulation this leads to

$$\frac{\tanh\alpha p_1 l_2}{\tanh\beta p_1 l_2} = \frac{\alpha}{\beta}\left(\frac{\beta^2 - \mathscr{A}}{\alpha^2 - \mathscr{A}}\right), \tag{6.3.79}$$

where

$$\mathscr{A} = \frac{\mathscr{A}^1_{01111}[\mathscr{A}^1_{02121}\mathscr{A}^1_{01212} - (\mathscr{A}^1_{02112})^2]}{\mathscr{A}^1_{02121}[\mathscr{A}^1_{01111}\mathscr{A}^1_{02222} - (\mathscr{A}^1_{01122})^2]} \tag{6.3.80}$$

and α, β are given in terms of the elastic moduli by (6.3.57) with (6.3.32).

For a given value of λ_3 and for given mode number p_1 equation (6.3.79) determines the set of points in (λ_1, λ_2)-space for which a non-trivial solution with $B_1 = D_1 = B_2 = D_2 = 0$ can occur. Such points define eigenconfigurations for the considered boundary-value problem and mode number. The corresponding eigenmodes are

$$\left.\begin{aligned} v_1 &= -\sin p_1 x_1(A_1\cosh\alpha p_1 x_2 + C_1\cosh\beta p_1 x_2),\\ v_2 &= \cos p_1 x_1(A_2\sinh\alpha p_1 x_2 + C_2\sinh\beta p_1 x_2) \end{aligned}\right\} \tag{6.3.81}$$

for mode numbers given by (6.3.54), and

$$\left.\begin{aligned} v_1 &= \cos p_1 x_1 (A_1 \cosh \alpha p_1 x_2 + C_1 \cosh \beta p_1 x_2),\\ v_2 &= \sin p_1 x_1 (A_2 \sinh \alpha p_1 x_2 + C_2 \sinh \beta p_1 x_2) \end{aligned}\right\} \tag{6.3.82}$$

for mode numbers (6.3.56), the ratios of the constants being given by either equation in (6.3.77) together with (6.3.60)

With $A_1 = C_1 = A_2 = C_2 = 0$, on the other hand, the bifurcation criterion is easily seen to be

$$\frac{\tanh \beta p_1 l_2}{\tanh \alpha p_1 l_2} = \frac{\alpha}{\beta}\left(\frac{\beta^2 - \mathscr{A}}{\alpha^2 - \mathscr{A}}\right), \tag{6.3.83}$$

and the eigenmodes are

$$\left.\begin{aligned} v_1 &= \begin{Bmatrix} -\sin p_1 x_1 \\ \cos p_1 x_1 \end{Bmatrix} (B_1 \sinh \alpha p_1 x_2 + D_1 \sinh \beta p_1 x_2),\\ v_2 &= \begin{Bmatrix} \cos p_1 x_1 \\ \sin p_1 x_1 \end{Bmatrix} (B_2 \cosh \alpha p_1 x_2 + D_2 \cosh \beta p_1 x_2) \end{aligned}\right\} \tag{6.3.84}$$

for p_1 given by $\begin{Bmatrix} (6.3.54) \\ (6.3.56) \end{Bmatrix}$.

Whether or not there exist real pairs (λ_1, λ_2) satisfying (6.3.79) or (6.3.83) for any p_1 and λ_3 depends on the form of constitutive law, and this enters these equations through α, β and $\mathscr{A}$. Note that since Case 1 excludes $\alpha^2 = \beta^2$ equations (6.3.79) and (6.3.83) cannot be satisfied simultaneously.

In Case 2 substitution of (6.3.62) into (6.3.76) and use of (6.3.63)–(6.3.65) leads to the bifurcation criterion

$$\frac{\sinh 2\alpha p_1 l_2}{2\alpha p_1 l_2} = -\frac{\alpha^2 - \mathscr{A}}{\alpha^2 + \mathscr{A}} \tag{6.3.85}$$

for $B_1 = D_1 = B_2 = D_2 = 0$, where $\mathscr{A}$ is again given by (6.3.80), and

$$\frac{\sin 2\alpha p_1 l_2}{2\alpha p_1 l_2} = \frac{\alpha^2 - \mathscr{A}}{\alpha^2 + \mathscr{A}}. \tag{6.3.86}$$

for $A_1 = C_1 = A_2 = C_2 = 0$. The details, which involve some tedious algebraic manipulations, are left as an exercise for the reader.

The bifurcation criterion for Case 3 can be obtained directly from that for Case 1 by replacing α, β by $\gamma + i\delta, \gamma - i\delta$ respectively, where γ and δ are given by

(6.3.66). Equations (6.3.79) and (6.3.83) respectively yield

$$\frac{\sinh 2\gamma p_1 l_2}{\sin 2\delta p_1 l_2} = \frac{\gamma(\mathscr{A} - \gamma^2 - \delta^2)}{\delta(\mathscr{A} + \gamma^2 + \delta^2)} \tag{6.3.87}$$

and

$$\frac{\sinh 2\gamma p_1 l_2}{\sin 2\delta p_1 l_2} = -\frac{\gamma(\mathscr{A} - \gamma^2 - \delta^2)}{\delta(\mathscr{A} + \gamma^2 + \delta^2)}. \tag{6.3.88}$$

For Case 4 the bifurcation criterion is obtained from either (6.3.85) or (6.3.86) by setting $\alpha = \beta = \mathrm{i}\delta$, where $\delta = (a/c)^{1/4}$.

In Case 5 only one non-trivial mode is possible, this corresponding to $A_1 = C_1 = A_2 = 0$ in (6.3.67). The bifurcation criterion is

$$\begin{aligned}\frac{\tanh \alpha p_1 l_2}{\alpha p_1 l_2} = {} & (\mathscr{A}^1_{01122})^2 \mathscr{A}^1_{01212} \mathscr{A}^1_{02121} [\mathscr{A}^1_{02121} \mathscr{A}^1_{01212} \\ & - \mathscr{A}^1_{02112}(\mathscr{A}^1_{01122} + \mathscr{A}^1_{02112})]^{-2}\end{aligned} \tag{6.3.89}$$

and can be obtained either directly by substitution of (6.3.67) into (6.3.76) or by a limiting process letting $\beta \to 0$ in (6.3.83) bearing in mind that $\mathscr{A} \to 0$ simultaneously with β.

It is a straightforward matter to show that in Case 6 there is no non-trivial solution to the incremental problem.

Substitution of (6.3.43) into the boundary conditions (6.3.52) subject to (6.3.72) and with $\mathscr{A}^1_{01111} = 0$ shows that in Case 7 $\psi = \kappa = 0$, but non-trivial modes with

$$\phi = A \begin{cases} -\sin p_1 x_1 \\ \cos p_1 x_1 \end{cases}, \qquad \chi = B \begin{cases} \cos p_1 x_1 \\ \sin p_1 x_1 \end{cases}$$

and

$$p_1 l_1 = \begin{cases} n\pi \\ (n - \frac{1}{2})\pi \end{cases} \quad n = 1, 2, \ldots$$

are possible provided

$$\tfrac{1}{3} p_1^2 l_2^2 \mathscr{A}^1_{01212} \mathscr{A}^1_{01122} = \mathscr{A}^1_{02222}(\mathscr{A}^1_{01122} + 2\mathscr{A}^1_{02112}). \tag{6.3.90}$$

For a given p_1 for which (6.3.90) is satisfied, the ratio of the constants A and

B is given by

$$B/A = \{\tfrac{1}{2}p_1^2 l_2^2 \mathscr{A}^1_{01122}(\mathscr{A}^1_{01122} + \mathscr{A}^1_{02112}) \\ + \mathscr{A}^1_{02121}\mathscr{A}^1_{02222}\}/p_1\mathscr{A}^1_{01122}.$$

The result (6.3.90) may also be obtained from (6.3.89) by taking the limit as $\alpha \to 0$.

Case 8 will occur, if at all, only at isolated points in $(\lambda_1, \lambda_2, \lambda_3)$-space. The boundary conditions (6.3.52), coupled with (6.3.73), reduce the solution (6.3.44) to $v_1 = 0$, $v_2 =$ arbitrary function of x_2. Similarly, (6.3.46) reduces to $v_1 = 0$, $v_2 =$ arbitrary function of x_1, while (6.3.48) yields $v_1 = 0$, $v_2 =$ constant.

The above discussion has not exhausted all the possibilities. For example, since the boundary conditions are not symmetric with respect to interchange of x_1 and x_2, different results can be obtained in Cases 5 and 7 if the conditions $a = 0$, $c > 0$ are replaced by $a > 0$, $c = 0$. Different boundary conditions will affect the details of the solutions but not their general structure so we do not examine alternative boundary conditions here.

In each of the regimes considered above the existence of critical values of (λ_1, λ_2) satisfying the relevant bifurcation criterion for a given λ_3 depends, in particular, on the signs of the stress components t_1, t_2 given by (6.3.18). In the stable domain, of course, bifurcation cannot occur since $\mathscr{A}^1_0$ is positive definite, but it is clear from our discussion in Section 6.2.7 that stability may fail where strong ellipticity holds; in particular, this may happen where the stresses are compressive. In this connection we recall from (6.2.101) that for *plane* incremental deformations necessary and sufficient conditions for the exclusion condition (6.2.133) to hold are

$$\left.\begin{aligned} &\frac{\partial t_1^{(1)}}{\partial \lambda_1} > 0, \quad \frac{\partial t_1^{(1)}}{\partial \lambda_1}\frac{\partial t_2^{(1)}}{\partial \lambda_2} - \frac{\partial t_1^{(1)}}{\partial \lambda_2}\frac{\partial t_2^{(1)}}{\partial \lambda_1} > 0, \\ &(t_1^{(1)} - t_2^{(1)})/(\lambda_1 - \lambda_2) > 0, \; t_1^{(1)} + t_2^{(1)} > 0. \end{aligned}\right\} \tag{6.3.91}$$

This does not take account of possible instabilities due to out-of-plane displacements. Corresponding necessary and sufficient conditions for strong ellipticity are, from (6.2.152)–(6.2.154),

$$\left.\begin{aligned} &\frac{\partial t_1^{(1)}}{\partial \lambda_1} > 0, \quad \frac{\partial t_2^{(1)}}{\partial \lambda_2} > 0, \quad \frac{(\lambda_1 t_1^{(1)} - \lambda_2 t_2^{(1)})}{\lambda_1^2 - \lambda_2^2} > 0, \\ &\frac{\partial t_1^{(1)}}{\partial \lambda_1}\frac{\partial t_2^{(1)}}{\partial \lambda_2} + \left(\frac{\lambda_1 t_1^{(1)} - \lambda_2 t_2^{(1)}}{\lambda_1^2 - \lambda_2^2}\right)^2 - \left(\frac{\partial t_1^{(1)}}{\partial \lambda_2} + \frac{\lambda_2 t_1^{(1)} - \lambda_1 t_2^{(1)}}{\lambda_1^2 - \lambda_2^2}\right)^2 \\ &\quad + 2\left(\frac{\lambda_1 t_1^{(1)} - \lambda_2 t_2^{(1)}}{\lambda_1^2 - \lambda_2^2}\right)\left(\frac{\partial t_1^{(1)}}{\partial \lambda_1}\frac{\partial t_2^{(1)}}{\partial \lambda_2}\right)^{1/2} > 0. \end{aligned}\right\} \tag{6.3.92}$$

The choice of constitutive law also has an important influence on the bifurcation criteria. In general, however, the constitutive law enters the bifurcation criteria in a complicated way and it is therefore difficult to extract meaningful information without selecting a relatively simple form of strain-energy function and using numerical calculations. The situation is different in the corresponding problem for incompressible materials; for such materials some general conclusions can be drawn, as we see shortly. For this reason we use the results for incompressible materials, rather than those for unconstrained materials, to illustrate the consequences of bifurcation criteria.

As far as the writer is aware the bifurcation criteria derived are new results in that they apply to a general isotropic elastic strain-energy function. Special cases of some of these results have appeared previously in the literature, but attention has been restricted to specific constitutive laws from the outset. We cite, in particular, Burgess and Levinson (1972) who examined Case 1 for two choices of strain-energy function; further references are contained in their paper.

At this point we observe that in Cases 1–5 and 7 the mode number and geometry appear in the bifurcation criteria only through a single parameter, namely $p_1 l_2$. Thus, an increase (respectively decrease) in the mode number at fixed aspect ratio l_2/l_1 is equivalent to an increase (respectively decrease) in l_2/l_1 at fixed mode number so the two quantities need not be varied independently in any numerical calculations. In Case 8 the bifurcation criterion (the simultaneous vanishing of certain material parameters) involves no mode number and is independent of the geometry.

Apart from the content of the problems given below our discussion of unconstrained materials is now complete.

Problem 6.3.1 For the strain-energy function

$$W = \tfrac{1}{2}\mu(\lambda_1^2 + \lambda_2^2 + \lambda_3^2 - 3 - 2\ln J) + \tfrac{1}{2}\mu'(J-1)^2$$

defined in (4.4.1) show that α^2 and β^2 are λ_1^2/λ_2^2 and

$$\frac{\mu(\lambda_1^2 + 1) + \mu' J^2}{\mu(\lambda_2^2 + 1) + \mu' J^2}.$$

Assuming that $\mu > 0, \mu' < 0, |\mu/\mu'| > 1$ and $\lambda_3 = 1$, find the sets of points in the (λ_1, λ_2)-plane for which each of Cases 1, 2, 5, 7, 8 pertains. Analyse the bifurcation criterion for the case when $\lambda_1 = \lambda_2$.

Problem 6.3.2 For a pure dilatation with $\lambda_1 = \lambda_2 = \lambda_3 = \lambda$ and corresponding Cauchy stress $t_1 = t_2 = t_3 = t$ we have

$$\mathscr{A}^1_{0ijkl} = \mu_1 \delta_{ij}\delta_{kl} + (\mu_2 + t)\delta_{ik}\delta_{jl} + \mu_2 \delta_{il}\delta_{jk},$$

where μ_1 and μ_2 are defined in Problem 6.1.5. Deduce that $\alpha^2=\beta^2=1$ and show that the bifurcation criterion is

$$\frac{\sinh 2p_1l_2}{2p_1l_2}=\pm\frac{(t+2\mu_2)(\mu_1+\mu_2)}{[2t^2+t(3\mu_1+5\mu_2)+2\mu_2(\mu_1+\mu_2)]}.$$

More generally, show that if $\lambda_1=\lambda_2$ and $t_1=t_2=t$ then $\alpha^2=\beta^2=1$ and

$$\frac{\sinh 2p_1l_2}{2p_1l_2}=\pm\frac{\mathscr{A}^1_{01212}(\mathscr{A}^1_{01111}+\mathscr{A}^1_{01122})-t\mathscr{A}^1_{01111}}{\mathscr{A}^1_{01212}(\mathscr{A}^1_{01111}+\mathscr{A}^1_{01122})+t\mathscr{A}^1_{01111}}.$$

Show that the associated modes (v_1, v_2) satisfy the biharmonic equation

$$v_{\alpha,1111}+2v_{\alpha,1122}+v_{\alpha,2222}=0.$$

Problem 6.3.3 Examine the asymptotic forms of the bifurcation criteria (6.3.79) and (6.3.83) for $l_2/l_1 \ll 1$. In particular, consider the (plane) simple compression test in which $t_1=0$ and $t_2<0$ for both $\mu'>0$ and $\mu'<0$ in respect of the strain-energy function used in Problem 6.3.1.

Problem 6.3.4 A half-space of isotropic elastic material is defined by $x_2 \le 0$ relative to rectangular Cartesian coordinates in some current configuration. The underlying homogeneous deformation is a pure strain with stretches $\lambda_1, \lambda_2, \lambda_3$ corresponding to the x_1, x_2, x_3 axes maintained by Cauchy stresses t_1, t_2, t_3.

Assuming that the incremental traction on $x_2=0$ vanishes and that the incremental deformation has the form (6.3.53) with $u_\alpha(x_2)\to 0$ as $x_2\to-\infty$ ($\alpha=1,2$), show that, in Case 1,

$$u_1=A_1\,\mathrm{e}^{\alpha p_1x_2}+C_1\,\mathrm{e}^{\beta p_1x_2},$$
$$u_2=A_2\,\mathrm{e}^{\alpha p_1x_2}+C_2\,\mathrm{e}^{\beta p_1x_2},$$

where $A_1/A_2, C_1/C_2$ are given by (6.3.60), $\alpha>0$ and $\beta>0$.

Obtain the bifurcation criterion

$$\mathscr{A}^1_{01111}\{(\mathscr{A}^1_{02112})^2-\mathscr{A}^1_{02121}\mathscr{A}^1_{01212}\}$$
$$+\alpha\beta\mathscr{A}^1_{02121}\{\mathscr{A}^1_{01111}\mathscr{A}^1_{02222}-(\mathscr{A}^1_{01122})^2\}=0$$

provided $\mathscr{A}^1_{01122}+\mathscr{A}^1_{02112}\neq 0$ and show that it reduces to

$$\mathscr{A}^1_{01111}-\mathscr{A}^1_{01122}=0$$

when $\lambda_1=\lambda_2$.

Derive corresponding results for Cases 2 and 3.

Problem 6.3.5 If $\lambda_1 = \lambda_2$ show that the solution of equations (6.3.23) may be represented in the forms

$$v_1 = \frac{\partial \phi}{\partial x_1} + \frac{\partial \psi}{\partial x_2}, \qquad v_2 = \frac{\partial \phi}{\partial x_2} - \frac{\partial \psi}{\partial x_1},$$

where ϕ, ψ are functions of (x_1, x_2) satisfying the biharmonic equation, $\mathscr{A}^1_{01111}\nabla^2\phi$ and $\mathscr{A}^1_{02121}\nabla^2\psi$ are conjugate harmonic functions and ∇^2 is the two-dimensional Laplacean operator.

Problem 6.3.6 Show that for Cases 1–4 the bifurcation criteria may be written

$$\frac{\sinh(\alpha+\beta)\eta}{(\alpha+\beta)}[\alpha\beta+\mathscr{A}] = \pm\frac{\sinh(\alpha-\beta)\eta}{(\alpha-\beta)}[\alpha\beta-\mathscr{A}]$$

with

$$\alpha^2\beta^2 = a/c,$$

$$(\alpha+\beta)^2 = 2(b+\sqrt{ac})/c,$$

$$(\alpha-\beta)^2 = 2(b-\sqrt{ac})/c,$$

and $[\sinh(\alpha-\beta)\eta]/(\alpha-\beta)$ replaced by η when $\beta = \alpha$, where $\eta = p_1 l_2$.

Analysis for incompressible materials For an incompressible material the underlying principal stretches are such that

$$\lambda_1\lambda_2\lambda_3 = 1, \tag{6.3.93}$$

and the principal Cauchy stresses are given by (6.3.19). The incremental governing equations (6.3.24) are coupled with the incremental incompressibility constraint (6.3.25), while the other conditions of the problem are just as for unconstrained materials. The bifurcation results are very similar to those obtained for unconstrained materials, but, as we see below, the material properties enter the bifurcation criteria in a much simpler form in the case of incompressibility.

From (6.3.25) we deduce the existence of a scalar function $\psi(x_1, x_2)$ such that

$$v_1 = \frac{\partial \psi}{\partial x_2}, \qquad v_2 = -\frac{\partial \psi}{\partial x_1}. \tag{6.3.94}$$

Substitution of (6.3.94) into (6.3.24) and elimination of $\dot{q}$ leads to an equation for ψ, namely

$$\mathscr{A}^1_{01212}\psi_{,1111}+(\mathscr{A}^1_{01111}+\mathscr{A}^1_{02222}-2\mathscr{A}^1_{01122}-2\mathscr{A}^1_{02112})\times \\ \times\psi_{,1122}+\mathscr{A}^1_{02121}\psi_{,2222}=0. \tag{6.3.95}$$

With

$$\psi=f(p_1x_1+p_2x_2),$$

analogously to (6.3.27), inserted into (6.3.95) we obtain

$$a'p_1^4+2b'p_1^2p_2^2+c'p_2^4=0, \tag{6.3.96}$$

where

$$\left.\begin{aligned} a'=\mathscr{A}^1_{01212},\quad c'=\mathscr{A}^1_{02121},\\ 2b'=\mathscr{A}^1_{01111}+\mathscr{A}^1_{02222}-2\mathscr{A}^1_{01122}-2\mathscr{A}^1_{02112}.\end{aligned}\right\} \tag{6.3.97}$$

Necessary and sufficient conditions for the (two-dimensional) strong ellipticity condition to hold are

$$a'>0,\quad c'>0,\quad b'>-\sqrt{a'c'}, \tag{6.3.98}$$

the first two of these inequalities being equivalent since, from (6.3.15), $a'/c'=\lambda_1^2/\lambda_2^2$. By contrast we note that (6.3.33) are necessary but not sufficient for strong elliplicity in the case of unconstrained materials. The inequalities (6.3.98) may be deduced from the results of Problem 6.2.14.

Just as for unconstrained materials the results depend on the location in the strongly-elliptic domain or on its boundary. The various possibilities are obtained by replacing a, b, c in Cases 1–4, 6, 8 by a', b', c', Cases 5 and 7 not arising for incompressible materials.

We now write

$$\psi=u(x_2)\sin p_1x_1 \tag{6.3.99}$$

so that

$$v_1=u'(x_2)\sin p_1x_1,\qquad v_2=-p_1u(x_2)\cos p_1x_1 \tag{6.3.100}$$

and the boundary conditions (6.3.50) and (6.3.51) are satisfied if p_1 is given by (6.3.54), or if p_1 is given by (6.3.56) when $\sin p_1x_1$ is replaced by $\cos p_1x_1$ in (6.3.99).

With the help of (6.3.3), (6.3.15) and (6.3.19) we see that the boundary conditions on $x_2 = \pm l_2$ take the forms

$$\dot{S}_{021} \equiv \mathscr{A}^1_{02121}(v_{1,2} + v_{2,1}) - t_2 v_{2,1} = 0,$$
$$\dot{S}_{022} \equiv \mathscr{A}^1_{01122} v_{1,1} + (\mathscr{A}^1_{02222} - q) v_{2,2} + \dot{q} = 0.$$

By writing $\dot{q} = w(x_2)\cos p_1 x_1$ and eliminating $w(x_2)$ by means of the first equation in (6.3.24) these boundary conditions can be rearranged as

$$\left.\begin{aligned} &c'u''(\pm l_2) + (c' - t_2)p_1^2 u(\pm l_2) = 0, \\ &c'u'''(\pm l_2) - (2b' + c' - t_2)p_1^2 u'(\pm l_2) = 0, \end{aligned}\right\} \tag{6.3.101}$$

while (6.3.95) yields the equation

$$c'u''''(x_2) - 2b'p_1^2 u''(x_2) + a'p_1^4 u(x_2) = 0. \tag{6.3.102}$$

For the following discussion we use the notation

$$\alpha^2 = (b' + \sqrt{b'^2 - a'c'})/c', \quad \beta^2 = (b' - \sqrt{b'^2 - a'c'})/c'. \tag{6.3.103}$$

Case 1: $a' > 0,\ c' > 0,\ b' > \sqrt{a'c'}$; α, β real and distinct

We write the general solution of (6.3.102) as

$$\begin{aligned} u(x_2) = {} & A_1 \cosh \alpha p_1 x_2 + A_2 \sinh \alpha p_1 x_2 \\ & + B_1 \cosh \beta p_1 x_2 + B_2 \sinh \beta p_1 x_2. \end{aligned} \tag{6.3.104}$$

Substitution of this into (6.3.101), followed by some rearrangement, leads to

$$\left.\begin{aligned} &(c'\alpha^2 + c' - t_2)A_1 \cosh \alpha p_1 l_2 \\ &\quad + (c'\beta^2 + c' - t_2)B_1 \cosh \beta p_1 l_2 = 0, \\ &(c'\alpha^2 - 2b' - c' + t_2)\alpha A_1 \sinh \alpha p_1 l_2 \\ &\quad + (c'\beta^2 - 2b' - c' + t_2)\beta B_1 \sinh \beta p_1 l_2 = 0, \end{aligned}\right\} \tag{6.3.105}$$

$$\left.\begin{aligned} &(c'\alpha^2 + c' - t_2)A_2 \sinh \alpha p_1 l_2 \\ &\quad + (c'\beta^2 + c' - t_2)B_2 \sinh \beta p_1 l_2 = 0, \\ &(c'\alpha^2 - 2b' - c' + t_2)\alpha A_2 \cosh \alpha p_1 l_2 \\ &\quad + (c'\beta^2 - 2b' - c' + t_2)\beta B_2 \cosh \beta p_1 l_2 = 0. \end{aligned}\right\} \tag{6.3.106}$$

Two independent modes are possible. Firstly, if $A_2 = B_2 = 0$, a non-trivial solution

$$u(x_2) = A_1 \cosh \alpha p_1 x_2 + B_1 \cosh \beta p_1 x_2 \tag{6.3.107}$$

is possible provided

$$\frac{\tanh \alpha p_1 l_2}{\tanh \beta p_1 l_2} = \frac{\beta}{\alpha}\left(\frac{c'\alpha^2 + c' - t_2}{c'\beta^2 + c' - t_2}\right)^2. \tag{6.3.108}$$

Secondly, a solution

$$u(x_2) = A_2 \sinh \alpha p_1 x_2 + B_2 \sinh \beta p_1 x_2 \tag{6.3.109}$$

is possible if

$$\frac{\tanh \alpha p_1 l_2}{\tanh \beta p_1 l_2} = \frac{\alpha}{\beta}\left(\frac{c'\beta^2 + c' - t_2}{c'\alpha^2 + c' - t_2}\right)^2. \tag{6.3.110}$$

In deriving the above formulae we have made use of the connection

$$c'\alpha^2 + c' - t_2 = -(c'\beta^2 - 2b' - c' + t_2).$$

Since (6.3.107) yields incremental displacements v_1 and v_2 which are respectively odd and even functions of x_2 the mode of deformation is said to be *flexural*. On the other hand, (6.3.109) corresponds to a *barreling* mode of deformation since v_1 and v_2 are even and odd functions of x_2 respectively.

When the underlying state of stress is such that $t_2 = 0$, the bifurcation criterion (6.3.108) reduces to

$$\frac{\tanh \alpha p_1 l_2}{\tanh \beta p_1 l_2} = \frac{\beta}{\alpha}\left(\frac{\alpha^2 + 1}{\beta^2 + 1}\right)^2 \tag{6.3.111}$$

and similarly (6.3.110) becomes

$$\frac{\tanh \alpha p_1 l_2}{\tanh \beta p_1 l_2} = \frac{\alpha}{\beta}\left(\frac{\beta^2 + 1}{\alpha^2 + 1}\right)^2. \tag{6.3.112}$$

These latter two bifurcation criteria have been examined in detail in the literature, notably by Sawyers and Rivlin (1974) and Sawyers (1977), and we refer to these papers for further details. Equations (6.3.108) and (6.3.110), however, have not previously appeared in the literature as far as the author is aware.

The implications of the bifurcation criteria will be discussed after the list of special cases is complete.

Case 2: $a'>0, c'>0, b'=\sqrt{a'c'}; \alpha^2=\beta^2=\lambda_1/\lambda_2$

The solution (6.3.104) is replaced by

$$\begin{aligned} u(x_1) = {} & A_1 \cosh \alpha p_1 x_2 + A_2 \sinh \alpha p_1 x_2 \\ & + B_1 \alpha p_1 x_2 \sinh \alpha p_1 x_2 + B_2 \alpha p_1 x_2 \cosh \alpha p_1 x_2 \end{aligned} \tag{6.3.113}$$

and the boundary conditions yield

$$\begin{aligned} & (b'+c'-t_2)A_1 \cosh \alpha p_1 l_2 + [2b' \cosh \alpha p_1 l_2 \\ & \quad + (b'+c'-t_2)\alpha p_1 l_2 \sinh \alpha p_1 l_2]B_1 = 0, \\ & \quad -(b'+c'-t_2)A_1 \sinh \alpha p_1 l_2 + [(b'-c'+t_2)\times \\ & \quad \times \sinh \alpha p_1 l_2 - (b'+c'-t_2)\alpha p_1 l_2 \cosh \alpha p_1 l_2]B_1 = 0, \\ & (b'+c'-t_2)A_2 \sinh \alpha p_1 l_2 + [2b' \sinh \alpha p_1 l_2 \\ & \quad + (b'+c'-t_2)\alpha p_1 l_2 \cosh \alpha p_1 l_2]B_2 = 0, \\ & -(b'+c'-t_2)A_2 \cosh \alpha p_1 l_2 + [(b'-c'+t_2)\times \\ & \quad \times \cosh \alpha p_1 l_2 - (b'+c'-t_2)\alpha p_1 l_2 \sinh \alpha p_1 l_2]B_2 = 0. \end{aligned}$$

The bifurcation criterion is *either*

$$b'+c'-t_2 \equiv \frac{t_1\lambda_2 - t_2\lambda_1}{\lambda_1-\lambda_2} = 0 \tag{6.3.114}$$

in which case $B_1 = B_2 = 0$ and barreling and flexural modes with $A_1 \neq 0$, $A_2 \neq 0$ can occur simultaneously, *or*

$$\begin{aligned} \frac{\sinh 2\alpha p_1 l_2}{2\alpha p_1 l_2} & = \pm \frac{b'+c'-t_2}{3b'-c'+t_2} \\ & \equiv \pm \frac{(t_1\lambda_2 - t_2\lambda_1)(\lambda_1+\lambda_2)}{t_1\lambda_2(3\lambda_1-\lambda_2) - t_2\lambda_1(3\lambda_2-\lambda_1)}, \end{aligned} \tag{6.3.115}$$

in which the plus (minus) prefix corresponds to the flexural (barreling) mode. When $t_2 = 0$, equation (6.3.115) reduces to

$$\frac{\sinh 2\alpha p_1 l_2}{2\alpha p_1 l_2} = \pm \frac{\lambda_1+\lambda_2}{3\lambda_1-\lambda_2}, \tag{6.3.116}$$

a result first given by Sawyers and Rivlin (1974).

Clearly, since the left-hand side of (6.3.116) is positive, flexural modes can only occur if $\lambda_2 < 3\lambda_1$ and barreling modes if $\lambda_2 > 3\lambda_1$. Moreover, in the flexural case it is necessary that $\lambda_2 > \lambda_1$ for (6.3.116) to have a real solution (λ_1, λ_2) for given $p_1 l_2$.

Case 3: $a' > 0, \quad c' > 0, \quad b'^2 < a'c'; \quad \alpha = \gamma + i\delta, \quad \beta = \gamma - i\delta$

Here

$$\gamma = \left(\frac{b' + \sqrt{a'c'}}{2c'}\right)^{1/2}, \qquad \delta = \left(\frac{\sqrt{a'c'} - b'}{2c'}\right)^{1/2} \tag{6.3.117}$$

and the solution for $u(x_2)$ again has the representation (6.3.104).

It is left as an exercise to show that the bifurcation criterion can be written

$$\frac{\sinh 2\gamma p_1 l_2}{\sin 2\delta p_1 l_2} = \pm \frac{\gamma\,[2c'\lambda_1\lambda_2(b' + c' - t_2) - c'^2\lambda_1^2 + (c' - t_2)^2\lambda_2^2]}{\delta\,[2c'\lambda_1\lambda_2(b' + c' - t_2) + c'^2\lambda_1^2 - (c' - t_2)^2\lambda_2^2]}, \tag{6.3.118}$$

with the plus (minus) sign corresponding to flexural (barreling) modes, and to deduce that when $t_2 = 0$ this becomes

$$\frac{\sinh 2\gamma p_1 l_2}{\sin 2\delta p_1 l_2} = \pm \frac{\gamma\,[2\lambda_1(b' + c') - \lambda_2 t_1]}{\delta\,[2\lambda_1(b' + c') + \lambda_2 t_1]}. \tag{6.3.119}$$

Case 4: $a' > 0,\ c' > 0,\ b' = -\sqrt{a'c'};\ \alpha^2 = \beta^2 = -\lambda_1/\lambda_2$

The bifurcation criterion for this case is obtained from that for Case 3 in the limit $\gamma = 0$ with $\delta^2 = \lambda_1/\lambda_2$. In particular, (6.3.119) becomes

$$\frac{\sin 2\delta p_1 l_2}{2\delta p_1 l_2} = \pm \frac{\lambda_1 - \lambda_2}{3\lambda_1 + \lambda_2}. \tag{6.3.120}$$

Note that this may also be obtained from (6.3.115) by putting $\alpha = i\delta$ and replacing λ_2 by $-\lambda_2$.

As we mentioned earlier, Cases 5 and 7 do not arise for incompressible materials. Case 6 yields the trivial solution $v_1 = v_2 = 0$. In Case 8, ψ is an arbitrary function of x_1 and x_2 subject to the restrictions

$$\psi_{,2} = \psi_{,11} = 0 \quad \text{on} \quad x_1 = \pm l_1$$
$$\psi_{,11} = \psi_{,112} = 0 \quad \text{on} \quad x_2 = \pm l_2.$$

Discussion of the bifurcation criteria To assist our discussion we first express a', b' and c' in terms of the strain-energy function. For this purpose we introduce the notation

$$\lambda = \lambda_1 \lambda_3^{1/2}, \qquad \lambda^{-1} = \lambda_2 \lambda_3^{1/2}, \tag{6.3.121}$$

bearing in mind (6.3.93), and define

$$\hat{W}(\lambda) = W(\lambda_1, \lambda_2, \lambda_3). \tag{6.3.122}$$

In (6.3.122) we regard λ and λ_3 as independent but suppress the appearance of λ_3 for convenience since it is fixed.

After a little algebra it follows from (6.3.15) and (6.3.97) that

$$a' = \lambda^4 c', \quad b' = \tfrac{1}{2}\lambda^2 \hat{W}'' - c', \quad c' = \lambda \hat{W}'/(\lambda^4 - 1), \tag{6.3.123}$$

where $\hat{W}' = \mathrm{d}\hat{W}(\lambda)/\mathrm{d}\lambda$. *The strong ellipticity condition* (6.3.98) *therefore amounts to*

$$\frac{\lambda \hat{W}'}{\lambda^2 - 1} > 0, \qquad \lambda^2 \hat{W}'' + \frac{2\lambda \hat{W}'}{\lambda^2 + 1} > 0. \tag{6.3.124}$$

Since

$$t_1 = \lambda \hat{W}' \tag{6.3.125}$$

when $t_2 = 0$, we take

$$\hat{W}'(\lambda) \gtreqless 0 \quad \text{according as} \quad \lambda \gtreqless 1 \tag{6.3.126}$$

so that plane extension (contraction) accompanies tension (compression).

The conditions to be satisfied by $\hat{W}$ in the various cases are as follows

$$\text{Case 1:} \quad \lambda^2 \hat{W}'' > \frac{2\lambda \hat{W}'}{\lambda^2 - 1} > 0. \tag{6.3.127}$$

$$\text{Case 2:} \quad \lambda^2 \hat{W}'' = \frac{2\lambda \hat{W}'}{\lambda^2 - 1} > 0. \tag{6.3.128}$$

$$\text{Case 3:} \quad -\frac{2\lambda \hat{W}'}{\lambda^2 + 1} < \lambda^2 \hat{W}'' < \frac{2\lambda \hat{W}'}{\lambda^2 - 1}, \quad \frac{\lambda \hat{W}'}{\lambda^2 - 1} > 0. \tag{6.3.129}$$

$$\text{Case 4:} \quad -\frac{2\lambda \hat{W}'}{\lambda^2 + 1} = \lambda^2 \hat{W}'', \quad \frac{\lambda \hat{W}'}{\lambda^2 - 1} > 0. \tag{6.3.130}$$

Case 6: $\hat{W}''>0, \quad \hat{W}'=0.$ (6.3.131)

Case 8. $\hat{W}''=0, \quad \hat{W}'=0.$ (6.3.132)

Since in the limit as $\lambda \to 1$, $\hat{W}'(\lambda)/(\lambda-1) \to \hat{W}''(1)$ we observe that $\lambda = 1$ is excluded except by Cases 2 and 8.

To illustrate the distribution of these cases in the range $0<\lambda<\infty$ we now consider a single-term strain-energy function of the form (4.3.66) with the notation $\mu_1=\mu$, $\alpha_1=m$. Thus, on use of (6.3.121), we obtain

$$\hat{W}(\lambda)=\mu\{(\lambda^m+\lambda^{-m})\lambda_3^{-(1/2)m}+\lambda_3^m-3\}/m. \tag{6.3.133}$$

The first inequality in (6.3.124) is satisfied provided $\mu m>0$, so, for convenience, we take $\mu>0$ and $m>0$. The following arguments are unaffected if (μ, m) are replaced by $(-\mu, -m)$ throughout.

We write

$$\lambda^2\hat{W}''-\frac{2\lambda\hat{W}'}{\lambda^2-1}=\mu\lambda_3^{-(1/2)m}f_m(\lambda),$$

$$\lambda^2\hat{W}''+\frac{2\lambda\hat{W}'}{\lambda^2+1}=\mu\lambda_3^{-(1/2)m}g_m(\lambda), \tag{6.3.134}$$

where

$$\left.\begin{aligned} f_m(\lambda)&=(m-1)\lambda^m+(m+1)\lambda^{-m}-2(\lambda^m-\lambda^{-m})/(\lambda^2-1),\\ g_m(\lambda)&=(m-1)\lambda^m+(m+1)\lambda^{-m}+2(\lambda^m-\lambda^{-m})/(\lambda^2+1),\end{aligned}\right\} \tag{6.3.135}$$

noting that

$$\mu f_m(\lambda)=-\mu f_{-m}(\lambda), \qquad \mu g_m(\lambda)=-\mu g_{-m}(\lambda).$$

The first point to note is that

$$g_m(\lambda)-f_m(\lambda)>0 \qquad \text{for all } \lambda>0, \quad m>0. \tag{6.3.136}$$

The properties of $f_m(\lambda)$ and $g_m(\lambda)$, which we now discuss, are shown in Fig. 6.5.

From (6.3.135) it is straightforward to show that

$$f_m(\lambda)>0, \qquad \lambda\neq 1, \qquad f_m(1)=0 \qquad \text{for } m>1,$$

$$f_1(\lambda)\equiv 0,$$

$$f_m(\lambda)<0, \qquad \lambda\neq 1, \qquad f_m(1)=0 \qquad \text{for } 0<m<1.$$

Thus, it follows from (6.3.136) that

$$g_m(\lambda)>0, \qquad 0<\lambda<\infty, \qquad \text{for } m\geq 1$$

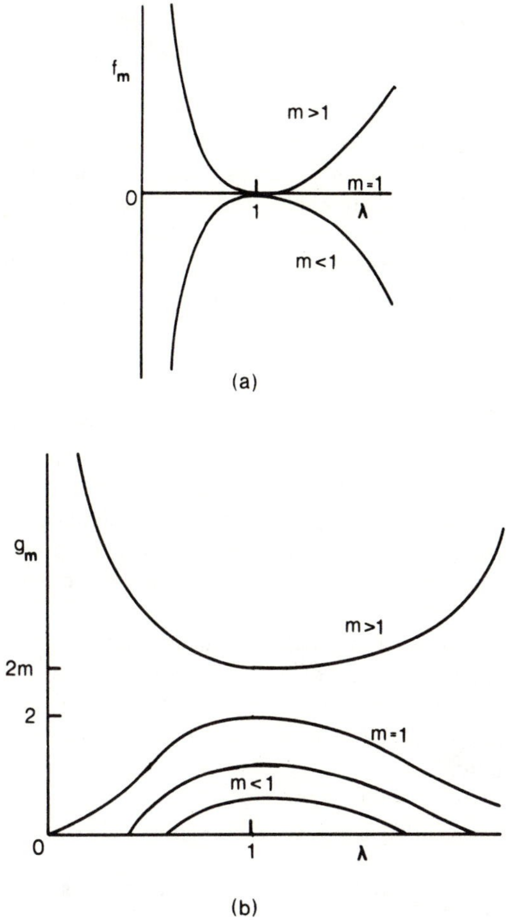

Fig. 6.5 Properties of the functions (a) $f_m(\lambda)$ and (b) $g_m(\lambda)$ defined by (6.3.135) for $m>1$, $m=1$ and $m<1$. In (b) the strongly-elliptic regime is characterized by $g_m(\lambda)>0$.

and the governing equations are strongly elliptic for all λ. If $m>1$, Case 1 applies for $\lambda\neq 1$, while Case 2 applies when $\lambda=1$; if $m=1$, Case 2 applies for all λ. Note that $m=2$ corresponds to the neo-Hookean strain-energy function.

If $0<m<1$, the equation

$$g_m(\lambda)=0$$

has a pair of roots, λ_m, λ_m' say, such that

$$0<\lambda_m<1<\lambda_m'<\infty, \qquad \lambda_m\lambda_m'=1,$$

λ_m increases and λ'_m decreases as m decreases, and $\lambda_0 = \lambda'_0 = 1$ in the limit $m \to 0$. Moreover,

$$g_m(\lambda) > 0 \qquad \text{for} \qquad \lambda_m < \lambda < \lambda'_m,$$

and for this range of values Case 3 applies if $\lambda \neq 1$ and Case 2 when $\lambda = 1$. For $\lambda = \lambda_m$ or λ'_m, Case 4 applies. Case 6 cannot hold in respect of (6.3.133) and Case 8 can only hold trivially for $\mu = 0$ or $m = 0$.

It is worth nothing here that in the plane tension test governed by (6.3.125) with $t_1 > 0$ the (nominal) traction $t_1^{(1)} = \lambda_1^{-1} t_1 = \lambda_3^{1/2} \hat{W}'$ has a maximum where $\hat{W}'' = 0$. For (6.3.133) this occurs at a value of λ given by

$$\lambda^{2m} = \frac{1+m}{1-m} \tag{6.3.137}$$

provided $0 < m < 1$ (or $-1 < m < 0$). At this value of λ, $g_m(\lambda)$ is positive so that strong ellipticity fails beyond the load maximum as λ increases. This point was made by Hill and Hutchinson (1975) who discussed the plane tension problem for a general class of incrementally linear solids. We refer to their paper for more details, some of which are relevant to elastic materials. The corresponding plane compression problem was examined by Young (1976).

With this background in mind we now examine the bifurcation criteria qualitatively for a general form of $\hat{W}$ and quantitatively for particular choices of $\hat{W}$. Unless stated otherwise we restrict attention to the plane tension/compression test so that $t_2 = 0$ with $t_1 > 0$ $(\lambda > 1)$ for tension and $t_1 < 0$ $(\lambda < 1)$ for compression.

Case 1: After a a little algebra (6.3.111) and (6.3.112) can be rearranged as

$$(\lambda^3 \hat{W}'' + \hat{W}') \frac{\sinh \eta(\alpha+\beta)}{(\alpha+\beta)} = \pm (\lambda^3 \hat{W}'' - \hat{W}') \frac{\sinh \eta(\alpha-\beta)}{(\alpha-\beta)}, \tag{6.3.138}$$

where

$$\eta = p_1 l_2 \tag{6.3.139}$$

with p_1 given by (6.3.54) or (6.3.56), and the plus (respectively minus) sign on the right-hand side corresponds to flexural (respectively barreling) modes of deformation. In terms of $\hat{W}$ we also have

$$\left.\begin{aligned} (\alpha+\beta)^2 &= \frac{(\lambda^4-1)}{\hat{W}'}\left(\lambda \hat{W}'' + \frac{2\hat{W}'}{\lambda^2+1}\right), \\ (\alpha-\beta)^2 &= \frac{(\lambda^4-1)}{\hat{W}'}\left(\lambda \hat{W}'' - \frac{2\hat{W}'}{\lambda^2-1}\right), \end{aligned}\right\} \tag{6.3.140}$$

and $\alpha\beta = \lambda^2$.

For $\lambda > 1$ it follows from (6.3.126) and (6.3.127) that

$$\lambda^3 \hat{W}'' + \hat{W}' > \lambda^3 \hat{W}'' - \hat{W}' > 0.$$

Therefore, since $x^{-1} \sinh x$ is a monotonic increasing function of x for $x > 0$, and $\alpha + \beta > \alpha - \beta$, the left-hand side of (6.3.138) is greater than the right-hand side for each sign. Thus, *neither flexural nor barreling bifurcation modes can occur in tension.*

In compression, $\lambda < 1$ and hence

$$\lambda^3 \hat{W}'' - \hat{W}' > 0, \qquad \lambda^3 \hat{W}'' + \hat{W}' > \frac{(3\lambda^2 - 1)\lambda \hat{W}'}{(\lambda^2 - 1)}.$$

The sign of $\lambda^3 \hat{W}'' + \hat{W}'$ is critical here. If $\lambda^2 > \frac{1}{3}$ then $\lambda^3 \hat{W}'' + \hat{W}' > 0$ and barreling modes are not possible. A necessary condition for barreling modes to occur is $\lambda^3 \hat{W}'' + \hat{W}' < 0$, and hence $\lambda^2 < \frac{1}{3}$. On the other hand, if $\lambda^3 \hat{W}'' + \hat{W}' < 0$ flexural modes are not possible; they are certainly possible for $\lambda^2 \geq \frac{1}{3}$ and may also exist for $\lambda_c^2 \leq \lambda^2 \leq \frac{1}{3}$, where λ_c is the first (and possibly only) value of λ for which $\lambda^3 \hat{W}'' + \hat{W}'$ vanishes as λ decreases.

Thus, under increasing compressive load, with λ decreasing from unity, flexural modes (if they occur at all) will be initiated for λ in the interval $\lambda_c < \lambda < 1$, while barreling modes can only appear subsequently (if at all) for $0 < \lambda < \lambda_c$. Whether or not solutions to (6.3.138) actually exist depends on η and on the properties of $\hat{W}$. In particular, for slender specimens, corresponding to $\eta \ll 1$, flexural modes certainly exist, as will be seen shortly from an asymptotic analysis of (6.3.138).

In order to extract more information from (6.3.136), we consider the class of strain-energy functions (6.3.133), recalling that Case 1 applies for $m > 1$ and $\lambda \neq 1$. We have

$$\lambda^3 \hat{W}'' + \hat{W}' = \mu \lambda_3^{-(1/2)m} h_m(\lambda)$$

where

$$h_m(\lambda) = (m-1)\lambda^{m+1} + \lambda^{m-1} + (m+1)\lambda^{-m+1} - \lambda^{-m-1}.$$

The characteristics of $h_m(\lambda)$ are depicted in Fig. 6.6 for $m > 1$, $m = 1$ and $0 < m < 1$.

For $m = 1$, $h_m(\lambda)$ vanishes where $\lambda^2 = \frac{1}{3}$. As m increases from unity the value of λ^2 for which $h_m(\lambda)$ vanishes in compression decreases from $\frac{1}{3}$ and the range of values of λ for which barreling is possible shrinks. For $m > 1$, $h_m(\lambda)$ is also a monotonic increasing function of λ. The range of values $0 < m < 1$ is covered in Cases 3 and 4.

When $\lambda = 1$ the difference between the left- and positive right-hand sides of (6.3.138) is

$$\tfrac{1}{2}\hat{W}''(1)(\sinh 2\eta - 2\eta)$$

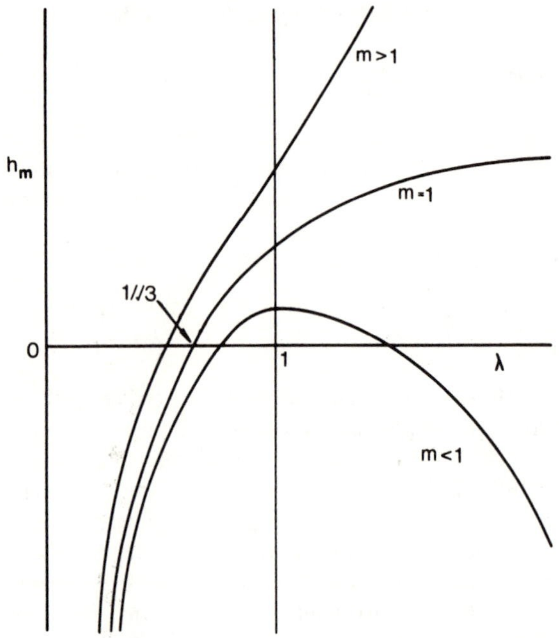

Fig. 6.6 Properties of the function $h_m(\lambda)$ for $m > 1$, $m = 1$ and $m < 1$.

which is positive for $\eta > 0$. In view of the properties of $h_m(\lambda)$ described above, it is clear that (6.3.138) must have a flexural solution for every $\eta > 0$ in compression. This argument is reinforced by our numerical calculations for $m = 2$. These are illustrated in Fig. 6.7, in which η_0 is plotted against the value of λ satisfying (6.3.138), where $\eta_0 \equiv \lambda^2\eta$ is independent of λ.

Results for other values of $m \geq 1$ are similar, with λ_c, the value of λ at which $h_m(\lambda)$ vanishes, decreasing with increasing m. For the Mooney–Rivlin material

$$\hat{W}' = (\mu_1\lambda_3^{-1} - \mu_2\lambda_3)(\lambda - \lambda^{-3}),$$

with $\mu_1 > 0$, $\mu_2 < 0$, and the results are identical to those for $m = 2$.

For given aspect ratio L_2/L_1 it is clear from Fig. 6.7 that flexural bifurcation becomes possible first for the lowest mode number and for the value of λ corresponding to $\eta_0 = \pi L_2/2L_1$ on the flexural curve. Other modes, corresponding to $\eta_0 = \pi L_2/L_1$, $3\pi L_2/2L_1, \ldots$, come into play at successively lower values of λ. For barreling, on the other hand, higher-order modes occur first, and the lowest-order modes may not be possible.

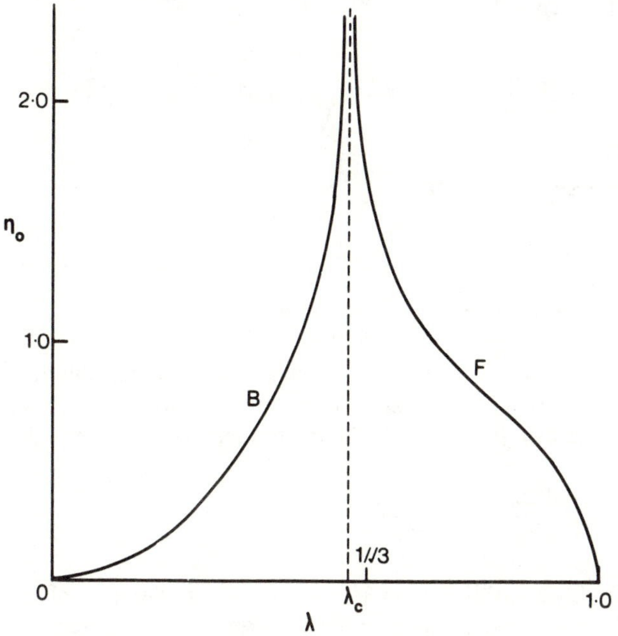

Fig. 6.7 Solution of (6.3.138) for the neo-Hookean strain-energy function ($m = 2$), showing critical values of η_0 plotted against λ for barreling (B) and flexural (F) modes of bifurcation in simple compression.

To complete our discussion of Case 1 we consider the asymptotic expansion of (6.3.138) for small η, corresponding to a slender specimen with $l_2/l_1 \ll 1$ and lowest mode number. Bifurcation into a flexural mode occurs for λ close to unity, and we write

$$\lambda = 1 + c_1\eta^2 + c_2\eta^4 + c_3\eta^6 + \cdots.$$

After some tedious algebra and use of the result $\hat{W}'''(1) = -3\hat{W}''(1)$, we obtain

$$c_1 = -\tfrac{1}{3}, \qquad c_2 = \tfrac{13}{30}, \qquad c_3 = -\tfrac{25}{42} - \frac{1}{81}\left\{\frac{\hat{W}''''(1)}{\hat{W}''(1)}\right\},$$

the first two of which are independent of $\hat{W}$ and λ_3. The details are left to the reader.

The stress t_1 associated with this critical value of λ and calculated from

(6.3.125) is given by†

$$t_1/\hat{W}''(1) = -\tfrac{1}{3}\eta^2 + \tfrac{17}{45}\eta^4 - \left[\tfrac{83}{210} + \frac{1}{54}\left\{\frac{\hat{W}''''(1)}{\hat{W}''(1)}\right\}\right]\eta^6 + \cdots,$$

and, in respect of (6.3.133),

$$\hat{W}''''(1)/\hat{W}''(1) = m^2 + 11$$

so that $-t_1$ increases with m^2.

Case 2: We now have $\alpha = \beta = \lambda$. First we consider the situation in which $\lambda \neq 1$ and $t_2 = 0$. The bifurcation criterion (6.3.114) yields $t_1 = 0$, which is incompatible with (6.3.126). The bifurcation criterion (6.3.116) gives

$$\frac{\sinh 2\lambda\eta}{2\lambda\eta} = \pm\frac{\lambda^2 + 1}{3\lambda^2 - 1}, \tag{6.3.141}$$

and, as in Case 1, this has no solution for $\lambda > 1$. The plot of η_0 against λ arising from (6.3.141) is similar to that shown in Fig. 6.7, except that $\lambda_c^2 = \frac{1}{3}$, and is independent of the form of $\hat{W}$.

However, (6.3.141) is coupled with (6.3.128) and, in general, both equations hold simultaneously for at most isolated values of λ. Indeed, for the strain-energy functions (6.3.133) with $m \neq 1$, Case 2 applies if and only if $\lambda = 1$, and (6.3.141) cannot be satisfied for $\eta > 0$. In fact, Case 2 applies at $\lambda = 1$ for an arbitrary form of $\hat{W}$ for which $\hat{W}''(1) > 0$. Exceptionally, for $m = 1$, (6.3.128) holds for all λ and Fig. 6.7 with $\lambda_c^2 = \frac{1}{3}$ then represents the bifurcation criterion.

We now turn to the equibiaxial stress test in which $\lambda = 1$ and $t_1 = t_2$. In the limit as $\lambda \to 1$, use of (6.3.19) and (6.3.15) shows that (6.3.114) becomes

$$\mathscr{A}^1_{01111} - \mathscr{A}^1_{01122} - q = 0$$

or, in terms of $\hat{W}$, simply

$$2t_1 = \hat{W}''(1) > 0. \tag{6.3.142}$$

†The first term yields the classical formula for the Euler buckling stress, $\frac{1}{4}\hat{W}''(1)$ being equal to the shear modulus in the ground state. For discussion of related experimental work see Beatty (1977).

Similarly, (6.3.115) becomes

$$\frac{\sinh 2\eta}{2\eta} = \pm \frac{\hat{W}''(1) - 2t_1}{\hat{W}''(1) + 2t_1}. \tag{6.3.143}$$

Clearly, the stress state $t_1 = t_2 = 0$ is excluded by (6.3.142) and by (6.3.143) for $\eta > 0$. In other words, bifurcation (of the type considered here) cannot happen if $t_1 = t_2 = 0$ except (in a flexural mode) in the limit $\eta \to 0$, which is of no practical interest.

Equation (6.3.143) has no solution for $t_1 > 0$, but bifurcation modes (whose x_2-dependence is given by (6.3.113) with $B_1 = B_2 = 0$) can occur in *tension* when t_1 reaches the value given by (6.3.142). This result is independent of the parameter η and of the material properties embodied in $\hat{W}$. Note, however, that $\hat{W}''(1)$ depends on $\lambda_3(=\lambda_1^{-2})$.

In compression, on the other hand, (6.3.143) has flexural solutions for

$$-\hat{W}''(1) < 2t_1 < 0$$

and barreling solutions for

$$2t_1 < -\hat{W}''(1).$$

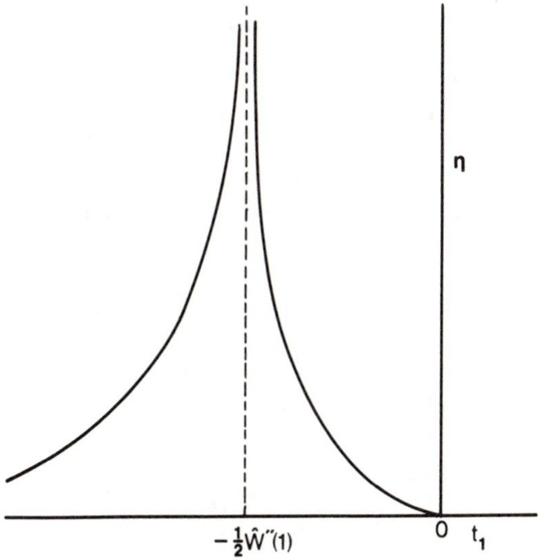

Fig. 6.8 Plot of η against critical stress t_1 in equibiaxial compression from equation (6.3.143).

Indeed, there is a one-to-one correspondence between the interval $0<\eta<\infty$ and each of these ranges of values of t_1. Thus, for each value of η equation (6.3.143) defines the critical stresses appropriate to flexural and barreling bifurcation. The relationship between η and t_1 is depicted in Fig. 6.8.

Since λ_3, and hence $\lambda_1=\lambda_3^{-1/2}$, is fixed the above criteria may alternatively be regarded as defining critical values of q through

$$q=t_1-\lambda_1\frac{\partial W}{\partial\lambda_1}$$

evaluated for $\lambda_1=\lambda_2$ and critical t_1.

The asymptotic solution of (6.3.143) for small η, analogous to that given in Case 1, is easily shown to be

$$t_1/\hat{W}''(1)=-\tfrac{1}{6}\eta^2+\tfrac{1}{45}\eta^4+\tfrac{1}{1890}\eta^6+\cdots.$$

Case 3: The bifurcation criterion for $t_2=0$ is obtained from (6.3.138) by writing $\alpha=\gamma+\mathrm{i}\delta$, $\beta=\gamma-\mathrm{i}\delta$, or directly from (6.3.119), to give

$$(\lambda^3\hat{W}''+\hat{W}')\frac{\sinh 2\gamma\eta}{\gamma}=\pm(\lambda^3\hat{W}''-\hat{W}')\frac{\sin 2\delta\eta}{\delta}, \tag{6.3.144}$$

where

$$4\gamma^2=\frac{(\lambda^4-1)}{\hat{W}'}\left(\lambda\hat{W}''+\frac{2\hat{W}'}{\lambda^2+1}\right)>0,$$

$$4\delta^2=\frac{(\lambda^4-1)}{\hat{W}'}\left(\frac{2\hat{W}'}{\lambda^2-1}-\lambda\hat{W}''\right)>0.$$

From our discussion of the strain-energy function (6.3.133) we recall that Case 3 requires $0<m<1$. In particular, this allows $\hat{W}''$ to be negative in tension for values of λ greater than that given by (6.3.137), and hence the possibility of bifurcation in tension as well as in compression. For the strain-energy function (6.3.133) with $m=\frac{1}{2}$ equation (6.3.144) becomes

$$\frac{(\lambda^2-4\lambda+2)}{\sqrt{-\lambda^2+4\lambda-1}}\sinh\left\{\eta(\lambda+1)\sqrt{\frac{-\lambda^2+4\lambda-1}{2}}\right\}$$
$$=\pm\frac{(\lambda^3-3\lambda^2+2\lambda-2)}{(\lambda-1)\sqrt{\lambda^2-1}}\sin\left\{\eta(\lambda-1)\sqrt{\frac{\lambda^2+1}{2}}\right\},$$

and the strong ellipticity condition holds for $2-\sqrt{3}<\lambda<2+\sqrt{3}$.

The solutions are plotted in Fig. 6.9. We note, in particular, that the

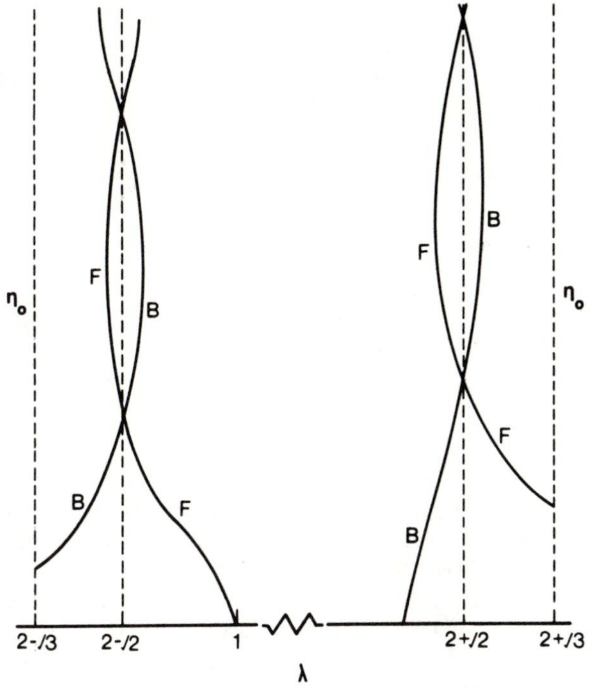

Fig. 6.9 Solution of (6.3.144) for (6.3.133) with $m = \frac{1}{2}$ in the strongly-elliptic regime $2-\sqrt{3} < \lambda < 2+\sqrt{3}$, with η_0 plotted against λ. Both barreling (B) and flexural (F) bifurcations occur in simple tension and in simple compression.

equations are satisfied for $\lambda = 2 \pm \sqrt{2}$ with η_0 then given by

$$\lambda^2\eta \equiv \eta_0 = \frac{\sqrt{2}\lambda^2 n\pi}{|\lambda - 1|\sqrt{\lambda^2+1}} \qquad n = 1, 2, \ldots .$$

In compression, the features displayed in Fig. 6.9 for

$$\eta_0 < \frac{\sqrt{2}\lambda^2\pi}{|\lambda - 1|\sqrt{\lambda^2+1}} \qquad (\text{with } \lambda = 2 - \sqrt{2})$$

are similar to those found in Case 1, but with a cut-off value of $\lambda = 2 - \sqrt{3}$ at the boundary of the strongly elliptic region. As η_0 increases, the oscillatory nature of the solution periodically interchanges the priorities of the flexural and barreling modes.

Unlike in Case 1, bifurcation, in both barreling and flexural modes, is also possible in tension. The bifurcation curves for tension are similar in shape to those for compression, as Fig. 6.9 shows, with a cut-off value of $\lambda = 2 + \sqrt{3}$. Similar results have been given by Sawyers (1977).

Case 4: Here we have $\gamma = 0$, $\delta = \lambda$, and (6.3.120) is written

$$\frac{\sin 2\lambda\eta}{2\lambda\eta} = \pm\frac{\lambda^2 - 1}{3\lambda^2 + 1}. \tag{6.3.145}$$

This is coupled with (6.3.130), so that for the strain-energy function considered in Case 3 we must solve (6.3.145) for $\lambda = 2 \pm \sqrt{3}$. To each of these values corresponds just one value of η_0 satisfying (6.3.145), as can be seen from Fig. 6.9.

Taken together the results for Cases 3 and 4 show that in either tension or compression bifurcation into either a flexural mode or a barreling mode of deformation is always possible prior to failure of ellipticity, for any value of η_0. Thus, such diffuse modes of bifurcation have priority over shear-band modes.

Finally in this section we remark that Case 8, which requires $\hat{W}' = \hat{W}'' = 0$, cannot hold for $\lambda = 1$ for a material which has non-zero ground-state shear modulus. Neither can it hold for any λ in respect of (6.3.133). It is possible to construct strain-energy functions for which these conditions do obtain for $\lambda \neq 1$, but they are of limited interest and we do not pursue the matter here.

Problem 6.3.7 For an unconstrained material the bifurcation problem for a half-space was outlined in Problem 6.3.4. Show that for the corresponding problem for an incompressible material the bifurcation criterion is simply

$$\lambda^3 \hat{W}'' + \hat{W}' = 0 \tag{6.3.146}$$

for *each* of Cases 1, 2, 3 when $t_2 = 0$, and that in Case 2 this is equivalent to $\lambda^2 = \frac{1}{3}$. Note that for the strain-energy function (6.3.133) the characteristics of $\lambda^3 \hat{W}'' + \hat{W}'$ at fixed λ_3 are those of $h_m(\lambda)$, illustrated in Fig. 6.6. Show also that Case 4 is incompatible with the above criterion, unlike the situation for a rectangular block of finite dimensions.

If $\lambda = 1$, $t_1 = t_2 \neq 0$ show that the bifurcation criterion is

$$t_1 = \pm\tfrac{1}{2}\hat{W}''(1). \tag{6.3.147}$$

Problem 6.3.8 Recalling from Section 6.2.5 that a shear-band mode of deformation is possible where ellipticity fails, show that for a strain-energy function for which Case 3 is appropriate the condition (6.3.146) always precedes

$$\lambda^2\hat{W}''+2\frac{\lambda\hat{W}'}{\lambda^2+1}=0$$

on a path of tensile or compressive loading from $\lambda=1$ ($t_2=0$).

Deduce that for the half-space problem considered above diffuse modes of bifurcation can always be activated prior to shear-band modes, and that, in tension, both types of mode occur after the maximum (nominal) load is passed. Further discussion of this problem is given by Reddy (1983).

For a rectangular block of finite dimensions show that, for certain values of η, both types of mode may occur simultaneously.

Problem 6.3.9 Show that, in respect of an incompressible material, the exclusion condition discussed in Section 6.2.5 may be written in the notation of the present section as

$$\int_{\mathscr{B}} \operatorname{tr}\{(\mathscr{A}_0^1\dot{\mathbf{A}}_0)\dot{\mathbf{A}}_0-q\dot{\mathbf{A}}_0^2\}\,\mathrm{d}v>0, \tag{6.3.148}$$

where $\dot{\mathbf{A}}_0=\operatorname{grad}\mathbf{v}$, for all $\mathbf{v}$ satisfying the (homogeneous) displacement boundary conditions on $\partial\mathscr{B}$, subject to $\operatorname{div}\mathbf{v}=0$ in $\mathscr{B}$.

For an isotropic material with strain-energy function W show that the integrand in (6.3.148), when referred to principal Eulerian axes, may be rearranged as

$$\begin{aligned}&\left\{\left(\lambda_1\frac{\partial}{\partial\lambda_1}-\lambda_3\frac{\partial}{\partial\lambda_3}\right)^2W-t_1-t_3\right\}v_{1,1}^2\\&+\left\{\left(\lambda_2\frac{\partial}{\partial\lambda_2}-\lambda_3\frac{\partial}{\partial\lambda_3}\right)^2W-t_2-t_3\right\}v_{2,2}^2\\&+\left\{\left(\lambda_1\frac{\partial}{\partial\lambda_1}-\lambda_3\frac{\partial}{\partial\lambda_3}\right)^2W+\left(\lambda_2\frac{\partial}{\partial\lambda_2}-\lambda_3\frac{\partial}{\partial\lambda_3}\right)^2W\right.\\&\left.-\left(\lambda_1\frac{\partial}{\partial\lambda_1}-\lambda_2\frac{\partial}{\partial\lambda_2}\right)^2W-2t_3\right\}v_{1,1}v_{2,2}\\&+\frac{1}{2}\sum_{i\neq j}\left\{\left(\frac{t_i-t_j}{\lambda_i^2-\lambda_j^2}\right)(\lambda_i^2v_{j,i}^2+\lambda_j^2v_{i,j}^2)+2\left(\frac{t_i\lambda_j^2-t_j\lambda_i^2}{\lambda_i^2-\lambda_j^2}\right)v_{i,j}v_{j,i}\right\},\end{aligned} \tag{6.3.149}$$

where $t_i=\lambda_i(\partial W/\partial\lambda_i)+q$. Deduce that necessary and sufficient conditions for this to be positive for all isochoric $\mathbf{v}$ are

$$A_1 - t_2 - t_3 > 0, \tag{6.3.150}$$

$$2(A_2A_3 + A_3A_1 + A_1A_2) - A_1^2 - A_2^2 - A_3^2 \\ - 4(A_1t_1 + A_2t_2 + A_3t_3) + 4(t_2t_3 + t_3t_1 + t_1t_2) > 0, \tag{6.3.151}$$

$$\frac{t_i^{(1)} - t_j^{(1)}}{\lambda_i - \lambda_j} > 0, \quad t_i^{(1)} + t_j^{(1)} > 0 \quad i \neq j \tag{6.3.152}$$

jointly, where

$$A_1 = \left(\lambda_2 \frac{\partial}{\partial \lambda_2} - \lambda_3 \frac{\partial}{\partial \lambda_3}\right)^2 W,$$

and similarly for A_2, A_3, and $t_i^{(1)} = \lambda_i^{-1} t_i$, $i = 1, 2, 3$. Note that the third of the above inequalities implies

$$\frac{t_i - t_j}{\lambda_i - \lambda_j} > 0 \qquad i \neq j.$$

Problem 6.3.10 Show that for the plane-strain problem with $v_3 = 0$ the integrand in (6.3.148) takes the form

$$\lambda^2 \hat{W}'' v_{1,1}^2 + \frac{\lambda \hat{W}'}{\lambda^4 - 1}(v_{1,2}^2 + 2v_{1,2}v_{2,1} + \lambda^4 v_{2,1}^2)$$

when $t_2 = 0$, and

$$\{\hat{W}''(1) - 2t_1\}\{v_{1,1}^2 + \tfrac{1}{4}(v_{1,2} + v_{2,1})^2\} + \tfrac{1}{2}t_1(v_{1,2} - v_{2,1})^2$$

when $\lambda_1 = \lambda_2$, where $\hat{W}$ is defined by (6.3.122) with (6.3.121).

Deduce that each of these is positive in an initial configuration in which $\lambda = 1$ and $t_1 = t_2 = 0$ for any non-vanishing (plane) displacement field satisfying the boundary condition (6.3.50). For the bifurcation problem examined in the present section, conclude that, on a path of deformation starting from this initial configuration, bifurcation becomes possible first in a *primary e-configuration* defined by the first value of λ met on this path which satisfies the relevant bifurcation criterion.

Problem 6.3.11 For an incompressible material body under *all-round dead load* show that bifurcation can occur where

$$\mathscr{A}_0^1 \dot{\mathbf{A}}_0 + \dot{q}\mathbf{I} - q\dot{\mathbf{A}}_0 = \mathbf{0}, \qquad \mathrm{tr}(\dot{\mathbf{A}}_0) = 0,$$

i.e. where the integrand in (6.3.148) is stationary. If the material is isotropic show that this is satisfied when either (a) the left-hand expression in (6.3.151) vanishes, (b) $(t_i^{(1)} - t_j^{(1)})/(\lambda_i - \lambda_j) = 0$ for any pair $i, j \neq i \in \{1, 2, 3\}$, or (c) $t_i^{(1)} + t_j^{(1)} = 0$ for any pair $i, j \neq i \in \{1, 2, 3\}$.

Deduce that the bifurcation modes are uniform and such that (a) $v_{i,j} = 0$ for every pair $i, j \neq i \in \{1, 2, 3\}$, (b) $\lambda_i v_{j,i} = \lambda_j v_{i,j} \neq 0$ with $v_{k,l} = 0$ for $kl \neq ij$ or $ji (\lambda_i \neq \lambda_j)$, (c) $\lambda_i v_{j,i} = -\lambda_j v_{i,j} \neq 0$ with $v_{k,l} = 0$ for $kl \neq ij$ or $ji (\lambda_i \neq \lambda_j)$, respectively. Show that when $\lambda_i = \lambda_j$ modes of type (a) and (b) can occur simultaneously where

$$\left(\lambda_1 \frac{\partial}{\partial \lambda_1} - \lambda_2 \frac{\partial}{\partial \lambda_2}\right)^2 W - 2t_1 = 0,$$

this being evaluated for $\lambda_1 = \lambda_2$.

Compare these with the corresponding results given in Section 6.2.3 for an unconstrained material, and show, in particular, that (b) and (c) respectively yield

$$\Omega_{ij} = \frac{\lambda_i - \lambda_j}{\lambda_i + \lambda_j} \Sigma_{ij} \quad \text{and} \quad \Omega_{ij} = \frac{\lambda_i + \lambda_j}{\lambda_i - \lambda_j} \Sigma_{ij}$$

as in Problem 6.2.1, where Ω_{ij} and Σ_{ij} are the components of the body spin and Eulerian strain-rate on the Eulerian principal axes.

Problem 6.3.12 For the plane-strain problem with $v_3 = 0$ and *in-plane dead loading* show that bifurcation can occur when either (a) $\lambda^2 \hat{W}'' - 2t_2 = 0$, (b) $(t_1^{(1)} - t_2^{(1)})/(\lambda_1 - \lambda_2) = 0$, or (c) $t_1^{(1)} + t_2^{(1)} = 0$ for $\lambda_1 \neq \lambda_2$, where $\hat{W}$ is defined by (6.3.122) with (6.3.121). Describe the bifurcation modes associated with each of (a), (b), (c). When $\lambda_1 = \lambda_2$ show that (a) and (b) coincide.

6.3.2 Global aspects of the plane-strain bifurcation of a rectangular block

The results of Problems 6.3.11 and 6.3.12 relate to the *local* invertibility of the stress-deformation relation

$$\mathbf{S} = \mathbf{H}(\mathbf{A}) + q\mathbf{B}^{\mathrm{T}}, \quad \det \mathbf{A} = 1 \tag{6.3.153}$$

and are analogous to corresponding results for unconstrained isotropic materials discussed in Section 6.2.3. The problem of the *global* invertibility of (6.3.153) was alluded to at the end of Section 6.2.2, following a detailed analysis of the same problem for unconstrained materials, and we now turn to its implications in the context of bifurcation theory.

We recall from Section 6.2.2 the polar decomposition $\mathbf{S} = \mathbf{T}^{(1)}\mathbf{R}^{\mathrm{T}}$ for isotropic materials and the possible combinations of signs of the principal values $t_1^{(1)}$, $t_2^{(1)}$, $t_3^{(1)}$ of $\mathbf{T}^{(1)}$, noting that at most one combination satisfies the dead-loading stability requirement

$$t_i^{(1)} + t_j^{(1)} \geq 0 \qquad i \neq j.$$

(For a homogeneously deformed incompressible material under all-round dead-loading, necessary and sufficient conditions for infinitesimal stability are (6.3.150)–(6.3.152) with $>$ replaced by $\geq$.)

Having obtained $\mathbf{T}^{(1)}$ and $\mathbf{R}$ the inversion of (6.3.153) is completed by obtaining $\mathbf{U}$ from

$$\mathbf{T}^{(1)} = \mathbf{H}(\mathbf{U}) + q\mathbf{U}^{-1}, \qquad \det \mathbf{U} = 1$$

or, equivalently, for an isotropic material possessing a strain-energy function W, by obtaining $\lambda_1, \lambda_2, \lambda_3$ from

$$t_i^{(1)} = \frac{\partial W}{\partial \lambda_i} + q\lambda_i^{-1}, \qquad \lambda_1\lambda_2\lambda_3 = 1, \tag{6.3.154}$$

where $t_i^{(1)}$ is prescribed $(i = 1, 2, 3)$. We thus have three equations for the determination of $\lambda_1, \lambda_2, \lambda_3$, namely

$$\lambda_1 t_1^{(1)} - \lambda_1 \frac{\partial W}{\partial \lambda_1} = \lambda_2 t_2^{(1)} - \lambda_2 \frac{\partial W}{\partial \lambda_2} = \lambda_3 t_3^{(1)} - \lambda_3 \frac{\partial W}{\partial \lambda_3}, \quad \lambda_1\lambda_2\lambda_3 = 1. \tag{6.3.155}$$

If required, q can be obtained subsequently from (6.3.154) in terms of $\lambda_1, \lambda_2, \lambda_3$, but we shall not need it in this section.

To illustrate the global results we consider the equibiaxial tension problem for which $t_1^{(1)} = t_2^{(1)}$ while λ_3 is fixed. With the notation defined in (6.3.121) and (6.3.122) together with

$$t^{(1)} = \lambda_3^{-1/2} t_1^{(1)} \tag{6.3.156}$$

the first equation in (6.3.155) yields *either* $\lambda_1 = \lambda_2$ (i.e. $\lambda = 1$) *or*

$$t^{(1)} = \frac{\lambda^2 \hat{W}'}{\lambda^2 - 1}. \tag{6.3.157}$$

By taking the limit $\lambda \to 1$ we see that (6.3.157) can hold for $\lambda = 1$ if and only if $t^{(1)} = \frac{1}{2}\hat{W}''(1)$. This corresponds precisely to the tensile bifurcation stress appropriate to the equibiaxial deformation path ($\lambda = 1$) discussed in Case 2 in Section 6.3.1 and alluded to in Problem 6.3.12.

Thus, under equibiaxial tensile loading starting from $t^{(1)}=0$ the deformation is such that $\lambda_1=\lambda_2=\lambda_3^{-1/2}$ up to $t^{(1)}=\frac{1}{2}\hat{W}''(1)$, beyond which value the deformation $\lambda_1=\lambda_2$ is unstable. The subsequent deformation path depends critically on the properties of $\hat{W}$, with λ and $t^{(1)}$ connected through (6.3.157). Since $t_1^{(1)}=t_2^{(1)}>0$ beyond the bifurcation point the deformation path is (neutrally) stable under plane strain conditions (i.e. λ_3 fixed) provided $A_3 \geq 2t_1>0$ (by specializing the results of Problem 6.3.9). With the help of the connection $t_1=\lambda_1 t_1^{(1)}=\lambda t^{(1)}$ and (6.3.157) we rewrite this requirement as

$$\lambda^2\hat{W}'' \geq \frac{2\lambda\hat{W}'}{\lambda^2-1}>0. \tag{6.3.158}$$

The deformation path with $\lambda \neq 1$ emanating from the bifurcation point $\lambda=1$, $t^{(1)}=\frac{1}{2}\hat{W}''(1)$ is therefore (neutrally) stable for materials whose strain-energy function satisfies (6.3.158). Note that (6.3.158) embraces only Cases 1, 2 amongst (6.3.127)–(6.3.132).

In respect of the strain-energy function (6.3.133), (6.3.158) is equivalent to $f_m(\lambda)\geq 0$, where $f_m(\lambda)$ is defined in (6.3.135). Thus, by reference to Fig. 6.5(a), we see that the path is (neutrally) stable for $m \geq 1$ and unstable for $m<1$. A bifurcation diagram, with λ plotted against $t^{(1)}$ for each of $m<1$, $m=1$ and $m>1$, is shown in Fig. 6.10. Two possible deformation branches emanate from the bifurcation point; one corresponds to $\lambda_1>\lambda_2$, the other to $\lambda_1<\lambda_2$, and, because of the loading symmetry, these are geometrically equivalent.

More complex bifurcation curves can be obtained for strain-energy

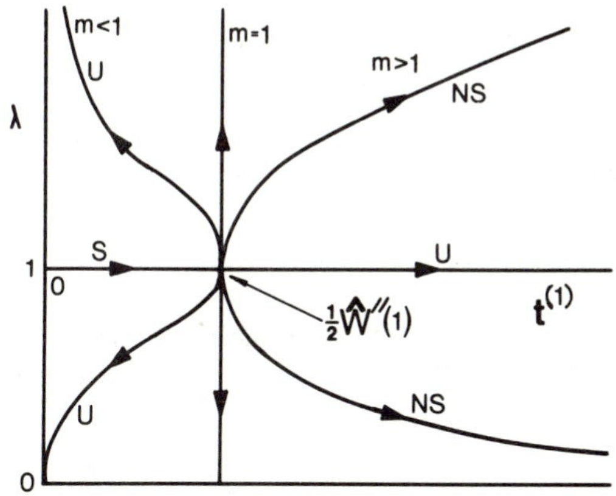

Fig. 6.10 Bifurcation diagram, showing λ plotted against $t^{(1)}$, for the equibiaxial tension of a rectangular block in respect of (6.3.133) with $m>1$, $m=1$ and $m<1$. The stability of the branches is labelled according to S = stable, NS = neutrally stable, U = unstable.

functions consisting of an $m>1$ and an $m<1$ term. For a general form of W the behaviour near $\lambda=1$ is governed by

$$\frac{\mathrm{d}t^{(1)}}{\mathrm{d}\lambda}=0, \qquad \frac{\mathrm{d}^2t^{(1)}}{\mathrm{d}\lambda^2}=\tfrac{1}{6}\hat{W}''''(1)-2\hat{W}''(1)$$

(both evaluated for $\lambda=1$), and, in particular, for (6.3.133) we have

$$\frac{\mathrm{d}^2t^{(1)}}{\mathrm{d}\lambda^2}=\tfrac{1}{6}(m^2-1)\hat{W}''(1)$$

at $\lambda=1$. Note that, for $m=1$, $t^{(1)}$ is independent of λ.

The related all-round dead-load problem has been examined in respect of the neo-Hookean strain-energy function ($m=2$) by Hill (1967), Rivlin (1974) and Sawyers (1976). Rivlin confined attention to the situation in which $t_1^{(1)}=t_2^{(1)}=t_3^{(1)}$, while Sawyers considered $t_1^{(1)}=t_2^{(1)}\neq t_3^{(1)}$. More recently, the first of these two cases has been discussed in detail by Ball and Schaeffer (1983) from the point of view of singularity theory. In particular, for the Mooney strain-energy function, they have shown that from a deformation branch on which $\lambda_1=\lambda_2\neq\lambda_3$ secondary bifurcation into a branch with no two stretches equal is possible.

Problem 6.3.13 For the neo-Hookean strain-energy function show that

$$\lambda_1\tau_1-\lambda_1^2=\lambda_2\tau_2-\lambda_2^2=\lambda_1^{-1}\lambda_2^{-1}\tau_3-\lambda_1^{-2}\lambda_2^{-2},$$

where $\tau_i=t_i^{(1)}/\mu$, $i=1,2,3$. If τ_1,τ_2,τ_3 are prescribed obtain an equation for λ_1.

Assuming that $\tau_1=\tau_2<\tau_3$ deduce that *either*

$$\text{(i)}\ \lambda_1=\lambda_2 \quad \text{and} \quad \lambda_1^6-\tau_1\lambda_1^5+\tau_3\lambda_1^2-1=0$$

or

$$\text{(ii)}\ \lambda_1+\lambda_2=\tau_1 \quad \text{and} \quad \lambda_1^3\lambda_2^3-\tau_3\lambda_1\lambda_2+1=0.$$

When $\tau_1=\tau_2=0$, and τ_3 is fixed at a value greater than $3(2)^{-2/3}$ show that $\lambda_1=\lambda_2=\lambda_3^{-1/2}$ and that, subject to $\lambda_3^3>2$, λ_3 is uniquely determined from the equation

$$\lambda_3^3-\tau_3\lambda_3^2-1=0.$$

Suppose that $\tau_1=\tau_2$ is now increased from zero. Show that the deformation follows the path described by (i) and is stable up to the point where $\lambda_1=\sqrt{\kappa}$,

$\tau_1 = 2\sqrt{\kappa}$, where κ is the unique positive root of

$$\kappa^3 - \tau_3\kappa + 1 = 0$$

satisfying $\kappa^3 < \frac{1}{2}$. Deduce that this is a bifurcation point and that for $\tau_1 > 2\sqrt{\kappa}$ the subsequent deformation follows the neutrally stable path governed by (ii) such that $\lambda_1\lambda_2 = \kappa$. Conclude that the bifurcation diagram is similar to that shown in Fig. 6.10 for $m > 1$.

Describe what happens when τ_1 reaches the value τ_3.

Problem 6.3.14 Assess the results of this section in the light of the theory of constitutive branching discussed in Section 6.2.8.

6.3.3 Other problems with underlying homogeneous deformation

In Section 6.3.1 we discussed the plane-strain bifurcation problem in some detail because it illustrates the features which are typical of local bifurcation analysis for problems with underlying homogeneous deformations. We do not, therefore, provide corresponding details here for other such problems; instead, we discuss briefly some relevant references but do not attempt to give an exhaustive list.

First we note that for a rectangular block under *uniaxial* stress (so that $t_2 = t_3 = 0$ and $\lambda_2 = \lambda_3$), with bifurcation restricted to the (1, 2)-plane, the bifurcation criteria are precisely those given in Section 6.3.1 appropriately specialized. In particular, for $\eta \ll 1$ the classical Euler formula for the buckling stress is obtained from the Case 1 results:

$$t_1 = -\tfrac{1}{3}\eta^2 \hat{W}''(1),$$

with $\hat{W}''(1) = 4\mu$ when $\lambda_3 = 1$, where μ is the classical shear modulus.

For a cylinder of uniform circular cross-section the incremental solution involves Bessel functions and, although explicit bifurcation criteria can be obtained for both barreling and flexural modes under suitable end conditions, useful information can only be obtained by numerical calculation. Exceptionally, when the radius of the cylinder is very much less than its length, the Euler buckling stress appropriate to a circular cross-section is obtained. For an axially loaded circular cylinder which is also rotating about its axis, details have been provided by Haughton and Ogden (1980a). Bending of a circular bar under various conditions has been analysed by Fosdick and Shield (1963). Further references are contained in these two papers.

Thus far our discussion has concerned problems with homogeneous incremental boundary data and the determination of bifurcation points. For problems in which the data are inhomogeneous the solutions are unique except at bifurcation points. At a bifurcation point that part of the solution normal (in an appropriate sense) to the e-mode is determined uniquely (recall Section 6.2.8). A number of problems involving inhomogeneous data have been solved, but, to the writer's knowledge, none of these solutions take proper account of the possibility of bifurcation.

Since the underlying deformation is homogeneous an incremental problem is similar in structure to the corresponding problem for an anisotropic linearly elastic solid with uniform elastic moduli. However, since $\dot{\mathbf{S}}_0$ is not in general symmetric the methods used in the linear theory do not in general carry over directly to the present circumstances. Exceptionally, for the problem of incremental deformations superposed on a homogeneous pure strain in which two principal stretches are equal, the general solution may be written down in terms of three scalar potential functions, as in the transversely isotropic linear theory (Green and Zerna, 1968). Some applications of this general theory are also given by Green and Zerna.

We complete this section with a number of problems related to the above discussion.

Problem 6.3.15 Specialize the results of Section 5.3.5 to the situation in which $\tau L \ll 1$ and discuss what happens when λ is a bifurcation point of the underlying uniaxial stress problem.

Problem 6.3.16 A solid sphere of incompressible isotropic elastic material is subjected to a radially-symmetric dead-load surface traction $t^{(1)}$. Show that, independently of the form of strain-energy function, when $t^{(1)}$ reaches the value 2μ, where μ is the ground-state shear modulus, bifurcation into a non-spherical shape is possible.

Confirm this result by use of a local bifurcation analysis, working along the following lines. Firstly, assuming that the incremental deformation is axisymmetric, obtain the incompressibility equation

$$\frac{\partial u}{\partial r}+\frac{2u}{r}+\frac{1}{r}\left(\frac{\partial v}{\partial \theta}+v\cot\theta\right)=0,$$

where u, v are the r, θ components of displacement in spherical polar coordinates.

Secondly, use the above equation and the results of Section 1.5.4 to show that the equilibrium equations can be reduced to

$$\frac{\partial^2 u}{\partial r^2}+\frac{4}{r}\frac{\partial u}{\partial r}+\frac{2u}{r^2}+\frac{1}{r^2}\left(\frac{\partial^2 u}{\partial \theta^2}+\cot\theta\frac{\partial u}{\partial \theta}\right)=-\frac{\partial \dot{q}^*}{\partial r},$$

$$r\frac{\partial^2 v}{\partial r^2}+2\frac{\partial v}{\partial r}-\frac{\partial^2 u}{\partial r\partial \theta}=-\frac{\partial \dot{q}^*}{\partial \theta},$$

and that the (homogeneous) boundary conditions become

$$(2 - t^{(1)*})\frac{\partial u}{\partial r} = -\dot{q}^* \qquad \text{on } r = a,$$

$$r\frac{\partial v}{\partial r} + (1 - t^{(1)*})\left(\frac{\partial u}{\partial \theta} - v\right) = 0 \qquad \text{on } r = a,$$

where

$$\dot{q}^* = 2\dot{q}/(\mathscr{A}^1_{01111} - \mathscr{A}^1_{01122} + t^{(1)} - q),$$
$$t^{(1)*} = 2t^{(1)}/(\mathscr{A}^1_{01111} - \mathscr{A}^1_{01122} + t^{(1)} - q)$$

evaluated for $\lambda_1 = \lambda_2 = \lambda_3 = 1$, and a is the radius of the sphere.

Thirdly, write

$$u = f_n(r)P_n(\cos\theta), \qquad v = g_n(r)\frac{\mathrm{d}}{\mathrm{d}\theta}P_n(\cos\theta), \qquad \dot{q}^* = h_n(r)P_n(\cos\theta),$$

where $P_n(\cos\theta)$ is the Legendre polynomial of order n, and express the equations and boundary conditions in terms of $f_n(r)$. Finally, find $f_n(r)$ and show that the bifurcation criterion for tension is $t^{(1)*} = 2$ (independent of n). Is bifurcation in compression possible?

Problem 6.3.17 Repeat the analysis of Problem 6.3.17 for a compressible material. If dead load is replaced by hydrostatic load, can bifurcation occur?

Problem 6.3.18 For the simple shear deformation

$$\mathbf{A} = \mathbf{I} + \gamma \mathbf{e}_1 \otimes \mathbf{e}_2$$

discussed in Sections 2.2.6 and 4.4, show that the shear traction in the direction $\mathbf{e}_1$ on a plane normal to $\mathbf{e}_2$ is expressible in the form

$$\tfrac{1}{2}\gamma\left(\frac{t_1^{(1)} - t_2^{(1)}}{\lambda_1 - \lambda_2} + \frac{t_1^{(1)} + t_2^{(1)}}{\lambda_1 + \lambda_2}\right),$$

where $\lambda_1\lambda_2 = 1$ and $\lambda_1 - \lambda_2 = \gamma(>0)$.

By writing

$$\tilde{W}(\gamma) = W(\lambda_1, \lambda_1^{-1}, 1)$$

deduce that the above traction may also be written as $\tilde{W}'(\gamma)$.

For this deformation show that the (plane) strong ellipticity condition

(6.3.124) is equivalent to

$$\tilde{W}'(\gamma)/\gamma > 0, \qquad \tilde{W}''(\gamma) > 0$$

in respect of an incompressible material. Write down the corresponding inequalities for an unconstrained material.

By considering incremental deformations superposed on the above simple shear, with zero displacements prescribed on the boundaries $x_2 = 0, h$ and $x_1 = \gamma x_2,\ \gamma x_2 + 1 (0 \le x_2 \le h)$, conclude that as γ increases bifurcation first occurs, if at all, where $W''(\gamma) = 0$ and corresponds to shear band formation. (It will be helpful to recall the result stated at the end of the first paragraph of Section 6.2.7.)

Analyse the bifurcation problem in which the above boundary conditions are replaced by $v_2 = 0$, $\dot{S}_{021} = 0$ on $x_2 = 0, h$ with zero (incremental) traction on $x_1 = \gamma x_2,\ \gamma x_2 + 1\ (0 \le x_2 \le h)$.

Problem 6.3.19 The current configuration of a circular cylinder rotating with angular speed ω about its axis and subject to axial load is as described in Section 5.3.7. Obtain the equations governing incremental displacements having components u, v, w relative to cylindrical polar coordinates r, θ, z.

If $w = 0$ and u, v are independent of z, show that the equilibrium equations can be reduced to

$$\frac{\partial^2 u}{\partial r^2} + \frac{2}{r}\frac{\partial u}{\partial r} + \frac{1}{r^2}\frac{\partial^2 u}{\partial \theta^2} - \frac{1}{r^2}\frac{\partial v}{\partial \theta} + \frac{\partial \dot{q}^*}{\partial r} = -\omega^{*2}\left(r\frac{\partial u}{\partial r} + u\right),$$

$$r\frac{\partial^2 v}{\partial r^2} + \frac{\partial v}{\partial r} - \frac{v}{r} - \frac{\partial^2 u}{\partial r \partial \theta} + \frac{1}{r}\frac{\partial u}{\partial \theta} + \frac{\partial \dot{q}^*}{\partial \theta} = -\omega^{*2} r\frac{\partial u}{\partial \theta},$$

where

$$\dot{q}^* = \dot{q}/\mathscr{A}^1_{01212}, \qquad \omega^{*2} = \rho\omega^2/\mathscr{A}^1_{01212}$$

with $\mathscr{A}^1_{01212}$ evaluated for $\lambda_1 = \lambda_2$.

From the incompressibility condition deduce the existence of a scalar function ψ such that

$$u = \frac{1}{r}\frac{\partial \psi}{\partial \theta}, \qquad v = -\frac{\partial \psi}{\partial r}$$

and show that ψ satisfies the (two-dimensional) biharmonic equation

$$\left(\frac{\partial^2}{\partial r^2} + \frac{1}{r}\frac{\partial}{\partial r} + \frac{1}{r^2}\frac{\partial^2}{\partial \theta^2}\right)^2 \psi = 0.$$

If the lateral surface of the cylinder is free of traction, show that

$$r\frac{\partial v}{\partial r} - v + \frac{\partial u}{\partial \theta} = 0 \quad \text{on } r = a,$$

$$\frac{\partial u}{\partial r} + \dot{q}^* = 0 \quad \text{on } r = a$$

and express these boundary conditions in terms of ψ.

By taking

$$\psi = f_n(r) \sin n\theta \qquad n = 1, 2, \ldots$$

show that bifurcation can occur when ω reaches the value given by

$$a^2 \omega^{*2} = 2(n - n^{-1}) \qquad n \geq 2.$$

Explain how the critical values of ω depend on the axial load.

Problem 6.3.20 If $\omega = 0$ in Problem 6.3.19 and the lateral surface is subjected to a cylindrically symmetric dead-load traction $t^{(1)}$, show that bifurcation can occur for $t^{(1)} = \pm 2\mathscr{A}^1_{01212}$, where $\mathscr{A}^1_{01212}$ is evaluated for $\lambda_1 = \lambda_2$.

6.3.4 Bifurcation of a pressurized spherical shell

The spherically-symmetric deformation of a spherical shell of incompressible isotropic elastic material was discussed in Section 5.3.2. In the present section we examine the possibility of bifurcation from such a deformation into an aspherical configuration for a shell subjected to internal pressure.

First, we recall some notation from Section 5.3.2. The principal stretches $\lambda_1, \lambda_2, \lambda_3$ correspond to the spherical polar coordinates r, θ, ϕ respectively, and we have

$$\lambda_1 = \lambda^{-2}, \qquad \lambda_2 = \lambda_3 = \lambda, \tag{6.3.159}$$

where

$$\lambda = \left(1 - \frac{a^3 - A^3}{r^3}\right)^{-1/3} \qquad a \leq r \leq b. \tag{6.3.160}$$

The values of λ at the interior and exterior surfaces are

$$\lambda_a = a/A, \qquad \lambda_b = b/B \tag{6.3.161}$$

respectively, and these are connected by (5.3.13).

With zero traction on $r=b$ and pressure P on $r=a$ we have

$$t_1 \equiv \lambda_1 \frac{\partial W}{\partial \lambda_1} + q = \begin{cases} -P & r=a \\ 0 & r=b, \end{cases} \tag{6.3.162}$$

and, from (5.3.21), P is given by

$$P = \int_{\lambda_b}^{\lambda_a} \frac{\hat{W}'(\lambda)\mathrm{d}\lambda}{\lambda^3 - 1}, \tag{6.3.163}$$

where

$$\hat{W}(\lambda) = W(\lambda^{-2}, \lambda, \lambda). \tag{6.3.164}$$

The principal Cauchy stress difference is given by

$$t_1 - t_2 = \tfrac{1}{2}\lambda\hat{W}', \tag{6.3.165}$$

for

$$1 < \lambda_b \le \lambda \le \lambda_a, \tag{6.3.166}$$

it being assumed that

$$\frac{\lambda\hat{W}'}{2(\lambda^6 - 1)} \equiv \mathscr{A}^1_{01212} > 0 \tag{6.3.167}$$

(and $P > 0$).

In view of the underlying symmetry, the components of $\boldsymbol{\mathscr{A}}^1_0$ on the principal axes are, from (6.3.15)–(6.3.17),

$$\left.\begin{aligned} &\mathscr{A}^1_{01111}, \mathscr{A}^1_{01122} = \mathscr{A}^1_{01133}, \mathscr{A}^1_{02222} = \mathscr{A}^1_{03333}, \mathscr{A}^1_{02233}, \\ &\mathscr{A}^1_{01212} = \mathscr{A}^1_{01313}, \mathscr{A}^1_{01221} = \mathscr{A}^1_{01331}, \mathscr{A}^1_{02121} = \mathscr{A}^1_{03131}, \\ &\mathscr{A}^1_{02323} = \mathscr{A}^1_{03232} = \tfrac{1}{2}\left\{\mathscr{A}^1_{02222} - \mathscr{A}^1_{02233} + \lambda_2 \frac{\partial W}{\partial \lambda_2}\right\} \\ &\qquad\qquad = \mathscr{A}^1_{02332} + \lambda_2 \frac{\partial W}{\partial \lambda_2}, \end{aligned}\right\} \tag{6.3.168}$$

subject to the symmetries appropriate to Green elasticity. Clearly, these components depend on r through λ.

With the help of (6.3.3)–(6.3.5) the equilibrium equation (6.3.1) takes the form

$$\operatorname{div}(\mathscr{A}_0^1\dot{\mathbf{A}}_0) + \operatorname{grad}\dot{q} - \dot{\mathbf{A}}_0^{\mathrm{T}}\operatorname{grad} q = \mathbf{0}, \tag{6.3.169}$$

with

$$\dot{\mathbf{A}}_0 = \operatorname{grad}\mathbf{v}, \qquad \operatorname{div}\mathbf{v} = 0, \tag{6.3.170}$$

in the absence of body forces. By means of (6.3.3), (6.3.8) and (6.3.162) the boundary conditions can be put as

$$\left\{\mathscr{A}_0^1\dot{\mathbf{A}}_0 + \dot{q}\mathbf{I} + \lambda_1\frac{\partial W}{\partial\lambda_1}\dot{\mathbf{A}}_0\right\}^{\mathrm{T}}\mathbf{n} = \begin{cases} -\dot{P}\mathbf{n} & \text{on } r = a \\ \mathbf{0} & \text{on } r = b, \end{cases} \tag{6.3.171}$$

where $\dot{P}$ is the incremental applied pressure.

For simplicity we restrict attention to axisymmetric bifurcations and let $\mathbf{v}$ have (r, θ, ϕ)-components $(u, v, 0)$ respectively, with u, v independent of ϕ. From the results of Section 1.5.4, the matrix of components of $\dot{\mathbf{A}}_0$ on the principal axes is

$$\begin{bmatrix} \dfrac{\partial u}{\partial r} & \dfrac{1}{r}\left(\dfrac{\partial u}{\partial\theta} - v\right) & 0 \\ \dfrac{\partial v}{\partial r} & \dfrac{1}{r}\left(u + \dfrac{\partial v}{\partial\theta}\right) & 0 \\ 0 & 0 & \dfrac{1}{r}(u + v\cot\theta) \end{bmatrix}$$

and the incompressibility condition (6.3.170) is therefore

$$r\frac{\partial u}{\partial r} + 2u + \frac{\partial v}{\partial\theta} + v\cot\theta = 0. \tag{6.3.172}$$

Again by reference to Section 1.5.4, and with the help of the underlying equilibrium equation

$$r\frac{\mathrm{d}t_1}{\mathrm{d}r} + 2(t_1 - t_2) = 0,$$

we find that the two components of (6.3.169) which are not satisfied identically are expressible as

$$\left(\mathscr{A}^1_{01111}-\mathscr{A}^1_{01122}+\lambda_1\frac{\partial W}{\partial\lambda_1}-\mathscr{A}^1_{01212}\right)\frac{\partial^2 u}{\partial r^2}$$
$$+\mathscr{A}^1_{02121}\frac{1}{r^2}\left(\frac{\partial^2 u}{\partial\theta^2}+\frac{\partial u}{\partial\theta}\cot\theta+2u\right)$$
$$+\left\{r\frac{\mathrm{d}}{\mathrm{d}r}\left(\mathscr{A}^1_{01111}-\mathscr{A}^1_{01122}+\lambda_1\frac{\partial W}{\partial\lambda_1}\right)\right.$$
$$+2\left(\mathscr{A}^1_{01111}-\mathscr{A}^1_{01122}+\lambda_1\frac{\partial W}{\partial\lambda_1}\right)-2\mathscr{A}^1_{01212}$$
$$+2\left(\lambda_1\frac{\partial W}{\partial\lambda_1}-\lambda_2\frac{\partial W}{\partial\lambda_2}\right)+\mathscr{A}^1_{02222}+\mathscr{A}^1_{02233}$$
$$\left.-2\mathscr{A}^1_{01122}+\lambda_2\frac{\partial W}{\partial\lambda_2}\right\}\frac{1}{r}\frac{\partial u}{\partial r}=-\frac{\partial\dot{q}}{\partial r}, \tag{6.3.173}$$

$$\mathscr{A}^1_{01212}\frac{\partial^2 v}{\partial r^2}+\left(r\frac{\mathrm{d}}{\mathrm{d}r}\mathscr{A}^1_{01212}+2\mathscr{A}^1_{01212}\right)\frac{1}{r}\frac{\partial v}{\partial r}$$
$$-\left(\mathscr{A}^1_{02222}-\mathscr{A}^1_{01122}-\mathscr{A}^1_{01212}+\lambda_1\frac{\partial W}{\partial\lambda_1}\right)\frac{1}{r}\frac{\partial^2 u}{\partial r\partial\theta}$$
$$+\left(r\frac{\mathrm{d}}{\mathrm{d}r}\mathscr{A}^1_{01212}+2\mathscr{A}^1_{01212}-\mathscr{A}^1_{02222}+\mathscr{A}^1_{02233}-\lambda_2\frac{\partial W}{\partial\lambda_2}\right)\times$$
$$\times\frac{1}{r^2}\left(\frac{\partial u}{\partial\theta}-v\right)=-\frac{1}{r}\frac{\partial\dot{q}}{\partial\theta}. \tag{6.3.174}$$

Correspondingly, the boundary conditions (6.3.171) reduce to

$$\left(\mathscr{A}^1_{01111}-\mathscr{A}^1_{01122}+\lambda_1\frac{\partial W}{\partial\lambda_1}\right)\frac{\partial u}{\partial r}+\dot{q}=\begin{cases}-\dot{P} & \text{on} \quad r=a\\ 0 & \text{on} \quad r=b,\end{cases} \tag{6.3.175}$$

$$\frac{\partial v}{\partial r}+\frac{1}{r}\left(\frac{\partial u}{\partial\theta}-v\right)=0 \qquad \text{on } r=a,b. \tag{6.3.176}$$

In a similar manner to that indicated in Problem 6.3.16, we aim to solve

(6.3.172)–(6.3.174) by setting

$$u = f_n(r)P_n(\cos\theta), \quad v = g_n(r)\frac{\mathrm{d}}{\mathrm{d}\theta}P_n(\cos\theta),$$
$$\dot{q} = h_n(r)P_n(\cos\theta) \tag{6.3.177}$$

for $n = 0, 1, 2, \ldots$. For $n = 0$ in (6.3.177), equation (6.3.172) yields

$$f_0(r) = a^3\dot{\lambda}_a/r^2\lambda_a,$$

where $\dot{\lambda}_a$ in the increment in λ_a associated with $\dot{P}$, i.e. $\dot{P} = (\mathrm{d}P/\mathrm{d}\lambda_a)\dot{\lambda}_a$ with $\mathrm{d}P/\mathrm{d}\lambda_a$ given by (5.3.22). This solution corresponds to continuing spherical symmetry. For $n \geq 1$ elimination of g_n and h_n between (6.3.172)–(6.3.174) and use of the equation

$$\frac{\mathrm{d}^2}{\mathrm{d}\theta^2}P_n(\cos\theta) + \cot\theta\frac{\mathrm{d}}{\mathrm{d}\theta}P_n(\cos\theta) + n(n+1)P_n(\cos\theta) = 0$$

leads to an equation for f_n, namely

$$\begin{aligned}
&\frac{\mathrm{d}}{\mathrm{d}r}\left\{r^4\mathscr{A}^1_{01212}f_n''' + r^3\left(r\frac{\mathrm{d}}{\mathrm{d}r}\mathscr{A}^1_{01212} + 4\mathscr{A}^1_{01212}\right)f_n''\right.\\
&\quad + r^3\frac{\mathrm{d}}{\mathrm{d}r}(2\mathscr{A}^1_{01212} + t_1)f_n'\\
&\quad + [n(n+1)(2\mathscr{A}^1_{01221} + 2\mathscr{A}^1_{01122} - \mathscr{A}^1_{01111} - \mathscr{A}^1_{02222})\\
&\quad \left. - (\mathscr{A}^1_{01221} + \mathscr{A}^1_{02121} + \mathscr{A}^1_{02233} - \mathscr{A}^1_{02222})]r^2f_n'\right\}\\
&\quad + (n^2 + n - 2)\left\{r^2\frac{\mathrm{d}^2}{\mathrm{d}r^2}(\mathscr{A}^1_{01212} - t_1)\right.\\
&\qquad\qquad + r^2\frac{\mathrm{d}}{\mathrm{d}r}[r^{-1}(\mathscr{A}^1_{01221} + \mathscr{A}^1_{02121} + \mathscr{A}^1_{02233}\\
&\qquad\qquad \left. - \mathscr{A}^1_{02222})] + n(n+1)\mathscr{A}^1_{02121}\right\}f_n = 0,
\end{aligned} \tag{6.3.178}$$

where $f_n' \equiv \mathrm{d}f_n/\mathrm{d}r$, $f_n'' \equiv \mathrm{d}^2f_n/\mathrm{d}r^2$ etc.

In terms of f_n the boundary conditions (6.3.175) and (6.3.176) become

$$r^3\mathscr{A}^1_{01212}f'''_n + r^2\left(r\frac{\mathrm{d}}{\mathrm{d}r}\mathscr{A}^1_{01212} + 6\mathscr{A}^1_{01212}\right)f''_n$$

$$+\left\{r\frac{\mathrm{d}}{\mathrm{d}r}(2\mathscr{A}^1_{01212} + t_1) + n(n+1)(2\mathscr{A}^1_{01221} + 2\mathscr{A}^1_{01122}\right.$$

$$- \mathscr{A}^1_{01111} - \mathscr{A}^1_{02222} - \mathscr{A}^1_{01212})$$

$$\left. - (\mathscr{A}^1_{01221} + \mathscr{A}^1_{02121} + \mathscr{A}^1_{02233} - \mathscr{A}^1_{02222}) + 6\mathscr{A}^1_{01212}\right\}rf'_n$$

$$+ (n^2+n-2)\left\{r\frac{\mathrm{d}}{\mathrm{d}r}(\mathscr{A}^1_{01212} - t_1) + \mathscr{A}^1_{01221} + \mathscr{A}^1_{02121}\right.$$

$$\left. + \mathscr{A}^1_{02233} - \mathscr{A}^1_{02222}\right\}f_n = 0 \qquad \text{on } r = a, b, \tag{6.3.179}$$

$$r^2 f''_n + 2rf'_n + (n^2+n-2)f_n = 0 \qquad \text{on } r = a, b \tag{6.3.180}$$

for $n \geq 1$, $\dot{P}$ having been taken care of by $n = 0$.

The solution for $n = 1$ The system (6.3.178)–(6.3.180) simplifies when $n = 1$ to the extent that a closed-from solution can be obtained, as we now show.

On putting $n = 1$ and making use of the connections

$$\lambda\frac{\mathrm{d}}{\mathrm{d}\lambda}\left(\lambda_1\frac{\partial W}{\partial\lambda_1}\right) = -2\left(\mathscr{A}^1_{01111} - \mathscr{A}^1_{01122} + \lambda_1\frac{\partial W}{\partial\lambda_1}\right),$$

$$\lambda\frac{\mathrm{d}}{\mathrm{d}\lambda}\left(\lambda_2\frac{\partial W}{\partial\lambda_2}\right) = \mathscr{A}^1_{02222} + \mathscr{A}^1_{02233} - 2\mathscr{A}^1_{01122} + \lambda_2\frac{\partial W}{\partial\lambda_2},$$

we integrate equation (6.3.178) to give

$$r^2\mathscr{A}^1_{01212}f'''_1 + r\left(r\frac{\mathrm{d}}{\mathrm{d}r}\mathscr{A}^1_{01212} + 4\mathscr{A}^1_{01212}\right)f''_1$$

$$+\left\{(\lambda^3+3)r\frac{\mathrm{d}}{\mathrm{d}r}\mathscr{A}^1_{01212} - 4\lambda^6\mathscr{A}^1_{01212}\right\}f'_1 = 0 \tag{6.3.181}$$

in conjunction with the boundary conditions (6.3.179) and (6.3.180), which are easily seen to reduce together to

$$rf''_1 + 2f'_1 = 0 \qquad \text{on } r = a, b. \tag{6.3.182}$$

Next, with use of the result

$$r\frac{d\lambda}{dr} = \lambda(1 - \lambda^3), \tag{6.3.183}$$

obtainable from (6.3.160), equation (6.3.181) can be integrated to give

$$rf_1'' + (3 + \lambda^3)f_1' = c_1 r/\lambda \mathscr{A}^1_{01212}, \tag{6.3.184}$$

where c_1 is a constant. This may also be integrated, with the help of (6.3.183), to give

$$r^4 f_1' = 2\lambda c_1 \int \frac{(\lambda^3 - \lambda^{-3})}{\hat{W}'} r^4 dr + \lambda c_2, \tag{6.3.185}$$

where use has been made of (6.3.167), and c_2 is a constant.

Substitution of (6.3.184) and (6.3.185) into (6.3.182), followed by elimination of c_2, an integration by parts, use of (6.3.160) and (6.3.183), and some algebraic manipulation leads to

$$\int_{\lambda_b}^{\lambda_a} (\lambda^3 - 1)^{-2/3}(\lambda^2 \hat{W}'' - \lambda \hat{W}')(\hat{W}')^{-2} d\lambda = 0 \tag{6.3.186}$$

(or $c_1 = 0$).

Thus, bifurcation into an $n = 1$ mode of deformation is possible at values of λ_a for which (6.3.186) is satisfied, λ_b being given in terms of λ_a by

$$\lambda_b = \{1 + A^3(\lambda_a^3 - 1)/B^3\}^{1/3}. \tag{6.3.187}$$

For a thin-walled shell, with $(B - A)/A = \frac{1}{3}\varepsilon \ll 1$, equation (6.3.186) approximates to

$$\lambda \hat{W}'' - \hat{W}' = 0 \tag{6.3.188}$$

since, from (6.3.187),

$$\lambda_a - \lambda_b = \tfrac{1}{3}\varepsilon(\lambda_a - \lambda_a^{-2}) + O(\varepsilon^2),$$

where $\lambda_b < \lambda < \lambda_a$. As far as (6.3.188) is concerned it is not necessary to distinguish between λ, λ_a, λ_b provided terms of order ε are neglected.

Before proceeding further it is necessary to examine the properties of $\hat{W}$ and, more particularly, the sign of $\lambda \hat{W}'' - \hat{W}'$ for $1 \le \lambda < \infty$. In this regard we take as our starting point the inequalities

$$\left.\begin{array}{l}\hat{W}' \gtreqless 0 \text{ according as } \lambda \gtreqless 1,\\ \hat{W}''(1) > 0,\end{array}\right\} \tag{6.3.189}$$

these together being equivalent to (6.3.167).

In connection with (6.3.188), we recall from Section 5.3.2 that the surface tension T in a thin-walled shell is given by

$$T = \tfrac{1}{6}\varepsilon A \lambda^{-1} \hat{W}'. \tag{6.3.190}$$

so that (6.3.188) is equivalent to $\mathrm{d}T/\mathrm{d}\lambda = 0$. Correspondingly, from (5.3.23), the internal pressure P is given by

$$P = \tfrac{1}{3}\varepsilon \lambda^{-2} \hat{W}'. \tag{6.3.191}$$

It follows immediately from (6.3.189)–(6.3.191) that

$$\frac{\mathrm{d}P}{\mathrm{d}\lambda} > 0 \Rightarrow \frac{\mathrm{d}T}{\mathrm{d}\lambda} > 0 \qquad (\lambda \geq 1).$$

We conclude that a necessary condition for bifurcation to occur is that P reaches a local maximum as λ increases from unity (i.e. as the shell radius increases from its initial value).

To help illustrate the possibilities we depict some typical characteristics of the functions $T^* \equiv \lambda^{-1}\hat{W}'$ and $P^* \equiv \lambda^{-2}\hat{W}'$ in Fig. 6.11. Also included for comparison is the graph of $\hat{W}'$ (which, in essence, represents the in-surface nominal stress). Firstly, in Fig. 6.11(a), we assume that P^* has a maximum at λ_0, followed by a minimum at λ_0', after which P^* increases monotonically with λ. Likewise, T^* is shown to have a maximum at λ_1 and a minimum at λ_1' (these corresponding to a pair of $n = 1$-mode bifurcation points). By contrast $\hat{W}'$ is monotonic increasing for $\lambda > 1$.

Since it is easy to show that

$$\frac{\mathrm{d}T^*}{\mathrm{d}\lambda} = \lambda \frac{\mathrm{d}P^*}{\mathrm{d}\lambda} + P^* \geq \frac{\mathrm{d}P^*}{\mathrm{d}\lambda} \quad (\lambda \geq 1)$$

it follows that, with the characteristics of P^* shown in Fig. 6.11(a), bifurcation points necessarily occur in pairs if they occur at all. Moreover, such points occur between the pressure maximum and minimum.

For the class of strain-energy functions (4.3.66), we have

$$\hat{W} = \sum_{p=1}^{N} \mu_p (2\lambda^{\alpha_p} + \lambda^{-2\alpha_p} - 3)/\alpha_p. \tag{6.3.192}$$

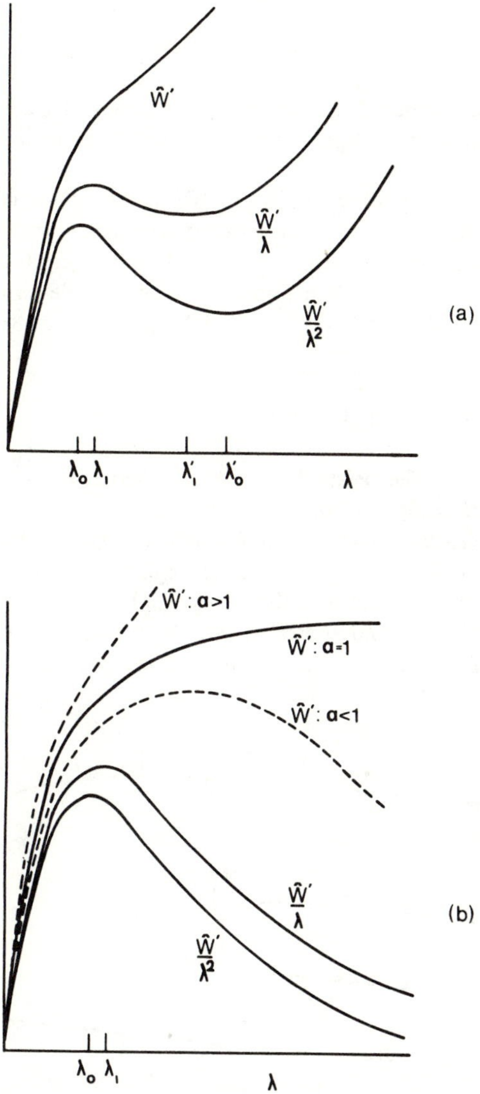

Fig. 6.11 Plots of $\hat{W}'(\lambda)$, $\lambda^{-1}\hat{W}'(\lambda)$, $\lambda^{-2}\hat{W}'(\lambda)$ against λ illustrating (a) typical maximum–minimum characteristics with two bifurcation points λ_1, λ_1', (b) case with no minimum and a single bifurcation point λ_1 in respect of (6.3.192) with $N = 1$.

Use of this in (6.3.188) shows that a necessary condition for ($n = 1$) bifurcation is

$$-1 < \alpha_p < 2 \qquad \text{some } p \in \{1, \ldots, N\}. \tag{6.3.193}$$

Note, in particular, that this rules out the Mooney strain-energy function ($\alpha_1 = 2, \alpha_2 = -2$) even though it possesses the P^* behaviour displayed in Fig. 6.11(a).

A second possibility, shown in Fig. 6.11(b), allows P^* to have a maximum (at λ_0), after which P^* decays monotonically to zero as $\lambda \to \infty$, as in the case of a neo-Hookean strain-energy function. With T^* following the same pattern, there is just a single bifurcation point (λ_1). For a single-term strain-energy function, corresponding to $N = 1$ in (6.3.192), equation (6.3.188) is satisfied for λ given by

$$\lambda^{3\alpha_1} = 2(\alpha_1 + 1)/(2 - \alpha_1), \tag{6.3.194}$$

subject to (6.3.193). The behaviour of $\hat{W}'$ is different for $\alpha_1 > 1$, $\alpha_1 = 1$ and $\alpha_1 < 1$, as shown in Fig. 6.11(b).

We now return to the bifurcation criterion (6.3.186). Clearly, a necessary condition for (6.3.186) to be satisfied is that λ_a has reached a value such that (6.3.188) holds for some λ in the interval (λ_b, λ_a). It is clear from (6.3.187) that the length of this interval increases with the shell thickness ratio B/A. Therefore, when the material characteristics are as shown in Fig. 6.11(a),

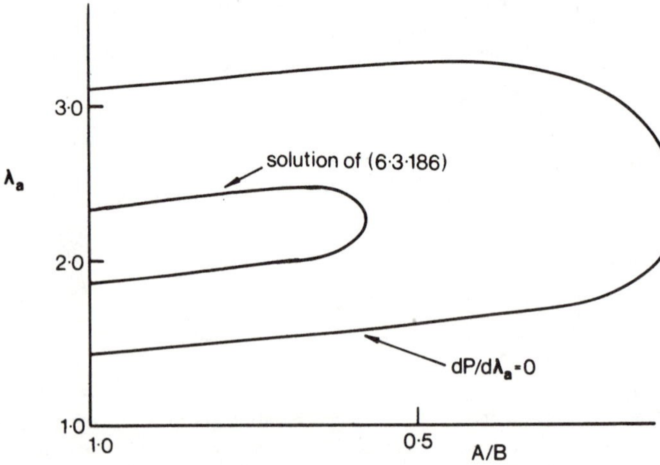

Fig. 6.12 Plot of critical values of λ_a against A/B corresponding to $n = 1$ mode bifurcation, compared with the solution of $dP/d\lambda_a = 0$.

bifurcation is ruled out for values of B/A larger than some critical value because the integrand in (6.3.186) is then negative for only a relatively small part of the interval (λ_b, λ_a). Furthermore, beyond a second critical value of B/A, P becomes a monotonic increasing function of $\lambda_a(>1)$. This behaviour is illustrated in Fig. 6.12 where the dependence on A/B of the critical values λ_a satisfying (6.3.186) is shown along with the solution of $\mathrm{d}P/\mathrm{d}\lambda_a = 0$. Results for $A/B = 1$ are those appropriate to the membrane theory limit.

In respect of the characteristics shown in Fig. 6.11(b), on the other hand, bifurcation is possible for arbitrary shell thicknesses. The details of this case are left for the reader to study.

In the membrane limit two independent modes of incremental deformation are available at an $n = 1$ bifurcation point. The first, for which $u = 0$, $v = -B \sin\theta$, where B is a constant, corresponds to thinning of the material at one pole of the spherical membrane and thickening at the other pole, the spherical shape being maintained. In the second, we have $v = 0$, $u = A\cos\theta$, where A is a constant, the equatorial radius is maintained, but the shape becomes aspherical (slightly egg-shaped) with respect to the equatorial plane. For a shell of finite thickness such modes interact through their dependence on r.

Results for $n \geq 2$ Apart from notational differences and some other minor amendments the analysis so far in this section follows closely that given by Haughton and Ogden (1978). We refer to this paper for further details and discussion of the $n = 1$ results and for references to earlier work. For mode numbers $n \geq 2$, corresponding results have also been given by Haughton and Ogden, and here we summarize briefly their conclusions.

First, in respect of a membrane, Haughton and Ogden (1978) provided a bifurcation criterion for a general value of n. In particular, for the constitutive properties illustrated in Fig. 6.11(a), they showed that, associated with each mode number n are pairs of bifurcation points, (λ_n, λ_n') say, nested according to

$$\lambda_0 < \lambda_1 < \lambda_2 < \cdots < \lambda_n < \cdots < \lambda_n' < \cdots < \lambda_2' < \lambda_1' < \lambda_0',$$

where (λ_0, λ_0') define the pressure turning points. In general, only a finite number of such pairs exist. Indeed, for a large class of strain-energy functions, there is at most one pair of bifurcation points (corresponding to $n = 1$). However, for certain strain-energy functions bifurcation points of all orders exist, and, specifically, in respect of (6.3.192), a necessary condition for this is

$$-1 < \alpha_p < 1 \qquad \text{some } p \in \{1, \ldots, N\}. \tag{6.3.195}$$

This condition is also sufficient when $N = 1$.

To show how the existence of bifurcation points for different mode

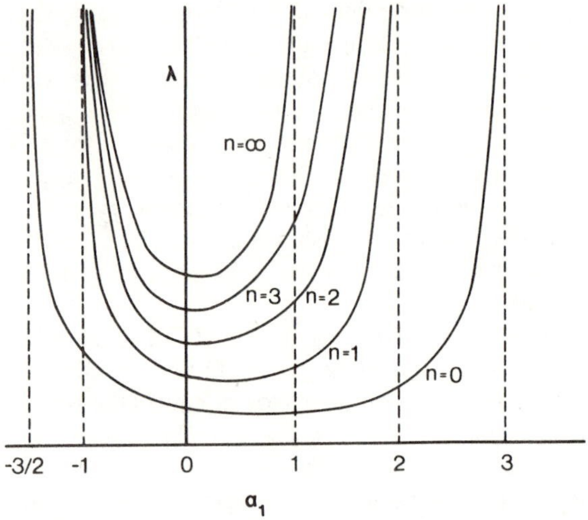

Fig. 6.13 Plot of critical values of λ against α_1 for different mode numbers n in respect of bifurcation of a spherical membrane shell.

numbers depends on the form of constitutive law we have calculated the critical values of λ appropriate to a membrane in respect of a single term strain-energy function and for a range of values of the exponent α_1. The results are shown in Fig. 6.13, where λ is plotted against α_1 for $n=0,1,2,3,\infty$, $n=0$ defining the pressure maximum points. We note that, in accordance with (6.3.193), no bifurcation points exist if $\alpha_1 \geq 2$ or $\alpha_1 \leq -1$, while for $-1<\alpha_1<2$ the constitutive properties are as shown in Fig. 6.11(b). Thus, for a given α_1, each mode number n possesses at most one bifurcation point, λ_n say, and $\lambda_0<\lambda_1<\lambda_2<\cdots<\lambda_n<\cdots$, there being no pressure minimum.

For thick-walled shells closed-form bifurcation criteria cannot be obtained from (6.3.178)–(6.3.180) for $n\geq 2$ and numerical calculations are needed to extract useful information. Such calculations have been carried out by Haughton and Ogden (1978). Provided B/A is not too large they found similar results to those shown in Fig. 6.13 for a single-term strain-energy function, and they gave details for $A/B=0.85$.

In order to illustrate how the shell thickness affects the existence of bifurcation points in respect of a single-term strain-energy function satisfying (6.3.195), we plot critical values of λ_a against A/B in Fig. 6.14 for a series of mode numbers. The left-hand end of the horizontal axis corresponds to the membrane limit $A/B=1$. The effect of increasing the shell thickness is to progressively phase out the bifurcation points corresponding to the largest mode numbers, until, in the limit of a spherical cavity in an infinite medium ($A/B=0$), the pressure maximum disappears.

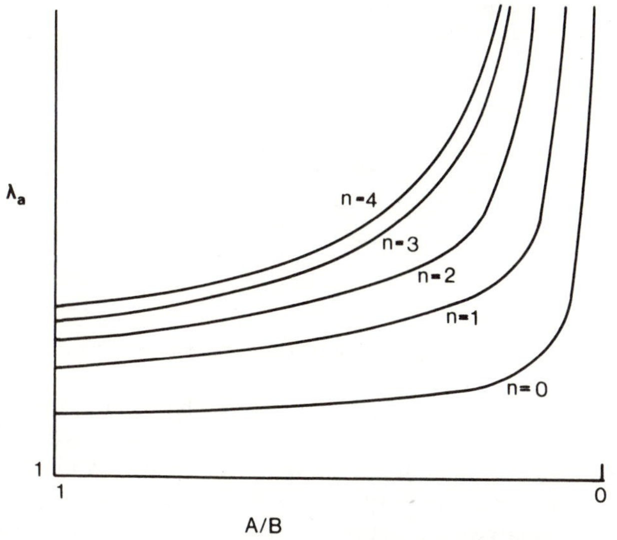

Fig. 6.14 Plot of critical values of λ_a against A/B for different mode numbers n and for a single-term strain-energy function satisfying (6.3.195).

By contrast, for strain-energy functions with the properties exhibited in Fig. 6.11(a), bifurcation points corresponding to $n \geq 2$ can only occur if at all, inside the closed curve defining $n = 1$ bifurcation points shown in Fig. 6.12. For each n (≥ 2) for which bifurcation points exist, these points aggregate into a curve in the (A/B, λ_a)-plane similar in shape to the $n = 1$ curve but with a higher 'cut-off' value of A/B.

Global interpretation of the results Our analysis has been restricted to the local behaviour in the neighbourhood of a spherically symmetric configuration, with the incremental modes at the bifurcation points defining, in essence, the *initial* tangents to the aspherical deformation paths emanating from the points. In order to trace the global development of the aspherical paths it is necessary to return from the linearized equations to the fully non-linear equations. However, even in the case of a membrane, analysis of these equations is difficult and little information of a qualitative nature has been forthcoming. The equations have been solved numerically by Haughton (1980), who has shown that, for a certain form of strain-energy function, the bifurcation points λ_1, λ_1' are linked.

Fig. 6.15 shows a global bifurcation diagram in which some measure of asphericity (which need not be specified here) is plotted against λ. For the present discussion, attention is confined to strain-energy functions with the

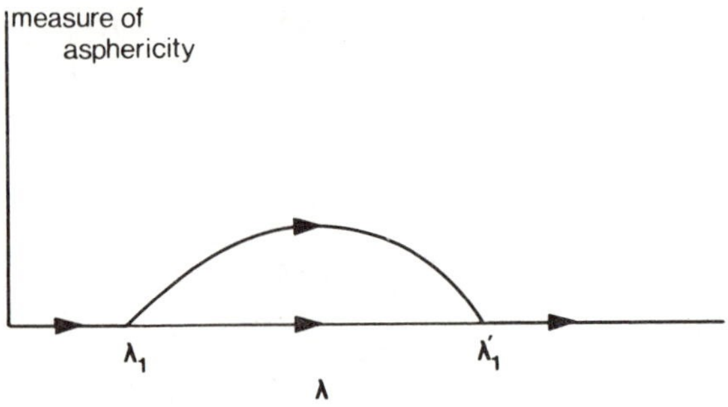

Fig. 6.15 Global bifurcation diagram for an internally pressurized spherical membrane shell.

properties shown in Fig. 6.11(a) and to membrane shells (according to Fig. 6.12 results for thick-walled shells for which bifurcation is possible follow the same pattern as those for a membrane). In Fig. 6.15 the vertical axis measures the departure from spherical symmetry while the horizontal axis, on which are marked the bifurcation points λ_1, λ_1', corresponds to the spherically symmetric path of deformation.

As λ increases from unity the spherical shape is maintained up to the value λ_1, after which point the shape becomes aspherical. The asphericity grows with further increase in λ, reaches a maximum, decreases and then fades out, the spherical shape being regained at the second bifurcation point λ_1' and thereafter retained. Since P is measured at all stages of the inflation, we *define* the parameter λ on the aspherical part of the deformation path as the (unique) inverse of (6.3.191) on the interval (λ_1, λ_1'), even though λ does not then have its original physical interpretation.

The behaviour depicted in Fig. 6.15 reflects what is observed in experiments on meteorological balloons, and references to such experiments are given in Haughton and Ogden (1978).

The spherically symmetric path of deformation is clearly unstable for $\lambda_1 < \lambda < \lambda_1'$. A stability analysis, based on the exclusion functional (6.2.139) appropriately specialized, is left for the reader to carry out. We emphasize, however, that, on the primary deformation path, stability is lost at λ_1, *not* at the pressure maximum point λ_0.

Finally, we remark that the effect of small inhomogeneities on the above results has been examined by Needleman (1977) and Haughton (1980).

Problem 6.3.21 Analyse the buckling of a spherical shell under *external* pressure and show that the results can be obtained by setting $P < 0$ and $\lambda_a < 1$ in the equations of this section. (Some numerical results for this problem have been given by Haughton and Ogden (1978, p. 135).)

Problem 6.3.22 The current configuration of a circular cylindrical tube subject to internal pressure and axial loading is described in Section 5.3.3. Analyse the problem of bifurcation from such a configuration under suitably chosen incremental boundary conditions. (This problem has been discussed in detail by Haughton and Ogden (1979a) for a membrane tube and (1979b) for a thick-walled tube. Results for rotating tubes under axial loading are given by Haughton and Ogden (1980b, c).)

6.4 WAVES AND VIBRATIONS

Although this book is not primarily concerned with the discussion of time-dependent problems it is nevertheless instructive at this point to give a brief account of the theory of infinitesimal waves and vibrations in a finitely-deformed elastic solid. Specifically, the intention here is to show the connection between this theory and the theory of quasi-static incremental deformations considered so far in Chapter 6.

With the incremental displacement $\boldsymbol{\chi} \equiv \mathbf{v}$ now dependent on time t, the incremental equation of motion in the absence of body forces is written

$$\operatorname{div} \dot{\mathbf{S}}_0 = \rho \frac{\partial^2 \mathbf{v}}{\partial t^2} \tag{6.4.1}$$

when referred to the current configuration, where ρ is the current density and a superposed dot indicates an incremental quantity (not a time derivative).

Henceforth we restrict attention to homogeneous underlying deformations. The Cartesian component form of (6.4.1) is then

$$\mathscr{A}^1_{0ijkl} v_{l,ik} = \rho \frac{\partial^2 v_j}{\partial t^2} \tag{6.4.2}$$

for unconstrained materials, and

$$\mathscr{A}^1_{0ijkl} v_{l,ik} + \dot{q}_{,j} = \rho \frac{\partial^2 v_j}{\partial t^2} \tag{6.4.3}$$

together with

$$v_{i,i} = 0 \tag{6.4.4}$$

for incompressible materials.

First we consider the propagation of plane waves in a material without boundary. We write

$$\mathbf{v} = \mathbf{m} f(\mathbf{n}\cdot\mathbf{x} - ct), \qquad \dot{q} = q_0 f'(\mathbf{n}\cdot\mathbf{x} - ct), \tag{6.4.5}$$

where q_0 is a constant and f is a twice-continuously differentiable function. In (6.4.5), c is the *wave speed*, the unit vector $\mathbf{n}$ defines the *direction of propagation* of the wave and the unit vector $\mathbf{m}$ is called the *polarization* of the wave.

Substitution of (6.4.5) into (6.4.2) and (6.4.3) yields

$$\mathscr{A}^1_{0ijkl} n_i n_k m_l = \rho c^2 m_j \tag{6.4.6}$$

and

$$\mathscr{A}^1_{0ijkl} n_i n_k m_l + q_0 n_j = \rho c^2 m_j \tag{6.4.7}$$

respectively, the latter being coupled with

$$m_i n_i = 0. \tag{6.4.8}$$

It follows from (6.4.7) and (6.4.8) that

$$q_0 = -\mathscr{A}^1_{0ijkl} n_i n_j n_k m_l. \tag{6.4.9}$$

Returning to the definition (6.2.155) of the acoustic tensor $\mathbf{Q}_0(\mathbf{n})$ we obtain the component representation

$$Q_{0ij} = \mathscr{A}^1_{0piqj} n_p n_q. \tag{6.4.10}$$

With the help of (6.4.9) equations (6.4.6) and (6.4.7) respectively may be written

$$\mathbf{Q}_0(\mathbf{n})\mathbf{m} = \rho c^2 \mathbf{m} \tag{6.4.11}$$

and

$$\mathbf{Q}_0^*(\mathbf{n})\mathbf{m} = \rho c^2 \mathbf{m}, \tag{6.4.12}$$

where

$$\mathbf{Q}_0^*(\mathbf{n}) = \mathbf{Q}_0(\mathbf{n}) - \mathbf{n} \otimes [\mathbf{Q}_0^{\mathrm{T}}(\mathbf{n})\mathbf{n}]. \tag{6.4.13}$$

For a Green-elastic material $\mathbf{Q}_0(\mathbf{n})$ is symmetric but, in general, $\mathbf{Q}_0^*(\mathbf{n})$ is not.

Equation (6.4.11) is called the *propagation condition*. It determines possible wave speeds and polarizations for any given direction of propagation in an unconstrained material, as does the propagation condition (6.4.12) for an incompressible material. When $\mathbf{n}$ is given and $\mathbf{m}$ is known, subject to $\mathbf{m}\cdot\mathbf{n}=0$ for an incompressible material, the wave speed is obtained from (6.4.11) or (6.4.12) with (6.4.13) in the form

$$\rho c^2 = [\mathbf{Q}_0(\mathbf{n})\mathbf{m}]\cdot\mathbf{m}. \tag{6.4.14}$$

Since $\mathbf{Q}_0(\mathbf{n})$ is symmetric for a Green-elastic material the eigenvalues ρc^2 of (6.4.11) are real and there exists a triad of mutually orthogonal eigenvectors $\mathbf{m}$ for each $\mathbf{n}$. Moreover, by reference to Section 6.2.7 and to equation (6.4.14), we see that the wave speeds are real for all $\mathbf{n}$ if and only if the strong ellipticity condition holds. The wave speeds are determined as functions of $\mathbf{n}$ from the *characteristic equation*

$$\det\{\mathbf{Q}_0(\mathbf{n}) - \rho c^2\mathbf{I}\} = 0. \tag{6.4.15}$$

For incompressible materials we deduce from (6.4.13) that

$$\mathbf{Q}_0^{*\mathrm{T}}(\mathbf{n})\mathbf{n} = \mathbf{0}$$

and hence that $\mathbf{Q}_0^*(\mathbf{n})$ has rank two. Equation (6.4.12) possesses just two eigenvectors, orthogonal to $\mathbf{n}$, but, since $\mathbf{Q}_0^*(\mathbf{n})$ is not in general symmetric, not necessarily mutually orthogonal. The characteristic equation

$$\det\{\mathbf{Q}_0^*(\mathbf{n}) - \rho c^2\mathbf{I}\} = 0 \tag{6.4.16}$$

correspondingly has two solutions for ρc^2 (recall the results of Problem 6.2.14).

If the material is isotropic the components of $\mathbf{Q}_0(\mathbf{n})$ on the Eulerian principal axes $\mathbf{v}^{(1)}$, $\mathbf{v}^{(2)}$, $\mathbf{v}^{(3)}$ are given in Problem 6.2.13. For a Green-elastic material these are such that

$$JQ_{0ii} = \lambda_i^2\frac{\partial^2 W}{\partial\lambda_i^2}n_i^2 + \left(\frac{\lambda_i(\partial W/\partial\lambda_i) - \lambda_j(\partial W/\partial\lambda_j)}{\lambda_i^2 - \lambda_j^2}\right)\lambda_j^2 n_j^2$$
$$+\left(\frac{\lambda_i(\partial W/\partial\lambda_i) - \lambda_k(\partial W/\partial\lambda_k)}{\lambda_i^2 - \lambda_k^2}\right)\lambda_k^2 n_k^2, \tag{6.4.17}$$

$$JQ_{0ij} = \left(\frac{\partial^2 W}{\partial\lambda_i\partial\lambda_j} + \frac{\lambda_j(\partial W/\partial\lambda_i) - \lambda_i(\partial W/\partial\lambda_j)}{\lambda_i^2 - \lambda_j^2}\right)\lambda_i\lambda_j n_i n_j, \tag{6.4.18}$$

where (i, j, k) is a cyclic permutation of $(1, 2, 3)$. For an unconstrained material $J = \lambda_1\lambda_2\lambda_3$, while $J = 1$ for an incompressible material.

When the direction of propagation is along a principal direction, $\mathbf{n} = \mathbf{v}^{(1)}$ say, it is easy to see that for an unconstrained material the polarization vectors satisfying (6.4.11) are simply $\mathbf{v}^{(1)}$, $\mathbf{v}^{(2)}$, $\mathbf{v}^{(3)}$. The first corresponds to a *longitudinal wave* ($\mathbf{m} = \mathbf{n}$) with

$$\rho c^2 = \lambda_1 \frac{\partial t_1}{\partial \lambda_1} \equiv J^{-1} \frac{\partial^2 W}{\partial \lambda_1^2}.$$

The second and third give *transverse waves* ($\mathbf{m}\cdot\mathbf{n} = 0$) with

$$\rho c^2 = \frac{t_1 - t_2}{\lambda_1^2 - \lambda_2^2} \lambda_1^2, \qquad \rho c^2 = \frac{t_1 - t_3}{\lambda_1^2 - \lambda_3^2} \lambda_1^2$$

respectively. For an incompressible material the solutions of (6.4.12) are just these two transverse waves.

In general, longitudinal waves exist in an unconstrained material only for certain isolated directions $\mathbf{n}$; the existence of at least one such $\mathbf{n}$ satisfying

$$\mathbf{Q}_0(\mathbf{n})\mathbf{n} = \rho c^2 \mathbf{n}$$

with $\rho c^2 > 0$ has been established by Truesdell (1966) using the assumption

$$[\mathbf{Q}_0(\mathbf{n})\mathbf{n}]\cdot\mathbf{n} > 0 \qquad \text{all } \mathbf{n} \tag{6.4.19}$$

but otherwise irrespective of the underlying state of deformation.

The result applies to a Cauchy-elastic material; for a Green-elastic material the longitudinal wave is coupled with a pair orthogonal transverse waves if the strong ellipticity condition holds. Exceptionally, longitudinal waves exist for all $\mathbf{n}$ when the underlying deformation is a pure dilatation or when the constitutive law has a special form, again under the assumption that (6.4.19) holds. For a discussion of the latter situation for both Cauchy- and Green-elastic materials see Ogden (1970b, c) and the references therein.

For transverse waves propagating in a Cauchy-elastic material existence results have been given by Chadwick and Currie (1975). Under certain conditions a pair of orthogonal transverse waves can propagate for every direction $\mathbf{n}$ (Ogden, 1970b, c). For more background information see Truesdell and Noll (1965) and Wang and Truesdell (1973).

The propagation condition for infinitesimal plane waves is formally identical to the propagation condition for plane acceleration waves (i.e. propagating planes carrying discontinuities in the acceleration and in the velocity gradient), and it was in the context of acceleration waves that the work of Truesdell (1966) and of Chadwick and Currie (1975) was set. For a detailed account of acceleration waves in unconstrained materials, we cite the article by Chen (1973) and the review by McCarthy (1975), each of which contains a comprehensive list of references. Papers by Ogden (1974b), Scott (1975, 1976) and Borejko and Chadwick (1980) deal with various aspects of acceleration waves in constrained materials. When the wave speed is zero the acceleration wave is referred to as a *stationary discontinuity*. Such discontinuities arise in the quasi-static context as discontinuities in the displacement gradient, as in the case of the shear-band deformation discussed in connection with the failure of ellipticity in Sections 6.2.5 and 6.2.7. Recently, a number of authors have studied these discontinuities in relation to boundary-value problems involving cracks. A concise and readable survey of this work has been given by Knowles (1982).

Thus far attention has been restricted to waves propagating in unbounded material. For material bodies with boundaries appropriate incremental boundary conditions need to be applied. For example, in considering the propagation of waves in a half-space the quasi-static formulation outlined in Problems 6.3.4 and 6.3.7, may be amended to include time dependence with the given boundary conditions applying for all $t \geq 0$. In the general (plane-strain) problem (6.3.27) is replaced by

$$v_\alpha = m_\alpha f(p_1 x_1 + p_2 x_2 - ct),$$

with f specialized to exponential/trigonometric form as appropriate.

Wave speeds for possible waves through the material are obtained by solution of the *secular equation* (or *characteristic equation* for consistency with the terminology used in respect of (6.4.15)). This results when the general solution of the equations is introduced into the boundary conditions. The point to emphasize here is that when $c = 0$ the secular equation reduces to the bifurcation criterion for the corresponding quasi-static problem. This limiting case may be interpreted as a standing wave, and is analogous to a stationary discontinuity in the context of acceleration waves.

For an excellent account of surface waves in a finitely-deformed elastic material see Chadwick and Jarvis (1979). The literature contains relatively little work which deals with the vibration of deformed bodies of finite size, and, considering the algebraic complexity involved in simple quasi-static problems (as evidenced in Section 6.3.1), this is not altogether surprising. A notable exception is the paper by Haughton (1982) which embodies some earlier work. This paper examines various modes of vibration of an axially-loaded circular cylinder rotating about its axis and includes discussion of the special cases in which either the axial load or the angular velocity vanishes. A feature of problems of this type is that, in general, numerical procedures have to be adopted at an early stage in the analysis if results of practical interest are required. A summary of other work concerned with time-dependent problems in finite elasticity is contained in the book by

Eringen and Suhubi (1974). This includes discussion of radial oscillations of spherical shells and circular cylindrical tubes without the restriction to small deformations.

Problem 6.4.1 Obtain the propagation condition for a homogeneously deformed elastic material which is inextensible in a single direction.

Problem 6.4.2 For a longitudinal wave propagating in the $(\mathbf{v}^{(1)}, \mathbf{v}^{(2)})$ principal plane show that the propagation condition (6.4.11) reduces to

$$\mathscr{A}^1_{01111} n_1^2 + (\mathscr{A}^1_{01122} + \mathscr{A}^1_{01212} + \mathscr{A}^1_{02121}) n_2^2 = \rho c^2,$$
$$(\mathscr{A}^1_{01122} + \mathscr{A}^1_{01212} + \mathscr{A}^1_{02121}) n_1^2 + \mathscr{A}^1_{02222} n_2^2 = \rho c^2,$$

in respect of an isotropic Green-elastic material. Obtain necessary and sufficient conditions for the existence of such a wave.

Problem 6.4.3 In an incompressible isotropic Green-elastic material show that a pair of transverse waves exists when the direction of propagation lies in a principal plane. If $n_3 = 0$ show that the wave speeds are given by

$$\rho c^2 = \mathscr{A}^1_{01212} n_1^4 + (\mathscr{A}^1_{01111} + \mathscr{A}^1_{02222} - 2\mathscr{A}^1_{01122} - 2\mathscr{A}^1_{02112}) \times$$
$$\times n_1^2 n_2^2 + \mathscr{A}^1_{02121} n_2^4$$

and

$$\rho c^2 = \frac{t_1 - t_3}{\lambda_1^2 - \lambda_3^2} \lambda_1^2 n_1^2 + \frac{t_2 - t_3}{\lambda_2^2 - \lambda_3^2} \lambda_2^2 n_2^2,$$

corresponding to polarizations with $m_1 = n_2$, $m_2 = -n_1$, $m_3 = 0$ and $m_1 = m_2 = 0$, $m_3 = 1$ respectively.

For the same direction of propagation find necessary and sufficient conditions for the existence of a transverse wave with $m_3 = 0$ in an unconstrained material.

REFERENCES

J.M. Ball, Convexity Conditions and Existence Theorems in Non-linear Elasticity, *Archive for Rational Mechanics and Analysis*, Vol. 63, pp. 337–403 (1977).

J.M. Ball, Discontinuous Equilibrium Solutions and Cavitation in Non-linear Elasticity, *Philosophical Transactions of the Royal Society of London, Series A*, Vol. 306, pp. 557–611 (1982).

J.M. Ball and D.G. Schaeffer, Bifurcation and Stability of Homogeneous Equilibrium Configurations of an Elastic Body under Dead-load Tractions, *Mathematical Proceedings of the Cambridge Philosophical Society*, Vol. 94, pp. 315–340 (1983).

M.F. Beatty, Elastic Stability of Rubber Bodies in Compression, in *Finite Elasticity*, Applied Mechanics Symposia Series Vol. 27 (Ed. R.S. Rivlin), American Society of Mechanical Engineers, pp. 125–150 (1977).

P. Borejko and P. Chadwick, Elastic Relations for Acceleration Waves in Elastic Materials, *Wave Motion*, Vol. 2, pp. 361–374 (1980).

I.W. Burgess and M. Levinson, The Instability of Slightly Compressible Rectangular Rubberlike Solids under Biaxial Loadings, *International Journal of Solids and Structures*, Vol. 8, pp. 133–148 (1972).

P. Chadwick and P.K. Currie, On the Existence of Transverse Elastic Acceleration Waves, *Mathematical Proceedings of the Cambridge Philosophical Society*, Vol. 77, pp. 405–413 (1975).

P. Chadwick and D.A. Jarvis, Surface Waves in a Pre-stressed Elastic Body, *Proceedings of the Royal Society of London, Series A*, Vol. 366, pp. 517–536 (1979).

P. Chadwick and R.W. Ogden, On the Definition of Elastic Moduli, *Archive for Rational Mechanics and Analysis*, Vol. 44, pp. 41–53 (1971a).

P. Chadwick and R.W. Ogden, A Theorem of Tensor Calculus and its Application to Isotropic Elasticity, *Archive for Rational Mechanics and Analysis*, Vol. 44, pp. 54–68 (1971b).

P.J. Chen, Growth and Decay of Waves in Solids, in *Handbuch der Physik*, Vol. VIa/3 (Ed. C.A. Truesdell), pp. 303–402, Springer (1973).

D.R.J. Chillingworth, J.E. Marsden and Y.H. Wan, Symmetry and Bifurcation in Three-dimensional Elasticity: Part I, *Archive for Rational Mechanics and Analysis*, Vol. 80, pp. 295–331 (1982).

J.L. Ericksen, Special Topics in Elastostatics, *Advances in Applied Mechanics*, Vol. 17, pp. 189–244 (1977).

A.C. Eringen and E.S. Suhubi, *Elastodynamics*, Vol. I, Academic Press (1974).

R.L. Fosdick and R.T. Shield, Small Bending of a Circular Bar Superimposed on Finite Extension or Compression, *Archive for Rational Mechanics and Analysis*, Vol. 12, pp. 223–248 (1963).

A.E. Green and W. Zerna, *Theoretical Elasticity*, 2nd edition, Oxford University Press (1968).

M.E. Gurtin, The Linear Theory of Elasticity, in *Handbuch der Physik*, Vol. VIa/2 (Ed. C.A. Truesdell), pp. 1–295, Springer (1973).

M.E. Gurtin, On Uniqueness in Finite Elasticity, in *Finite Elasticity* (Eds. D.E. Carlson and R.T. Shield), Proceedings of the IUTAM Symposium, Lehigh University 1980, pp. 191–199, Martinus Nijhoff (1982).

D.M. Haughton, Post Bifurcation of Perfect and Imperfect Spherical Elastic Membranes, *International Journal of Solids and Structures*, Vol. 16, pp. 1123–1133 (1980).

D.M. Haughton, Wave Speeds in Rotating Elastic Cylinders at Finite Deformation, *Quarterly Journal of Mechanics and Applied Mathematics*, Vol. 35, pp. 125–139 (1982).

D.M. Haughton and R.W. Ogden, On the Incremental Equations in Non-linear Elasticity II. Bifurcation of Pressurized Spherical Shells, *Journal of the Mechanics and Physics of Solids*, Vol. 26, pp. 111–138 (1978).

D.M. Haughton and R.W. Ogden, Bifurcation of Inflated Circular Cylinders of Elastic Material under Axial Loading—I. Membrane Theory for Thin-walled Tubes, *Journal of the Mechanics and Physics of Solids*, Vol. 27, pp. 179–212 (1979a).

D.M. Haughton and R.W. Ogden, Bifurcation of Inflated Circular Cylinders of Elastic Material under Axial Loading—II. Exact Theory of Thick-walled Tubes, *Journal of the Mechanics and Physics of Solids*, Vol. 27, pp. 489–512 (1979b).

D.M. Haughton and R.W. Ogden, Bifurcation of Finitely-deformed Rotating Elastic Cylinders, *Quarterly Journal of Mechanics and Applied Mathematics*, Vol. 33, pp. 251–265 (1980a).

D.M. Haughton and R.W. Ogden, Bifurcation of Rotating Circular Cylindrical Elastic Membranes, *Mathematical Proceedings of the Cambridge Philosophical Society*, Vol. 87, pp. 357–376 (1980b).

D.M. Haughton and R.W. Ogden, Bifurcation of Rotating Thick-walled Elastic Tubes, *Journal of the Mechanics and Physics of Solids*, Vol. 28, pp. 59–74 (1980c).

R. Hill, On Uniqueness and Stability in the Theory of Finite Elastic Strain, *Journal of the Mechanics and Physics of Solids*, Vol. 5, pp. 229–241 (1957).

R. Hill, Eigenmodal Deformations in Elastic/Plastic Continua, *Journal of the Mechanics and Physics of Solids*, Vol. 15, pp. 371–386 (1967).

R. Hill, Constitutive Inequalities for Isotropic Elastic Solids under Finite Strain, *Proceedings of the Royal Society of London, Series A*, Vol. 314, pp. 457–472 (1970).

R. Hill, Aspects of Invariance in Solid Mechanics, *Advances in Applied Mechanics*, Vol. 18, pp. 1–75 (1978).

R. Hill, On the Theory of Plane Strain in Finitely Deformed Compressible Materials, *Mathematical Proceedings of the Cambridge Philosophical Society*, Vol. 86, pp. 161–178 (1979).

R. Hill, Constitutive Branching in Elastic Materials, *Mathematical Proceedings of the Cambridge Philosophical Society*, Vol. 92, pp. 167–181 (1982).

R. Hill and J.W. Hutchinson, Bifurcation Phenomena in the Plane Tension Test, *Journal of the Mechanics and Physics of Solids*, Vol. 23, pp. 239–264 (1975).

R.J. Knops and E.W. Wilkes, Theory of Elastic Stability, in *Handbuch der Physik*, Vol. VIa/3 (Ed. C.A. Truesdell), pp. 125–302, Springer (1973).

J.K. Knowles, Localized Shear near the Tip of a Crack in Finite Elastostatics, in *Finite Elasticity* (Eds. D.E. Carlson and R.T. Shield), Proceedings of the IUTAM Symposium, Lehigh University 1980, pp. 257–268, Martinus Nijhoff (1982).

J.K. Knowles and E. Sternberg, On the Ellipticity of the Equations of Non-linear Elastostatics for a Special Material, *Journal of Elasticity*, Vol. 5, pp. 341–361 (1975).

J.K. Knowles and E. Sternberg, On the Failure of Ellipticity of the Equations for Finite Elastostatic Plane Strain, *Archive for Rational Mechanics and Analysis*, Vol. 63, pp. 321–336 (1977).

M.F. McCarthy, Singular Surfaces and Waves, in *Continuum Physics*, Vol. 2 (Ed. A.C. Eringen), pp. 450–521, Academic Press (1975).

A. Needleman, Inflation of Spherical Rubber Balloons, *International Journal of Solids and Structures*, Vol. 13, pp. 409–421 (1977).

R.W. Ogden, Compressible Isotropic Elastic Solids under Finite Strain: Constitutive Inequalities, *Quarterly Journal of Mechanics and Applied Mathematics*, Vol. 23, pp. 457–468 (1970a).

R.W. Ogden, *On Constitutive Relations for Elastic and Plastic Materials*, Ph.D. Dissertation, Cambridge University (1970b).

R.W. Ogden, Waves in Isotropic Elastic Materials of Hadamard, Green or Harmonic Type, *Journal of the Mechanics and Physics of Solids*, Vol. 18, pp. 149–164 (1970c).

R.W. Ogden, Large Deformation Isotropic Elasticity I: on the Correlation of Theory and Experiment for Incompressible Rubberlike Solids, *Proceedings of the Royal Society of London, Series A*, Vol. 326, pp. 565–584 (1972a).

R.W. Ogden, Large Deformation Isotropic Elasticity II: on the Correlation of Theory and Experiment for Compressible Rubberlike Solids, *Proceedings of the Royal Society of London, Series A*, Vol. 328, pp. 567–583 (1972b).

R.W. Ogden, On Isotropic Tensors and Elastic Moduli, *Proceedings of the Cambridge Philosophical Society*, Vol. 75, pp. 303–319 (1974a).

R.W. Ogden, Growth and Decay of Acceleration Waves in Incompressible Elastic Solids, *Quarterly Journal of Mechanics and Applied Mathematics*, Vol. 27, pp. 451–464 (1974b).

R.W. Ogden, *Elastic Deformations of Rubberlike Solids*, in Mechanics of Solids, The Rodney Hill 60th Anniversary Volume (Eds. H.G. Hopkins and M.J. Sewell), pp. 499–537, Pergamon Press (1982).

G.P. Parry, On the Relative Strengths of Intrinsic Stability Criteria, *Quarterly Journal of Mechanics and Applied Mathematics*, Vol. 31, pp. 1–7 (1978).

B.D. Reddy, The Occurrence of Surface Instabilities and Shear Bands in Plane Strain Deformation of an Elastic Half-space, *Quarterly Journal of Mechanics and Applied Mathematics*, Vol. 36, pp. 337–350 (1983).

R.S. Rivlin, Stability of Pure Homogeneous Deformations of an Elastic Cube under Dead Loading, *Quarterly of Applied Mathematics*, Vol. 32, pp. 265–271 (1974).

K.N. Sawyers, Stability of an Elastic Cube under Dead Load: Two Equal Forces, *International Journal of Non-linear Mechanics*, Vol. 11, pp. 11–23 (1976).

K.N. Sawyers, Material Stability and Bifurcation in Finite Elasticity, in *Finite Elasticity*, Applied Mechanics Symposia Series Vol. 27 (Ed. R.S. Rivlin), American Society of Mechanical Engineers, pp. 103–123 (1977).

K.N. Sawyers and R.S. Rivlin, Bifurcation Conditions for a Thick Elastic Plate under Thrust, *International Journal of Solids and Structures*, Vol. 10, pp. 483–501 (1974).

K.N. Sawyers and R.S. Rivlin, On the Speed of Propagation of Waves in a Deformed Elastic Material, *Journal of Applied Mathematics and Physics* (*ZAMP*), Vol. 28, pp. 1045–1057 (1977).

K.N. Sawyers and R.S. Rivlin, On the Speed of Propagation of Waves in Deformed Compressible Elastic Materials, *Journal of Applied Mathematics and Physics* (*ZAMP*), Vol. 29, pp. 245–251 (1978).

N.H. Scott, Acceleration Waves in Constrained Elastic Materials, *Archive for Rational Mechanics and Analysis*, Vol. 58, pp. 57–75 (1975).

N.H. Scott, Acceleration Waves in Incompressible Elastic Solids, *Quarterly Journal of Mechanics and Applied Mathematics*, Vol. 29, pp. 295–310 (1976).

R.T. Shield and R.L. Fosdick, Extremum Principles in the Theory of Small Elastic Deformations Superposed on Large Elastic Deformations, in *Modern Developments in the Mechanics of Continua* (Ed. S. Eskinazi), pp. 107–125 (1966).

C.A. Truesdell, Existence of Longitudinal Waves, *Journal of the Acoustical Society of America*, Vol. 40, pp. 729–730 (1966).

C.A. Truesdell and W. Noll, The Non-linear Field Theories of Mechanics, in *Handbuch der Physik*, Vol. III/3 (Ed. S. Flügge), Springer (1965).

C.-C. Wang and C.A. Truesdell, *Introduction to Rational Elasticity*, Noordhoff (1973).

N.J.B. Young, Bifurcation Phenomena in the Plane Compression Test, *Journal of the Mechanics and Physics of Solids*, Vol. 24, pp. 77–91 (1976).

L. Zee and E. Sternberg, Ordinary and Strong Ellipticity in the Equilibrium Theory of Incompressible Hyperelastic Solids, *Technical Report* No. 51, Division of Engineering and Applied Science, California Institute of Technology (1982).

CHAPTER 7

Elastic Properties of Solid Materials

7.1 PHENOMENOLOGICAL THEORY

The phenomenological theory of elasticity is concerned with the *description* of the observed elastic behaviour of real materials. More specifically, it deals with the mathematical representation of the macroscopic elastic properties found from experimental tests, and *not* with the explanation of how these properties arise from the underlying microscopic structure.[†]

The theory assumes that the material is perfectly elastic. This, of course, is an idealization but, in practice, it provides a very good approximation to actual material behaviour in many circumstances. The assumption is valid if time-dependent effects such as creep and relaxation are negligible over the time scale of the experimental procedures. Once the property of elasticity has been established in this sense, it is usual to assume that it can be described in terms of a strain-energy function. This requires of the material the absence of hysteresis effects[‡], again an idealization. Although such effects can be accommodated by Cauchy elasticity, the simplicity gained by working with a single scalar function (the strain-energy function) is an overriding consideration. It is therefore assumed (not always justifiably) that hysteresis is negligible.

[†]The macroscopic property of elasticity arises from two distinct mechanisms at the microscopic level. Firstly, in the case of polymeric materials such as rubber it is the 'uncoiling' of the long-chain molecules which is responsible for the elasticity as the bulk material is stretched. Such molecules are interconnected by means of chemical cross-links to form a three-dimensional network. For summaries of the molecular theory and its connection with the phenomenological theory see Treloar (1975) and Ogden (1982). The second mechanism is due to the interatomic forces present in materials with a regular atomic structure, i.e. crystals. The macroscopic description of crystal elasticity has been summarized in Section 4.2.5. For discussion of the relationship between interatomic forces and macroscopic elasticity we refer to Hill (1975), Ericksen (1977), Milstein and Hill (1977, 1978, 1979) and Parry (1980) and the references cited in these papers.

[‡]Some hysteresis effects, however, can be predicted by using a non-convex strain-energy function.

With the strain-energy function W, per unit reference volume, taken in its most general form (subject to the restriction imposed by objectivity) the objective is to compare the predictions of the resulting theoretical stress-deformation relations with experimental data over as wide a range of different deformations as possible. We therefore regard W as a function of the right stretch tensor $\mathbf{U}$, this having six independent components. To simplify the problem, some general properties are assessed first, with a view to reducing the range of experimental tests needed.

(a) Material homogeneity Since elasticity is a local theory (in the sense described in Section 4.1.2), the elastic properties of any material may be determined from experiments carried out on homogeneous specimens (see Section 4.1.3 for a definition of material homogeneity). In order to do this it is necessary to prepare test specimens which are as close as possible to being homogeneously constituted. This can be achieved reasonably well in practice, and preliminary tests on samples cut from a representative specimen can be used to check the accuracy with which the requirement is met. *For a homogeneous material, W can be determined from experimental tests involving homogeneous deformations.* The properties of an inhomogeneous or composite elastic material can then, in principle, be described in terms of the properties of W, known locally for each constituent material.

(b) Material symmetry Simple experimental tests serve to detect any anisotropy in the material properties. The degree of anisotropy can be assessed by further tests designed to highlight material symmetries such as those appropriate to the crystal classes (Sections 4.2.5 and 4.3.2). If the material is isotropic then the number of experimental tests required to determine W is relatively small and the accompanying mathematics is correspondingly simplified.

(c) Internal constraints Preliminary tests may also be used to expose evidence on which to base the imposition of certain internal constraints. For example, the compressibility of the material can be assessed by measuring volume changes associated with a number of simple mechanical tests. If the dilatation is very small then the incompressibility constraint is a good approximation. Such constraints, like material symmetries, reduce the numbers of experiments needed and simplify the mathematical structure.

As far as anisotropic materials are concerned both the experimental procedures and the mathematics involved are more complicated than for isotropic materials, as we have indicated above. We refer to Green and Adkins (1970) for a discussion of anisotropy, with the remark that although there is a considerable body of theory concerned with anisotropic materials there is very little in the literature which relates this to the properties of real materials, at least for large deformations. Indeed, there is a lack of relevant experimental data for this purpose. In this chapter, therefore, we illustrate the general principles and methods of the phenomenological theory by describing in detail its application to isotropic materials.

For an isotropic material, W depends on $\mathbf{U}$ through the principal stretches $\lambda_1, \lambda_2, \lambda_3$. Thus, an experimental set-up in which *pure strains* are imposed on a rectangular block of material is sufficiently general for the determination of W (although, in practice, such deformations are difficult to achieve in general). We refer to Section 4.3.5 for a limited account of isotropic forms of W for both unconstrained and incompressible materials, and to Treloar (1975) and Ogden (1982) for detailed discussions aimed specifically at rubberlike materials.

In Section 7.2 we discuss some deformations associated with simple experimental tests, attention being concentrated on homogeneous deformations. We deal first with the theory for incompressible isotropic materials and then relate this to the properties of rubberlike materials. Rubber is the prime example of a material displaying non-linear elastic response. The corresponding basis for the discussion of compressible materials is also examined. In passing, we remark that, in general, non-homogeneous deformations result from the application of boundary tractions. Thus, once W has been found from tests involving homogeneous deformations it can be used to solve problems (such as those discussed in Sections 5.2 and 5.3) in which the deformations are not homogeneous, and so predict results for particular materials. Such results can be assessed against subsequent experimental findings and be held as a further check on the validity of the considered form of W for the materials in question.

In Section 7.3 the effect of small changes in material properties is investigated. For a given set of material constants appearing in W, it is assumed that the solution of a certain boundary-value problem is known. The perturbation of this solution due to small changes in the material constants is then governed by an incremental boundary-value problem similar to those discussed in Chapter 6. In Section 7.4 this theory is applied to a 'slightly compressible' material, and, in particular, it enables volume changes to be calculated in terms of known isochoric deformations for an arbitrary form of strain-energy function. Finally, the results are discussed in relation to rubberlike materials.

7.2 ISOTROPIC MATERIALS

7.2.1 Homogeneous pure strain of an incompressible material

We consider the homogeneous pure strain of a rectangular block with dimensions L_1, L_2, L_3 into a rectangular block with dimensions l_1, l_2, l_3. The deformation is described by the principal stretches $\lambda_1, \lambda_2, \lambda_3$, where $l_i = \lambda_i L_i$

$(i=1,2,3)$. For an incompressible material, to which attention is confined here, we have

$$\lambda_1\lambda_2\lambda_3=1, \tag{7.2.1}$$

leaving just two independent stretches.

Since we are dealing with isotropic materials, we recall from (4.3.39) that the strain-energy function is subject to the symmetries

$$W(\lambda_1,\lambda_2,\lambda_3)=W(\lambda_1,\lambda_3,\lambda_2)=W(\lambda_3,\lambda_1,\lambda_2), \tag{7.2.2}$$

coupled with (7.2.1).

From (4.3.49) the principal Cauchy stresses are given by

$$t_i=\lambda_i\frac{\partial W}{\partial\lambda_i}-p \qquad i=1,2,3. \tag{7.2.3}$$

Because of (7.2.1) we are able to eliminate λ_3 from explicit consideration and introduce the notation

$$\tilde{W}(\lambda_1,\lambda_2)\equiv\tilde{W}(\lambda_2,\lambda_1)=W(\lambda_1,\lambda_2,\lambda_1^{-1}\lambda_2^{-1}). \tag{7.2.4}$$

It then follows from (7.2.3) that

$$t_1-t_3=\lambda_1\frac{\partial\tilde{W}}{\partial\lambda_1}, \qquad t_2-t_3=\lambda_2\frac{\partial\tilde{W}}{\partial\lambda_2}. \tag{7.2.5}$$

For consistency with the classical theory W must satisfy

$$\left.\begin{aligned}&\tilde{W}(1,1)=0, \qquad \frac{\partial\tilde{W}}{\partial\lambda_\alpha}(1,1)=0 \quad \alpha=1,2,\\ &\frac{\partial^2\tilde{W}}{\partial\lambda_1\partial\lambda_2}(1,1)=2\mu, \qquad \frac{\partial^2\tilde{W}}{\partial\lambda_\alpha^2}(1,1)=4\mu \quad \alpha=1,2,\end{aligned}\right\} \tag{7.2.6}$$

to which (6.1.88) reduce when the substitution (7.2.4) is made, μ being the ground state shear modulus.

The two stress-deformation relations (7.2.5) provide the theoretical basis for relating experimentally determined values of the stress components and the two independently assignable principal stretches. In practice, $\tilde{W}$ can be found from experiments on plane rectangular sheets of material in plane stress ($t_3 = 0$). Equations (7.2.5) then reduce to

$$t_1 = \lambda_1 \frac{\partial \tilde{W}}{\partial \lambda_1}, \qquad t_2 = \lambda_2 \frac{\partial \tilde{W}}{\partial \lambda_2}. \tag{7.2.7}$$

We note in passing that equations (7.2.5) are unaffected by the superimposition of an arbitrary hydrostatic stress and that p has been eliminated.

Biaxial deformations of the type governed by (7.2.7) can provide sufficient data from which to determine $\tilde{W}$ for a wide range of values of λ_1 and λ_2. An alternative experimental arrangement which provides equivalent information involves the extension and inflation of a thin-walled circular cylindrical tube. The ends of the tube are closed by fixing the tube over circular disks, and the extension is achieved by hanging weights from one end when the tube is suspended vertically. Pressure is then applied through a hole in one of the disks until the circular cylindrical shape is regained. This is done for a series of disks of different radii. The theoretical basis for this experiment is now described.

Let λ_1 and λ_2 be the axial and circumferential stretches. Then the reference and current lengths, L and l, of the cylinder are related by $l = \lambda_1 L$. Since the tube has thin walls we have the approximations $a = \lambda_2 A$, $h = \lambda_3 H$, where A and a are the reference and current radii of the cylinder, and H and h are the corresponding wall thicknesses. The quantity ε, defined by $\varepsilon = H/A$, is such that $\varepsilon \ll 1$.

On conversion to the notation of the present section the inflating pressure P, given in Section 5.3.3, becomes

$$P = \varepsilon \lambda_1^{-1} \lambda_2^{-1} \frac{\partial \tilde{W}}{\partial \lambda_2}. \tag{7.2.8}$$

When the pressure on the end faces of the cylinder is taken into account, it is easy to see that the axial load N applied to the ends is given by $N = 2\pi A^2 F$, where

$$F = \varepsilon \frac{\partial \tilde{W}}{\partial \lambda_1} - \tfrac{1}{2} \varepsilon \lambda_2 \lambda_1^{-1} \frac{\partial \tilde{W}}{\partial \lambda_2}, \tag{7.2.9}$$

thus modifying the formula of Section 5.3.3 which applied to an open-ended tube. Measured values of λ_1, λ_2, F (or, equivalently, N) and P enable $\tilde{W}$ to be determined from (7.2.8) and (7.2.9), ε being a known geometrical quantity.

Some important special cases of (7.2.5) arise when λ_1 and λ_2 are interdependent, and we now examine these separately. In connection with the first two we refer to the deformations discussed at the beginning of Section 2.2.6.

(a) Simple tension In simple tension $t_2 = t_3 = 0$ and we write $t_1 = t$, $\lambda_1 = \lambda$. By symmetry $\lambda_2 = \lambda_3 = \lambda^{-1/2}$ and the stress-deformation relations (7.2.5) combine to give

$$t = \lambda \frac{\mathrm{d}\hat{W}}{\mathrm{d}\lambda}, \tag{7.2.10}$$

where

$$\hat{W}(\lambda) = \tilde{W}(\lambda, \lambda^{-1/2}). \tag{7.2.11}$$

(b) Pure shear For this deformation $\lambda_3 = 1$ and we set $\lambda_1 = \lambda$, $\lambda_2 = \lambda^{-1}$ with $t_1 = t$, $t_2 = 0$. A non-vanishing stress t_3 is required to maintain $\lambda_3 = 1$. On elimination of t_3 from (7.2.5) we obtain

$$t = \lambda \frac{\mathrm{d}\hat{W}}{\mathrm{d}\lambda}, \tag{7.2.12}$$

where

$$\hat{W}(\lambda) = \tilde{W}(\lambda, \lambda^{-1}). \tag{7.2.13}$$

(c) Equibiaxial tension In equibiaxial tension we have $t_1 = t_2 = t$ and $t_3 = 0$ coupled with $\lambda_1 = \lambda_2 = \lambda$, and hence (7.2.5) gives

$$t = \tfrac{1}{2}\lambda \frac{\mathrm{d}\hat{W}}{\mathrm{d}\lambda}, \tag{7.2.14}$$

where

$$\hat{W}(\lambda) = \tilde{W}(\lambda, \lambda). \tag{7.2.15}$$

Each of (a)–(c) plays a role in the determination of material properties, but, since only a single disposable parameter, λ, is involved, neither is sufficient on its own to characterize the strain-energy function. This latter task falls to (7.2.5). However, in practice, larger values of λ_1 and λ_2 can be achieved by means of (a)–(c) than by the biaxial deformation tests governed by (7.2.5). So, once $\tilde{W}(\lambda_1, \lambda_2)$ has been found for a certain range of values of λ_1 and λ_2 information from tests governed by (a)–(c) provides an additional check on $\tilde{W}(\lambda_1, \lambda_2)$ for that range and also enables the coverage of $\tilde{W}(\lambda_1, \lambda_2)$ to be extended beyond that range. These points will be brought out in Section 7.2.2 where the theory is compared with experimental results for rubberlike materials.

Since, as we have already noted, the addition of an arbitrary hydrostatic pressure to the stress components does not affect equations (7.2.5) and their specializations, it follows that the equation governing equibiaxial tension is equivalent to that governing uniaxial compression. Similarly, uniaxial tension is equivalent to equibiaxial compression. This means that the properties of $\tilde{W}$ can be assessed from *tension* tests alone ($t_1 > 0$, $t_2 \geq 0$, $t_3 \geq 0$, say), and avoids the need to apply compressive stresses (which raise difficulties associated with instability).

Finally in this section, we note that equibiaxial tension results may be obtained from experiments which involve the inflation of a thin-walled spherical shell. When the wall thickness is very much less than the radius of the shell the deformation in the shell surface is equibiaxial and approximately uniform. We recall from equation (5.3.23) that if $\lambda_1 = \lambda_2 = \lambda$ are the in-surface stretches and P is the inflating pressure then

$$P = \tfrac{1}{3}\varepsilon\lambda^{-2}\frac{\mathrm{d}\hat{W}}{\mathrm{d}\lambda}, \tag{7.2.16}$$

where $\hat{W}$ is defined by (7.2.15). A full description of the problem is given in Section 5.3.2, but the results of Section 6.3.4 must also be accounted for since (7.2.16) assumes that spherical symmetry is maintained.

7.2.2 Application to rubberlike materials

In this section we relate the theory discussed in Section 7.2.1 to experimental data for rubberlike materials. Firstly, we note that rubber is the most common example of a material whose mechanical response can be regarded as non-linearly elastic. Sheets or tubes of rubberlike material with substantially homogeneous properties can be prepared. The elastic response is essentially isotropic, although a degree of anisotropy can be introduced in certain preparation processes. In the case of a plane sheet, for example, the isotropy can be checked by reversing the roles of t_1 and t_2 in any series of tests. Rubberlike materials can also be regarded as incompressible since volume changes are very small (the dilatation $J - 1$ being of order 10^{-4}) in standard uniaxial or biaxial tension tests. Of course, quite large volume changes can be achieved in hydrostatic pressure tests, but this is a reflection of the fact that the ratio μ/κ of shear to bulk modulus is very small. We shall consider this point in detail in Section 7.4.2 in connection with the theory of slightly compressible materials.

The assumptions underlying the theory of Section 7.2.1 are therefore applicable to rubberlike materials, at least as a first approximation. In Sections 7.4.2 and 7.4.3 we shall discuss the effect of relaxing the incompressibility constraint. In the present section we confine attention to the theory of Section 7.2.1, specialized as the need arises, and take representative experimental data from biaxial tests. A detailed review of the elastic properties of rubberlike solids has recently been provided by Ogden (1982) and we refer to this paper for background information and references. It includes a historical sketch of experimental work and a discussion of different formulations of the underlying theory.

Essentially, the objective is to determine $\tilde{W}(\lambda_1, \lambda_2)$ for as large as possible a domain in the positive quadrant of (λ_1, λ_2)-space. One way of doing this is to fix one of the stretches, λ_2 say, and vary λ_1 while recording the accompanying values of t_1 and t_2. The data can then be represented by plotting $t_1 - t_2$ against λ_1 for each fixed λ_2 and setting the results against the theoretical relation

$$t_1 - t_2 = \lambda_1 \frac{\partial \tilde{W}}{\partial \lambda_1} - \lambda_2 \frac{\partial \tilde{W}}{\partial \lambda_2}, \tag{7.2.17}$$

obtained from (7.2.7).

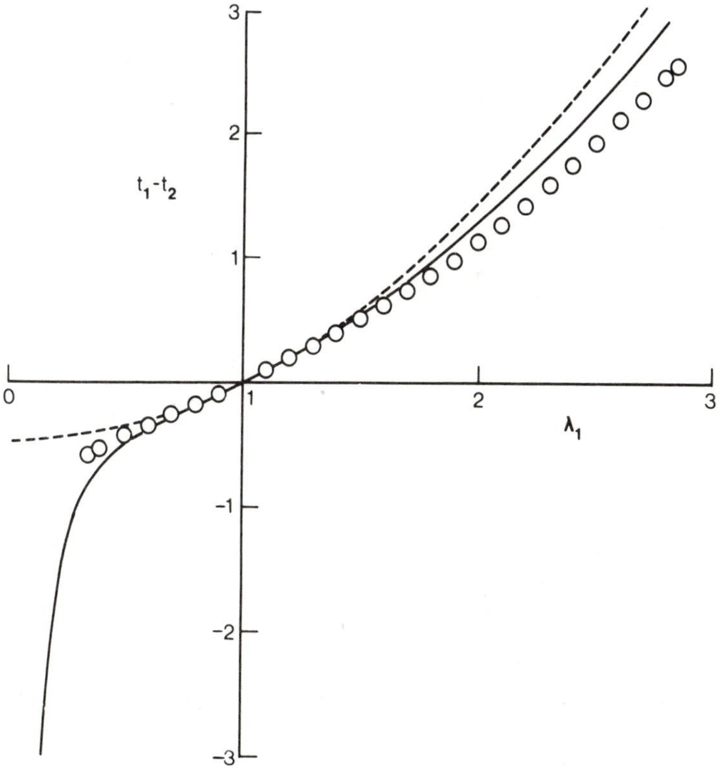

(a)

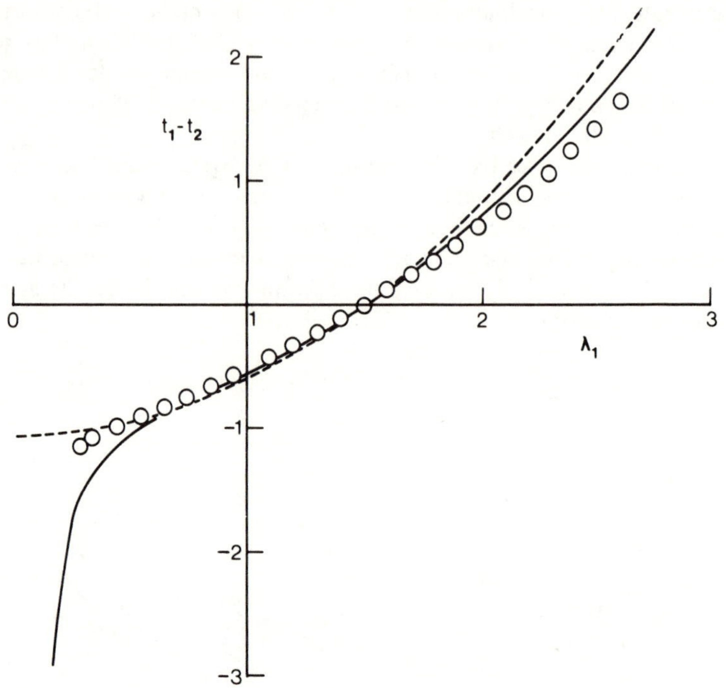

(b)

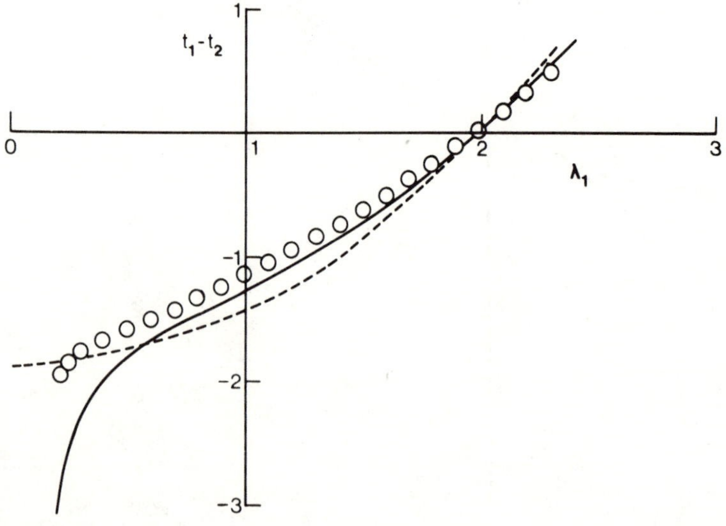

(c)

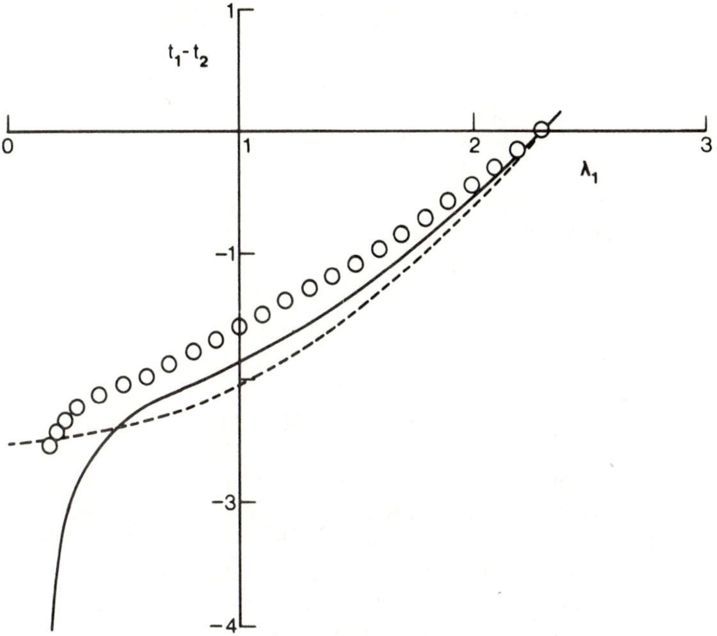

(d)

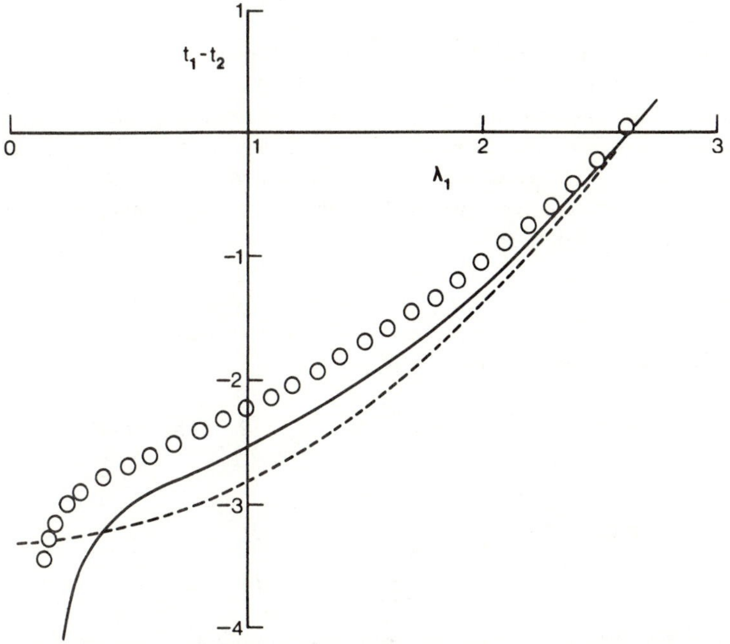

(e)

Fig. 7.1 Biaxial deformation of a rectangular sheet. Plot of $t_1 - t_2$ against λ_1: data of Jones and Treloar (○) compared with neo-Hookean (broken curves) and Mooney (continuous curves) theories; (a) $\lambda_2 = 1.0$, (b) $\lambda_2 = 1.502$, (c) $\lambda_2 = 1.984$, (d) $\lambda_2 = 2.295$, (e) $\lambda_2 = 2.623$. The units of $t_1 - t_2$ are $\mathrm{N\,mm}^{-2}$.

We intimated in Section 7.2.1 that all the required data can be obtained if attention is restricted to tension tests. Clearly, (7.2.17) will cover the domain defined by $\lambda_1 \geq 1$, $\lambda_2 \geq 1$. It will also provide data for $\lambda_1 > 1$, $\lambda_2 < 1$ (or, equivalently, $\lambda_1 < 1$, $\lambda_2 > 1$, in view of the symmetry (7.2.4)) since pre-extension to $\lambda_1 > 1$ with $t_2 = 0$ will set $\lambda_2 < 1$; further increases in λ_1 can then be imposed after fixing λ_2. In the latter case, $\lambda_1\lambda_2 > 1$ and hence $\lambda_3 = \lambda_1^{-1}\lambda_2^{-1} < 1$; from measured values of $\lambda_2 < 1$ and calculated values of $\lambda_3 < 1$ we are able to deduce data for the remaining part of the positive (λ_1, λ_2)-quadrant—that defined by $0 < \lambda_1 < 1$, $0 < \lambda_2 < 1$—by feeding the information about λ_3, with $t_3 = 0$, into (7.2.17), i.e. setting $t_1 = 0$ and letting λ_1 take on the numerical values of λ_3.

To properly assess the properties of a given material, all the experimental tests should, ideally, be performed on a single material specimen or, as near as possible, on identical specimens cut from a single sample (a sheet of rubber for example). This was not always done in some of the early experimental work, but has been the practice more recently. With this point in mind we take our experimental data from the comprehensive biaxial tests carried out by Jones and Treloar (1975) on small rectangular rubber sheets. Their results are representative of those available from biaxial experiments (for references to other work see Ogden, 1982) and are provided in a form suitable for representation by (7.2.17).

Jones and Treloar (1975) carried out experiments with λ_2 fixed at each of the values 1, 1.502, 1.984, 2.295, 2.623 and their data, with $t_1 - t_2$ plotted against λ_1, are shown in Fig. 7.1 ((a)–(e) respectively). They did not perform experiments with λ_2 fixed at values less than unity, but they obtained results for $\lambda_1 < 1$, $\lambda_2 < 1$ by reversing the roles of λ_1 and λ_3, as described above. Thus, their data are restricted in that they leave the domain $0 < \lambda_1 < 1$, $0 < \lambda_2 < 1$ in question. Nevertheless, a great deal of useful information is provided.

Along with the data in Fig. 7.1 are shown the theoretical curves calculated for the neo-Hookean and Mooney strain-energy functions, (4.3.63) and (4.3.62), which we now write

$$\tilde{W}(\lambda_1, \lambda_2) = \tfrac{1}{2}\mu_1(\lambda_1^2 + \lambda_2^2 + \lambda_1^{-2}\lambda_2^{-2} - 3), \qquad \mu_1 = \mu, \tag{7.2.18}$$

and

$$\begin{aligned}\tilde{W}(\lambda_1, \lambda_2) = {} & \tfrac{1}{2}\mu_1(\lambda_1^2 + \lambda_2^2 + \lambda_1^{-2}\lambda_2^{-2} - 3) \\ & - \tfrac{1}{2}\mu_2(\lambda_1^{-2} + \lambda_2^{-2} + \lambda_1^2\lambda_2^2 - 3), \qquad \mu_1 - \mu_2 = \mu,\end{aligned} \tag{7.2.19}$$

respectively, where μ is the ground-state shear modulus.

These are prototype strain-energy functions for rubberlike materials. The neo-Hookean strain-energy function arises from the Gaussian molecular statistical theory of rubber elasticity, but, as Fig. 7.1 shows, its predictions agree with observation only for values of (λ_1, λ_2) near $(1, 1)$ and depart markedly from the data away from $(1, 1)$. The status of the neo-Hookean strain-energy function for large deformations is therefore somewhat limited, but it provides a good working model of the elastic response of rubberlike materials for small deformations. For a discussion of the relationship between the statistical theory and the phenomenological theory of rubber elasticity we refer to Treloar (1975) and Ogden (1982). The Mooney strain-energy function provides a (limited) measure of the departure of observed behaviour from the predictions of the Gaussian theory since, to a certain extent, it rectifies the discrepancy between the neo-Hookean theory and the data. However, its predictions also differ significantly from the data at values of (λ_1, λ_2) sufficiently distant from $(1, 1)$, although they are qualitatively correct.

The *shapes* of the neo-Hookean and Mooney curves shown in Fig. 7.1 are not affected by a change in λ_2. The curves themselves are merely subjected to a vertical translation by such a change. This important shape invariant property is also possessed by the strain-energy functions

$$W(\lambda_1, \lambda_2, \lambda_3) = \sum_{p=1}^{N} \mu_p(\lambda_1^{\alpha_p} + \lambda_2^{\alpha_p} + \lambda_3^{\alpha_p} - 3)/\alpha_p \tag{7.2.20}$$

or, equivalently,

$$\tilde{W}(\lambda_1, \lambda_2) = \sum_{p=1}^{N} \mu_p(\lambda_1^{\alpha_p} + \lambda_2^{\alpha_p} + \lambda_1^{-\alpha_p}\lambda_2^{-\alpha_p} - 3)/\alpha_p, \tag{7.2.21}$$

with the ground-state shear modulus given by

$$\sum_{p=1}^{N} \mu_p \alpha_p = 2\mu, \tag{7.2.22}$$

and

$$W(\lambda_1, \lambda_2, \lambda_3) = w(\lambda_1) + w(\lambda_2) + w(\lambda_3), \tag{7.2.23}$$

where the function w is arbitrary, subject to

$$w(1) = 0, \qquad w'(1) + w''(1) = 2\mu. \tag{7.2.24}$$

The *separable* form of strain-energy function (7.2.23) incorporates (7.2.18)–(7.2.20) as special cases. The assumption that the strain-energy function is expressible in the form (7.2.23) for rubberlike materials is known as the *Valanis-Landel hypothesis*, following the paper by Valanis and Landel (1967). Some theoretical justification for use of (7.2.23) is provided by the

Taylor expansion of $W(\lambda_1, \lambda_2, \lambda_3)$ near $\lambda_1 = \lambda_2 = \lambda_3 = 1$, which, as shown by Ogden (1974), is of separable form up to the fifth order in $\lambda_i - 1$ $(i = 1, 2, 3)$. There are four independent constants associated with this expansion, one each for the second to fifth orders. We note that (7.2.19) involves only two constants.

On close inspection it can be seen that the experimental data shown in Fig. 7.1 possess the shape invariant property referred to above. This becomes clearer in Fig. 7.2 where the data for all five values of λ_2 are plotted on a single diagram. The curves through the data points for different values of λ_2 are indistinguishable in shape, as may be confirmed by superposition (Jones and Treloar, 1975). It follows that the biaxial data, as shown, for the rubber

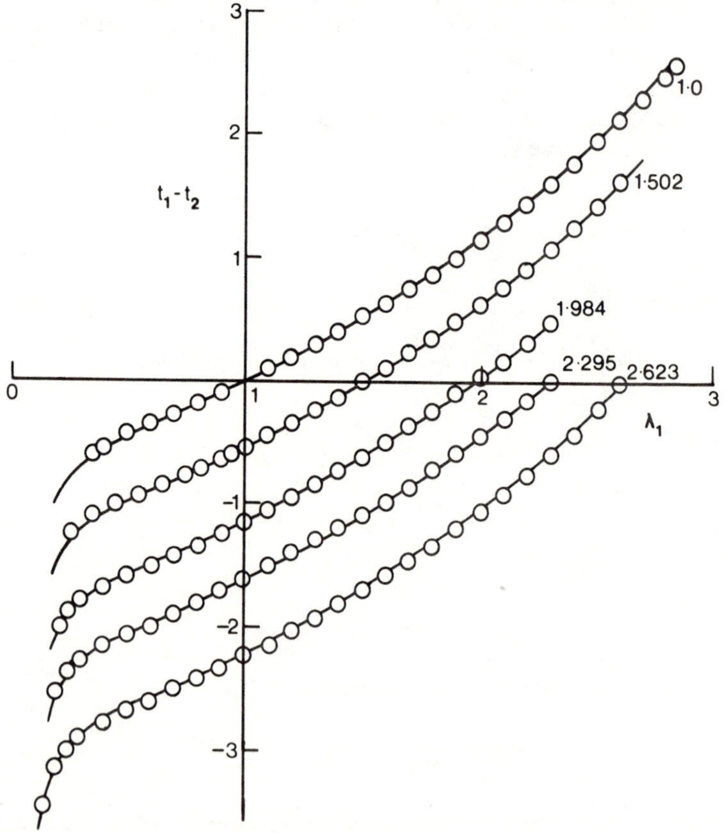

Fig. 7.2 Biaxial deformation of a rectangular sheet. Plot of $t_1 - t_2$ against λ_1: data of Jones and Treloar (○) for $\lambda_2 = 1.0, 1.502, 1.984, 2.295, 2.623$ compared with the predictions (———) of the strain-energy function (7.2.20) with $N = 3$ and material constants (7.2.28). The units of $t_1 - t_2$ are $\mathrm{N\,mm^{-2}}$.

in question tested by Jones and Treloar may be represented by a strain-energy function of the separable form (7.2.23), at least for the range of deformations they achieved. The most general form of $\tilde{W}(\lambda_1, \lambda_2)$ possessing the required shape invariant property is

$$\tilde{W}(\lambda_1, \lambda_2) = w(\lambda_1) + w(\lambda_2) + \bar{w}(\lambda_1 \lambda_2),$$

where $\bar{w}$ is a function independent of w. But, by reversing the roles of, say, λ_1 and λ_3 in the experiments one can deduce that $\bar{w}(\lambda_1\lambda_2) = w(\lambda_1^{-1}\lambda_2^{-1}) \equiv w(\lambda_3)$, so that (7.2.23) is recovered.

In view of the above results, further discussion of (7.2.23) is desirable. From (7.2.17) we obtain

$$t_1 - t_2 = \lambda_1 w'(\lambda_1) - \lambda_2 w'(\lambda_2). \tag{7.2.25}$$

Clearly, the shape of the curves may be determined by restricting attention to $\lambda_2 = 1$, so that (7.2.25) becomes

$$t_1 - t_2 = \lambda_1 w'(\lambda_1) - w'(1), \tag{7.2.26}$$

the equation corresponding to pure shear. The vertical shift due to a change from $\lambda_2 = 1$ to $\lambda_2 \neq 1$ is then given by the difference between the right-hand sides of (7.2.25) and (7.2.26), namely

$$w'(1) - \lambda_2 w'(\lambda_2). \tag{7.2.27}$$

As Figure 7.1 illustrates, the curves predicted by the Mooney strain-energy function have broadly the correct shape, but the predicted vertical shift departs from experimentally determined values, and increasingly so as λ_2 increases. Any strain-energy function of the form (7.2.23) must account for the shape of the data curve, through (7.2.26), and the vertical shift, through (7.2.27). The specific strain-energy function constructed by Jones and Treloar to fit their data is of the form (7.2.20) with $N = 3$, the values of the constants being given by

$$\begin{aligned} &\alpha_1 = 1.3, \qquad \alpha_2 = 4.0, \qquad \alpha_3 = -2.0, \\ &\mu_1 = 0.69, \qquad \mu_2 = 0.01, \qquad \mu_3 = -0.0122\,\mathrm{N\,mm^{-2}}. \end{aligned} \tag{7.2.28}$$

The theoretical curves calculated for (7.2.28) are shown in Fig. 7.2 along with the data, and it can be seen that the agreement between theory and experiment is excellent.

At this point some general remarks about the shape of the curves in Fig. 7.2 in relation to equation (7.2.26) are called for. Specifically, we use (7.2.20) for this purpose, in which case (7.2.26) becomes

$$t_1 - t_2 = \sum_{p=1}^{N} \mu_p(\lambda_1^{\alpha_p} - 1). \tag{7.2.29}$$

In view of the discussion following (7.2.17), this equation is applicable for $\lambda_1 < 1$ as well as $\lambda_1 \geq 1$. The vertical shift (7.2.27) now becomes

$$\sum_{p=1}^{N} \mu_p(1 - \lambda_2^{\alpha_p}). \tag{7.2.30}$$

For pure shear deformations with λ_1 in the range 0.8 to 1.2, say, the data can be represented by a strain-energy function of the form (7.2.20) with $N = 1$ and α_1 taking any value between about 1 and 2, $\mu_1 > 0$ being chosen so that $\mu_1\alpha_1 = 2\mu$, as pointed out by Ogden (1982). However, a single-term strain-energy function is not sufficient to cover also values of λ_1 up to about 3.0, and a second term must be introduced for this purpose. To provide the correct trend for large λ_1 this term must have α_2 and μ_2 positive. The marked downturn of the data for λ_1 less than about 0.8, on the other hand, cannot be accounted for by such a two-term strain-energy function. Its description requires a negative exponent, and we therefore introduce a third term with $\alpha_3 < 0$ and $\mu_3 < 0$. Thus, in general, at least three terms are needed in (7.2.20) to describe the stress-deformation characteristics embodied in the experimental data. The constants (7.2.28) satisfy the above criteria. The Mooney strain-energy function is also consistent with these criteria, except that it contains only two pairs of constants.

Amongst other results which provide confirmation of the separability property are those obtained by James, Green and Simpson (1975) and Kawabata and Kawai (1977) from experiments with plane sheets, and by Vangerko and Treloar (1978) from experiments on thin rubber tubes, the latter being based on the theory described by (7.2.8) and (7.2.9). Vangerko and Treloar achieved larger values of λ_1 and λ_2 in their experiments than did Jones and Treloar (1975), and an interesting feature of their results is that some departure from separability was found near the largest values of λ_1 and λ_2 reached (where $\lambda_1\lambda_2$ exceeded about 12).

In connection with the tube experiments it is worth noting here that there is a certain non-uniqueness in the relation between the axial loading F and the axial extension λ_1 at fixed internal pressure P, as demonstrated by Haughton and Ogden (1979). More particularly, a given F may be associated with two different values of λ_1 which are quite close in numerical value, and conversely. This is illustrated in Fig. 7.3 where $F^* \equiv \varepsilon^{-1}F$ is plotted against

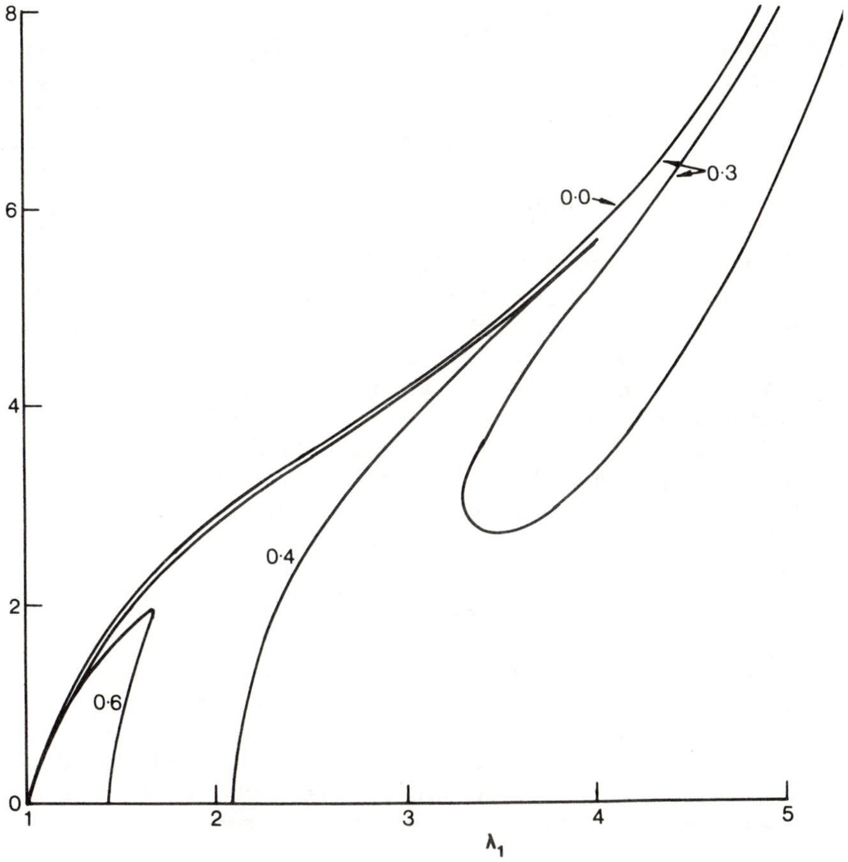

Fig. 7.3 Plot of the axial stress $F^* \equiv \varepsilon^{-1}F$ against λ_1 for an axially-loaded, internally-pressurized circular cylindrical membrane at fixed values 0.0, 0.3, 0.4, 0.6 of $P^* \equiv \varepsilon^{-1}P$. The units of F^* and P^* are $\mathrm{N\,mm^{-2}}$.

λ_1 for certain fixed values of $P^* \equiv \varepsilon^{-1}P$. Clearly, care must be exercised in interpreting the experimental results in this situation. Note that non-uniqueness is not found for $P = 0$ (simple tension).

Once it has been established on the basis of biaxial tests that the elastic properties of certain rubberlike materials can be characterized by means of a separable form of strain-energy function, it remains to determine the detailed form of the scalar function w occurring in (7.2.23). This has been done by Jones and Treloar (1975) in relation to their own biaxial data, resulting in the three-term strain-energy function with constants (7.2.28). Jones and Treloar also obtained simple tension data (for λ up to nearly 3.0) and these were found to be consistent with the predictions of this strain-energy function. More generally, we remark here that data from separate simple tension, pure shear and equibiaxial tension tests can provide valuable additional information and also confirm the form of w constructed from general biaxial data.

Earlier experiments by Treloar (1944), with a different rubber from that used by Jones and Treloar, produced data for simple tension, pure shear and equibiaxial tension up to quite large values of the stretch λ (about 7.0 in simple tension). These data, though not coupled with general biaxial data, have formed the basis for much subsequent work. In particular, using (7.2.20), but in the absence of prior justification from biaxial tests, Ogden (1972a) constructed a strain-energy function to fit Treloar's data. As in the case of (7.2.28) this has three terms, the constants being given by

$$\begin{aligned} &\alpha_1 = 1.3, \qquad \alpha_2 = 5.0, \qquad \alpha_3 = -2.0, \\ &\mu_1^* = 1.491, \qquad \mu_2^* = 0.003, \qquad \mu_3^* = -0.0237 \end{aligned} \tag{7.2.31}$$

in dimensionless form, where $\mu_p^* = \mu_p/\mu$ $(p = 1, 2, 3)$ and $\mu = 4.225\,\mathrm{kg\,cm^{-2}}$.

The closeness of fit to Treloar's data is illustrated in Fig. 7.4, separately for (a) simple tension, (b) pure shear, and (c) equibiaxial tension, with the nominal stress $t^{(1)} = t/\lambda$ plotted against λ. The accompanying theoretical stress-deformation relations, based on (7.2.20), are respectively

$$t^{(1)} = \sum_{p=1}^{N} \mu_p(\lambda^{\alpha_p - 1} - \lambda^{-(1/2)\alpha_p - 1}), \tag{7.2.32}$$

$$t^{(1)} = \sum_{p=1}^{N} \mu_p(\lambda^{\alpha_p - 1} - \lambda^{-\alpha_p - 1}), \tag{7.2.33}$$

$$t^{(1)} = \sum_{p=1}^{N} \mu_p(\lambda^{\alpha_p - 1} - \lambda^{-2\alpha_p - 1}) \tag{7.2.34}$$

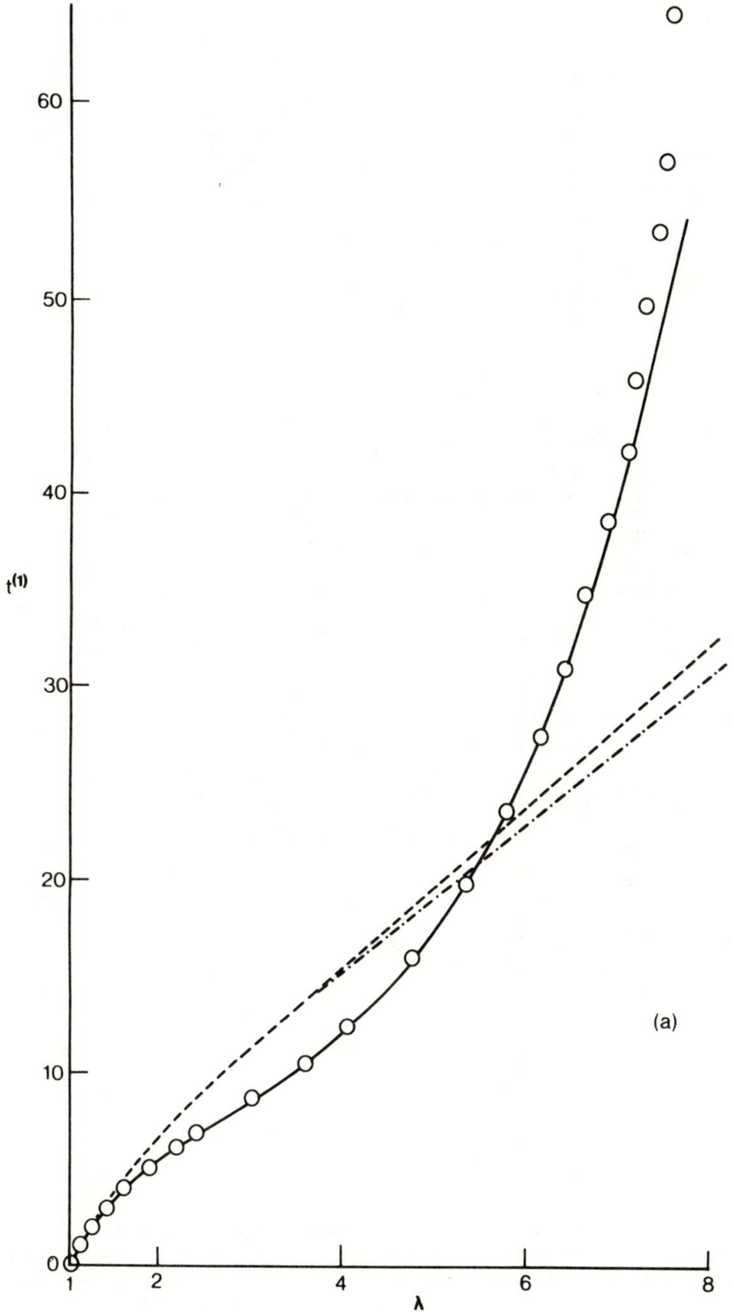
60
50
40
$t^{(1)}$
30
20
10
0
1
2
4
6
8
λ
(a)

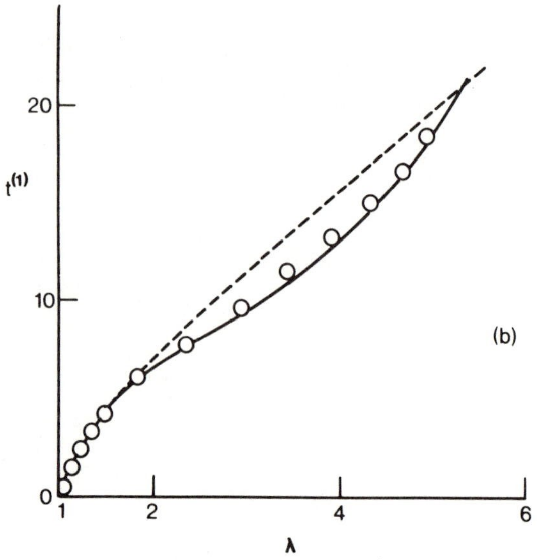

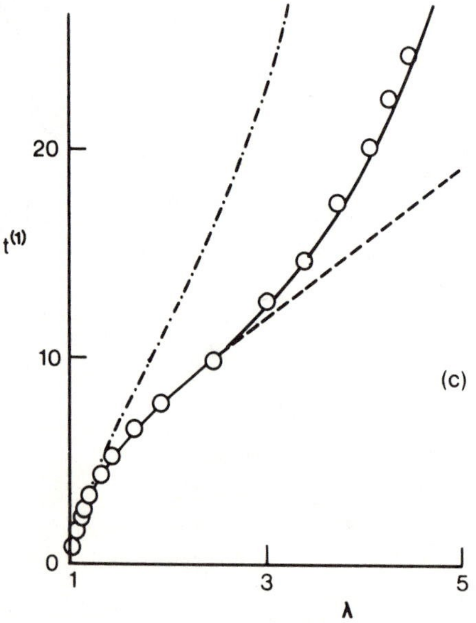

Fig. 7.4 Plot of $t^{(1)}$ against λ. Treloar's data (○) compared with predictions of neo-Hookean (-----), Mooney (-.-.-.) and three-term (———) strain-energy functions: (a) simple tension, (b) pure shear, (c) equibiaxial tension. In (b) the neo-Hookean and Mooney curves coincide. The units of $t^{(1)}$ are $\mathrm{kg\,cm^{-2}}$.

from (7.2.10), (7.2.12) and (7.2.14). In retrospect, it is the separable form of the strain-energy function which underlies its success. For further discussion we refer to Ogden (1972a, 1982). More recently, Treloar's data have been used together with (7.2.20) for $N = 3$ and $N = 4$ by Twizell and Ogden (1983) in a systematic curve-fitting exercise based on a non-linear optimization procedure.

Also plotted in Fig. 7.4 are the theoretical curves calculated for the neo-Hookean and Mooney strain-energy functions. These emphasize the discrepancy between theory and experiment already discussed in relation to the biaxial data. Note, in particular, that the exponent $\alpha_1 = 2$ is not large enough to account for the upward trend of the data at large values of λ in simple tension.

Clearly, separability is an important feature of rubber elasticity. However, as Vangerko and Treloar (1978) have shown, departure from separability is evident at sufficiently large values of λ_1 and λ_2. Since biaxial data are only available for a limited number of different rubber compounds it is therefore premature to conclude that the extent of separability is so great for other rubberlike materials. A systematic series of experiments on a wide range of different rubberlike materials is therefore called for in order to ascertain how separability depends on, for example, the chemical constitution of the material.

Throughout this section the deformation has been described in terms of the fundamental variables λ_1 and λ_2. However, in most of the early theoretical work and in the theoretical interpretation of experimental results the symmetrical invariants

$$I_1 \equiv \lambda_1^2 + \lambda_2^2 + \lambda_3^2, \qquad I_2 \equiv \lambda_2^2\lambda_3^2 + \lambda_3^2\lambda_1^2 + \lambda_1^2\lambda_2^2$$

were preferred. Since stress-deformation equations based on I_1 and I_2 are not suitable for discussing separability (as explained by Ogden, 1982) we do not examine them here. For a full account of this approach and its disadvantages we refer to the book by Treloar (1975) and the review by Ogden (1982). The latter paper also contains a summary of the underlying molecular theories of rubber elasticity and their relationship to the macroscopic property of separability.

Finally in this section we mention some experimental work which involves non-homogeneous deformations. Rivlin and Saunders (1951) carried out pure torsion and combined extension/torsion tests on solid circular cylinders of rubber. On the basis of the theory, described in Section 5.3.5, the data from these tests were found by Ogden and Chadwick (1972) to meet the predictions of the strain-energy function (7.2.20) for $N = 3$, with constants (7.2.31). Gent and Rivlin (1952) obtained corresponding data for the torsion of a circular cylindrical tube under internal pressure (the results of Problem 5.3.10 provide the required theory). For a discussion of other experimental results for circular cylindrical tubes we refer to Varga (1966). A summary of some experimental arrangements of engineering relevance is given by Lindley (1974).

7.2.3 Homogeneous pure strain of a compressible material

For a compressible isotropic elastic material we have (7.2.2) but, in general, the principal stretches do not satisfy (7.2.1). We write

$$J = \lambda_1 \lambda_2 \lambda_3 \tag{7.2.35}$$

and the required stress-deformation relations are

$$Jt_i = \lambda_i \frac{\partial W}{\partial \lambda_i} \qquad i = 1, 2, 3. \tag{7.2.36}$$

As in Section 7.2.2 we restrict attention here to the homogeneous pure strain of a rectangular block.

In order to develop the theory in close parallel with that described in Section 7.2.2 for incompressible materials we decompose the deformation into its isochoric and dilatational parts. Thus, as in (2.2.69) we write

$$\lambda_i^* = J^{-1/3} \lambda_i \qquad i = 1, 2, 3, \tag{7.2.37}$$

so that

$$\lambda_1^* \lambda_2^* \lambda_3^* = 1, \tag{7.2.38}$$

and we refer to $\lambda_1^*, \lambda_2^*, \lambda_3^*$, as the *modified stretches.* We can now take J and any pair of the modified stretches as independent variables. For an isochoric deformation $J = 1$ while for a pure dilatation $\lambda_1^* = \lambda_2^* = \lambda_3^* = 1$.

We now introduce the notation

$$W^*(\lambda_1^*, \lambda_2^*, \lambda_3^*, J) \equiv W(\lambda_1, \lambda_2, \lambda_3), \tag{7.2.39}$$

where the symmetry (7.2.2) of W is carried over to W^*. With the help of (7.2.37) equation (7.2.36) becomes

$$Jt_i = \lambda_i^* \frac{\partial W^*}{\partial \lambda_i^*} - \left(\tfrac{1}{3} \sum_{j=1}^{3} \lambda_j^* \frac{\partial W^*}{\partial \lambda_j^*} - J \frac{\partial W^*}{\partial J} \right) \quad i = 1, 2, 3. \tag{7.2.40}$$

Note that this is analogous to (7.2.3) and reduces to it in the incompressible limit. It follows that

$$J(t_i - t_j) = \lambda_i^* \frac{\partial W^*}{\partial \lambda_i^*} - \lambda_j^* \frac{\partial W^*}{\partial \lambda_j^*} \tag{7.2.41}$$

for each pair $i, j \in \{1, 2, 3\}$ together with

$$\tfrac{1}{3}(t_1 + t_2 + t_3) = \frac{\partial W^*}{\partial J}. \tag{7.2.42}$$

As in the incompressible theory, we retain just two of the modified stretches as independent variables and, accordingly, we write

$$\tilde{W}^*(\lambda_1^*, \lambda_2^*, J) \equiv W^*(\lambda_1^*, \lambda_2^*, \lambda_1^{*-1}\lambda_2^{*-1}, J). \tag{7.2.43}$$

Equations (7.2.41) and (7.2.42) are now replaced by

$$J(t_1 - t_3) = \lambda_1^* \frac{\partial \tilde{W}^*}{\partial \lambda_1^*}, \qquad J(t_2 - t_3) = \lambda_2^* \frac{\partial \tilde{W}^*}{\partial \lambda_2^*}, \tag{7.2.44}$$

$$\tfrac{1}{3}(t_1 + t_2 + t_3) = \frac{\partial \tilde{W}^*}{\partial J}. \tag{7.2.45}$$

Equations (7.2.44) should be compared with (7.2.5).

For consistency with the classical theory $\tilde{W}^*$ must satisfy

$$\left.\begin{aligned}
&\tilde{W}^*(1,1,1) = \frac{\partial \tilde{W}^*}{\partial \lambda_1^*}(1,1,1) = \frac{\partial \tilde{W}^*}{\partial \lambda_2^*}(1,1,1) = \frac{\partial \tilde{W}^*}{\partial J}(1,1,1) = 0,\\
&\frac{\partial^2 \tilde{W}^*}{\partial \lambda_1^{*2}}(1,1,1) = \frac{\partial^2 \tilde{W}^*}{\partial \lambda_2^{*2}}(1,1,1) = 2\frac{\partial^2 \tilde{W}^*}{\partial \lambda_1^* \partial \lambda_2^*}(1,1,1) = 4\mu,\\
&\frac{\partial^2 \tilde{W}^*}{\partial J^2}(1,1,1) = \kappa,
\end{aligned}\right\} \tag{7.2.46}$$

where μ and κ are the classical shear and bulk moduli. These are equivalent to (6.1.87).

In general, the dilatational and distortional parts of the deformation are coupled through (7.2.44) and (7.2.45) but, in essence, the distortional response of the material is governed by (7.2.44) and the dilatational response by (7.2.45). For a pure dilatation $\lambda_1^* = \lambda_2^* = 1$ and equations (7.2.44) reduce to

$$\frac{\partial \tilde{W}^*}{\partial \lambda_1^*}(1,1,J) = \frac{\partial \tilde{W}^*}{\partial \lambda_2^*}(1,1,J) = 0 \tag{7.2.47}$$

and (7.2.45) becomes

$$t = \frac{\partial \tilde{W}^*}{\partial J}(1,1,J), \tag{7.2.48}$$

where $t_1 = t_2 = t_3 = t$ is the accompanying hydrostatic stress. For a pure distortion $J = 1$ and (7.2.44) simplify to

$$\left.\begin{aligned} t_1 - t_3 &= \lambda_1^* \frac{\partial \tilde{W}^*}{\partial \lambda_1^*}(\lambda_1^*, \lambda_2^*, 1), \\ t_2 - t_3 &= \lambda_2^* \frac{\partial \tilde{W}^*}{\partial \lambda_2^*}(\lambda_1^*, \lambda_2^*, 1), \end{aligned}\right\} \tag{7.2.49}$$

while the hydrostatic part of the stress is given by

$$\tfrac{1}{3}(t_1 + t_2 + t_3) = \frac{\partial \tilde{W}^*}{\partial J}(\lambda_1^*, \lambda_2^*, 1). \tag{7.2.50}$$

We note that (7.2.44) is unaffected by the superposition of a hydrostatic pressure P, except through a change in the value of J, but (7.2.45) is modified to

$$\tfrac{1}{3}(t_1 + t_2 + t_3) - P = \frac{\partial \tilde{W}^*}{\partial J}. \tag{7.2.51}$$

The properties of $\tilde{W}^*$ may be determined from biaxial experiments in an environment of hydrostatic pressure in which λ_1^*, λ_2^* and J are controlled independently. The required governing equations are obtained by setting $t_3 = 0$ in (7.2.44) and (7.2.51) to give

$$Jt_1 = \lambda_1^* \frac{\partial \tilde{W}^*}{\partial \lambda_1^*}, \quad Jt_2 = \lambda_2^* \frac{\partial \tilde{W}^*}{\partial \lambda_2^*} \tag{7.2.52}$$

and

$$\tfrac{1}{3}(t_1 + t_2) - P = \frac{\partial \tilde{W}^*}{\partial J}. \tag{7.2.53}$$

the latter of which may be rearranged as

$$JP = \tfrac{1}{3}\left(\lambda_1^* \frac{\partial \tilde{W}^*}{\partial \lambda_1^*} + \lambda_2^* \frac{\partial \tilde{W}^*}{\partial \lambda_2^*}\right) - J\frac{\partial \tilde{W}^*}{\partial J}. \tag{7.2.54}$$

In simple tension $t_2 = 0$ and we write $\lambda_1^* = \lambda^*$, $\lambda_2^* = \lambda^{*-1/2}$, $t_1 = t$ and

$$\hat{W}^*(\lambda^*, J) = \tilde{W}^*(\lambda^*, \lambda^{*-1/2}, J). \tag{7.2.55}$$

Equations (7.2.52) and (7.2.54) then simplify to

$$\tfrac{1}{3}Jt = \tfrac{1}{3}\lambda^* \frac{\partial \hat{W}^*}{\partial \lambda^*} = J\frac{\partial \hat{W}^*}{\partial J} + JP, \tag{7.2.56}$$

the first of which is analogous to (7.2.10). It is left for the reader to obtain corresponding equations for pure shear and equibiaxial tension.

When $P = 0$, equation (7.2.54) serves to determine J in terms of λ_1^* and λ_2^* or, equivalently, in terms of λ_1 and λ_2, and the second equation in (7.2.56) has a similar interpretation. This information will be drawn on in Section 7.4.3 where the present theory is applied to slightly compressible materials. We refer to Section 4.3.5 for a discussion of strain-energy functions for compressible materials.

Problem 7.2.1 Show that for a thin-walled spherical shell of compressible isotropic elastic material the internal pressure P, given by (7.2.16) for an incompressible material, is

$$P = \tfrac{1}{3}\varepsilon J^{-1}\lambda^{*-2}\frac{\partial \hat{W}^*}{\partial \lambda^*},$$

where

$$\hat{W}^*(\lambda^*, J) = \tilde{W}^*(\lambda^*, \lambda^*, J),$$

and that this is coupled with the equation

$$\tfrac{1}{3}\lambda^* \frac{\partial \hat{W}^*}{\partial \lambda^*} = J\frac{\partial \hat{W}^*}{\partial J}.$$

Problem 7.2.2 A thin-walled circular cylindrical tube of compressible isotropic elastic material is subjected to axial extension and internal pressure, the ends of the tube being closed. If P is the pressure and $2\pi A^2 F$ the axial load, where A is the reference radius of the tube, show that

$$P = \varepsilon J^{-1}\lambda_1^{*-1}\lambda_2^{*-1}\frac{\partial \tilde{W}^*}{\partial \lambda_2^*},$$

$$F = \varepsilon J^{-1/3}\frac{\partial \tilde{W}^*}{\partial \lambda_1^*} - \tfrac{1}{2}\varepsilon J^{-1/3}\lambda_1^{*-1}\lambda_2^*\frac{\partial \tilde{W}^*}{\partial \lambda_2^*},$$

where λ_1^* and λ_2^* are the modified stretches corresponding to the axial and azimuthal directions respectively and ε is as defined in Section 5.3.3.

7.3 THE EFFECT OF SMALL CHANGES IN MATERIAL PROPERTIES

Suppose we identify a (finite) number of dimensionless material constants, $\eta_1, \eta_2, \ldots$, say, occurring in the elastic constitutive law. We represent these jointly by the vector $\boldsymbol{\eta}$ and incorporate them explicitly into the constitutive law by writing the strain-energy function as

$$W(\mathbf{A}, \boldsymbol{\eta}), \tag{7.3.1}$$

$\mathbf{A}$ being the deformation gradient. The nominal stress-deformation relation is then

$$\mathbf{S} = \frac{\partial W}{\partial \mathbf{A}}(\mathbf{A}, \boldsymbol{\eta}) \tag{7.3.2}$$

for an unconstrained material.

In (7.3.1) we regard $\mathbf{A}$ and $\boldsymbol{\eta}$ as independent variables, but for a specific boundary-value problem the deformation $\boldsymbol{\chi}$, with $\mathbf{A} = \operatorname{Grad} \boldsymbol{\chi}$, will in general depend on the values of $\boldsymbol{\eta}$. Any change in $\boldsymbol{\eta}$ for such a problem will be reflected in a change in $\mathbf{A}$.

When there are no body forces the basic boundary-value problem is described by

$$\operatorname{Div} \mathbf{S} = \mathbf{0} \quad \text{in} \quad \mathscr{B}_0, \tag{7.3.3}$$

$$\boldsymbol{\chi} = \boldsymbol{\xi} \quad \text{on} \quad \partial\mathscr{B}_0^x, \tag{7.3.4}$$

$$\mathbf{S}^{\mathrm{T}}\mathrm{N} = \boldsymbol{\sigma} \quad \text{on} \quad \partial\mathscr{B}_0^\sigma, \tag{7.3.5}$$

together with (7.3.2).

Suppose that the solution $\boldsymbol{\chi}$ to this problem is known. Now let the material properties be changed by means of an increment $\delta\boldsymbol{\eta}$ in the value of $\boldsymbol{\eta}$ and assume that for the same boundary-value problem the solution $\boldsymbol{\chi}$ is adjusted by a corresponding increment $\delta\boldsymbol{\chi}$. Let $\operatorname{Grad} \delta\boldsymbol{\chi}$ be denoted by $\delta\mathbf{A}$. Then, to the first order in incremental quantities, the strain-energy function (7.3.1) is replaced by

$$W(\mathbf{A}, \boldsymbol{\eta}) + \left\{\frac{\partial W}{\partial \boldsymbol{\eta}}(\mathbf{A}, \boldsymbol{\eta})\right\} \cdot \delta\boldsymbol{\eta} + \operatorname{tr}\left\{\frac{\partial W}{\partial \mathbf{A}}(\mathbf{A}, \boldsymbol{\eta})\, \delta\mathbf{A}\right\} \tag{7.3.6}$$

and the nominal stress by

$$\mathbf{S} + \left(\delta\boldsymbol{\eta} \cdot \frac{\partial}{\partial \boldsymbol{\eta}}\right)\mathbf{S} + \mathscr{A}^1 \delta\mathbf{A}, \tag{7.3.7}$$

where $\mathscr{A}^1 = \partial \mathbf{S}/\partial \mathbf{A}$ is the tensor of first-order moduli defined in Section 6.1.2.

For equilibrium to be maintained under this change the incremental deformation must satisfy the equation

$$\mathrm{Div}(\mathscr{A}^1\,\delta\mathbf{A}) + \mathrm{Div}\left\{\left(\delta\boldsymbol{\eta}\cdot\frac{\partial}{\partial\boldsymbol{\eta}}\right)\mathbf{S}\right\} = \mathbf{0} \quad \text{in } \mathscr{B}_0 \tag{7.3.8}$$

and the boundary conditions

$$\delta\boldsymbol{\chi} = \mathbf{0} \quad \text{on} \quad \partial\mathscr{B}_0^x, \tag{7.3.9}$$

$$\delta\mathbf{S}^{\mathrm{T}}\mathrm{N} = \delta\boldsymbol{\sigma} \quad \text{on} \quad \partial\mathscr{B}_0^\sigma, \tag{7.3.10}$$

where $\delta\boldsymbol{\sigma}$ reflects the change $\delta\boldsymbol{\chi}$ (but does not depend directly on $\delta\boldsymbol{\eta}$). In the case of dead loading $\delta\boldsymbol{\sigma} = \mathbf{0}$.

Equation (7.3.8) is identical in form to the equation governing incremental elastic deformations in the presence of a body force, as discussed in Section 6.2.1. The body-force term in (7.3.8) is a known function of the underlying deformation $\boldsymbol{\chi}$ and the prescribed increment $\delta\boldsymbol{\eta}$. In this sense the problems governed by (7.3.8)–(7.3.10) are more difficult than those exemplified in Section 6.3, since a particular solution associated with the body force needs to be found. If $\boldsymbol{\chi}$ is an eigenconfiguration for the underlying problem an eigensolution needs to be added to the particular solution. An unstable configuration may be stabilized by the change in material properties; it is left to the reader to analyse the effect of $\delta\boldsymbol{\eta}$ on the stability criteria discussed in Section 6.2.3.

For an incompressible material, det $\mathbf{A} = 1$, (7.3.2) is replaced by

$$\mathbf{S} = \frac{\partial W}{\partial \mathbf{A}}(\mathbf{A}, \boldsymbol{\eta}) - p\mathbf{B}^{\mathrm{T}}$$

and (7.3.7) is correspondingly modified, with $\mathrm{tr}(\mathbf{B}^{\mathrm{T}}\,\delta\mathbf{A}) = 0$ provided the constraint is maintained under the change $\delta\boldsymbol{\eta}$. Similar equations can be obtained in respect of other constraints, such as inextensibility.

The transition from the constitutive law of a constrained material to that of an unconstrained material (or conversely) can also be achieved by means of a change in an appropriately defined $\boldsymbol{\eta}$. Such a transition is covered by the above theory but, because of the nature of the material constants involved (for each constraint one such constant becomes infinite in the limit as the constraint is approached), care must be exercised in assessing the relative orders of magnitude of the terms in the Taylor expansion of the constitutive law with respect to $\delta\boldsymbol{\eta}$. In Section 7.4 this is illustrated in detail for the important special case in which the incompressibility constraint is relaxed but volume changes are small. Applications of the theory are also discussed.

7.4 NEARLY INCOMPRESSIBLE MATERIALS

For an incompressible material we let the deformation be denoted by χ_0 and its gradient Grad χ_0 by $\mathbf{A}_0$, so that

$$\det \mathbf{A}_0 = 1. \tag{7.4.1}$$

Let $W_0(\mathbf{A}_0)$ denote the strain-energy function. The nominal stress-deformation relation is then

$$\mathbf{S}_0 = \frac{\partial W_0}{\partial \mathbf{A}_0} - p_0 \mathbf{B}_0^{\mathrm{T}}, \tag{7.4.2}$$

where $\mathbf{B}_0^{\mathrm{T}} = \mathbf{A}_0^{-1}$ and p_0 is arbitrary. If $\mathbf{T}_0$ denotes the Cauchy stress then from (7.4.2) we obtain

$$\tfrac{1}{3}\operatorname{tr}(\mathbf{T}_0) = \tfrac{1}{3}\operatorname{tr}\left(\mathbf{A}_0 \frac{\partial W}{\partial \mathbf{A}_0}\right) - p_0, \tag{7.4.3}$$

the hydrostatic part of the stress.

As we shall see in what follows these formulae are recovered from corresponding equations in the theory for unconstrained materials by taking the limit of vanishing dilatation.

7.4.1 Compressible materials and the incompressible limit

For a compressible material let χ and $\mathbf{A} = \operatorname{Grad} \chi$ denote the deformation and its gradient. Then, in the notation introduced in (2.2.67), we decompose $\mathbf{A}$ into its dilatational and distortional parts,

$$J = \det \mathbf{A}, \tag{7.4.4}$$

and

$$\mathbf{A}^* = J^{-1/3} \mathbf{A} \tag{7.4.5}$$

respectively, where

$$\det \mathbf{A}^* = 1, \tag{7.4.6}$$

generalizing (7.2.37).

Subject to (7.4.6) we regard $\mathbf{A}^*$ and J as independent variables and write the strain-energy function as

$$W^*(\mathbf{A}^*, J) \equiv W(\mathbf{A}). \tag{7.4.7}$$

The nominal stress-deformation relation is then obtained in the form

$$\mathbf{S} = \frac{\partial W}{\partial \mathbf{A}} = J^{-1/3}\left(\frac{\partial W^*}{\partial \mathbf{A}^*} - p^*\mathbf{B}^{*\mathrm{T}}\right), \tag{7.4.8}$$

where

$$p^* = \tfrac{1}{3}\,\mathrm{tr}\left(\mathbf{A}^*\frac{\partial W^*}{\partial \mathbf{A}^*}\right) - J\frac{\partial W^*}{\partial J} \tag{7.4.9}$$

and $\mathbf{B}^*$ is the inverse of $\mathbf{A}^{*\mathrm{T}}$. The derivation of (7.4.8) requires an expression for $\partial \mathbf{A}^*/\partial \mathbf{A}$, the Cartesian component form of which is obtained from (7.4.5) as

$$\partial A^*_{ij}/\partial A_{kl} = J^{-1/3}(\delta_{ik}\delta_{jl} - \tfrac{1}{3}A^*_{ij}B^*_{kl}).$$

The Cauchy stress is given by $\mathbf{T} = J^{-1}\mathbf{AS}$, and it follows from (7.4.8) and (7.4.9) that the hydrostatic part of the stress is simply

$$\tfrac{1}{3}\,\mathrm{tr}(\mathbf{T}) = \frac{\partial W^*}{\partial J}. \tag{7.4.10}$$

For the strain energy to vanish in the reference configuration we must have

$$W^*(\mathbf{I}, 1) = 0, \tag{7.4.11}$$

where $\mathbf{I}$ is the identity, and, by reference to (7.4.8)–(7.4.10), we deduce that the condition

$$\frac{\partial W^*}{\partial J}(\mathbf{I}, 1) = 0 \tag{7.4.12}$$

ensures that the reference configuration is stress free. Equations (7.4.11) and (7.4.12) generalize formulae given for isotropic materials in (7.2.46). By generalizing the final equation in (7.2.46) we define the ground-state bulk modulus κ as

$$\kappa = \frac{\partial^2 W^*}{\partial J^2}(\mathbf{I}, 1). \tag{7.4.13}$$

Suppose that $\boldsymbol{\mu} \equiv (\mu_1, \mu_2, \ldots)$ represents the remaining ground-state elastic moduli (just a single shear modulus in the case of an isotropic material), and write $\boldsymbol{\eta} = \boldsymbol{\mu}/\kappa, \eta = |\boldsymbol{\eta}|$.

For an incompressible material $\eta = 0$ and the equations (7.4.1)–(7.4.3) are recovered from (7.4.6), (7.4.8) and (7.4.10) respectively in the limit $J \to 1$ by identifying χ with χ_0, $W^*(\mathbf{A}_0, 1)$ with $W_0(\mathbf{A}_0)$ and p^* with p_0. In the limiting process the term $\partial W^*/\partial J$, being the hydrostatic part of the stress, becomes arbitrary as $J \to 1$. This can be seen by considering the dilatation $\varepsilon \equiv J - 1$ to be small and expanding (7.4.10) as a Taylor series in the neighbourhood of $J = 1$. Thus

$$\tfrac{1}{3}\operatorname{tr}(\mathbf{T}) = \frac{\partial W^*}{\partial J}(\mathbf{A}^*, 1) + \varepsilon \frac{\partial^2 W^*}{\partial J^2}(\mathbf{A}^*, 1) + \cdots, \tag{7.4.14}$$

and for a pure dilatation ($\mathbf{A}^* = \mathbf{I}$) use of (7.4.12) and (7.4.13) reduces this to

$$\tfrac{1}{3}\operatorname{tr}(\mathbf{T}) = \kappa\varepsilon + \cdots.$$

In the limit let $\varepsilon \to 0$ and $\kappa \to \infty$ in such a way that their product remains finite. (It is assumed that higher-order terms in (7.4.14) are successively of smaller orders of magnitude and give no contribution to $\tfrac{1}{3}\operatorname{tr}(\mathbf{T})$ in the limit.)

Suppose that $\boldsymbol{\mu}$ is fixed and that the order of magnitude of $W^*(\mathbf{A}^*, 1)$ is $\mu = |\boldsymbol{\mu}|$. Since the term $\partial W^*(\mathbf{A}^*, 1)/\partial J$ in (7.4.14) is not involved in the limiting process we can regard it also as being of order μ. In general $\partial^2 W^*(\mathbf{A}^*, 1)/\partial J^2$ will depend on κ as well as on $\boldsymbol{\mu}$, but, in view of (7.4.13), it will be of order κ when η is small. We assume that $\partial^3 W^*(\mathbf{A}^*, 1)/\partial J^3$, $\partial^4 W^*(\mathbf{A}^*, 1)/\partial J^4$, etc., are of order κ at most. It then follows from (7.4.14) that the dilatation is of order η provided that the order of magnitude of the hydrostatic part of the stress does not exceed μ. Thus, when η is small, volume changes are correspondingly small unless the hydrostatic part of the stress approaches order κ.

When η is small and ε is of order η the strain-energy function may be expanded analogously to (7.4.14) as

$$W^*(\mathbf{A}^*, J) = W^*(\mathbf{A}^*, 1) + \varepsilon \frac{\partial W^*}{\partial J}(\mathbf{A}^*, 1) + \tfrac{1}{2}\varepsilon^2 \frac{\partial^2 W^*}{\partial J^2}(\mathbf{A}^*, 1) + \cdots \tag{7.4.15}$$

The second and third terms on the right-hand side of (7.4.15) are of the same order of magnitude and of order η times the first term. Under the assumptions we have made, the subsequent terms are of order $\eta^2, \eta^3, \ldots$ times the first term.

7.4.2 Nearly incompressible materials

A *nearly incompressible* (or *slightly compressible*) material is defined to be a compressible material which $\eta \ll 1$, such that χ approaches χ_0 as η approaches zero and

$$W^*(\mathbf{A}_0, 1) = W_0(\mathbf{A}_0). \tag{7.4.16}$$

In what follows we assume $\eta \ll 1$.

Let the deformation in the nearly incompressible material with strain-energy function $W^*(\mathbf{A}^*, J)$ be given by

$$\chi = \chi_0 + \eta \mathbf{u}, \tag{7.4.17}$$

where χ_0 is the deformation in the incompressible material with strain-energy function $W_0(\mathbf{A}_0)$ defined by (7.4.16), to the first order in η. The vector function $\mathbf{u}$ is such that Grad $\mathbf{u}$ is of order unity or less. It follows from (7.4.17) that

$$\mathbf{A} = \mathbf{A}_0 + \eta \operatorname{Grad} \mathbf{u} \tag{7.4.18}$$

and hence

$$\varepsilon \equiv J - 1 = \eta \operatorname{tr}\{(\operatorname{Grad} \mathbf{u})\mathbf{B}_0^{\mathrm{T}}\} \equiv \eta \operatorname{div} \mathbf{u} \tag{7.4.19}$$

to the first order in η, where div represents the divergence taken with respect to $\mathbf{x}_0 \equiv \chi_0(\mathbf{X})$. From (7.4.5), (7.4.18) and (7.4.19) we obtain

$$\mathbf{A}^* = \mathbf{A}_0 + \eta\{\operatorname{Grad} \mathbf{u} - \tfrac{1}{3}(\operatorname{div} \mathbf{u})\mathbf{A}_0\}, \tag{7.4.20}$$

also to the first order in η.

With the strain-energy function expanded in the form (7.4.15) the nominal stress (7.4.8) becomes

$$\begin{aligned} \mathbf{S} = \frac{\partial W^*}{\partial \mathbf{A}^*}(\mathbf{A}^*, 1) - \Bigg\{ \tfrac{1}{3} \operatorname{tr}\left(\mathbf{A}^* \frac{\partial W^*}{\partial \mathbf{A}^*}(\mathbf{A}^*, 1) \right) \\ - \frac{\partial W^*}{\partial J}(\mathbf{A}^*, 1) - \varepsilon \frac{\partial^2 W^*}{\partial J^2}(\mathbf{A}^*, 1) \Bigg\} \mathbf{B}^{*\mathrm{T}} + \mu O(\eta). \end{aligned} \tag{7.4.21}$$

If the order η terms in (7.4.21) are neglected then $\mathbf{A}^*$ may be replaced by $\mathbf{A}_0$ and we obtain the zero-order approximation

$$\begin{aligned} \mathbf{S}_0 = \frac{\partial W^*}{\partial \mathbf{A}_0}(\mathbf{A}_0, 1) - \Bigg\{ \tfrac{1}{3} \operatorname{tr}\left(\mathbf{A}_0 \frac{\partial W^*}{\partial \mathbf{A}_0}(\mathbf{A}_0, 1) \right) \\ - \frac{\partial W^*}{\partial J}(\mathbf{A}_0, 1) - \varepsilon \frac{\partial^2 W^*}{\partial J^2}(\mathbf{A}_0, 1) \Bigg\} \mathbf{B}_0^{\mathrm{T}} \end{aligned} \tag{7.4.22}$$

for the stress. Use of (7.4.16) and comparison of (7.4.22) with (7.4.2) establishes the connection

$$p_0 = \tfrac{1}{3}\operatorname{tr}\left\{ \mathbf{A}_0 \frac{\partial W_0}{\partial \mathbf{A}_0}(\mathbf{A}_0) \right\} - \frac{\partial W^*}{\partial J}(\mathbf{A}_0, 1) - \varepsilon \frac{\partial^2 W^*}{\partial J^2}(\mathbf{A}_0, 1)$$

or, equivalently, from (7.4.3),

$$\varepsilon \frac{\partial^2 W^*}{\partial J^2}(\mathbf{A}_0, 1) = \tfrac{1}{3}\operatorname{tr}(\mathbf{T}_0) - \frac{\partial W^*}{\partial J}(\mathbf{A}_0, 1). \tag{7.4.23}$$

Thus, the nearly incompressible theory with the zero-order approximation $\chi = \chi_0$ is equivalent to the incompressible theory provided the first-order deformation (7.4.17) yields the dilatation ε determined by (7.4.23). To the first order in η equation (7.4.23) provides an expression for $\varepsilon = \operatorname{div}\mathbf{u}$ in terms of the known deformation gradient $\mathbf{A}_0$ from the incompressible theory when $\partial W^*(\mathbf{A}_0, 1)/\partial J$ and $\partial^2 W^*(\mathbf{A}_0, 1)/\partial J^2$ are prescribed. In order to determine $\mathbf{u}$ completely the first-order equilibrium equation needs to be considered. This is simply

$$\operatorname{Div}(\mathbf{S} - \mathbf{S}_0) = \mathbf{0}, \tag{7.4.24}$$

where, to the first-order in η, (7.4.8), (7.4.15) and (7.4.20) yield

$$\begin{aligned} \mathbf{S} - \mathbf{S}_0 = {} & \mathscr{A}^1\,\delta\mathbf{A} - \tfrac{1}{3}\varepsilon\mathscr{A}^1\mathbf{A}_0 - \tfrac{1}{3}\varepsilon\mathbf{S}_0 + \varepsilon\frac{\partial^2 W^*}{\partial \mathbf{A}_0 \partial J}(\mathbf{A}_0, 1) + \tfrac{1}{2}\varepsilon^2 \frac{\partial^3 W^*}{\partial \mathbf{A}_0 \partial J^2}(\mathbf{A}_0, 1) \\ & - \tfrac{1}{3}\varepsilon \operatorname{tr}\left\{ \mathbf{A}_0 \frac{\partial^2 W^*}{\partial \mathbf{A}_0 \partial J}(\mathbf{A}_0, 1) + \tfrac{1}{2}\varepsilon \mathbf{A}_0 \frac{\partial^3 W^*}{\partial \mathbf{A}_0 \partial J^2}(\mathbf{A}_0, 1) \right\} \mathbf{B}_0^{\mathrm{T}} \\ & + \varepsilon \left\{ \frac{\partial W^*}{\partial J}(\mathbf{A}_0, 1) + \varepsilon \frac{\partial^2 W^*}{\partial J^2}(\mathbf{A}_0, 1) + \tfrac{1}{2}\varepsilon \frac{\partial^3 W^*}{\partial J^3}(\mathbf{A}_0, 1) \right\} \mathbf{B}_0^{\mathrm{T}}, \end{aligned} \tag{7.4.25}$$

where $\delta\mathbf{A} = \eta \operatorname{Grad}\mathbf{u}$ and $\mathscr{A}^1 = \partial \mathbf{S}_0/\partial \mathbf{A}_0$ with $\mathbf{S}_0$ given by (7.4.22).

Since ε and $\mathbf{A}_0$ are known equation (7.4.24) can now be written

$$\operatorname{Div}(\mathscr{A}^1\,\delta\mathbf{A}) + \mathbf{F} = \mathbf{0},$$

where $\mathbf{F}$ is a known body force of order $\eta\mu$. The present theory is now clearly seen to be embodied in that described in Section 7.3. Higher-order approximations may also be considered but we do not discuss them here. Notice, in particular, that while the zero-order theory requires use of the term $\partial^2 W^*(\mathbf{A}_0, 1)/\partial J^2$ the first-order theory uses $\partial^3 W^*(\mathbf{A}_0, 1)/\partial J^3$, and so on for higher orders.

For a more detailed account of the above theory we refer to Ogden (1978). This paper also discusses the application of the zero-order theory to the calculation of volume changes accompanying the extension and torsion of a circular cylinder of isotropic material. For a different approach to the theory see Spencer (1959, 1964) and for other applications see Spencer (1959) and Ogden (1979). Further references are contained in Ogden (1978). In Section 7.4.3 we specialize the results of this section to the case of pure homogeneous strain of an isotropic material with a view to its application to some simple problems.

Problem 7.4.1 Specialize the formula (7.4.23) for an isotropic material and obtain an expression for ε for the combined axial and torsional shear deformation described in Section 5.3.6.

If the strain-energy function is given by

$$W(\lambda_1, \lambda_2, \lambda_3) = \tfrac{1}{2}\mu(\lambda_1^2 + \lambda_2^2 + \lambda_3^2 - 3 - 2\ln J) + \tfrac{1}{2}\kappa(J-1)^2$$

show that for the plane strain torsional shear of a circular annulus of radii A and $B(> A)$ the dilatation is given by

$$\kappa\varepsilon = \sigma^2\left(\frac{1}{A^2B^2} - \frac{1}{R^4}\right)\bigg/ 16\pi^2\mu, \qquad A \le R \le B,$$

where σ is defined in Section 5.3.6.

7.4.3 Pure homogeneous strain of a nearly incompressible isotropic material

For a compressible isotropic elastic material we recall from (7.2.44) and (7.2.45) that the stress-deformation relations may be expressed as

$$J(t_1 - t_3) = \lambda_1^* \frac{\partial \tilde{W}^*}{\partial \lambda_1^*}, \qquad J(t_2 - t_3) = \lambda_2^* \frac{\partial \tilde{W}^*}{\partial \lambda_2^*}, \tag{7.4.26}$$

$$\tfrac{1}{3}(t_1 + t_2 + t_3) = \frac{\partial \tilde{W}^*}{\partial J}, \tag{7.4.27}$$

where the strain-energy function $\tilde{W}^*(\lambda_1^*, \lambda_2^*, J)$ is defined by (7.2.39) and (7.2.43), and is subject to (7.2.46).

For a nearly incompressible material the expansion (7.4.15) becomes

$$\tilde{W}^*(\lambda_1^*, \lambda_2^*, J) = \tilde{W}^*(\lambda_1^*, \lambda_2^*, 1) + \varepsilon \frac{\partial \tilde{W}^*}{\partial J}(\lambda_1^*, \lambda_2^*, 1)$$
$$+ \tfrac{1}{2}\varepsilon^2 \frac{\partial^2 \tilde{W}^*}{\partial J^2}(\lambda_1^*, \lambda_2^*, 1) + \cdots \tag{7.4.28}$$

and the formula (7.4.23) simplifies to

$$\varepsilon\frac{\partial^2\tilde{W}^*}{\partial J^2}(\lambda_{10},\lambda_{20},1)=\tfrac{1}{3}(t_{10}+t_{20}+t_{30})-\frac{\partial\tilde{W}^*}{\partial J}(\lambda_{10},\lambda_{20},1), \tag{7.4.29}$$

where $\lambda_{10},\lambda_{20}$ are the principal stretches associated with the incompressible (or zero-order) theory and t_{10},t_{20},t_{30} are the principal components of $\mathbf{T}_0$.

Equation (7.4.29) forms the zero-order approximation to (7.4.27). The corresponding approximation to (7.4.26) is

$$\left.\begin{aligned} t_1-t_3&=\lambda_{10}\frac{\partial\tilde{W}}{\partial\lambda_{10}}(\lambda_{10},\lambda_{20}),\\ t_2-t_3&=\lambda_{20}\frac{\partial\tilde{W}}{\partial\lambda_{20}}(\lambda_{10},\lambda_{20}),\end{aligned}\right\} \tag{7.4.30}$$

where

$$\tilde{W}(\lambda_{10},\lambda_{20})=\tilde{W}^*(\lambda_{10},\lambda_{20},1) \tag{7.4.31}$$

is the strain-energy function arising in the incompressible theory discussed in Section 7.2.1. The terms omitted from the right-hand side of (7.4.30) are of order η relative to those retained.

Clearly, the primary stress-deformation characteristics of a nearly incompressible material are determined on the basis of the incompressible theory through (7.4.30) and (7.4.31), and the neglected order η terms may be ignored for practical purposes. At the zero-order level the remaining constitutive functions $\partial\tilde{W}^*(\lambda_{10},\lambda_{20},1)/\partial J$ and $\partial^2\tilde{W}^*(\lambda_{10},\lambda_{20},1)/\partial J^2$ appear only in (7.4.29), and measured values of ε in biaxial tension tests serve to provide some information about these functions. If, on the other hand, these functions are prescribed then (7.4.29) simply determines ε as a function of the deformation $\lambda_{10},\lambda_{20}$. In order to determine λ_1^*,λ_2^* the first-order approximation must be used. The resulting first-order equations also serve to provide further information about the above two constitutive functions, as we now illustrate.

Let

$$\lambda_i=\lambda_{io}(1+\varepsilon_i)\qquad \mathrm{i}=1,2,3, \tag{7.4.32}$$

so that, correct to first order,

$$\varepsilon=\varepsilon_1+\varepsilon_2+\varepsilon_3 \tag{7.4.33}$$

and

$$\lambda_i^*=\lambda_{io}(1+\varepsilon_i-\tfrac{1}{3}\varepsilon)\qquad i=1,2,3. \tag{7.4.34}$$

Then the first-order approximation to the first equation in (7.4.26), with (7.4.28), is easily shown to be

$$\begin{aligned} t_1 - t_3 = t_{10} - t_{30} &+ (\varepsilon_1 - \tfrac{1}{3}\varepsilon)\times \\ &\times \left\{ \lambda_{10}^2 \frac{\partial^2 \tilde{W}^*}{\partial \lambda_{10}^2}(\lambda_{10}, \lambda_{20}, 1) + \lambda_{10} \frac{\partial \tilde{W}^*}{\partial \lambda_{10}}(\lambda_{10}, \lambda_{20}, 1) \right\} \\ &+ (\varepsilon_2 - \tfrac{1}{3}\varepsilon)\lambda_{10}\lambda_{20} \frac{\partial^2 \tilde{W}^*}{\partial \lambda_{10} \partial \lambda_{20}}(\lambda_{10}, \lambda_{20}, 1) \\ &- \varepsilon \left\{ \lambda_{10} \frac{\partial \tilde{W}^*}{\partial \lambda_{10}}(\lambda_{10}, \lambda_{20}, 1) + \lambda_{10} \frac{\partial^2 \tilde{W}^*}{\partial \lambda_{10} \partial J}(\lambda_{10}, \lambda_{20}, 1) \right. \\ &\left. + \tfrac{1}{2}\varepsilon \lambda_{10} \frac{\partial^3 \tilde{W}^*}{\partial \lambda_{10} \partial J^2}(\lambda_{10}, \lambda_{20}, 1) \right\}, \end{aligned} \tag{7.4.35}$$

and similarly for $t_2 - t_3$.

For a problem in which the stresses are prescribed, for example, $t_i = t_{i0} (i = 1, 2, 3)$, and then (7.4.35) together with the corresponding equation for $t_2 - t_3$ provides a pair of equations for ε_1 and ε_2 involving the derivatives of $\partial \tilde{W}^*(\lambda_{10}, \lambda_{20}, 1)/\partial J$ and $\partial^2 \tilde{W}^*(\lambda_{10}, \lambda_{20}, 1)/\partial J^2$ with respect to λ_{10} and λ_{20}. Since ε is known from (7.4.29) equation (7.4.33) then serves to determine ε_3.

Problem 7.4.2 For the homogeneous biaxial stress of a rectangular block with $t_3 = 0$ show that the implicit relationship

$$J \frac{\partial \tilde{W}^*}{\partial J} = \frac{1}{3}\left(\lambda_1^* \frac{\partial \tilde{W}^*}{\partial \lambda_1^*} + \lambda_2^* \frac{\partial \tilde{W}^*}{\partial \lambda_2^*} \right)$$

between $J, \lambda_1^*, \lambda_2^*$ is obtained from (7.4.26) and (7.4.27). By differentiating this equation with respect to λ_α^* $(\alpha = 1, 2)$, show that

$$\begin{aligned} \frac{\partial J}{\partial \lambda_\alpha^*} = &\left[\tfrac{1}{3}\lambda_\beta^* \frac{\partial^2 \tilde{W}^*}{\partial \lambda_\beta^* \partial \lambda_\alpha^*} + \frac{1}{3} \frac{\partial \tilde{W}^*}{\partial \lambda_\alpha^*} - J \frac{\partial^2 \tilde{W}^*}{\partial \lambda_\alpha^* \partial J} \right] \times \\ &\times \left[J \frac{\partial^2 \tilde{W}^*}{\partial J^2} + \frac{\partial \tilde{W}^*}{\partial J} - \tfrac{1}{3}\lambda_\gamma^* \frac{\partial^2 \tilde{W}^*}{\partial \lambda_\gamma^* \partial J} \right]^{-1}, \end{aligned}$$

where there is summation over β and γ from 1 to 2, and use (7.2.46) to deduce that this reduces to $2\mu/\kappa \equiv 2\eta$ in the reference configuration.

Problem 7.4.3 Obtain the specialization of the formulae in Problem 7.4.2 to the case of simple tension by writing $\lambda_1^* = \lambda^*$, $\lambda_2^* = \lambda^{*-1/2}$ and

$$\hat{W}^*(\lambda^*, J) \equiv \tilde{W}^*(\lambda^*, \lambda^{*-1/2}, J)$$

as in (7.2.55), and deduce that $\mathrm{d}J/\mathrm{d}\lambda^* = \eta$ in the reference configuration. Show that the corresponding specialization of (7.4.29) is

$$\varepsilon \frac{\partial^2 \hat{W}^*}{\partial J^2}(\lambda, 1) = \tfrac{1}{3}t - \frac{\partial \hat{W}^*}{\partial J}(\lambda, 1), \tag{7.4.36}$$

where $\lambda = \lambda_{10}, t = t_{10}$ with

$$t = \lambda \frac{\mathrm{d}\hat{W}}{\mathrm{d}\lambda}, \qquad \hat{W}(\lambda) \equiv \hat{W}^*(\lambda, 1). \tag{7.4.37}$$

7.4.4 Application to rubberlike materials

Here we give a brief account of the application of the theory described in Section 7.4.3, taking rubberlike materials, for which η is typically of order 10^{-4}, as an illustrative example. The discussion follows closely that by Ogden (1982, Section 3.4).

With the strain-energy function given by (7.4.28) it is assumed that the first term $\tilde{W}^*(\lambda_1^*, \lambda_2^*, 1)$, bearing in mind (7.4.31), is known from the work on incompressible materials described in Section 7.2.2. Equation (7.4.29) with $t_{30} = 0$ then relates the dilatation to the two independent functions $\partial\tilde{W}^*(\lambda_{10}, \lambda_{20}, 1)/\partial J$ and $\partial^2\tilde{W}^*(\lambda_{10}, \lambda_{20}, 1)/\partial J^2$ in biaxial stress. When a hydrostatic pressure P is superposed on the biaxial stress equation (7.4.29) becomes

$$\varepsilon \frac{\partial^2 \tilde{W}^*}{\partial J^2}(\lambda_{10}, \lambda_{20}, 1) + P = \tfrac{1}{3}(t_{10} + t_{20}) - \frac{\partial \tilde{W}^*}{\partial J}(\lambda_{10}, \lambda_{20}, 1), \tag{7.4.38}$$

which specializes (7.2.54) provided P is not too large. For each fixed pair $(\lambda_{10}, \lambda_{20})$ and the associated pair (t_{10}, t_{20}), measured values of ε and P provide information about both constitutive functions from pure homogeneous biaxial deformation experiments. Further properties of these functions can be found from first-order equations such as (7.4.35).

Recalling (7.4.27) we define the *bulk modulus* for a general state of deformation as

$$\frac{1}{3}\frac{\partial}{\partial J}(t_1 + t_2 + t_3) \equiv \frac{\partial^2 \tilde{W}^*}{\partial J^2}. \tag{7.4.39}$$

This approximates to $\partial^2\tilde{W}^*(\lambda_1^*, \lambda_2^*, 1)/\partial J^2$ in circumstances where ε is of order η and reduces to the ground-state bulk modulus κ in pure dilatation ($\lambda_1^* = \lambda_2^* = 1$). Evidence from experiments which measure volume changes in simple tension indicates that the bulk modulus is 'isotropic', i.e. independent of λ_1^* and

λ_2^*, and therefore equal to κ. For discussion of this point and citation of references to experimental work see Ogden (1982). Although justification for adopting the assumption is as yet somewhat limited it is convenient here to set the bulk modulus equal to κ so that, when $P=0$, equation (7.4.38) becomes

$$\kappa\varepsilon = \tfrac{1}{3}(t_{10}+t_{20}) - \frac{\partial \tilde{W}^*}{\partial J}(\lambda_{10}, \lambda_{20}, 1). \tag{7.4.40}$$

Equation (7.4.40) now serves to determine the form of the function $\partial\tilde{W}^*(\lambda_{10}, \lambda_{20}, 1)/\partial J$.

The only deformation for which relevant volume-change data are available is that in simple tension, and we therefore specialize (7.4.40) accordingly. With the notation as used in Problem 7.4.3, we obtain

$$\kappa\varepsilon = \tfrac{1}{3}t - \frac{\partial \hat{W}^*}{\partial J}(\lambda, 1) \tag{7.4.41}$$

and we now confine attention to this special case.

In Fig. 7.5 we plot $\kappa\varepsilon/\mu$ against λ. The continuous curve was constructed to

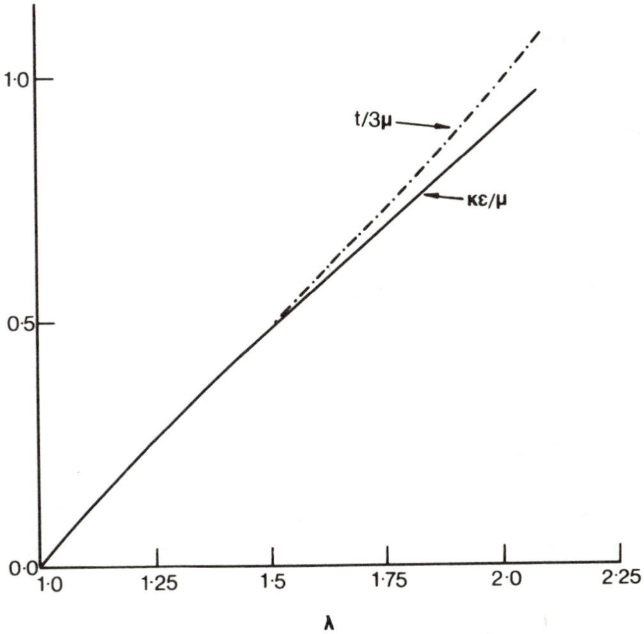

Fig. 7.5 Plot of $\kappa\varepsilon/\mu$ against λ for simple tension. The continuous curve represents Penn's data and the broken curve the corresponding plot of $t/3\mu$.

fit the volume-change data of Penn (1970) and is representative of the available data for simple tension. For references to other data see Ogden (1982). The broken curve is a plot of $t/3\mu$ against λ, also from Penn's data. In view of (7.4.41) the difference between the two curves is a measure of the function $\partial\hat{W}^*(\lambda, 1)/\partial J$. The separation of the curves for λ greater than about 1.5 establishes that such a function should be included in the strain-energy function. For full discussion of its properties and significance we refer to Ogden (1982).

We now turn our attention to circumstances in which the dilatation ε is not restricted to being of order η, so that the hydrostatic part of the stress exceeds order μ in magnitude and the expansion (7.4.28) is no longer appropriate. For our purposes it suffices to restrict attention to the class of strain-energy functions defined by

$$\tilde{W}^*(\lambda_1^*, \lambda_2^*, J) = \tilde{W}^*(\lambda_1^*, \lambda_2^*, 1) + (J-1)\frac{\partial\tilde{W}^*}{\partial J}(\lambda_1^*, \lambda_2^*, 1) + \kappa g(J), \tag{7.4.42}$$

where, in view of (7.2.46),

$$g(1) = g'(1) = 0, \qquad g''(1) = 1. \tag{7.4.43}$$

The assumption that the bulk modulus is 'isotropic' is embodied in (7.4.42).

Substitution of (7.4.42) into (7.4.26) and (7.4.27) shows that

$$t_i - t_j = \mu O(1) \qquad i, j \in \{1, 2, 3\}$$

and

$$\tfrac{1}{3}(t_1 + t_2 + t_3) = \kappa g'(J) + \mu O(1)$$

since $\tilde{W}^*(\lambda_1^*, \lambda_2^*, 1)$ and $\partial\tilde{W}^*(\lambda_1^*, \lambda_2^*, 1)/\partial J$ are typically of order μ. We deduce that

$$t_i = \kappa g'(J) + \mu O(1) \qquad i \in \{1, 2, 3\},$$

so that when $J-1$ exceeds $O(\eta)$ the first term on the right-hand side is dominant and the stress is essentially hydrostatic and equal to $\kappa g'(J)$. We therefore write

$$P = -\kappa g'(J), \tag{7.4.44}$$

P being the applied hydrostatic pressure.

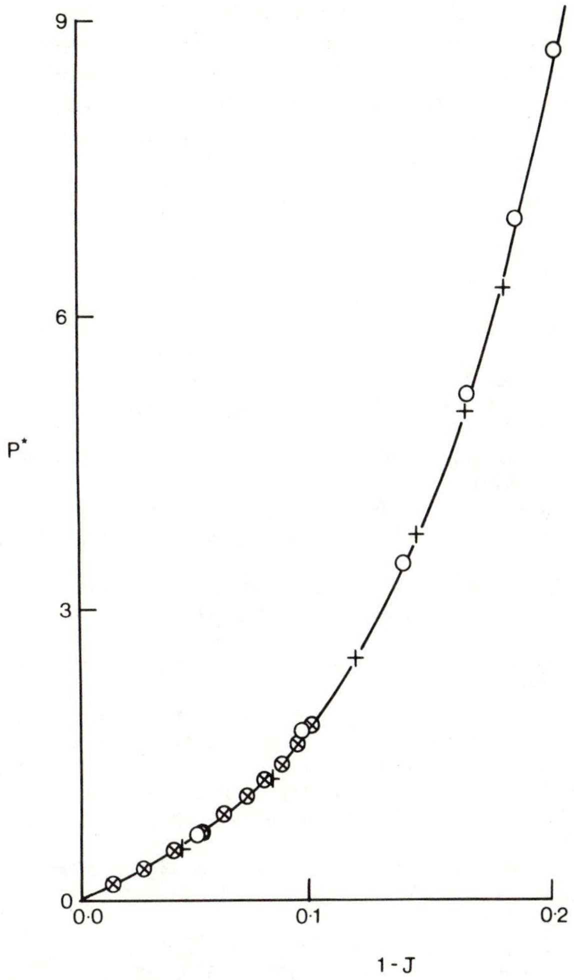

Fig. 7.6 Plot of the dimensionless pressure P^* against $1-J$. Theory (———) compared with the data of Adams and Gibson (⊗) and Bridgman (○, +).

In hydrostatic compression tests values of J down to about 0.8 have been achieved by a number of experiments, and data from the work of Adams and Gibson (1930) and Bridgman (1945) is shown in Fig. 7.6 with the dimensionless pressure P^* plotted against $1-J$. The continuous curve, shown for comparison, is calculated from the theoretical form

$$g'(J)=\tfrac{1}{9}(J^{-1}-J^{-10})$$

with $P^* = 9P/\kappa$, and this illustrates the good agreement obtainable between theory and experiment. For full details and for references to other experimental work see Ogden (1972b). Because of problems associated with the formation of cavities only relatively small volume changes have been achieved in hydrostatic tension (Gent and Lindley, 1959) and we do not discuss this case here.

REFERENCES

L.H. Adams and R.E. Gibson, The Compressibility of Rubber, *Journal of the Academy of Sciences, Washington*, Vol. 20, pp. 213–223 (1930).

P.W. Bridgman, The compression of 61 Substances to 25,000 kg/cm^2 Determined by a New Rapid Method, *Proceedings of the American Academy of Arts and Sciences*, Vol. 76, pp. 9–24 (1945).

J.L. Ericksen, Special Topics in Elastostatics, *Advances in Applied Mechanics*, Vol. 17, pp. 189–244 (1977).

A.N. Gent and P.B. Lindley, Internal Rupture of Bonded Rubber Cylinders in Tension, *Proceedings of the Royal Society of London, Series A*, Vol. 249, pp. 195–205 (1959).

A.N. Gent and R.S. Rivlin, Experiments on the Mechanics of Rubber—II. The Torsion, Inflation and Extension of a Tube, *Proceedings of the Physical Society, Series B*, Vol. 65, pp. 487–501 (1952).

A.E. Green and J.E. Adkins, *Large Elastic Deformations*, 2nd edition, Oxford University Press (1970).

D.M. Haughton and R.W. Ogden, Bifurcation of Inflated Circular Cylinders of Elastic Material under Axial Loading—I. Membrane Theory for Thin-walled Tubes, *Journal of the Mechanics and Physics of Solids*, Vol. 27, pp. 179–212 (1979).

R. Hill, On the Elasticity and Stability of Perfect Crystals at Finite Strain, *Mathematical Proceedings of the Cambridge Philosophical Society*, Vol. 77, pp. 225–240 (1975).

A.G. James, A. Green and G.M. Simpson, Strain-energy Functions of Rubber—I. Characterization of Gum Vulcanizates, *Journal of Applied Polymer Science*, Vol. 19, pp. 2033–2058 (1975).

D.F. Jones and L.R.G. Treloar, The Properties of Rubber in Pure Homogeneous Strain, *Journal of Physics D: Applied Physics*, Vol. 8, pp. 1285–1304 (1975).

S. Kawabata and H. Kawai, Strain-energy Density Functions of Rubber Vulcanizates from Biaxial Extension, *Advances in Polymer Science*, Vol. 24, pp. 89–124 (1977).

P.B. Lindley, *Engineering Design with Natural Rubber*, 4th edition, Technical Bulletin, The Malaysian Rubber Producers' Research Association (1974).

F. Milstein and R. Hill, Theoretical Properties of Cubic Crystals at Arbitrary Pressure—I. Density and Bulk Modulus, *Journal of the Mechanics and Physics of Solids*, Vol. 25, pp. 457–477 (1977).

F. Milstein and R. Hill, Theoretical Properties of Cubic Crystals at Arbitrary Pressure—II. Shear Moduli, *Journal of the Mechanics of Physics and Solids*, Vol. 26, pp. 213–239 (1978).

F. Milstein and R. Hill, Theoretical Properties of Cubic Crystals at Arbitrary Pressure—III. Stability, *Journal of the Mechanics and Physics of Solids*, Vol. 27. pp. 255–279 (1979).

R.W. Ogden, Large Deformation Isotropic Elasticity: on the Correlation of Theory and Experiment for Incompressible Rubberlike Solids, *Proceedings of the Royal Society of London, Series A*, Vol. 326, pp. 565–584 (1972a).

R.W. Ogden, Large Deformation Isotropic Elasticity: on the Correlation of Theory and Experiment for Compressible Rubberlike Solids, *Proceedings of the Royal Society of London, Series A*, Vol. 328, pp. 567–583 (1972b).

R.W. Ogden, On Isotropic Tensors and Elastic Moduli, *Proceedings of the Cambridge Philosophical Society*, Vol. 75, pp. 427–436 (1974).

R.W. Ogden, *Nearly Isochoric Elastic Deformations: Application to Rubberlike Solids, Journal of the Mechanics and Physics of Solids*, Vol. 26, pp. 37–57 (1978).

R.W. Ogden, Nearly Isochoric Elastic Deformations: Volume Changes in Plane Strain, *Quarterly of Applied Mathematics*, Vol. 36, pp. 337–345 (1979).

R.W. Ogden, Elastic Deformations of Rubberlike Solids, in *Mechanics of Solids*, The Rodney Hill 60th Anniversary Volume (Eds. H.G. Hopkins and M.J. Sewell), pp. 499–537, Pergamon Press (1982)

R.W. Ogden and P. Chadwick, On the Deformation of Solid and Tubular Cylinders of Incompressible Isotropic Elastic Material, *Journal of the Mechanics and Physics of Solids*, Vol. 20, pp. 77–90 (1972).

G.P. Parry, Continuum Models of Non-linearly Elastic Perfect Crystals, in *Continuum Models of Discrete Systems* (Eds. E. Kröner and K.-H. Anthony), pp. 81–96, University of Waterloo Press (1980).

R.W. Penn, Volume Changes Accompanying the Extension of Rubber, *Transactions of the Society of Rheology*, Vol. 14, pp. 509–517 (1970).

R.S. Rivlin and D.W. Saunders, Large Elastic Deformations of Isotropic Materials—VII. Experiments on the Deformation of Rubber, *Philosophical Transactions of the Royal Society of London Series A*, Vol. 243, pp. 251–288 (1951).

A.J.M. Spencer, On Finite Elastic Deformations with a Perturbed Strain-energy Function, *Quarterly Journal of Mechanics and Applied Mathematics*, Vol. 12, pp. 129–145 (1959).

A.J.M. Spencer, Finite Deformation of an almost Incompressible Elastic Solid, in *Second-order Effects in Elasticity, Plasticity and Fluid Dynamics* (Eds. M. Reiner and D. Abir), pp. 200–216, Pergamon Press (1964).

L.R.G. Treloar, Stress-strain Data for Vulcanized Rubber under Various Types of Deformation, *Transactions of the Faraday Society* Vol. 40, pp. 59–70 (1944).

L.R.G. Treloar, *The Physics of Rubber Elasticity*, 3rd edition, Oxford University Press (1975).

E.H. Twizell and R.W. Ogden, Non-linear Optimization of the Material Constants in Ogden's Stress-deformation Function for Incompressible Isotropic Elastic Materials, *Journal of the Australian Mathematical Society, Series B*, Vol. 24, pp. 424–434 (1983).

K.C. Valanis and R.F. Landel, The Strain-energy Function of a Hyperelastic Material in Terms of the Extension Ratios, *Journal of Applied Physics*, Vol. 38, pp. 2997–3002 (1967).

H. Vangerko and L.R.G. Treloar, The Inflation and Extension of Rubber Tube for Biaxial Strain Studies, *Journal of Physics D: Applied Physics*, Vol. 11, pp. 1969–1978 (1978).

O.H. Varga, *Stress-Strain Behavior of Elastic Materials*, Interscience (1966).

APPENDIX 1

Convex Functions

Suppose $\mathbb{E}$ is a finite-dimensional vector-space and let the domain $\mathscr{D}$ be an open subset of $\mathbb{E}$. $\mathscr{D}$ is said to be a *convex* domain if it contains the line segment $t\mathbf{u} + (1-t)\mathbf{v}$ for all $\mathbf{u}, \mathbf{v} \in \mathscr{D}$ and $0 < t < 1$.

Let $\mathbf{f}\colon \mathscr{D} \to \mathbb{E}$ be continuous. Then $\mathbf{f}$ is said to be *convex* on $\mathscr{D}$ if

$$\{\mathbf{f}(\mathbf{u}) - \mathbf{f}(\mathbf{v})\}\cdot(\mathbf{u} - \mathbf{v}) \geq 0$$

for all $\mathbf{u}, \mathbf{v} \in \mathscr{D}$ and *strictly convex* on $\mathscr{D}$ if

$$\{\mathbf{f}(\mathbf{u}) - \mathbf{f}(\mathbf{v})\}\cdot(\mathbf{u} - \mathbf{v}) > 0 \tag{A.1}$$

for all $\mathbf{u}, \mathbf{v} \in \mathscr{D}$ with$\mathbf{v} \neq \mathbf{u}$. In the latter case $\mathbf{f}$ is also said to be *strictly monotone* on $\mathscr{D}$.

Since $\mathbf{f}(\mathbf{u}) = \mathbf{f}(\mathbf{v})$ violates (A.1) unless $\mathbf{u} = \mathbf{v}$ it follows that (A.1) implies $\mathbf{f}$ is *one-to-one* on $\mathscr{D}$. Its inverse $\mathbf{f}^{-1}$ is then strictly convex on $\mathbf{f}(\mathscr{D})$.

If $\mathbf{f}$ is continuously differentiable a consequence of (A.1) is that the linear mapping $\mathbf{L} \equiv \partial\mathbf{f}(\mathbf{u})/\partial\mathbf{u}\colon \mathbb{E} \to \mathbb{E}$ is positive semi-definite, i.e.

$$(\mathbf{L}\mathbf{w})\cdot\mathbf{w} \geq 0 \qquad \text{for all } \mathbf{w} \in \mathbb{E}, \tag{A.2}$$

at each point $\mathbf{u} \in \mathscr{D}$, and is positive definite, i.e.

$$(\mathbf{L}\mathbf{w})\cdot\mathbf{w} > 0 \qquad \text{for all } \mathbf{w} \neq \mathbf{0} \text{ in } \mathbb{E}, \tag{A.3}$$

in $\mathscr{D}$ *except possibly on a nowhere dense subset*. For a proof of this result see Bernstein and Toupin (1962). The converse is untrue in general, but if $\mathscr{D}$ is a convex domain and (A.3) holds at each point in $\mathscr{D}$ then (A.1) follows. In other words, strict *local* convexity implies strict *global* convexity in a convex domain.

The latter result is proved by writing $\mathbf{w} = \mathrm{d}\mathbf{u}$, so that (A.3) becomes $\mathrm{d}\mathbf{f}\cdot\mathrm{d}\mathbf{u} > 0$, and taking $\mathrm{d}\mathbf{u}$ to be $(\mathbf{v} - \mathbf{u})\mathrm{d}t$ ($\mathrm{d}t > 0$), i.e. tangential to the straight line segment $(1-t)\mathbf{u}+t\mathbf{v}$ joining $\mathbf{u}$ to $\mathbf{v}$. Thus $\mathrm{d}\mathbf{f}\cdot(\mathbf{v}-\mathbf{u})>0$ and integration along the line then yields (A.1)

Suppose next that $\phi:\mathscr{D}\to\mathbb{R}$ is a continuously differentiable function. Then ϕ is said to be *strictly (globally) convex* on $\mathscr{D}$ if

$$\phi(\mathbf{v})-\phi(\mathbf{u})-(\mathbf{v}-\mathbf{u})\cdot\frac{\partial\phi}{\partial\mathbf{u}}(\mathbf{u})>0 \tag{A.4}$$

for all $\mathbf{u},\mathbf{v}\in\mathscr{D}$ such that $\mathbf{v}\neq\mathbf{u}$. By reversing the roles of $\mathbf{u}$ and $\mathbf{v}$ in (A.4) and adding the resulting inequality to (A.4) we obtain

$$\left\{\frac{\partial\phi}{\partial\mathbf{u}}(\mathbf{u})-\frac{\partial\phi}{\partial\mathbf{v}}(\mathbf{v})\right\}\cdot(\mathbf{u}-\mathbf{v})>0 \tag{A.5}$$

for all $\mathbf{u},\mathbf{v}\in\mathscr{D}$ with $\mathbf{v}\neq\mathbf{u}$, which is equivalent to (A.1) with $\mathbf{f}=\partial\phi/\partial\mathbf{u}$. In general (A.5) does not imply (A.4) but if $\mathscr{D}$ is convex a variant of the line segment argument used above shows that (A.4) is necessary and sufficient for (A.5). (The word strictly can be removed from the above if $>$ is replaced by $\geq$ in (A.4).)

If ϕ is twice continuously differentiable then (A.5) implies that the Hessian $\partial^2\phi(\mathbf{u})/\partial\mathbf{u}\partial\mathbf{u}$ is positive definite except on a nowhere dense set. Since the Hessian is symmetric it follows from (A.2) and an extremum property of quadratic forms that equality holds for $\mathbf{w}\neq\mathbf{0}$ in (A.2) if and only if $\mathbf{Lw}=\mathbf{0}$, i.e. where $\det\mathbf{L}=0$. Thus (A.5) ensures that $\partial^2\phi(\mathbf{u})/\partial\mathbf{u}\partial\mathbf{u}$ is positive definite except for those $\mathbf{u}$ for which $\det(\partial^2\phi(\mathbf{u})/\partial\mathbf{u}\partial\mathbf{u})=0$, this equation defining a nowhere dense set in $\mathscr{D}$. On the other hand, if ϕ is strictly locally convex on $\mathscr{D}$ (i.e. (A.3) holds with $\mathbf{L}\equiv\partial^2\phi(\mathbf{u})/\partial\mathbf{u}\partial\mathbf{u}$ for each $\mathbf{u}\in\mathscr{D}$) and $\mathscr{D}$ is a convex domain it follows that ϕ is strictly globally convex on $\mathscr{D}$.

Now suppose that $\mathbb{E}$ represents the vector space of second-order tensors. We modify our notation accordingly so that $\mathbf{u},\mathbf{v}$ are replaced by $\mathbf{E},\mathbf{E}^*$, $\mathbf{f}$ by $\mathbf{G}$, $\mathbf{L}$ by $\mathscr{L}$ and $\mathbf{w}$ by $\dot{\mathbf{E}}$. The inequalities (A.1), (A.3) and (A.4) then become

$$\operatorname{tr}\{(\mathbf{T}^*-\mathbf{T})(\mathbf{E}^*-\mathbf{E})\}>0 \qquad \mathbf{E}^*\neq\mathbf{E}, \tag{A.6}$$

$$\operatorname{tr}\{(\mathscr{L}\dot{\mathbf{E}})\dot{\mathbf{E}}\}>0 \qquad \dot{\mathbf{E}}\neq\mathbf{0}, \tag{A.7}$$

and

$$\phi(\mathbf{E}^*)-\phi(\mathbf{E})-\operatorname{tr}\{\mathbf{T}(\mathbf{E}^*-\mathbf{E})\}>0 \qquad \mathbf{E}^*\neq\mathbf{E} \tag{A.8}$$

respectively, where $\mathbf{T}=\mathbf{G}(\mathbf{E})$, $\mathbf{T}^*=\mathbf{G}(\mathbf{E}^*)$, with $\mathbf{G}=\partial\phi/\partial\mathbf{E}$ in (A.8).

Henceforth we restrict attention to symmetric tensors $\mathbf{E}$ in three physical dimensions, with $\mathbf{G}$ correspondingly symmetric and $\mathbb{E}$ therefore six-dimensional. In particular, we are concerned with *isotropic* tensor functions ϕ and $\mathbf{G}$.

Let e_1, e_2, e_3 denote the principal values of $\mathbf{E}$, ordered according to $e_1 \geq e_2 \geq e_3$, and let t_1, t_2, t_3 be those of $\mathbf{T} = \mathbf{G}(\mathbf{E})$, $\mathbf{T}$ and $\mathbf{E}$ being coaxial. Likewise let e_1^*, e_2^*, e_3^* and t_1^*, t_2^*, t_3^* be the principal values of $\mathbf{E}^*$ and $\mathbf{T}^* \equiv \mathbf{G}(\mathbf{E}^*)$ respectively.

We consider first the inequality

$$\sum_{i=1}^{3} (t_i^* - t_i)(e_i^* - e_i) > 0 \tag{A.9}$$

with $e_i^* \neq e_i$ for at least one $i \in \{1, 2, 3\}$. We may choose $e_i^* = e_j$ and $e_k^* = e_k$, where (i, j, k) is a permutation of $(1, 2, 3)$, so that, because of the isotropy, $t_i^* = t_j$, $t_k^* = t_k$ and (A.9) reduces to

$$(t_i - t_j)(e_i - e_j) > 0 \qquad i \neq j, \quad e_i \neq e_j. \tag{A.10}$$

This implies that t_1, t_2, t_3 are in the same algebraic order as e_1, e_2, e_3, i.e. $t_1 \geq t_2 \geq t_3$.

Next we prove that (A.9) is necessary and sufficient for (A.6) to hold. Necessity is established simply by choosing $\mathbf{E}^*$ in (A.6) to be coaxial with $\mathbf{E}$. To show sufficiency we note that

$$\mathrm{tr}\{(\mathbf{T}^* - \mathbf{T})(\mathbf{E}^* - \mathbf{E})\} = \sum_{i=1}^{3} (t_i^* e_i^* + t_i e_i) - \mathrm{tr}(\mathbf{T}^*\mathbf{E} + \mathbf{T}\mathbf{E}^*).$$

However, from a result of Hill (1968, 1970) we have that $\mathrm{tr}(\mathbf{T}\mathbf{E}^*)$ is stationary when $\mathbf{E}^*$ is coaxial with $\mathbf{T}$ (for fixed principal values) and the maximum value is obtained when e_1^*, e_2^*, e_3^* are in same algebraic order as t_1, t_2, t_3. In view of (A.10) the choice $e_1^* \geq e_2^* \geq e_3^*$ shows that $\mathrm{tr}(\mathbf{T}\mathbf{E}^*) \leq \sum_{i=1}^{3} t_i e_i^*$ and similarly $\mathrm{tr}(\mathbf{T}^*\mathbf{E}) \leq \sum_{i=1}^{3} t_i^* e_i$. Thus

$$\mathrm{tr}\{(\mathbf{T}^* - \mathbf{T})(\mathbf{E}^* - \mathbf{E})\} \geq \sum_{i=1}^{3} (t_i^* - t_i)(e_i^* - e_i)$$

and (A.6) follows from (A.9).

By adapting the results (6.2.211) – (6.2.213) to the present notation we see that (A.7) is equivalent to

(a) matrix $(\partial t_i / \partial e_j)$ is positive definite, (A.11)

(b) $(t_i - t_j)/(e_i - e_j) > 0 \qquad i \neq j$ (A.12)

together. If (A.11) holds on an open convex subset of (e_1, e_2, e_3)-space (a subspace of $\mathbb{E}$) then, again by the line segment argument, it follows that (A.9) also holds on that subset, and this in turn implies (A.12).

Because of the isotropy we may write $\phi(\mathbf{E}) \equiv \hat{\phi}(e_1, e_2, e_3)$, where $\hat{\phi}$ is symmetric with respect to pairwise interchange of e_1, e_2, e_3. Use of the extremal property of the trace of a tensor product then shows that (A.8) is equivalent to

$$\hat{\phi}(e_1^*, e_2^*, e_3^*) - \hat{\phi}(e_1, e_2, e_3) - \sum_{i=1}^{3} t_i(e_i^* - e_i) > 0 \tag{A.13}$$

with $e_i^* \neq e_i$ for at least one $i \in \{1, 2, 3\}$, where

$$t_i = \frac{\partial \hat{\phi}}{\partial e_i}(e_1, e_2, e_3) \tag{A.14}$$

(Hill, 1968).

REFERENCES

B. Bernstein and R.A. Toupin, Some Properties of the Hessian Matrix of a Strictly Convex Function, *Journal für reine und angewandte Mathematik*, Vol. 210, pp. 65–72 (1962).
R. Hill, On Constitutive Inequalities for Simple Materials—I, *Journal of the Mechanics and Physics of Solids*, Vol. 16, pp.229–242 (1968).
R. Hill, Constitutive Inequalities for Isotropic Elastic Solids under Finite Strain, *Proceedings of the Royal Society of London, Series A*, Vol. 314, pp. 457–472 (1970).

APPENDIX 2

Glossary of Symbols

Symbol	Description
$\mathbf{A}$	deformation gradient tensor
A	matrix of components of $\mathbf{A}$
A_{ij}	components of $\mathbf{A}$, elements of A
$\mathbf{A}^*$	distortional part of $\mathbf{A}$
$\dot{\mathbf{A}}, \delta\mathbf{A}$	increment in $\mathbf{A}$
$\delta\mathbf{A}, \delta\dot{\mathbf{A}}$	variation in $\mathbf{A}, \dot{\mathbf{A}}$
$\mathscr{A}^1$	fourth-order tensor of first-order elastic moduli
$\mathscr{A}^\nu$	$(2\nu+2)$th-order tensor of νth-order elastic moduli, $\nu = 1, 2, \ldots$.
$\mathscr{A}_0^\nu$	tensor of νth-order instantaneous elastic moduli
$\mathscr{A}^\nu_{\alpha i \beta j \ldots}$	components of $\mathscr{A}^\nu$
$\mathscr{A}^\nu_{0ijkl\ldots}$	components of $\mathscr{A}_0^\nu$
$\mathbf{B} = (\mathbf{A}^T)^{-1}$	inverse deformation gradient tensor
B	body
$\mathscr{B}_0$	place occupied by body in reference configuration
$\mathscr{B}, \mathscr{B}_t$	place occupied by body in current configuration
$\partial\mathscr{B}_0$	boundary of $\mathscr{B}_0$
$\partial\mathscr{B}, \partial\mathscr{B}_t$	boundary of $\mathscr{B}$, $\mathscr{B}_t$
$\mathbf{b}$	body force vector (per unit mass)
$\mathbf{b}_0$	body force vector—Lagrangean description
$\dot{\mathbf{b}}, \delta\mathbf{b}$	increment in $\mathbf{b}$
$\mathscr{D}$	subset of $\mathscr{E}$
$\mathscr{E}$	Euclidean point space
$\mathbb{E}$	Euclidean vector space
$\mathbf{E}, \mathbf{E}^{(2)}$	Green strain tensor
$\mathbf{E}$	infinitesimal strain tensor
$\mathbf{E}^{(m)}$	Lagrangean strain tensor
$E\{\cdot\}$	energy functional
$E_c\{\cdot\}$	complementary energy functional
$\mathbf{F}(\mathbf{U})$	general Lagrangean strain tensor, coaxial with $\mathbf{U}$

$\mathbf{G}, \mathbf{G}^{(m)}, \mathbf{G}^{\mathsf{F}}$	response functions of a Cauchy elastic material
$\mathbf{H}$	response function of a Cauchy elastic material
$\mathbf{I}$	second-order identity tensor
$I_1(\mathbf{T}), I_2(\mathbf{T}), I_3(\mathbf{T})$	principal invariants of the second-order tensor $\mathbf{T}$
J	local ratio of current to reference volume
$\mathscr{L}^1$	fourth-order tensor $\partial\mathbf{F}/\partial\mathbf{U}$
$\mathscr{L}^2$	sixth-order tensor $\partial^2\mathbf{F}/\partial\mathbf{U}^2$
$\mathscr{L}^{(m)1}$	fourth-order tensor of first-order elastic moduli
$\mathscr{L}^{(m)\nu}$	$(2\nu+2)$th-order tensor of νth-order elastic moduli
$\mathscr{L}_0^{(m)\nu}$	tensor of νth-order instantaneous elastic moduli
$\mathscr{M}^1$	inverse of $\mathscr{L}^1$
$\mathbf{M}$	energy momentum tensor
M	torsional couple
N	axial load
$\mathbf{n}$	unit outward normal to $\partial\mathscr{B}$
$\mathbf{N}$	unit outward normal to $\partial\mathscr{B}_0$
$\mathbf{N}$	inward normal to eigensurface
O	observer
$\mathbf{O}$	origin
p	hydrostatic part of stress in an incompressible elastic material
q	same as $-p$
$\mathbf{Q}$	orthogonal second-order tensor
Q	component matrix of $\mathbf{Q}$
Q_{ij}	components of $\mathbf{Q}$
$\mathbf{Q}(\mathbf{N}), \mathbf{Q}_0(\mathbf{n})$	acoustic tensors
$\mathbb{R}$	the set of real numbers
$\mathbf{R}$	second-order rotation tensor
R	matrix of components of $\mathbf{R}$
$\mathbf{S}$	nominal stress tensor
$\dot{\mathbf{S}}, \delta\mathbf{S}$	increment in $\mathbf{S}$
$\delta\mathbf{S}, \delta\dot{\mathbf{S}}$	variation in $\mathbf{S}$, $\dot{\mathbf{S}}$
$\mathbf{T}$	Cauchy stress tensor
T, T'	matrices of components of $\mathbf{T}$
$\hat{\mathbf{T}}$	Kirchhoff stress tensor
$\mathbf{T}^{(1)}$	symmetrized Biot stress tensor
$\mathbf{T}^{(2)}$	second Piola–Kirchhoff stress tensor
$\mathbf{T}^{(m)}$	stress tensor conjugate to $\mathbf{E}^{(m)}$
$\mathbf{T}^{\mathsf{F}}$	stress tensor conjugate to $\mathbf{F}(\mathbf{U})$
$\mathbf{t}, \mathbf{t}(\mathbf{n})$	stress vector
$t_i, \hat{t}_i, t_i^{(m)}$	principal components of $\mathbf{T}, \hat{\mathbf{T}}, \mathbf{T}^{(m)}$
$\mathbf{U}$	right stretch tensor
U	component matrix of $\mathbf{U}$

$\mathbf{u}^{(i)}$	Lagrangean principal axes
$\mathbf{u}$	displacement vector
$\mathbf{V}$	left stretch tensor
V	component matrix of $\mathbf{V}$
V	vector space
$\mathbf{v}^{(i)}$	Eulerian principal axes
$\mathbf{V}$	velocity—Lagrangean description
$\mathbf{v}$	velocity—Eulerian description
$\mathbf{v}$	incremental displacement
W	strain-energy function
W_c	complementary energy function
X	particle
$\mathbf{X}$	particle or position vector in reference configuration
$\mathbf{x}$	position vector in current configuration
x_i, X_α	Cartesian components of $\mathbf{x}, \mathbf{X}$
α_p	material constant
$\boldsymbol{\Gamma}$	velocity gradient
γ	amount of shear in simple shear deformation
δ_{ij}	Kronecker delta
ε_{ijk}	alternating symbol
κ	bulk modulus
$\lambda_1, \lambda_2, \lambda_3$	principal stretches
$\lambda_1^*, \lambda_2^*, \lambda_3^*$	modified principal stretches
λ	Lame modulus, principal stretch
μ	material constant, in particular shear modulus
μ_p	material constant
$\boldsymbol{\xi}$	prescribed placement vector on $\partial\mathscr{B}_o$
ρ_0, ρ	reference, current density
$\boldsymbol{\Sigma}$	Eulerian strain-rate
$\boldsymbol{\sigma}$	prescribed traction on $\partial\mathscr{B}_o$
σ	magnitude of $\boldsymbol{\sigma}$
$\dot{\boldsymbol{\sigma}}, \delta\boldsymbol{\sigma}$	increment in $\boldsymbol{\sigma}$
ϕ_0, ϕ_1, ϕ_2	constitutive functions
χ_0, χ	reference, current configuration
$\boldsymbol{\chi}$	deformation
$\dot{\boldsymbol{\chi}}, \delta\boldsymbol{\chi}$	increment in $\boldsymbol{\chi}$
$\delta\boldsymbol{\chi}, \delta\dot{\boldsymbol{\chi}}$	variation in $\boldsymbol{\chi}$, $\dot{\boldsymbol{\chi}}$
$\boldsymbol{\Omega}$	body spin
$\boldsymbol{\Omega}^{(L)}, \boldsymbol{\Omega}^{(E)}$	spin of Lagrangean, Eulerian principal axes
$\mathbf{0}$	zero vector
$\mathbf{0}$	zero tensor

Index

A CATALOG OF SELECTED

DOVER BOOKS

IN SCIENCE AND MATHEMATICS

A CATALOG OF SELECTED
DOVER BOOKS
IN SCIENCE AND MATHEMATICS

ROTARY-WING AERODYNAMICS, W.Z. Stepniewski. Clear, concise text covers aerodynamic phenomena of the rotor and offers guidelines for helicopter performance evaluation. Originally prepared for NASA. 537 figures. 640pp. 6⅛ × 9¼.
64647-5 Pa. $15.95

DIFFERENTIAL GEOMETRY, Heinrich W. Guggenheimer. Local differential geometry as an application of advanced calculus and linear algebra. Curvature, transformation groups, surfaces, more. Exercises. 62 figures. 378pp. 5⅜ × 8½.
63433-7 Pa. $9.95

INTRODUCTION TO SPACE DYNAMICS, William Tyrrell Thomson. Comprehensive, classic introduction to space-flight engineering for advanced undergraduate and graduate students. Includes vector algebra, kinematics, transformation of coordinates. Bibliography. Index. 352pp. 5⅜ × 8½. 65113-4 Pa. $9.95

A SURVEY OF MINIMAL SURFACES, Robert Osserman. Up-to-date, in-depth discussion of the field for advanced students. Corrected and enlarged edition covers new developments. Includes numerous problems. 192pp. 5⅜ × 8½.
64998-9 Pa. $8.95

ANALYTICAL MECHANICS OF GEARS, Earle Buckingham. Indispensable reference for modern gear manufacture covers conjugate gear-tooth action, gear-tooth profiles of various gears, many other topics. 263 figures. 102 tables. 546pp. 5⅜ × 8½. 65712-4 Pa. $14.95

SET THEORY AND LOGIC, Robert R. Stoll. Lucid introduction to unified theory of mathematical concepts. Set theory and logic seen as tools for conceptual understanding of real number system. 496pp. 5⅜ × 8¼. 63829-4 Pa. $12.95

A HISTORY OF MECHANICS, René Dugas. Monumental study of mechanical principles from antiquity to quantum mechanics. Contributions of ancient Greeks, Galileo, Leonardo, Kepler, Lagrange, many others. 671pp. 5⅜ × 8½.
65632-2 Pa. $14.95

FAMOUS PROBLEMS OF GEOMETRY AND HOW TO SOLVE THEM, Benjamin Bold. Squaring the circle, trisecting the angle, duplicating the cube: learn their history, why they are impossible to solve, then solve them yourself. 128pp. 5⅜ × 8½. 24297-8 Pa. $4.95

MECHANICAL VIBRATIONS, J.P. Den Hartog. Classic textbook offers lucid explanations and illustrative models, applying theories of vibrations to a variety of practical industrial engineering problems. Numerous figures. 233 problems, solutions. Appendix. Index. Preface. 436pp. 5⅜ × 8½. 64785-4 Pa. $11.95

CURVATURE AND HOMOLOGY, Samuel I. Goldberg. Thorough treatment of specialized branch of differential geometry. Covers Riemannian manifolds, topology of differentiable manifolds, compact Lie groups, other topics. Exercises. 315pp. 5⅜ × 8½. 64314-X Pa. $9.95

HISTORY OF STRENGTH OF MATERIALS, Stephen P. Timoshenko. Excellent historical survey of the strength of materials with many references to the theories of elasticity and structure. 245 figures. 452pp. 5⅜ × 8½. 61187-6 Pa. $12.95